Foam Films and Foams
Fundamentals and Applications

PROGRESS IN COLLOID AND INTERFACE SCIENCE

Series Editors
Reinhard Miller and Libero Liggieri

The *Progress in Colloid and Interface Science* series consists of edited volumes and monographs dedicated to relevant topics in the field of colloid and interface science. The series aims to present the most recent research developments and progress made on particular topics. It covers fundamental and/or applied subjects including any type of interface, their theoretical description and experimental analysis; formation and characterization of disperse systems; bulk properties of solutions; experimental methodologies; self-assembling materials at interfaces; and applications in various areas including agriculture, biological systems, chemical engineering, coatings, cosmetics, flotation processes, food processing, materials processing, and household products.

Interfacial Rheology
Reinhard Miller and Libero Liggieri

Bubble and Drop Interfaces
Reinhard Miller and Libero Liggieri

Drops and Bubbles in Contact with Solid Surfaces
Michele Ferrari, Libero Liggieri and Reinhard Miller

Colloid and Interface Chemistry for Nanotechnology
Peter Kralchevsky, Reinhard Miller and Francesca Ravera

Computational Methods for Complex Liquid–Fluid Interfaces
Mohammad Taeibi Rahni, Mohsen Karbaschi and Reinhard Miller

Physical-Chemical Mechanics of Disperse Systems and Materials
Eugene D. Shchukin and Andrei S. Zelenev

Foam Films and Foams
Fundamentals and Applications
Dotchi Exerowa, Georgi Gochev, Dimo Platikanov, Libero Liggieri and Reinhard Miller

For more information about this series, please visit: https://www.crcpress.com/Progress-in-Colloid-and-Interface-Science/book-series/CRCPROCOLINT

Foam Films and Foams
Fundamentals and Applications

Edited by
Dotchi Exerowa
Georgi Gochev
Dimo Platikanov
Libero Liggieri
Reinhard Miller

CRC Press
Taylor & Francis Group
Boca Raton London New York

CRC Press is an imprint of the
Taylor & Francis Group, an **informa** business

CRC Press
Taylor & Francis Group
6000 Broken Sound Parkway NW, Suite 300
Boca Raton, FL 33487-2742

First issued in paperback 2021

ISBN 13: 978-1-03-223584-4 (pbk)
ISBN 13: 978-1-4665-8772-4 (hbk)

DOI: 10.1201/9781351117746

Visit the Taylor & Francis Web site at
http://www.taylorandfrancis.com

and the CRC Press Web site at
http://www.crcpress.com

Dedication

DOTCHI EXEROWA (1935–2017)

Dotchi Exerowa passed away on October 2, 2017, in Sofia. The Bulgarian and international colloid and interface community lost a scientist of high professional authority and a prominent representative of the Bulgarian School of Colloid and Interface Science.

Dotchi Exerowa was born on May 20, 1935, in the town of Varna, Bulgaria. In 1958, she graduated in chemistry at Sofia University "St. Kliment Ohridski" with honors and joined the newly founded Institute of Physical Chemistry at the Bulgarian Academy of Sciences (IPC-BAS), where the Department of Interfaces and Colloids was headed by Professor Alexei Scheludko, founder of the Bulgarian School of Colloid and Interface Science. Dotchi Exerowa completed her PhD thesis ("Free thin liquid films and foams") in 1969 and obtained habilitation in 1971. In 1987, she completed her Doctor of Science Thesis ("Formation and stability of black foam films"). In 1983, she succeeded Professor Scheludko as Head of the Department of Interfaces and Colloids in IPC-BAS and led the department until 2005. Parallel with her duties at IPC-BAS, she also lectured on "Physicochemical methods in biology," "Foams and emulsions," and "Physical chemistry of liquid interfaces" at Sofia University. In 1988, she became Professor in Physical Chemistry and in 2004 was elected as a Full Member of the Bulgarian Academy of Sciences.

The start of the scientific career of Dotchi Exerowa coincided with the brilliant period when the basis of surface forces studies in the area of fluid interfaces was being researched all over the world. Her research activity actually began in 1958 when she elaborated her MSc thesis entitled "Electrostatic disjoining pressure in foam films" under the supervision of Alexei Scheludko. This study was published soon after in *Kolloid Zeitschrift* [1]. Three conceptual papers that followed may be considered as benchmarks for the further development of the Bulgarian School of Colloids and Interface Science [2–4]. Her subsequent scientific achievements were related to the study of surface forces in thin liquid films, as well as to adjoining aspects in the physics and chemistry of interfacial phenomena. They might be grouped into three major trends: (i) new scientific instrumentation; (ii) pioneering new research results; and (iii) innovative applications of the new scientific results.

New scientific instrumentation: Over the course of several decades, microscopic foam films have been one of the most actively investigated research topics of colloids and interfaces. Their specific kinetic and thermodynamic properties promote them as a simple and comprehensible model for the investigation of surface forces and the stability of dispersive systems. They also have an important independent role as a vital constituent of biological systems, thus bridging the fundamental physicochemical knowledge among biology, biomedicine, biotechnologies, and life sciences. The progress in this research field is based on the effectiveness of various experimental methods developed for the formation and the investigation of thin liquid films. One of the most successful techniques for the study of these films is the microinterferometric setup for the formation and investigation of thin films, put forward in the 1960s by Dotchi Exerowa and Alexei Scheludko. The key element of this setup is the so-called Scheludko–Exerowa cell. Using this cell, the first experimental verification of the classical theory of Derjaguin, Landau, Verwey, and Overbeek (DLVO theory) was accomplished [1,5]. Later on, this methodology was modified and further developed by Exerowa and her research group and applied to the investigation of a variety of fluid systems. Particularly important is the novel modification—Thin Liquid Film Pressure Balance Technique (TLF-PBT)—which allows the direct measurement of interaction forces in the microscopic liquid films, gaining the disjoining pressure/

film thickness isotherms ($\Pi(h)$) for a wide range of thickness values and pressures, including the transition region of common black/Newton black films (CBF/NBF).

The Scheludko–Exerowa microinterferometric technique turned out to be a very reliable research instrumentation and is now used worldwide in most cases where surface forces at fluid interfaces are investigated. Variations of these setups are working in a number of laboratories: University of California at Berkley (USA), Technical University of Virginia (USA), Max-Planck Institute of Colloid and Interface Research in Potsdam (Germany), Moscow State University (Russia), Institute of Surface Chemistry, Stockholm (Sweden), Ecole Normale, Paris (France), Institute of Food Research, Norwich (UK), and so on. In 2004–2005, Dotchi Exerowa initiated the design of the latest version of the microinterferometric technique with the additional option for electrochemical investigations of thin liquid films. It was constructed by Exerowa's collaborators [6,7] in partnership with Canadian colleagues (University of Alberta and CanmetENERGY, Edmonton). The technique is applied to the investigation of emulsion films stabilized with natural surfactants. These thin liquid films are of crucial importance for understanding the properties of dispersive systems, which are obtained in the oil, pharmaceutical, cosmetic, and food industries.

Pioneering new research results: Historically, the microscopic foam film was the first simple model of two interacting fluid interfaces. Using the microscopic foam film model, Exerowa and collaborators investigated the fundamental aspects of the formation and stability of foam films. It was verified for the first time that there are two types of equilibrium films: common black films (CBF) and Newton (bilayer) black films (NBF). One of the basic research trends to which Exerowa had a significant contribution was the detailed investigation of the NBFs. A new theoretical approach to explain their formation and properties was proposed with NBFs regarded as a 2D-ordered system. The rupture of the bilayers was related to a fluctuation mechanism of nanoscale hole formation. The hole formation was treated as a nucleation process of a new phase in the 2D-system (NBF). The leading role of the amphiphile concentration and the short-range intermolecular forces for the formation and stabilization of these films was precisely verified. One of the basic achievements of Exerowa and her group was the promotion of the NBF as an efficient model for the study of interaction forces in biological membrane configurations.

Innovative applications of the new scientific results: Although the primary accomplishments of Dotchi Exerowa concerned the fundamental studies of fluid interfacial properties and interactions, in many cases she transferred the results into innovative advances and applications. One indicative example was the development of a new method for the estimation of the lung maturity of neonates. As is well-known, the deficiency of the lung alveolar surfactant causes lung immaturity and leads to so-called Respiratory Distress Syndrome (RDS) in newborn infants. Through the investigation of bilayer films formed from phospholipids and the components of alveolar surfactant, Exerowa and collaborators demonstrated that NBFs may be applied successfully for the exploration of alveolar surfactant insufficiency. The microscopic foam film was proposed as a novel *in vitro* model for the study of the alveolar interface and the estimation for the stability of the alveoli. Based on this model, a new method for early diagnostics of lung maturity of neonates was proposed and a new methodology for the design and optimization of therapeutic pulmonary preparations was developed. The method was often called the Exerowa Black Film method [8] and is implemented in a number of medical institutions in Bulgaria, Europe, and the USA. In recognition of these accomplishments, Dotchi Exerowa was elected as a member of the International Association for Lung Surfactant System and of the EURAIL Commission ("EURope Against Immature Lung"). The very good correspondence of the clinical results and the parameters of the *in vitro* model suggested a new hypothesis for the structure of the alveolar surfactant layer. This is the so-called "bilayer-monolayer model", which allows deeper understanding of the major physiological process of breathing [9].

Another significant application trend was the study of the properties and the stability of foams. Starting from their experience on foam films, Dotchi Exerowa and her group succeeded in relating their peculiarities to the stability of foam systems. The basic success here was the elucidation of the role of the foam film type (common thin, common black, or bilayer) for the lifetime and the syneresis (drainage) of foams [10]. The obtained results were very useful for the optimization of real foam-producing formulations, to be used, for example, in firefighting, food, and cosmetic industries.

Many of Exerowa's research results were included in the monograph *"Foam and Foam Films,"* first published in 1990 (Khimiya, Moscow). The second edition, extended and revised, was published by Elsevier in 1998. She had also authored a number of chapters in the encyclopedic books edited by: Hans Lyklema (*"Fundamentals of Interface and Colloid Science,"* Elsevier, 2005); Johan Sjöblom (*"Emulsions and Emulsion Stability,"* CRC Taylor and Francis, 2005); Tharwat Tadros (*"Emulsion Science and Technology,"* Wiley-VCH, 2009); Victor Starov (*"Nanoscience: Colloidal and Interfacial Aspects,"* CRC Press, Taylor and Francis, 2010); Hiroyuki Ohshima and Kimiko Makino (*"Colloid and Interface Science in Pharmaceutical Research and Development,"* Elsevier, 2014). In 2007, the book *"Colloid Stability—The Role of Surface Forces"* (edited by Th. Tadros, Wiley-VCH), was published in two volumes and dedicated to the 70th birthdays of Dotchi Exerowa and Dimo Platikanov. In 2009, another book, *"Highlights in Colloid Science"* (Wiley-VCH), was published, with Dotchi Exerowa as a co-editor.

During her long and successful scientific career, Dotchi Exerowa initiated and developed numerous research contacts and scientific collaborations with leading scientists and research laboratories in the field of physical chemistry of interfaces and colloids (France, Great Britain, Germany, Spain, Poland, Russia, Canada, Italy, Japan, etc.). She was elected to the scientific committees of many international conferences and symposia. She was member of the editorial boards of leading colloid chemistry journals (*Current Opinion in Colloid and Interface Science, Advances in Colloid and Interface Science, Colloids and Surfaces A, Colloid and Polymer Science, Colloid Journal* (in Russian)). In 1997, she was a co-chair (together with Dimo Platikanov) of the 9th International Conference on Surface and Colloid Chemistry, organized under the aegis of the International Association of Colloid and Interface Scientists (IACIS). Professor Exerowa had also been elected twice as a member of the IACIS Council and was closely engaged in the development and the organizational advance of the international colloid and interface scientific community. A tribute to the high quality of her research achievements and fruitful international collaborations was the international symposium (Dotchi Exerowa Symposium—Smart and Green Interfaces: Fundamentals and Diagnostics) held in Sofia, October 29–31, 2015, and dedicated to her 80th birthday. More than 110 researchers from 23 countries participated in this event and a special issue with selected papers was published recently in *Colloids and Surfaces A: Physicochemical and Engineering Aspect*, vol. 519, 2017.

The death of Dotchi Exerowa is a great loss to the colloid and interface community. She will be remembered by her colleagues, collaborators, and former students for her deep passion for science, the wide horizon of her research interests, and her keen intuition to address challenging problems, thus advancing truly basic and pioneering studies. The legacy of the important results of her scientific, teaching, and applied activities constituted a vital contribution to the productive development of the Bulgarian School of Colloid and Interface Science and were key pillars supporting the progress and international reputation of this school.

Elena Mileva
Sofia, Bulgaria

BIBLIOGRAPHY

1. Scheludko, A., Exerowa, D. Über den elektrostatischen Druck in Schaumfilmen aus wässringen Ektrolytlösungen. *Koll Z* 1959; 165: 148.
2. Scheludko, A., Exerowa, D. Instrument for interferometric measuring of the thickness of microscopic foam layers. *Comm Dept Chem Bulg Acad Sci* 1959; 7: 123.
3. Exerowa, D., Scheludko, A. Taches noires et stabilité des mousses. In: Overbeek, J.Th.G. (Ed.), *Chemistry, Physics and Application of Surface Active Substances*, Vol. 2. Gordon & Breach Sci. Publ., London, 1964, p. 1097.

4. Exerowa, D., Scheludko, A. Porous plate method for studying microscopic foam and emulsion films. *Compt Rend Acad Bulg Sci* 1971; 24: 47.

5. Scheludko, A., Exerowa, D. Über den elektostatischen und van der Waalsschen zusätzlichen Druck in wässeringen Schaumfilmen. *Kolloid Zeitschrift* 1960; 168: 24.

6. Panchev, N., Khristov, K., Czarnecki, J., Exerowa, D., Bhattacharjee, S., Masliyah, J. A new method for water-in-oil emulsion film studies. *Colloids Surf A* 2008; 315: 74.

7. Mostowfi, F., Khristov, K., Czarnecki, J., Masliyah, J., Bhattacharjee, S. Electric field mediated breakdown of thin liquid films separating microscopic emulsion droplets. *Appl Phys Lett* 2007; 90: 184102.

8. Cordova, M., Mautone, A.J., Scarpelli, E.M. Rapid in vitro tests of surfactant film formation: Advantages of the Exerowa black film method. *Pediatr Pulmonol* 1996; 21: 373.

9. Kashchiev, D., Exerowa, D. Structure and Surface Energy of the Surfactant Layer on the Alveolar Surface. *Eur Biophys J* 2001; 30: 34.

10. Khristov, K., Exerowa, D. Influence of the Foam Film Type on the Foam Drainage Process. *Colloids Surf* 1995; 94: 303.

Dedication

DIMO PLATIKANOV (1936—2017)

Dimo Platikanov passed away on October 21, 2017, in Sofia. He enjoyed a long and distinguished career at the University of Sofia, Bulgaria.

Dimo Platikanov was Professor of Physical Chemistry and a former Head of the Department of Physical Chemistry of the University of Sofia. This department was established in 1925 by Professor Iwan Stranski (1897–1979) and was headed by Professor Rostislav Kaishev (1908–2002) and Professor Alexei Scheludko (1920–1995).

Dimo Nikolov Platikanov was born on February 2, 1936, in Sofia. His mother was Lyuba Zaharieva Trifonova (1908–1989), a philologist—Germanist, and his father was Nikola Dimov Platikanov (1898–1984) who was a well-known specialist in cattle breeding and professor in the University of Sofia and Fellow of the Bulgarian Academy of Sciences. Dimo's sister, Vesselina Platikanova, is also a physical chemist, and Professor Boyan Mutaftchiev (France)—a well-known specialist in crystal growth, is his cousin.

Dimo Platikanov graduated from the University of Sofia in 1958. Regarded by his professors as an unusually gifted student, Dimo was immediately employed as a lecturer in the Department of Physical Chemistry of the University of Sofia. In that department, Dimo elaborated his PhD thesis, then later presented his DSc thesis; there he became first an associate professor and later, a professor. During that time, he worked for Professor Scheludko in his new Bulgaria colloid science research area and taught physical chemistry to the students in chemistry.

Dimo was a brilliant experimenter; but he contributed to pure theory as well: "Disjoining Pressure of Curved Films," 1985 (with B.V. Toshev) [1], "Line Tension in Three-Phase Equilibrium Systems," 1988 (with B. V. Toshev and A. Scheludko) [2], "Disjoining Pressure, Contact Angle and Line Tension in Free Thin Liquid Films," 1992 (with B.V. Toshev) [3], and "Wetting: Gibbs' Superficial Tension Revisited," 2006 (with B.V. Toshev) [4].

The books on the modern development of colloid science as an extension to nano science appeared under the editorship of Dimo Platikanov and Dochi Exerowa, his wife who passed away weeks before him, are much appreciated by the international colloid science community: "*Highlights in Colloid Science*" (Wiley, 2008) [5].

Two basic thermodynamic quantities that characterize the three-phase contact with a thin liquid film are contact angle and line tension. Line tension, first introduced by Gibbs, is similar to surface tension, but in contrast, it could be either positive or negative. It is extremely important to provide reliable methodologies for its experimental measurement. The contact (wetting) angles measurment and the description of the wetting phenomena have a prolonged history. In these systems, thin liquid films usually exist. The disjoining pressure with van der Waals, electrostatic, steric, and other components is the basic thermodynamic quantity that characterizes the behavior of such systems. Symmetrical as well as asymmetrical thin liquid films exist. There are black films of two types: thicker common films and thinner Newton films. All these systems have been examined carefully by Professor Platikanov and his co-workers and more than 20 unique experimental techniques have been elaborated. This was not a surprise because Dr. Platikanov had the talent to practice science with accuracy and responsibility and it marked his long scientific career.

Anyone in the study of thin films knows the Sceludko–Exerowa microinterferometric method. Platikanov used this method with several improvements to determine the shape of the three-phase contact in the presence of films under dynamic and as well as equilibrium conditions. The initial stages of forming plane-parallel liquid films were studied and the so-called "dimpling" effect was recognized. This paper, broadly cited by other researchers, was published in 1964 in the *Journal of Physical Chemistry* [6]. Later, the equilibrium profile of the transition zone between a thin wetting film and its adjacent meniscus had been experimentally obtained (with Zorin and Kolarov) [7,8]; in this study, the evidence of the theoretical evidence predicted by Martynov, Ivanov, and Toshev of "contact thickness" (instead of "contact angle") [9] had been approved.

In Platikanov's research group, three methods for experimental determination of the line tension of the contact line perimeter between a Newton black film and a bulk solution were developed: the method of the diminishing bubble, the method of the critical bubble, and the method of the porous plate (with Nedyalkov and others). These methods were based on the dependence of the contact angle on the radius of the circular contact line perimeter. The contact angle increases when the radius decreases if the line tension is positive; the contact angle decreases when the radius decreases if the line tension is negative. Platikanov's data experimentally confirmed the existence of positive and negative values of the line tension. The change of the sign of the line tension can be realized by changing the content of the bulk solution or by a change of temperature [10,11].

The thinnest Newton black films are solid bodies rather than liquid objects. Thus, defects (holes) in the black film could appear spontaneously. One may expect an increase of gas permeability through such films in comparison with the gas permeability through thicker films. Dimo's group succeeded in proving this fact (12–14]. The gas permeability coefficients of Newton black films strongly depend on the surfactant concentration. It was an important result. Thus, the theory by Kashchiev and Exerowa [15] for fluctuational formation of nanoscopic holes within films and biomembranes was experimentally grounded.

Professor Dimo Platicanov was a socially responsible person. He was always ready to support and help his colleagues. Up until the last time he was President of the International Foundation "St. Cyril and St. Methodius"—many young boys and girls were happy to develop their professional identities with the support of that foundation. Professor Platikanov was a President of the Hunboldt-Union in Bulgaria some time ago. Dimo's successful Presidency of the International Association of Colloids and Interface Sciences (IACIS) 2002–2006 should also be mentioned.

Professor Platikanov possessed a strong historical sensitivity. In recent years, I have published numerous texts on the history of Bulgarian society in the past—many historical testimonies—forgotten, distorted, or forbidden in the second part of the twentieth century, were revived [16]. Dimo read these texts critically and with great interest. Every day in our offices at the Department of Physical Chemistry, we two met each other to talk and to discuss these literary documents. Thus, our friendship was developed and strengthened.

Dimo was a great man and a distinguished scholar. With his scientific achievements and personal features, Professor Platikanov will be always remembered and honored.

B. V. Toshev
Sofia, Bulgaria

REFERENCES

1. Toshev, B.V., Platikanovm, D. Disjoining pressure of curved films. *Compt R Acad Bulgare Sci* 1985; 38: 703–705.
2. Toshev, B.V., Platikanov, D., Scheludko, A. Line tension in three-phase equilibrium systems. *Langmuir* 1988; 4: 489–499.
3. Toshev, B.V., Platikanov, D. Disjoining pressure, contact angle and line tension in free thin liquid films Adv. *Colloid Interface Sci* 1992; 40: 157–189.
4. Toshev, B.V., Platikanov, D. Wetting: Gibbs' superficial tension revisited. *Colloids Surf A* 2006; 291: 177–180.
5. Platikanov, D., Exerowa, D. (Eds.) *Highlights in Colloid Science.* Wiley: Weinheim, 2008.

6. Platikanov, D. Experimental investigation on the "dimpling" pf thin liquid films. *J Phys Chem* 1964; 68: 3619–3624.

7. Zorin, Z., Platikanov, D., Kolarov, T. The transition region between aqueous wetting films on quartz and the adjacent meniscus. *Colloids Surf* 1987; 22: 133–145.

8. Kolarov, T., Zorin, Z., Platikanov, D. Profile of the transition region between aqueous wetting films on quartz and the adjacent meniscus. *Colloids Surf* 1990; 51: 37–47.

9. Martynov, G.A., Ivanov, I.B., Toshev, B.V. On the mechanical equilibrium of the free liquid film with the meniscus. *Kolloidn Zh* 1976; 38: 474–479.

10. Platikanov, D., Nedyalkov, M., Scheludko, A. Line tension of Newton black films. I. Determination by the critical bubble method. *J Colloid Interface Sci* 1980; 75: 612–619.

11. Platikanov, D., Nedyalkov, N., Nasteva, V. Line tension of Newton black films. II. Determination by the diminishing bubble method. *J Colloid Interface Sci* 1980; 75: 620–628.

12. Nedyalkov, M., Krustev, R., Stankova, A., Platikanov, D. On the mechanism of permeation of gas though Newton black films at different temperatures. *Langmuir* 1992; 8: 3142–3144.

13. Krustev, R., Platikanov, D., Nedyalkov, M. Permeability of common black foam films to gas I. *Colloids Surf A* 1993; 79: 129–136.

14. Krustev, R., Platikanov, D., Stankova, A., Nedyalkov, M. Permeation of gas through Newton black films at different chain length of the surfactant. *J Disp Sci Tech* 1997; 18: 789–800.

15. Exerowa, D., Kaschiev, D. Hole-mediated stability and permeability of bilayers. *Contemp Phys* 1986; 27: 429–461.

16. https://bvtoshev.wordpress.com/

Contents

SECTION I Adsorption Layers

SECTION II Foam Films

SECTION III Foams

SECTION IV   Structure of Foams and Antifoaming

SECTION V   Applications

Content

Preface

The present book *"Foam Films and Foams: Fundamentals and Applications,"* is Volume 7 of the book series "Progress in Colloid and Interface Science." It is dedicated to one of the most complex topics in colloid and interface science—liquid foams. From a practical point of view, foams are well known by everyone, and we are in contact with foam every day. However, complete understanding of foam formation and its subsequent stabilization or destabilization is not yet fully understood.

As will be shown in this book, foams are hierarchical systems. A top-down approach shows that it consists of many foam bubbles in contact with each other, forming liquid films in between. On either side of such a foam film, there is an adsorption layer self-assembled by surfactants or polymers or their mixtures. Hence, the properties of these adsorption layers and the properties of the respective liquid films control the formation and stability of the resulting foam.

This book represents the state of the art of the fundamentals of foam films and foams together with a comprehensive description of the corresponding adsorption layers, and discusses important aspects in various practical applications. The starting point is adsorption layers formed by surfactants, polymers, and their mixtures. They are the basic elements of foam films formed at the two film sides. Today, we cannot yet say which of the many adsorption layer properties are the most decisive ones for stabilizing a foam film. However, what is very clear is the need of an adsorption layer for the stabilization of a foam film, which otherwise would immediately rupture. In turn, a foam would never exist if the bubbles, their building blocks, were not stabilized against coalescence.

The present book aims at presenting the entire philosophy of the formation and stabilization of foam and discusses the most recent state of the art of each level of its hierarchy. The correlations of properties between adsorption layers and films and also between films and foam are the center of interest. The final list of applications demonstrates how important this fundamental knowledge is for the production of foams for various modern technologies or practical fields.

The present book appears 20 years after a book with a similar title, *"Foam and Foam Films"*, written by D. Exerowa and P. Kruglyakov, published by Elsevier in 1998. In a sense, the present book is its successor; however, it should be underlined that the present book is a completely new book, written by a number of authors.

Dotchi Exerowa
Georgi Gochev
Dimo Platikanov
Libero Liggieri
Reinhard Miller

Introduction

Foams are omnipresent in any moment of life. The appearance of foams in everyday life is connected to foams in many situations, such as at the seaside or in rivers, on top of a beer, and during dishwashing. Food and beverages provide extra pleasure when foamed. Frothed milk, whipped cream, meringue and mousse are popular examples. The appearance of foams is advantageous or even vital in technologies like flotation, while on the other hand, it reduces the efficiency in detergents. New materials with extraordinary properties are based on foams of which ceramic or metallic foams are prominent examples.

Foams are hierarchical systems, consisting of foam bubbles in certain contact with each other, resulting in the formation of foam films. This is an important structural element of the foam and plays a dominating role in its existence. Together with the Plateau borders and vertexes, the foam films form a unified capillary system. All the most important foam processes, including gas bubble expansions and their lifetimes, are determined by the thickness, structure, and physicochemical properties of foam films.

The two sides of a foam film (lamella) are two solution/air surfaces which are separated by a thin layer of the liquid solution. This is usually an aqueous solution of stabilizing surfactants or polymers or their mixtures (called foaming agents). These amphiphilic substances form two adsorption layers on both film surfaces, which are actually responsible for the stability of the foam film and thus of the foam. In foam films, there are normal forces in addition to the lateral forces along the surface which usually act in the adsorption layers. The normal forces appear when the foam film becomes sufficiently thin owing to the surface forces—Van der Waals forces, electrostatic (double layer) forces, steric forces, structural forces, and so on. Their resultant force per unit film area is the disjoining pressure which, together with the lateral forces, determines the foam film stability as well as the foam film thickness. Depending on the total thickness of the foam films, the resulting foams are called wet (relatively thick films) or dry—in the case of extremely thin foam films where there is no liquid solution inside the film, so-called Newton black films.

This book represents the state of the art of the fundamentals of foam films and foams together with a comprehensive description of the corresponding adsorption layers, and discusses important aspects in various practical applications of foams. The book aims at presenting the entire philosophy of foam formation and foam stability.

A separate section (Chapter 1) provides a presentation of the basic principles of adsorption layers which is necessary for a quantitative understanding of the main phenomena in foam films and foams. An introduction is dedicated to the dynamics and thermodynamics of adsorption layers. This comprises properties of amphiphilic molecules such as surfactants, polymers, proteins, and their mixtures, adsorbed at the solutions/air interface. The mechanical properties of adsorption layers, the dilatational and shear viscoelasticities, are also discussed. Interfacial peculiarities caused by micelle dynamics or the presence of molecular aggregates in the solution bulk are analyzed.

The next two extensive Sections II and III present and discuss the most important knowledge about foam films, and, respectively, about foams. Each of these two sections starts with a historical survey which represents unique material on the evolution of the understanding of films (Chapter 2) and foams (Chapter 8). It starts from pure admiration of the beauty of soap bubbles and foams, goes through the attempt to quantify the shape of lamellae and of foams by scientific capacities such as Newton, Plateau, and Gibbs, and finally leads to a discussion of the role of foaming and defoaming agents. Basics of the physical chemistry of a foam film, together with experimental methods for studying foam films are presented in Chapter 3. Considerable progress in the study of foam film properties and the reasons for their stability is due to the single microscopic model foam film obtained and studied by a special technique and proved to be adequate for the films in a polyhedral foam. A large amount of experimental data are presented in Chapters 4–7 for various types of foaming agents, such as surfactants, polymers, proteins, polyelectrolytes, and their mixtures as well as solid particles.

Then, in Chapters 9–15, an extensive presentation follows on the formation and stabilization/destabilization of foams which includes an analysis of the underlying mechanisms and scientific principles for regulating foam properties. Production of foams with indispensable technological and exploitation qualities involves determining the quantitative relationships between foam structure parameters (expansion ratio, dispersity, bubble shape, film thickness, and capillary pressure) and the kinetics of the processes controlling foam stability (drainage, gas diffusion transfer, coalescence and film rupture, and foam collapse).

Section IV is dedicated to the destruction of foams and antifoaming, to the physics of foam structures, and to the rheology of foams, although these themes could each be explored in separate books.

Section V is dedicated to the most important fields of applications, such as foams in food, foams in detergents, foams in flotation, foams in pharmacy and cosmetics, foam fractionation, and foams in firefighting, and finally with interesting insights into metallic foams and ceramic foams as well as recent attempts for foam studies in microgravity in space conditions.

A general trend has recently been established: Close contacts and the exchange of the results and ideas on foams between the physicists on one side and the chemists (physical chemists, colloid chemists) on the other side. Thus, the emergence of new areas of scientific research on foam films and foams can be expected. On the other hand, the development of new materials and products based on nanoparticles (actually colloid particles!) increasingly involves foam films and foams. The use of foam films and foams in biology and medicine should also be taken into account. Hence, significant developments of this scientific area can be expected.

Editors

Dotchi Exerowa graduated in chemistry at Sofia University "St. Kliment Ohridski" and researched her PhD thesis as well as habilitation at the Institute of Physical Chemistry at the Bulgarian Academy of Sciences (IPC-BAS). She was head of the Department of Interfaces and Colloids at the IPC-BAS until 2005. In 2004, she was elected as a full member of the Bulgarian Academy of Sciences. Her scientific interests were focused on thin liquid films and foams and the mechanisms of lung surfactants.

Georgi Gochev studied chemistry at Sofia University "St. Kliment Ohridski" and earned his PhD in the Bulgarian Academy of Sciences, Sofia, 2009. He later spent several years as a postdoctoral researcher in the Max Plank Institute of Colloids and Interfaces in Potsdam and at the University Münster (Germany). His scientific interests are in surface phenomena at soft interfaces, adsorption dynamics, surface rheology, thin liquid films and foam, and emulsion stability with particular emphasis on polymers and proteins.

Dimo Platikanov graduated in chemistry at the University of Sofia "St. Kliment Ohridski." Here, in the Department of Physical Chemistry, he defended his PhD thesis and his habilitation, and became a full professor and head of the department. He was very much involved in the organization of colloid and interface sciences, in particular in the International Association of Colloid and Interface Scientists. Dimo Platikanov's main scientific interests are liquid films, in particular wetting films.

Libero Liggieri is senior researcher and group leader at the CNR Institute of Condensed Matter Chemistry and Energy Technologies in Genoa (Italy). His research interests are mainly in the field of dynamics of surfactants and particles layers at the liquid interface, with applications for emulsions, foams, and materials. He is particularly involved in surface science studies of microgravity.

Reinhard Miller studied mathematics in Rostock and colloid chemistry in Dresden. After his PhD and habilitation at the Academy of Sciences in Berlin (Germany), he spent postdoctoral time at the University of Toronto. Since 1992, he has been a group leader at the Max Planck Institute of Colloids and Interfaces in Potsdam, Germany. His main fields of interests are thermodynamics and kinetics of surfactants and proteins at liquid interfaces, interfacial rheology, foams, and emulsions.

Contributors

Dirk Blunk
Department of Chemistry
University of Cologne
Cologne, Germany

Jan Cilliers
Department of Earth Science and Engineering
Imperial College London
London, UK

Jason N. Connor
Future Industries Institute
University of South Australia
Mawson Lakes
and
Department of Chemical Engineering
University of Adelaide
Adelaide, South Australia, Australia

F. A. Costa Oliveira
LNEG- National Laboratory of Energy and
    Geology, I.P.
UER - Renewable Energies and Energy
    Systems Integration Unit
Estrada do Paço do Lumiar
Lisboa, Portugal

Wiebke Drenckhan
Laboratory of Physics of Solids
Paris-Sud University
Orsay
and
Institute Charles Sadron
CNRS UPR, Strasbourg, France

Jan Engmann
Institute of Materials Science
Nestlé Research Center Lausanne
Vers-chez-les-Blanc
Lausanne
and
Nestlé Product Technology Center
Orbe, Switzerland

Dotchi Exerowa
Institute of Physical Chemistry
Bulgarian Academy of Science
Sofia, Bulgaria

F. García-Moreno
Helmholtz Centre Berlin
Institute of Applied Materials
Berlin, Germany

P. R. Garrett
School of Chemical Engineering and
    Analytical Science
The Mill, University of Manchester
Manchester, United Kingdom

Cécile Gehin-Delval
Institute of Materials Science
Nestlé Research Center Lausanne
Vers-chez-les-Blanc
Lausanne
and
Nestlé Product Technology Center
Orbe, Switzerland

Georgi Gochev
Institute of Physical Chemistry
Bulgarian Academy of Science
Sofia, Bulgaria

and

Max-Planck Institute of Colloids and
    Interfaces
Potsdam-Golm, Germany

Urs T. Gonzenbach
de Cavis AG
Dübendorf, ZH, Switzerland

Deniz Z. Gunes
Institute of Materials Science
Nestlé Research Center Lausanne
Vers-chez-les-Blanc
Lausanne
and
Nestlé Product Technology Center
Orbe, Switzerland

Stefan Hutzler
School of Physics
Trinity College Dublin
The University of Dublin
Ireland

Khristo Khristov
Institute of Physical Chemistry
Bulgarian Academy of Science
Sofia, Bulgaria

Nora Kristen-Hochrein
Berlin-Brandenburg Academy of Sciences
Berlin, Germany

Marcel Krzan
Jerzy Haber Institute of Catalysis and Surface
   Chemistry
Polish Academy of Sciences
Krakow, Poland

Agnieszka Kulawik-Pióro
Institute of Organic Chemistry and
   Technology
Department of Chemical Engineering and
   Technology
Cracow University of Technology
Krakow, Poland

D. Langevin
Laboratory of Physics of Solids
Paris-Sud University
Orsay, France

Martin E. Leser
Institute of Materials Science
Nestlé Research Center Lausanne
Vers-chez-les-Blanc
Lausanne
and
Nestlé Product Technology Center
Orbe, Switzerland

Meike Lexis
Karlsruhe Institute of Technology
Institute for Mechanical Process Engineering
   and Mechanics
Gotthard-Franz-Straße
Karlsruhe, Germany

Libero Liggieri
Institute of Condensed Matter Chemistry and
   Technologies for Energy
Genoa, Italy

Elena Mileva
Institute of Physical Chemistry
Bulgarian Academy of Sciences
Sofia, Bulgaria

Reinhard Miller
Max-Planck Institute of Colloids and
   Interfaces
Potsdam-Golm, Germany

Gareth Morris
Department of Earth Science and
   Engineering
Imperial College London
United Kingdom

Dimo Platikanov
Department of Physical Chemistry
Sofia University
Sofia, Bulgaria

Arnaud Saint-Jalmes
Institute of Physics of Rennes
Rennes, France

Christophe Schmitt
Institute of Materials Science
Nestlé Research Center Lausanne
Vers-chez-les-Blanc
Lausanne
and
Nestlé Product Technology Center
Orbe, Switzerland

Rossen Sedev
Future Industries Institute
University of South Australia
Mawson Lakes, South Australia
and
WA School of Mines: Minerals, Energy &
   Chemical Engineering
Curtin University
Bentley, Western Australia, Australia

Philip N. Sturzenegger
de Cavis AG
Dübendorf, ZH, Switzerland

Aouatef Testouri
Laboratory of Physics of Solids
Paris-Sud University
Orsay
and
Institut Charles Sadron
CNRS UPR
Strasbourg, France

Roumen Todorov
Institute of Physical Chemistry
Bulgarian Academy of Science
Sofia, Bulgaria

Bożena Tyliszczak
Department of Chemistry and Technology of
   Polymers
Cracow University of Technology
Krakow, Poland

Vamseekrishna Ulaganathan
Future Industries Institute
University of South Australia
Adelaide, Australia

M. Vignes-Adler
Laboratory of Modeling and Simulation Multi
   Scale
University Paris-Est Marne-la-Vallée
Marne la Vallée, France

Norbert Willenbacher
Karlsruhe Institute of Technology
Institute for Mechanical Process Engineering
   and Mechanics
Gotthard-Franz-Straße
Karlsruhe, Germany

Section I

Adsorption Layers

1 The Surface Layer as the Basis for Foam Formation and Stability

Libero Liggieri, Elena Mileva, and Reinhard Miller

CONTENTS

1.1   THERMODYNAMICS OF ADSORPTION

The thermodynamic description of adsorption layers is an important prerequisite for a quantitative understanding of any processes going on at liquid interfaces. It provides equations of state which express the surface pressure as the function of surface layer composition, and the corresponding adsorption isotherms which determine the dependence of the adsorption of each dissolved component on their bulk concentrations. These equations can also provide an easier measurable surface tension or surface pressure isotherms.

The description of experimental data, mainly equilibrium surface tension isotherms, can be easily done by the Langmuir adsorption model or the corresponding von Szyszkowski surface tension equation. However, significant deviations of the experimental tensions from the model values are often observed, so that more sophisticated models, such as the Frumkin model or the only recently derived reorientation and surface aggregation models, have to be applied.

1.1.1   GENERAL BASIS OF THE THERMODYNAMICS OF ADSORPTION

A comprehensive overview of the thermodynamics of soluble adsorption layers at liquid interfaces was presented in a recent book on surfactants [1]. It was shown that the most efficient starting point is the equation of Butler [2]:

$$\mu_i^s = \mu_i^{0s} + RT \ln f_i^s x_i^s - \gamma \omega_i \tag{1.1}$$

Here, $\mu_i^{0s}(T,P)$ is the standard chemical potential of component i which, in contrast to all other approaches, depends only on temperature T and pressure P and not on surface tension γ. f_i, x_i, and ω_i are the activity coefficients, mole fractions, and molar surface areas of component i, and the superscript "s" refers to the surface.

Equivalent to Equation 1.1 for the surface, we get the chemical potential for the bulk phase α in the following form

$$\mu_i^\alpha = \mu_i^{0\alpha} + RT \ln f_i^\alpha x_i^\alpha \tag{1.2}$$

In equilibrium, both potentials must be equal, so Equations 1.1 and 1.2 lead to

$$\mu_i^{0s} + RT \ln f_i^s x_i^s - \gamma \omega_i = \mu_i^{0\alpha} + RT \ln f_i^\alpha x_i^\alpha \tag{1.3}$$

By defining a respective standard state (infinite dilution of the solution), denoted by a subscript 0, one can obtain the following general relationships for the solvent

$$\ln \frac{f_0^s x_0^s}{f_0^\alpha x_0^\alpha} = -\frac{(\gamma_0 - \gamma)\omega_0}{RT} \tag{1.4}$$

and for the surfactants

$$\ln \frac{f_i^s x_i^s / f_{(0)i}^s}{K_i f_i^\alpha x_i^\alpha / f_{(0)i}^\alpha} = -\frac{(\gamma_0 - \gamma)\omega_i}{RT} \tag{1.5}$$

with $K_i = (x_i^s / x_i^\alpha)_{x_i^\alpha \to 0}$ being the distribution coefficients of component i at infinite dilution. From these generic equations, various special models can be derived, as was shown in detail in [1]. In the next paragraphs, we first summarize some classical adsorption models without any details of their derivation, and then, in the subsequent paragraph, give some details about two new models derived recently as refinements of the classical models.

1.1.2 CLASSICAL ADSORPTION MODELS

More than 200 years ago, William Henry found a linear dependence for gas solubility in a liquid on its pressure [3]. Such a linear relationship is also applicable to the adsorption Γ of a surfactant at a liquid surface as a function of its bulk concentration c:

$$\Gamma = Kc. \tag{1.6}$$

K is the Henry constant, which can be used as a simple measure for the surface activity of a surfactant. Such a linear adsorption model allows, of course, only a rather rough description of experimental data, mainly in the range of very low surface coverage. However, for some estimations, it is quite suitable. By using the Gibbs fundamental adsorption equation [4] in its simplest form

$$\Gamma = -\frac{1}{RT} \frac{d\gamma}{d\ln c} \tag{1.7}$$

we get a generic equation demonstrating how the two main quantities Γ and γ and the bulk concentration c are interrelated. However, it does not allow a clear physical picture of the adsorption layer to be identified. Only when linked with a certain adsorption model does it help in finding relationships suitable for the analysis and interpretation of experimental data. If we combine, for example, the adsorption isotherm of Equation 1.6 with the Gibbs Equation 1.7, we end up with a surface tension isotherm in the following form

$$\gamma = \gamma_0 - RTKc \tag{1.8}$$

where γ_0 is the surface tension of pure water. Experiments show that this linear equation has, of course, only a narrow range of validity.

In his studies on fatty acids, von Szyszkowski [5] tried to find some regularities in the measured surface tension isotherms and he found that an equation of the form

$$\frac{\gamma_0 - \gamma}{\gamma_0} = B \ln\left[1 + \frac{c}{A}\right] \tag{1.9}$$

describes the values rather well. Initially, the parameters A and B in Equation 1.9 were empirical constants in a purely empirical equation. However, later Langmuir derived an adsorption isotherm on the basis of physical principles having the following form [6]:

$$\Gamma = \Gamma_\infty \frac{c/a}{1 + c/a} \tag{1.10}$$

which, by comparison, allowed him to give a physical meaning to these empirical coefficients. The coefficient a in Equation 1.10 is called the Langmuir equilibrium adsorption constant. It is often used as a simple measure of the surface activity of a surfactant and corresponds to the concentration at which half of the surface is covered by surfactant molecules $\Gamma = \Gamma_\infty/2$, while Γ_∞ is the maximum number of adsorbed molecules. When we set $B = RT\Gamma_\infty/\gamma_0$ and $A = a$, from Equation 1.9, we obtain an equation

$$\gamma = \gamma_0 - RT\Gamma_\infty \ln(1 + c/a). \tag{1.11}$$

We obtain exactly the same equation when we combine Equation 1.10 with the Gibbs fundamental adsorption equation Equation 1.7. In this way, the empirical parameters in von Szyszkowski's equation attain a physical meaning. At small surfactant concentrations, we can approximate $\ln(1 + x)$ by x, so that the Henry constant can be defined by $K = \Gamma_\infty/a$ using the quantities of the Langmuir model given by Equation 1.11. Therefore, a as well as K are measures for the surface activity of the respective surfactant.

Although Gibbs' Equation 1.7 cannot be used for a quantitative analysis of experimental data, it plays a central role in the thermodynamics of adsorption layers. With its help, theoretical relationships can be transferred into each other. For example, Equation 1.11 can be turned into an equivalent equation of state, that is, a relationship between surface tension or surface pressure $\Pi = \gamma_0 - \gamma$:

$$\Pi = -RT\Gamma_\infty \ln\left[1 - \frac{\Gamma}{\Gamma_\infty}\right]. \tag{1.12}$$

From the analysis of surface tension data for aqueous fatty acid solutions $\gamma(c)$, Frumkin found out that with an increase in the alkyl chain length, the deviation from the Langmuir adsorption model increases [7]. To decrease this deviation, he proposed to add an interaction term to Equation 1.12 in order to take this effect into consideration:

$$\Pi = -RT\Gamma_\infty \ln\left[1 - \frac{\Gamma}{\Gamma_\infty}\right] - a_F \left[\frac{\Gamma}{\Gamma_\infty}\right]^2. \tag{1.13}$$

The coefficient a_F accounts for the interaction between the alkyl chains of the adsorbed molecules. Although Frumkin added this term empirically, Lucassen-Reynders later demonstrated that it can be also derived on a thermodynamic basis [8].

The adsorption models of Frumkin, Langmuir, and Henry are of increasing complexity, however, they transfer into each other when the concentration of the adsorbing surfactant molecules in the solution is reduced step by step. The recently developed models called reorientation and surface aggregation models, respectively, are again a bit more complex, but also reduce to the Frumkin or Langmuir models under certain simplifications.

1.1.3 New Adsorption Models

Less than 20 years ago, two new models were developed, the Reorientation Model [9] and the Aggregation Model [10]. The Reorientation Model contains two molar areas for each adsorbing molecule, while ω_2 corresponds to the interfacial area required by a single adsorbed molecule at an empty interface, and the molar area ω_1 is the minimum value of the required area. As schematically shown in Figure 1.1, this model can physically be understood in terms of different orientations of the adsorbed molecule to the surface.

This physical picture reflects the transition from a more tilted orientation in a diluted adsorption layer to an upright orientation with a smaller molar area for a closely packed adsorption monolayer. For surfactants of the type C_nEO_m (oxyethylated alcohols), another interpretation was given in the literature [11]. Here are discussions on the increase of surface coverage (or surface pressure), whereby the EO part occupies less and less area at the interface.

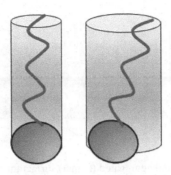

FIGURE 1.1 Schematic for the different orientation modes of a surfactant at the solution surface as one of the possible physical pictures for the reorientation model.

The Reorientation Model was further refined in [11,12] by taking into consideration the nonideality of entropy and enthalpy, and the intrinsic compressibility ε of the adsorbed layer. The basic equations developed for this model are:

For the surface equation of state -

$$-\frac{\Pi\omega_0}{RT} = \ln(1-\Gamma\omega) + \Gamma(\omega-\omega_0) + a(\Gamma\omega)^2 \tag{1.14}$$

and for the adsorption isotherms for states 1 and 2, respectively -

$$b_1 c = \frac{\Gamma_1\omega_0}{(1-\Gamma\omega)^{\omega_1/\omega_0}} \exp\left(-2a\Gamma\omega\frac{\omega_1}{\omega_0}\right) \tag{1.15}$$

$$b_2 c = \frac{\Gamma_2\omega_0}{(\omega_2/\omega_1)^\alpha (1-\Gamma\omega)^{\omega_2/\omega_0}} \exp\left(-2a\Gamma\omega\frac{\omega_2}{\omega_0}\right). \tag{1.16}$$

In this set of equations, we use $\omega = (\omega_1\Gamma_1 + \omega_2\Gamma_2)/\Gamma$ as the mean molar area, $\theta = \omega\Gamma = \omega_1\Gamma_1 + \omega_2\Gamma_2$ as the surface coverage, $\Gamma = \Gamma_1 + \Gamma_2$ as the total adsorption, $\omega_1 = \omega_0(1-\varepsilon\Pi\theta)$ with $\omega_2 > \omega_1$, b_i is the adsorption equilibrium constant of component i, and a is the intermolecular interaction constant. It is also assumed that the molar area ω_0 of a surfactant molecule at $\Pi = 0$ is identical to the molar area of the solvent. The coefficient α is the power law exponent. The ratio of adsorptions in the two possible states of the adsorbed molecules is expressed by a relationship which follows from Equations 1.15 and 1.16 for $b_1 = b_2$:

$$\frac{\Gamma_1}{\Gamma_2} = \frac{(\omega_1/\omega_2)^\alpha}{(1-\Gamma\omega)^{(\omega_2-\omega_1)/\omega_0}} \exp\left(-2a\Gamma\omega\frac{(\omega_2-\omega_1)}{\omega_0}\right). \tag{1.17}$$

The second new so called Aggregation Model [10] was also developed only recently and includes only one molar area for the adsorbed molecules ω_1, identical to the Langmuir and Frumkin models. Additionally, a critical surface aggregation concentration Γ_c and the surface aggregation number n were introduced. Figure 1.2 schematically shows the formation of trimers at the interface, which of course require less space than three single surfactant molecules.

The basic equations for the aggregation model are:

$$\text{Equation of state:} \frac{\Pi\omega}{RT} = -\ln\left\{1-\Gamma_1\omega\left[1+(\Gamma_1/\Gamma_c)^{n-1}\right]\right\} \tag{1.18}$$

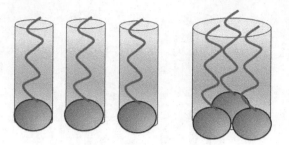

FIGURE 1.2 Schematic of single and aggregated (trimer) surfactants adsorbed at the surface.

$$\text{Adsorption isotherm}: bc = \frac{\Gamma_1\omega}{\left\{1 - \Gamma_1\omega\left[1 + (\Gamma_1/\Gamma_c)^{n-1}\right]\right\}^{\omega_1/\omega}} \tag{1.19}$$

while the mean molar area is again obtained from a weighted average

$$\omega = \omega_1 \frac{1 + n(\Gamma_1/\Gamma_c)^{n-1}}{1 + (\Gamma_1/\Gamma_c)^{n-1}}. \tag{1.20}$$

The Aggregation Model is a specific type of the Frumkin model where the interaction between the chains of the adsorbed molecules leads to the formation of small aggregates.

1.1.4 ADSORPTION MODELS FOR SURFACTANT MIXTURES

In many practical situations, mixtures of surfactants are used because solutions of single surfactants cannot provide the desired properties. A generalization of the given models is not trivial although we assume some simplifications. If we assume, for example, that the minimum molar area $1/\Gamma_\infty$ is identical for all adsorbing species, we can easily find the generalized Langmuir adsorption model

$$\Pi = -RT\Gamma_\infty \ln\left[1 - \sum_{i=1}^{r} \frac{\Gamma_i}{\Gamma_\infty}\right]. \tag{1.21}$$

However, the equal molar area at the surface is acceptable only for very few surfactants. A much more general understanding of mixed adsorption layers was possible only with the first quantitative model based on the known properties of all components in the mixture. In [12,13], a set of equations was derived for binary surfactant mixtures which corresponds to a generalized Frumkin model:

$$\Pi = -\frac{RT}{\omega_0}\left[\ln\left(1 - \theta_1 - \theta_2\right) + \theta_1\left(1 - \frac{1}{n_1}\right) + \theta_2\left(1 - \frac{1}{n_2}\right) + a_1\theta_1^2 + a_2\theta_2^2 + 2a_{12}\theta_1\theta_2\right] \tag{1.22}$$

Here, θ_i is the relative surface coverage by component i. With some simplifications, a relationship can be obtained that can be easily applied to experimental data

$$\exp\overline{\Pi} = \exp\overline{\Pi}_1 + \exp\overline{\Pi}_2 - 1 \tag{1.23}$$

The dimensionless surface pressure for the mixture $\overline{\Pi} = \Pi\omega/RT$ and for the two components $\overline{\Pi}_1 = \Pi_1\omega/RT$ and $\overline{\Pi}_2 = \Pi_2\omega/RT$, respectively, are defined such that they correspond to the same concentrations c_i in the individual and mixed solutions, while the average molar area ω is given again as in Equation 1.14.

1.1.5 ADSORPTION MODELS FOR PROTEINS AND PROTEIN/SURFACTANT MIXTURES

The formation and properties of protein adsorption layers at the solution/air surface are quite different from those of surfactants. Proteins require a much larger space which can also decrease with the increasing surface coverage. In a review [14], the state of the art of adsorption models for proteins has been summarized.

The main feature in protein adsorption models is that we have to assume different conformations of adsorbed protein molecules which correspond to different molar areas. At low surface coverage or surface pressure, the protein molecule occupies the maximum area ω_{max}, while in a fully packed layer, the minimum area ω_{min} is required. With this physical picture, the following equation of state was derived [14]:

$$-\frac{\Pi\omega_0}{RT} = \ln(1 - \theta_P) + \theta_P(1 - \omega_0/\omega_P) + a_P\theta_P^2, \tag{1.24}$$

where a_P is the intermolecular interaction parameter and ω_0 is the molar area of a solvent molecule or a protein segment at the surface. The total adsorption of protein molecules in all n possible states $\Gamma_P = \Sigma_{i=1}^n \Gamma_{Pi}$ leads then to the surface coverage of $\theta_P = \omega_P\Gamma_P = \Sigma_{i-1}^n \omega_i\Gamma_{Pi}$. ω_P is the average molar area of adsorbed proteins in a surface monolayer. With a further increase in the protein concentration, for many proteins, the formation of bilayers (or multilayers) is observed. Theoretical models for a multilayer adsorption can be derived as was shown in [14], but it is not discussed in detail here.

For mixtures of a protein with surfactants, we must discriminate between ionic and nonionic surfactants. While nonionic surfactants can interact with the protein only via their hydrophobic chains, and therefore, adsorb at an interface mainly only in a competitive way to the protein molecules, ionic surfactants can interact with the protein via electrostatic and hydrophobic interactions [15]. The electrostatic interaction results in the formation of complexes which have a stronger surface active (more hydrophobic) than the original protein. Hence, the situation at the surface of a mixed solution is such that the complexes adsorb in competition to free surfactant molecules.

An equation of state for protein/nonionic surfactant mixtures can be easily derived when we assume $\omega_0 \cong \omega_S$ [16]:

$$-\frac{\Pi\omega_0^*}{RT} = \ln(1 - \theta_P - \theta_S) + \theta_P(1 - \omega_0/\omega_P) + a_P\theta_P^2 + a_S\theta_S^2 + 2a_{PS}\theta_P\theta_S, \tag{1.25}$$

where the indices S and P refer to the surfactant and protein, respectively. As one can see, all parameters in this theoretical model refer to the single component except for the parameter a_{PS} which describes the interaction between the protein and surfactant molecules. Note that the small differences between ω_0 and ω_S can be considered by using the averaged molar area

$$\omega_0^* = \frac{\omega_0\theta_P + \omega_{S0}\theta_S}{\theta_P + \theta_S}, \tag{1.26}$$

where ω_{S0} is the real molar area of the surfactant. More details on this model were discussed in [16].

For a mixture of a protein with an ionic surfactant, we finally end up with an equation of state identical to Equation 1.25 and obtain the subscript PS instead of P [17]. The corresponding adsorption isotherm for protein/surfactant complexes in state 1 reads [17]:

$$b_{PS}\left(c_P^m c_S\right)^{1/(1+m)} = b_{PS} c_P^{m/(1+m)} c_S^{1/(1+m)} = \frac{\omega \Gamma_1}{\left(1 - \theta_{PS} - \theta_S\right)^{\omega_1/\omega}} \exp\left[-2a_{PS}(\omega_1/\omega)\theta_{PS} - 2a_{SPS}\theta_S\right]. \qquad (1.27)$$

Similar isotherm equations can be obtained for any of the possible i states.

1.1.6 SELF-ASSEMBLY OF PARTICLES AT LIQUID INTERFACES

Partly hydrophobic particles also self-assemble at liquid interfaces. The pioneering work for such systems was done by Binks et al. [18]. Aveyard et al. [19] demonstrated that emulsions can be stabilized solely by particles. Interacting with surfactants, particles can be even tailored to self-assemble at interfaces and hinder any droplet/bubble coalescence. Thus, particle/surfactant mixtures are most suitable for the stabilization of liquid films and liquid dispersed systems such as emulsions and foams [20,21].

The fact that particles can stabilize liquid dispersed systems has been known for more than a hundred years since Pickering produced emulsions stabilized by mustard seed particles [22]. The use of particles as a stabilizer became important almost a century after the findings of Pickering when particles, in particular nanoparticles, at interfaces, turned out to be extremely interesting from a scientific and technological point of view. The special attention of scientists and engineers was caused by an increasing application in oil recovery, food processing, cosmetics, and membrane-based separation and purification techniques.

The attachment of particles at liquid/air or liquid/liquid interfaces is determined by their hydrophilic/hydrophobic balance [23]. The gain in free energy ΔE of a particle of radius R attaching at a planar interface, is given by [24]:

$$\Delta E = -\pi R^2 \gamma (1 - \cos\theta)^2, \qquad (1.28)$$

where θ is the contact angle of the particle at the interface and γ is the interfacial tension between the two fluids.

The gain in free energy ΔE depends strongly on the wettability of the particle. Thus, the contact angle θ and the radius of the particle R are key parameters, leading to very strong attachment of the particles at the interface, sometimes called irreversibly absorbed. The attachment energy can exceed several orders of magnitude of the thermal energy kT [25].

It is worth noting here that the behavior of liquid interfaces covered by particles is essentially different from those covered by surfactant molecules. Surfactants form a dynamic adsorption layer and are subjected to fast exchange processes between the bulk phases and the fluid interface, in contrast to nanoparticles [23,24].

An overview on particle loaded liquid interfaces and the effect of added surfactants on the thermodynamic and nonequilibrium properties was given very recently in [26] so that we do not need to discuss further details.

1.1.7 SURFACE TENSION ISOTHERMS FOR SELECTED SURFACTANTS, PROTEINS, AND THEIR MIXTURES

This paragraph is dedicated to the demonstration of the suitability of the discussed adsorption models for the quantitative description of experimental data. The first example is based on the historical data of Frumkin from 1925 [7] concerning medium chain fatty acids at the water/air interface. These data were the reason for Frumkin to search for an improvement of the Langmuir adsorption model in order to better describe the measured isotherms. Figure 1.3 shows the measured equilibrium surface tensions for hexanoic acid solutions in a broad concentration range. These data were compared by the Langmuir and Frumkin model and one can easily see that the Frumkin isotherm is in perfect agreement with the experimental data.

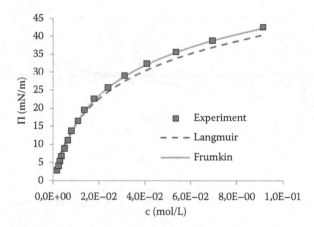

FIGURE 1.3 Surface tension isotherm of hexanoic acid in 0.07 n HCl; symbols ■—measured values taken from [7], dotted line—Langmuir model, solid line—Frumkin model.

In [1], a large number of additional examples were presented and it was shown that the Langmuir isotherm, although rather simple in its structure, agrees very well with the experimental data measured, for example, for the members of the homologous series of alkyl dimethyl phosphine oxides. For another class of nonionic surfactants, the oxethylated alcohols, the Reorientation Model, given by the set of Equations 1.14 through 1.17, allows a perfect description of experimental isotherms.

Figure 1.4 shows as an example of the data for a mixed surfactant system. The experimental equilibrium surface tension isotherms for the individual $C_{12}EO_5$ and $C_{14}EO_8$ and their mixture taken at a mixing ratio of 3:1 [27] is shown. The theoretical isotherms were calculated from the equations given above by Equations 1.14 through 1.17, using the following physical values [27]:

$C_{12}EO_5$ (component 1): $\omega_{10} = 4.2 \cdot 10^5$ m^2/mol, $\omega_{12} = 1.0 \cdot 10^6$ m^2/mol,
$\quad \alpha_1 = 2.5$, $\varepsilon_1 = 0.006$ m/mN and $b_1 = 5.1 \cdot 10^3$ m^3/mol;
$C_{14}EO_8$ (component 2): $\omega_{20} = 5.7 \cdot 10^5$ m^2/mol, $\omega_{22} = 1.0 \cdot 10^6$ m^2/mol,
$\quad \alpha_2 = 2.8$, $\varepsilon_2 = 0.007$ m/mN and $b_2 = 1.0 \cdot 10^5$ m^3/mol.

As one can see, the model provides a perfect agreement with the experimental data.

Another example for a mixed system is shown in Figure 1.5 [28]. This case corresponds to a mixture of the protein lysozyme with the anionic surfactant sodium dodecyl sulfate (SDS). The

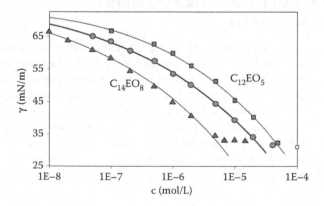

FIGURE 1.4 Equilibrium surface tension isotherms for solutions of $C_{12}EO_5$ (■) and $C_{14}EO_8$ (▲), and for a 3:1 mixture of $C_{12}EO_5/C_{14}EO_8$ (●); the theoretical curves were calculated with the proposed model using the parameters given in the text.

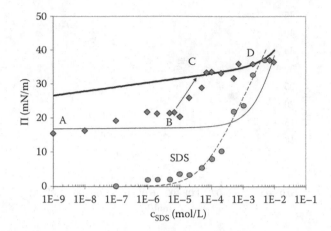

FIGURE 1.5   Surface pressure isotherm for aqueous solutions of SDS (●) and lysozyme/SDS mixtures at a fixed lysozyme concentration of 7×10^{-7} mol/L (◆), lines are theoretical dependencies explained in the text.

isotherm of the pure SDS is also shown in this figure. At pH 7, one lysozyme molecule has 8 positive net charges. Due to electrostatic interactions between the protein and surfactant molecules, complexes are formed which are more hydrophobic than the original protein molecules. Therefore, these complexes have higher surface activity than the native lysozyme. The curves were calculated from the model given above by Equations 1.25 through 1.27 [17]. The thin solid line corresponds to a model without any complex formation, that is, assuming that the surface activity of the protein remains the same in the mixed solutions. The bold solid line corresponds to the adsorption of complexes formed by one lysozyme molecule and eight SDS molecules having a surface activity ten times higher than that of the native protein. The shown theoretical curves in Figure 1.5 reproduce the experimental findings very well. The letters given in the graph mark some important changes in the composition of the adsorption layer formed at the solution/air interface. In the concentration range between points A and B, the effect of electrostatic bonding of SDS to the protein is more or less negligible. When the SDS concentrations c_S exceeds 10^{-6} mol/l, the interionic bonding of SDS becomes significant and the hydrophobicity of the complex increases (range between the points B and C). This finally results in an increased adsorption activity (range between points C and D).

1.2   EXPERIMENTAL METHODS FOR ADSORPTION LAYER CHARACTERIZATION

The knowledge of adsorption layer properties at the air/solution interface is of key importance for understanding the formation and stability of foams and foam films. The major applied experimental methods may be systematized as providing equilibrium and dynamic mean field characteristics (e.g., tensiometric methods) or elucidating details of the structural features and the adjoining structure-properties relationships (advanced optical instrumentation, neutron, and x-ray scattering techniques). In particular, for liquid/fluid interfaces, there are quite a number of tensiometry tools.

This section gives only a brief overview of the most frequently used or most easily applied methods based on single drop and bubble surfaces which was summarized recently in [29] and on more general methods in [30]. In addition, some selected optical methods are reviewed here briefly.

1.2.1   SURFACE AND INTERFACIAL TENSION

For understanding the formation of foams and foam films, dynamic surface tensions are essential and provide important insight about the effect of surface active compounds in view of their potential as foaming agents. Although these measurements do not provide a direct measure of the number of amphiphilic molecules at the solution surface, they allow to determine accurately the amount of adsorbed surfactants as a function of concentration and time. This is due to Equation 1.7, which gives a relationship between surface tension and surface concentration (adsorption).

Classical methods, such as ring and plate tensiometry, are widespread in many laboratories. However, these methods actually have quite a number of drawbacks, however, due to the unavailability of international norms, they are still often used. The capillary rise method and the drop volume tensiometry are suitable techniques for the analysis of the surface behavior of aqueous solutions. In many places, however, these methods have been or are about to be replaced by much more efficient, powerful, and very accurate techniques, which we want to introduce here briefly. Here, we consider the maximum bubble pressure (MBPT) and capillary pressure tensiometry (CPT), and the drop/bubble profile analysis tensiometry (PAT) for which detailed descriptions can be found in [29].

The MBPT counts as the most reliable methodology for dynamic surface tension data at extra short adsorption times. Using a suitable experimental protocol, it is possible to measure values for adsorption times even shorter than 10^{-3} s [31]. On the other hand, due to the specific construction for short adsorption times, most of the MBPT instruments are not suitable for long adsorption times so that this method typically has to be complemented by a second technique more suitable for much longer adsorption times. From this point of view, the MBPT and PAT complement each other perfectly and provide a total adsorption time range from milliseconds up to hours [32].

The scheme given in Figure 1.6 shows the main elements of a modern MBPT. The air is pumped through a filter, a gas flow capillary, and a gas reservoir into a capillary at the tip of which the bubbles are formed. The flow capillary is required to measure and control the gas flow rate. A second pressure sensor is mounted to the gas reservoir to measure the change of pressure in the growing gas bubble. All parts of the measuring procedure are computer controlled and the experimental data are also registered and analyzed by the PC. The various details and the optimum parameters of the many elements in the instruments are also described in a book chapter [33].

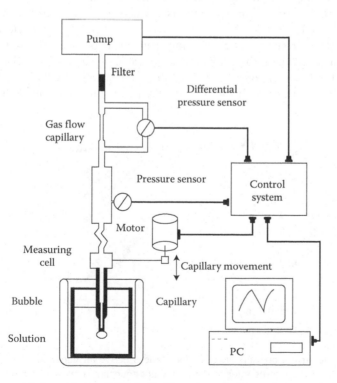

FIGURE 1.6 Scheme with the main elements of a modern MBPT.

The pressure P measured with the pressure sensor is a direct measure for the surface tension γ via the simplest form of the Laplace equation

$$\gamma = \frac{r_0}{2} Pf. \tag{1.29}$$

Here, r_0 is the known capillary radius and f is a correction factor which accounts for the deviation of the bubble from a spherical shape. This correction factor f is needed for capillaries with a radius $r_0 > 0.1$ mm when the deviation of the bubble from sphericity becomes larger than the accuracy of the pressure measurement. For a sufficiently slow bubble formation process, P contains only the capillary and hydrostatic pressure and, therefore, the surface tension can be easily obtained.

However, for fast growing bubbles, the measured total pressure P contains more components. The main part is the pressure in the measuring system P_s, which is proportional to the capillary pressure $P_c = 2\gamma/r_0$. In addition, we have the hydrostatic liquid pressure $P_H = \Delta\rho gH$ and the excess pressure P_d caused by dynamic effects. In [34], it was demonstrated that this pressure component includes the aerodynamic resistance of the capillary and viscous and inertia effects in the liquid. Note, $\Delta\rho$ here is the difference between the densities of liquid and gas, g is the gravity constant, and H is the immersion depth of the capillary into the liquid. Thus, we get the respective equation for calculating the dynamic surface tension from the measured pressure

$$\gamma = f \frac{r_0 (P_s - P_H)}{2} - \Delta\gamma_a - \Delta\gamma_v. \tag{1.30}$$

The term $\Delta\gamma_a$ describes the aerodynamic resistance of the capillary, while $\Delta\gamma_v$ reflects the viscous and inertia effects in the liquid. Both terms are negligible for slow bubble formation but have significant values in measurements at very short adsorption times.

An example for the application of the MBPT is given below in Figure 1.7 showing the dynamic surface tensions for different $C_{14}EO_8$ solutions all above the critical micellar concentration (CMC = 1.1×10^{-5} mol/L) as a function of the effective surface age. This effective or adsorption time has been calculated from the bubble life time using an algorithm discussed in [35].

From the data, it becomes clear that the presence of micelles significantly accelerates the adsorption process, although the monomer concentration remains almost constant. The accelerated adsorption provides information on the micelle formation/dissolution rate constant, as was quantitatively analyzed in [31].

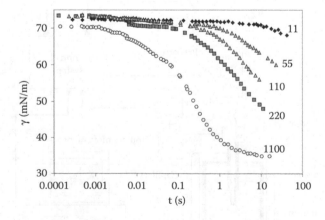

FIGURE 1.7 Dynamic surface pressure as a function of the effective surface lifetime for five different $C_{14}EO_8$ solutions; the numbers at the curves refer to the surfactant concentrations given in µmol/L. (According to Fainerman, V.B. et al. *J Colloid Interface Sci* 2006; 302: 40–46 [31].)

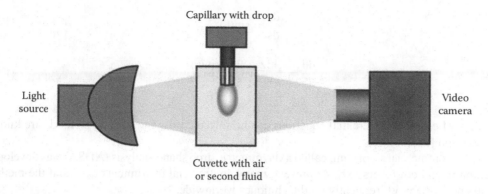

Capillary with drop

Light
source

Video
camera

Cuvette with air
or second fluid

FIGURE 1.8 Schematic of the drop and bubble profile analysis tensiometry (PAT).

The method that best complements the MBPT is the drop and bubble profile analysis tensiometry (PAT), which provides tension data for the time interval from a few seconds up to hours or even days.

The main principle consists of acquiring the image of a drop or bubble, extracting the shape coordinates from the image, and fitting it with the Gauss Laplace equation of capillarity (Figure 1.8).

This equation has the general form

$$\gamma\left(\frac{1}{R_1}+\frac{1}{R_2}\right)=\Delta P \tag{1.31}$$

where R_1 and R_2 are the two principal radii of interface curvature and ΔP is the pressure difference across the interface. Equation 1.31 can be transferred into a set of three first-order differential equations expressed by the geometric parameters of the drop/bubble profile, that is, the arc length, s, and the normal angle, θ, between the drop radius and the z-axis, as shown in Figure 1.9:

$$\frac{dx}{ds}=\cos(\theta) \tag{1.32}$$

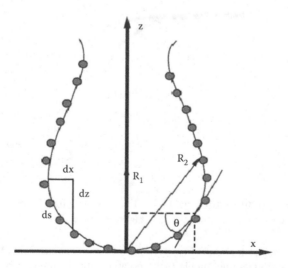

FIGURE 1.9 Definition of coordinates of a drop/bubble profile to be used for solving the Gauss Laplace equation.

$$\frac{dz}{ds} = \sin(\theta) \tag{1.33}$$

$$\frac{d\theta}{ds} = \pm \frac{\beta z}{b^2} + \frac{2}{b} - \frac{\sin(\theta)}{x}. \tag{1.34}$$

This set of ordinary differential equations can be solved easily, and various methods are known for its solution [36].

The first commercial algorithm, called axisymmetric drop shape analysis (ADSA) was developed by Neumann and coworkers [37]. At present, many commercial instruments exist and the method became one of the most frequently used techniques worldwide.

In order to demonstrate that these two methods, MBPT and PAT, are complementary, we show the surface tension data for two aqueous Triton X-45 solutions. Except in the range of short times for the PAT, the measured data are lower than expected. The reason for this discrepancy is the remarkable initial load of a drop in PAT when starting the experiment at time t = 0. The solid curves, calculated for a diffusion controlled adsorption based on a Reorientation Model, clearly demonstrate an excellent fitting, that is, an excellent correspondence of the data measured with the two methods.

The third instrumentation we want to briefly discuss here is the capillary pressure tensiometry (CPT). This method is similar to the MBPT, except that the pressure is recorded over the whole period of time for a single bubble or drop in contrast to only the maximum value used in the MBPT. The CPT can be applied according to various protocols as was explained in [38] (Figure 1.10).

The measuring set ups for the CPT are based on two generally different strategies—an open or a closed measuring cell. The closed cell geometry is shown schematically in Figure 1.11. The scheme of the MBPT, in contrast, has an open cell geometry, as one can see in the scheme of Figure 1.6.

The CPT is a method with a major advantage—it works for any liquid/fluid interface and also for systems composed of two liquids with the same density. It is also the method of choice for experiments under microgravity [41].

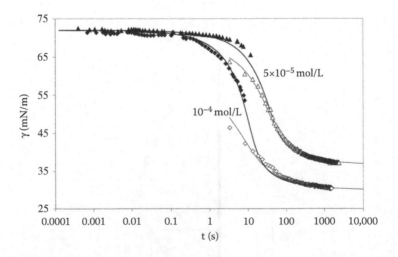

FIGURE 1.10 Dynamic surface tensions of two Triton X-45 solutions measured by the maximum bubble pressure (filled symbols) and emerging bubble (open points) methods; the curves are calculated from the Reorientation Model given by Equations 1.14 through 1.16. (According to Fainerman, V.B. et al. *Colloids Surfaces A* 2009; 334: 8–15 [39].)

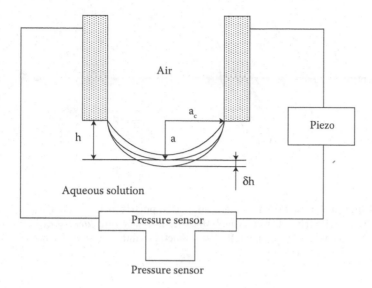

FIGURE 1.11 Schematic of a capillary pressure tensiometer based on a so called closed measuring cell. (Adapted from Kovalchuk, V.I. et al. In: Miller, R., Liggieri, L. (Eds.), *Bubble and Drop Interfaces*, Vol. 2, Progress in Colloid and Interface Science, Brill Publ: Leiden, 2011, pp. 143–178 [40].)

1.2.2 FURTHER METHODS TO CHARACTERIZE LIQUID INTERFACES

Foam systems stabilized by complex surfactant/surfactant (surfactant-mixture), polymer/surfactant, protein/surfactant or nanoparticles/polymer aqueous solutions have peculiar properties and stability behavior due to the increasing importance of the specific adsorption layer structure at the air-solution interface. In such cases, there is a great potential for complementing the tensiometric and rheological studies with matching investigations on the time evolution of these structures and the adjoining bulk/interface property relationships. During the last two decades, several new methodologies for characterization of adsorption layers air/solution interfaces have been developed. They are based on advanced optical, neutron, and x-ray scattering techniques.

The most important optical methods include linear (ellipsometry) and second order nonlinear optical spectroscopies (vibrational sum frequency generation and second harmonic generation) [42,43].

Ellipsometry has been widely applied for the study of adsorption of materials at the solid/fluid interface. However, it is only recently being applied more extensively for the investigation of fluid/fluid interfaces [43–46]. It is based on the fact that reflection of polarized light at an interface depends on its state of polarization. Light polarized perpendicular to the plane of reflection (s-polarized) is reflected differently from light polarized in the plane of reflection (p-polarization). The instrumentation provides information about some aspects of the structure at liquid interfaces involving interpretation in terms of interfacial thickness and refractive index [47].

A typical scheme of the instrumentation is presented in Figure 1.12 [43]. In this PCSA (polarizer, compensator, sample, analyzer) configuration, the orientation of the angles of polarizer and compensator are chosen in such a way that the elliptically polarized light is completely linearly polarized after it is reflected off the sample. Ellipsometric measurements provide the changes in the reflectance through the amplitude angle Ψ and the phase difference Δ between the parallel and perpendicular components of a polarized light beam upon reflection from the sample (Figure 1.12).

These quantities are functions of the complex reflectivity coefficients r_p and r_s for $\hat{p}$- and $\hat{s}$-polarized light beams, with amplitudes E^i, E^r, and phases δ^i, δ^r, respectively, and $\Delta = \left(\delta_p^r - \delta_s^r \right) - \left(\delta_p^i - \delta_s^i \right)$.

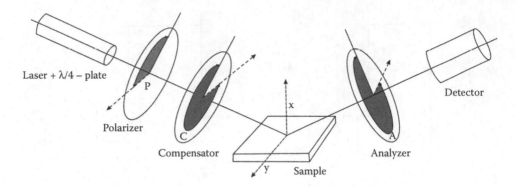

FIGURE 1.12 Ellipsometer in PCSA configuration (P—polarizer, C—compensator, S—sample, A—analyzer). According to (Reprinted from *Novel Methods to Study Interfacial Layers*, Vol. 11, Studies in Interface Science. Motschmann, H., Teppner, R., Ellipsometry in interface science, pp. 1–42, Copyright 2001, with permission from Elsevier [43].)

$$\tan \Psi = \frac{\left| E_p^r \right| / \left| E_p^i \right|}{\left| E_s^r \right| / \left| E_s^i \right|} \qquad (1.35)$$

$$\tan \Psi e^{-i\Delta} = \frac{r_p}{r_s} \qquad (1.36)$$

Examples of typical results from ellipsometric studies are shown in Figure 1.13 [46].

From the above measurements and using appropriate theoretical model assumptions, important adsorption layer characteristics may be calculated as presented in Figure 1.14a and b.

Usually, ellipsometry applied to adsorption layers of nonionic soluble surfactants directly yields the surface excess and is often used as an alternative and/or supplement of surface tension measurements [43,48]. Ellipsometry applied to adsorption layers of ionic soluble surfactants, however, does not determine the surface excess. The ellipsometric signal may show a nonmonotonicbehavior which

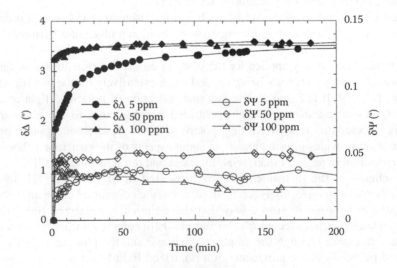

FIGURE 1.13 Time evolution of ellipsometric angles, $\delta\Delta = \Delta - \Delta_{\text{bare substance}}$ and $\delta\Psi = \Psi - \Psi_{\text{bare substance}}$ during the adsorption of block copolymer $E_{106}B_{16}$ at air/water interface. $E_{106}B_{16}$ is added at t = 0. (According to Blomqvist, B.R. et al. *Langmuir* 2005; 21: 5061–5068 [46].)

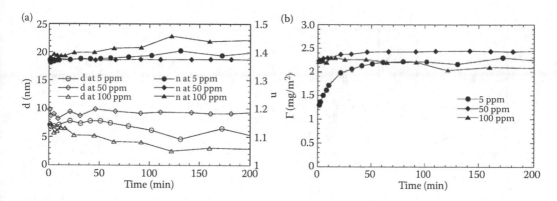

FIGURE 1.14 (a) Average thickness (d) and average refractive index (n) of the polymer film, at bulk solution concentrations of 5, 50, and 100 ppm of $E_{106}B_{16}$, as calculated from the values of $\delta\Delta$ and $\delta\Psi$ on Figure 1.13. (b) Adsorbed amounts (Γ) of $E_{106}B_{16}$, as calculated using the n and d values on panel (a). (According to Blomqvist, B.R. et al. *Langmuir* 2005; 21: 5061–5068 [46].)

is caused by a redistribution of the ions between the Stern layer and the diffuse double layer. The data analysis within the classical model of a charged double layer thus results in an estimation of the prevailing ion distribution at the interface [49].

The major advantages of the ellipsometric approach are: (i) it does not require reference measurements; (ii) it is extremely sensitive to very thin layers (less than the thickness of a monolayer); and (iii) it is fast and gets the full spectrum in a few seconds. The major limitation is that it does not directly give the physical parameters of the sample (adsorption layer thickness and refractive index), but requires a theoretical model-based analysis. The flexibility of the fluid boundary also complicates the model scheme. Additionally, the adsorbed layers are usually very thin, transparent, lie on an interface between nonadsorbing media (air and water), and the refractive index contrast between the adsorbed layer and the aqueous substrate is generally much weaker than at the solid/liquid interface. This requires the ellipsometry to usually be combined with other interface sensitive techniques [48,50,51].

Another option is the application of second order nonlinear optical (NLO) spectroscopy, based on the nonlinear response of the dielectric polarization of media to the light electric field [42,52–55]. For example, vibrational sum frequency generation spectroscopy (VSFG) is an NLO technique where two laser beams mix at the interface and generate an output beam with a frequency equal to the sum of the two input frequencies. The two input beams access the surface and the output beam is picked up by a detector. Usually, one of the beams is held at a constant frequency and the other is a laser. When the latter is adjusted, the coincidence between the photon energy and the energy of the molecular vibrational mode results in a resonant enhancement of the VSFG response, because at the interface of two isotropic phases, the inversion symmetry is broken and the VSFG signal is produced within the interfacial region. The inherent surface sensitivity makes this approach a suitable method for studying the vibrational spectroscopy of molecules at interfaces. In the graphical analysis of the output intensity against the wave number, this is represented by peaks. Thus, the VSFG signal contains information about molecular orientation and arrangement. By using different combinations of polarization for the input beams, important features about the distribution of the adsorbed species and their orientation can be retrieved. Since VSFG detects vibrational modes which are rather localized to specific groups of atoms within the molecules, the relative orientation of different groups within the same molecule can be obtained as well.

Second harmonic generation (SHG, also called frequency doubling) is an NLO technique in which photons with the same frequency interacting with the substrate are combined to generate new photons with twice the energy, twice the frequency, and half the wavelength of the initial photons. SHG, as an even order nonlinear optical effect, is only allowed in media without inversion symmetry. It is a special case of SFG.

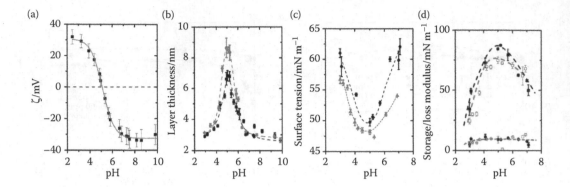

FIGURE 1.15 (a) Zeta potential of BLG as a function of solution pH. (b) Thickness of BLG layers adsorbed to the air—water interface against the solution pH, determined by ellipsometry. The concentrations of BLG solutions are 15 (squares) and 54 μM (circles). (c) Surface tension for 10 (circles) and 50 μM (triangles) BLG aqueous solutions as a function of pH. Each data point corresponds to a measured value after 30 min. (d) Interfacial dilational elasticity E′ (squares) and viscosity E (circles) of BLG. Filled symbols correspond to a BLG concentration of 10 μM while open symbols correspond to 50 μM. (According to Engelhardt, K. et al. *Langmuir* 2013; 29: 11646–11655 [54].)

Molecular orientation can be probed by detecting the polarization of the second harmonic signal, generated from a polarized beam. For substrates with bulk inversion symmetry, SHG is only allowed at interfaces. SHG provides an intrinsic surface specificity and the analysis of polarization-dependent SHG measurements yields the symmetry of the interface, the number density, and the orientation of the amphiphile. These data can be used to assess some peculiar features of adsorption layers of soluble surfactants [42]. In the case of embedded interfaces, this technique can also be applied, particularly when an intense laser is used which is capable of penetrating deep into the material and no direct contact with the sample is needed.

The above discussed experimental methodologies may be applied to complement the properties of foams and interfaces with the information on several length scales [54]. The macroscopic properties of aqueous β lactoglobulin (BLG) foams are juxtaposed to the molecular structure and interactions at the solution/air interface. In Figures 1.15 and 1.16, the results of an example for the use of such a combined approach are presented. Through the combination of interfacial sensitive methods (sum frequency generation, ellipsometry, bubble profile analysis tensiometry, and surface dilational rheology) and foam rheology measurements, three regions of different pH are identified, where both macroscopic and microscopic properties change and are interconnected. At the isoelectric point of the interface (pH~5, Figure 1.15a), the interfacial layers of BLG carry no net charge and the formation of disordered and agglomerated BLG multilayers is presumed, resulting in a maximum in the surface coverage of BLG.

For alkaline and acidic pH conditions, the protein—protein interactions transform from attractive to a highly repulsive regime which leads to the formation of BLG monolayers and to highly polar ordered water molecules. Changes in the layer thickness (Figure 1.15b) and the protein—protein interaction (attractive or repulsive) are shown to modify the interfacial dilational elasticity E′ which also exhibits a maximum at pH~5 (Figure 1.15d). It is established that the thick and disordered adsorption layers result in the highest foam stability because the gas bubbles are protected by the proteins arranged at the air/solution interface, while thin and more ordered layers with strong electrostatic repulsions result in a lower resistance to mechanical disturbances and less stable foams.

One of the most effective techniques for examining complex aqueous systems and interfaces is specular neutron reflectometry [48,56–59]. This instrumentation is based on the determination of the specular neutron reflectivity, defined as the ratio of the intensities between reflected and incoming beams:

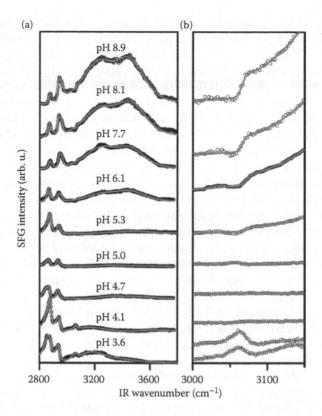

FIGURE 1.16 (a) Vibrational SFG spectra in the region of CH (2800–3100 cm^{-1}) and OH stretching vibrations (3000–3800 cm^{-1}) for BLG adsorbed to the air–water interface. Spectra are recorded at different pH. (b) Magnification of the spectra in (a) showing changes in the polarity of aromatic CH stretching band at 3050 cm^{-1} in more detail. (According to Engelhardt, K. et al. *Langmuir* 2013; 29: 11646–11655 [54].)

$$R = I_{ref}/I_{inc} \approx \left(16\pi^2/Q^2\right)\left|\varrho(Q)\right|^2. \tag{1.37}$$

R is a function of the wave vector normal to the interface ($Q = (4\pi/\lambda)\sin\theta$, where λ is the neutron wavelength and θ is the grazing angle of incidence), and it is related to the density profile in the direction normal to the interface:

$$\varrho(Q) = \int_{-\infty}^{\infty} e^{-iQz}\rho(z)\,dz; \quad \rho(z) = \sum_{i=1}^{m} b_i n_i(z), \tag{1.38}$$

where b_i and n_i are the scattering lengths and the number densities of the i-th element in a given volume containing m various components.

The basic advantage of this approach is that the scattering length density at the interface can be manipulated using hydrogen and deuterium isotopic substitution. Neutrons are scattered by atomic nuclei and different nuclei have different scattering properties. Particularly important is the fact that hydrogen (H) and deuterium (D) have neutron scattering lengths that are different in sign and magnitude: $b_H = -3.74 \times 10^{-5}$ Å and $b_D = +6.67 \times 10^{-5}$ Å. The large difference between the scattering lengths of H and D leads to wide differences in the scattering lengths of protonated and deuterated substrates. Many hydrogen containing materials, including water and hydrocarbons, have negative scattering length densities. Thus, it is possible to add D_2O to H_2O so that its scattering length density is adjusted to be the same as that of air, that is, zero. There is no reflection from the air/water interface for water of this adjusted composition (0.081 mol fraction of D_2O) and it is called null reflecting water (NRW). For a solution of deuterated species in NRW, reflection will only occur if there is adsorption of the deuterated material at the interface. It is, therefore, possible by varying the isotopic composition, to obtain information about the adsorption of particular surface active agent in a mixed system. For instance, in the case of an aqueous solution containing more than one species adsorbed at the interface, as in polymer/surfactant mixtures, the coverages of the individual components can be determined by measuring the reflectivity for the deuterated surfactant in NRW in the presence of protonated polymer and repeating the measurement with deuterated polymer and protonated surfactant. The obtained reflectivities come almost entirely from the deuterated species since the scattering length density of the protonated species is negligible. Thus, the adsorbed amounts of both polymer and surfactant can be obtained separately.

The neutron reflectivity measurements give more information than just adsorbed amounts, providing access to the structure of the surface layer. Using a direct method of analysis based on a suitable theoretical approximation, the volume fraction distributions of individually deuterium labelled components can be obtained. Besides, neutrons are highly penetrating and typically nonperturbing. This allows for great flexibility in sample environments and the use of delicate sample materials (e.g., biological specimens).

The disadvantages of neutron reflectometry include: (i) the high cost of the required infrastructure; (ii) some materials may become radioactive upon exposure to the beam; and (iii) the relatively lower flux and higher background of the technique (as compared to x-ray reflectivity, for example) and hence reduce the measurement resolution.

The x-ray scattering approach contains an inherent combination of much higher intensity and much lower background scattering. This makes the signal-to-noise ratio in x-ray reflection several orders of magnitude better than for neutron refractometry. But because it lacks the selectivity afforded by contrast variation, it has been less widely used for the study of soluble species adsorbed at the air/water interface [48]. However, when there is an in-plane order of the monolayer, as is common for insoluble monolayers, x-ray grazing incidence diffraction becomes a uniquely powerful technique [60–62]. Thus, it can give, for example, the symmetry of the 2-D lattice and determine the surface lattice parameters.

X-ray radiation has a generally rather large penetration depth as compared to electrons. Due to this property, x-ray diffraction is not surface sensitive. So, for the studies of surface structure at air/solution interfaces, the solution is to use Grazing Incidence Diffraction (GIXD) [61,62]. GIXD measurements are performed at very low incident angles to maximize the signal from the thin adsorption layers. Thus, for the GIXD, the incident and diffracted beams are made nearly parallel. The stationary incident beam makes a very small angle with the sample surface (typically 0.3–3°), which increases the path length of the x-ray beam through the interfacial film. This helps to amplify the diffracted intensity, while at the same time, reducing the diffracted intensity from the substrate. Overall, there is a considerable increase in the film signal-to-background ratio. Since the path length increases when the grazing incidence angle is used, the diffracting volume increases proportionally. This is the reason for the increased signal strength. The approach is particularly useful for complex systems containing large synthetic or natural polymers.

1.3   ADSORPTION DYNAMICS

This chapter aims at describing the fundamentals of adsorption kinetics of surfactants and proteins at liquid interfaces. The starting point of almost all old and new models is the work of Ward and Tordai [63] published in 1946. It proposes that the formation is mainly controlled by the diffusion of surfactant molecules toward the interface. Specific aspects for surfactants or proteins come into play when defining the equilibrium state, that is, defining the condition at an infinite time in form of the corresponding adsorption isotherm. In addition, some experimental techniques can be adequately described only if specific initial and boundary conditions are considered. This is true, for example, for the maximum bubble pressure or drop volume tensiometry, where the area of the interface is increasing during the whole measurement process. This section gives first a short general introduction into the physical picture of the adsorption layer formation. Then, the classical model of Ward and Tordai is described, including additional mechanisms for the transfer of molecules from the solution bulk into the adsorbed state. Then the situation for mixtures of a surfactant with other surfactants or with a protein is discussed. Finally, some characteristic examples are given.

1.3.1 GENERAL PHYSICAL PICTURE

Adsorption layers of surfactants are very dynamic objects with continuous fluxes of surfactants to and from the interface. Depending on the adsorption state, there is a net adsorption or a net desorption flux to or from the interface. Only in the adsorption equilibrium, that is, when the adsorption Γ reaches the equilibrium value Γ_0, are both fluxes the same (see Figure 1.17). Thus, we should call the situation dynamic equilibrium.

For freshly formed interfaces, the adsorbed amount is typically zero, $\Gamma(t = 0) = 0$, that is, $\Gamma < \Gamma_0$ and the adsorption process starts until the value of Γ_0 is reached. In the opposite case, that is, when more molecules are at the interface than are required for the adsorption equilibrium, molecules have to leave the interface, that is, they have to desorb.

In the past, there were generally two ideas to describe the dynamics of adsorption layer formation, depending on which of the two processes is the slower: the diffusion as the transport process of molecules from the bulk to the interface, or the step of molecules from the dissolved to the adsorbed state. The diffusion controlled model assumes that the diffusional transport of interfacial active

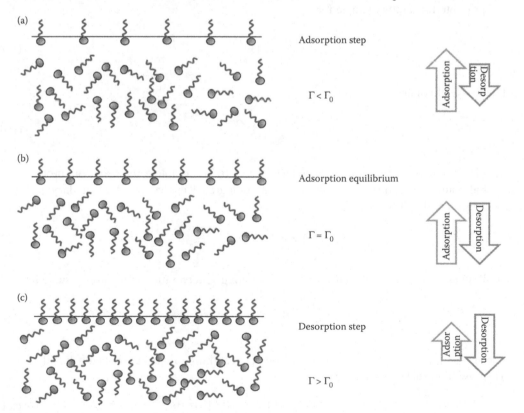

FIGURE 1.17 Adsorption fluxes to and from the interface, depending on the adsorption state; (a) for $\Gamma < \Gamma_0$, there is a net adsorption flux to the interface; (b) when the adsorption layer is at equilibrium, the adsorption and desorption fluxes are identical; (c) for $\Gamma > \Gamma_0$, there is a net desorption flux from the interface.

molecules from the bulk to the interface is the rate-controlling process. In contrast, the so called kinetic controlled model assumes that the transfer mechanisms of molecules from the solution bulk to the adsorption layer is slow and hence is the rate determining mechanism.

Figure 1.17 shows adsorption fluxes to and from the interface, depending on the adsorption state; (a) for $\Gamma < \Gamma_0$ there is a net adsorption flux to the interface; (b) when the adsorption layer is at equilibrium the adsorption and desorption fluxes are identical; and (c) for $\Gamma > \Gamma_0$, there is a net desorption flux from the interface.

Transport in the solution bulk is controlled by the diffusion of surfactant molecules in absence of any liquid flow or convection. The transfer of molecules from the subsurface, the liquid layer adjacent to the interface, to the interface itself happens without transport and is based just on molecular steps, such as rotations or flip-flops.

1.3.2 CLASSICAL ADSORPTION KINETICS MODELS

As mentioned above, since the work of Ward and Tordai in 1946 [63], all adsorption kinetics models are directly or indirectly linked to this basic work. Below we show the main features of this model and how more complicated models are derived from it.

1.3.2.1 The Diffusion Controlled Adsorption Model

The diffusion controlled adsorption model of Ward and Tordai [63] assumes that the transfer of molecules from the subsurface to the interface is much faster than the transport from the bulk to the subsurface by diffusion. In the absence of any flow in the liquid bulk, Fick's second diffusion law can be presented in a rather simple form

$$\frac{\partial c}{\partial t} = D \frac{\partial^2 c}{\partial x^2}. \tag{1.39}$$

The first Fickian law

$$\frac{\partial \Gamma}{\partial t} = D \frac{\partial c}{\partial x}, \tag{1.40}$$

serves as a boundary condition at the interface $x = 0$. To complete the transport problem, one additional boundary condition and an initial condition are necessary. Typical boundary conditions are: an infinite bulk phase

$$\lim_{x \to \infty} c(x,t) = c_0 \text{ at } t > 0 \tag{1.41}$$

or a bulk phases of limited depth h where a diminishing concentration gradient is assumed

$$\frac{\partial c}{\partial x} = 0 \text{ at } x = h; \quad t > 0. \tag{1.42}$$

The usual initial condition is a homogenous concentration distribution and a freshly formed interface with no surfactant adsorption

$$c(x,t) = c_0 \text{ at } t = 0 \text{ and } \Gamma(0) = 0. \tag{1.43}$$

Any other initial condition according to the actual experimental conditions can be chosen, however, a simple analytical solution is not possible then. In many cases, an initial load of the interface is more realistic than a completely free interface. Hence, it is suitable to assume $\Gamma(0) = \Gamma_d \neq 0$, which leads to a modified version of the Ward and Tordai equation:

$$\Gamma(t) = \Gamma_d + 2\sqrt{\frac{D}{\pi}}\left[c_o\sqrt{t} - \int_0^{\sqrt{t}} c(0, t-\tau)d\sqrt{\tau}\right]. \qquad (1.44)$$

This equation was derived for flat interfaces. If the interface is curved, as it is the case in experiments using small single drops or bubbles, an additional term has to be added to the right-hand side [64]

$$\Gamma(t) = \Gamma_d + 2\sqrt{\frac{D}{\pi}}\left[c_0\sqrt{t} - \int_0^{\sqrt{t}} c_S(0, t-\tau)d\left(\sqrt{\tau}\right)\right] \pm \frac{c_0 D}{R}t \qquad (1.45)$$

When the radius of curvature R is sufficiently large, this additional term does not need to be considered.

1.3.2.2 Impact of the Subsurface Interface Transfer Mechanism

Besides the transport of molecules by diffusion in the solution bulk, an additional time step could be the transfer of molecules from the solution bulk into the adsorbed state at the interface. The most frequently used is the following rate equation

$$\frac{d\Gamma}{dt} = k_a c_0\left(1 - \frac{\Gamma}{\Gamma_\infty}\right) - k_d\frac{\Gamma}{\Gamma_\infty}, \qquad (1.46)$$

which in equilibrium corresponds to the Langmuir adsorption model. With a zero net flux ($d\Gamma/dt$) = 0, the Langmuir isotherm is obtained. The constants k_a and k_d are the adsorption and desorption rate constants. Other adsorption models, for example the Frumkin model, have been used instead of the simple Langmuir model, as was proposed by Fainerman [65] or MacLeod and Radke [66].

A more general model results when the transport by diffusion and the transfer of molecules from the solution to the interface and back are considered simultaneously. The first author to propose this model was Baret [67]. The coupling between the transfer mechanisms such as Equation 1.46 with the Ward and Tordai Equation 1.44 can be managed when assuming that the concentration in the subsurface $c(0,t)$ is the one that controls the transfer mechanism, that is, c_0 in Equation 1.46 has to be replaced by $c(0,t)$:

$$\frac{d\Gamma}{dt} = k_a c(0, t)\left(1 - \frac{\Gamma}{\Gamma_\infty}\right) - k_d\frac{\Gamma}{\Gamma_\infty}. \qquad (1.47)$$

The set of Equations 1.44 and 1.47 represents an integro-differential equation for which an analytical solution cannot be found. This set of equations has to be solved numerically as was shown in [68].

1.3.2.3 Molecular Reorientation or Aggregation within the Interfacial Layer

After Nikolov et al. [69] had discussed the possibility of coexisting states in surfactant adsorption layers, more publications were dedicated to his subject [70,71]. In 1996, Fainerman et al. proposed new thermodynamic models for adsorption layers comprised of surfactant molecules that can undergo molecular changes such as a change in orientation [9] or a two dimensional aggregation at liquid interfaces [72]. These new ideas have an immense impact on the adsorption dynamics. In the book published in 2001, groups of surfactants are discussed which follow one or the other adsorption mechanism [73].

In contrast to what was discussed above, any changes in orientation or formation of small aggregates (see Figures 1.1 and 1.2) are assumed to proceed much faster than the diffusion of surfactants in the bulk so that models based on such molecular processes are still diffusion controlled and no additional rate constant must be introduced. As shown in [74], any two-dimensional aggregation at the interface leads to an acceleration of the adsorption process as compared to the situation where no aggregates are formed, while the process of orientation changes decelerates the rate of adsorption.

1.3.2.4 Approximate Solutions

The Ward and Tordai Equation 1.37 requires some numerical efforts to be solved and compared to experimental data. For rough estimations, however, simple asymptotic solutions or approximations can be applied. In 1952, Sutherland [75] derived a first approximation, which is valid only for the linear adsorption isotherms. Later, Hansen [76] presented simple equations where the surface concentration $\Gamma(t)$ was expressed as a function of the dimensionless time $\Theta = Dt/(\Gamma_0/c_0)^2$. For short times, $\sqrt{\Theta} \leq 0.2$ was obtained

$$\frac{\Gamma(t)}{\Gamma_0} \approx 2\sqrt{\frac{\Theta}{\pi}} \left(1 - \frac{\sqrt{\pi\Theta}}{2} \pm \cdots\right), \tag{1.48}$$

while for sufficiently large times for $\sqrt{\Theta} \geq 5.0$, he got

$$\frac{\Gamma(t)}{\Gamma_0} \approx 1 - \frac{1}{\sqrt{\pi\Theta}} \left(1 - \frac{1}{2\Theta} \pm \cdots\right). \tag{1.49}$$

The simplest approximation is obtained when only the first term of Equation 1.48 is used

$$\Gamma(t) = 2c_0 \sqrt{\frac{Dt}{\pi}}. \tag{1.50}$$

This equation described, of course, only very roughly the course of the adsorption, that is, for a short time range. The limits of this and other approximate solutions were discussed in detail by Makievski et al. in [77].

1.3.2.5 Adsorption Kinetics from Micellar Solutions

Above the CMC of a surfactant, one could assume that the adsorption kinetics from such a solution remains identical to that right below the CMC. However, although the monomer concentration of the surfactant remains almost constant and micelles are not surface active and hence do not adsorb, the rate of adsorption is remarkably enhanced in the presence of micelles in the bulk. The reason is that because micelles are in a local equilibrium with the monomers and whenever surfactant molecules adsorb at the surface, micelles close to the surface tend to disintegrate so as to reestablish the local monomer-micelle equilibrium [78]. The more micelles that exist in the bulk, the faster is this local equilibration and the stronger is the support for the flux of monomers to the surface.

Seen from the other side, we can assume that measuring the adsorption kinetics of micellar solutions would allow understanding of not only the transport of monomers to the surface required to let them adsorb, but also to analyze which mechanism and with which rate constant the micelles disintegrate. This fact can be used to study the micelle kinetics of surfactants in solutions above the CMC. As an example, in [79] the rates of formation/dissolution of micelle of sodium dodecyl sulfate (SDS), various oxethylated alcohols (C_nEO_m), and their mixtures were studied.

Above the CMC of surfactant solutions, hydrophobic molecules can be solubilized in the core of the micelles. A prominent example is the kinetics of SDS solutions. This anionic surfactant tends to hydrolyzation in aqueous solution so that part of it turns into the homologous alcohol dodecanol. In solution, this long chain alcohol is extremely surface active and adsorbs at the air/solution surface about 500 times stronger than the SDS molecules. This leads to an adsorption kinetics which is essentially governed by the "impurity" dodecanol rather than by the main surfactant SDS. Above the CMC, however, SDS forms micelles and solubilizes the dodecanol, thus the latter is depleted from the solution and hence from the interface so that the adsorption layer properties are mainly controlled now by the SDS molecules. A remaining indirect effect of the dodecanol comes from its impact on the kinetics of the SDS micelles [80] slowing down the micellar disintegration process. In [78] and other overview articles [81,82], more details are given on the quantification of the impact of micelles on the dynamic surface properties.

1.3.2.6 Effect of Electric Charge on the Adsorption Kinetics of Ionic Surfactant

When surfactants are charged, the ions have to be transported by diffusion within an electric field the extension of which is determined by the surface charge. This surface charge and the electric potential distribution in the solution bulk are functions of time due to the process of adsorption at the surface. Surface charge/surface potential and potential distribution have to be taken into account when describing the adsorption process. Instead of the simple 2nd Fickian law, a modified equation must be used which considers the electric field in which the surfactant ions must be transported by diffusion [83]

$$\frac{\partial c}{\partial t} = D \frac{\partial}{\partial x} \left(\frac{\partial c}{\partial x} - c \frac{e}{kT} \frac{\partial \psi}{\partial x} \right), \tag{1.51}$$

where k is the Boltzmann constant, T is the absolute temperature, and e is the electric charge of an electron. The potential distribution $\psi(x)$ depends on the model of the electric double layer, as was shown in much detail in [84] or [85].

1.3.3 MECHANISMS FOR SURFACTANT MIXTURES

In contrast to the fundamental studies on surfactants, in practical applications, technical surfactants and their mixtures are used. Therefore, knowledge of the competitive adsorption of surfactant mixtures at liquid interfaces is required. There are different experimental conditions to form mixed surfactant layers, and the respective theories have to consider the corresponding initial and boundary conditions.

1.3.3.1 Mixed Nonionic Surfactant Solutions

The general model also typically starts from the Ward and Tordai equation. However, instead of Equation 1.44 for a single surfactant, a set of these equations is required, one for each of the r adsorbing compounds, while the initial and boundary conditions are taken with respect to the same physical reasoning for each component. As a result, we end up with a system of r integral equations which have to be solved simultaneously [86]. For a flat interface, when R is much larger than the diffusion layer thickness $R \gg \Gamma_0/c_0$, we can use the set of r equations of the following format

$$\Gamma_i(t) = \Gamma_{d,i} + 2\sqrt{\frac{D_i}{\pi}} \left(c_{oi}\sqrt{t} - \int_0^{\sqrt{t}} c_i(0, t-\tau)d\sqrt{\tau} \right), \quad i = 1,...,r \tag{1.52}$$

where the $\Gamma_{d,i}$ are the initial adsorption values for the components i and D_i are the corresponding diffusion coefficients.

The set of these r integral equations is interrelated via a multi-component adsorption isotherm, such as given by Equation 1.22. For the simplest case of a linear adsorption isotherm for each compound in Equation 1.6, instead of the set of r interrelated Equation 1.52, we get a system of independent equations of the type:

$$\Gamma_i(t)=\Gamma_{0,i}\left(1-\exp(D_i t/K_i^2)\,\mathrm{erfc}(\sqrt{D_i t}/K_i)\right) \tag{1.53}$$

where erfc(x) is the complementary error function, $\Gamma_{0,i}$ are the equilibrium adsorption values for component i, and K_i are the corresponding Henry constants as given in Equation 1.6. As the adsorption model of Henry does not consider any interaction of adsorbed molecules with each other, these r equations can be solved independently and the sum of adsorptions yields the total adsorption as a function of time $\Gamma_{tot}(t) = \Sigma_{i=1}^{r}\Gamma_i(t)$.

A theoretical simulation of the dependencies $\gamma(t)$ for some mixtures of the two nonionic surfactants decyl and tetradecyl dimethyl phosphine oxides (C_{10}DMPO and C_{14}DMPO) is shown in Figure 1.18. According to the data published by Lunkenheimer et al. [88], the surface activities of these two surfactants differ by almost two orders of magnitude, as expected from the Traube rule. As we can see, up to an adsorption time of about 100 s, the values of all $\gamma(t)$ dependencies are almost identical and correspond to the adsorption of the pure C_{10}DMPO. At about this adsorption time, the component C_{14}DMPO with the longer alkyl chain and hence higher surface activity and much lower bulk concentration starts to adsorb and competes with the C_{10}DMPO molecules adsorbed at the surface. This competition leads to a certain displacement of C_{10}DMPO. The larger the C_{14}DMPO concentration, the lower is the final surface tension of the mixed C_{10}DMPO and C_{14}DMPO solutions. The typical course of the dynamic surface tensions $\gamma(t)$ shows two steps, one corresponding to the moment of equilibration of the less surface active compound C_{10}DMPO and the second that of the second component C_{14}DMPO. These characteristics of the two compounds are very visible because of the remarkable difference in surface activity, while for mixtures of surface active compounds with smaller differences in the surface activity, the steps are less or even not at all visible.

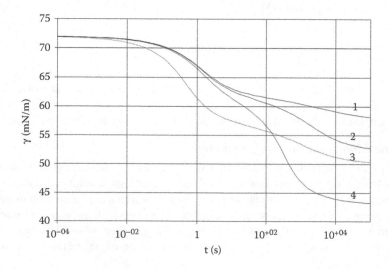

FIGURE 1.18 Calculated dynamic surface tension $\gamma(t)$ using Equation 1.8 to simulate the simultaneous adsorption of mixtures containing C_{10}DMPO and C_{14}DMPO: $c_{10} = 10^{-7}$ (1, 2, 4), 2×10^{-7} mol/cm³ (3), $c_{14} = 10^{-9}$ (1, 3), 3×10^{-9} (2), 10^{-8} mol/cm³ (4). (According to Miller, R. et al. *Tenside Surfactants Detergents* 2003; 40: 256–259 [57].)

1.3.3.2 Mixtures of Ionic Surfactant of Opposite Sign

For mixtures of nonionic surfactants or ionic surfactants of the same sign of the charge, the molecules adsorb in competition at the surface. When, however, a cationic and an anionic surfactant are mixed, ion pairs are formed [89]. These resulting double chain molecules behave completely different from what the single parts of the ion pairs did. These new entities in solution are very hydrophobic and thus have extremely high surface activity. Some authors call this a synergistic effect, however, it does not have much to do with synergism when two ions of opposite charge form a "lipid-like" highly surface active ion pair. For equimolar mixtures, the solutions essentially contain one type of surfactant and the surface properties can be described as for standard surfactants. Note that the surface properties cannot be deduced from those of the single components. For nonequimolar mixtures, the excess quantity of the higher concentration compound remains in competition with the ion pair component so that the adsorption layer properties have to be described by theories developed for mixtures, where one component is the ion pair and the other is the excess of the higher concentrated compound.

There are not many studies about this type of surfactant mixtures. For water/oil systems, one could speculate that the very hydrophobic ion pairs could get transported to the interface and dissolved in the oil phase once formed in the aqueous phase. Alternatively, and in particular above a certain concentration, the ion pair molecules could form larger aggregates which then would precipitate.

1.3.4 Adsorption Mechanisms for Proteins and Protein/Surfactant Mixtures

In contrast to mixed surfactant solutions, mixtures of surfactants with macromolecules or proteins show particular adsorption behavior. In mixed solutions containing nonionic polymers and surfactants, the interaction is mainly hydrophobic in nature. For ionic polymers mixed with oppositely charged surfactants, complexes are formed. The generally not surface active ionic macromolecules get hydrophobized via the electrostatic binding of oppositely charged surfactant ions and then adsorb at the interface [90,91].

In a recent work, the effect of alkyl chain length of the anionic surfactants sodium alkyl sulfate on the complex formation with the cationic polymer polyallylamine hydrochloride (PAH) was systematically investigated [92] and it was found that for the shorter alkyl chain surfactants (sodium decyl and dodecyl sulfate), the ionic interaction dominates and complexes are formed which are increasingly hydrophobic. For the longer chains (sodium tetradecyl and hexadecyl sulfate), the electrostatic binding to the PAH is only initially effective, followed by a hydroprobic interaction, leading to decreasing surface activity of the formed complexes.

A new type of experiments can be performed with drop profile analysis tensiometry using a special designed double capillary [93]. With this equipment, the volume of a drop can be exchanged by another liquid, that is, a protein solution drop can "*in vivo*" be replaced by a pure buffer solution and this via another exchange of the drop bulk into a surfactant solution drop. During the first step, protein molecules adsorb at the drop surface, in the second step the nonadsorbed protein is replaced, and during the third step, surfactant is injected into the drop, starting to adsorb in competition to the pre-adsorbed proteins. This experimental route for forming a mixed protein/surfactant layer is called sequential adsorption, in contrast to the traditional simultaneous adsorption routes where the two components adsorb from a mixed solution. In the first case, protein/surfactant complexes are formed at the drop surface only, while in the second case, the complexes form in the solution prior to adsorption. This methodology was successfully applied to the solution/air [94] and solution/oil interface [95]. Recently, a summary was given concerning the differences observed for an adsorption layer formed in one way or the other [15].

1.3.5 Examples for Selected Surfactant and Protein Systems

Figure 1.19 presents dynamic surface tensions for diluted solutions of Triton X-100 as measured by bubble profile analysis tensiometry. The dotted lines are theoretical dependencies calculated for a

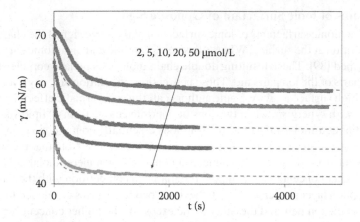

FIGURE 1.19 Dynamic surface tensions of Triton X-100 solutions at different concentrations studied by bubble profile analysis tensiometry; symbols—experimental data; dotted lines—calculated from the reorientation model. (According to Fainerman, V.B. et al. *Colloids Surfaces A* 2009; 334: 8–15 [39].)

diffusion controlled adsorption based on the Reorientation Model given by Equations 1.14 through 1.16. Note the model calculations were made using the parameters obtained from the analysis of the equilibrium surface tension isotherms. The only adjustable parameter was the diffusion coefficient. The analysis of the data shows that with increasing concentration, the optimum apparent diffusion coefficient becomes smaller.

Similar results were obtained in [39] for the other studied Triton X-n surfactants. The reason for this concentration dependence of the obtained diffusion coefficients can be the inhomogeneity of the Triton samples. Additionally, simplifications in the theory can lead to some systematic dependencies such as this. A more sophisticated theory, however, is less easy to handle and requires numerical solutions of the complete set of equations, including the transport by diffusion and the hydrodynamics. This discussion will be made in depth in a forthcoming book of this series dealing with particular computational fluid dynamics simulations demonstrating how far the accuracy of a theoretical interpretation can go in the limits of available computer time.

As discussed in respect to the data in Figure 1.19 and also in general above, most technical surfactants are mixtures of different compounds. Thus, it is essential to have efficient tools to also analyze the interfacial behavior of mixed surfactant solutions. The dynamic surface tensions for three mixed solutions of the two homologous surfactants $C_{10}DMPO$ and $C_{14}DMPO$ are presented in Figure 1.20. The theoretical curves were calculated for diffusion coefficients between 10^{-6} cm²/s and 3×10^{-6} cm²/s, as mentioned in the legend of Figure 1.20. It is evident that the experiments can be described very well by a diffusion controlled adsorption as was discussed in more detail in [87].

For protein solutions, the interpretation of experimental data is more complicated due to the much more complex equation of state, as given by Equation 1.24. Moreover, as proteins are extremely highly surface active and have a rather large molecular weight, they adsorb at very low molar concentrations. In [96], dynamic surface tension measurements were for β-casein (BCS) solutions under two different experimental conditions. One of the protocols was the classical one, that is, a bubble was formed in the solution and then kept constant while the surface tension was continuously measured from the profile of the bubble. In the second protocol, a continuous flow of the solution was arranged around the bubble so that no transport of protein molecules from a longer distance toward the bubble surface was required. Instead, the adsorption of the protein molecules happens by diffusion only within a thin hydrodynamic boundary layer of thickness δ. In Figure 1.21, the data are obtained for three different BCS concentrations in the presence of a forced convection. While the data shown in Figure 1.21 indicate that equilibrium for the three concentrations is established

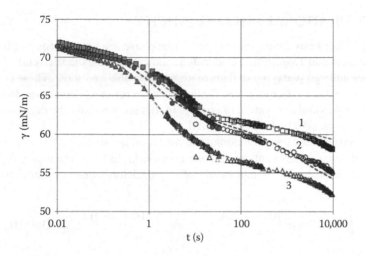

FIGURE 1.20 Dynamic surface tensions of a mixture of $C_{10}DMPO$ and $C_{14}DMPO$; concentration ratio $c_{10}|c_{14} = 10^{-4}$ mol/L$|10^{-6}$ mol/L (1), 10^{-4} mol/L$|3 \times 10^{-6}$ mol/L (2), 2×10^{-4} mol/L$|3 \times 10^{-6}$ mol/L (3); dotted lines—calculated for $D = 10^{-6}$ (1), 3×10^{-6} (2) 2×10^{-6} (3) cm^2/s using the diffusion model given by Equations 1.22 and 1.52, data for t < 10 s – BPA1, data for t > 10 s – PAT1. (According to Miller, R. et al. *Tenside Surfactants Detergents* 2003; 40: 256–259 [87].)

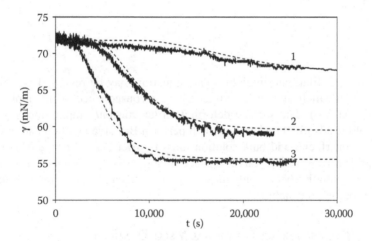

FIGURE 1.21 Dynamic surface tension of BCS solution with forced convection at three concentrations 2×10^{-9} mol/L (1), 5×10^{-9} mol/L (2), and 10^{-8} mol/L (3); solid lines calculated from a simplified diffusion model. (According to Fainerman, V.B. et al. *Colloids Surfaces A* 2006; 282–283: 217–221 [96].)

within less than 30.000 s, the adsorption without convection at the same bulk concentrations takes more than one order of magnitude longer time (see details in [96]).

The theoretical curves calculated for a forced convection, that is, diffusion only within a thin layer δ around the bubble surface, are also shown in Figure 1.21. For a diffusion coefficient $D = 2.7 \times 10^{-10}$ m^2/s and a layer thickness of δ = 0.27 mm, an excellent agreement with the experimental data at all three concentrations can be achieved.

Recently, systematic studies were performed for ß-lactoglobulin at the water/air and water/tetradecane interfaces, including investigations of the equation of state [97,98], dynamics of adsorption layer formation [99,100], and dilational viscoelasticity [101,102]. The studies included the effects of protein concentration, pH, and the ionic strength of the solution.

1.4   SURFACE DILATIONAL VISCO-ELASTICITY

The presence of surface active substances in the liquid phase is a key prerequisite for the formation and stability of foams and foam films. Under dynamic conditions, the surfactant layers at the air/solution interfaces undergo two types of deformations: dilational and shear. These deformations lead to the onset of two types of surface stresses: elastic, related to reversible processes, and viscous, linked to irreversible processes and dissipation of energy. In this section, only the dilational deformations, both elastic and viscous, will be briefly reviewed.

Due to the expansion and/or contraction of the interface, the surface area A is a function of time t. Let us assume that the experiments start from a state with initial surface area A_0 and that A(t) is known. The surface deformation $\alpha(t)$ and the rate of surface deformation $\dot{\alpha}(t)$ are defined as [103,104]:

$$\alpha(t) = \int_0^t \dot{\alpha}(t)dt = \ln\frac{A(t)}{A_0} \approx \frac{\Delta A}{A_0}, \quad \dot{\alpha}(t) \equiv \frac{1}{A}\frac{dA}{dt} = \frac{d\alpha(t)}{dt}, \Delta A(t) = A(t) - A_0. \quad (1.54)$$

The surface stress is presented as the deviation of the surface tension with time t, namely the difference between the instantaneous value $\gamma(t)$ and the initial value γ_0: $\Delta\gamma(t) = \gamma(t) - \gamma_0$.

The relationship of the surface stress $\Delta\gamma(t)$, the surface deformation, $\alpha(t)$, and the rate of surface deformation $\dot{\alpha}(t)$ is the rheological constitutive equation and it characterizes the surface dilational viscoelasticity of the adsorbed interfacial layer. For small ΔA, it is presented by the following expression [103,104]:

$$E = \frac{\Delta\gamma}{\Delta A/A_0} \quad (1.55)$$

where E is the surface dilational modulus. This equation may be regarded as a time dependent response function, in which the surface stress $\Delta\gamma$ is a consequence of the applied compression (surface deformation $\alpha(t)$). The surface deformation (expansion/compression) results in various rearrangements of the adsorbed molecules which perform Brownian motion and interact with each other. This causes interfacial and bulk solution mass transfer and exchange of the surface active species. A comprehensive overview on the surface dilational viscoelasticity may be found in [103–108]. Here, we preferentially specify only those elements and cases which are most relevant to foam and foam film properties and stability.

1.4.1   GENERAL BASIS OF SURFACE DILATIONAL VISCO-ELASTICITY

As can be seen from Equation 1.55, the surface dilational viscoelasticity of liquid interfaces is a direct consequence of the surface tension concept and is related to the reaction of the interfacial layer to the imposed external disturbances. Thus, by probing the reaction of the air/solution interface to well defined perturbations, data about the structure and the interactions in the surface layer and about the mechanisms of the adsorption process in a particular system may be extracted. The disturbances, relevant to foam film and foam properties, are most often mechanical: the dilation/contraction of the adsorption layer is achieved through harmonic, trapezoidal or other sequential disturbances, performed in a systematic manner.

Let us assume that the unperturbed system is characterized by an equilibrium distribution of the surface active species between the bulk and the interfacial regions. Any perturbation of the equilibrium surfactant adsorption layer imposes the onset of surface tension gradients. In the case of small deviations from the equilibrium state, the surface dilational modulus Equation 1.55 may be expressed as a product of a thermodynamic factor (frequency independent) and a term related to the mechanism of the dynamic response of the interface (frequency dependent) [105]:

$$E = -E_0 \frac{d\Gamma}{d\ln A}.$$ (1.56)

Here

$$E_0 = -\left(\frac{d\gamma}{d\ln\Gamma}\right)_{eq} = \left(\frac{d\Pi}{d\ln\Gamma}\right)_{eq}, \quad \Pi = \gamma_0 - \gamma_{eq}$$ (1.57)

is a thermodynamic quantity, the so called Gibbs elasticity of the interfacial layer (high frequency limit of the dilational elasticity), Γ is the surface concentration of the surfactant (adsorption), and Π is the equilibrium surface pressure. Depending on the specificity of the investigated system, any of the adsorption models already discussed in Section 1.1 may be applied for the calculation of E_0. The dynamic response to the external disturbance is embedded in the second term of Equation 1.57—$d\Gamma/d\ln A$. The latter refers to the transport adsorption mechanisms and mass conservation in that particular system. By changing the frequency of the applied perturbation within a wide range of magnitudes, different time scales characteristic of various possible adsorption and mass transfer mechanisms may be identified and investigated.

In the specific case of small amplitudes of the applied disturbance, the systems are very well described by the linear constitutive model [105,108,109]. If, for example, the perturbation is sinusoidal and can be reduced to a small change of the surface area with a constant frequency f, all interfacial quantities influenced by this disturbance will vary as harmonic functions of the same frequency. Thus, the interfacial deformation with amplitude $\tilde{A}$ may be presented as $\Delta A = \tilde{A}e^{i2\pi ft}$. It will cause a surface stress $\Delta\gamma = \tilde{\gamma}e^{i(2\pi ft+\phi)}$ with amplitude $\tilde{\gamma}$, and phase shift ϕ, due to the viscous dissipative losses during the relaxation process. The surface dilational modulus defined by Equation 1.56 acquires the form of a complex number:

$$E(f) = E_r(f) + iE_i(f) = \varepsilon_d + i2\pi f\eta_d$$ (1.58)

with E(f) having two contributions: a real part, the surface dilational elasticity ε_d, which accounts for the elastic response of the interface, and an imaginary part, related to the surface dilational viscosity, η_d, which is a measure of the energy dissipation during the mass transfer of the surface active species in the relaxation processes recovering the initial equilibrium state of the perturbed system. The real and imaginary parts of the complex dynamic surface viscoelasticity can be calculated from the amplitudes of the surface tension and surface area, and the phase shift between the oscillations of these two quantities [110]:

$$E(f) = A_0 \frac{\tilde{\gamma}}{\tilde{A}}\cos\phi + iA_0 \frac{\tilde{\gamma}}{\tilde{A}}\sin\phi.$$ (1.59)

Thus, the dilational viscoelasticity depends on the specific structure and interactions in the surfactant adsorption layer, and on the exchange mechanisms of matter between bulk and interfacial regions in response to the applied perturbations.

1.4.2 EXPERIMENTAL METHODS

All experimental methods for investigating the surface dilational rheology are based on a perturbation of the mechanical equilibrium at the interface and the subsequent measurement of the system's response [105,108,110–112]. The approach is usually based on measuring the time course of the surface tension and the evaluation of surface dilational viscoelasticity using Equation 1.55. There are two major types of experimental techniques which operate in different frequency ranges and allow the determination of surface viscoelasticity for frequency values of $\sim 10^{-3}$ Hz to ~ 2 MHz: (i) setups based on application of Langmuir trough and operating with a (nearly) flat liquid interface; (ii) instrumentation based on the investigation of curved menisci in the oscillating drop and bubble methods.

In the case of a flat liquid interface one can induce almost uniform surface compression and expansion in the center of a Langmuir trough [108,111,112]. The measurement of the surface tension is usually performed by a Wilhelmy plate. Stress relaxation studies may be performed following a sudden uniaxial in-plane compression of the Langmuir film, induced by the barriers of the trough. This approach is very appropriate to study a system presenting a single slow relaxation process.

When more response mechanisms are present with different values of the characteristic times, techniques allowing frequency tuning are preferable [110]. For example, the perturbation may be invoked by barrier or ring oscillations causing harmonic expansions and compressions of the flat liquid surface at a given amplitude. The major difference consists in the way of invoking the surface deformation. In the oscillation barrier method, the barriers are subjected to a sinusoidal motion at a constant frequency f and small amplitude. Typical experiments within the linear regime involves compression-expansion cycles with area amplitudes of less than 10% of the initial surface area A_0. The applied perturbation should be small enough, otherwise the in-plane shear components cannot be disregarded and the viscoelastic response of the interface becomes nonlinear. Therefore, the method of oscillation barrier is most reliable at f < 1 Hz, when the length of the surface longitudinal waves exceeds the length of the trough and the surface tension oscillations may be considered homogeneous. A comprehensive overview on this type of instrumentation is given in [110–112].

The oscillating ring method [110,111,113] uses a cylindrical glass ring with a sandblasted inner surface to ensure complete wetting. The oscillations of the ring along its axis result in oscillations of the liquid surface area inside the ring due to periodic changes of the meniscus height and shape of the internal ring surface. The main advantage of the method is the almost pure dilational deformations of the liquid surface and thereby a negligible contribution of the shear stresses.

There are also more sophisticated methods based on Langmuir trough application. For example, surface wave methods may excite capillary waves (ripples) [112]. One approach is to probe the light scattered by thermally excited capillary waves (Surface Quasi-Elastic Light Scattering). The light scattering arises from fluctuations of the refractive index due to random capillary roughness. Another method initiates the capillary waves through an external electrical field (electrocapillary waves). More specific details may be found in [110–112,114,115].

The methods based on curved liquid menisci are the most widely used in the last two decades [105,110,114]. Among the important advantages of these setups are: small liquid volumes (microliters), wide commercialization at reasonable prices, easy to operate, and well developed software packages for registration and processing of the experimental data. There are two basic variants: drop and bubble profile analysis tensiometry (PAT), and capillary pressure tensiometry (CPT).

In recent years, the drop and bubble profile analysis tensiometry has developed into one of the major instrumentations for studies in the field of interfacial research. The method is based on the shape analysis of a hanging, sessile or buoyant drop or bubble. The principle is based on the fact that the specific shape results from the competition of two forces, surface tension and gravity. The radius of curvature of the liquid interface is related to the surface tension. The surface tension can be determined by fitting the Gauss-Laplace equation to the profile coordinates of the bubble/drop, using γ as a fitting parameter [104,105,108,110,116]:

$$\gamma \left(\frac{1}{R_1} + \frac{1}{R_2} \right) = \Delta P_0 + \Delta \rho g z. \tag{1.60}$$

Here, R_1 and R_2 are the principle radii of the interfacial curvature, ΔP_0 is the pressure difference across the interface, $\Delta \rho$ is the density difference, g is the local gravitational constant, and z is the vertical height measured from the reference plane. A major criterion for the applicability of the method is that the dimensionless shape factor $\beta = \Delta\rho gb^2/\gamma$ fulfills the inequality $|\beta| > 0.1$, with b being the curvature radius at the drop/bubble apex.

Through a video camera, the image of the drop/bubble is transferred to a computer, digitized, and by a fitting algorithm, the surface tension γ is determined. Using a precise dosing system, the size of the drop or bubble cannot be kept constant, but also allows generating harmonic volume changes, which in turn lead to a periodic sinusoidal area compression and expansion. Thus, harmonic surface tension changes are induced from which the surface viscoelasticity can be determined.

The method has some limitations as well [115]. The major issues are: sometimes shear surface properties must also be taken into account [110,111,115]; the oscillation can have nonradial components, and the meniscus shape then could not be described by the Gauss-Laplace equation. These disadvantages limit the frequency range assessable to the oscillating drop and bubble method in PAT from above by up to a frequency of ~ 1 Hz. Depending on the systems, the operation of the source of mechanical oscillations, and for the oscillating bubbles in particular, this limit can be even lower. The three-phase contact line might also influence the meniscus oscillations [117], therefore, the material of the capillary can also affect the surface elasticity. Thus, the PAT techniques are usually applicable at sufficiently low deformation frequencies, usually between 0.001 and 0.2 Hz.

For the determination of interfacial rheological properties at higher frequency values, smaller and spherical drops/bubbles are required. The experiments are performed by capillary pressure tensiometry (CPT) [105,108,110]. The data are obtained by the direct measurement of the pressure difference across the two phases $P_c(t)$ separated by a spherical interface of radius $R(t)$, thus giving the surface tension as a function of time $\gamma(t)$.

$$\gamma(t) = \frac{P_c(t)R(t)}{2} \tag{1.61}$$

The surface dilational elasticity is obtained during a controlled variation of the drop/bubble volume. The measurements are provided for the frequency range between 0.1 and 100 Hz.

In both methods, PAT and CPT, the surface area variation is obtained by varying in a controlled way the volume of a drop or a bubble, while the surface tension response is directly measured from either the bubble profile or is calculated by the capillary pressure. These methods are particularly appropriate to study systems presenting single or multiple but slow relaxation processes. When more processes are present and they have significantly different values of the characteristic times, techniques allowing frequency tuning within a broader frequency range are preferable [110,117,118].

1.4.3 CLASSICAL APPROACH

The simplest case for the calculation of surface viscoelasticity is that of a solution with only one soluble low molecular mass surfactant. If the only relaxation process controlling the adsorption kinetics is molecular diffusion from the bulk of the solution to the interface [103–105,108,117], then Equation 1.56 is transformed into the well known van den Tempel-Lucassen formula:

$$E = E_r + iE_i = E_0 \frac{1+\zeta+i\zeta}{1+2\zeta+2\zeta^2}, \tag{1.62}$$

where $\zeta = \sqrt{f_0/f}$, and $E_0 = -(d\gamma/d\ln\Gamma)_{eq}$ are the high frequency limiting (Gibbs) elasticity. Here, $f_0 = (D/4\pi)(dc/d\Gamma)_{eq}^2$ is a characteristic frequency corresponding to the specific time of the relaxation diffusion mechanism. The two parameters, E_0 and f_0, do not depend on the applied frequency, but are innate properties of the particular interfacial layer [104,105,110]. Equation 1.62 is widely used and is usually coupled with one of the existing thermodynamic models, as briefly reviewed in Section 1.4.1. The course of E_r and E_i against the perturbation frequency are presented in Figure 1.22 for the particular case of a diffusion controlled relaxation process [105,118].

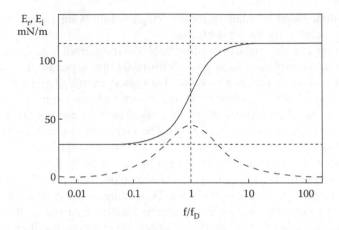

FIGURE 1.22 Example of real (solid line) and imaginary (dashed line) parts of the surface dilational modulus E. (According to Ravera, F., Liggieri, L., Loglio, G. In: Miller, R., Liggieri, L. (Eds.), *Progress in Colloid and Interface Science*, Vol. 1, Brill: London–Boston, 2009, pp. 137–177, Ravera, F., Ferrari, M., Liggieri, L. *Colloids and Surfaces A* 2006; 282–283: 210–216 [103,118].)

The course of the imaginary part E_i of the surface dilational modulus against the disturbance frequency is marked by the onset of a maximum corresponding to the characteristic frequency of the particular relaxation process, while an inflection point is registered in the real part E_r at the same frequency value. This course of the curves has a general validity because it comes directly from the consideration of the relaxation phenomenon as a linear kinetic process.

Note that Equation 1.62 describes the dilation rheology of the surfactant adsorbed from solution onto a flat surface. In the case of curved interfaces, for low oscillation frequencies and small sized oscillation drops and bubbles (e.g., the radii 1.5–2 mm), the actual geometry of the oscillating object can play a significant role. The respective equations have been derived by Joos [35]. Thus, the complex elasticity of spherical objects may be presented as in Equation 1.63 for the adsorption onto a bubble and onto a drop surface, respectively [35,119]:

$$E = \frac{E_0}{1 + (D/i2\pi fR)(dc/d\Gamma)(1 + nR)},$$

$$E = \frac{E_0}{1 + (D/i2\pi fR)(dc/d\Gamma)[nR\coth(nR) - 1]}$$

(1.63)

where $n^2 = i2\pi f/D$, and R is the bubble/drop radius.

In complex fluids, when other relaxation processes besides the molecular diffusion of the surfactant are possible both in the interfacial and/or bulk phases, more sophisticated models have to be applied. The classical approach of Lucassen-van den Tempel may be extended, and additional characteristic frequencies could be defined. This is possible if the characteristic times (and frequencies) for each of the relaxation processes are different enough so that they could be detected by the surface rheological measurements. Such processes are, for example, molecular reorientation, change of conformation of the surface active substance, aggregation, chemical reactions, formation of specific complexes, and so on. The detailed development of this concept may be found, for example, in [118–120].

A simpler case is that of one soluble surfactant in the system, but the diffusion from the solution bulk is not the only relaxation process counterbalancing the perturbation of the equilibrium state of the adsorption layer. For example, if different surface states for the adsorbed molecules are possible, a kinetic mechanism to pass from one state to the other should also be considered. In practice, a new variable X is included which related to the second relaxation process. As such a parameter, the average surface area per molecule may be chosen when reorientation of the adsorbed molecules has to be accounted for. Therefore, the surface dilational modulus as defined by Equation 1.56 may be presented as [105,118]:

$$E = -E_{0\Gamma} \frac{d\ln\Gamma}{d\ln A} - E_{0X} \frac{d\ln X}{d\ln A}, \tag{1.64}$$

where $E_{0\Gamma} = -(d\gamma/d\ln\Gamma)_{eq}$ and $E_{0X} = -(d\gamma/d\ln X)_{eq}$ are equilibrium thermodynamic quantities, calculated for the initial state. They can be obtained using the respective equilibrium adsorption isotherm relevant to a particular system. The second factors in each of the two terms of the sum are the frequency dependent contributions. If one may assume a linear behavior of the system for the second relaxation process as well, and the perturbation is harmonic, the following expression for the surface dilational modulus is obtained:

$$E = \frac{E_{0\Gamma} - i\lambda E_{0G}}{-(1 + \zeta - i\zeta)i\lambda + (1 + G + \zeta - i\zeta/(1 + G))}, \tag{1.65}$$

where $E_{0G} = E_{0\Gamma} + E_{0X}(d\ln X^0/d\ln\Gamma)$ with X^0 being the initial value of the additional variable X; $\lambda = (f_k/f)$, with f_k standing for the characteristic frequency of the second relaxation process; $G = (1 - (dc_s/d\Gamma)/(\partial c_s/\partial\Gamma)^0)$, where c_s is the sublayer surfactant concentration. The presence of a second relaxation process modifies the shape of the frequency curves of the surface dilational elasticity at high frequency values (Figure 1.23). Thus, the extra relaxation process causes the onset of a second maximum in the imaginary part $E_i(f)$ and a second inflexion point in the $E_r(f)$.

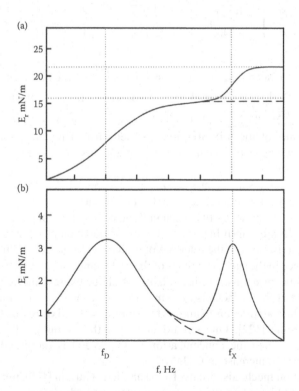

FIGURE 1.23 A representative sketch of real (a) and imaginary (b) parts of the surface dilational modulus E for the case of additional relaxation process (solid lines) compared to the Lucassen-van den Tempel model (dashed lines). (According to Ravera, F., Liggieri, L., Loglio, G. In: Miller, R., Liggieri, L. (Eds.), *Progress in Colloid and Interface Science*, Vol. 1, Brill: London–Boston, 2009, pp. 137–177, Ravera, F., Ferrari, M., Liggieri, L. *Colloids and Surfaces A* 2006; 282–283: 210–216 [103,118].)

Generally, Equation 1.65 is obtained only assuming that the variation of X is in phase with the variation of surface stress, without assuming a particular mechanism for the dynamic aspect of the average molar area. Equations 1.64 through 1.65 are also in agreement with the expression of the viscoelasticity obtained in the framework of the two state model, for reorientable surfactants as well. More examples of its applicability may be found in [80,105,120–124].

If ζ vanishes, then the surface dilational modulus for an insoluble adsorption layer is obtained:

$$E = E_{0G} + (E_{0\Gamma} - E_{0G}) \frac{if / f_k}{1 + if / f_k}. \tag{1.66}$$

1.4.4 Mechanisms for Surfactant Mixtures

The cases of surface dilational rheology of surfactant mixtures have been investigated in detail, for example, in [106,117,120,122,125–130]. Let us deal with a mixture of two soluble surfactants. Assuming the only relaxation process is molecular diffusion, and in the case of a harmonic perturbation of frequency f, the complex surface dilational modulus (Equation 1.56) for the mixture is expressed as [120,122]:

$$
\begin{aligned}
E = &-\frac{1}{B}\left(\frac{\partial \gamma}{\partial \Gamma_1}\right)_{\Gamma_2}\left[\sqrt{\frac{i2\pi f}{D_1}}a_{11} + \sqrt{\frac{i2\pi f}{D_2}}a_{12}\frac{\Gamma_1}{\Gamma_2} + \frac{i2\pi f}{\sqrt{D_1 D_2}}(a_{11}a_{22} - a_{12}a_{21})\right] \\
&-\frac{1}{B}\left(\frac{\partial \gamma}{\partial \Gamma_2}\right)_{\Gamma_1}\left[\sqrt{\frac{i2\pi f}{D_1}}a_{21}\frac{\Gamma_1}{\Gamma_2} + \sqrt{\frac{i2\pi f}{D_2}}a_{22} + \frac{i2\pi f}{\sqrt{D_1 D_2}}(a_{11}a_{22} - a_{12}a_{21})\right]
\end{aligned}
\tag{1.67}
$$

Here, Γ_j, c_j, D_j are the adsorption, the bulk concentration, and the diffusion coefficient of the j-th component; $a_{ij} = (\partial \Gamma_i / \partial c_j)\big|_{c_{k \neq j}}$ are partial derivatives, determined from the adsorption isotherm; $B = 1 + \sqrt{i2\pi f/D_1}a_{11} + \sqrt{i2\pi f/D_2}a_{22} + i2\pi f / \sqrt{D_1 D_2}(a_{11}a_{22} - a_{12}a_{21})$.

The elasticity modulus of the mixtures may be determined for any model of the interfacial layer provided the dependencies of the surface tension and adsorption on the concentrations of the components are accessible.

For example, if the binary surfactant mixture corresponds to the generalized Frumkin model Equation 1.22, the average molar area ω will be defined as $\omega_0 = (\omega_1\theta_1 + \omega_2\theta_2)/(\theta_1 + \theta_2)$, where $\theta_i = \omega_i \times \Gamma_i$ is the surface coverage by the surfactant molecules of component i. The model involves only the parameters which are determined from the results obtained in the studies of surface tension, adsorption, and dilational rheology of the solutions of individual surfactants. Thus, it assumes that there is no substantial difference in the surface activity of the surfactants and all variations might be ascribed only to the different concentrations. So, it is also possible to express the limiting high frequency elasticity for the mixture of the surfactants in terms of viscoelasticities of the individual components.

The procedure is illustrated as an example for solutions of mixtures of $C_{14}DMPO/C_{10}DMPO$ at different concentrations [122]. One should note that the modulus of the surface dilational viscoelasticity increases when the concentration of $C_{14}DMPO$ is increased, while the opposite tendency is observed for the increase of $C_{10}DMPO$.

More refined adsorption mechanisms can also be applied. Thus, in [130], mixtures of oxyethylated surfactant are investigated based on the equation of state and adsorption isotherms for surfactants exhibiting reorientation and intrinsic compressibility of the interfacial layer, as described in Section 1.1.3. The general tendency is that at the initial stage of the adsorption process, the surfactant of lower surface activity is adsorbed first. It is then continuously displaced by the surfactant with higher surface activity (Figure 1.24).

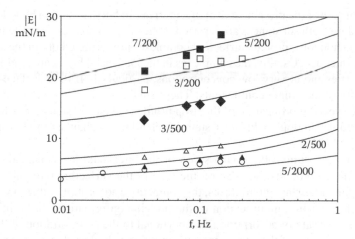

FIGURE 1.24 Viscoelastic modulus |E| as a function of frequency (f) for mixed $C_{14}DMPO/C_{10}DMPO$ solutions at different concentrations; labels refer to the $C_{14}DMPO/C_{10}DMPO$ concentrations in the mixture in μM; symbols are experimental points. (According to Aksenenko, E.V. et al. *J Phys Chem C* 2007; 111: 14713–14719 [122].)

1.4.5 MECHANISMS FOR PROTEINS AND PROTEIN/SURFACTANT MIXTURES

The surface dilational properties of proteins at the air/solution interface show specific dependencies on the bulk concentration c and the adsorption Γ, as compared to low molecular mass surfactants [131–136]. These characteristics also have variations, depending on the natural conformation of the protein in the bulk—globular or linear flexible. Thus, during the adsorption process, the surface tension values usually pass through an "induction period" when they remain close to the pure water value, followed by a steep decrease within a narrow range of time. This effect is more pronounced for globular proteins as compared to linear flexible species. Following the sharp drop of the surface tension values, the complete equilibrium values are usually achieved very slowly. Additionally, within a narrow protein concentration range, the thickness of the adsorption layer increases considerably, with the effect being more pronounced for linear flexible proteins. Passing over a certain bulk concentration value, the further increase of c has no effect on the interfacial pressure and the overall Π vs. Γ-curve acquires an S-shape. The net possible increase of the interfacial pressure (Π < 30 mN/m), as compared to the pure water/air interface, is generally lower than the values accessible for several low molecular mass surfactants (Π > 40 mN/m).

These peculiarities are closely related to the intrinsic compressibility of the protein adsorption layer: the conformation distribution of the protein at the interface depends on the surface coverage and multiple states of the protein molecules may be acquired. The states may change during the compression/expansion process as well. The molecular area becomes smaller with increasing protein concentration.

The protein/surfactant mixed adsorption layers also have several distinct features [15,128,137–141]. The interactions of proteins and surfactants are different in the solution bulk and in the interfacial layers and depend on the natural conformation of the protein (linear flexible or globular), the type of surfactant (ionic, nonionic), and other properties of the solution (pH, ionic strength, additives). Usually, bulk and interfacial complexes are formed which differ in solubility and surface activity and determine the specific rheological properties at the air/solution interface. The complexes are formed due to the action of hydrophobic and electrostatic forces, very often resulting in substantial conformation changes of the protein [142].

The adsorption from mixed protein/surfactant solutions is investigated by two major experimental approaches: sequential or simultaneous routes. In the sequential adsorption route, the protein is adsorbed at the interface first and then the surfactant is introduced into the system. Therefore, the protein/surfactant interactions and the formation of protein/surfactant mixed layers are affected by the initial conformation of the protein, both at the interface and in the solution bulk. In a simultaneous adsorption procedure, both substances are inserted in the system and they compete for the interfacial area. Due to the interactions, protein/surfactant complexes are formed, both in the solution bulk and at the air/solution interface. Usually, the complexes in the interfacial layers are readily detected in the rheological properties and at very low concentrations, particularly if their surface activity exceeds the surface activity of the single components. These studies have only recently begun to develop, mainly due to the special modification of the drop profile analysis tensiometer—the coaxial double capillary combined with a double dosing system which allows exchange of the drop volume during the experiments [114,139,143].

At simultaneous adsorption, the composition of the interfacial layer depends not only on the concentration of the ingredients, but also on the type of the surfactant (nonionic or ionic). In the case of protein/nonionic surfactant mixtures, the competitive adsorption is due to weak hydrophobic interactions which result in a modification of the protein conformation. In the initial stage of the adsorption process, the rate of adsorption is proportional to the concentration of the surface active species in the system. Thus, if the surfactant concentration is higher, it is adsorbed first and then it could be replaced by the more surface active protein. With a substantial increase of the surfactant concentration in the system, the surfactant molecules compete with the surface active protein/surfactant complexes and gradually displace them from the adsorption layer. This change in the adsorption layer composition, related to the overall change in the composition of the system, is usually very sensitive to pH and to the ionic strength of the solution.

In the case of protein/ionic surfactant mixtures, the mechanism is different. Upon increasing the surfactant concentration, electrostatic interactions dominate and the charges of the protein molecule are compensated by forming electro-neutral complexes. The latter are higher surface active as compared to the native protein molecule. Upon a further increase of the surfactant concentrations, hydrophobic interactions begin to dominate, thus making the complexes more hydrophilic and less surface active. Thus, the surface active molecules compete with the complexes for the areas in the interfacial layer and gradually replace them (particularly at concentrations above CMC).

1.4.6   Adsorption Mechanisms for Selected Surfactant and Protein Systems

Here, several examples are presented which illustrate the application of the above discussed principles and conceptions for the characterization (diagnostics) of complex fluid systems typically used in foam and foam film studies.

1.4.6.1   Surfactants

The viscoelasticity modulus as a function of the surfactant concentration has been investigated at low frequencies using drop/bubble profile tensiometry. An example is shown in Figure 1.25 [79,108].

The experimental results demonstrate the onset of characteristic maxima at certain surfactant concentrations for both $E_i(C_{C12DMPO})$ and $E_r(C_{C12DMPO})$. The decrease of the surface dilational elasticity and surface dilational viscosity at higher surfactant concentrations is related to the effect of micelles (above the critical micelle concentration (CMC)) [79,103,108]. The onset of a surface tension gradient brings about a disturbance of the equilibrium between the monomers and the micelles in the system. This initiates a consecutive disintegration of micellar aggregates. The result is a release of additional monomers which also participate in the equilibration processes, following the perturbation of the adsorbed layer during the surface rheology experiments. Upon an increase of the number of micelles in the solution, the dilational viscoelasticity of the adsorption layer at the air/solution interface is further decreased.

Similar effects have been registered, limited to the surface dilational elasticity, in the case of certain low-molecular mass surfactants (LMM), both for ionic and non-ionic surfactants, in the premicellar concentration domain [144–146] (Figure 1.26).

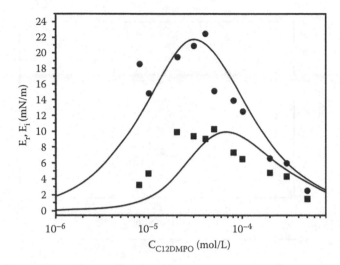

FIGURE 1.25 Real and imaginary parts of the complex viscoelasticity as a function of C_{12}DMPO concentration at f = 0.1 Hz; E_i—filled squares and E_r—filled circles. (According to Miller, R. et al. *Colloid and Polymer Science* 2010; 288: 937–950 [108].)

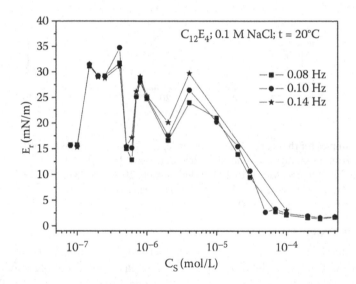

FIGURE 1.26 Surface dilational elasticities of aqueous $C_{12}E_4$ solutions against the surfactant concentration. CMC is 5.8×10^{-5} M. (According to Arabadzhieva, D. et al. *Ukr J Physics* 2011; 58: 801–810, Arabadzhieva, D. et al. *Colloids and Surfaces A* 2011; 392: 233–241 [144,145].)

The sequential change of minima and maxima are experimental evidences for the presence of premicellar aggregates in the solution bulk at concentrations in the intermediate concentration range (below CMC, but higher than those from the Henri region). The above example concerns the nonionic surfactant tetraethyleneglycol monododecyl ether ($C_{12}E_4$), but similar results have been obtained for ionic surfactants as well. The effect is related to the structure of the LMM surfactant and to the specific solution conditions (e.g., presence of salt, etc.) which ensure the manifestation of well balanced hydrophilicity-hydrophobicity of the amphiphilic molecule. It is strong enough to have an impact on both dynamic and equilibrium surface tension values at the solution/air interface, and on the drainage kinetics of foam film.

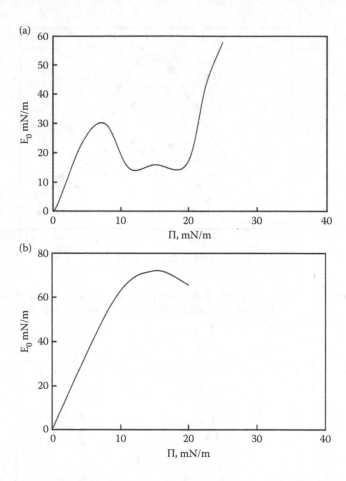

FIGURE 1.27 Examples for the shape of $E_0(\Pi)$ dependencies for (a) linear flexible (BCS); (b) globular (BSA) proteins. (Adapted from Benjamins, J., Lucassen-Reynders, E.H. In: Miller, R., Liggieri, L. (Eds.), *Progress in Colloid and Interface Science*, Vol. 1, Brill: London-Boston, 2009, pp. 257–305, Noskov, B.A. *Adv Colloid Interface Sci* 2014; 206: 22–238, Lucassen-Reynders, E.H., Benjamins, J., Fainerman, V.B. *Curr Opin Colloid Interface Sci* 2010; 15: 264–270 [131–133].)

1.4.6.2 Polymer Systems

Depending on the native bulk conformation of polymer molecules in aqueous solutions (globular or flexible), the limiting elasticity E_0 has a specific course against the surface pressure at the air/solution interface. For example, the surface dilational elasticity of linear flexible proteins usually passes through a maximum, followed by a minimum, and then increases again upon raise of the surface pressure (Figure 1.27a). In the case of globular proteins, however, a monotonic raise of the surface dilational elasticity up to a maximum is observed with increasing the adsorption and the protein concentration (Figure 1.27b). The value of this maximum is higher than the respective data for linear flexible proteins.

1.4.6.3 Polymer/Surfactant Systems

Although the surface pressure isotherms of mixtures obtained on different routes (sequential and simultaneous adsorption of the components) do not vary significantly, the interfacial dilational properties reveal important dissimilarities. In the case of nonionic surfactants, the surface dilational moduli against the surfactant concentration change in a similar manner, both at simultaneous and sequential adsorption. In the case of ionic surfactants, the courses of these moduli are quite different (Figure 1.28). The formation of surfactant/protein complexes in the bulk definitely affects the interfacial behavior and the sequentially adsorbed surfactant is always more effective in displacing

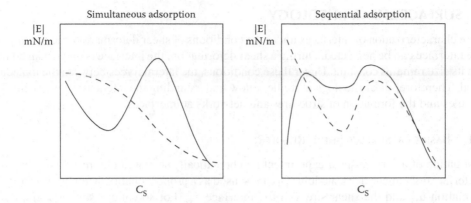

FIGURE 1.28 Dilatational viscoelastic modulus of polymer/surfactant mixtures: β-casein/C$_{12}$DMPO (continuous line) and β-casein/DoTAB (dashed line) solutions at a fixed β-casein concentration (1 × 10^{-6} mol/L) formed by simultaneous and sequential adsorption as a function of ionic (DoTAB) and non-ionic (C$_{12}$DMPO) surfactant concentration (C$_S$). (According to Maldonado-Valderrama, J., Rodríguez-Patino, J.M. *Curr Opin Colloid Interface Sci* 2010; 15: 271–282 [139].)

the adsorbed protein [147–149]. These conceptions are illustrated here by the data obtained for mixtures of the linear flexible protein BCS with the nonionic surfactant dodecyl dimethyl phosphine oxide (C$_{12}$DMPO), and the ionic surfactant dodecyltrimethyl ammonium bromide (DoTAB).

1.4.6.4 Medical Applications

The last example illustrates the option of specific application of surface dilational rheology in medicine [150]. It is well known that Neurosyphilis (NS) diagnostics may be complicated because the syphilitic process can either be asymptomatic or does not exhibit clear clinical indications. Therefore, more sensitive detection methods for this pathology need to be developed. One prospective option is related to the fact that NS results in specific changes of the liquor cerebrospinalis (CSF) which may be captured by surface dilational rheology measurements (Figure 1.29). As we can see from Figure 1.29, the real values E$_r$ for the patients without clinical signs of accompanying disease are essentially higher than those for syphilitic patients with expressed neurological disturbances.

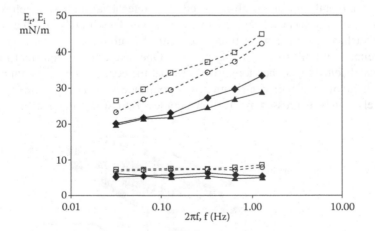

FIGURE 1.29 Dependencies of the real (E$_r$, upper points, bold curves) and imaginary (E$_i$, lower points, thin curves) viscoelasticity components. The filled symbols designate the data for patients with syphilis accompanied by neurologic diseases; the empty symbols are the data for patients with syphilis but without accompanying neurologic diseases. (According to Kazakov, V.N. et al. *Colloids Surfaces A* 2011; 391: 190–194 [150].)

1.5 SURFACE SHEAR RHEOLOGY

For the characterization of interfaces under the conditions of shear deformation, the compressibility of the interface can be neglected. During a shear deformation, the interface is only changed in shape while its size remains constant. Under these conditions, the lateral forces between the molecules are probed. Therefore, shear rheological studies allow understanding the interactions between adsorbed molecules and the formation of structures and networks at interfaces.

1.5.1 Basics of Surface Shear Rheology

The situation of a surface shear experiment is schematically shown in Figure 1.30. In the xy-plane, the interfacial shear elasticity modulus G_i can be used as a proportionality factor between the applied deformation u_{xy} and the shear stress in the interface τ_{xy}. For so called "solid" layers, an elastic (Hookian) behavior is observed, given by $\tau_{xy} = G_i u_{xy}$. On the contrary, for ideal "liquid" layers, a pure viscous (Newtonian) behavior is observed with $\tau_{xy} = \eta_i du_{xy}/dt = \eta \hat{u}_{xy}$, where $\hat{u}_{xy}$ is the rate of shear and η_i is the interfacial shear viscosity.

Attractive interactions between elements at an interface, such as adsorbed or spread molecules (surfactants, lipids, proteins, polymers) can lead to an increase in G_i and η_i. In order to overcome the interaction forces, energy is required to allow the interface elements to flow. Due to the complexity in the composition and the interactions at the interface, most layers have both an elastic and viscous behavior, that is, they behave viscoelastically. The viscoelastic response of these layers is equivalent to that in a complex bulk rheology, that is, the interfacial shear modulus (G^*) consists of a real (storage) G' and an imaginary (loss) component G'':

$$Gi^* (\omega) = Gi'(\omega) + iGi''(\omega) \equiv Gi' + i\omega\eta i \tag{1.68}$$

where ω is the circular frequency. For oscillations at a given frequency and very small amplitudes, the shear viscosity η_i can be calculated from the loss modulus $G_i'' = \omega\eta_i$.

Changes in the mechanical behavior of liquid interfaces can be caused by adsorption of surface active molecules from the bulk phase or by spreading an insoluble monolayer. Surface layers formed by typical surfactants show a dilational viscoelastic behavior, but the shear rheological behavior is very weak. Typically, polymers or proteins adsorbed or spread at an interface form layers with a measurable shear viscoelasticity.

In any case, the liquid subphase is always needed for interfacial studies, therefore, the properties of the monomolecular interfacial layers cannot be studied separately. As the measuring body located in the surface layer is always in contact with the subphase bulk, any measured quantity is partly influenced by bulk rheological quantities. One of the most important requirements for interfacial shear rheometers is, therefore, to ensure that the contribution of the interface dominates the measured signal over the bulk contribution. The contribution of the interface (X_i) and bulk phase (X_v), respectively, can be expressed by the dimensionless Boussinesq number Bo

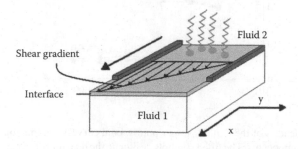

FIGURE 1.30 Schematic of the shear gradient in the slit of a shear rheometer. (See Krägel, J., Derkatch, S.R., In: Miller, R., Liggieri, L. (Eds.), *Interfacial Rheology*, Brill, Leiden: 2009, p. 372–428 [151].)

$$Bo = X_i/lX_v \tag{1.69}$$

where l is a characteristic length. In classical shear rheometers, this length is the contact line of the measuring body in the interface. For high values, $Bo \gg 1$, the measured quantities are mainly governed by interfacial properties, while for small values, $Bo \ll 1$, the instrument would not be suitable for interfacial studies. In many cases, however, we can have $Bo \approx 1$, which entails a quantitative analysis of the impact of the liquid flow pattern on the measured signals in order to separate the interfacial effects. Many commercial instruments provide the necessary algorithms for this often very complicated analysis.

1.5.2 Instruments for Measuring Interfacial Shear Rheology

The available shear rheology instruments can be divided into direct and indirect methods. In the indirect methods, the deformed is generated externally, for example, by the rotation of the measurement vessel or by displacement of a body in contact with the interface. Canal viscometers are examples for the indirect methods. Changes in the flow behavior of the liquid bulk and interface provide information on the shear viscoelasticity of the interfacial layer. This group of instruments is mainly applied to liquid/gas interfaces.

In contrast, in the direct methods, the interface is sheared by means of a suitable measuring body which is in direct contact with the interface and creates a shear gradient in the interface, which can be applied in a steady (rotational deformation) or dynamic (oscillation deformation) way to air/water and oil/water interfaces. Geometries of the measuring body with a small contact length to the interface and a low moment of inertia (torsion pendulum with a sharp edge or rings or thin needles) are most suitable for sensitive instruments. In Figure 1.31, the torsion pendulum is shown as an example for a direct interfacial shear rheometer. The geometry is essentially given by the two radii r_1 and r_2 of the ring and measuring vessel, respectively. The shear stress S can be determined from the transferred torque via the relationship

$$S = \frac{M}{2\pi}\left[\frac{1}{r_1^2} + \frac{1}{r_2^2}\right]. \tag{1.70}$$

The theoretical background for the data interpretation was derived by Tschoegl [152], who brought together a mechanical equivalent to the various rheological compounds in the experimental set up, such as the elasticity of the torsion wire, the inertia of the measuring body, the friction of the liquid

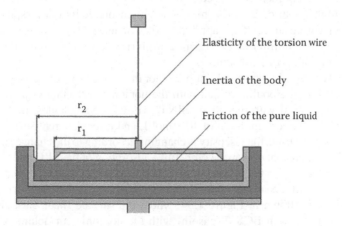

Elasticity of the torsion wire

Inertia of the body

Friction of the pure liquid

r_2

r_1

FIGURE 1.31 Scheme of a torsion pendulum rheometer.

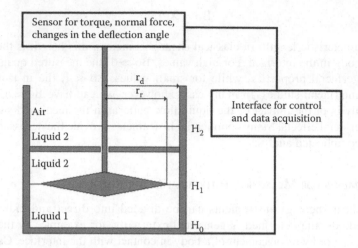

FIGURE 1.32 Scheme of a conventional rheometer with a directly coupled biconical disc for interfacial studies. (See Krägel, J., Derkatch, S.R., Miller, R. *Adv Colloid Interface Sci* 2008; 144: 38–53, Erni, P. et al. *Rev Sci Instrum* 2003; 74: 4916–4924 [153,154].)

subphase(s), and the shear viscosity and elasticity of the interfacial layer. More details, including the data acquisition and interpretation and the required calibration processes in order to determine the system constants, have been given elsewhere by Krägel and Derkatch in [153].

Measuring geometries with small contact lines to the interface and a small inertia of the measuring body, (e.g., the torsion pendulum as shown in Figure 1.31) allow accurate studies even of very low viscoelasticity values of interfacial layers. In contrast, with the use of biconical disks with traditional but very sensitive modern rheometers, (see Figure 1.32) only layers at the air/water or oil/water interface can be studied with a sufficiently high accuracy.

1.5.3 SHEAR VISCO-ELASTICITY OF SELECTED INTERFACIAL LAYERS

The shear rheology can be studied for quite a number of systems, adsorbed layers as well as spread monolayers containing polymers, proteins, and surfactants. For soluble adsorption layers of classical surfactants, however, the mentioned rheometers are not suitable and very special methods have to be applied. For example, Petkov et al. [155] described a method based on the drag coefficient of a small spherical particle floating at the interface under the action of an external capillary force. The determined drag coefficient then gives access to the shear viscosity of adsorbed layers (e.g., of SDS) with a sensitivity of the order of 10^{-2} μNs/m. For spread insoluble lipid layers, the torsion pendulum rheometer, however, can be successfully applied [156]. Condensed layers of phospholipids like DPPC (dipalmitoylphosphatidylcholine) or DMPE (1,2-dimyristoyl-sn-glycero-3-phosphoethanolamine) show shear viscosities of up to 1 mNs/m.

An experimental example is shown in Figure 1.33 for the surface shear viscosity of aqueous solutions of the protein BLG (ß-lactoglobulin) mixed with the nonionic surfactant $C_{10}DMPO$ (decyl dimethyl phosphinoxide), as measured with the ISR-1 (SINTERFACE Technologies, Berlin). The shown data correspond to a fixed BLG concentration of 10^{-6} mol/L and three different surfactant concentrations. With increasing adsorption time, the viscosity is changing according to the changes in the composition of the surface layer. For the present system, a competitive adsorption of the BLG and $C_{10}DMPO$ molecules has been found. The decrease in the viscosity with time is evidence for the fact that BLG molecules are step by step displaced from the surface due to an increasing adsorption of the surfactants [153].

A second example is shown in Figure 1.34, where the surface shear elasticity is measured for mixed solutions of the protein BCS (ß-casein) with the nonionic surfactants $C_{12}DMPO$ (dodecyl dimethyl phosphinoxide) and the cationic surfactant $C_{12}TAB$ (dodecyl trimethyl ammonium bromide)

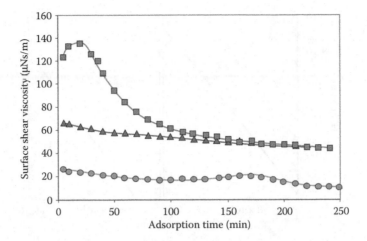

FIGURE 1.33 Change of the surface shear viscosity of mixed ß-lactoglobulin/C_{10}DMPO mixtures at a fixed protein concentration of 1×10^{-6} mol/L and different C_{10}DMPO concentrations: 10^{-5} mol/L (■), 2×10^{-5} mol/L (▲), 3×10^{-5} mol/L (●) in dependence on the adsorption time. (According to Krägel, J., Derkatch, S.R., Miller, R. *Adv Colloid Interface Sci* 2008; 144: 38–53 [153].)

as a function of the surfactant concentration. For both surfactants, a decrease in the shear elasticity is observed, however, with C_{12}DMPO it is much steeper. Hence, the C_{12}DMPO displaces the protein molecules more efficiently from the surface.

The experiments presented in both Figures 1.33 and 1.34 were performed with the torsion pendulum rheometer. As mentioned above, recently commercial bulk rheometers have become so sensitive that they can be also applied for interfacial studies. An example is the rheometer MCR 301 (produced by Anton Paar). For such interfacial measurements, a special measuring body in form of a biconus is provided, and the respective data analysis software for separating bulk from interfacial effects is also provided.

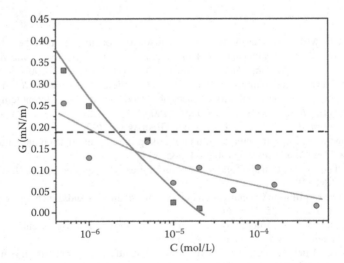

FIGURE 1.34 Surface shear elasticity of mixed adsorption layers formed after 500 min from solutions of BCS/C_{12}DMPO (■) and BCS/C_{12}TAB (●) at a fixed protein concentration of 10^{-6} mol/L; measurements were performed with the torsion pendulum rheometer ISR-1 (SINTERFACE Technologies, Berlin); the curves are a guide for the eye; the horizontal line stands for the 10^{-6} mol/L BCS solution without added surfactant. (According to Kotsmar, C. et al. *J Phys Chem B* 2009; 113: 103–113 [157].)

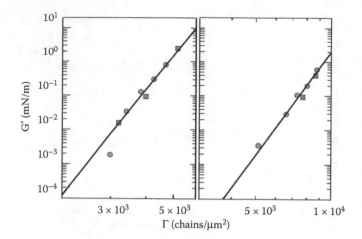

FIGURE 1.35 Shear elasticity G′ as a function of the polymer concentration in the semidilute regime; the two graphs corresponds to samples with different molecular weights of 2.708×10^5 g/mol (left) and 1.59×10^5 g/mol (right); measurements were made with two different rheometers: (●) ISR-1 from SINTERFACE Technologies and (■) MCR-301 from Anton Paar; the straight lines are power law fits.

In order to demonstrate that a new instrument measures correct physical values, a study was performed with two interfacial rheometers, working with different principles, such as the torsion pendulum ISR-1 and the modified bulk rheometer MCR301-ISR [158]. Figure 1.35 shows the interfacial shear storage modulus G′ as function of the surface concentration for two polymers of different molecular weights. The values of G′ are found in the range between 10^{-4} and 15 mN/m. The data measured with the two rheometers also show excellent agreement for the G″ values. This convincingly demonstrates that both techniques complement each other perfectly.

Note the straight lines represent power law fits of the shear modulus against the polymer concentration, $G' \sim \Gamma^{y'}$, with an exponent of 10 ± 1, as was discussed in detail in [158].

REFERENCES

1. Fainerman, V.B., Miller, R. Thermodynamics of adsorption of surfactants at the solution-fluid interface. In: Möbius, D., Miller, R., Fainerman, V.B. (Eds.), *Surfactants—Chemistry, Interfacial Properties and Application, Studies in Interface Science*, Vol. 13, Elsevier: Amsterdam, 2001, pp. 99–188.
2. Butler, J.A.V. The thermodynamics of the surfaces of solutions. *Proc Roy Soc Ser A* 1932; 138: 348–375.
3. Henry, W. Experiments on the quantity of gases absorbed by water, at different temperatures, and under different pressures. *Phil Trans Roy Soc London* 1803; 93: 29–42.
4. Gibbs, J.W. *Collected Works*, Vol 1. Longmans, Green & Co, 1928.
5. von Szyszkowski, B. Experimentelle Studien über kapillare Eigenschaften der wäßrigen Lösungen von Fettsäuren. *Z Phys Chem (Leipzig)* 1908; 64: 385–414.
6. Langmuir, I. The constitution and fundamental properties of solids and liquids. II. Liquids. *J Amer Chem Soc* 1917; 39: 1848–1907.
7. Frumkin, A. Die Kapillarkurve der höheren Fettsäuren und die Zustandsgleichung der Oberflächenschicht. *Z Phys Chem (Leipzig)* 1925; 116: 466–484.
8. Lucassen-Reynders, E.H. Interactions in mixed monolayers I. Assessment of interaction between surfactants. *J Colloid Interface Sci* 1973; 42: 554–562.
9. Fainerman, V.B., Miller, R., Wüstneck, R., Makievski, A.V. Adsorption isotherm and surface tension equation for a surfactant with changing partial molar area. 1. Ideal surface layer. *J Phys Chem* 1996; 100: 7669–7675.
10. Fainerman, V.B., Miller, R. Surface tension isotherms for surfactant adsorption layers including surface aggregation. *Langmuir* 1996; 12: 6011–6014.
11. Fainerman, V.B., Lylyk, S.V., Aksenenko, E.V., Makievski, A.V., Petkov, J.T., Yorke, J., Miller, R. Adsorption layer characteristics of triton surfactants 1. Surface tension and adsorption isotherms. *Colloids Surfaces A* 3341–3347.

12. Fainerman, V.B., Miller, R., Adsorption isotherms at liquid interfaces. In: Somasundaran, P., Hubbard, A. (Eds.), *Encyclopedia of Surface and Colloid Science*, 2nd. Edition, Vol. 1, Taylor & Francis: Boca Raton, 2009, pp. 1–15.

13. Fainerman, V.B., Wüstneck, R., Miller, R. Surface tension of mixed surfactant solutions. *Tenside Surfactants Detergents* 2001; 38: 224–229.

14. Fainerman, V.B., Lucassen-Reynders, E.H., Miller, R. Description of the adsorption behaviour of proteins at water/fluid interfaces in the framework of a two-dimensional solution model. *Adv Colloid Interface Sci* 2003; 106: 237–259.

15. Dan, A., Gochev, G., Krägel, J., Aksenenko, E.V., Fainerman, V.B., Miller, R. Interfacial rheology of mixed layers of food proteins and surfactants. *Current Opinion in Colloid Interface Sci* 2013; 18: 302–310.

16. Miller, R., Fainerman, V.B., Leser, M.E., Michel, M. Surface tension of mixed non-ionic surfactant/ protein solutions: Comparison of a simple theoretical model with experiments. *Colloids Surfaces A* 2004; 233: 39–42.

17. Fainerman, V.B., Zholob, S.A., Leser, M.E., Michel, M., Miller, R. Adsorption from mixed ionic surfactant/protein solutions—analysis of ion binding. *J Phys Chem* 2004; 108: 16780–16785.

18. Binks, B.P. Particles as surfactants - similarities and differences. *Current Opinion in Coll Interface Sci* 2002; 7: 21–41.

19. Aveyard, R., Binks, B.P., Clint, J.H. Emulsions stabilised solely by colloidal particles. *Adv Colloid Interface Sci* 2003; 100: 503–546.

20. Limage, S., Krägel, J., Schmitt, M., Dominici, C., Miller, R., Antoni, M. Rheology and structure formation in diluted mixed particle-surfactant systems. *Langmuir* 2001; 26: 16754–16761.

21. Stocco, A., Rio, E., Binks, B.P., Langevin, D. Aqucous foams stabilized solely by particles. *Soft Matter* 2011; 7: 1260–1267.

22. Pickering, S.U. Emulsions. *J Chem Soc Trans* 1907; 91: 2001–2021.

23. Binks, B.P., Horozov, T.S. *Colloidal Particles at Liquid Interfaces*. Cambridge University Press: Cambridge, 2006.

24. Levine, S., Bown, B.D., Partridge, S.J. Stabilization of emulsions by fine particles I. Partitioning of particles between continuous phase and oil/water interface. *Colloids Surf* 1989; 38: 325–43.

25. Binks, B.P. Particles as surfactants: Similarities and differences. *Curr Opin Colloid Interface Sci* 2002; 7: 1–41.

26. Guzman, E.K., Santini, E., Liggieri, L., Ravera, F., Loglio, G., Krägel, J., Maestro, A., Rubio, R.G., Grigoriev, D.O., Miller, R., Particle-surfactant interaction at liquid interfaces. In: Kralchevsky, P., Miller, R., Ravera, F. (Eds.), *Progress in Colloid Interface Science*, Vol. 4, Colloid Chemistry in Nanotechnology, CRC Press: Boca Raton, 2013, pp. 77–109.

27. Fainerman, V.B., Aksenenko, E.V., Petkov, J.T., Miller, R., Adsorption layer characteristics of mixed oxyethylated surfactant solutions. *J Phys Chem B* 2001; 114: 4503–4508.

28. Miller, R., Alahverdjieva, V.S., Fainerman, V.B. Thermodynamics and rheology of mixed protein/ surfactant adsorption layers. *Soft Matter* 2008; 4: 1141–1146.

29. Miller, R., Liggieri, L. (Eds.), Bubble and drop interfaces. In: *Progress in Colloid and Interface Science*, Vol. 2, Brill Publ.: Leiden, 2011.

30. Möbius, D., Miller, R. (Eds.), Novel Methods to Study Interfacial Layers. In: *Studies in Interface Science*, Vol. 11, Elsevier: Amsterdam, 2001.

31. Fainerman, V.B., Mys, V.D., Makievski, A.V., Petkov, J.T., Miller, R. Dynamic surface tension of micellar solutions in the millisecond and sub-millisecond time range. *J Colloid Interface Sci* 2006; 302: 40–46.

32. Schulze-Schlarmann, J., Stubenrauch, C., Miller, R. Dynamic surface tensions of C10EO4 solutions measured by bubble pressure tensiometry and drop profile analysis. *Tenside Surfactants Detergents* 2005; 42: 307–312.

33. Fainerman, V.B., Miller, R., Maximum bubble pressure tensiometry: Theory, analysis of experimental constrains and applications. In: Miller, R., Liggieri, L. (Eds.), *Bubble and Drop Interfaces*, Vol. 2, Progress in Colloid and Interface Science, Brill Publ: Leiden, 2011, pp. 75–118.

34. Kovalchuk, V.I., Dukhin, S.S. Dynamic effects in maximum bubble pressure experiments. *Colloids Surfaces A* 2001; 192: 131–155.

35. Joos, P. *Dynamic Surface Phenomena*. VSP: Utrecht, 1999.

36. Maze, C., Burnet, G. A non-linear regression method for calculating surface tension and contact angle from the shape of a sessile drop. *Surface Science* 1969; 13: 451–470.

37. Rotenberg, Y., Boruvka, L., Neumann, A.W. Determination of surface tension and contact angle from the shapes of axisymmetric fluid interfaces. *J Colloid Interface Sci* 1983; 93: 169–183.

38. Javadi, A., Krägel, J., Pandolfini, P., Loglio, G., Kovalchuk, V.I., Aksenenko, E.V., Ravera, F., Liggieri, L., Miller, R. Short time dynamic interfacial tension as studied by the growing drop technique. *Colloids Surfaces A* 2010; 365: 62–69.

39. Fainerman, V.B., Lylyk, S.V., Aksenenko, E.V., Liggieri, L., Makievski, A.V., Petkov, J.T., Yorke, J., Miller, R. Adsorption layer characteristics of Triton surfactants 2. Dynamic surface tensions and adsorption dynamics. *Colloids Surfaces A* 2009; 334: 8–15.

40. Kovalchuk, V.I., Ravera, F., Liggieri, L., Loglio, G., Javadi, A., Kovalchuk, N.M., Krägel, J., Studies in capillary pressure tensiometry and interfacial dilational rheology. In: Miller, R., Liggieri, L. (Eds.), *Bubble and Drop Interfaces*, Vol. 2, Progress in Colloid and Interface Science, Brill Publ: Leiden, 2011, pp. 143–178.

41. Banhart, J., García-Moreno, F., Hutzler, S., Langevin, D., Liggieri, L., Miller, R., Saint-Jalmesand, A., Weaire, D. Foams and emulsions in space. *Europhysics News* 2008; 39: 26–28.

42. Motschmann, H., Teppner, R., Bae, S., Haageand, K., Wantke, D. What do linear and non-linear optical techniques have to offer for the investigation of adsorption layers of soluble surfactants? *Colloid Polym Sci* 2000; 278: 425–433.

43. Motschmann, H., Teppner, R., Ellipsometry in interface science. In: Möbius, D., Miller, R. (Eds.), *Novel Methods to Study Interfacial Layers*, Vol. 11, Studies in Interface Science, Elsevier: Amsterdam, 2001, pp. 1–42.

44. Harke, M., Teppner, R., Schulz, O., Orendi, H., Motschmann, H. Description of a single modular optical setup for ellipsometry, surface plasmons, waveguide modes, and their corresponding imaging techniques including Brewster angle microscopy. *Rev Sci Instrum* 1997; 68: 3130–3134.

45. Walsh, C.B., Wen, X., Franses, E.I. Ellipsometry and infrared reflection absorption spectroscopy of adsorbed layers of soluble surfactants at the air-solution interface. *J Colloid Interface Sci* 2001; 233: 295–305.

46. Blomqvist, B.R., Benjamins, J.W., Nylander, T., Arnebrant, T. Ellipsometric characterization of ethylene-oxide-butylene oxide diblock copolymer adsoerption at the air-water interface. *Langmuir* 2005; 21: 5061–5068.

47. Mora, M.F., Wehmeyer, J.L., Synowicki, R., Garcia, C.D., Investigationg protein adsorption via spectroscopic ellipsometry. In: Puleo, D.A., Bizios, R. (Eds.), *Biological Interactions on Material Surfaces. Understanding and Controlling Protein, Cell, and Tissue Responses*, Springer-Verlag: New York, 2009, pp. 19–41.

48. Taylor, D.J.F., Thomas, R.K., Penfold, J. Polymer/surfactant interactions at the air/water interface. *Adv Colloid Interface Sci* 2007; 132: 69–110.

49. Koelsch, P., Motschmann, H. A method for direct determination of the prevailing counterion distribution at a charged surface. *J Phys Chem B* 2004; 108: 18659–18664.

50. Penfold, J. The structure of the surface of pure liquids. *Rep Prog Phys* 2001; 64: 777–814.

51. Teppner, R., Bae, S., Haage, K., Motschmann, H. On the analysis of ellipsometric measurements of adsorption layers at fluid interfaces. *Langmuir* 1999; 15: 7002–7007.

52. Jubb, A.M., Hua, W., Allen, H.C. Environmental chemistry at vapour/water interfaces: Insights from vibrational sum frequency generation spectroscopy. *Ann Rev Phys Chem* 2012; 63: 107–130.

53. Engelhardt, K., Peukert, W., Braunschweig, B. Vibrational sum-frequency generation at protein modified air–water interfaces: Effects of molecular structure and surface charging. *Curr Opin Colloid & Interface Science* 2014; 19: 207–215.

54. Engelhardt, K., Lexis, M., Gochev, G., Konnerth, C., Miller, R., Willenbacher, N., Peukert, W., Braunschweig, B. pH effects on the molecular structure of β-lactoglobulin modified air-water interfaces and its impact on foam rheology. *Langmuir* 2013; 29: 11646–11655.

55. Ulaganathan, V., Gochev, G., Gehin-Delval, C., Leser, M.E., Gunes, D.Z., Miller, R., Effect of pH and salt concentration on rising air bubbles in ß-lactoglobulin solutions. *Colloids Surfaces A* 2016; 505:165–170.

56. Penfold, J., Thomas, R.K. Mixed surfactants at the air-water interface. *Ann Rep Prog Chem Sect C* 2010; 106: 14–35.

57. Zhao, X.B., Pan, F., Lu, J.R. Interfacial assembly of proteins and peptides: Recent examples studied by neutron reflection. *J R Soc Interface* 2009; 6: S659–S670.

58. Penfold, J., Thomas, R.K. Neutron reflectivity and small angle neutron scattering: An introduction and perspective on recent progress. *Curr Opin Colloid & Interface Science* 2014; 19: 198–206.

59. Campbell, R.A., Tummino, A., Noskov, B.A., Varga, I. Polyelectrolyte/surfactant films spread from neutral aggregates. *Soft Matter* 2016; 12: 5304–5312.

60. Jensen, T.R., Kjaer, K., Structural properties and interactions of thin films at the air-liquid interface explored by synchrotron X-ray scattering, in *"Novel Methods to Study Interfacial Layers"*, Vol. 11, Studies in Interface Science, Möbius, D., Miller, R. (Eds.), Elsevier: Amsterdam, 2001, pp. 205–254.

61. Stefaniu, R.C., Brezesinski, G. X-ray investigation of monolayers formed at the soft air/water interface. *Curr Opin Colloid & Interface Science* 2014; 19: 216–227.
62. Brun, A.P., Clifton, L.A., Halbert, C.E., Lin, B., Meron, M., Holden, P.J., Lakey, J.H., Holt, S.A. Structural characterization of a model Gram-negative bacterial surface using lipopolysaccharides from rough strain of Escherichia coli. *Biomacromolecules* 2013; 14: 2014–2022.
63. Ward, A.F.H., Tordai, L. Time-dependence of boundary tensions of solutions. *J Phys Chem* 1946; 14: 453–461.
64. Miller, R., Fainerman, V.B., Aksenenko, E.V., Leser, M.E., Michel, M. Dynamic surface tension and adsorption kinetics of β casein at the solution/air interface. *Langmuir* 2004; 20: 771–777.
65. Fainerman, V.B. Adsorption kinetics of surfactants from solutions. *Kolloid Zh* 1977; 39: 106–112.
66. MacLeod, C.A., Radke, C.J. Surfactant exchange kinetics at the air/water interface from the dynamic tension of growing liquid drops. *J Colloid Interface Sci* 1994; 166: 73–88.
67. Baret, J.F. Theoretical model for an interface allowing a kinetic study of adsorption, *J Colloid Interface Sci* 1969; 30: 1–12.
68. Miller, R., Kretzschmar, G. Numerische Lösung für ein gemischtes Modell der diffusions- kinetik-kontrollierten Adsorption. *Colloid Polymer Sci* 1980; 258: 85–87.
69. Nikolov, A., Martynov, G., Exerowa, D. Associative interactions and surface tension in ionic surfactant solutions at concentrations much lower than the CMC. *J Colloid Interface Sci* 1981; 81: 116–124.
70. Lin, S.Y., McKeigue, K., Maldarelli, C. Diffusion-limited nterpretation of the induction period in the relaxation in surface tension due to adsorption. *Langmuir* 1991; 7: 1055–1066.
71. Hirte, R., Lunkenheimer, K. Surface equation of state and transitional behavior of adsorption layers of soluble amphiphiles at fluid interfaces. *J Phys Chem* 1996; 100: 13786–13793.
72. Aksenenko, E.V., Fainerman, V.B., Miller, R. Dynamics of surfactant adsorption from solution considering aggregation within the adsorption layer. *J Phys Chem* 1998; 102: 6025–6028.
73. Fainerman, V.B., Möbius, D., Miller, R. (Eds.), Surfactants—Chemistry, Interfacial Properties and Application, *Studies in Interface Science*, Vol. 13, Elsevier: Amsterdam, 2001.
74. Fainerman, V.B., Miller, R., Aksenenko, E.V., Makievski, A.V., Krägel, J., Loglio, G., Liggieri, L. Effect of surfactant interfacial orientation/aggregation on adsorption dynamics. *Adv Colloid Interface Sci* 2000; 86: 83–101.
75. Sutherland, K.L. The kinetics of adsorption at liquid surfaces. *Austr J Sci Res* 1952; A5: 683–696.
76. Hansen, R.S. The theory of diffusion controlled adsorption kinetics with accompanying evaporation. *J Phys Chem* 1960; 64: 637–641.
77. Makievski, A.V., Fainerman, V.B., Miller, R., Bree, M., Liggieri, L., Ravera, F. Determination of equilibrium surface tension values by extrapolation via long time approximations. *Colloids Surfaces A* 1997; 122: 269–273.
78. Miller, R., Noskov, B.A., Fainerman, V.B., Petkov, J.G., Impact of micellar kinetics on dynamic interfacial properties of surfactant solutions, in *Highlights in Colloid Science*, D. Platikanov, D. Exerowa (Eds.), Wiley-VHC: Weinheim, 2008, pp. 247–259.
79. Fainerman, V.B., Aksenenko, E.V., Mys, A.V., Petkov, J.T., Yorke, J., Miller, R. Adsorption layer characteristics of mixed SDS/CnEOm solutions. 3. Dynamics of adsorption and surface dilational rheology of micellar solutions. *Langmuir* 2010; 26: 2424–2429.
80. Fainerman, V.B., Aksenenko, E.V., Petkov, J.T., Miller, R. Influence of solubilized dodecane on the dynamic surface tension and dilational rheology of micellar Triton X-45 and SDS solutions. *Colloids Surfaces A* 2012; 413: 125–129.
81. Eastoe, J., Dalton, J.S. Dynamic surface tension and adsorption mechanisms of surfactants at the air-water interface. *Adv Colloid Interface Sci*, 2000; 85: 103–144.
82. Danov, K.D., Kralchevsky, P.A., Denkov, N.D., Ananthapadmanabhan, K.P., Lips, A. Mass transport in micellar surfactant solutions: 1. Relaxation of micelle concentration, aggregation number and polydispersity. *Adv Colloid Interface Sci* 2006; 119: 1–16.
83. Warszyński, P., Wantkeand, K.D., Fruhner, H. Theoretical description of surface elasticity of ionic surfactants. *Colloids Surfaces A* 2001; 189: 29–53.
84. Kalinin, V.V., Radke, C.J. An ion-binding model for ionic surfactant adsorption at aqueous-fluid interfaces. *Colloids Surfaces A* 1996; 114: 337–350.
85. Danov, K.D., Kralchevsky, P.A., Ananthapadmanabhan, K.P., Lips, A. Micellar surfactant solutions: Dynamics of adsorption at fluid interfaces subjected to stationary expansion. *Colloids Surfaces A* 2006; 282–283: 143–161.
86. Miller, R., Lunkenheimer, K., Kretzschmar, G. Ein modell für die diffusionskontrollierte adsorption von tensidgemischen an fluiden phasengrenzen. *Colloid Polymer Sci* 1979; 257: 1118–1120.

87. Miller, R., Makievski, A.V., Frese, C., Krägel, J., Aksenenko, E.V., Fainerman, V.B. Adsorption kinetics of surfactant mixtures at the aqueous solution—air interface. *Tenside Surfactants Detergents* 2003; 40: 256–259.

88. Lunkenheimer, K., Haage, K., Miller, R. On the adsorption properties of surface-chemically pure aqueous solutions of n-alkyl-dimethyl and n-alkyl-diethyl phosphine oxides. *Colloids Surfaces* 1987; 22: 215–224.

89. Frese, Ch., Ruppert, S., Schmidt-Lewerkühne, H., Wittern, K.P., Eggers, R., Fainerman, V.B., Miller, R., Analysis of dynamic surface tension data for SDS-DTAB mixed solutions. *PCCP* 6, 2004 1592–1596.

90. Aidarova, S., Sharipova, A., Krägel, J., Miller, R. Polyelectrolytes/surfactant mixtures in the bulk and at water/oil interfaces. *Adv Colloid Interface Sci* 2014; 205: 87–93.

91. Sharipova, A.A., Aidarova, S.B., Grigoriev, D., Mutalieva, B., Madibekova, G., Tleuova, A., Miller, R. Polymer-surfactant complexes for microencapsulation of vitamin E and its release. *Colloids Surfaces B* 2016; 137: 152–157.

92. Sharipova, A., Aidarova, S., Cernoch, P., Miller, R. Effect of surfactant hydrophobicity on the interfacial properties of polyallylamine hydrochloride/sodium alkyl sulphate at water/hexane interface. *Colloids Surfaces A* 2013; 438: 141–147.

93. Wege, H.A., Holgado-Terriza, J.A., Neumann, A.W., Cabrerizo-Vilchez, M.A. Axisymmetric drop shape analysis as penetration film balance applied at liquid-liquid interfaces. *Colloids Surf A* 1999; 156: 509–517.

94. Kotsmar, C., Grigoriev, D.O., Makievski, A.V., Ferri, J.K., Krägel, J., Miller, R., Möhwald, H. Drop profile analysis tensiometry with drop bulk exchange to study the sequential and simultaneous adsorption of a mixed β-casein /C12DMPO system. *Colloid Polymer Sci* 2008; 286: 1071–1077.

95. Dan, A., Kotsmar, C., Ferri, J.K., Javadi, A., Karbaschi, M., Krägel, J., Wüstneck, R., Miller, R. Mixed protein-surfactant adsorption layers formed in a sequential and simultaneous way at the water/air and water/oil interfaces. *Soft Matter* 2012; 8: 6057–6065.

96. Fainerman, V.B., Lylyk, S., Ferri, J.K., Miller, R., Watzke, H., Leser, M.E., Michel, M. Adsorption kinetics of proteins at the solution air interfaces with controlled bulk convection. *Colloids Surfaces A* 2006; 282–283: 217–221.

97. Gochev, G., Retzlaff, I., Aksenenko, E.V., Fainerman, V.B., Miller, R. Adsorption isotherms and equation of state for adsorbed β-Lactoglobulin layers at air/water surface. *Colloids Surfaces A* 2013; 422: 33–38.

98. Won, J.Y., Gochev, G.G., Ulaganathan, V., Krägel, J., Aksenenko, E.V., Fainerman, V.B., Miller, R. Effect of solution pH on the adsorption of BLG at the solution/tetradecane interface. *Colloids Surfaces A* 2017; 519: 161–167.

99. Ulaganathan, V., Retzlaff, I., Won, J.Y., Gochev, G., Gehin-Delval, C., Leser, M.E., Noskov, B.A., Miller, R., β-lactoglobulin adsorption layers at the water/air surface: 1. Adsorption kinetics and surface pressure isotherm: Effect of pH and ionic strength. *Colloids Surfaces A* 2017; 519: 153–160.

100. Won, J.Y., Gochev, G.G., Ulaganathan, V., Krägel, J., Aksenenko, E.V., Fainerman, V.B., Miller, R. Mixed Adsorption mechanism for the kinetics of BLG interfacial layer formation at the solution/tetradecane interface. *Colloids Surfaces A* 2017; 519: 146–152.

101. Ulaganathan, V., Retzlaff, I., Won, J.Y., Gochev, G., Gunes, D.Z., Gehin-Delval, C., Leser, M., Noskov, B.A., Miller, R., β-lactoglobulin adsorption layers at the water/air surface: 2. Dilational rheology: Effect of pH and ionic strength. *Colloids Surfaces A* 2017; 521: 167–176.

102. Won, J.Y., Gochev, G.G., Ulaganathan, V., Krägel, J., Aksenenko, E.V., Fainerman, V.B., Miller, R. Dilational viscoelasticity of BLG adsorption layers at the solution/tetradecane interface—effect of pH and ionic strength. *Colloids Surfaces A* 2017; 521: 204–210.

103. Lucassen-Reynders, E.H., Lucassen, J., History of dilational rheology, Chapter 2. In: Miller, R., Liggieri, L. (Eds.), *Progress in Colloid and Interface Science*, Vol. 1, Brill: London–Boston, 2009, pp. 40–78.

104. Zholob, S.A., Kovalchuk, V.I., Makievski, A.V., Krägel, J., Fainerman, V.B., Miller, R., Determination of the dilational elasticity and viscosity from the surface tension response to harmonic area perturbation, Ch. 3. In: Miller, R., Liggieri, L. (Eds.), *Progress in Colloid and Interface Science*, Vol. 1, Brill: London–Boston, 2009, pp. 80–105.

105. Ravera, F., Liggieri, L., Loglio, G., Dilational rheology of adsorbed layers by oscillating drops and bubbles, Ch. 5. In: Miller, R., Liggieri, L. (Eds.), *Progress in Colloid and Interface Science*, Vol. 1, Brill: London–Boston, 2009, pp. 137–177.

106. Ivanov, I.B., Danov, K.D., Ananthapadmanabhan, K.P., Lips, A. Interfacial rheology of adsorbed layers with surface reaction: On the origin of the dilational surface viscosity. *Adv Colloid Interface Sci* 2005; 114–115: 61–92.

107. Liggieri, L., Miller, R. Relaxation of surfactant adsorption layers at liquid interface. *Current Opinion in Colloid and Interface Sci* 2010; 15: 256–263.
108. Miller, R., Ferri, J.K., Javadi, A., Krägel, J., Mucic, N., Wüstneck, R. Rheology of interfacial layers. *Colloid and Polymer Science* 2010; 288: 937–950.
109. Sagis, L.M.C. Nonlinear rheological models for structured interfaces. *Physica A* 2010; 389: 1993–2006.
110. Ravera, F., Ferrari, M., Santini, E., Liggieri, L. Influence of surface processes on the dilational visco-elasticity of surfactant solutions. *Adv Colloid Interface Sci* 2005; 117: 75–100.
111. Noskov, B.A., Loglio, G., Miller, R. Dilational surface visco-elasticity of polyelectrolyte/surfactant solutions: Formation of heterogeneous adsorption layers. *Adv Colloid Interface Sci* 2011; 168: 179–197.
112. Noskov, B.A., Capillary waves in interfacial rheology, Ch. 4. In: Miller, R., Liggieri, L. (Eds.), *Progress in Colloid and Interface Science*, Vol. 1, Brill: London–Boston, 2009, pp. 103–136.
113. Kokelaar, J.J., Prins, A., de Gee, M.J. A new method for measuring the surface dilational modulus of a liquid. *J Colloid Interface Sci* 1991; 146: 507–511.
114. Monroy, F., Ortega, F., Rubio, R.G., Velarde, M.G. Surface rheology, equilibrium and dynamic features at interfaces, with emphasis on efficient tools for probing polymer dynamics at interfaces. *Adv Colloid Interface Sci* 2007; 134–135: 175–189.
115. Leser, M.E., Acquistapace, S., Cagna, A., Makievski, A.V., Miller, R. Limits of oscillation frequencies in drop and bubble shape tensiometry. *Colloids Surfaces A* 2005; 261: 25–28.
116. Cabrerizo-Vilchez, M.A., Wege, H.A., Holgado-Terriza, J.A., Neumann, A.W., Axisymmetric drop shape analysis as penetration Langmuir balance. *Rev Sci Instrum* 1999; 70: 2438–2444.
117. Noskov, B.A., Loglio, G. Dynamic surface elasticity of surfactant solutions. *Colloids Surfaces A* 1998; 143: 167–183.
118. Ravera, F., Ferrari, M., Liggieri, L. Modeling of dilational visco-elasticity of adsorbed layers with multiple kinetic processes. *Colloids and Surfaces A* 2006; 282–283: 210–216.
119. Fainerman, V.B., Aksenenko, E.V., Lylyk, S.V., Makievski, A.V., Ravera, F., Petkov, J.T., Yorke, J., Miller, R. Adsorption layer characteristics of Triton surfactants. 3. Dilational visco-elasticity. *Colloids Surfaces A* 2009; 334: 16–21.
120. Aksenenko, E.V., Kovalchuk, V.I., Fainerman, V.B., Miller, R. Surface dilational rheology of mixed adsorption layers at liquid interfaces. *Adv Colloid Interface Sci* 2006; 122: 57–66.
121. Kovalchuk, V.I., Miller, R., Fainerman, V.B., Loglio, G. Dilational rheology of adsorbed surfactant layers—role of the intrinsic two-dimensional compressibility. *Adv Colloid Interface Sci* 2005; 114–115: 303–313.
122. Aksenenko, E.V., Kovalchuk, V.I., Fainerman, V.B., Miller, R. Surface dilational rheology of mixed surfactant layers at liquid interfaces. *J Phys Chem C* 2007; 111: 14713–14719.
123. Fainerman, V.B., Aksenenko, E.V., Krägel, J., Miller, R. Visco-elasticity moduli of aqueous $C_{14}EO_8$ solutions as studied by drop and bubble shape methods. *Langmuir* 2013; 29: 6964–6968.
124. Fainerman, V.B., Kovalchuk, V.I., Aksenenko, E.V., Miller, R. Dilational viscoelasticity of adsorption layers measured by drop and bubble profile analysis: Reason for different results. *Langmuir* 2016; 32: 5500–5509.
125. Lucassen-Reynders, E.H., Interactions in mixed monolayers. III. Effect on dynamic surface properties. *J Colloid Interface Sci* 1973; 42: 573–580.
126. Garrett, P.R., Joos, P. Dynamic dilational surface properties of submicellar multicomponent surfactant solutions. Part I. Theoretical. *J Chem Soc Faraday Trans* 1976; 172(1): 2161–2173.
127. Jiang, Q., Valentini, J.E., Chiew, Y.C. Theoretical models for dynamic dilational surface properties of binary surfactant mixtures. *J Colloid Interface Sci* 1995; 174: 268–271.
128. Fainerman, V.B., Kovalchuk, V.I., Leser, M.S., Miller, R. Effect of the intrinsic compressibility on the dilational rheology of adsorption layers of surfactants, proteins and their mixtures. In: Tadros, Th. (Ed.), *Colloid and Interface Science*. Vol. 1, *Colloid Stability: The Role of Surface Forces, Part 1*, WILEY-VCH Verlag: Weinheim, 2007, pp. 307–333.
129. Aksenenko, E.V., Fainerman, V.B., Petkov, J.T., Miller, R. Dynamic surface tension of mixed oxyethylated surfactant solutions. *Colloids and Surfaces A* 2010; 365: 210–214.
130. Fainerman, V.B., Aksenenko, E.V., Zholob, S.A., Petkov, J.T., Yorke, J., Miller, R. Adsorption layer characteristics of mixed SDS/CnEOm solutions. 2. Dilational visco-elasticity. *Langmuir* 2010; 26: 1796–1801.
131. Benjamins, J., Lucassen-Reynders, E.H., Interfacial rheology of adsorbed protein layers, Ch. 7. In: Miller, R., Liggieri, L. (Eds.), *Progress in Colloid and Interface Science*, Vol. 1, Brill: London–Boston, 2009, pp. 257–305.

132. Noskov, B.A. Protein conformational transitions at the liquid–gas interface as studied by dilational surface rheology. *Adv Colloid Interface Sci* 2014; 206: 222–238.

133. Lucassen-Reynders, E.H., Benjamins, J., Fainerman, V.B., Dilational rheology of protein films adsorbed at fluid interfaces. *Curr Opin Colloid Interface Sci* 2010; 15: 264–270.

134. Wüstneck, R., Fainerman, V.B., Aksenenko, E.V., Kotsmar, C., Pradines, V., Krägel, J., Miller, R. Surface dilatational behavior of ß-casein at the solution/air interface at different pH values. *Colloids and Surfaces A* 2012; 404: 17–24.

135. Noskov, B.A., Mikhailovskaya, A.A., Lin, S.-Y., Loglio, G., Miller, R. Bovine serum albumin unfolding at the air/water interface. *Langmuir* 2010; 26: 17225–17231.

136. Mikhailovskaya, A.A., Noskov, B.A., Nikitin, E.A., Lin, S.-Y., Loglio, G., Miller, R. Dilational surface viscoelasticity of protein solutions. Impact of urea. *Food Hydrocolloids* 2014; 34: 98–103.

137. Kotsmar, C., Aksenenko, E.V., Fainerman, V.B., Pradines, V., Krägel, J., Miller, R. Equilibrium and dynamics of adsorption of mixed β-casein/surfactant solutions at the water/hexane interface. *Colloids Surfaces A* 2010; 354: 210–217.

138. Kovalchuk, V.I., Aksenenko, E.V., Miller, R., Fainerman, V.B., Surface dilational rheology of mixed adsorption layers of proteins and surfactants at liquid interfaces, Chapter 9. In: Miller, R., Liggieri, L. (Eds.), *Progress in Colloid and Interface Science*, Vol. 1, Brill: London–Boston, 2009, pp. 335–374.

139. Maldonado-Valderrama, J., Rodríguez-Patino, J.M. Interfacial rheology of protein-surfactant mixtures. *Curr Opin Colloid Interface Sci* 2010; 15: 271–282.

140. Miller, R., Leser, M.E., Michel, M., Fainerman, V.B. Surface dilational rheology of mixed ß-lactoglobulin/ surfactant adsorption layers at the air/water interface. *J Phys Chem* 2005; 109: 13327–13331.

141. Mikhailovskaya, A.A., Noskov, B.A., Lin, S.-Y., Loglio, G., Miller, R. Formation of protein/surfactant adsorption layer at the air/water interface as studied by dilational surface rheology. *J Phys Chem B* 2011; 115: 9971–9979.

142. Fainerman, V.B., Aksenenko, E.V., Krägel, J., Miller, R. Thermodynamics, interfacial pressure isotherms and dilational rheology of mixed protein-surfactant adsorption layers. *Adv. Colloid Interface Sci* 2016; 233: 200–222.

143. Kotsmar, C., Krägel, J., Kovalchuk, V.I., Aksenenko, E.V., Fainerman, V.B., Miller, R. Dilation and shear rheology of mixed β-casein/surfactant adsorption layers. *J Phys Chem B* 2009; 113: 103–113.

144. Arabadzhieva, D., Tchoukov, P., Mileva, E., Miller, R., Soklev, B. Impact of amphiphilic nanostructures on formation and rheology of adsorption layers and on foam film drainage. *Ukr J Physics* 2011; 58: 801–810.

145. Arabadzhieva, D., Mileva, E., Tchoukov, P., Miller, R., Ravera, F., Liggieri, L. Adsorption layer properties and foam film drainage of aqueous solutions of tetraethylene-glycol monododecyl ether. *Colloids and Surfaces A* 2011; 392: 233–241.

146. Arabadzhieva, D., Tchoukov, P., Mileva, E., Soklev, B. Interfacial layer properties and foam film drainage of aqueous solutions of hexadecyltrimethylammonium chloride. *Colloids and Surfaces A* 2014; 460: 28–37.

147. Maestro, A., Kotsmar, C., Javadi, A., Miller, R., Ortega, F., Rubio, R.G. Adsorption of β-casein - surfactant mixed layers at the air-water interface evaluated by interfacial rheology. *J Phys Chem B* 2012; 116: 4898–4907.

148. Dan, A., Wüstneck, R., Krägel, J., Aksenenko, E.V., Fainerman, V.B., Miller, R. Interfacial adsorption and rheological behavior of β-casein at the water/hexane interface at different pH. *Food Hydrocolloids* 2014; 34: 193–201.

149. Dan, A., Gochev, G., Miller, R. Tensiometry and dilational rheology of mixed β-lactoglobulin/ionic surfactant adsorption layers at water/air and water/hexane interfaces. *J Colloid Interface Sci* 2015; 449: 383–391.

150. Kazakov, V.N., Barkalova, E.L., Levchenko, L.A., Klimenko, T.M., Fainerman, V.B., Miller, R. Dilation rheology as medical diagnostics of human biological liquids. *Colloids Surfaces A* 2011; 391: 190–194.

151. Krägel, J., Derkatch, S.R., Interfacial Shear Rheology—An overview of measuring techniques and their applications. In: Miller, R., Liggieri, L. (Eds.), *Interfacial Rheology*, Brill: Leiden, 2009, pp. 372–428.

152. Tschoegl, N.W. The mathematical relations of the torsion pendulum in the study of surface films. *Kolloid Z* 1961; 181: 19–20.

153. Krägel, J., Derkatch, S.R., Miller, R. Interfacial shear rheology of protein-surfactant layers. *Adv Colloid Interface Sci* 2008; 144: 38–53.

154. Erni, P., Fischer, P., Windhab, E.J., Kusnezov, V., Stettin, H., Läuger, J. *Rev Sci Instrum* 2003; 74: 4916–4924.

155. Petkov, J.T., Danov, K.D., Denkov N.D., Precise method for measuring the shear surface viscosity of surfactant monolayers. *Langmuir* 1996; 12(11): 2650–2653.
156. Krägel, J., Kretzschmar, G., Li, J.B., Loglio, G., Miller, R., Möhwald, H. Surface rheology of monolayers. *Thin Solid Films* 1996; 284–285: 361–364.
157. Kotsmar, C., Krägel, J., Kovalchuk, V.I., Aksenenko, E.V., Fainerman, V.B., Miller, R. Dilation and shear rheology of mixed BCS/surfactant adsorption layers. *J Phys Chem B* 2009; 113: 103–113.
158. Maestro, A., Ortega, F., Monroy, F., Krägel, J., Miller, R. Surface Shear Rheology properties of poly(methylmetacrylate) Langmuir films: A comparison between two different surface shear rheometers. *Langmuir* 2009; 25: 7393–7400.

Section II

Foam Films

2 Historical Perspectives on Foam Films

Georgi Gochev, Dimo Platikanov, and Reinhard Miller

CONTENTS

Foam films are the subject of many studies in colloid and interface science. The history of this scientific topic dates back to the late 17th century, with pioneers such as Hooke [1], Newton [2], Plateau [3], and Gibbs [4,5] having created the basis of our present knowledge. Through their curiosity and systematic approach, they observed and were the first to explain many simple day-to-day phenomena. As the painting of Rembrandt below shows, soap bubbles were an attractive object for art even before these early scientific studies.

Rembrandt Harmenszoon van Rijn, Cupid with the Soap Bubble, 1634.
Liechtenstein Museum, Vienna. (Picture taken from an internet source http://www.liechtensteincollections.at)

The most intensive work on the fundamental properties of foam films was performed in the 20th century. This historical perspective, however, is focused mainly on the early findings for this topic. The recent state of the art of the many facets of foam films will be dealt with in the subchapters of Chapter 3.

2.1   EARLY TIMES: FROM HOOKE AND NEWTON TO PLATEAU AND GIBBS

In his book "*Statique experimentale et theoryque des liquides soumis aux seules forces moleculaires*" (*Experimental and Theoretical Statics of Liquids Subject to Only Molecular Forces*; hereafter simply referred to as *Statique*), published in 1873 [3], Plateau—the blind Belgian scientist—explained his observations and conclusions within a systematic study of various properties of foam films. In addition, along with his own achievements, he commented on the findings of scientists who preceded him in the period from the pioneering observations of Boyle, Hooke and Newton until 1869. We will discuss the early knowledge of foam films below mainly on the basis of this source [3].

Robert Boyle and Robert Hooke were the first to investigate foam films in the second half of the 17th century. According to §317 in *Statique* [3], Boyle (1627–1691), mostly known for his achievements in chemistry and physics, was the first to draw attention to the colors of foam bubbles. In 1663, he published his work entitled "*Experiments and observations upon colors*," from which we took the following excerpt [3]:

To show the chemists that one can make appear or disappear colors where there is neither increase nor change of the sulfurous, saline or mercurial principles of the bodies, I did not resort to the iris produced by the glass prism, nor to the colors which one sees, on a serene morning, in those dewdrops which reflect or refract suitably towards the eye the rays of the light; but I will point out to them what they can observe in their laboratories: because if a chemical essential oil or concentrated spirit of wine is shaken until bubbles develop at its surface, those offer brilliant and varied colors which disappear all at the moment when the liquid which constitutes the films falls down in the remainder of oil or spirit of wine; one can thus make it so that a colorless liquid shows various colors and loses them in one moment, without increase nor reduction in any of its hypostatic principles. And, to say in passing, it is worthy of remark that certain bodies, either colorless, or colored, being brought to a great thinness, acquire colors which they did not have before; indeed, besides the variety of colors that water made, viscous by soap acquires when it is inflated in spherical bubbles, terpentine, when air in a certain manner is insufflated there, provides bubbles variously colored, and, although these colors disappear as soon as the bubbles burst, those would probably continue to express varied nuances on their surface, if their texture were sufficiently durable.

Hooke (1635–1703) pursued broad scientific interests, ranging from physics to architecture. He came very close to reaching the conclusion that gravity follows an inverse square law, and that such a relation governs the motions of the planets. This idea was subsequently developed by Newton. Hooke was an assistant to Boyle, which may have been the reason for his special interest in the coloring of foam bubbles. In 1672, he communicated to the Royal Society his work "*On holes (Black Film) in Soap Bubbles*" [1,6], a facsimile of which is shown in Figure 2.1. We underlined some phrases to emphasize Hooke's astonishment about the results of his experiments on foam bubbles. For a fresh bubble, he observed that initially it was "*white and clear*," and after a certain time he noted that "*all variety of colors that may be observed in a rainbow*" appeared. These colors were grouped in several series and underwent several changes; however, after a certain period of time, the bubbles appeared again as white, but with large "holes." To the best of our knowledge, this is the first documented observation of the phenomenon of black spots in foam films.

Isaac Newton (1642–1727) studied the colors of film caps resting on a solution and attempted to find analogy with his earlier observations of colors of the air wedge between two glass plates, which were a "little convex" (i.e., Newton rings) [2]. As detailed in observations 17 and 18 in "*The Second Book of Optics*" [2], Newton covered a fresh bubble with clear glass, leaving it undisturbed by any air flow. In this way, he observed the appearance of colors in a very regular order as concentric rings encompassing the bubble apex as the color successions formed bands of several orders, as shown in Table 2.1. Subsequently, the color rings started dilating, flowing down, and spreading over the whole bubble area, becoming more distinguishable. Newton related this phenomenon to drainage in the foam film and film thinning. During the first stage, a small circular black spot emerged at the apex.

" paved with green tiles, as the houfes with divers colours.   The fteeples for the
" moft part like ours, but built of mud, except which are of fquare ftone."

Mr. HOOKE brought in his written account of an experiment, made March
13, upon a bubble of water and foap, which was ordered to be regiftered ',
and was as follows :

" By the help of a fmall glafs-pipe there were blown feveral fmall bubbles out
" of a mixture of foap and water; where it was obvious to obferve, that at the
" beginning to blow any of thefe bubbles, the orbicular film of water, which en-
" compalied a globe of air, appeared white and clear, without any appearance of
" colour ; but after fome time the film by degrees growing thinner, (part there-
" of falling down, and part thereof evaporating and wafting into air) there ap-
" peared upon the furface thereof all variety of colours, that may be obferved in
" a rainbow, beginning at firft with a pale yellow, then orange, red, purple,
" blue, green, and fo onward, with other the fame feri s or fucceffions of colours :
" in which it was farther notable, that the firft and laft feries of colours were
" very faint, and that the middlemoft order or feries was very bright and orien-
" tal.   After thefe colours had paffed over their feveral changes, the film of the
" bubble began to appear again white, and prefently up and down in this fecond
" white film there appear feveral holes, which by degrees increafe and grow big-
" ger, and feveral of them break into one another, till at length they become very
" confpicuous and big.   It is ftrange to obferve, how thofe holes will, by the
" blowing or moving of the ambient air, be carried up and down upon the en-
" compafled globe of air, and yet the bubble remain in its orbicular form with-
" out falling.   It is yet further ftrange, that after this, when the bubble breaks,
" its breaking is with a kind of impetus or crack, difperfing the parts in a kind
" of powder or mift.   It is yet further ftrange, that thofe parts of the bubble,
" which thus appear like holes through it, by the moving up and down upon
" the furface of the aerial globe will change its form, and from a circular
" be made eliptical, or any other undulated or waved form, in the fame man-
" ner as any of the colours, that are vifible on the bubble.   It is yet more ftrange,
" that though it is moft certain, that both the incompaffing and incompaffed
" air have furfaces, yet by no means, that I have yet made ufe of, will they
" afford either reflection or refraction, which all the other parts of the incom-
" paffed air do.   It is pretty hard to imagine, what curious net or invifible body
" it is, that fhould keep the form of the bubble, or what kind of magnetifm it
" is, that fhould keep the film of water from falling down, or the parts of in-
" cluded and including air from uniting.   The experiment, though at firft thought
" it may feem one of the moft trivial in nature, yet as to the finding out the na-
" ture and caufe of reflection, refraction, colours, congruity and incongruity,
" and feveral other properties of nature, I look upon it as one of the moft in-
" ftructive : of which more hereafter perhaps."

' Vol. iv. p. 128.

He

FIGURE 2.1   Documented observations of Hooke in 1672 (From Birch, T. A. *Millard London* 1757; 3: 29[6].)

TABLE 2.1
Newton's Color Bands

Order	Color Sequence
1	Black, Blue, White, Yellow, Red
2	Violet, Blue, Green, Yellow, Red,
3	Purple, Blue, Green, Yellow, Red,
4	Green, Red
5	Blue, Red
6	Blue, Red
7	...

First, he wondered why there was no reflection from the black portion of the film but, after more detailed inspection, he saw several smaller round spots which appeared even "blacker." Furthermore, he was able to see, albeit very faintly, the reflected images of the sun or a candle. Through these observations, he realized that there is some reflection from all parts of the film of different blackness. He also observed that some colored spots as well as small black spots moved chaotically; these spots were generated at the sides of the film and moved up toward the apex.

It is amazing that Newton succeeded in estimating the film thickness for a given color. He also noted, however, that when the medium in the wedge had a higher refractive index (e.g., water or oil), the diameters of the rings contracted, preserving the same colors. Newton found by exact measurements that the thicknesses of the interposed media were proportional to the inverse ratio of their refractive indices. Thus, because the thickness of the white film produced in a vacuum or in air is about 143 nm (see Reference 2 Book II, obs. 6), the white layer produced in water is a factor of 1/1.33 times that thickness (i.e., $\approx$107 nm). The Newton scale for liquid bubbles is summarized in Table 2.1.

The name of Johann Gottlob Leidenfrost (1715–1794) is mostly associated with the Leidenfrost effect. In 1756 (the same year he became a member of the Prussian Academy of Sciences), he published "*A Tract about Some Qualities of Common Water*" [7], in which, along with other findings, he reported for the first time on the "contractile force" of soap bubbles, that is, when a bubble is blown by a tube and the tube is left open, the bubble shrinks gradually and finally vanishes, expelling all the air it contains. Plateau critically commented on this work, stating that the bubble exists in a solid-like state; however, he conceded that Leidenfrost was the first to notice the contractile force (i.e., surface tension), saying "*Leidenfrost does not extend this contractile force to a general property of liquid surfaces, …, moreover, still according to him, the aqueous part of the film has a force of opposite nature, being an explosive force, which makes the bubble to burst.*" Leidenfrost also attempted to evaluate the film thickness in the moment when the bubble is formed by a simple method that was independent of the film color. Using a tube in which a soap solution spontaneously rises by capillarity, he inflated exactly the same amount of solution and obtained a bubble which did not carry any suspended drop. Then, by weighing the bubble, knowing the exact diameter, and assuming uniform thickness, he estimated the film thickness to be on the order of a micrometer. He also demonstrated that if one pierces the thickest part of a film with a needle and then withdraws the needle, the film does not rupture. However, if one touches the black portions with any other object, the film immediately ruptures.

In 1773, Wilke published a report in which he described the inflation of soap bubbles at a sufficiently low temperature to make them freeze [8]. Soon after bubble formation, small particles of snow in the form of small stars condensed and moved freely over the sides of the bubbles. Wilke wrote "Those who will repeat these experiments will discover there infinity of small curious and amusing details."

In 1782, the Italian physicist and natural philosopher Cavallo communicated on "the first airships"—he produced bubbles filled with hydrogen which rose in the air.

In 1820, the American inventor Samuel Morey reported on bubbles produced by melted resin which exhibited colors [9]. He observed the formation of a succession of bubbles attached to each other by thin "strings" and that some of them survived for up to eight months. It is a curious fact that one day a small girl came to Morey and showed him the similar but perfectly regular lines of more than twenty bubbles, each with a length of approximately 8 mm and a width of approximately 6 mm. These bubbles were divided from each other by intermediate "strings" of around 3 mm. Morey was unable to explain such a phenomenon.

In the period of 1836–1837, John William Draper published the results of fascinating experiments on the passage of gases through liquid films [10]. He observed that bubbles containing a certain gas that were present in an atmosphere of a different gas either decreased or increased gradually until reaching a certain size that corresponded to the equilibration of the gas composition inside and outside of the bubble. Draper also demonstrated gas permeability through soap films by another experiment in which he placed a plane film on the orifice of a bottle and put the bottle in an atmosphere of nitric oxide; after a few seconds, the plane film started to bulge and, within one to two minutes, a well-developed spherical bubble cap was established.

In the period of 1819–1844, the Italian scientist Ambrogio Fusinieri published remarkable findings [11–14]. Plateau commented that *"these memoirs contain, like that of Leidenfrost, excellent observations and a not easily acceptable theory."* Fusinieri, like others before him, paid attention to the effect of evaporation on the color evolution in soap film caps. Moreover, he was the first to produce and observe large plane films. In his experiments, he immersed the edge of a glass bell (around 15 cm in diameter) in solution; after he withdrew it and set it in a position such that a vertical black film formed, which he observed in daylight on a black screen. Similarly, he also observed the evolution of vertical plane films in metal rings of 5–7-cm diameter (Figure 2.2). Fusinieri made experiments either with soap caps or plane films for a number of liquids. He used the term "wedge-shaped films" to refer to vertical plane films and used the term "capillary corners" with reference to the small masses with transverse concave curvature that form at the border between the film and

FIGURE 2.2   A vertical film with white and black bands of the first order on top and colored bands of higher orders below. (Picture taken from an internet source Giorgio Carboni, Experiments on surface phenomena and colloids, https://www.pinterest.de/pin/430304939368927320/.)

either the solid frame (for plane films) or the solution surface (for film caps). He also attempted to explain the instantaneous disappearance of films upon bursting:

> ...when a film breaks, and thus the bond of the viscosity, which maintained the liquid in a film state, is removed, the expansive force, directed before in the direction of the surface of the film, acts in all directions, transforms the liquid into vapor, and gives the molecules violent movements of projection, accompanied by decomposition with releases of gas; one assures oneself of this release in what, after the rupture, one observes, on the solid edge to which the film was attached, a certain quantity of tiny hollow bubbles.

Fusinieri pointed out that one cannot attribute this kind of rupture to the compression of air entrapped in a bubble because the same phenomenon occurs for planar films.

Fusinieri observed a phenomenon that occurs during the evolution of a fresh vertical plane film, which becomes generally thinner owing to the gradual descent of the liquid; namely, that certain parts of the film became thinner than the surrounding film and exhibited colors. These colored spots immediately started to ascend, dragging a train (Figure 2.3). Later, Brewster also observed similar phenomena [15] although it is not clear if he was aware of the work of Fusinieri and he called these objects "tadpoles." In a vertical film, one can observe the color bands of several orders. With time, these bands dilate and decrease in number until the whole film is covered by white and black areas of the first order. The front gradually descends, with an irregular front casting black filaments down toward the white band; Dewar later referred to this process as "black fall." Different behavior is observed when the film is nearly horizontal; the changes in the pattern proceed more slowly and sometimes most of the film is covered with "tadpoles" such that the bands are not recognizable. In addition, black areas appear irregularly. Fusinieri performed small vertical oscillations of the vertical frame that holds the film and observed that the color bands oscillate like the strata of several liquids of different densities. This demonstrated that such banded (thick) films exhibit similar behavior to that of bulk liquids.

In the name of science, the 28 year old Joseph Antoine Ferdinand Plateau (1801–1883) exposed his eyes to the sun for 25 s, which seriously damaged his sight to the extent that by the age of 42 he was completely blind. Remarkably, Plateau continued his work, with the help of others, and performed a number of experiments with drops, bubbles, and films to gain fundamental insight into capillary action based on molecular forces. His work was finally summarized in *Statique* [3] in 1873. In the following year, Maxwell published a critical review of this book in *Nature* [16], calling its author *"a distinguished man of science who passes from phenomena to ideas, from experiment to theory."* Plateau's work is of great significance in surface science, particularly his experimental study of the action of surface tension, which determines the shape of liquid interfaces. In a beautiful and easily performed experiment, he formed an olive oil drop in a water/alcohol mixture of the same density

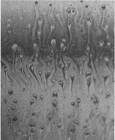

FIGURE 2.3 Photos of the "tadpoles." (Taken from an internet source Jane Thomas Soap Films, https://www.flickr.com/photos/jane_in_wales/sets/72157623594452058/.)

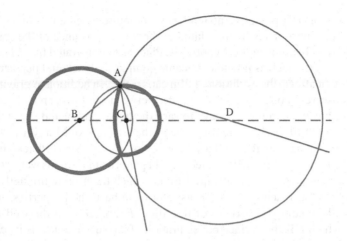

FIGURE 2.4 Two bubbles in contact with radii AB and AC; the angles BAC and DAC are equal to 60°.

as the oil; in this situation, the shape of the drop was independent of gravity and subject only to the action of surface tension, which makes it spherical (i.e., minimal area).

Plateau performed detailed systematic experiments on the persistence of bubbles, film caps, and plane films produced from model solutions containing water or water/glycerin mixtures of various ratios and different amounts of foaming agents (sodium oleate, saponin, albumin, etc.). In particular, he carried out a large number of experiments on solutions of sodium oleate in water/ glycerin (15:11), the so called "liquide glycerique." Plateau was interested in the forms adopted by liquids under different conditions and the surface of minimal area. Therefore, he performed numerous experiments with foam films attached to frames of different shapes and architectures. He determined the relation of the curvature of a film to the sizes of the two bubbles between which the film is formed (Figure 2.4). Only in the ideal case, when all other conditions are constant and the two bubbles have exactly the same size, will the film between them be flat.

Plateau also found that, at equilibrium, the pressure exerted by a spherical bubble of diameter d on the entrapped gas is inversely proportional to d. He proved the correctness of this relation experimentally by accurate measurement of the diameter of a bubble inflated by a pipe while the other end of the pipe was connected to a simple water manometer for measuring the pressure.

In §211 and 282 in *Statique* [3], Plateau concluded, on the basis of his experiments on rising bubbles and foam films, that the generation of such a peculiar system as a thin film is a necessary result of the cohesion and viscosity of the liquid. He also claimed that the cohesion of liquids is of the same order as that of solids, which was in agreement with the works of Henry [17] and Dupré [18]. Plateau also showed that the surfaces of some liquids have viscosities different from that of the bulk liquid, as expressed in the following quote:

> The surface layer of liquids has an intrinsic viscosity, independent of the viscosity of the interior of the mass; in some liquids, this surface viscosity is stronger than the interior viscosity, and often by much, as in water and especially in saponin solution; in others liquids it is, on the contrary, weaker than the interior viscosity, and often also by much, as in spirits of turpentine, alcohol, etc.

He presumed that surface tension is a destructive agent that disturbs the film persistence because the latter mainly depends on the degree to which the surface viscosity counteracts the surface tension. The generation and stability of liquid films is possible only when the surface tension is lower than the cohesion of the surface layers. Conversely, both Plateau in §161 in *Statique* [3] and Dupré [18] concluded that the surface tension does not depend on the film thickness "as long as this thickness is above a certain limit." Under such circumstances, the film should be equally persistent at any thickness as long as it possesses no tendency to rupture when it is thin. Lower persistence of films containing black portions is likely related to the lower resistance of the thinner parts toward external perturbations. To confirm this hypothesis, Plateau demonstrated that a large film of "liquide glycerique," protected as much as possible, became completely black and persisted for many days.

Another factor that affects the evolution of film caps is that, in addition to gravity effects, the "full mass," which is observed along the circular border between the film cap and the solution surface, exhibits a negative mean curvature (concave toward the liquid phase) and sucks the liquid from the film. These objects of "full masses" are the same as the "strings" of Morey, the "capillary corners" of Fusinieri, and the Gibbs triangles, and are now commonly known as Plateau borders.

Josiah Willard Gibbs (1839–1903) dedicated only 14 pages on his famous work "*On the Equilibrium of Heterogeneous Substances*" [4,5] to liquid films, which dealt predominantly with foam films. He considered a film "sufficiently thick for its interior to have the properties of matter in mass." Gibbs introduced the quantity "elasticity of the film," E, and derived the well-known expression $E = 4\Gamma^2(d\mu/dG)$, where Γ is the surface concentration of the surfactant, μ is its chemical potential, and G is its total amount of surfactant in the film per unit area. It is amazing that Gibbs came so close to the idea of the disjoining pressure but did not define it. In References 4 and 5 he wrote:

> Sooner or later, the interior will somewhere cease to have the properties of matter in mass. The film will then probably become unstable with respect to a flux of the interior …, the thinnest parts tending to become still more thin … very much as if there were an attraction between the surfaces of the film, insensible at greater distances, but becoming sensible when the thickness of the film is sufficiently reduced. We should expect this to determine the rupture of the film, and such is doubtless the case with most liquids.

In his book "*Soap Films: Studies of their Thinning*" [19], Karol Mysels wrote:

> It may be worth noting that while J. Willard Gibbs is generally considered as strictly a theoretical investigator, his insight into soap films was based on personal observations and experiment. … This is an excerpt from a letter [20] of a visitor to New Haven in 1878: "Willard was amusing himself examining soap bubbles, not blowing them in a pipe but dipping an ivory ring in the soap suds & standing it up under a box lined with black, but a glass side to look through, & then examined the bubble which was between the sides of the ring. It was quite interesting to see how different the bubbles would be with more or less soap in the water or with glycerin added.

2.2   THE LATE 19TH CENTURY AND THE EARLY 20TH CENTURY

Charles Vernon Boys was possibly the first to garner public interest in soap films and bubbles. He held numerous engaging lectures/demonstrations for audiences widely varying in age and scientific knowledge. His lectures were so popular that many were familiar with Boys' work before he published his book entitled "*Soap bubbles and the forces which mould them*" in 1890 [21].

Between 1843 and 1868, Plateau published many works in "Memoires of the Academy of Belgium" and summarized his results in *Statique* [3]. This, however, is not the right place to honor all of his contributions to the development of capillary theory and its application to thin liquid films. One of the most essential questions, dedicated to the present topic, is the suitability of experimental and theoretical methods to investigate the basic characteristics of thin liquid films, such as their thickness and surface tension, in order to elucidate the role of molecular interaction. Plateau explained that, as a part of the transition from a liquid to a vapor state, the liquid's energy changes not abruptly but in a continuous manner. A valuable contribution to this area was made by Johannes Diderik van der Waals. In his thesis, published in 1873 [22], the same year as *Statique* [1], he calculated the approximate thickness of the zone close to the surface, over which continuous change of energy is fulfilled, to be approximately 3 nm in the case of water. In other studies, Quincke concluded that the radius of molecular interactions is approximately 50 nm [23], and Plateau stated that it is less than 59 nm.

(a)   (b)

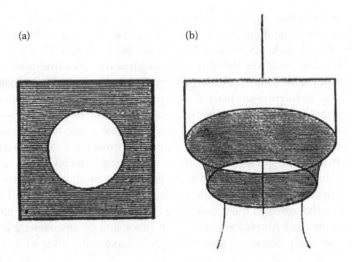

FIGURE 2.5 Experiments by van der Mensbrugghe: (a) Perforated plane film; (b) Catenoidal film with a suspended ring. (Pictures taken from van der Mensbrugghe, M.G. *Phil Magaz* 1867; 33: 270[24].)

The thin liquid film is an ideal object for studying the range of molecular interactions as the thickness of the thinnest black film is considered to be comparable to this range. In this direction, many attempts were made to measure the tension of a film and, in particular, the thickness of black films. Excellent examples demonstrating the action of surface tension are the experiments of van der Mensbrugghe [24], of which we consider two. The first example involves a perforated planar film in a frame (Figure 2.5a). The perforation is produced by means of a fine silk thread with connected ends, thus forming a closed contour, in the following manner: the wet thread is gently incorporated into the film such that it adopts an irregular shape; when the film portion inside the thread contour is destroyed, the thread quickly expands and covers the maximum area that can be surrounded by the given thread perimeter in accordance with the calculus of variations [3]; namely, a perfect circle. This circular shape is practically independent of gravity since turning the film from a horizontal to vertical position causes no changes in its configuration. The film occupies the minimal area under these circumstances, thus demonstrating the contractile force in a plane film due to the uniform tension acting over the film area irrespective of the shape. The second example is also striking: a film is formed in a ring and then comes into contact with a smaller ring that rests on a tripod; thus, two concentric films appear which are divided by the inner ring. By moving up the outer ring, a film with the shape of part of a catenoid (see §58, 80, 226, and 227 in *Statique* [3]) is formed, as shown in Figure 2.5b. If the smaller ring is light enough, it can be suspended in the film in equilibrium, which is defined by the balance between the weight (m) of the suspended system and the "contractile force" in the film, defined by its tension, γF. Under equilibrium, the tension of the film is $\gamma F = m/2\pi R\cos\alpha$, where R is the radius of the suspended ring and α is the angle between the tangent to the lowest point of the catenary contour and the vertical direction. If the radii of the two rings and the distance between them are known, the catenary curve can be found and the angle α calculated. If one blows onto the plane circular film, it moves down and carries away the smaller ring, thus elongating the catenoidal film. As soon as one stops blowing, the equilibrium state is immediately restored. Using both methods, van der Mensbrugghe came to very similar results and for films from "liquide glycerique," he found the surface tension to be around 58.8 mN/m; this value is twice the force exerted by the two faces of the film (i.e., the "superficial tension" of the liquid will be around 29.4 mN/m if no dependence on the tension on the film thickness is assumed).

In the period 1881–1906, Reinold and Rücker published results (Reference 25 and citations therein) on the thickness of black films obtained by either electrical or optical methods. They compared the data for several systems containing various amounts of different foaming agents and KNO₃ dissolved in water, as well as for salted "liquide glycerique." By measuring the resistance of the films, they evaluated their thicknesses and found that the specific electric resistance of films thicker than 374 nm is independent of the film thickness and equal to that of the bulk liquid. They also applied an optical method based on the interference of the film, assuming that the refractive index of the liquid in the film is equal to that in the bulk liquid. Finally, they reported a mean value obtained by both methods of about 12 nm for the thickness of the black film of an aqueous sodium oleate solution containing 3% KNO_3. From our current knowledge of surfaces and foam films, one should immediately recognize that such a high electrolyte concentration should greatly suppress the electrostatic disjoining pressure even in films with high double layer potential, thus enabling the formation of the thinnest film: the Newton black film. This film thickness gradually increases when the salt concentration is reduced and reaches about 22 nm in the absence of salt. For "liquide glycerique," the minimal thickness is 10 nm at 8.66% salt content and the maximal thickness is 24 nm in the absence of any added salt. These investigations were perhaps the first that revealed the dependence of the film thickness on the ionic strength of the solution from which the black film is formed. For salt-free solutions, any increase in the concentration of the soap agent leads to an increase in film thickness. Reinold and Rücker wrote [25]:

> It is difficult to assign a reason why the addition of salt to the liquid should produce so grate a difference in the results. In part, the better conducting salt probably masks effects which when soap alone is used become predominant....

In 1899 and 1906, Edwin Johonnott published two papers dealing with the thickness of black films [26,27]. He used two optical methods suggested by Albert Michelson: an interferometric and a photometric method. With the Michelson interferometer, he measured the thickness of the black films by putting a number of vertical films in frames in a row along the optical path of the interferometer, and adjusted them such that the displaced fringes appeared contiguous with those of the parallel light that did not pass through the films, as shown in Figure 2.6. The number of fringes displaced is determined by the central black fringe, which appeared broader in comparison to the neighboring color fringes (shown in Figure 2.6a in grayscale). For a fresh thick film, no fringes appear (Figure 2.6b). When the black film is formed at the top of the frame, it causes deflection in the vertical fringes owing to differences in optical path lengths of the reference beam and the number of films. The jump to a larger thickness at the front of the black film changes the direction of the fringes (Figure 2.6c,d). Finally, when the black film occupies the whole frame, all fringes become straight again (Figure 2.6e).

To calculate the mean film thickness (h), Johonnott introduced the formula $h = (l\lambda)/(2N(n-1))$, where l is the fringe displacement, λ is the sodium wavelength (589 nm), N is the number of films, and n is the refractive index of water (1.333) [26]. Johonnott measured the thickness of black films of aqueous sodium oleate solutions at different concentrations and with the addition of glycerin. A sealed cell was used in which the films on the frames were left to rest for a sufficient time in order to become black and for the deflection of the fringes to become detectable. He observed that the film thickness for a given solution depends on the humidity of the atmosphere around the film. When films are exposed to open air, they tend to thin to the smallest possible thickness, which is enhanced by evaporation

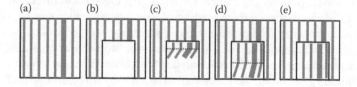

FIGURE 2.6 Vertical interferometer fringes of white light. (a) Normal view of the fringes. (b–e) Views after introduction of a frame with a film: (b) fresh thick film; (c,d) appearance of a black film at the top and its extension downward (the dotted line of the fringe breakpoints corresponds to the advancing front of the black film); and (e) a frame fully occupied with a black film. (Sketch reproduced from Johonnott, E.S. *Phil Magaz Series 5* 1899; 47: 501[26].)

of the liquid from the film; this final state is referred to as the "second black film" (Newton Black Film, NBF, according to the conventions of the International Union of Pure and Applied Chemistry, UIPAC). For films sealed in a cell at a given temperature, only the "first black film" (Common Black Film, CBF according to UIPAC) with a thickness of around 15 nm appeared for aqueous solutions of 1:40 and 1.70 sodium oleate. For more diluted solutions, even after two days, no NBF was detected. The formation of NBF was achieved by heating the whole system; after subsequent cooling, the CBF reformed. The results presented in Figure 2.7 [26] show the variation in film thickness with cyclic temperature changes. These results demonstrated for the first time the temperature-induced transition between the two states of black films. In these experiments, the cause of the transition from a CBF to an NBF can be attributed not only to increase of the temperature, but also to the respective increase in pressure upon warming the sealed chamber. Later, Johonnott invented a setup which allows variation of the ambient pressure around the film by which he studied the CBF-NBF transition. In all cases, the NBF had a constant thickness of around 6 nm.

Acting upon another suggestion from Michelson, Johonnott determined the film thickness by measuring the light reflected at different angles of incidence with respect to the film; this approach is still in use today. Measurements were made by comparing the images of two slits in the focus of the telescope of a spectrometer. The collimated light from the first slit is reflected by the film and the other slit was in front of a silvered glass photometer, as discussed in detail in the original paper [26]. Johonnott considered that both methods have their advantages and disadvantages. For example,

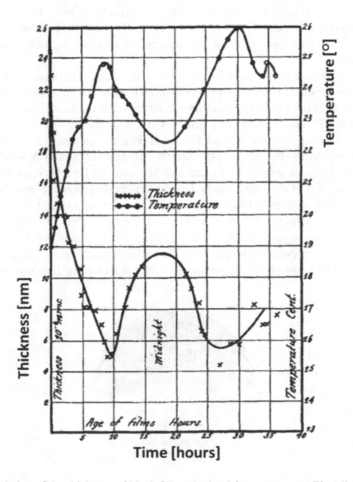

FIGURE 2.7 Evolution of the thickness of black films obtained from aqueous 1:70 sodium oleate solutions during cyclic temperature changes. (Graph taken from Johonnott, E.S. *Phil Magaz Series 5* 1899; 47: 501[26].)

the first one is better with respect to controlling temperature and humidity, but, on the other hand, a large number of films thin with variable rates, and therefore give only a mean value for the thickness. However, Johonnott demonstrated the feasibility of both methods and concluded that:

i. *the thickness of the black film of a soap solution is not constant, and may vary between 6 nm (second black film) and 40 nm (first black film);*
ii. *the film of a pure aqueous oleate solution may consist of two black films, the thickness of the second being about half the limiting thickness of the first, which is about 12 nm;*
iii. *the addition of glycerin to a pure oleate solution prevents the appearance of a second black film.*

Johonnott clearly distinguished between the NBF and CBF. He adopted thickness values of 6 and 12 nm, respectively, and made no allowance for the refractive index of the solution, which is different from that of water; recalculation with the refractive index of aqueous solutions gives values of 5.7 and 11.4 nm, respectively [30, p. 66]. Johonnott also discovered CBF-NBF inversion by varying the pressure of the atmosphere around the film by means of a special arrangement. He observed that it was possible to turn the CBF into an NBF by slightly increasing the pressure by pinching the soft tube connected to the inner compartment of the arrangement where the film rests. The NBF appeared as perfectly round spots within the CBF; after some time, the whole film turned into an NBF, while a subsequent small decrease in pressure caused the CBF to appear similarly as round spots within the NBF. Using a similar approach to that of van der Mensbrugghe [24], Johonnott measured the tension of approximately cylindrical films between a liquid surface and a ring suspended to a sensitive Oertling balance. He estimated the tension of the film to be 25–26 mN/m, which is more than two times smaller than the tension obtained by van der Mensbrugghe [24].

A striking phenomenon is Johonnott's so called "five blacks." He observed with a microscope five different states of black films which often appeared together in the field of view. The first three were "evanescent" and broke up as the film thinned, whereas the fourth and fifth corresponded to the "first black film" (CBF) and "second black film" (NBF), with thicknesses of approximately 12 and 6 m, respectively. Such multiplicity of black films was observed earlier by Brewster, who published an illustration of such films (Figure 2.8) [15], and later by Perrin [28] and Wells [29] in their works on stratification in films, as well as by Dewar and his assistant Lawrence, all of whom aimed to elucidate the constitution of thin liquid films. Lawrence provided an excellent summary of Dewar's works in a book published in 1929 [30] on the latest discoveries up to that time.

Jean Baptiste Perrin, who was honored in 1926 with the Nobel Prize for Physics for his work on Brownian motion, microscopically observed using reflected light a new kind of film produced from mixtures of fluorescein and potassium oleate [28]. The vertical film thinned normally until interference colors appeared and then, instead of the normal sequence of bands, the film was composed of flakes

FIGURE 2.8   Stratification in soap films: Illustration of the *second black* film and the *first black film*. (Picture taken from Brewster, D. *Trans Roy Soc Edinburgh* 1867; 24: 491[15].)

of different tints and every tint was uniformly colored, corresponding to a certain uniform thickness of the flake. These flakes appeared all over the film and grew irregularly until the whole film was covered by such patterns. By saturating the atmosphere around the film with solution vapor, thicker stratifications occurred as the carmine and green stages and were especially brilliant; in addition, color tints such as yellow and orange of the second order were surprisingly observed. This discovery led Perrin to the conclusion that stratification obeys some regular rule and he suggested that:

i. *in a stratified film, the thickness of each layer is an integer multiple of an elementary thickness which is of the order of 5 nm;*

ii. *the layers of a stratified film are formed by a superposition of identical elementary sheets of a suitable number.*

Based on Perrin's method for estimating the thickness of a black film, it is likely that the NBF (first order black) film corresponds to the elementary sheet. This method consists of counting the maximum number of stratifications between the NBF (the thinnest film) and a certain film stage of known thickness (obtained optically). For the latter, Perrin used the violet from the first order for which the thickness was around 210 nm (Reinold and Rücker reported 230 nm [25]) and counted 18–19 stages of black, gray, and white, 17–18 stages of yellow, orange, and red, and 1 stage of violet. This gives a thickness per stratification of 5.7 nm. This method is adequate only under the assumption that the stratification increment (elementary sheet thickness) is constant. Perrin modified this method and counted the number of stages between the dark and bright bands produced by monochromatic light. The difference in thickness between a dark and a bright band was found to be one quarter of the wavelength. He also divided the range of all stratifications into three regions of the film and counted the layers between the two stages for each region. Both the results obtained by subsequently observing the individual regions of the film and those obtained by observing the whole range between the NBF led to the same value of 5.4 nm. This confirmed that stratified films consist of elementary layers identical in thickness to the NBF. Perrin also considered the change in the refractive index of the soap solution compared to that of pure water to arrive at a corrected value of 5.2 nm. Wells [29] extended the work of Perrin and determined a value of 4.4 nm for the thickness of the NBF, which after the correction for the refractive index, was revised to 4.2 nm.

James Dewar is best known today for his invention of the Dewar flask and for his work with low temperature phenomena. However, he also conducted numerous experiments with bubbles and vertical plane films and described the phenomenon of the "black fall" as well as other features of thin films. For example, he kept a soap film in a bottle for more than three years [30]. The "black fall" was first described by Fusinieri [11–14]. This is the boundary between the black part and the thicker portions of the film. This boundary is quite sharp, which indicates a large and abrupt change in thickness (Figure 2.9a). For a vertical film or a bubble on a surface, this is a horizontal line or a

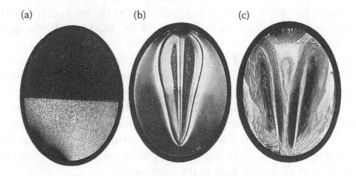

FIGURE 2.9 Photographs from Dewar's experiment with vertical plane films. (a) Critical black fall. (b) Film disturbed by an upward air jet. (c) Film disturbed by an electric spark. (Pictures taken from Lawrence, A.S.C. *Soap Films: A Study of Molecular Individuality.* Bell & Sons; 1929[30].)

circle, respectively. Dewar used the "black fall" as a measure of the rate of thinning of films and bubbles. In his experiment, bubbles were observed under different conditions; namely, (i) a bubble resting on the top of a capillary, (ii) a bubble suspended from a capillary, and (iii) a larger bubble suspended from a capillary with a smaller bubble attached. In the first case, the thinning was regular, as was observed for vertical films or film caps and the "black fall" gradually descended with time at a certain rate. In the second case, during thinning of the bubble, a black collar was formed in the region just below the point of contact with the capillary. This is a direct result of the suction of the "plateau border"; this suction is in the opposite direction to gravity drainage. Hence, the rate of the "black fall" is counterbalanced or enhanced, respectively, in these two conditions. In condition (iii), the "black fall" is strongly promoted by the suction from the "plateau border," which, in turn, is drained through the bubble attached to the main bubble under observation. Dewar also reported on "films as detectors"; that is, the film can drastically change its pattern and the way it develops in response to small perturbations. In Figure 2.9, vertical films disturbed by an upward air jet or electric sparkle are shown (panels B and C, respectively). The characteristic pattern in Figure 2.9b is further disturbed by sound (i.e., very small fluctuations of air pressure). All these factors lead to "abnormal development" of the film and to more peculiar behavior in response to stronger perturbations. In contrast to quiet films, the "black fall" is no longer a line and many black spots are formed all over the film, especially in the thicker parts. These small black spots ascend the film and coalesce into a black portion. Lawrence referred to this "tumultuous development" as "critical black fall" [30].

Eighty years after Drapper published his work *"Experiments on endosmosis"* [10], in which he considered the transfer of gasses through bubbles (see previous section), Dewar studied this phenomenon for "black bubbles" [30]. He found that the rate of gas penetration through the bubble is proportional to its internal excess pressure (the capillary pressure), $4\gamma/R$, where γ is the surface tension and R is the bubble radius. He also reported—as had Drapper [10]—that transfer of gas causes either expansion or contraction of the bubble. Dewar verified this experimentally by observing a bubble in a hydrogen atmosphere. He performed a series of experiments including a hanging chain of bubbles of gradually decreasing sizes obtained from different solutions. Dewar could not reach a definitive conclusion on the basis of these results, but pointed out that the variety of rates of observed gas effusion may be related to the individual characteristics of the black films obtained from the different solutions. Dewar gave two lectures on this subject: *"A Soap Bubble"* in 1878 and *"Soap Films as Detectors"* in 1923.

2.3   MID-20TH CENTURY

The knowledge gained in the past three centuries was the result not only of many various observations and experiments, but also theoretical works, including those of Plateau [3] and Gibbs [4,5]. However, in the early 20th century, some important questions remained to be answered. The phenomenology of foam films (i.e., the film lifetime, drainage and thinning, formation of a black film, and rupture) was qualitatively well studied; however, a general theory describing the formation of a film, its evolution, stability, and mechanism of rupture was lacking. Between the 1930s and 1960s, significant progress on thin liquid films was achieved.

Miles, Ross, and Schedlovsky studied the drainage of vertical films of different detergents in glass frames as well as the effects of detergent concentration and additives [31]. They observed the films in red light and followed the rate of descent of the bands. Two categories of drainage behavior were distinguished: rapid and slow. They attributed the different drainage rates to the surface viscosity of the film. For example, films formed from solutions of saponin, which forms very viscous and rigid film surfaces, drained very slowly. They found that for a certain solution composition, a transition temperature exists at which the slow drainage becomes faster and that this temperature is a function of the total concentration of all constituents in the solution.

Mysels, Shinoda, and Frankel extended this method by applying measurements of the film thickness from the intensity of reflected light and investigated the drainage and mechanisms of the thinning of vertical films of different solutions. These results were summarized in Reference 19.

These studies were later extended upon by Lyklema, Scholten, and Mysels [32], and three extreme types of films with respect to their drainage behavior were considered:

i. *Simple mobile film.* This is the most frequent case and was discussed above. Briefly, the bands are relatively regular and descend quickly in an ordinary manner as turbulence appears only along the borders. The black film is readily formed on top giving rise to a regular "black fall."

ii. *Irregular mobile film.* This film initially behaves like a simple mobile film, but after a certain time, "abnormal development" takes place after a black portion appears and starts to develop; this corresponds to Dewar's "critical black fall," as explained above.

iii. *Rigid film.* The surfaces of such films are highly resistant to any motion within the plane of the film. The drainage is very slow as the bands are closely spaced and irregularly curved. With time, the bands separate slowly becoming more horizontal and a black film appears along the border. Such films were those obtained from solutions containing sodium dodecyl sulfate around its critical micelle concentration and a considerable additional amount of dodecanol.

Such mobility or rigidity of the film surfaces affects not only the rate of drainage, but also the shape of the vertical film, that is, for mobile surfaces, the vertical cross section of the film has a wedged shape and drainage is rapid, whereas for rigid surfaces, it is parabolic and the drainage is slow. Lyklema, Scholten, and Mysels [19,32] evaluated the thickness of CBFs and NBFs by measuring the intensity of reflected light from the film, assuming the same refractive index as that of water. The uniform thickness of the NBF was pointed out earlier by Johonnot for films formed from aqueous sodium oleate solutions [26,27].

Another dynamic factor that affects film stability is the Gibbs elasticity, E. The importance of this characteristic was revealed by Gibbs [4,5]. However, quantification of the Gibbs elasticity suffers from difficulties with the determination of the chemical potential of self-assembling species like surfactants. Pioneering measurements were performed by Mysels, Cox, and Skewis in 1960 [33]. They detected a tenfold increase in E of rigid films in comparison to that of mobile films, thus experimentally proving the existence of film elasticity. Van den Tempel et al. [34,35] subsequently investigated the Gibbs elasticity theoretically and experimentally, and established that E increases as the film thins. Krotov and Rusanov developed a new experimental setup [36], which allowed simultaneous measurements of film tension, area, and thickness of horizontal, rectangular, and large foam films. They compared the experimental and theoretical values of E and studied its role in the stability of thick foam films.

Other aspects of foam films were also explored in this period. Drapper [10] and then Dewar [30] studied gas effusion through the foam film of a bubble. Brown, Thuman, and McBain [37] as well as Princen and Mason [38] measured the rate of bubble deflation owing to the permeability of the film to gasses by monitoring bubbles floating on solutions; it was shown that when fully covered by an NBF, bubbles are highly permeable. The penetration rates for several gasses were determined. The theoretical approach used for calculating the film permeability assumed that this is equal to the sum of the permeabilities of the water layer in the film and the two adsorption layers [39].

The concept of the disjoining pressure was introduced into the terminology of liquid films by Derjaguin and Obuchov in 1936 [40]. The action of disjoining pressure in a thin liquid layer defines it as a "thin liquid film," in contrast to a thicker layer where the disjoining pressure is negligible. In the 1940s, based on this concept, the theory of the stability of lyophobic colloids was independently created by Derjaguin and Landau [41], and Verwey and Overbeek [42]—the so called DLVO theory. This theory was a milestone in colloid and interface science. It explained the stability of disperse systems including foams and foam films. The disjoining pressure arises in foam films because of different types of surface forces acting in the film; for example, van der Waals molecular forces and electrostatic double layer forces (DLVO forces), as well as steric forces, structural forces, and so on (non-DLVO forces). The state of the art of experimental and theoretical works on colloid stability prior to the DLVO theory can be found in a historical review by Vincent published in 2012 [43].

Experimental verification of the action of the disjoining pressure in foam films was made by Derjaguin, Titievskaia et al. [44] as well as by Overbeek [45] and Duyvis [46]. Exploring very small, circular foam films, they measured the repulsive electrostatic component of the disjoining pressure neglecting the attractive van de Waals component and compared the results with those predicted by the theory. In 1959, studies of foam films gained momentum with the introduction of the microscopic foam film approach by Scheludko and Exerowa [47,48], who invented a special experimental cell (detailed description is given in Chapter 3.3). Using this approach, they experimentally proved the DLVO theory for foam films [47,48]. The same experimental approach was used also by Scheludko, Platikanov, and Manev [49] to measure the disjoining pressure of microscopic foam films from nonpolar liquids in which only molecular van der Waals forces of attraction operate. Large vertical foam films in a frame have also been studied simultaneously. Lyklema and Mysels [50] obtained experimental results for the electrostatic and van der Waals disjoining pressures and compared them with the DLVO theory. Jones, Mysels, and Scholten [51] performed detailed experiments with black foam films from aqueous solutions of sodium dodecyl sulfate. They examined the conditions under which common black films are stable. They completed a stability diagram of temperature (T) as a function of NaCl concentration (C_{NaCl}) and demonstrated that CBFs are stable at high T and low C_{NaCl}, whereas NBFs are stable at low T and high C_{NaCl}.

In the 1950s and 1960s, many powerful research groups under the guidance of prominent scientists such as Mysels and Princen (USA); Derjaguin and Rusanov (Soviet Union); Overbeek, Lyklema, Vrij, and van den Temple (Netherlands); Scheludko (Bulgaria); Hayden, Goodman, and Kitchener (England), worked in the field of liquid foam films and set the basis for our current understanding.

2.4 CONCLUSIONS

The chapter was not intended to describe the state of the art, but to illustrate how the gradual advance, made over the past three centuries, have led us to our current level of understanding. Famous names like Hooke, Newton, Plateau, and Gibbs were the pioneers in the field of foam films. Inspired by their observations and findings, others—often their co-workers—pushed the field forward and established a very solid basis in understanding foam films. Summaries of the state of the art throughout the development of this field have been published in numerous textbooks and monographs [52–57]; these summaries progress from mostly descriptive to theoretically founded. The most recent summaries published as book chapters appeared in 2005 [58,59]; however, these did not cover all facets of this rather complex topic. An extensive description of the subject was provided 17 years ago by Exerowa and Kruglyakov [60]. The present book intends to provide a systematic update and extension of this work. Detailed descriptions of the various subjects, which together represent a complete characterization of the exciting topic of foam films, will be available in the following chapters and subchapters and are, therefore, not discussed here.

ACKNOWLEDGMENTS

The authors are grateful to Adam Brotchie (Nature Publishing Group) for proofreading this manuscript.

REFERENCES

1. Hooke, R. On holes (Black Film) in soap bubbles. *Communic Roy Soc* 1672; Mar 28.
2. Newton, I. Optics. London; 1704. 4th ed., corrected; 1730. Dover; 1952.
3. Plateau, J.A.F. Statique experimentale et theoryque des liquides soumis aux seules forces moleculaires. Gautier-Villars, Trubner et cie, F. Clemm, 2 Vols; 1873.
4. Gibbs, J.W. On the equilibrium of heterogeneous substances. *Trans Conn Acad* 1887; 3: 108.
5. Gibbs, J.W. *Collected Works*. Longmans, Green & Co, New York; 1928. Vol. 1.
6. Birch, T. History of the royal society. *A Millard London* 1757; 3: 29.
7. Leidenfrost, J.G. Communis Nonnullis Qualitatibus Tractatus, 1756.

8. Wilke, Mr. Sur le forme de la neigre. *J de Phys de l'abbé Rozier* 1782; 1: 106.
9. Morey, S. Bubbles blown in melted rosin. *J Silliman 1st Series* 1820; 2: 179.
10. Draper, J.W. Experiments on endosmosis. *J Franklin Institute* 1836; 22: 27.
11. Fusinieri, A. Ricerche sui colon delle lamine sottili e sui loro rapporti coi colon prismatici. *J de Brugnatelli* 1819; 2: 319.
12. Fusinieri, A. Memoria copra i fenomeni chimici delle rolls sottili. *J de Brugnatelli* 1821; 4: 133,209, 287, 380 and 442.
13. Fusinieri, A. Della forza di repulsione che si sviluppa fra le parti dei corpi ridotti a minime dimensioni, ossia del calorico di spontanea espansione in lamine sottili. *J de Brugnatelli* 1823; 6: 34.
14. Fusinieri, A. Corne la forza repulsiva delta materia attenuata agisca all'atto della rottura di bolle o lamine piane di soluzione di sapone. *Ann delle scienze del Regno Lombardo-Veneto* 1844; 8: 213.
15. Brewster, D. On the colours of soap bubbles. *Trans Roy Soc Edinburgh* 1867; 24: 491.
16. Maxwell, J.C. Plateau on soap bubbles. *Nature* 1874; 10: 119.
17. Henry, W. Cohesion of liquids. *Phil Magaz* 1845; 26: 541.
18. Dupré, A.M. *Théorie mécanique de la chaleur.* Gauthier-Villars, Paris; 1869. Ch. 8.
19. Mysels, K.J., Shinoda, C., Frankel, S. *Soap Films: Studies on Their Thinning.* The University Press, Pergamon, Glasgow; 1959.
20. Wheeler, L.P. *Josiah Willard Gibbs.* Rev. ed., Yale University Press, New Haven; 1952; 260.
21. Boys, C.V. *Soap Bubbles and the Forces Which Mould Them.* Society for the promotion of Christian knowledge, London; E. & J.B. Young, New York; 1890.
22. van der Waals, J.D. On the continuity of the gas and liquid state. *Thesis,* University Leiden; 1873.
23. Quincke, G.H. Über die Entfernung in welcher die Molecularkräft der Capillarität noch wirksam sind. *Ann De M Poggendorff* 1869; 137: 402.
24. van der Mensbrugghe, M.G. On the tension of liquid films. *Phil Magaz* 1867; 33: 270.
25. Reinold, A.W., Rücker, A.W. On the thickness and electrical resistance of thin liquid films. *Phil Trans Roy Soc A* 1893; 184: 505.
26. Johonnott, E.S. Thickness of the black spot in liquid films. *Phil Magaz Series 5* 1899; 47: 501.
27. Johonnott, E.S. The black spot in thin liquid films. *Phil Magaz Series 6* 1906; 11: 746.
28. Perrin, J. The stratification of liquid lamellae. *Ann de Phys* 1918; 10: 165.
29. Wells, P.V. On the thickness of stratified lamellae. *Ann de Phys* 1921; 16: 79.
30. Lawrence, A.S.C. *Soap Films: A Study of Molecular Individuality.* Bell & Sons, London; 1929.
31. Miles, G.D., Ross, J., Schedlovsky, L. Film drainage: A study of the flow properties of films of solutions of detergents and the effect of added materials. *J Am Oil Chemist's Soc* 1950; 27: 268.
32. Lyklema, J., Scholten, P.C., Mysels, K.J. Flow in thin liquid films. *J Phys Chem* 1965; 69: 116.
33. Mysels, K.J., Cox, M.C., Skewis, J.D. The measurement of film elasticity. *J Phys Chem* 1961; 65: 1107.
34. Van der Temple, M., Lucassen, J., Lucassen-Reynders, E.H. Application of surface thermodynamics to Gibbs elasticity. *J Phys Chem* 1965; 69: 1798.
35. Prins, A., Arcuri, C., van den Tempel, M. Elasticity of thin liquid films. *J Colloid Interface Sci* 1967; 24: 84.
36. Krotov, V.V., Rusanov, A.I. Direct measurement of the Gibbs elasticity of liquid films (original title in Russian: Прямое измерение Гиббсовской упругости жидких пленок). *Doklady ANSSSR* 1970; 191: 866.
37. Brown, A.G., Thuman, W., McBain, J.W. Transfer of air through adsorbed surface films as a factor in foam stability. *J Colloid Sci* 1953; 8: 508.
38. Princen, H.M., Mason, S.G. The permeability of soap films to gasses. *J Colloid Sci* 1965; 20: 353.
39. Princen, H.M., Overbeek Mason, S.G. The permeability of soap films to gasses: II. A simple mechanism of monolayer permeability. *J Colloid Interface Sci* 1967; 24: 125.
40. Derjaguin, B.V., Obuchov, E. Anomalien dünner Flüssigkeitsschichten. III. Ultramikrometrische Untersuchungen der Solvathüllen und des "elementaren" Quellungsaktes. *Acta Physicochim URSS* 1936; 5: 1.
41. Derjaguin, B.V., Landau, L.D. Theory of the stability of strongly charged lyophobic sols and of the adhesion of strongly charged particles in solutions of electrolytes. *Acta Physicochim USSR* 1941; 14: 633.
42. Verwey, E.J.W., Overbeek, J.Th.G. *Theory of the Stability of Lyophobic Colloids.* Elsevier, Leiden; 1948.
43. Vincent, B. Early (pre-DLVO) studies of particle aggregation. *Adv Colloid Interface Sci* 2012; 170: 56.
44. Derjaguin, B.V., Titievskaia, A.S., Abricossova, I.I., Malkina, A.D. Investigations of the forces of interaction of surfaces in different media and their application to the problem of colloid stability. *Discus Faraday Soc* 1954; 18: 24.
45. Overbeek, J.T.G. Black soap films. *J Phys Chem* 1960; 64: 1178.

46. Duyvis, E.M. The equilibrium thickness of free liquid films. *Thesis*, University Utrecht, 1962.
47. Sheludko, A., Exerowa, D. Über den elektrostatischen Druck in Schaumfilmen aus wässerigen Elektrolytlösungen. *Kolloid-Z* 1959; 165: 148.
48. Sheludko, A., Exerowa, D. Über den elektrostatischen und van der Waalsschen zusätzlichen Druck in wässerigen Schaumfilmen. *Kolloid-Z* 1960; 168: 24.
49. Scheludko, A., Platikanov, D., Manev, E. Disjoining pressure in thin liquid films and the electro-magnetic retardation effect of the molecule dispersion interactions. *Faraday Discuss* 1965; 40: 253.
50. Lyklema, J., Mysels, K.J. A study of double layer repulsion and van der Waals attraction in soap films. *J Am Chem Soc* 1965; 87: 2539.
51. Jones, M.N., Mysels, K.J., Scholten, P.C. Stability and some properties of second black film. *Trans Faraday Soc* 1966; 62: 1336.
52. Scheludko, A. Thin liquid films. *Adv Colloid Interface Sci* 1967; 1: 391.
53. Bikerman, J.J. *Foams*. Springer-Verlag, Berlin; 1973.
54. Rusanov, A.I. *Phasengleichgewichte und Grenzflächenerscheinungen*. Akademie-Verlag, Berlin; 1978. (first published 1967 in Russian).
55. Isenberg, C. *The Science of Soap Films and Soap Bubbles*. Dover Publications, New York; 1992.
56. Ivanov, I.B. (Eds.). *Thin Liquid Films. Fundamentals and Applications*. Marcel Dekker, New York; 1988.
57. Mittal, K.L., Kumar, P. (Eds.). *Emulsions, Foams, and Thin Films*. Marcel Dekker, New York; 2000.
58. Platikanov, D., Exerowa, D. Thin liquid films. In: Lyklema, J. (Ed.), *Fundamentals of Interface and Colloid Science V; Soft Colloids*. Elsevier, Amsterdam; 2005, pp. 1–91.
59. Platikanov, D., Exerowa, D. Symmetric thin liquid films with fluid interfaces. In: Sjöblom, J. (Ed.), *Emulsions and Emulsion Stability*. CRC Taylor & Francis, Boca Raton; 2005, pp. 127–184.
60. Exerowa, D., Kruglyakov, P.M. Foam and foam films. In: Möbius, D., Miller, R. (Eds.), *Studies in Interface Science*. Elsevier, Amsterdam; 1998.

3 Fundamentals of Foam Films

Dimo Platikanov and Dotchi Exerowa

CONTENTS

In this chapter, the main fundamental knowledge of foam films is briefly reviewed. Further information about foam films can be found elsewhere [1–9]. The foam is a system of a gas phase (bubbles) dispersed in a liquid medium (usually aqueous solution). The liquid solution is contained in foam films which meet in Plateau borders and vertexes. The foam film is the most important structural element of foam since it determines foam stability and the foam's main properties. The foam film is a type of thin liquid film. Thin liquid films are always spontaneously formed when two particles of the dispersed phase (solid particles, liquid drops or gas bubbles) come close to each other. The liquid film is a symmetric film when both approaching particles have the same composition of substances, that is, the symmetric liquid film divides two identical microphases. Such symmetric liquid films are called foam films [10].

What exactly should we call a *thin liquid film*? Let us consider a foam film with parallel plane surfaces (Figure 3.1). The Cartesian coordinate system is oriented such that the x and y axes lay in the film plane, while the z-axis is normal to the film surfaces.

Obviously, the film dimension along the z-axis is much smaller than the dimensions along the x and y axes which determine the film area. The liquid phase from which the film is formed is denoted in Figure 3.1 by α and the adjacent gas phase—by β. The α/β interface is actually not a mathematical plane, but an interfacial layer with a finite thickness. Most of the physical quantities inside this interfacial layer vary along the z-axis due to the mutual influence of the phases α and β. When the distance between both interfaces is rather large (Figure 3.1a), the two interfacial layers are far away from each other. There is liquid between them with the same properties as the bulk liquid phase α. Such a liquid film is a *thick film* from the viewpoint of thermodynamics, nevertheless its thickness could actually only be a few micrometers.

When the film thickness, however, is small enough so that both interfacial layers overlap (Figure 3.1b), there is no liquid inside the film with the properties of the bulk liquid phase α. Just as in a single interfacial layer, most of the physical quantities vary along the z-axis over the entire liquid film from the bulk of one phase β up to the other phase β. This is the physical ground for the rise of the disjoining pressure in the film as well as for the change of the surface tension of the film interfaces. Some properties of the film become dependent on its thickness. Such a film is referred to as a *thin liquid film*.

It should be noted that this is a thermodynamic definition of the thin liquid film and it reflects the peculiar properties of the thin film and its thickness dependence. However, when kinetic properties (e.g., film drainage, rheology, dynamic elasticity, etc.) are considered, the liquid film is often called "thin" simply because of its very small thickness, not taking into account that it could be a thermodynamically "thick" film. For instance, Gibbs has considered the foam film elasticity assuming the same surface tension of the film surfaces to be as that of the bulk liquid.

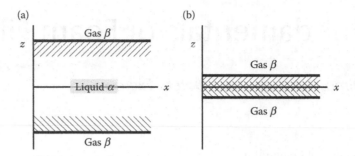

FIGURE 3.1 (a) Scheme of a "thick" foam film and (b) scheme of a thin foam film formed from the liquid phase α between two equal gas phases β.

The thermodynamic definition of the thin liquid film shows that the most important factor which determines its properties is the interaction between the two film interfaces, that is, the interactions due to different types of surface forces between the two adjacent phases across the liquid film. The total Gibbs energy of interaction $G(h)$ depends on the film thickness h. It determines a force per unit film area:

$$\Pi = -[\partial G(h)/\partial h]_{p,T} \tag{3.1}$$

The pressure Π is called *disjoining pressure*, which is positive in the case of repulsion between the two film interfaces, and negative in the case of attraction.

3.1 THERMODYNAMICS OF FOAM FILMS

There is no liquid inside the thin foam film with the properties of the liquid bulk phase α from which the film has been formed (Figure 3.1b) and most of the physical properties vary along the z-axis normal to the film. The thermodynamic description of such a system is based, according to Gibbs, on a simplified model. The differences between the values of physical quantities of the model and of the real system are introduced as excess quantities. The thermodynamic analysis presented here is for a symmetric, plane-parallel, horizontal thin foam film. This analysis, based only on the *model with two Gibbs dividing surfaces* [9], is developed according to [11]. It seems that this thermodynamic approach is most convenient for the interpretation of experimental results.

In the simplified foam film model (Figure 3.2), the entire thermodynamic system consists of: (i) the liquid phase α, in general a multicomponent solution; (ii) the gas phase β; and (iii) the small phase f, a thin foam film formed from and connected with the liquid phase α.

The foam film is drawn out in a solid frame; the solid material of the vessel walls and the frame is not soluble in phases α and β. The following simplifications are accepted in the film model. (a) The Plateau borders, respectively, the menisci, at the film contact with the solid walls are neglected. The transition film/meniscus/bulk liquid will be considered later. (b) Both interfaces between the film and gas phase β are replaced by two parallel, horizontal Gibbs dividing surfaces. The whole space outside the dividing surfaces is filled by gas with the bulk properties of phase β. (c) The space between the dividing surfaces is filled by liquid with the bulk properties of the reference liquid phase α. (d) The distance h between the dividing surfaces is defined as the thermodynamic thickness h of the foam film.

Initially a thick foam film is drawn by the frame from the liquid phase α. It can become thin only if $p^\alpha < p^\beta$; the pressure difference $\Delta p = p^\beta - p^\alpha$ is the driving force for the film thinning. Why can an equilibrium foam film be obtained at the end of the process of thinning if $\Delta p = $ constant? The pressure p in an isotropic bulk fluid phase (like phases α and β) is the same everywhere (Pascal's low). The following components of the pressure tensor are identical:

$$p_{xx} = p_{yy} = p_{zz} = p \tag{3.2}$$

This is not the case for a real interfacial layer as well as a real thin liquid film (Figure 3.1b). There, only the component p_{zz} of the pressure tensor normal to the film (or single interface) remains constant, while the components parallel to the film (or single interface) are functions of z:

$$p_{zz} = p_n = \text{constant} \tag{3.3}$$

$$p_{xx} = p_{yy} = p_t(z) \tag{3.4}$$

As a result, the *surface (interfacial) tension* γ arises at each interface between two bulk phases and its mechanical definition is called Bakker's equation:

$$\gamma = \int_{-\infty}^{+\infty} [p_n - p_t(z)] dz = \int_{-\infty}^{+\infty} [p^\beta - p_t(z)] dz. \tag{3.5}$$

The foam film, just as an interface, is anisotropic along the z-axis (Figure 3.1b) and by analogy, the *film tension* γ^f (Figure 3.2) can be defined by the same expression:

$$\gamma^f = \int_{-\infty}^{+\infty} \left[p^\beta - p_t(z) \right] dz. \tag{3.6}$$

Here the integration is carried out from the bulk of phase β over the entire film up to the bulk of phase β on the other film side. The film tension is (like the surface tension) a force per unit length acting tangential to a surface which is called the *surface of tension* of the foam film. In the case of a symmetric, plane-parallel foam film the surface of tension coincides with the mid-plane of the film. The property film tension requires that an external force per unit length γ^f has to pull the frame with the film at equilibrium (Figure 3.2).

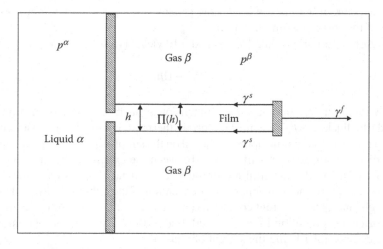

FIGURE 3.2 Scheme of a symmetric plane-parallel thin foam film in a frame formed from the liquid phase α and surrounded by the gas phase β.

It is also possible to introduce on the basis of the simplified film model a *surface tension of the film* γ^s. Its mechanical definition [12] is given by the equation

$$\gamma^s = \int_0^{h/2} [p^\alpha - p_t(z)]\mathrm{d}z + \int_{h/2}^{\infty} [p^\beta - p_t(z)]\mathrm{d}z. \tag{3.7}$$

In Equation 3.7, h is the *thermodynamic film thickness*, that is, the distance between the two Gibbs dividing surfaces. Hence γ^s depends on the location of the dividing surfaces, in contrast to the surface (interfacial) tension γ between two bulk phases as well as the film tension γ^f which are independent of the location of the single dividing surface. Both γ^s and γ^f determine the mechanical equilibrium of the film in the *tangential* direction.

In the simplified film model, the space between the two dividing surfaces is filled by a liquid with pressure p^α of the reference phase α (Figure 3.2). The film is thinning because of the pressure difference $\Delta p = p^\beta - p^\alpha$. Obviously, at equilibrium, Δp must be balanced by any additional pressure. This pressure, due to the action of surface forces in the film, has been introduced [13] as *disjoining pressure* Π (see Equation 3.1). Its mechanical definition [14] is given by

$$\Pi = p_n - p^\alpha, \tag{3.8}$$

that is, the disjoining pressure is the difference between the normal component of the film's pressure tensor (which is constant) and the pressure in the reference liquid phase α. Taking into account that

$$\Delta p = p^\beta - p^\alpha, \quad p^\beta = p_n, \tag{3.9}$$

it follows for symmetric, plane-parallel, horizontal foam films

$$\Pi = \Delta p = p^\beta - p^\alpha. \tag{3.10}$$

Equation 3.10 is the condition for mechanical equilibrium of the foam film in the *normal* direction. Positive Π means repulsion between the two film surfaces, that is, both surfaces are "disjoined" by the interactions due to surface forces. The resultant of all interactions per unit film area is actually Π. It balances the applied external pressure difference Δp. For a plane-parallel, horizontal, real foam film which contacts through a Plateau border the frame wall, Δp is the *capillary pressure* P_c at the curved surface of the corresponding meniscus.

The combination of Equations 3.6, 3.7, 3.9, and 3.10 yields [15,16]

$$\gamma^f = 2\gamma^s + \Pi h. \tag{3.11}$$

This is a very important relation between film tension γ^f, surface tension of the film γ^s, disjoining pressure Π, and film thickness h—the most important thermodynamic characteristics of a foam film.

According to the thermodynamic approach based on the simplified model of a foam film with two parallel Gibbs dividing surfaces, the differences between the extensive thermodynamic properties of the model and of the real foam film are introduced as surface excess quantities. All extensive surface excess properties of the film depend on the location of the Gibbs dividing surfaces. Just as in the case of a single interface, different conditions can be accepted to determine the dividing surface location. Most often, the condition $\Gamma_1^{sf} = 0$ is used, that is, the surface excess of component 1 in the film is zero, the component 1 being the solvent of phase α.

Obviously, the thermodynamic thickness h of the foam film, which is the distance between the two Gibbs dividing surfaces, also depends on their location, respectively, on the condition $\Gamma_1^{sf} = 0$.

Real foam films are usually stabilized by dense surfactant adsorption layers at both surfaces. That means the thermodynamic film thickness is closer to the thickness of the inner liquid layer of the film, rather than to the whole film thickness. However, the surfactant molecules can be very different as can the film structure. Hence, the comparison of the thermodynamic and the real film thickness should be done separately for each system.

Each total extensive quantity of the foam film can be presented as the sum of the doubled surface excess quantity plus the corresponding extensive quantity of the volume part of the film. Thus, the total internal energy U^f of a foam film is presented as the doubled surface excess internal energy $2U^{sf}$ plus the internal energy $U^{\alpha f}$ of the volume part of the foam film.

The fundamental thermodynamic equation for the surface excess of the internal energy U^{sf} of the foam film follows directly from the First and Second Laws of Thermodynamics:

$$2\,dU^{sf} = 2T\,dS^{sf} - \Pi\,dV^f + \gamma^f\,dA^f + 2\sum \mu_i\,dn_i^{sf}. \tag{3.12}$$

Here, the superscripts denote: f—film, s—surface excess; T is temperature, S—entropy, V—volume, A—area, μ_i—the chemical potential, and n_i—the moles of each component i. Following the usual procedure, the corresponding important Gibbs-Duhem relation of the Gibbs dividing surfaces of the foam film can be obtained, which when assigned to the unit film area reads

$$d\gamma^f = -2S_A^{sf}\,dT + h\,d\Pi - 2\sum_i \Gamma_i^{sf}\,d\mu_i. \tag{3.13}$$

Here, the subscript A denotes the surface excess extensive quantity related to the unit film area and $\Gamma_i^{sf} = n_i^{sf}/A^f$.

The temperature T and the chemical potential μ_i of each component i are considered constant in all three phases f, α, and β in the thermodynamic system examined (Figure 3.2). Hence from Equation 3.13 we obtain

$$\left(\frac{\partial \gamma^f}{\partial \Pi}\right)_{T,\mu_i} = h, \tag{3.14}$$

which after integration at constant temperature and all chemical potentials gives

$$\gamma^f = 2\gamma - \int_\infty^h \Pi\,dh + \Pi h \quad \text{or} \quad \gamma^f = 2\gamma + \Pi h + \Delta F(h). \tag{3.15}$$

The quantity $\Delta F(h)$ is usually called *interaction free (Helmholtz) energy of the foam film*. Obviously, this is the isothermal, reversible work per unit film area done against the disjoining pressure when thinning a thick film down to a small equilibrium thickness h. Equation 3.15 can be used to determine the quantity $\Delta F(h)$ from experimental data for γ and γ^f (the last two quantities can be directly measured). The product Πh is usually very small since h is extremely small for so called *black films*—the case when 2γ and γ^f noticeably differ. Then Πh is negligible and $\Delta F(h) \approx \gamma^f - 2\gamma$.

A foam film is always connected with a bulk liquid phase or with a solid wall as well as with other foam films in foam, through a *Plateau border*—a liquid body with concave surfaces like a meniscus. The Plateau border contains bulk liquid α from which the foam film has been formed. The contact film/Plateau border is schematically presented in Figure 3.3. On the left side, there is a part of the horizontal, plane-parallel, symmetric foam film and on the right side a part of the Plateau border. A *transition zone film/bulk* always exists between them [17]. At equilibrium, the disjoining

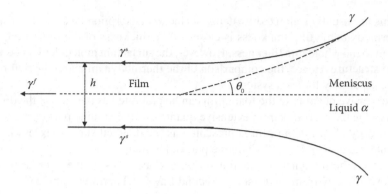

FIGURE 3.3 Scheme of part of a plane-parallel symmetric thin foam film (left side) connected through a transition zone with a meniscus of the liquid phase α.

pressure Π acts in the film and the surface tension of the film γ^s differs from the surface tension γ of the bulk liquid phase. Both Π and γ^s are due to the surface forces' interactions in the foam film. These interactions operate in the transition zone film/bulk as well, but they decrease with increasing thickness of the transition zone toward the meniscus. This means that both Π and $(\gamma - \gamma^s)$ gradually decrease along the transition zone in direction from the film to the Plateau border, becoming zero in the bulk liquid.

In a real system, the curved meniscus interface gradually becomes a flat film interface and γ gradually turns into γ^s. The simplified macroscopic thermodynamic description neglects the presence of a transition zone and assumes an abrupt change of γ (bulk liquid) into γ^s (film). This thermodynamic approach requires introduction of two other thermodynamic quantities: the *contact angle film/bulk* and the *line tension* of the contact line. Thus, the simplified thermodynamic description becomes equivalent to the real system.

In the Plateau border, the surface tension of the bulk liquid is γ and the shape of its concave interface is determined by the Laplace equation. If we also consider γ constant in the transition zone and extrapolate the meniscus interface according to the Laplace equation toward the film (the dotted curve in Figure 3.3), it intersects the mid-plane of the film (the horizontal dotted straight line). The line of intersection is the *contact line* of the film and the angle between both dotted lines is the *contact angle* θ_0 [18,19]. If we consider a cylindrical meniscus, that is, a straight contact line, the balance of forces in a tangential direction, taking into account Equation 3.15, is given by

$$\gamma^f = 2\gamma \cos\theta_0 = 2\gamma + \Pi h + \Delta F(h). \tag{3.16}$$

For a curved contact line (e.g., round film in a spherical meniscus), the balance of forces reads [20]

$$2\gamma \cos\theta_0 = \gamma^f + \frac{\kappa}{r^f}. \tag{3.17}$$

Here, κ is the line tension and r^f is the radius of curvature of the contact line. The term κ/r^f is actually a two-dimensional capillary pressure—a surface pressure difference on both sides of the curved contact line. However, this term is usually very small because of the very small κ-value; it should be taken into account at extremely small r^f only.

The contact angle θ_0 can be experimentally measured. From these experimental values with a measured γ-value, one can calculate, using Equation 3.16, values of the interaction free (Helmholtz) energy $\Delta F(h)$ of the film. As already noted, the term Πh is usually very small, that is, $\Pi h \approx 0$.

3.2 SURFACE FORCES IN FOAM FILMS

The most important factor which determines the properties of foam films is the interaction between the two film interfaces, as well as the interactions between the two adjacent phases across the liquid film. The thermodynamic quantity *disjoining pressure* Π is a result of these interactions due to different types of *surface forces* acting in the thin liquid films. The dependence of the disjoining pressure on the film thickness is called a *disjoining pressure isotherm*. Two categories of surface forces are usually distinguished: DLVO and non-DLVO surface forces [21,22]. The *Van der Waals molecular interactions* or *dispersion molecular forces* as well as the *electrostatic* or *double layer forces* are called DLVO forces (both are long range forces), the balance of which is the basis of the DLVO theory, that is, the theory of Deryaguin, Landau, Verwey, and Overbeek [23,24] of the stability of lyophobic colloids [25]. According to this theory, the disjoining pressure in a foam film is considered as a sum of an electrostatic component Π_{el} and a Van der Waals component Π_{vw}:

$$\Pi = \Pi_{el} + \Pi_{vw}. \tag{3.18}$$

The *electrostatic* or *double layer disjoining pressure* is given by the equation

$$\Pi_{el} = 2cRT(\cosh zy^m - 1), \tag{3.19}$$

with $m = h/2$ and $y^m = F\psi^m/RT$. Here, c is the concentration of a symmetric electrolyte of valence z, F is the Faraday's constant; ψ^m is the potential at a distance $h/2$ (at the central plane of the symmetric foam film) which is connected to the potential ψ^0 of the electric double layer at the film/adjacent gas phase surface through the relation

$$-\frac{\kappa h}{2} = \int_{y=yo}^{y=y^m} \frac{z\,dy}{\sqrt{2[\cosh(zy) - \cosh(zy^m)]}} \tag{3.20}$$

with

$$y = \frac{F\psi}{RT}, \quad y^0 = \frac{F\psi^0}{RT} \quad \text{and} \quad \kappa = \sqrt{\frac{F^2 z^2 c}{\varepsilon_0 \varepsilon RT}}, \tag{3.21}$$

where ε is the relative dielectric permittivity and $\varepsilon_0 = 8.854 \times 10^{-12}$ C^2 V^{-1} m^{-1} is the dielectric permittivity of vacuum.

An approximated expression of the electrostatic component of the disjoining pressure can be derived from Equation 3.19 at small values of ψ^m

$$\Pi_{el} = \frac{64cRT}{z^2} \left[\tanh\left(\frac{zy^0}{4}\right)\right]^2 e^{-\kappa h}. \tag{3.22}$$

Equation 3.22 is quite adequate for an easy calculation of Π_{el}, however, for a wider range of ψ^0 and c values, it is necessary to use the more general Equations 3.19 and 3.20.

The positive electrostatic disjoining pressure Π_{el} depends on the potential ψ^0 of the diffuse double layer at the foam film/gas interface. The relation between film thickness h and electrolyte concentration c obeys Equations 3.19 and 3.20. The potential ψ^0 can be calculated from the measured equilibrium film thickness h and known c value. This is the "method of the equilibrium thin liquid film" for determining the ψ^0 potential and for the study of the electric properties at the interface [26]. The method provides an equilibrium potential to be evaluated and all complications occurring at kinetic measurements are avoided. Detailed study with this method of the $h(c)$ and $h(pH)$ dependences in the absence of a surfactant gave $\psi^0 \approx 30$ mV at the aqueous electrolyte solution/air interface.

The values of the ψ^0-potential of the diffuse double layer at the aqueous solution surface can be used to calculate the surface charge density σ^0, according to

$$\sigma^0 = \left[\frac{2\varepsilon RTc}{\pi}\right]^{1/2}\left[1/2\cosh\left(\frac{F\psi^0}{RT}\right) - 1/2\cosh\left(\frac{F\psi^m}{RT}\right)\right]^{1/2}. \tag{3.23}$$

The calculation by the DLVO theory does not give an estimation of whether the values of ψ^0 and σ^0, are positive or negative. However, experimental measurements provide information about the ions adsorbed at the interface and it is possible to determine the potential sign. The $\psi^0(pH)$ dependence for aqueous solutions at constant ionic strength (HCl + KCl) shows that at pH > 5.5, the potential becomes constant and equal to about 30 mV. At pH < 5.5, the potential sharply decreases and becomes zero at pH $\approx$ 4.5, that is, an isoelectric state at the solution surface is reached. This means that the charge at the surface of the aqueous solutions is mainly due to the adsorption of H^+ and OH^- ions. The adsorption potential of the OH^- ions in the Stern layer is higher. It follows that the ψ^0-potential at a solution/air interface appears as a result of adsorption of OH^- ions [26].

Two theories, macroscopic and microscopic, are involved in the calculation of the *Van der Waals disjoining pressure* Π_{vw} in thin liquid films. According to the *microscopic theory*, the total interaction force in a flat gap between two semi-infinite phases decreases with distance much more slowly than the interaction force between two individual molecules. The following expression for Π_{vw} in a symmetric liquid film between two gas or condensed phases is obtained

$$\Pi_{vw} = -\frac{K_{vw}}{h^n} \tag{3.24}$$

where K_{vw} is the Van der Waals-Hamaker constant [27]); $K_{vw}(h) = A(h)/6\pi$, $A(h)$ is the Hamaker constant for symmetric films—a weak function of film thickness. In general, $n = 3$; however, if a correction for the electromagnetic retardation of dispersion forces is taken into account [28], we get $n = 4$.

A general formula for calculating the dispersion molecular interactions for any type of condensed phases has been derived in a *macroscopic theory* [29,30]. The attraction between bodies results from the existence of the fluctuational electromagnetic field of the substance. If this field is known for a thin liquid film, it is possible to determine the disjoining pressure in it. The stricter macroscopic theory avoids the approximations assumed in the microscopic theory, that is, additivity of forces, integration, and extrapolation of interactions of individual molecules in the gas to interactions in the condensed phase. The macroscopic theory is successfully used for the calculation of the molecular dispersion interactions, the $\Pi_{vw}(h)$. Several approximations have been derived since the exact equations are too complicated. A good approximation for symmetric flat films is

$$\Pi_{vw} = -\frac{\hbar}{8\pi^2 h^3}\int_0^\infty \left(\frac{x^2}{2} + x + 1\right)e^{-x}\frac{(\varepsilon_f - \varepsilon)^2}{(\varepsilon_f + \varepsilon)^2}d\omega \tag{3.25}$$

with $x = 2\omega h\varepsilon_f^{1/2}$, ε_f and ε are the relative dielectric permittivities (functions of the frequency ω) of the film and the adjacent phases, respectively, and $\hbar$ is Planck's constant divided by 2π. For small film thickness, a simpler approximation can be derived

$$\Pi_{VW} = -\frac{\hbar}{8\pi^2 h^3} \int_0^\infty \frac{(\varepsilon_f - \varepsilon)^2}{(\varepsilon_f + \varepsilon)^2} \, d\omega. \tag{3.26}$$

Another approximation for a foam film with relatively large thickness is

$$\Pi_{VW} = \frac{\hbar c}{h^4} \frac{\pi^2}{240\sqrt{\varepsilon_0}} \left(\frac{1-\varepsilon_0}{1+\varepsilon_0}\right)^2 \varphi\left(\frac{1}{\varepsilon_0}\right) \tag{3.27}$$

where ε_0 is the static value of the dielectric permittivity of the foam film; c is the speed of light. Here, Π_{VW} is inversely proportional to h^4, similar to Hamaker's formula which accounts for the electromagnetic retardation of dispersion forces [28].

For practical use of the macroscopic theory, equations with empirical constants are also appropriate for calculation of Π_{VW}, for instance [31]

$$\Pi_{VW} = h^{-3} \left[\frac{b + (a + ch)}{(1 + dh + eh^2)}\right] \tag{3.28}$$

where a, b, c, d, and e are empirical constants.

The non-DLVO forces [21,22] in foam films are: *steric surface forces*, due to the adsorption layers at films' surfaces of surfactant or polymer molecules; *structural (solvation, hydration) surface forces* which originate from modifications in the liquid structure adjacent to the film interfaces; as well as other surface forces, introduced by some authors. Most important for symmetric foam films are the *steric repulsion interactions*—long range for macromolecules, short range for small molecules.

In foam films stabilized by polymers, at low electrolyte concentrations c_{el}, the electrostatic repulsion dominates and decreases with increasing c_{el} until *steric repulsion* (which is independent of c_{el}) becomes operative at c_{cr}. Above c_{cr}, at lower pressure of the mushroom tails of the longer chains interact rather softly; at higher pressure, a brush-to-brush contact is realized and strong steric repulsion is expected. The transition from electrostatic to steric repulsion occurs at c_{cr} given by

$$\frac{1}{\kappa_{cr}} \equiv \left(\frac{\varepsilon kT}{8\pi e^2 N_A c_{cr}}\right)^{1/2} = R_F \tag{3.29}$$

where ε is the dielectric constant of the solvent; e is the elementary charge; N_A is Avogadro's number; $1/\kappa_{cr}$ is the Debye length at c_{cr}. R_F is the radius of gyration of a flexible neutral chain in a good solvent given by the Flory relation [32]:

$$R_F = aN^{3/5} \tag{3.30}$$

where a is an effective monomer size and N is the number of monomers.

If a macromolecule adsorbs at the interface as a separate coil, it should occupy an area of the order of the projected area of the molecule in the bulk solution, that is, R_F^2. However, the polymer chains are crowding the solution/air interface and are stretched, that is, they form a brush [33]. The thickness h_1 of the adsorption layer is determined by the conformation of the macromolecules at the interface and it can be calculated from the simple brush model:

$$h_1 = aN\left(\frac{a^2}{A_0}\right)^{1/3}, \tag{3.31}$$

with A_0 the area per molecule.

The "soft" steric repulsion determines a quite large film thickness, while a strong steric repulsion due to the brush-to-brush contact determines a considerably smaller thickness. Presumably, the polymer brushes repel each other. Under these conditions, the de Gennes scaling theory [33] for the interaction between two surfaces carrying polymer brushes applies. Accordingly, the steric disjoining pressure is given by

$$\Pi_{st} \cong \frac{kT}{D^3}(H^{-9/4} - H^{3/4}) \tag{3.32}$$

where $H = h/2h_1$ is the dimensionless film thickness and D is distance between two grafted sites. The first term is the osmotic pressure arising from the increased polymer concentration in the two compressed layers. The second one is an elastic restoring force (polymer molecules always tend to coil which is the origin of the negative sign of this term). De Gennes' theory gives a satisfactory description of the steric surface forces at pressures and film thicknesses where a brush-to-brush contact is realized.

3.3 EXPERIMENTAL METHODS FOR RESEARCH OF MICROSCOPIC FOAM FILMS

Original and unique methods have been developed for the investigation of foam films. The progress in the knowledge about foam films is due to a great extent on effective experimental techniques created for their investigation. Single foam films of different size and shape are the objects of these techniques: flat film in a frame in the mm/cm range, usually vertical, but also horizontal; bubbles blown at the orifice of a capillary tube; bubbles floating on the solution surface; films with cylindrical symmetry stretched between solid rings or cylinders; microscopic, round, horizontal foam films; and so on. A variety of physical methods is used for the film characterization: interference and intensity of light reflected by the film; microphotography; ellipsometry; Brewster angle measurement; infrared light absorption; x-ray reflectivity; electric conductivity; dynamometry; and so on. It is impossible to describe this great multitude of experimental techniques in a brief way (see e.g., [1–6]).

Therefore, only the methods of *microscopic foam films*, in conjunction with the *micro-interferometric technique* as well as the *pressure balance technique*, are considered. Small circular foam films, the radius of which is within the range of 10–500 μm, are considered to be microscopic films. The experimental technique for their study allows the measurement of thermodynamic quantities, the following of the kinetic behavior, the formation of black films, the realization of metastable states, and so on. An advantage is the possibility to work at very low surfactant concentrations.

The main details of the two most often used measuring cells of Scheludko and Exerowa [34,35] are presented in Figure 3.4. The microscopic film is formed in the middle of a biconcave liquid drop and is situated in a cylindrical glass holder of radius R by withdrawing liquid from it through the

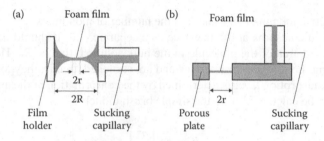

FIGURE 3.4 The main details of two measuring cells of Scheludko and Exerowa for the study of microscopic, horizontal, circular foam films: (a) foam film, formed in the middle of a biconcave liquid drop, situated in a cylindrical glass holder; (b) foam film, formed in a cylindrical hole in a porous sinter glass plate.

sucking capillary tube (Figure 3.4a). The suitable inner radius R of the film holder is 0.2–0.6 cm and the film radius ranges from 100 to 500 μm.

The periphery of the film is in contact with the solution from which the film is formed. In the case of aqueous foam films, the inner part of the holder carrying the biconcave drop is finely "furrowed" with vertical lines, closely situated to one another, which improve wetting [36]. In the case of foam films from nonpolar liquids ("oil" foam films), the film holder should be hydrophobized. A constant capillary pressure acts on the film formed in this cell (Figure 3.4a). It is determined by the radius of curvature of the meniscus. Two series of photographs of different states of microscopic foam films, taken under a microscope during the process of film thinning, are presented in Figure 3.5.

The film holder of the porous plate cell [35] is a cylindrical hole drilled in a porous sintered glass plate with the hole radius being considerably smaller, for instance <250 μm (Figure 3.4b). Plates of various pore radii can be used. If the meniscus penetrates into the pores, their radius determines the radius of curvature, that is, the small pore size allows increasing the capillary pressure until the gas phase can enter into them. The capillary pressure can be increased to more than 10^5 Pa, depending on the pores size and the surface tension of the solution. The film holders of both cells in Figure 3.4, respectively, the foam films, are situated in the closed space of the cell, saturated with the solution vapor.

The film thickness is the most important quantity to be experimentally determined. As already mentioned above, the definition of the film thickness is very difficult. The transition from the liquid film to the adjacent gas phase is not abrupt. There is an interfacial layer with finite thickness; the surfactant forms adsorption layers at both film interfaces. The thermodynamic description of an interface is based on the so called *Gibbs dividing surface*. The foam film thermodynamics involves two Gibbs dividing surfaces and the distance between them is defined as the *thermodynamic film thickness* [12,37]. Another approach is the mechanical definition: the film boundaries are two plane-parallel mathematical faces—the *surfaces of tension*, each of which is subject to a uniform, isotropic tension. The *mechanical film thickness* is defined as the distance h between the two surfaces of tension [38]. The big problem is that neither the thermodynamic nor the mechanical film thickness coincides exactly with the thickness which is experimentally determined using different physical methods. Most

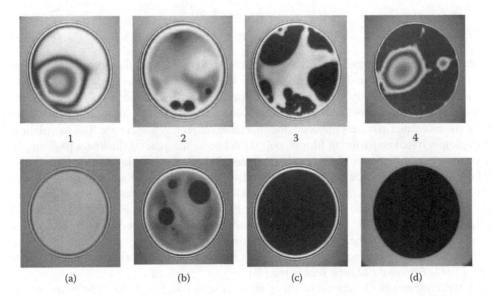

| 1 | 2 | 3 | 4 |
| (a) | (b) | (c) | (d) |

FIGURE 3.5 Formation of black spots and black films in a thicker unstable microscopic liquid film. The series (1–4)—foam film from a solution with complicated rheological properties which determine irregular film thinning. The series (a–d)—plane-parallel smooth thinning foam film (a), formation of black spots (b), common black film (c), and Newton black film (d)

of these methods are optical methods and the measurement of the film thickness is based on the abrupt change of a property, such as the refractive index, electron density, and so on at the film boundary.

The most widely used optical method—the interferometric (for microscopic foam films—microinterferometric) method is based on the measurement of the intensity of visible light reflected from the foam film. It can be registered as a dependence of photo-current versus time in which the extrema correspond to the interference maxima and minima, that is, film thickness is divisible to $\lambda/4n$ (where λ is the wavelength of light and n is the refractive index of the liquid in the film). Thus, knowing the order k of interference, it is easy to determine the film thickness at these points. Film thickness (between a maximum and a minimum) is calculated from the ratio between the measured intensities of the reflected monochromatic light I, corresponding to a certain thickness, and I_{max}, corresponding to the interference maximum, according to the formula [39–41]

$$h_w = \frac{\lambda}{2\pi n}\left(k\pi \pm \arcsin\sqrt{\frac{I/I_{max}}{1+[(n^2-1)/2n]^2(1-I/I_{max})}}\right), \tag{3.33}$$

where h_w is called the equivalent film thickness, that is, the thickness of a foam film with uniform refractive index n of the bulk solution from which the film is formed. The accuracy of thickness measurements with the microinterferometric technique is ± 0.2 nm.

It is clear that h_w is not equal to the real physical film thickness. Although h_w does not coincide with the thermodynamic thickness h, h_w is close to it if the last is defined through the condition of zero interfacial excess of the solvent in the film. The difference between h_w and h could be neglected for any thickness larger than 30 nm. However, for thinner liquid films, it is necessary to also account for the film structure. The three layer film model ("sandwich model") with an aqueous core of thickness h_2 and refractive index n_2 and two homogenous layers of hydrocarbon chains of the adsorbed surfactant of thickness h_1 each and refractive index n_1 is often used for foam films. The thickness of the aqueous core h_2 is most often determined according to the formula [42]

$$h_2 = h_w - 2h_1\frac{n_1^2-1}{n_2^2-1}. \tag{3.34}$$

Since, however, each model involves some assumptions, the calculation of h_2 always renders a certain inaccuracy. The most important problem in the three layer model concerns the position of the plane that divides the hydrophobic and hydrophilic parts of the adsorbed surfactant molecule. In some cases, it seems reasonable to have this plane passing through the middle of the hydrophilic head of the molecule, in others the head does not enter into the aqueous core. The calculations of the film thickness based on different film models do not solve the general thickness problem; however, they provide possibilities for a reasonable interpretation of experimental results.

The microinterferometric technique is usually used for determination of disjoining pressure isotherms $\Pi(h)$ as well. At equilibrium, Π is equal to the pressure difference Δp applied from outside to the film. For microscopic horizontal circular foam films (Figures 3.4 and 3.5), $\Delta p = P_c$ is the capillary pressure of the concave meniscus around the film, determined by the meniscus curvature and the surface tension of the bulk solution. The capillary pressure is experimentally accessible and at equilibrium $\Pi = P_c$ can be determined; this is the base of the experimental method called *Thin Liquid Film—Pressure Balance Technique* [43].

The experimental block scheme of this method is shown in Figure 3.6. The films are formed in the porous plate measuring cell (Figure 3.4b). The hydrodynamic resistance in the porous plate is sufficiently small and the maximum capillary pressure which can be applied to the film is determined by the pore size and the surface tension γ of the solution. When the maximum pore size is 0.5 μm, the capillary pressure is $\sim 3 \times 10^5$ Pa at $\gamma = 70$ mN/m.

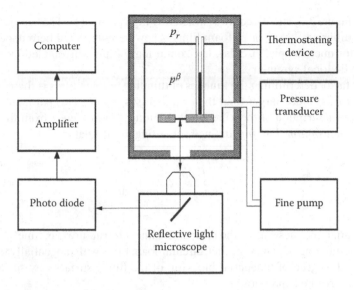

FIGURE 3.6 Block scheme of the Thin Liquid Film—Pressure Balance Technique.

The cell is placed in a temperature controlled device and mounted on a microscopic table. Thus, the film can be monitored and measured photometrically in reflected light. The reflected light enters the photodiode and its signal is amplified and registered by a computer. The regulation of the capillary pressure is achieved by a special membrane pump which allows a gradual and reversible change in the gas pressure p^β in the closed cell. Values of Π less than 100 Pa prove to be much more difficult to measure, so there should be an entire conformity with the equation giving the balance of pressures acting in the film [44] and the geometry of the measuring cell:

$$\Pi = p^\beta - p_r + \frac{2\gamma}{r} - \Delta\rho g h_c, \tag{3.35}$$

where p_r is the external reference pressure, usually atmospheric pressure; r is radius of the sucking capillary tube; $\Delta\rho$ is density difference between the gas and the surfactant solution; h_c is the height of the solution in the sucking capillary tube above the film.

The microscopic foam film experimental technique is also used for determining two other important thermodynamic characteristics of the foam films: the *contact angle* θ_o appearing at the contact of the film with the bulk solution from which it is formed, and the *film tension* γ^f related to it. Two methods for the θ_o measurement have been developed: the topographic method and the film expansion method [45,46].

Besides many variants of the methods reviewed above, there are also other original methods for studying specific aspects of the behavior and properties of foam films: kinetics of film thinning, film stability, film electroconductivity, film permeability to gas, film viscoelasticity, and so on.

3.4 NONEQUILIBRIUM BEHAVIOR OF FOAM FILMS

When a foam film is formed from a bulk liquid phase α in a surrounding gas β, it is initially a thick liquid film. It becomes a thin liquid film as the result of the *process of thinning* under the driving force $\Delta p = p^\beta - p^\alpha$. This pressure difference can be: the capillary pressure P_c of the Plateau border (meniscus), a hydrostatic pressure difference, a Δp created by experimental device, and so on. An equilibrium foam film is obtained only if Δp is counterbalanced by a positive disjoining pressure Π. However, when Π remains lower than Δp or is even negative, no equilibrium can be established. The process of thinning leads to either *film rupture* or to the *jump-like formation* of a much thinner *black foam film* (Figure 3.5) stabilized by a sufficiently high positive Π. These processes are the most important nonequilibrium behavior of foam films. They are considered here under strictly defined conditions only for microscopic, circular, horizontal foam films, surrounded by a double concave meniscus with cylindrical symmetry.

An important factor determining the kinetics of thinning of foam films is their anisodiametricity, that is, their radii r are much larger than their thicknesses, $r \gg h$. Two parameters are introduced for the quantitative description of the kinetics of thinning: the *lifetime* τ from film formation until rupture and the *rate v of film thinning*, defined according to the relations

$$\tau = \int_{h_0}^{h_{cr}} \frac{dh}{v}; \quad v = -\frac{dh}{dt} \tag{3.36}$$

where h_0 is the initial thickness; h_{cr} is the critical thickness of rupture; t is time.

The velocity of thinning of plane-parallel, circular foam films with tangentially immobile surfaces because of the high degree of tangential blocking of the films' surfaces by surfactant adsorption layers, is given [47] by the expression

$$\frac{dh^{-2}}{dt} = a(P_c - \Pi), \quad a = \frac{4}{3\eta r^2}. \tag{3.37}$$

Here, the pressure term represents the difference between the capillary pressure P_c of the meniscus and the disjoining pressure Π in the foam film, where η is the viscosity. This is actually an application of the Stephan-Reynolds relation [48] (see Reynolds' lubrication theory [49]) for the rate v_{Re} of thinning/thickening of a liquid layer between two solid circular plates. The applicability of Eq. (3.37) imposes the following four requirements: (1) the surfaces should be tangentially immobile (zero surface velocity); (2) the film surfaces should remain plane-parallel; (3) the capillary pressure P_c of the meniscus should not be affected by film thinning; (4) the viscosity η should not depend on film thickness.

Requirement (1) in some cases can deviate considerably. The suppression of the tangential mobility of the foam film surfaces is affected by the presence of a surfactant through the Marangoni effect [50]. When the liquid drains from a foam film, a gradient of surface tension is created at its surfaces counterbalancing the viscous stress. This gradient, equivalent to a respective surfactant adsorption gradient, causes a surfactant mass transfer: diffusion flow from the bulk to the film surface, and surface flow in the direction of the adsorption gradient. The surfactant might effectively block the fluid surfaces of foam films. The deviation from Equation 3.37 resulting from the Marangoni effect has been experimentally observed and theoretically predicted [51] by the expression

$$\frac{v}{v_{Re}} = 1 - \left[1 - \frac{2D_s}{Dh}\left(\frac{\partial \Gamma}{\partial c}\right)\right]\frac{3D\eta}{\Gamma(\partial \gamma/\partial c)} \tag{3.38}$$

where D, D_s are, respectively, the bulk and surface diffusion coefficients of the surfactant; γ is the surface tension. The quantity $Ma = |\Gamma(\partial \gamma/\partial c)|/D\eta$ is called the Marangoni number and plays an important role in all transport processes at phase interfaces. Other factors (e.g., surface viscosity) affecting surface mobility of films and, hence, kinetics of their thinning are also analyzed [52]. Equation 3.38 describes correctly the foam film's hydrodynamics only when the factor *Ma* has large values—equivalent to a small deviation from zero surface velocity. The general analysis of the role of surface mobility on the rate of thinning indicates that at high surface mobility ($Ma < 1$), the hydrodynamics of film thinning changes strongly, thus leading to a deviation from Equation 3.37.

Equation 3.37 requires further that the liquid drainage from the foam film is strictly axial symmetric between the parallel walls. However, nonequilibrium foam films are in fact never plane-parallel. This is determined by the balance between hydrodynamic and capillary pressure. Only microscopic films of radii less than 100 μm retain their quasiparallel surfaces during thinning, which makes them particularly suitable for model studies. Films of larger radii exhibit significant deviations from the plane-parallel shape which affect both the kinetics of thinning and their stability [53]. Slightly larger films ($r \geq 100$ μm) keep the axial symmetry of thinning, but they lose their plane-parallel shape. In the center, a typical thickening can be formed, known as *dimple* (a lens-like formation) whose periphery is a thinner *barrier ring*. The dimple forms spontaneously as a result of the hydrodynamic resistance to thinning in the periphery of the circular liquid film [54,55]. Experiments proved that the rate of thinning is practically equal in both the dimple's center and barrier ring. This leads to an increase in the nonuniformity of thickness. This nonuniformity increases with increasing film radius as well.

During drainage of larger circular, horizontal films ($r > 200$ μm), more complex phenomena have been observed. One or more thicker domains form in such films, dividing them into parts [53]. During film thinning, these "channels" move and sometimes separate from one end and sink into the meniscus. The axial symmetry of drainage is disturbed. Such a film drains faster than expected for a homogenous symmetrically draining film of the same size. An attempt for theoretical explanation of these effects was based on the idea [56] that the transition from symmetric to asymmetric drainage depends on the presence and the properties of the stabilizing surfactant.

It seems that the deviations from the plane-parallel shape during film thinning are the main reason for considerable differences between the time of thinning measured and calculated from Equation 3.37. The experimental results clearly indicate that Equation 3.37 can be applied only to sufficiently small films ($r < 100$ μm), that is, to films of uniform thickness. A theory of the dynamics of large, circular, horizontal films with developed subdomains in them derived an equation [57] about thinning of films of nonhomogeneous thickness:

$$v = \frac{1}{6\eta} \sqrt[5]{\frac{h^{12}\Delta p^8}{4\gamma^3 r^4}}. \tag{3.39}$$

The experimental results are in good agreement with the theoretical prediction. Equation 3.39 holds for large films with strongly expressed thickness inhomogeneities.

The stability of a foam film plays a decisive role for the stability of foams. The study of processes leading to film rupture is useful for understanding the reasons for their stability. The simplest explanation of film rupture involves reaching a thermodynamically unstable state. A typical example of thermodynamically unstable systems is asymmetric foam film in which the Van der Waals contribution to the disjoining pressure obeys Hamaker's relation, Equation 3.24. These are films made from some aqueous surfactant solutions containing sufficient amounts of an electrolyte to suppress the electrostatic component of the disjoining pressure as well as films formed from nonaqueous solutions.

During thinning, the thermodynamically unstable foam films keep their shape in a large range of thickness until a rather small thickness is approached, at which the film ruptures. This thickness is called *critical thickness of rupture* h_{cr}. Therefore, the thermodynamic instability is a necessary but not a sufficient condition for film instability. Two processes govern the film instability—film thinning with retaining film shape, and film rupture. Contemporary understanding of foam film rupture is based on the concept of the existence of fluctuational waves on liquid surfaces [58]. According to this approach, the film is ruptured by unstable waves, that is, waves the amplitudes of which increase with time. The rupture occurs at the moment when the amplitude Δh or its root mean square value $\sqrt{(\Delta h)^2}$ of a certain unstable wave grows up to the order of the film thickness

$$\sqrt{(\Delta h)^2} \approx h_{cr}. \tag{3.40}$$

The basis of this model has been developed in [2,59]. It was shown that the condition of increase in amplitude of a wave is equivalent to the condition of increase in local pressure

$$\delta(\Delta p_c + \Pi) > 0 \qquad (3.41)$$

$$\delta \Delta p_c = -\gamma k^2 \Delta h \qquad (3.42)$$

Δp_c is the capillary pressure corresponding to wavelength $\lambda = 2\pi/k$ and amplitude Δh; $\delta \Pi = (d\Pi/dh)\Delta h$ is the respective perturbation of disjoining pressure Π. Hence, Equation 3.41 yields

$$k^2 < \frac{[d\Pi/dh]}{\gamma} = k_{cr}^2. \qquad (3.43)$$

The upper limit k_{cr} of the unstable spectrum range is related to the *Scheludko number* [60].

In thick films ($h > 0.5\ \mu m$), only capillary forces act against surface deformations, that is, $\delta \Delta p_c \gg \delta \Pi$ and fluctuation waves are practically stable for the whole wavelength spectrum determined by Equation 3.40. Moreover, the steady state amplitudes of the capillary waves determined from the equipartial law $\sqrt{\overline{(\Delta h)^2}} \approx \sqrt{kT/\gamma}$ at typical conditions ($\gamma \sim 50\ mN\ m^{-1}$; $kT = 4 \times 10^{-21}\ J$) have values of the order of $\sqrt{\overline{(\Delta h)^2}} \sim 0.1\ nm$, that is, thick films are not only stable, but remain practically unaffected by thermal fluctuations.

In the process of film thinning, the interactions due to surface forces in the foam film become stronger and the attractive components corresponding to the negative disjoining pressure have a destabilizing effect (deepening of amplitude). When only Van der Waals forces act in the film do Equations 3.43 and 3.24 give [59] for k_{cr}

$$k_{cr} = \sqrt{\frac{3K_{VW}}{\gamma h^4}}. \qquad (3.44)$$

Once formed, the unstable waves grow until one of them (the fastest) conforms with Equation 3.40 and then the film ruptures. During this time, the film thins more depending on the conditions under which it is produced. This kinetic part of the theory of film rupture at a critical thickness has been formulated and partially solved in [61]. An important feature of the kinetics of film rupture is the random character of the process. Here, the question is about the correct description of the effect of fluctuations on the evolution of single waves. The further development of the theory leads to an expression [62] which seems suitable from an experimental point of view:

$$h_{cr} = \frac{(kT)^{1/10} K_{VW}^{2/5}}{(v\eta)^{1/5} \gamma^{3/10}}. \qquad (3.45)$$

Since rupture is a process with a clearly pronounced random character, reliable measurements of h_{cr} are possible only with microscopic circular foam films in which nonfluctuation disturbances are eliminated. The experimental data are in good agreement with Equation 3.45. The K_{VW} value recalculated from the experimental data of h_{cr} is very close to the one theoretically calculated according to the Lifshitz theory $K_{VW} = 10^{-21}\ J$. Later some corrections were also introduced in the theory of film rupture.

Another approach [63] to the rupture of thin liquid films was based on stochastic modeling of this critical transition. Autocorrelation functions for steady state and for thinning liquid films are obtained. A method for the calculation of the lifetime τ and h_{cr} of films was introduced. It accounts for the effect of the spatial correlation of waves. The existence of subdomains leads to decrease in

τ and increase in h_{cr}, that is, an increase in the probability for film rupture. Coupling of surface waves' dynamics and rate of drainage v leading to stabilization of thinning films are also accounted for in [64].

Foam films lose their stability during thinning when they reach a certain critical thickness h_{cr}. There are two possibilities: either the film ruptures or a local jump-like thinning in the film occurs. Since at very small thickness a foam film looks black in reflected light, the jump-like local thinning appearance is called formation of *black spots*. A black spot is a very small round area of a *black film* which is much thinner than the surrounding area of the unstable foam film. The appearance of black spots, their expansion, merging, and formation of black foam films are shown in Figure 3.5. The first series (1–4) of consecutive photos illustrates this process in a foam film from a solution with complicated rheological properties which determine irregular film thinning. In contrast, the second series in Figure 3.5 shows the formation of black spots (b) and black films (c and d) in a smooth thinning plane-parallel foam film (a).

It appears that the theory of foam films' rupture is applicable not only to the process of rupture by local thinning, but also to the formation of black spots. Hence, black spots can serve to detect the mechanism of local flexion in the film which allows to roughly estimate the fluctuation wavelength ($\lambda/2$ is ca. 1 μm). Such general treatment of instability including the formation of black spots can be employed as an additional tool to verify the theory of rupture. An important result is that rupture of unstable films as well as formation of black spots occurs at the same critical thickness h_{cr}. It is ca. 30 nm for foam films from aqueous surfactant solutions and the color of the films is gray at this thickness.

3.5   BLACK FOAM FILMS

Black foam films can reach an extremely small thickness. When observed under a microscope, they are black since they reflect minimal light when their thickness is below 20 nm. Therefore, they could be called nanofilms as well. The IUPAC nomenclature [8] distinguishes two equilibrium phase states of black films: *common black film*, CBF (Figure 3.5c) and *Newton black film*, NBF (Figure 3.5d). There is a pronounced transition between them, that is, CBF can transform into NBF or vice versa [4].

The CBF, just as the thicker foam films, can be described by the three layer film model or "sandwich model"—a liquid core between two adsorption layers of surfactant molecules. The NBF, however, has a bilayer structure without a free liquid core between the two layers of surfactant molecules, that is, a bilayer of amphiphilic molecules [4]. In the behavior of the latter, the short range molecular interactions prove to be of major importance. The definition "liquid film" is hardly valid for bilayers. They possess a higher degree of ordering similar to that of liquid crystals. It has been proved, however, by infrared spectroscopy [65] and electrical conductivity measurements [66], that there is water in the NBF. It is most probable that the adsorption layers contain a certain quantity of water but are not separated by an aqueous core. This is also confirmed by ellipsometric measurements [67] and by precise x-ray reflectivity measurements with CBF and NBF [68].

The dependence of disjoining pressure versus thickness for relatively large h values of a foam film is consistent with the DLVO theory. However, black films exhibit a diversion from the DLVO theory which is expressed in the specific course of the $\Pi(h)$ isotherm. A $\Pi(h)$ isotherm (in an arbitrary scale) of a foam film from surfactant + aqueous electrolyte solution is shown in Figure 3.7. The two types of black films, CBF and NBF, are clearly distinguished as two thermodynamic phase states, the black films being stabilized by long- and short-range surface forces, respectively.

In the right hand side of the isotherm (Figure 3.7), the curve passes a shallow minimum, after which the disjoining pressure becomes positive and increases up to a maximum. In this range, foam films exist, their equilibrium being described by the DLVO theory. If $h < h_{cr}$, the film is a common black film, CBF, schematically presented in the figure. At the equilibrium film thickness h_1, the disjoining pressure equals the external (capillary) pressure, $\Pi = P_c$. The equilibrium of CBF is also described by the DLVO theory, although discrepancies between the theoretical and experimentally

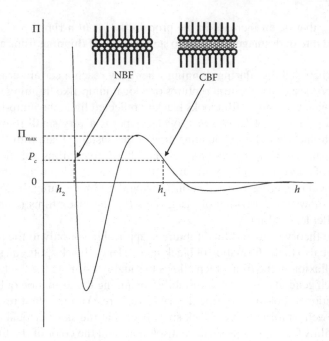

FIGURE 3.7 Schematic presentation of a disjoining pressure Π vs. film thickness h isotherm of a symmetric thin foam film in arbitrary scale.

obtained $\Pi(h)$ isotherms at film thicknesses below 20 nm are established [4,69]. The pressure difference, $\Pi_{max}-P_c$, is the barrier which hinders the transition to a film of smaller thickness.

According to the DLVO theory, the disjoining pressure should decrease infinitely to the left of Π_{max}. However, experimental results [46,70] show the existence of a second minimum in the $\Pi(h)$ isotherm after which the disjoining pressure sharply ascends. Another equilibrium is established on the rising left hand side of the isotherm, again under the condition $\Pi = P_c$. This black film with a smaller equilibrium thickness h_2 is actually the bilayer Newton black film (NBF), schematically presented in the figure as well. The left branch of the $\Pi(h)$ curve is not described by the DLVO theory, unlike the preceding minimum. Obviously, at this extremely small film thickness short range non-DLVO surface forces, for instance steric interactions, determine this part of the $\Pi(h)$ isotherm.

The two types of black films, CBF and NBF, show rather different properties due to their different structures. This difference determines the CBF/NBF transition as well. A number of thermodynamic parameters determine if CBF or NBF is in equilibrium: temperature T, electrolyte concentration c_{el}, surfactant concentration c_s, pH, and capillary pressure P_c. Besides the values of these parameters, the kind of the surfactant (including amphiphilic polymers) is also decisive for the existence of either CBF or NBF. The temperature is a very important parameter: systematic studies [71, 72] of black foam films have shown that at given c_{el}, c_s, and P_c the change of temperature causes a CBF/NBF transition; at high temperatures, the equilibrium black films are CBF and at low temperatures—NBF. Measurements [66] of the longitudinal specific electrical conductivity κ_f of black foam films have shown that the temperature dependence $\kappa_f\ (T)$ is different for CBF and NBF, respectively, and the activation energy of the ions' mobility is also rather different.

Thermodynamic quantities which show very different values for CBF and NBF are the contact angle θ_o and film tension γ^f. Systematic measurements of θ_o [46,69,73] established a very sharp jump in the $\theta_o(c)$ curve at the CBF/NBF transition. A critical electrolyte concentration c_{cr} at which the transition occurs has been determined. The film tension γ^f which is related through Equation 3.16 with the contact angle θ_o, shows a similar behavior. Its value for CBF ($\theta_o < 1°$) is close to 2γ of the bulk solution, while for NBF (θ_o of the order of $10°$) it is essentially different. Accordingly, the values of the interaction Helmholtz energy $\Delta F(h)$ of the film are rather different for CBF and NBF.

The stability in respect to rupture and permeability to gas of the NBF, which are bilayers of amphiphilic molecules, can be considered from a unified point of view [74,75]. The NBF bilayer can be regarded as consisting of two monolayers of amphiphilic molecules mutually adsorbed on each other. Each of the monolayers can be filled with a maximum of $N_m = 1/A_o$ molecules, but their thermal motion reduces their density below N_m. This means that vacancies of amphiphilic molecules (i.e., molecule-free sites) exist in the bilayer. The vacancies cluster together to form holes and if they are sufficient in number and size, they can make the bilayer permeable to molecular species. When $c_s < c_e$, nucleus holes can form and, by irreversible overgrowth, cause the rupture of the bilayer; c_e is a certain concentration of monomer amphiphilic molecules in the solution, called bilayer equilibrium concentration. If $c_s = c_e$, the bilayer is truly stable (and not metastable) in respect to rupture by hole nucleation. It must be emphasized that the bilayer also retains infinite stability for $c_s > c_e$. The bilayer cannot rupture despite the presence of a certain population of holes in it.

For steady state nucleation, the nucleation theory derived an explicit $\tau(c_s)$ dependence of the bilayer mean lifetime τ on the bulk concentration c_s

$$\tau(c_s) = A \exp\left[\frac{B}{\ln(c_e/c_s)}\right],\tag{3.46}$$

with A and B being constants. In some cases, τ can be so short that experimental observation of the bilayer after its formation is possible only with a certain probability W depending on the resolution time t_r of the particular equipment used. In direct visual observations of the bilayer rupture, for instance, $t_r \approx 0.5$ s, which is the reaction time of the eye. Since observation of the bilayer is possible only if the bilayer has ruptured during the time $t > t_r$, the nucleation theory yields W as a function of $c_s \leq c_e$

$$W(c_s) = \exp\left[-t_r/A \exp\left[\frac{B}{\ln(c_e/c_s)}\right]\right].\tag{3.47}$$

Equations 3.46 and 3.47 show that both τ and W sharply increase with the bulk surfactant concentration c_s in a relatively narrow range. These theoretical dependences, $\tau(c_s)$ and $W(c_s)$, can be easily checked since τ, W, and c_s are measurable quantities. The mean lifetime τ is measured as the time elapses from the moment of formation of a bilayer with a given radius until the moment of its rupture. Due to the fluctuation character of the film rupture, the film lifetime τ and probability W are random parameters, hence they should be determined by averaging data from a great number of measurements.

REFERENCES

1. Mysels, K.J., Shinoda, K., Frankel, S. *Soap Films*. Pergamon Press: New York, 1959.
2. Scheludko, A. Thin liquid films. *Adv Colloid Interface Sci* 1967; 1: 391.
3. Clunie, I.S., Goodman, J.F., Ingram, B.T. Thin liquid films. In: Matijevic, E. (Ed.), *Surface and Colloid Science*, Vol. 3. Wiley: New York, 1971.
4. Exerowa, D., Kruglyakov, P.M. *Foam and Foam Films*. Elsevier: Amsterdam, 1998.
5. Ivanov, I. *Thin Liquid Films*. Marcel Dekker: New York, 1988.
6. Platikanov, D., Exerowa, D. Thin liquid films. In: Lyklema, J. (Ed.), *Fundamentals of Interface and Colloid Science*, Vol. 5, Chapter 6. Elsevier: Amsterdam, 2005.
7. Exerowa, D., Kashchiev, D., Platikanov, D. Stability and permeability of amphiphile bilayers. *Adv Colloid Polymer Sci* 1992; 40: 201.
8. Ter Minassian-Saraga, L. Thin films including layers: Terminology in relation to their preparation and characterization. *Pure & Appl Chem* 1994; 66: 1667.
9. Gibbs, J.W. *Collected Works*. Longmans Green: London, 1928.
10. Gochev, G., Platikanov, D., Miller, R. Chronicles of foam films. *Adv Colloid Interface Sci* 2016; 233: 115–125.

11. de Feijter, J.A. Thermodynamics of thin liquid films. Chapter 1. In: Ivanov, I. (Ed.), *Thin Liquid Films*. Marcel Dekker: New York, 1988.

12. Toshev, B.V., Ivanov, I.B. Thermodynamics of thin liquid films. *Colloid Polymer Sci* 1975: 253: I. Basic relations and conditions of equilibrium, 558; II. Film thickness and its relation to the surface tension and the contact angle, 593.

13. Deryaguin, B.V., Obukhov, E.V. Anomalien dünner Flüssigkeitsschichten. III. Ultramikrometrische Untersuchungen der Solvathüllen und des "elementaren" Quellungsaktes [Anomalies of thin liquid layers. III. Ultramicroscopic studies on the solvation shell and the "elenmentary" act of swelling]. *Acta Physicochim URSS* 1936; 5: 1.

14. Deryagin, B.V., Churaev, N.V. К вопросу об определении понятия расклинивающего давления и его роли в равновесии и течении тонких пленок [On the problem of determining the concept of the disjoining pressure and its role in the equilibrium and the flow of thin films] *Kolloidny Zhur* 1976; 38: 438.

15. Rusanov, A.I. К термодинамике пленок. 1. Об упругости толстых пленок [To the thermodynamics of films. 1. On the elasticity of thick films] *Kolloidny Zhur* 1966; 28: 551.

16. Rusanov, A.I. *Phasengleichgewichte und Grenzflächenerscheinungen* [Phase equilibria and interfacial phenomena] Akademie Verlag: Berlin, 1978.

17. Kolarov, T., Zorin, Z., Platikanov, D. Profile of the transition region between aqueous wetting films on quartz and the adjacent meniscus. *Colloids Surfaces* 1990; 51: 37.

18. Princen, H.M., Mason, S.G. Shape of a fluid drop at a fluid-liquid interface. I. Extension and test of two-phase theory. *J Colloid Sci* 1965; 20: 156.

19. Deryagin, B.V., Martynov, G.A., Gutop, Y.V. Термодинамика и устойчивость свободных пленок [Thermodynamics and stability of free-standing films] *Kolloidny Zhur* 1965; 27: 357.

20. Veselovskij, V.S., Pertsov, V.N. Адгезия капель к твердым поверхностним [Adhesion of droplets to solid surfaces] *Zhur Fiz Khim* 1936; 8: 245.

21. Derjaguin, B.V., Churaev, N.V., Muller, V.M. *Surface Forces*. Consult Bureau: New York, 1987.

22. Israelachvili, J.N. *Intermolecular and Surface Forces*. Academic Press: New York, 1991.

23. Derjaguin, B.V., Landau, L.D. Theory of the stability of strongly charged lyophobic sols and of the adhesion of strongly charged particles in solutions of electrolytes. *Acta Physicochim USSR* 1941; 14: 633.

24. Verwey, E.J.W., Overbeek, J.T.G. *Theory of Stability of Lyophobic Colloids*. Elsevier: Amsterdam, 1948.

25. Vincent, B. Early (pre-DLVO) studies of particle aggregation. *Adv Colloid Interface Sci* 2012; 170: 56.

26. Exerowa, D. Effect of adsorption, ionic strength and pH on the potential of the diffuse electric layer. *Kolloid-Z* 1969; 232: 703.

27. Scheludko, A., Exerowa, D. Über den elektrostatischen und van der Waalsschen zusätzlichen Druck in wässerigen Schaumfilmen. [On the electrostatic and van der Waals extra pressure in aqueous foam films] *Kolloid-Z* 1960; 168: 24.

28. Casimir, H.B.G., Polder, D. The influence of retardation on the London-van der Waals forces. *Phys Rev* 1948; 73: 360.

29. Lifshitz, E.M. The theory of molecular attractive forces between solids. *Zhur Exper Teoret Fiz* 1956; 2: 73.

30. Dzyaloshinskii, I.E., Lifshitz, E.M., Pitevskii, L.P. The general theory of van der Waals forces. *Adv Phys* 1961; 10: 165.

31. Donners, W.A. PhD Thesis. University Utrecht, 1976.

32. Flory, P. *Principles of Polymer Chemistry*. Cornell University Press: New York, 1956.

33. de Gennes, P.G. The general theory of van der Waals forces. *Macromolecules* 1980; 13: 1069.

34. Scheludko, A., Exerowa, D. Setup for interferometric measurement of the thickness of microscopic foam films. *Comm Dept Chem Bulg Acad Sci* 1959; 7: 123.

35. Exerowa, D., Scheludko, A. Porous plate method for studying microscopic foam and emulsion films. *Compt Rend Acad Bulg Sci* 1971; 24: 47.

36. Exerowa, D., Zacharieva, M., Cohen, R., Platikanov, D. Dependence of the equilibrium thickness and double layer potential of foam films on the surfactant concentration. *Colloid Polymer Sci* 1979; 257: 1089.

37. Sheludko, A. Black Films. *Annuaire Univ Sofia Fac Chimie* 1967/1968; 62: 47.

38. Eriksson, J.C., Toshev, B.V. Disjoining pressure in soap film thermodynamics. *Colloids Surfaces* 1982; 5: 241.

39. Vašiček, C.J. *Optics of Thin Films*. North Holland Publ Co: Amsterdam, 1960.

40. Scheludko, A., Platikanov, D. Untersuchung dünner flüssiger Schichten auf Quecksilber [Investigation on thin liquid layers on mercury] *Kolloid-Z* 1961; 175: 150.

41. Scheludko, A. Sur certaines particularités des lames mousseuses. I. Formation, amincissement et pression complémentaire [On some peculiarities of foam lamellas. I. Formation, thinning and extra pressure] *Proc Koninkl Nederl Akad Wetenschap* 1962; B65: 76.

42. Duyvis, E.M. PhD Thesis. University Utrecht, 1962.

43. Exerowa, D., Kolarov, T., Khristov, K. Direct measurement of disjoining pressure in black foam films. I. Films from an ionic surfactant. *Colloids Surfaces* 1987; 22: 171.

44. Bergeron, V., Radke, C.J. Equilibrium Measurements of Oscillatory Disjoining Pressures in Aqueous Foam Films. *Langmuir* 1992; 8: 3020.

45. Scheludko, A., Radoev, B., Kolarov, T. Tension of liquid films and contact angles between film and bulk liquid. *Trans Faraday Soc* 1968; 64: 2213.

46. Kolarov, T., Scheludko, A., Exerowa, D. Contact angle between black film and bulk liquid. *Trans Faraday Soc* 1968; 64: 2864.

47. Scheludko, A. Über das Ausfließen der Lösung aus Schaumfilmen. *Kolloid-Z* 1957; 155: 39.

48. Stephan, J.S. Versuche über die scheinbare Adhäsion [Experiments on the apparent adhesion] *Math Natur Akad Wiss* 1884; 69: 713.

49. Reynolds, O. On the theory of lubrication and its application to Mr. Beauchamp tower's experiments, including an experimental determination of the viscosity of olive oil. *Phil Trans Royal Soc* 1886; A177: 157.

50. Marangoni, C. Sul principio della viscosita' superficiale dei liquidi stabilito dal sig. J. Plateau [On the principle of superficial viscosity of liquids by Mr. J. Plateau] *Nuovo Cimento Ser* 1872; 2: 239.

51. Radoev, B., Manev, E., Ivanov, I. Geschwindigkeit der Verdünnung flüssiger Filme [Velocity of thinning of liquid films] *Kolloid-Z* 1969; 234: 1037.

52. Barber, A., Hartland, S. The effects of surface viscosity on the axisymmetric drainage of planar liquid films. *Canad J Chem Eng* 1976; 54: 279.

53. Scheludko, A. Thin liquid films. *Annuaire Univ Sofia Fac Chim* 1964/65; 59: 263.

54. Frankel, S., Mysels, K.J. On the "dimpling" during the approach of two interfaces. *J Phys Chem* 1962; 66: 190.

55. Platikanov, D. Experimental investigation on the "dimpling" of thin liquid films. *J Phys Chem* 1964; 68: 3619.

56. Joye, J., Hirasaki, G., Miller, C. Asymmetric Drainage in Foam Films. *Langmuir* 1994; 10: 3174.

57. Manev, E., Tsekov, R., Radoev, B. Effect of thickness non-homogeneity on the kinetic behaviour of microscopic foam films. *J Dispers Sci Technol* 1997; 18: 769.

58. Mandelstam, L. Effect of thickness non-homogeneity on the kinetic behaviour of microscopic foam films. *Ann Physik* 1913; 346: 609.

59. Scheludko, A. Sur certaines particularités des lames mousseuses. II. Stabilité cinétique, épaisseur critique et épaisseur d'équilibre [On some peculiarities of foam lamellas. II. Kinetic stability, critical thickness and equilibrium thickness] *Proc Koninkl Nederl Akad Wetenschap* 1962; B65: 87.

60. Patzer, J.F., Homsy, G.M. Hydrodynamic stability of thin spherically concentric fluid shells. *J Colloid Interface Sci* 1975; 51: 499.

61. Vrij, A. Possible mechanism for the spontaneous rupture of thin, free liquid films. *Faraday Discussion Chem Soc* 1966; 42: 23.

62. Radoev, B., Scheludko, A., Manev, E. Critical thickness of thin liquid films. Theory and experiment. *J Colloid Interface Sci* 1983; 95: 254.

63. Tsekov, R., Radoev, B. Life time of nonthinning liquid films - influence of the surface waves spatial correlations. *Adv Colloid Interface Sci* 1992; 38: 353.

64. Sharma, A., Ruckenstein, E. Stability, critical thickness, and the time of rupture of thinning foam and emulsion films. *Langmuir* 1987; 3: 760.

65. Corkill, J., Goodman, J., Orgden, C., Tate, J. The structure and stability of black foam films. *Proc Roy Soc* 1963; A273: 84.

66. Platikanov, D., Rangelova, N. Electroconductivity of Black Foam Films. In: Deryagin, B.V. (Ed.), *Research in Surface Forces*, Vol. 4. Consultants Bureau: New York, 1972, p. 246.

67. den Engelsen, D., Frens, G. Ellipsometric investigation of black soap films. *J Chem Soc Faraday Trans I* 1974; 70: 237.

68. Belorgey, O., Benattar, J.J. Structural properties of soap black films investigated by x-ray reflectivity. *Phys Rev Lett* 1991; 66: 313.

69. de Feijter, J., Vrij, A. Contact angles in thin liquid films: III. Interaction forces in Newton black soap films. *J Colloid Interface Sci* 1979; 70: 456.

70. Huisman, F., Mysels, K.J. Contact angles in thin liquid films: III. Interaction forces in Newton black soap films. *J Phys Chem* 1969; 73: 489.

71. Jones, M., Mysels, K., Scholten, P. Stability and some properties of second black film. *Trans Faraday Soc* 1966; 62: 1336.
72. Platikanov, D., Nedyalkov, M. *Annuaire Univ Sofia Fac Chim* 1969/70; 64: 353.
73. Exerowa, D., Khristov, K., Zacharieva, M. Метастабильные черные пленки [Metastable black films]. In: Deryagin, B.V. (Ed.), *Poverkhnostnye sily v tonkikh plenok*. Proc. VI confer. Surface Forces, Nauka: Moscow, 1979, pp. 186.
74. Exerowa, D., Kashchiev, D. Nucleation mechanism of rupture of newtonian black films. I. Theory. *J Colloid Interface Sci* 1980; 77: 501.
75. Exerowa, D., Kashchiev, D. Hole-mediated stability and permeability of bilayers. *Contemp Phys* 1986; 27: 429.

4 Surfactant Stabilized Foam Films

Elena Mileva

CONTENTS

This chapter is focused on the properties of microscopic foam films from aqueous solutions containing low-molecular mass surfactants (LMMS). Two basic aspects of the surfactant effect on the formation and stability of foam films can be identified: (i) dynamic phenomena related to the onset of flow-induced surface tension gradients at the film interfaces (Marangoni effects) and their influence on the film drainage behavior; and (ii) effects of bulk self-assembled surfactant nanostructures on film drainage and stability. All effects are modified by the specific confinement conditions of the film geometry. Although these factors are intermingled, depending on the specific conditions in the system they have distinct features and one or another may prevail and determine the performance of the foam films. The particular manifestation of these aspects depends on the type of LMMS (ionic, nonionic, zwitterionic; soluble, insoluble), and on other circumstances which modify the hydrophilic-lipophilic balance (HLB) in the system (e.g., added electrolyte, temperature, pH, etc.). Insofar as the foam films are important structural components of the foams, the fine tuning of their properties affects the overall foam formation and properties thus bringing about new possibilities for the design of particular foam formulations aimed at targeted applications.

The key aspect of the surfactant stabilization role is the relationship between the structure and properties of the adsorption layers on one hand, and the dynamic and stability characteristics of the foam films on the other hand. The impact of surfactants on the drainage kinetics and the stability of foam films has been a matter of attention in colloid and interface science for more than half a century (e.g., [1–9]). The recent accomplishments in the field are based on the development of some new trends and ideas. One important aspect is that during the film formation and drainage, LMMS not only trigger the onset of interfacial tension gradients (Marangoni effects), but may give rise to considerable surface dilational and shear viscoelasticities. The relation of these characteristics has been extensively commented on in literature (e.g., [8,9] and references therein). Recently, owing to the rapid improvement of the techniques investigating interfacial structure and rheology (Chapter 1, this book), and the development of combined experimental protocols, including advanced tensiometry (e.g., [10]) and microscopic thin liquid film instrumentation [2,4], additional phenomena and properties have been explored. This field of research has a high potential to lead to better understanding the role of the LMMS adsorption layers for the properties of foam films and much effort has been made to relate the newly gained interfacial characteristics to their drainage and stability performance.

4.1 BASICS OF MARANGONI EFFECTS IN FOAM FILMS

The kinetic stabilization of microscopic foam films by LMMS is due to the coupling of the specific film hydrodynamics and the mass transfer of the surfactant species. The classic approach is based on the concepts of physicochemical hydrodynamics [11] and has been extensively reviewed in other sources (e.g., [5–7,12,13] and references therein). Here, we shall point out only the basic principles that rule the initial formation and thinning of the films.

The essence of the surfactant effect on the drainage behavior of the foam films may be illustrated by the scaling analysis of the governing dynamic equations, specified for horizontal microscopic films. Such a model system originates from the experimental conditions for formation and drainage of foam films using the microinterferometric setup of Scheludko-Exerowa [1,2]. It has been established that the drainage behavior of thin films is governed primarily by the tangential mobility of the interfaces. In the course of film formation, the close approach of the two air/solution surfaces at small gap widths results in overall retardation of the fluid motion and the appearance of an effective zone of interaction, where the major viscous dissipation of energy is concentrated [14,15]. This substantiates the notion that the theoretical modeling of the film outflow is based on the creeping-flow hydrodynamic equations:

$$\nabla p = \eta^f \nabla^2 \mathbf{v}, \quad \nabla \cdot \mathbf{v} = 0 \tag{4.1}$$

The following notations are used: p, $\mathbf{v}$—for the pressure and the velocity fields; ρ^f, η^f—for the density and the solution bulk viscosity. The governing hydrodynamic Equation 4.1 is further analyzed through introducing a number of scaling parameters: H^f, R^f—the characteristic dimensions of the film; U^f, U^s—the characteristic radial velocities of the film and interfacial outflow; $\Delta U = U^f - U^s$—difference between the characteristic velocities in the film bulk and at its interfaces; V^f is the characteristic drainage velocity of the foam film (Figure 4.1) [12,14,15].

In the case of limited tangential mobility of the film interfaces and due to the strong anisodiametry of the film flow region, that is, $\varepsilon = H^f/R^f \ll 1$, additional simplification of Equation 4.1 is possible. Thus if

$$\frac{(U^f - U^s)}{U^f} = \frac{\Delta U}{U^f} > (H^f/R^f)^2, \quad V^f = -\left(\frac{dh}{dt}\right), \quad U^f \sim \frac{V^f R^f}{H^f} \tag{4.2}$$

the film outflow may be modelled through the lubrication-flow mode (also called Reynolds or Stephan-Reynolds film flow model [1,16–18]). In a cylindrical coordinate system (r, φ, z) (Figure 4.1), the Reynolds equation acquires the form of:

$$\frac{\partial^2 v_r}{\partial^2 z} = \frac{1}{\eta^f} \frac{\partial p}{\partial z}, \quad \frac{\partial p}{\partial z} = 0, \quad \frac{1}{r} \frac{\partial}{\partial r}(r v_r) + \frac{\partial v_z}{\partial z} = 0 \tag{4.3}$$

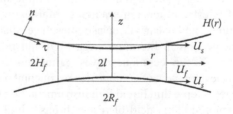

FIGURE 4.1 Schematic presentation of the foam film hydrodynamics. H^f, R^f—the characteristic length scales of the film; U^f, U^s—the characteristic scales of the radial velocity in the film and of the film interfaces. According to (Reprinted from *Emulsions: Structure, Stability and Interactions, Vol. 4 of Interfacial Science and Technology*. Mileva, E., Radoev, B., Hydrodynamic interactions and stability of emulsion films, Copyright 2004, with permission from Elsevier [12]).

It should be noted that in the absence of LMMS, the Reynolds flow model is not the proper asymptotic of the film hydrodynamics. In such a case $\Delta U/U^f \sim (H^f/R^f)^2$ and, instead of Equation 4.3, the scaling analysis results in the following set of flow equations [7,19,20]:

$$\frac{\partial}{\partial r}\frac{1}{r}(rv_r)+\frac{\partial^2 v_r}{\partial z^2}=\frac{1}{\eta^f}\frac{\partial p}{\partial z}, \quad \frac{1}{\eta^f}\frac{\partial p}{\partial z}=\frac{\partial^2 v_z}{\partial^2 z}, \quad \frac{1}{r}\frac{\partial}{\partial r}(rv_r)+\frac{\partial v_z}{\partial z}=0. \qquad (4.4)$$

As already mentioned, the surfactants have a profound influence on the tangential mobility of liquid interfaces [7,8,12,21] and the film drainage time is usually increased upon the raise of the surfactant concentration in the system (e.g., [2,4,22]). The experimental findings reveal that upon increase of the surfactant quantity, the mean lifetime of microscopic horizontal foam films increases systematically [22]. One such example is shown on Figure 4.2:

In order to explain the essence of this phenomenon, the mass transfer of the stabilizing LMMS has to be clarified. The surfactant mass transfer is usually presented through the convective diffusion equation [7,12,17,21]:

$$(\mathbf{v}\cdot\nabla)c = D^f\nabla^2 c \qquad (4.5)$$

with D^f being the bulk diffusion coefficient.

The flow-induced surfactant transfer enforces a decrease of the interfacial mobility of the film surfaces through the well known Marangoni effects [5,6,11]: In the course of the film drainage, the flow sweeps the surfactant molecules outside the film region toward the neighboring meniscus. The uneven distribution of the amphiphilic molecules on the interfaces results in the emergence of local surface tension gradients. The latter cause the onset of tangential forces which act in a direction opposite to the film fluid outflow. Toward the interfacial regions of surfactant deficiency, diffusion fluxes are directed. If LMMS is insoluble, only a surface (2D) diffusion process strives to restore the surface tension gradients [5,8,23]. In the case of soluble LMMS, there is also a surfactant flux from the film bulk. This coupling of the surfactant mass transfer and the film flow is reflected in the respective boundary conditions, which may be expressed in a local surface coordinate system (n,τ):

$$P^f_{n\tau} = \nabla_\tau\gamma+\eta_s\nabla^2_\tau v_\tau, \quad \nabla_\tau\left(\Gamma v_\tau\right)=D^s\nabla^2_\tau\,\Gamma+\left(j^f_n\right)_{z=h(r)}. \qquad (4.6)$$

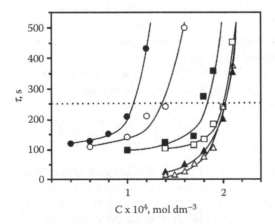

FIGURE 4.2 Dependence of mean lifetime τ on surfactant concentration for black foam films, obtained from aqueous solutions of sodium dodecyl sulfate, at various electrolyte concentrations: (●) 0.1 M NaCl; (○) 0.13 M NaCl; (■) 0.145 M NaCl; (□) 0.2 M NaCl; (▲) 0.29 M NaCl; (△) 0.5 M NaCl. The temperature is 22°C, according to (Nikolova, A., Exerowa, D. *Colloids Surfaces A* 1999; 149: 185 [22]).

The following notations are used: $\Delta_\tau = \partial/\partial\tau$; $h(r)$ stands for the generant curve of the film interface. Equation 4.6 is the condition for the continuity of the tangential component of the stress tensor at the film interfaces; γ is the surface tension. Moreover η_s designates the surface viscosity; in most cases, for surfactant stabilized films this term may be neglected because it results only in a slight modification of the interfacial mobility [6,17,24,25]. Equation 4.6 is related to the flow-induced surface tension gradients due to deflection of the adsorption layer coverage from its equilibrium values during the film drainage. These gradients evoke mass transfer fluxes, directed towards the places of deficiency of surfactants at the film interfaces: a surface diffusion flux $D^s\nabla_\tau^2\Gamma$, and a bulk mass flux j_n^f. Here, Γ is the surface concentration of the surfactant; D^s is the surface diffusion coefficient.

The geometrical anisodiametry of the foam films modifies not only the fluid flow, but has an effect on the surfactant mass transfer [6,7,17,26,27]. Commonly, the bulk surfactant flux toward the film interface includes two consecutive stages: diffusion from the bulk to the subsurface layer, and adsorption from the subsurface to the interface [11]. As has been shown in [6,7,17], the diffusion limited case is usually the rate-determining stage for LMMS in a microscopic liquid film, and

$$\left(j_n^f\right)_{z=h(r)} = -D^f\left(\frac{\partial c}{\partial n}\right)_{z=h(r)}. \tag{4.7}$$

Additional scaling parameters may be introduced [12,21,26,27]: $\Delta c_r^f, \Delta c_z^f$—characteristic changes of the surfactant concentration in radial and normal directions, respectively; δ^f—the scale parameter of the concentration diffusion layer inside the film; $Pe^f = U^f R^f/D^f$—the film Peclet number. Thus, the scaling of the separate terms in Equation 4.6 is:

$$\left(j_n^f\right)_{z=h(r)} \sim D^f\frac{\Delta c_z^f}{H^f}, \quad \nabla_\tau\gamma \sim \frac{\partial\gamma}{\partial r} = \left|\frac{\partial\gamma}{\partial c}\right|_0\frac{\partial c}{\partial r} \sim \left|\frac{\partial\gamma}{\partial c}\right|_0\frac{\Delta c_r^f}{R^f}. \tag{4.8}$$

Using Equation 4.5, the following interrelation between the scaling quantities Δc_r^f and Δc_z^f may be formulated [12,26,27]:

$$\Delta c_r^f = \frac{Pe^f + (R^f/\delta^f)^2}{Pe^f + 1}\Delta c_z^f. \tag{4.9}$$

If $Pe^f < 1$, due to the geometric confinement of the film flow, the concentration diffusion layer scales as $\delta^f \sim H^f \ll R^f$. Besides, $\Delta c_r^f \sim c^m - c^f$ and $\Delta c_z^f \sim c^m - c_s^f$, where c^m scales the surfactant concentration in the bulk of the film meniscus region, c^f is the film bulk concentration, and c_s^f is the subsurface concentration (see Figure 4.3).

Due to the film anisodiametry $\varepsilon = H^f/R^f \ll 1$, the scalings of the surfactant fluxes in the normal and tangential directions to the film interface are linked as:

$$j_n^f \sim D^f\frac{\Delta c_z^f}{H^f} \sim D^f\varepsilon\frac{\Delta c_r^f}{R^f} \sim \varepsilon j_\tau^f \tag{4.10}$$

The relationship (4.10) illustrates the fact that because of the film confinement conditions, the onset of a surface tension gradient causes the emergence of a bulk concentration gradient between the film bulk and the adjacent meniscus region as well. In order to compensate this bulk surfactant concentration gradient and to ensure the further supply of the interfacial regions with LMMS, new quantities of surfactant molecules enter the film from the adjacent meniscus (in counterflow to the film drainage outflow). This flux, however, has to cover a longer distance until reaching the interfacial domains depleted of surfactant molecules. This circumstance gives additional time to the emerged tangential forces, acting opposite to the outflow, to be maintained for a while and the interfacial

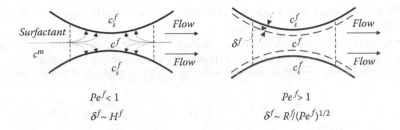

FIGURE 4.3 Schematic presentation of the mutual influence of the film hydrodynamics and the surfactant mass transfer. δ^f stands for the thickness of the concentration diffusion layer inside the film; c_s^f, c^f and c^m are the concentration characteristic scales of the subsurface layer, inside the film bulk, and in the film meniscus, respectively. According to (Reprinted from *Emulsions: Structure, Stability and Interactions, Vol. 4 of Interfacial Science and Technology*. Mileva, E., Radoev, B., Hydrodynamic interactions and stability of emulsion films, Copyright 2004, with permission from Elsevier [12]).

outflow to be retarded. The result is an effective immobilization of the film interfaces and a delay of the overall film drainage process. The tangential mobility of the film interfaces may be appraised through the following correlation [12]:

$$U^s = \frac{U^f}{1+f_h+f_c}, \quad \frac{\Delta U}{U^f} = \frac{f_h+f_c}{1+f_h+f_c}. \tag{4.11}$$

Here, $f_h = (H^f/R^f)^2 = \varepsilon^2$ is the hydrodynamic factor; f_c is the concentration factor related to the Marangoni effect:

$$f_c = \frac{\dfrac{\Gamma_0}{\eta^f D^f}\varepsilon\dfrac{\delta^f}{R^f}\left|\dfrac{\partial\gamma}{\partial c}\right|_0\dfrac{(R^f/\delta^f)^2+Pe^f}{1+Pe^f}}{1+\dfrac{D^s}{D^f R^f}\left(\dfrac{\delta^f}{R^f}\right)\left|\dfrac{\partial\Gamma}{\partial c}\right|_0\dfrac{(R^f/\delta^f)^2+Pe^f}{1+Pe^f}}, \quad \text{and} \quad f_c = \frac{\dfrac{\Gamma_0}{\eta^f D^f}\left|\dfrac{\partial\gamma}{\partial c}\right|_0}{1+\dfrac{D^s}{D^f R^f}\left|\dfrac{\partial\Gamma}{\partial c}\right|_0} \quad \text{for } Pe^f < 1 \tag{4.12}$$

All quantities denoted by $|\,|_0$ refer to the respective equilibrium values and are taken from independent surface tension measurements [7,12,17]. Following Levich [6,11,12], it is assumed that the surface concentration may be presented as $\Gamma = \Gamma_0 + \Gamma'$, with Γ_0 the equilibrium value in the absence of fluid motion, and Γ' is the perturbation due to the film outflow.

The above described scaling scheme leads to asymptotic models for the major cases coupling the film flow and the surfactant mass transfer. For example, the validity criterion for the Reynolds flow model (4.2) may be reformulated in terms of tangential mobility of the film interfaces [12,21,26]:

$$\frac{\Delta U}{U^f} = \frac{f_h+f_c}{1+f_h+f_c} > \varepsilon^2. \tag{4.13}$$

If $Pe^f < 1$, the convective diffusion Equation 4.5 is transformed into the Laplace equation for the surfactant concentration:

$$\Delta c = 0 = D^f\left(\frac{1}{r}\frac{\partial}{\partial r}r\frac{\partial c}{\partial r}+\frac{\partial^2 c}{\partial z^2}\right) \tag{4.14}$$

Its solution may be presented as a single layer potential [12,27]. The coupling problem is then reduced to a Fredholm integral equation for the tangential mobility of the air/solution interface, which may be treated according to known procedures. More details may be found in the original papers [12,27].

The estimation of the mean film drainage velocity $V^f = -(dh/dt)$ (as defined by Equation 3.36) is linked to the flow and surfactant scaling parameters by the following relationship:

$$\frac{V^f}{V^{imob}} \sim 1 + \frac{1}{f_h + f_c} \approx 1 + \left(\left(\frac{H^f}{R^f}\right)^2 + \frac{\frac{\Gamma_0}{\eta^f D^f}\left|\frac{\partial\gamma}{\partial c}\right|_0}{1 + \frac{D^s}{D^f R^f}\left|\frac{\partial\Gamma}{\partial c}\right|_0}\right)^{-1}, \tag{4.15}$$

where $V^{imob} \approx \Delta P(H^f)^3/\eta^f(R^f)^2$ is the characteristic drainage velocity for tangentially immobile interfaces (Reynolds flow model); $\Delta P = P_c - \Pi$ is the difference between the capillary pressure in the meniscus region and the disjoining pressure in the foam film. If Equation 4.15 is inserted into Equation 3.36 ($\tau = \int_{h_0}^{h_{cr}} dh/V^f$), an expression for the evaluation of the mean drainage time of the film is obtained. The value of h_{cr} is usually extracted from drainage experiments on microscopic foam films (Chapter 3, this book). Note that the already cited expression Equation 3.38 is a variant of Equation 4.15 [17,18,28].

Other model developments have also been advanced. For example, one of the well known facts from the experimental investigation of the microscopic foam films is the onset of asymmetric foam film drainage [2]. It has been observed that in films of diameters more than 200 μm, and in the conditions of higher tangential mobility of the interfaces, thicker portions in the form of "channels" are usually observed in the draining film. The latter disturb the symmetric thinning outflow and lead to quicker overall drainage of the films. For the first time, an attempt to relate this effect to the presence of LMMS is proposed in [29,30]. The key idea of the proposed model is that the transition from symmetric to asymmetric drainage depends on the properties of the stabilizing surfactant. The disturbance growth factor in [29,30], however, is particularly related only to surface viscosity and surface diffusion and thus concerns primarily insoluble LMMS. The treatment in [30] has been developed further in [31] so as to account for the effect of soluble film-stabilizing surfactant, and a new growth factor is defined:

$$\frac{D^s H_0^f}{V^f R^f}\left(\frac{I_1}{I_2}\right)^2\left[1 + \frac{6\pi\eta^f}{I_1}\frac{D^f H_0^f}{\Gamma_0 V^f |\partial\gamma/\partial c|_0}\left(\frac{2\pi}{\lambda}\right)^2\left(\frac{R^d}{H_0^f}\right)^2\right]$$

$$+ \frac{3\pi}{I_2}\frac{\eta^f D^f R^f}{\Gamma_0 H_0^f |\partial\gamma/\partial c|_0} \sim \begin{cases} <1 & unstable \\ >1 & stable \\ \sim 0 & marginal \end{cases} \tag{4.16}$$

Here, H_0^f is the characteristic film thickness before the onset of an asymmetric drainage; λ is the wavelength of the thickness perturbation. $R^d \sim \lambda$ is the minimum radial circumference where the perturbation in the film thickness might appear. An asymmetric outflow of this type may be observed only for larger films ($R^f > R^d$). Insofar as the radial outflow in the film continues after the onset of the thickness perturbation, the disturbance should be carried radially outward in the course of film drainage. Such a situation has often been observed experimentally [2,4]. Thus, the wave front acquires a form of a spiral wave.

In other studies, the deformability of the film interfaces and presence of multiple thickness nonhomogeneities in the films have explicitly been taken into account (e.g., [13,32]). However, the essence of the surfactant stabilization remains the same: there is always an onset of global or local Marangoni effect, retarding the tangential mobility of the interfaces and leading to a delay of the overall film drainage process.

4.2   FOAM FILMS WITH SURFACTANT NANOSTRUCTURES

At higher surfactant concentrations, the foam film stabilization includes additional peculiarities which are related to the presence of self-assembled nanostructures.

The onset and growth of self-assembled aggregates in LMMS aqueous solutions is a consequence of the balance of weaker interactions operating in the investigated systems (van derWaals, hydrophobic, screened electrostatic, etc.) They manifest themselves in two major peculiarities: (1) in bulk solutions the nanostructures are continuously and reversibly exchanging portions with one another; (2) any alteration in the solution's conditions (electrolyte concentration, total quantity of the amphiphile, temperature, etc.) immediately affects both the intra- and interaggregate interactions and, therefore, influences the entire size distribution of the amphiphilic entities in the system. Upon formation of thin liquid films, the immediate environment of any surfactant or surfactant aggregate changes dramatically. Thus, it should not be expected that the self-assembled structures will remain the same as within the bulk solution. The diphilic character of the surfactant molecules results in the respective aggregation on the interfaces as well. The latter phenomenon significantly affects the adsorption properties of the surfaces. Thus, the foam films are regarded as being obtained from a solution that contains surfactant nanostructures with a definite size distribution. It is very convenient to view the foam film as a three-layered model: the film bulk phase and two 2D surface phases (air/liquid interfaces). It is assumed that the self-assembly is possible in all the layers and they are in thermal, mechanical, and chemical equilibrium. The details may be found in [33,34]. Here, only a synopsis of the general scheme is presented.

If it may be assumed that the volume fraction of the surfactant nanostructures is relatively low, the size distribution curve of the self-assembled species for the film bulk is presented as mole fractions of the respective components. All quantities characterizing the bulk solution are denoted with a superscript (^{b}):

$$X_n^b = X_n^{0,b} X_n^{int,b} X_n^{ad,b} X_n^{f,b} \tag{4.17}$$

where

$X_n^b = \dfrac{N_n^b}{N_w^b + \sum_i i N_i^b}$ is the mole fraction of the bulk solution self-assemblies containing n surfactant molecules; $X_w^b = \dfrac{N_w^b}{N_w^b + \sum_i i N_i^b}$ is the mole fraction of water; and

$$X_n^{0,b} = \left(X_1^b \right)^n \exp\left[\frac{n\mu_1^{0,b} - \mu_n^{0,b}}{kT} \right] \tag{4.18}$$

$$X_n^{int,b} = \exp\left[\frac{\sum_j \left(n u_{1j}^b - u_{nj}^b \right) N_j^b}{kTV_f} \right] \tag{4.19}$$

$$X_n^{ad,b} = \exp\left[\frac{-2\gamma_b}{kT} \right] \tag{4.20}$$

$$X_n^{f,b} = \exp\left[\frac{2}{kT}\left\{2\gamma_f(h) + \Pi(h)h\right\}\left\{n\frac{\partial A_f}{\partial N_1^b} - \frac{\partial A_f}{\partial N_n^b}\right\}\right]. \tag{4.21}$$

In the above expressions, $\mu_1^{0,b}$, $\mu_n^{0,b}$ are the standard chemical potentials of a monomer and the n-mer. In the above equations u_{nj}^b stands for the bulk interaction potential between i-mer and j-mer. It is assumed that the interaggregate interactions do not explicitly depend on the film thickness. That is acceptable for smaller aggregates and not very long range interaction forces. In this model, the specific conditions in the foam films are accounted for by the dependence of the disjoining pressure on the film thickness. The notation γ_b stands for the surface tension at the air/solution interface; A_f is the area of the film surface; N_i^b, N_w^b are the numbers of i-mers and of the water molecules in the film bulk. The advantage of the presentation Equations 4.17 through 4.21 is that the bulk micelle formation and the interspecies interaction influence are effectively decoupled from the adsorption and specific film effects.

The adsorption layers at the film interfaces are regarded as nonautonomous 2D phases (e.g., [35,36]). The size distribution curve of the 2D nanostructures is obtained in complete analogy to the bulk micellization. All quantities characterizing the interfaces are denoted with a superscript (^{s}). The result is:

$$X_n^s = X_n^{0,s} X_n^{int,s} X_n^{f,s} \tag{4.22}$$

where $X_n^s = \dfrac{N_n^s}{N_w^s + \sum_i i N_i^s}$ is the mole fraction of the interfacial self-assemblies containing n surfactant molecules; and

$$X_n^{0,s} = \left(X_1^s\right)^n \exp\left[\frac{n\mu_1^{0,s} - \mu_n^{0,s}}{kT}\right] \tag{4.23}$$

$$X_n^{int,s} = \exp\left[\frac{\sum_j \left(nu_{1j}^s - u_{nj}^s\right)N_j^s}{kTA_f} - \frac{\Delta a_n^s}{2kTA_f^2}\sum_i N_i^s \sum_j N_j^s u_{ij}^s\right], \quad \Delta a_n^s = na_1^s - a_n^s, \tag{4.24}$$

$$X_n^{f,s} = \exp\left[-\frac{\Delta a_n^s}{kT}\gamma_f(h)\right]\exp\left[\frac{v_a}{kT}\Pi(h)\left(n\frac{\partial N_1^b}{\partial N_1^s} - \frac{\partial N_n^b}{\partial N_n^s}\right)\right]\exp\left[-\frac{\Delta a_n^s}{kT}\int_\infty^h \left(\frac{\partial\Pi(h')}{\partial h}\right)h' \, dh'\right] \tag{4.25}$$

Here, $\mu_1^{0,s}$, $\mu_n^{0,s}$ are the standard chemical potentials of a monomer and an n-aggregate in the surface layer, they contain information about the intrinsic surface self-assembling properties of the surfactant at the air-solution interface before foam films are formed; a_i^s, a_n^s are the effective areas assigned to every surfactant monomer and an aggregate. It is assumed that the 2D aggregate can take only one position on the interface. In analogy to Equation 4.21, the film tension $\gamma_f(h)$ (see Equation 3.15) and the disjoining pressure $\Pi(h)$ both enter the size distribution curve. However, the latter become important only if $\Delta a_n^s \neq 0$ and $n(\partial N_1^b/\partial N_1^s) - (\partial N_n^b/\partial N_n^s) \neq 0$. Provided that larger premicellar aggregates prevail and $\Delta a_n^s > 0$, the disjoining pressure may play a significant role in the size distribution of the interfacial self-assemblies.

The basic assumption of the theoretical scheme is that the volume fraction of the surfactant nanostructures is relatively low. It has been established that if the films are thicker and the disjoining pressure may be neglected, the rearrangement of the surfactant species upon thinning results in additional self-assembling of the surfactant molecules. If the foam films are so thin that the disjoining pressure becomes operative, its effect on the bulk structure reorganization may be observed only if $i(\partial A_f / \partial N_1^b) - (\partial A_f / \partial N_i^b) \neq 0$. If the concentration of added electrolyte is high enough so that the electrostatic component of the surface forces is effectively suppressed, the leading term is the van der Waals constituent Π_{vw}. The disjoining pressure then acts in the direction of enhanced destruction of the existing aggregates, both in the film bulk and at the film interfaces. As the film drainage proceeds, the number of the free monomers in the film region increases. Thus, the self-assemblies are acting as "reservoirs" of surfactant molecules. The availability of these "reservoirs" influences the drainage process and the results are routinely registered by the investigation of foam film thinning kinetics.

Specific film characteristics are distinguished depending on whether the system is in the premicellar concentration domain (PMC) or above the respective critical micelle concentration (CMC).

4.2.1 PREMICELLAR CONCENTRATION DOMAIN

At intermediate surfactant concentrations which are about one to two orders of magnitude lower than the respective CMC, the hydrophilic-hydrophobic balance in LMMS aqueous solutions may result in crumbly self-assembled structures, which contain smaller numbers of surfactant molecules than in the micellar aggregates above CMC. These nanostructures are named premicelles [37,38]. The first experimental evidences for the presence of bulk amphiphilic nanostructures in aqueous solutions, at concentrations below CMC, date from 1958. Mukerjee et al. [39] measured the electrical conductivity of water solutions of sodium dodecyl sulfate. For the explanation of the results, they presume that the surfactant molecules form dimers, thus minimizing the contact between water and the hydrophobic tails. Later experimental data linked to premicellar aggregation have been reported by other authors [40–43].

The existence of a plateau section in the surface tension isotherm is one of the major signs evidencing the micellar self-assembly in surfactant solutions above CMC [44]. However, in studies on aqueous solutions from the ionic surfactant sodium dodecyl sulfate, it has been found that the surface tension isotherms contain kink and plateau portions for intermediate concentrations [37]. The unusual concentration courses of the adsorption layer properties have been interpreted as possible effects of premicellar self-assembly. Similar results have been found for aqueous solutions of n-heptanol in the presence of added electrolyte [45]. Recent studies on aqueous solutions of nonionic surfactants, namely tri- [46] tetra- [47,48] and penta-ethylene [46,48,49] glycol monododecyl ethers (C_mE_n) and hexadecyltrimethylammonium chloride [50], found that other interfacial properties, particularly dilational surface rheology, may also be related to the presence of premicelles.

The presence of premicelles is important and they may have an appreciable influence on the properties of the microscopic foam films. This notion has been substantiated through systematic experimental studies on the drainage kinetics of foam films [51–55]. The investigations identify specific features of the routinely measured drainage characteristics against the surfactant quantity within the same (intermediate) concentration range where atypical courses of surface tension isotherms are registered. All microscopic films drain quickly and rupture in a minute without reaching equilibrium thickness. The drainage itself is in a regime of enhanced tangential mobility of the film interfaces. However, particular courses of the drainage characteristics against the surfactant concentration have been registered which are related to the presence of premicelles: (i) the formation of "unstable" black formations; (ii) concentration synchronies of the odds in foam film drainage parameters and the occurrence of low concentration kinks and plateau portions of the respective surface tension isotherms.

Usually, during the drainage of surfactant-stabilized films, thickness irregularities—black patterns—are observed [1,2]. These are thinner portions, visualized as black spots, which emerge within the background plane parallel film, expand, and evolve into common black films (CBF) or Newton black films (NBF). These black films survive for longer time intervals (minutes, hours) [2].

A common feature of the indicated systems is that the adsorption layer at the liquid/air interface is (almost) closely packed (see e.g., Figure 3.5a–d). In the surfactant solutions from the premicellar concentration domain, however, similar but "unstable" black patterns are observed (Figure 4.4) [33,55–57]. The term "unstable" is used because in such cases all of the resulting films are rupturing films. The unstable black formations themselves have lifetimes of not more than several seconds. The respective foam films drain quickly and survive for at most about a minute.

Particularly interesting objects are the black dots (Figure 4.4a). They mark the concentration values where the kink and plateau portions of the adsorption isotherms at lower concentration are situated (see Figure 4.5a), live for 3–10 s, and do not grow in size. The emergence of these black patterns is related to the initial onset of the premicelles and the black dots are viewed as detectors for the presence of amphiphilic nanostructures in the initial aqueous LMMS solutions [56–59]. The "unstable" spots (Figure 4.4b) live for about a second, quickly grow in size, and the film ruptures before they embrace the whole area of the background foam film. These spots are characteristic features of the concentration range of the premicellar plateau portion in the surface tension isotherms (see Figure 4.5a). Just as in the case of black spots in films obtained at higher surfactant quantities, the rate expansion of these patterns is related to the specific combination of surface tension, kinetics of drainage, and surface forces. What is established by now is that the rate of expansion of the "unstable" spots is always linear with the time [57].

The experiments show that the onset of "unstable" black patterns is closely connected to the sharp increase of drainage time with the concentration (Figure 4.5b). The general theoretical background of this mechanism has been explained in [33,34,55]. As already stated, within the intermediate (premicellar) concentration domain the microscopic foam films drain in a regime of high interfacial mobility and rupture in a minute or two. The observed unstable black patterns are a clear sign of the thickness inhomogeneities. The latter create an extra option for the onset of local differences in the general coupling mechanism of the film hydrodynamics and the surfactant mass transfer. In the thinner portions of the film, the black pattern regions, the local flow is retarded. The mechanism of this retardation is closely related to the fact that the surfactant nanostructures (premicelles) in

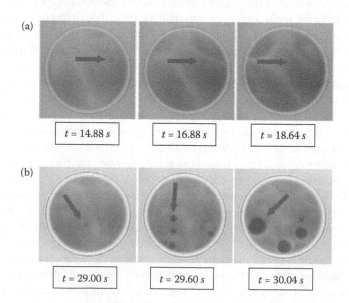

(a)

| $t = 14.88\ s$ | $t = 16.88\ s$ | $t = 18.64\ s$ |

(b)

| $t = 29.00\ s$ | $t = 29.60\ s$ | $t = 30.04\ s$ |

FIGURE 4.4 Representative snapshots of films with "unstable" black patterns for aqueous solutions of hexadecyltrimethyl ammonium chloride (CTAC) of the intermediate (premicellar) surfactant concentration range and in the presence of 0.1 M NaCl: (a) black dots, $C_S = 2.5 \times 10^{-7}\,M$ CTAC; (b) "unstable" black spots, $C_S = 2.5 \times 10^{-6}\,M$ CTAC. The foam film experiments are performed with the film microinterferometric technique of Scheludko-Exerowa, at temperature 20°C.

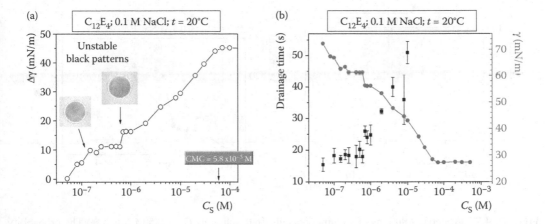

FIGURE 4.5 Aqueous solutions of tetra-ethylene [46,48,49] glycol monododecyl ether; $C_{el} = 0.1$ M NaCl; $t = 20°C$. (a) "Unstable" black patterns: snapshots of black dots and juxtaposition to the peculiarities of the surface tension isotherm. The foam film experiments are performed with the foam film microinterferometric technique of Scheludko-Exerowa. (b) Mean drainage time of foam films against the surfactant concentration. $\Delta\gamma = \gamma_0 - \gamma$; γ_0 is the equilibrium surface tension of the aqueous solution of the electrolyte. According to (Arabadzhieva, D. *Colloids and Surfaces A* 2011; 392: 233 [47]).

the film bulk serve as an extra reservoir for surfactant molecules. The flow sweeps the surfactant molecules outside the microfilm, locally creating a surface tension gradient. The latter evokes the emergence of tangential forces acting in a direction opposite to the fluid outflow and resulting in effective immobilization of the black pattern interfaces (Figure 4.6).

The existing nanostructures in the black pattern bulk are already destroyed and it is depleted of surfactant molecules. In the neighboring thicker regions, there are still self-assemblies. Upon drainage, they are destroyed further on releasing an additional number of monomers. The latter may participate in the feed-up of the film interfaces within the black pattern region. This surfactant extra flow provides additional time for the emerged local surface tension gradient to be kept for a while and the interface flow to be retarded. A respective retardation of the thinning within the black pattern domain is obtained. The increase of the overall concentration of the initial solution leads to the onset of more of these black formations. The larger their number is, the sharper is the rise-up of the overall mean drainage time of the film.

At higher overall surfactant quantity, the number and the probability of the onset of these black patterns are increased (Figure 4.7a). The mean drainage time of the microscopic foam films against the surfactant concentration also runs in synchrony with the peculiarities of the interfacial properties [55].

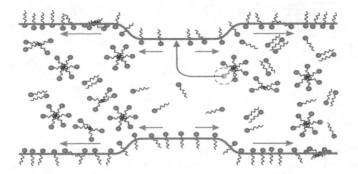

FIGURE 4.6 A schematic sketch presenting the impact of premicellar structures on the drainage kinetics of the foam film. According to (Arabadzhieva, D. *Colloids and Surfaces A* 2011; 392: 233 [47]).

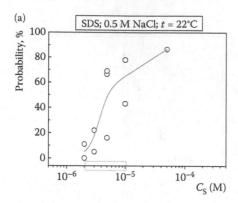
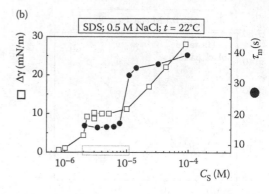

FIGURE 4.7 Aqueous solutions of sodium dodecylsulfate solutions $C_{el} = 0.5$ M NaCl. (a) The probability to observe "unstable" black patterns against the surfactant concentration; (b) Mean drainage times of foam films against the surfactant concentration. According to (Mileva, E., Tchoukov, P., Tadros, Th (Ed.): Surfactant nanostructures in foam films. *Colloid Stability: The Role of Surface Forces –Part 1; Vol. 1. Colloids and Interface Science Series.* 2007. Ch. 8. Copyright Wiley-VCH Verlag GmbH & Co. KGaA. Reproduced with permission [55]).

It increases sharply within the same concentration range where the irregularities of the adsorption layer properties are observed [55]. This slow down mechanism combines the presence of amphiphilic nanostructures with the specific film hydrodynamics and the mass transfer of the surfactant molecules in thinning foam films. The coordinated manifestation of interfacial and thin film peculiarities may be considered a characteristic feature of the premicellar (intermediate) surfactant concentration domain.

A detailed examination of the rheological properties of the air-water interfacial layers of aqueous solutions of nonionic surfactants C_mE_n have been performed and some additional structure properties interrelations are determined [46–49]. The equilibrated air/solution surface is subjected to area disturbances in a Profile Analysis Tensiometer (within the frequency range of 0.002–0.2 Hz), and in a Capillary Pressure Tensiometer (up to 100 Hz). The basic result is that the run of the surface dilational elasticities against the surfactant concentration exhibits a sequence of maxima (Figure 4.8).

It has been established that the maxima are typical for the concentration domain where the odds in the surface tension isotherms are observed and they are registered at both low and high frequencies of the imposed interfacial disturbances. These surface rheological data support the premicellar notion. During the imposed interfacial disturbances, the surface area of the bubble is periodically increased and compressed, resulting in the appearance of surface tension gradients. The surface

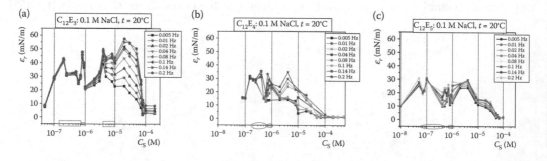

FIGURE 4.8 Surface dilational elasticities of aqueous solutions of (a) tri-, (b) tetra- and (c) pentaethyleneglycol monododecyl ethers against the surfactant concentration. The experiments are performed with PAT-1 within the frequency range of 0.005–0.2 Hz; $C_{el}=0.1$ M NaCl; $t = 20°C$. According to (Arabadzhieva, D., Soklev, B., Mileva, E. *Colloids and Surfaces A* 2013; 419: 194; Arabadzhieva, D. et al. *Colloids and Surfaces A* 2011; 392: 233; Arabadzhieva, D. et al. *Ukr J Physics* 2011; 56: 801 [46–48]).

dilational elasticity is a measure of the resistance against the creation of a surface tension gradient at the interface and of the rate at which this gradient disappears once the system is again left to itself [60,61]. It is indicative of both the scale of surface tension changes and of the system's capacity to restore the initial equilibrium values when the disturbance is released. If there are no premicelles, the usual course of the dilational elasticity against the amphiphile concentration encompasses a maximum and a fall down at the CMC value. However, the surface rheology results as presented in Figure 4.8 demonstrate that a phenomenon of this type is repeatedly observed within the intermediate concentration range. Several sequential minima and maxima are registered in the curves presenting the relationship between the surface dilational elasticity and the surfactant concentration. These maxima could be looked at as signs for the formation of different bulk premicelles. The combination of these effects is considered as an experimental indication for the chronological onset of different self-assembled nanostructures upon the successive raise of the surfactant concentration. These interrelations between adsorption layer properties and microscopic foam film drainage have also been registered for cationic surfactant hexadecyl trimethylammonium chloride, as well [50].

Generally, the time and length scales of the performed film drainage and rheological phenomena are of the same order of magnitude. Both the film drainage and the surface dilational studies are related to processes that comprise the onset of surface tension gradients and the consecutive relaxations of these gradients due to mass transfer of the surfactant molecules. Therefore, the specific peculiarities in the runs of the interfacial layer properties and of the drainage parameters have common origin, related to dynamic phenomena (specific manifestation of the Marangoni effect) and modified by the presence of bulk premicellar self-assemblies.

These results show that the interfacial properties at the air/solution interface are key factors that influence the behavior and stability of microscopic foam films. The nature of this influence depends on the surfactant concentration. In the premicellar concentration range, the dilatational rheology and surface dilational elasticities in particular, are most directly related to the foam film drainage characteristics. The enhanced coupling of the interfacial rheology and the film drainage kinetics is due to the presence of premicelles in the solution bulk. In the course of film thinning, the premicellar self-assemblies can be decomposed and thus they serve as an additional source of amphiphilic molecules which have a sensitive impact on the foam film drainage kinetics.

Premicellar aggregates have been detected in similar aqueous systems by other experimental methods as well. For example, in [62] evidences for the existence of premicellar multimers in the bulk of aqueous solutions of tetraethylene glycol monodecyl ether ($C_{10}E_4$) have been presented. The scattered light intensity and the normalized time autocorrelation functions extracted from dynamic light scattering studies show that within a concentration domain ($C_S = 0.78 - 0.82$ μmol/mL), which is below CMC, both monomers and bulk premicelles coexists; while at higher concentrations only signals from the genuine micelles are detected.

4.2.2 FOAM FILMS FROM MICELLAR SOLUTIONS

There are micelles in aqueous surfactant solutions at concentrations above CMC. Depending on the structure of the LMMS molecule, presence of salts, and so on, these self-assemblies are either monodisperse (spherical) or polydisperse (cylindrical, disc-like). In the previous subsections, the case of relatively low volume fraction of the surfactant nanostructures is discussed. At higher LMMS concentrations, the volume fraction of the surfactant nanostructures (micelles) may be considerable. Applying the microscopic foam film technique, the so called "stratification phenomenon" is registered, namely a consecutive stepwise thinning of the films. The earliest evidences of the foam film stratification phenomena have been described by Johonnott [63] and Perrin [64]. Particularly interesting are the studies of Bergeron et al. [65,66]. They investigated foam films stabilized by sodium dodecyl sulfate. Precise measurements of the disjoining pressure isotherms employing the Thin Liquid Film-Pressure Balance Technique (TLF-PBT) and the dynamic method of Scheludko-Exerowa have been performed. It has been established that upon drainage, the film goes in a stepwise manner through several metastable stages until the final equilibrium thickness is reached. At each step, black spots emerge within the thicker background plane-parallel film. The spots expand until the whole area of the film is covered and a film of transitional thickness is obtained. Then again, new black spots emerge and expand within the metastable background film and another (lower) thickness value is reached, and so on.

The interpretation of these experiments is based on the following vision of the structure of the film: Due to the high surfactant concentration, the film interfaces are covered with densely packed adsorption layers. Near the interfaces, there are narrow regions of bilayer-like (disc-like) structures, while in the film bulk there are micelles as in the bulk of the initial solution (Figure 4.9).

Two key suggestions came out of this study. First, the height of the observed thickness steps may be associated with the structural parameters of the existing micellar aggregates and is a function of the surfactant concentration, the quantity of the added electrolyte, and the temperature. Second, there is a gradual transition from diffuse micelle-containing domains in relatively thick films to bilayer-like structures in thinner films. The stepwise sheeting in the course of the film drainage is presumed to reflect the consecutive removal of the arrays of ordered micelles (layer by layer) from the film bulk toward the meniscus and is related to the onset of oscillatory disjoining pressure [66–68]. Recently, a theoretical approach has been developed which allows from the experimental step-wise dependence of the film thickness on time to determine the micelle aggregation number and the micelle charge [69,70]. The methodology is also applied for the case of mixed micelles obtained in solutions of ionic and nonionic LMMS [70].

The film stratification phenomenon is a common phenomenon in the high concentration surfactant domain and is observed for foam films from aqueous solutions of all types of amphiphiles: ionic, nonionic, as well as in the case of solutions of surfactant and surfactant/polyelectrolyte mixtures [71–74]. In order to better understand the essence of this phenomenon, it should be mentioned that similar stratification phenomena have been registered in films stabilized with nanoparticles [75] as well. The treatment of thin liquid films from the viewpoint of statistical mechanics has revealed the common background of these phenomena [75,76]. It is related to the structure and layering of the

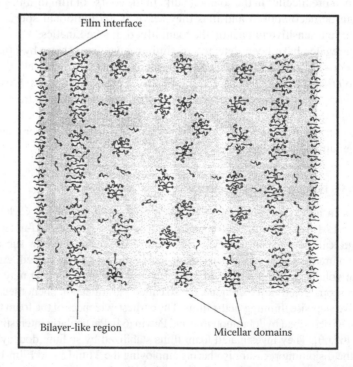

FIGURE 4.9   Schematic representation of possible surfactant structuring in thin liquid foam films from micellar solutions. According to (Bergeron, V., Radke, C. *Langmuir* 1992; 8: 3020 [65]).

fluids in the vicinity of large interfaces. In the confined space of the films, these effects are amplified. In [84] it is shown that the layering and stratification in thin liquid films could be well-conceived if the discrete (molecular) nature of the solution (including the solvent) is taken into account. Concise formalism is developed based on the generalization of Ornstein-Zernike equations for fluid mixtures and a simple multicomponent hard core model. It is found that a hard sphere colloidal suspension, comprised of larger entities (nanostructures) and smaller species (solvent), tends to be ordered in a monolayer structure next to the film interface (Figure 4.10).

The layering is driven by excluded volume interactions [76]. It is also established that the solvent molecules enhance the structural forces, both among the larger particles themselves, and between them and the film interfaces. This effect, however, may be observed only at a higher volume fraction of the larger species. Such an organization of the film fluid promotes oscillatory structural interactions between the film interfaces. Introducing a considerable size polydispersity at a fixed volume fraction of the components results in a decrease of the stratification effect. An extension of the analysis for nonspherical particles is also provided (dumbbells, flexible linear chains and network-forming fluid). The model studies show that the stratification is more pronounced when the fluid is composed of spherical species.

To summarize, the onset of the stepwise thinning of foam films from micellar solutions (at high surfactant concentrations) is not a particular sign of the presence of surfactant micellar aggregates, but comes out always when a considerable volume fraction of any type of nanostructures is present. The more stable these structures are (e.g., nanoparticles, micelles, liposomes, vesicles, etc.), the more pronounced the stratification phenomena in foam films will be. However, the specific role of the surfactant nanostructures is associated with the following peculiarities: (i) their labile character usually results in a variety of form and size distributions at the film interfaces, which are different from the respective self-assembled species in the film bulk; (ii) other film properties, different from stratification, may also become important. Thus, upon formation and drainage of the films, there is often a reorganization of the initially existing nanostructures and they may be destroyed as well. Therefore, the foam films can not only serve as an instrumentation for the investigation of

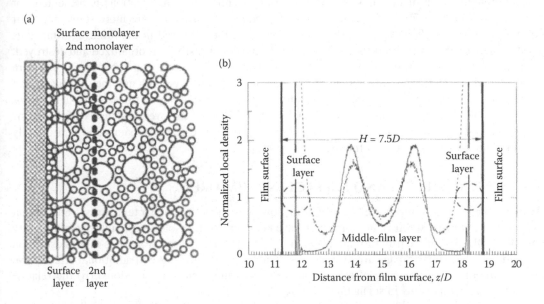

FIGURE 4.10 (a) Schematic 2D presentation of local density distribution of a binary suspension of hard core colloids near the film surface; (b) Monte Carlo data for local density distribution of macroions in a film. According to (Reprinted from *Emulsions: Structure, Stability and Interactions.* Henderson, D., Trokhymchuk, A.D., Wasan, D. Structure and layering of fluids in thin films. Chapter 7. pp. 259–311, Copyright 2004, with permission from Elsevier [75]).

LMMS self-assemblies, but the surfactant aggregates themselves can have a significant impact on the properties and the stability of the foam films.

The most distinctive characteristic property of foam films from micellar solutions is the onset of thickness inhomogeneities in the form of black patterns. From a structural point of view, these formations are more like the usual black spots (precursors of CBF and/or NBF). The time evolution (expansion) of these black spots, appearing within the background gray film at each consecutive step of the film sheeting, has been analyzed in detail [66,71]. It is obvious that the parameters of the expanding spots contain valuable information both about the structural peculiarities of the micellar aggregates and about the foam film drainage mechanism and stability. Bergeron et al. [66] have found a linear law of spot radius expansion with time, quite similar to the case of the premicellar concentration [57]. The drainage itself is reported to occur by rim expansion and by thinning of the thicker portion of the film outside the rim due to the combined actions of surface tension, disjoining, viscous, and capillary suction forces. Langevin, however, reported a somewhat different relationship: spot radius expansion scaled with $t^{1/2}$ [71]. This study concerns expansion dynamics of domains in stratifying foam films containing micelles, colloidal particles or polymer surfactant complexes. There is still no clear explanation of the difference between [66] and [71]. One possible explanation of these differences may be the specific stability of the obtained self-assembled species. Thus, at lower stability of the aggregates they release surfactant molecules upon film drainage. This results in smoother expansion dynamics of the respective black patterns and the rate of expansion of the spots is linearly dependent on time. In the case of more stable aggregates and nanoparticles, the expansion is somewhat delayed and the spots' radii scale with $t^{1/2}$.

Evidences for these general arguments might be found in the investigations of micelle-stabilized foams [77,78]. No doubt, the presence of micellar nanostructures has an impact not only on the properties and the stability of thin liquid films, but on the foam systems which contain them as key ingredients as well. As already mentioned, this aspect is also closely related to the stability and lifetime of the existing micelles, and to the possible reorganization of micellar aggregates in the bulk of the film. An overview on such effects and on their relationships to the kinetics of formation and the disintegration of micelles in view of various technological applications is presented, for example, in Reference 79. If, for example, the surfactant nanostructures are more stable and their volume fraction is sufficiently high, they might be trapped and layered in the thinning films, thus additionally stabilizing them. If they are more labile and could easily be reorganized and destroyed, their stabilizing role is diminished. Nonetheless, even at these high surfactant concentrations, there are intermediate cases when the micelles release monomers slowly enough and the stabilization of the films is achieved due to the smooth upholding of the adsorption layers on the interfaces. Thus, the films typically drain down to Newton black films (NBF) and the overall stability of the foams is essentially determined by the lifetimes of these films. In these cases, the foam stability is directly related to the type of the films in the foam systems [2].

4.3   SPECIFIC EFFECTS AND SURFACTANT MIXTURES

By now, the general coupling effects of film flow dynamics and surfactant mass transfer that determine the kinetic stability of foam films have been outlined. Insofar as the interfacial mobility has a key influence on the film formation and drainage, the state and the properties of the adsorption layers during the film drainage are also of considerable importance. Several additional effects have been detailed: (i) the presence of simple electrolytes in the surfactant solutions; (ii) the relative content of the components in surfactant mixtures.

The presence of electrolyte modifies the adsorption behavior of the surfactant at the fluid interface. In the lower surfactant concentration range, the increase of the electrolyte concentration at constant surfactant concentration results in an enhanced adsorption at the interface [2,4,22,80]. Besides, there may be a reorganization of the adsorption layer coverage related to the modification of the effective head group area (hydrophilic portion) of the amphiphile [81]. In the early studies of Exerowa and

Scheludko [1–3,82], it was experimentally established that these effects have a strong impact on the film drainage time and stability. It has also been found that from a certain critical electrolyte concentration ($C_{el,cr}$) on, there is no further change in the film properties [1,2]. This concentration is the threshold value which marks the transition from CBF to NBF.

It is notable that the presence of electrolyte is important also for the case of nonionic surfactants as well. Thus in [83,84], it is shown that the stability depends on the concentration of the added salt. The effect is related to the specific structure of the air/water interface and the so called "preferential" adsorption of OH^- [2,85].

Studies are performed with aqueous solutions of nonionic surfactants: n-dodecyl-β-D-maltoside (β-$C_{12}G_2$) and tetraethylene-glycol-monodecylether ($C_{10}E_4$) [84]. Insofar as the surfactant concentration is high, common black films (CBF) are formed. Unlike the case of kinetic stabilization of a rupturing film in the premicellar concentration domain, however, the adsorption layers have an immediate influence on the stability of foam films. Combined studies on the dilational elasticities of the interfacial layers at various disturbance frequencies reveal more details about this relationship. It is established that at low values of the disturbance frequencies, the higher the surfactant concentration is, the lower the surface dilational elasticity of the adsorption layer. Conversely, at high values of the disturbance frequencies, the interfacial monolayers tend to behave as insoluble and the higher the concentration is, the higher the surface dilational elasticity. So, in this case there is a clear correlation between the film stability and the high frequency limit of the surface elasticity and there is a minimum value of the dilational elasticity that is required to stabilize the film.

Specific effects are registered for aqueous solution of LMMS mixtures. For example, in systems containing nonionics, there is a competitive adsorption of the components. Therefore, the fine tuning of the interfacial layer properties, as well as of the drainage behavior of the foam films, may be ensured by the proper choice of the surfactants in view of their structure and surface activities.

Thus, in the case of foam films stabilized by mixtures of nonionic surfactants n-dodecyl-β-D-mhaltoside (β-$C_{12}G_2$) and hexaethylene-glycol-monododecyl-ether ($C_{12}E_6$), the peculiarities may be related to the specific properties of the head groups [86,87]. It may be assumed that the glucose group is comparable to ~4 ethylene oxide units. The flexibility of the hydrophilic portions of these surfactants is also quite different: the maltoside unit behaves like a hard disc, while the ethylene oxide units perform here more like short polymer (oligomer) chains. So, the hydration of the ethylene-based surfactant is one order of magnitude higher than that of the sugar-based surfactant. Another point of interest is the pH dependence of the surface charge: the maltoside is pH insensitive down to the isoelectric point, while that of ethylene oxide changes linearly with pH value. Therefore, in the ethylene oxide case, the adsorbed molecules change both their hydration degree (easy uptake and release of water) and their conformation (high flexibility). It is also established that at a lower total surfactant concentration, the aqueous solutions of 1:1 surfactant mixture behaves like pure $C_{12}E_6$ solution, both with regard to interfacial properties (surface tension), and to the foam films' performance. Additionally, at the critical micelle concentration, the films stabilized by this surfactant mixture drain as films containing only a solution of pure $C_{12}E_6$. However, in the presence of traces of $C_{12}E_6$, for example in a 50:1 mixture (β-$C_{12}G_2$:$C_{12}E_6$), the interfacial properties are dominated by the major component.

These studies may be put into a broader context of complex interrelations of various bulk and interfacial properties of aqueous solution of a mixture (1:1) of the nonionic surfactants, including both foam films and foams [88]. The runs of the disjoining pressure isotherms for various total surfactant concentrations have been compared. While below CMC values, the course of the curves resembles those of single nonionic surfactants, the results above CMC show peculiarities: (i) common black films are obtained at high pressures (>100 Pa); (ii) the transition from common black film to Newton black film lasts for hours; (iii) a clear irreproducibility of the disjoining pressure isotherms has also been observed. Some of these special features might again be related to the different flexibilities of the head groups and higher surface activity of $C_{12}E_6$. But the exact mechanism of coupling of interfacial structural and rheological properties and the foam film drainage and foam stability is still not completely clear.

In view of the role of the hydrophilic and hydrophobic portions of the LMMS molecule on the tunability of its surface activity on the air/solution interface, interesting model studies are performed for the so called Lennard-Jones surfactants. These are model systems that may be viewed as primitive models of nonionic surfactants. The cases of a single surfactant [89] and a binary mixture of two surfactants with different lengths of the solvophobic tails [90] are investigated. The numerical experiments show that the tail is preferably stretched, while the head group, consisting of several "hydrophilic" blobs, is a random coil. Therefore, it is the tail attraction that acts as a specific tuning parameter in these systems: small tail–tail attraction energy increases the adsorption and decreases the critical aggregation concentration (CAC) as compared to the same surfactant structure with no tail–tail attraction. The conclusion is that the surfactant with stronger tail–tail attractions strives to abandon the aqueous solvent and to adsorb relatively closer to the vapor phase. These results add to a better comprehension of the factors that determine the detailed picture of the interfacial coverage in the case of nonionic surfactants. They have to be accounted for and might be used in the interpretation of foam film performance in the case of aqueous nonionic LMMS.

More complex cases have been considered as well. For example, aqueous foam films stabilized by mixtures of a nonionic (n-dodecyl-β-D-maltoside (β-$C_{12}G_2$) or dodecyldimethyl phosphine oxide ($C_{12}DMPO$)), and a cationic surfactant (dodecyl trimethylammonium bromide ($C_{12}TAB$)) are investigated in [91,92]. The basic result is that the nonionic surfactant dominates both the mixed monolayer and the mixed micelles due to its higher surface activity. It is also responsible for the observed minimum in the surface tension isotherm of the 1:50 (nonionic:ionic) mixture. The behavior of the respective foam films is studied at various mixing ratios of the surfactants. Both charge neutralization and charge reversal are observed. Thus, for 1:1 mixtures and at 0.5 CMC, the foam films are unstable and rupture, while at higher total concentrations, NBF are formed. When the composition of the mixture is 50:1 (nonionic:ionic) and below the critical micellar concentration, either CBF or NBF are formed. In the opposite case of a surfactant ratio 1:50, when the cationic surfactant dominates in the mixture, a steep increase in the thickness of the films is observed and only CBF are obtained. The comparison of results for β-$C_{12}G_2$/$C_{12}TAB$ and $C_{12}DMPO$/$C_{12}TAB$ mixtures shows similar results. Insofar as the length of the tails of the nonionic surfactants is the same, this outcome means that the type of the nonionic surfactant seems irrelevant to tuning the charge densities of the foam film interfaces. The exact mechanism of stabilization, however, still remains a matter of discussion and further elaboration.

REFERENCES

1. Scheludko, A. Thin liquid films. *Adv Colloid Interface Sci* 1967; 1: 391.
2. Exerowa, D., Kruglyakov, P.M. *Foam and Foam Films*. Elsevier: Amsterdam, 1998.
3. Exerowa, D., Scheludko, A. Taches noires et stabilité des mousses. In: Overbeek, JThG, editor. *Chemistry, Physics and Application of Surface Active Substances*; Vol. 2. Gordon & Breach Sci Publ: London, 1964. p. 1097.
4. Platikanov, D., Exerowa, D. Thin liquid films. In: Lyklema, J. (Ed.), *Findamentals of Interface and Colloid Science*, Vol. 5. Elsevier: Amsterdam, 2005, Ch. 6.
5. Lee, J., Hodgson, T. Film flow and coalescence-I. Basic relations, film shape and criteria for interface mobility. *Chem Eng Sci* 1968; 23: 1375.
6. Ivanov, I.B. Effect of surface mobility on the dynamic behavior of thin liquid films. *Pure Appl Chem* 1980; 52: 1241.
7. Ivanov, I.B., Dimitrov, D.S. Thin film drainage. In: Ivanov, I.B. (Ed.), *Thin Liquid Films*. Marcel Dekker: New York, 1988, p. 379.
8. Lucassen, J. Dynamic properties of free liquid films and foams. In: Lucassen-Reynders, E.H. (Ed.), *Anionic Surfactants: Physical Chemistry of Surfactant Action*, Marcel Dekker: New York, 1981, Ch. 6.
9. Kovalchuk, V.I., Krägel, J., Pandolfini, P., Loglio, G., Liggieri, L., Ravera, F., Makievski, A.V., Miller, R. Dilational rheology of thin liquid films. In: Miller, R., Liggieri, L. (Eds.), *Interfacial Rheology*. Brill: London-Boston, 2009, Ch. 12.

10. Loglio, G., Pandolfini, P., Miller, R., Makievski, A.V., Ravera, F., Ferrari, M., Liggieri, L. Drop and bubble shape analysis as a tool for dilational rheological studies of interfacial layers. In: Möbius, D., Miller, R. (Eds.), *Novel Methods to Study Interfacial Layers*. Elsevier: Amsterdam, 2001, p. 439.

11. Levich, V. *Physicochemical Hydrodynamics*. Prentice Hall: Engelwood Cliffs, 1962.

12. Mileva, E., Radoev, B. Hydrodynamic interactions and stability of emulsion films. In: Petsev, D.N. (Ed.), *Emulsions: Structure, Stability and Interactions, Vol. 4 of Interfacial Science and Technology*. Elsevier: London, 2004, Ch. 6.

13. Kralchevsky, P., Danov, K., Denkov, N. Chemical physics of colloid systems and interfaces. In: Birdi, K.S. (Ed.), *Handbook of Surface and Colloid Chemistry*. CRC Press: Boca Raton, 2008, Ch. 7.

14. Mileva, E., Radoev, B. Effective zones of hydrodynamic interaction of fluid particles at small separations. *Coll Polym Sci* 1986; 264: 823.

15. Mileva, E., Radoev, B. Hydrodynamic interaction of two emulsion droplets at small separations. *Coll Polym Sci* 1985; 263: 587.

16. Scheludko, A. Sur certaines particularités des lames mousseuses. II. Stabilité cinétique, épaisseur critique et épaisseur d'équilibre. *Proc Konikl Ned Akad Wetenschap* 1962; B65: 87.

17. Ivanov, I.B., Dimitrov, D.S., Somasundaran, P., Jain, R.K. Thinning of films with deformable surfaces: Diffusion-controlled surfactant transfer. *Chem Eng Sci* 1985; 40: 137.

18. Radoev, B., Dimitrov, D., Ivanov, I. Hydrodynamics of thin liquid films. Effect of the surfactant on the rate of thinning. *Coll Polym Sci* 1974; 252: 50.

19. Dimitrov, D., Radoev, B. *Compt Rend de l'Academie bulgare des Sciences* 1976; 29: 1649.

20. Nikolov, L., Mileva, E. Model study on emulsion systems with high interfacial mobility. *Compt Rend de l'Academie bulgare des Sciences* 2010; 63: 1435.

21. Mileva, E., Radoev, B. The influence of surfactants on the hydrodynamic interactions in emulsion systems. *Coll Polym Sci* 1986; 264: 965.

22. Nikolova, A., Exerowa, D. Rupture of common black films: Experimental study. *Colloids Surfaces A* 1999; 149: 185.

23. Mileva, E., Radoev, B. Tangential mobility of a thin layer in the presence of insoluble surfactant. *Commun Dept Chem Bulg Acad Sci* 1991; 24: 513.

24. Scriven, L. Dynamics of a fluid interface. Equation of motion for Newtonian surface fluids. *Chem Eng Sci* 1960; 12: 98.

25. Sørensen, T. Instabilities induced by mass transfer, low surface tension and gravity at isothermal and deformable fluid interfaces. In: Sørensen, T. (Ed.), *Dynamics and Instability of Fluid Interfaces*. Lecture Notes in Physics, vol. 105. Springer-Verlag: Berlin, 1979, Ch.1, pp. 1–74.

26. Mileva, E., Radoev, B. Mass transfer and hydrodynamic interaction in emulsion systems. I. Scaling concepts. *Colloids Surfaces A* 1993; 74: 259.

27. Mileva, E., Nikolov, L. Mass transfer and hydrodynamic interaction in emulsion systems. II. Potential theory. *Colloids Surfaces A* 1993; 74: 267.

28. Vassilieff, C., Manev, E., Ivanov, I.B., VI. Internationale Tangung uber Grennzflachen-aktive Stoffe Abh. Akad Wiss DDR, Abtl Math naturwiss. Technik, No.1; 1987; p. 465.

29. Joye, J., Hirasaki, G., Miller, C. Asymmetric drainage in foam films. *Langmuir* 1994; 10: 3174.

30. Joye, J., Hirasaki, G., Miller, C. Numerical simulation of instability causing asymmetric drainage in foam films. *J Coll Interface Sci* 1996; 177: 542.

31. Mileva, E., Exerowa, D. A model for the asymmetric drainage of thin liquid films. In: *Proc. Second World Congress on Emulsion*, Bordeaux, France, 1997, Vol. 2, p. 2-2-235/1-10.

32. Manev, E., Tsekov, R., Radoev, B. Effect of thickness non-homogeneity on the kinetic behaviour of microscopic foam films. *J Dispersion Science Technology* 1997; 18: 769.

33. Mileva, E., Exerowa, D. Self-assembled structures in thin liquid films. *Colloids Surf A* 1999; 149: 207.

34. Mileva, E., Exerowa, D. Self-assembly of amphiphilic molecules in foam films. In: Mittal, K., Kumar, P. (Eds.), *Emulsions, Foams and Thin Films*. Marcel Dekker: New York, 2000, Ch. 15.

35. Rusanov, A. *Phasengleichgewichte und Grenzflaechenerscheinungen*. Akademie: Berlin, 1978.

36. Lopatkin, A. *Theoretical Basis of Physical Adsorption*. Moscow University Press: Moscow, 1983.

37. Nikolov, A., Martynov, G., Exerowa, D. Associative interactions and surface tension in ionic surfactant solutions at concentrations much lower than the CMC. *J Colloid Interface Sci* 1981; 81: 116.

38. Vold, M. Micellization process with emphasis on premicelles. *Langmuir* 1992; 8: 1082.

39. Mukerjee, P., Mysels, K., Dulin, C. Dilute solutions of amphipathic ions. I. Conductivity of strong salts and dimerization. *J Phys Chem* 1958; 62: 1390.

40. Kanicky, J.R., Shah, D.O. Effect of premicellar aggregates on the pKa of fatty acid soap solutions. *Langmuir* 2003; 19: 2034.

41. Neumann, M., Schmitt, C., Iamazaki, E. A fluorescence emission study of the formation of induced premicelles in solutions of polyelectrolytes and ionic surfactants. *J Colloid Interface Sci* 2003; 264: 490.

42. del Gutierrez-Hijar, D.P., Becerra, F., Puig, J., Soltero-Martinez, J.F.A., Sierra, M.B., Schulz, P. Properties of two polymerizable surfactant aqueous solutions: Dodecyl-ethylmethacrylate-dimethylammonium bromide and hexadecylethylmethacrylate dimethylamonium bromide. I. Critical micelle concenration. *Coll Polym Sci* 2004; 283: 74.

43. Lopez-Fontan, J.L., Gonzalez-Perez, A., Costa, J., Ruso, J.M., Prieto, G., Schulz, P., Sarmiento, F. The critical micelle concentration of tetraethylammonium perfluoro-octylsulfonate in water. *J Colloid Interface Sci* 2006; 294: 458.

44. Mukerjee, P., Mysels, K. *Critical Micelle Concentrations of Aqueous Surfactant Systems.* NSRDS: Washington, 1971.

45. Arabadzhieva, D., Tchoukov, P., Mileva, E., Exerowa, D. Experimental investigation of foam films stabilized with n-heptanol. In: Dragieva, J., Balabanova, E. (Eds.), *Nanoscience & Nanotechnology*. Heron Press: Sofia, 2006, pp. 230–232.

46. Arabadzhieva, D., Soklev, B., Mileva, E. Amphiphilic nanostructures in aqueous solutions of triethyleneglycol monododecyl ether. *Colloids and Surfaces A* 2013; 419: 194.

47. Arabadzhieva, D., Mileva, E., Tchoukov, P., Miller, R., Ravera, F., Liggieri, L. Adsorption layer properties and foam film drainage of aqueous solutions of tetra-ethylene glycol monododecyl ether. *Colloids and Surfaces A* 2011; 392: 233.

48. Arabadzhieva, D., Tchoukov, P., Mileva, E., Miller, R., Soklev, B. Impact of amphiphilicnanostructures on formationand rheology of interfaciallayers and on foam film drainage. *Ukr J Physics* 2011; 56: 801.

49. Arabadzhieva, D., Mileva, E., Tchoukov, P., Miller, R. Investigations of adsorption layers stabilized with nonionic surfactant pentaethyleneglycol monododecyl ether. In: Kashchiev, D. (Ed.), *Nanoscale Phenomena and Structures*. Prof. Marin Drinov Publishing House: Sofia, 2008, p. 179.

50. Arabadzhieva, D., Tchoukov, P., Soklev, B., Mileva, E. Surface rheology of adsorption layers and foam film drainage kinetics of aqueous solutions of hexadecyltrimethylammonium chloride. *Colloids and Surfaces A* 2014; 460: 28.

51. Tchoukov, P., Mileva, E., Exerowa, D. Experimental evidences of self-assembly in foam films from amphiphilic solutions. *Langmuir* 2003; 19: 1215.

52. Mileva, E., Exerowa, D. Foam films as instrumentation in the study of amphiphile self-assembly. *Adv Colloid Interface Sci* 2003; 100–102: 547.

53. Tchoukov, P., Mileva, E., Exerowa, D. Drainage time peculiarities of foam films from amphiphilic solutions. *Colloids Surf A* 2004; 238: 19–25.

54. Mileva, E., Tchoukov, P., Exerowa, D. Amphiphilic nanostructures in thin liquid films. *Adv Colloid Interface Sci* 2005; 114–115: 47–52.

55. Mileva, E., Tchoukov, P. Surfactant nanostructures in foam films. In: Tadros, Th (Ed.), *Colloid Stability: The Role of Surface Forces—Part 1; Vol. 1. Colloids and Interface Science Series*. WILEY-VCH: Weinheim, 2007, Ch. 8.

56. Mileva, E., Exerowa, D., Tchoukov, P. Black dots as a detector of self-assembly in thin liquid films. *Colloids Surf A* 2001; 186: 83.

57. Tchoukov, P., Mileva, E. Expansion rate of "unstable" black spots in microscopic foam films. *Compt Rend de l'Academie bulgare des Sciences* 2011; 64: 1273.

58. Mileva, E., Exerowa, D. Amphiphilic nanostructures in foam films. *Curr Opin Coll Interface Sci* 2008; 13: 120.

59. Mileva, E. Impact of adsorption layers on thin liquid films. *Curr Opin Coll Interface Sci* 2010; 15: 315.

60. Liggieri, L., Miller, R. Relaxation of surfactants adsorption layers at liquid interfaces. *Curr Opin Coll Interface Sci* 2010; 15: 256.

61. Miller, R., Ferri, J.K., Javadi, A., Krägel, J., Mucic, N., Wüstneck, R. Rheology of interfacial layers. *Coll Polym Sci* 2010; 288: 937.

62. Lee, Y.C., Lui, H.S., Lin, S.Y., Huang, H.F., Wang, Y.Y., Chou, L.W. An observation of the coexistence of multimers and micelles in the nonionic surfactant C10E4 solution by dynamic light scattering. *J Chinese Inst Chem Eng* 2008; 39: 75.

63. Johonnott, E.S. The black spots in thin liquid films. *Phil Mag* 1906; 11: 746.

64. Perrin, J. La statification des lames liquides. *Ann Phys (Paris)* 1918; 10: 160.

65. Bergeron, V., Radke, C. Equilibrium measurements of oscillatory disjoining pressures in aqueous foam films. *Langmuir* 1992; 8: 3020.

66. Bergeron, V., Jimenez-Laguna, A.I., Radke, C. Hole formation and sheeting in the drainage of thin liquid films. *Langmuir* 1992; 8: 3027.

67. Khristov, Khr, Exerowa, D., Kruglyakov, P. Multilayer foam films and foams from surfactant solutions with high solubilizing ability. *Colloids Surf A* 1993; 78: 221.

68. Exerowa, D., Lalchev, Z. Bilayer and multilayer foam films: Model for study of the alveolar surface and stability. *Langmuir* 1986; 2: 668.

69. Danov, K., Basheva, E., Kralchevsky, P., Ananthapadmanabhan, K.P., Lips, A. The metastable states of foam films containing electrically charged micelles or particles: Experiment and quantitative interpretation. *Adv Colloid and Interface Science* 2011; 168: 50.

70. Kralchevsky, P., Danov, K., Anachkov, S. Micellar solutions of ionic surfactants and their mixtures with nonionic surfactants: Theoretical modeling vs. experiment. *Colloid Journal* 2014; 76: 255.

71. Heinhg, P., Belttran, C., Langevin, D. Domain growth dynamics and local viscosity in stratifying foam films. *Phys Rev E* 2006; 73: 051607-1–8.

72. Langevin, D. Polyelectrolyte and surfactant mixed solutions. Behaviour at surfaces and in thin films. *Adv Colloid Interface Sci* 2001; 89–90: 467.

73. Stubenrauch, C., von Klitzing, R. Disjoining pressure in thin liquid foam and emulsion films—new concepts and perspectives. *J Phys: Condens Matter,* 2003; 15: R1197.

74. Taylor, S., Czarnecki, J., Masliyah, J. Aqueous foam films stabilized by sodium naphtanates. *J Colloid Interface Sci* 2006; 299: 283.

75. Henderson, D., Trokhymchuk, A.D., Wasan, D. Structure and layering of fluids in thin films. Chapter 7. In: Petsev, D. (Ed.), *Emulsions: Structure, Stability and Interactions.* Elsevier: New York, 2004, pp. 259–311.

76. Trokhymchuk, A.D., Henderson, D., Nikolov, A., Wasan, D. Entropically driven ordering in a binary colloidal suspension near a planar wall. *Phys Rev E* 2001; 64: 012401–012404.

77. Dushkin, C., Stoichev, T., Horozov, T., Mehreteab, A., Bronze, G. Dynamics of foams of ethoxylated ionic surfactant in the presence of micelles and multivalent ions. *Colloid Polym Sci* 2003; 281: 130.

78. Rusanov, A., Krotov, V., Nekrasov, A. Extremes of some foam properties and elasticity of thin foam films near critical micelle concentration. *Langmuir* 2004; 20: 1511.

79. Patist, A., Kanicky, J.R., Shukla, P.K., Shah, D.O. Importance of micellar kinetics in relation to technological processes. *J Colloid Interface Sci* 2002; 245: 1.

80. Exerowa, D., Kolarov, T., Khristov, Khr. Direct measurement of disjoining pressure in black foam films. I. Films from an ionic surfactant. *Colloids & Surfaces* 1987; 22: 171.

81. Exerowa, D., Nikolov, A., Zacharieva, M. Common black and Newton film formation. *J Colloid Interface Sci* 1981; 81: 419.

82. Exerowa, D. Effect of adsorption, ionic strength and pH on the potential of the diffuse electric layer. *Kolloid-Zeitschrift* 1969; 232: 703.

83. Belttran, C., Langevin, D. Electtrostatic effects in films stabilized by non-ionic surfactants. *J Colloid Interface Sci,* 2007; 312: 47.

84. Santini, E., Ravera, F., Ferrari, M., Stubenrauch, C., Makievski, A., Krägel, J. A surface rheological study of non-ionic surfactants at the water-air interface and the stability of the corresponding thin foam films. *Colloids Surf A* 2007; 298: 12.

85. Exerowa, D., Zacharieva, M. Investigation of the isoelectric points at the solution/air interface. In: Derjaguin, B. (Ed.), *Research in Surface Forces,* Vol. 4. Consultants Bureau: New York, 1972, p. 234.

86. Angarska, J., Stubenrauch, C., Manev, E. Drainage of foam films stabilized with mixtures of non-ionic surfactants. *Colloids Surf A* 2007; 309: 189.

87. Simulescu, V., Stubenrauch, C., Manev, E. Drainage and critical thickness of foam films from aqueous solutions of mixed nonionic surfactants. *Colloids Surf A* 2008; 319: 21.

88. Stubenrauch, C., Claesson, P.M., Rutland, M., Manev, E., Johansson, I., Pedersen, J.S., Langevin, D. Mixtures of n-dodecyl-β-D-maltoside and hexaoxyethylene dodecyl ether—Surface properties, bulk properties, foam films, and foams. *Adv Colloid Interfcae Sci* 2010; 155: 5.

89. Howes, A.J., Radke, C.J. Monte Carlo simulation of Lennard-Jones nonionic surfactant adsorption at liquid/vapour interface. *Langmuir* 2007; 23: 1835.

90. Howes, A.J., Radke, C.J. Monte Carlo simulation of mixed Lennard-Jones nonionic surfactant adsorption at liquid/vapour interface. *Langmuir* 2007; 23: 11580.

91. Buchavzov, N., Stubenrauch, C. A disjoining pressure study of foam films stabilized by mixtures of nonionic and ionic surfactants. *Langmuir* 2007; 19: 5315.

92. Carey, E., Stubenrauch, C. A disjoining pressure study of foam films stabilized by mixtures of non-ionic ($C_{12}DMPO$) and an ionic surfactant ($C_{12}TAB$). *J Colloid Interface Sci* 2010; 343: 314.

5 Foam Films Stabilized by Polymers and Proteins

Georgi Gochev and Nora Kristen-Hochrein

CONTENTS

5.1 INTRODUCTION

Amphiphilic macromolecules adsorb at liquid/fluid interfaces and form thicker layers compared to those formed by low molecular weight surfactants. Such layers have specific structure such as brush-like and loop-like types in the case of polymeric surfactants or a rigid viscoelastic network in the case of proteins. In a foam film, the physical contact between two such surfaces gives rise to the action of an efficient repulsive surface force, that is, steric stabilization of thin liquid films and colloid systems in general [1–4]. On the other hand, polymeric surfactants and especially proteins are capable of significantly modifying the rheological characteristics of liquid/fluid interfaces [5–7] and significantly disturbing the mobility of these interfaces. Foam films with less mobile surfaces exhibit slower drainage and these were classified by Mysels as rigid films [8].

Investigations on foam films stabilized by macromolecules dates back several decades ago with the pioneering works of Musselwhite and Kitchener [9] and Yampolskaya et al. [10] on protein foam films, and of Lyklema and van Vliet on polymer foam films [11]. The microinterferometric method [1,12], explained in detail in Chapter 3.3, has been employed in these studies. Later on, the same method was continuously involved into investigations on foam films stabilized by block copolymers [1,3,13–16], by charged [17–19] and neutral [16,20,21] graft copolymers, and by proteins [22–42].

Addition of surfactants to protein solutions is able to affect, sometimes dramatically, the characteristics of protein foam films [35–39,42]. This is another case of polyelectrolytes of weak amphiphilic character which generally do not stabilize thin liquid films. Polyelectrolytes are capable of forming equilibrium foam films in the presence of an ionic low molecular weight surfactant. Such polyelectrolyte/surfactant foam films may differ from the surfactant alone foam films in respect to drainage, thickness, and stability [43–46]. Another way to involve polyelectrolytes in interfacial processes is by applying chemical amphiphilization of the polyelectrolyte chain. Such ionic block and graft copolymers can be efficient stabilizers of foam films [17–19]. X-ray reflectivity experiments with gravity-drained vertical foam films drawn in a frame from solutions proved to be feasible in revealing the structure of foam films from polymer [18,19], protein [47–49] or mixed polymer/protein [49] solutions. The diminishing bubble method was also applied for investigating protein foam films [47,48,50].

We review and discuss below some main findings on thin foam films divided into three subsections, namely Polymer Foam Films, Polyelectrolyte/Surfactant Foam Films, and Protein Foam Films.

5.1.1   POLYMER FOAM FILMS

Adsorbing polymers produce interfacial layers with various architectures and thicknesses. The physical contact between such two layers in a thin liquid film gives rise to an additional component of the film disjoining pressure Π, denoted steric component Π_{st}, acts independently of the classical DLVO forces (electrostatic Π_{el} and van der Waals Π_{vw}), and originates from the steric interaction between the macromolecular interfacial layers. It should be noted that these types of forces can differ from the short-range forces in the films from low molecular weight surfactants [1] because of the much larger size and higher conformational entropy for the macromolecules. Assuming the action of only these three types of surface forces, the disjoining pressure in foam films reads: $\Pi(h) = \Pi_{el}(h) + \Pi_{vw}(h) + \Pi_{st}(h)$, with h the film thickness. Further below we discuss the properties of foam films stabilized by polymers in terms of disjoining pressure and film thickness*. The theoretical backgrounds of the DLVO theory and the de Gennes scaling theory for steric interactions [51–53] are explained in Chapter 3.2.

Lyklema and van Vliet investigated for the first time, to our best knowledge, foam films from polymeric surfactants [11]. They interferometrically measured the equilibrium thickness of horizontal foam films with radius 1.5–3 mm stabilized either by poly(vinyl alcohol) PVA or by partially esterified poly(methacrylic acid) PMA. The PVA samples differ in their molecular weight (M_w) and chain architecture (blocky or random) which both determine pronounced differences in film characteristics. Film stability was discussed on the basis of the experimental disjoining pressure isotherm $\Pi(h)$ for films from the different samples. The films from the two blocky PVAs show higher stability than the films from the random PVA (weaker steric repulsion Π_{st}). The film thickness h was found to be dependent of M_w and only weakly affected by the polymer concentration C_p. The authors attributed the fact that h is larger than twice the thickness (δ) of a single layer (measured by ellipsometry) to the existence of a few longer tails present in the adsorbed layer. Such terminally anchored isolated tails might be invisible in ellipsometry, but are capable of generating steric repulsion in the foam films. For the case of foam films from PMA, the difference ($h - 2\delta$) is less than that for PVA films, which leads to the conclusion that adsorbing polyelectrolytes are less capable of forming long tails and this determines the smaller film thickness.

5.1.1.1   Foam Films Stabilized by Block Copolymers

Foam films obtained from solutions of triblock copolymers of the $E_xP_yE_x$ type (Pluronics, E stands for an ethylene oxide monomer and P for a propylene oxide monomer) were extensively studied by Sedev et al. [1,13,14,54]. The action of electrostatic forces Π_{el} in these foam films is demonstrated by the variation of the equivalent film thickness h_w with changing C_{el} or/and pH at constant capillary pressure [13,55]; note that similar behavior was also observed for oil-in-water emulsion films obtained from Pluronic polymer solutions [56]. The experimental $h_w(C_{el})_{pH}$ and $h_w(pH)_{C_{el}}$ dependences for foam films stabilized by Pluronic F108 ($E_{122}P_{56}E_{122}$) are plotted in Figure 5.1.

---

* Note that the term "film thickness" can have different definitions depending on the context – with the interferometric method the equivalent film thickness (h_w) is measured, which differs from the real physical thickness (h) of the film. These terms are quantitatively related and h can be calculated from the measured h_w; however, at film thicknesses larger than 30 nm, it is generally assumed that $h \approx h_w$, for details see Chapter 3.3.

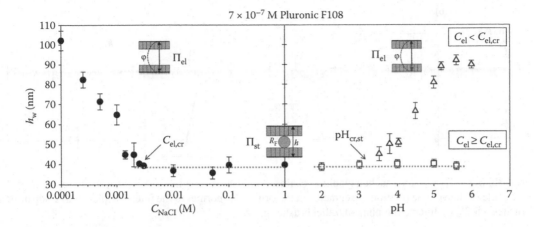

FIGURE 5.1 Equivalent film thickness h_w of foam films from Pluronic F108 as a function of: (left) the electrolyte concentration in native polymer solutions (pH 5.5–6); (right) pH at two electrolyte concentrations below and above $C_{el,cr}$. (Plots adapted from the data set reported in Sedev R, Exerowa D. *Adv Colloid Interface Sci* 1999; 83: 111 [13].)

The screening effect of the increasing electrolyte concentration on Π_{el} causes a gradual reduction in h_w until the critical concentration $C_{el,cr}$. At a given double layer potential φ_0, the film thickness is predominantly determined by the thickness of the diffuse part of the double layer (Debye length) $1/\kappa$ which in turn depends on C_{el}, for details see Chapter 3.2. Under other conditions, an increase in C_{el} results in a reduction in film thickness and beyond the critical electrolyte concentration $C_{el,cr}$, the films are stabilized by brush-to-brush steric repulsion [1,3,13]. The film thickness in the h_w(pH) dependence was measured at different C_{el} [13] and Figure 5.1 (right) shows two examples chosen to be, respectively, below and above $C_{el,cr}$. The origin of the negative charging of liquid/fluid interfaces and the corresponding electrostatic interaction in foam and emulsion films is continuously debated in the literature, but however, it is commonly accepted to be due to adsorption of hydroxyl ions at the interface and their incorporation into the hydration shell of soluble moieties of the adsorbed molecules [1,3,57–60]. This hypothesis is valid as well for foam [1,13,55] and emulsion films [56] from block copolymers. The final h_w-values at $C_{el} \geq C_{el,cr}$ are virtually independent of C_{el} and pH. At $C_{el} \leq C_{el,cr}$, the film thickness is pH-dependent and decreases with reduction of pH until a certain value is reached below which pH does not affect the film thickness (such critical pH value is denoted $pH_{cr,st}$ [1,13]). At $pH_{cr,st}$, Π_{el} vanishes because of obliteration of the diffuse double layer potential due to electroneutralization of the surface charge determining hydroxyl ions in an acidic environment.

In the absence of Π_{el}, the foam films are stabilized by additional force of steric origin Π_{st}. Quantitative description of Π_{st} in foam films from block copolymers [1,3,13] is feasible on the basis of the scaling concept of Alexander [51] and de Gennes [52,53]. This theory considers the two cases of *i)* a brush layer of terminally attached polymer chains to an interface under good solvent conditions; and *ii)* the repulsive interaction between such two layers, for details see Chapter 3.2. Figure 5.2 shows schematic representations of the structure of a single brush layer (a–c) and two brushes approaching each other (d,e); the notations are D—distance between the anchoring points, which defines the area per molecule, δ—thickness of the adsorbed layer, R_F—radius of gyration (Flory radius) of a macromolecule in bulk phase, h—distance of separation between the two interfaces (film thickness). The steric disjoining pressure between two brushes can be evaluated by use of Equation 3.32 (see Chapter 3.2).

The range of the steric interaction in foam films from block copolymer solutions is continuously discussed in the literature [1,3,13]. For example, Equation 3.32 was fitted to the experimental Π-h isotherms for foam films from solutions of the $E_{122}P_{56}E_{122}$ tri- and the $E_{106}B_{16}$ di-block copolymers with the brush thickness δ as a fitting parameter yielding $\delta = 11.1$ nm and 15.5 nm, respectively

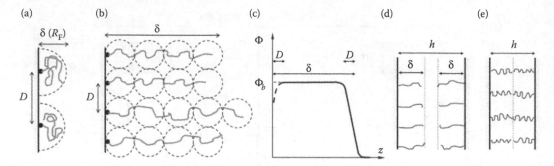

FIGURE 5.2 Terminally attached lyophilic chains. (a) Mushroom regime $D > R_F$; (b) Brush regime $D < R_F$; (c) Profile of the volume fraction of segments in the brush (Φ_b) normal to the interface plane; (d) Undisturbed brushes $h > 2\delta$; (e) Interacting plane-parallel brushes $h \le 2\delta$.

(B stands for a butylene oxide monomer). The range of such steric interaction between polymer layers is several times larger than that measured for Newton black films from low molecular weight surfactants. Further studies in this direction concluded that the brush thickness increases linearly on increasing the number of E_x segments of the polymeric chain [14]. Similar behavior is observed for foam films stabilized by copolymers of the $E_x B_y E_x$ and $E_x P_y$ types [3,15]. Different branches in the disjoining pressure isotherm Π-h correspond to films stabilized by brushes with different thicknesses as the brush thickness decreases when reducing the length of the soluble E_x chain(s) for every copolymer. A similar trend was found for emulsion films from Pluronics solutions [56]. In another case—that of charge amphiphilic polymers (amphiphilic polyelectrolytes)—one also has to take into an account changes in brush thickness and structure due to gradual chain neutralization under the screening effect of increasing ionic strength as discussed on the basis of x-ray reflectivity measurements in References 17–19.

5.1.1.2 Foam Films Stabilized by Graft Copolymers

Amphiphilic graft copolymers of the AB_m type are also able to provide stabilization of foam and emulsion films. Water soluble copolymers are synthesized by random grafting of m hydrophobic alkyl chains B onto a hydrophilic backbone A. For a given A-chain length, m determines the degree of grafting $S_{graft}(m)$. The degree of grafting is a major factor for the properties and stability of foam films stabilized by graft copolymers [17,18,20]. The reason for this is the specific architecture of the adsorption layers at the polymer solution interface as studied by measurements of the surface forces acting between solid particles [61] and in foam [17,18,20] and emulsion [62] films. Figure 5.3 shows a schematic representation of the structure of layers from graft copolymers at film surfaces. The copolymer molecules are anchored at the solution/air(oil) interface by multipoint attachments

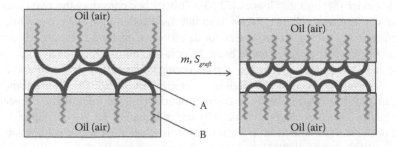

FIGURE 5.3 Schematic representation of the steric stabilization of foam and emulsion films from AB_m graft copolymers with different degrees of grafting.

of the alkyl chains thus forming loops and tails protruding into the solution. Higher S_{graft} leads to reduction of the loop size.

As discussed above, the critical electrolyte concentration $C_{el,cr}$ represents the transition from electrostatic to steric stabilization in foam films from nonionic block copolymers [1,3,13,15]. A similar behavior is observed for foam films from hydrophobically modified Inulin (HMI) graft copolymers and black films are obtained at $C_{el} \geq C_{el,cr}$ [20]. In this study, four copolymers were used, hereafter designated A, B, C, and D, with increasing S_{graft}, where S_{graft} for every sample obeys the proportion D = 3C/2 = 3B = 6A and the loop size correspondingly decreases in the order A > B > C > D. The disjoining pressure isotherm shows remarkably higher stability for the black films from the two copolymers with the larger loops (A and B) compared to the data for the copolymers C and especially D which form smaller loops. This provides evidence that the loop size controls the magnitude of the steric interaction in these films.

The curves for the films from copolymers A, B, and C in Figure 5.4 virtually coincide up to a pressure of several hundred Pa. Within this low pressure region, a kind of "soft" interaction on [11,13] between isolated longer tails and larger loops could be assumed to determine the film thickness and such an interaction is practically independent of the degree of grafting. A remarkable exception is the case of the much less stable films from copolymer D. These films readily thin down with a small pressure increase and rupture at several hundred Pa which implies only a weak Π_{st} due to very thin loop layers at film surfaces. At higher compressions, such "soft" interaction is overcome and the films are stabilized by the actual loop-to-loop repulsion. The curves for copolymers A and B in Figure 5.4 split as the shift in the equivalent film thickness h_w is attributed to the different loop size. This was further demonstrated by model calculations based on Equation 3.32 showing a difference of 1 nm between the h_w-values of the films from these two copolymers [20].

X-ray reflectivity measurements were performed for vertical foam films drawn in a frame from solutions of hydrophobically modified poly(acrylic acid) Na salt, HMPAH [17,18]. The film thickness was studied as a function of the polymer concentration C_p and for various formulations of the chemical structure of HMPAH which differ in (*i*) the degree of grafting; (*ii*) the alkyl chain length; and (*iii*) the molecular weight M_w. It was found for copolymers with equal M_w that higher S_{graft} causes a sharp increase in the film thickness with increasing C_p, while lower S_{graft} results in an initial plateau region where the film thickness is almost independent of C_p until a threshold C_p value [18]. On the other hand, at same M_w and S_{graft}, reduction of the alkyl chain length from C_{18} to C_{12} results in shifting of the dependence with one order of magnitude toward higher polymer concentrations. However, it is

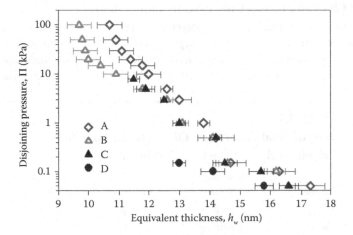

FIGURE 5.4   Disjoining pressure isotherm for foam films obtained from solutions of 10^{-4} M hydrophobically modified inulin polymeric surfactants with different degrees of grafting A < B < C < D; for details see the text. (Plots adapted from the data set reported in Gochev G et al., *Colloids Surfaces A* 2011; 391: 101 [20].)

evident that a decrease in either S_{graft} or alkyl chain length causes a reduction of the surface activity of the copolymers. Measurements of foam films from copolymers of different M_w revealed that in the concentration region below the threshold C_p the film thickness scales with $M_w^{1/2}$ [17]. Results on black foam films from HMPAH [18] show that film thickness decreases with increasing C_{el} which accounts for electrostatic effects in the films as in the case of charged di-block copolymers [18]. Beyond a certain C_{el}, the film thickness remains constant and independent of C_{el} which could be interpreted as an indication for formation of films of the NBF type.

So far, we have discussed the "brush-to-brush" and "loop-to-loop" steric interaction in foam films from block and graft copolymers. It is easy to conclude that such polymers can be efficient in the stabilization of foam films and that the mechanisms of steric stabilization are related to the structure and thickness of the interfacial polymer layers. Recently, some reports on foam films stabilized by star-like copolymers appeared [63]. These films are characterized by specific parameters, such as film thickness, critical electrolyte concentration, surfactant concentration for obtaining a stable film, and so on. The interfacial conformation of such molecules seems to be complicated and the mechanisms of stabilization of foam films from such species need further elucidation.

5.1.2 POLYMER/SURFACTANT FOAM FILMS

Polyelectrolyte/surfactant mixtures in foams are of great relevance in many practical applications, like personal care or cleaning. The combination of the components in the mixtures has a great impact on the final state before film rupture. Depending on the charge of the used polyelectrolytes and surfactants, either a CBF or a NBF is formed before rupture. Table 5.1 gives an overview about the type of foam films that are formed at different polyelectrolyte/surfactant charge combinations.

In foam film studies of these systems, it is crucial to choose the concentrations of the components carefully. In order to get homogenous and continuously thinning films, the surfactant concentration should be below the critical micelle concentration (CMC) and the polyelectrolyte concentration below the overlap concentration (C'). The CMC determines the onset of micelles formation in the bulk solution, while C' corresponds to the polyelectrolyte concentration where polyelectrolytes start to overlap and form a network. Both phenomena lead to structural forces in foam films. Furthermore, to avoid aggregates in the bulk solution that prevent the formation of homogeneous films, it is important to choose the composition of both components such that the critical aggregation concentration (CAC) is not exceeded [43].

5.1.2.1 Likely Charged Polyelectrolyte/Surfactant Systems
The influence of mixtures of likely charged polyelectrolytes and surfactants on the surface tension is negligible. It is assumed that due to the like charge, the adsorbed surfactant molecules at the interface repel the polyelectrolytes, so that no surface active complexes are formed that could lower the surface tension. However, depending on the length of the alkyl chain of the surfactant, interactions between

TABLE 5.1
Type of Final Foam Film, Observed for Specific Polyelectrolyte/Surfactant Combinations

Polymer	Nonionic Surfactant	Cationic Surfactant	Anionic Surfactant
Anionic	CBF	CBF	CBF
Cationic	NBF	CBF	–
Nonionic	CBF	–	CBF

Source: Kolaric B et al., *J Phys Chem B* 2003; 107: 8152 [64].

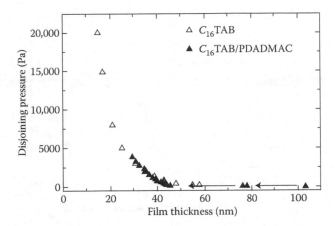

FIGURE 5.5 Π-h_w isotherm for foam films obtained from pure $C_{16}TAB$ (0.1 mM) and 0.1 mM $C_{16}TAB$ + 5 mM PDADMAC solutions. (Kristen N, von Klitzing R. Effect of polyelectrolyte/surfactant combinations on the stability of foam films. Soft Matter 2010; 6: 849. Reproduced by permission of The Royal Society of Chemistry [43].)

the hydrophobic backbone of the polyelectrolyte and the surfactant can occur, which results in a depletion of the surfactant from the surface. This is, for example, the case in mixtures of $C_{16}TAB$ and PDADMAC (poly-diallyldimethylammonium-chloride) shown in Figure 5.5 [43].

Foam film studies of this system with and without polyelectrolytes show formation of CBFs. The films are very stable and no transition to an NBF occurs in the investigated capillary pressure range. The disjoining pressure isotherms Π-h for foam films from the mixed system show lower film stability but similar film thickness. This indicates that the polyelectrolyte does not act like a simple salt, which would lead to electrostatic screening between the film surfaces but does not increase the osmotic pressure in the film either. However, the addition of polyelectrolytes leads to a reduced stability of the foam films, which is an indication of the significant influence of the motion of the polyelectrolyte chains in the film bulk on the film stability.

5.1.2.2 Oppositely Charged Polyelectrolyte/Surfactant Systems

In the case of oppositely charged polyelectrolyte/surfactant systems, so far only formation of CBFs has been observed. Since these films are stabilized by the repulsion of the two opposing interfaces, the adsorption of the surfactant and/or the polyelectrolyte should influence the foam film stability. In the low concentration regime of the surfactant (below 10^{-4} M), the addition of polyelectrolytes often leads to the formation of surface active complexes. As a result, the surface tension is lower for polyelectrolyte/surfactant mixtures compared to that of the pure surfactant. This happens at the critical surface aggregation concentration (CSAC) [58,65]. The driving force for formation of complexes between oppositely charged polyelectrolytes and surfactants is the action of electrostatic and hydrophobic forces. The charged polymer segments attract the oppositely charged head groups of the surfactant molecules. Additionally, interactions between the hydrophobic surfactant tails and the hydrophobic backbone of the polyelectrolyte can occur [66].

In this low concentration regime, only loosely packed monolayers cover the air/water interface [67–69]. The distance between the surfactant molecules depends on the degree of charge of the polymer and can be calculated from the Gibbs equation that is applied to the surface tension isotherms. The charged monomer units of the polymer form complexes with the surfactant molecules at the interface while the uncharged parts of the polyelectrolyte chain dangle into the bulk solution. The higher the degree of charge, the smaller the distance between the surfactant molecules is. This is due to the shorter distance between the charged monomer units [68,70,71]. The thickness of the interfacial layer is also influenced by the degree of charge of the polymer. Highly charged polyelectrolytes such as PSS (poly-styrene-sulfonate-sodium salt) adsorb flatly at

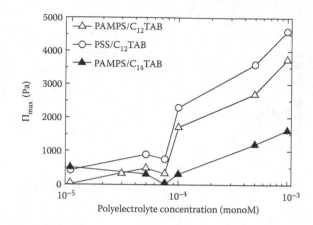

FIGURE 5.6 Stability of polyelectrolyte/surfactant films; maximum pressure before film rupture versus polyelectrolyte concentration at a fixed surfactant concentration of 10^{-4} M. (Reprinted with permission from Kristen N et al. Films from oppositely charged polyelectrolytes/surfactant mixtures: Effect of Surfactant and polyelectrolyte hydrophobicity. Langmuir 2010; 26: 9321. Copyright 2010 American Chemical Society [77].)

the interface while polyelectrolytes with a lower degree of charge form thicker layers due to the loops that are extended into the solution [72,73].

Foam film studies with varied polyelectrolyte concentration and fixed surfactant concentration have shown that the foam films are very sensitive to the polymer/ surfactant concentration ratio and to the type and hydrophobicity of the two components [74–77]. Various C_nTAB/PAMPS (poly-acrylamido-methyl-propanesulfonate-sodium salt) and C_nTAB/PSS systems have been investigated and, in most cases, a minimum in stability has been observed close to the nominal isoelectric point (pI), see Figure 5.6. However, the exact shape of the foam stability curve is strongly affected by the used surfactant. In the case of $C_{14}TAB$ and a highly charged polyelectrolyte, the foam film stability is reduced towards pI and no stable foam films could be formed at the point of equal charges. Once this point is crossed, the stability of the films increases again with increasing polyelectrolyte concentration. On the other hand, in the case of $C_{12}TAB$, the foam film stability first increases upon polyelectrolyte addition and only a minimum stability is observed around pI. In the concentration range above pI, very stable foam films are observed. In general, no stable foam films can be obtained from pure $C_{12}TAB$ solutions, nevertheless, the addition of all types of negatively charged additives as hydrophobic and hydrophilic polyelectrolytes, with a long or short chain, always leads to the formation of stable foam films. Depending on the systems, the concentration regime in which the initial stabilization appears is different.

The general properties of foam films formed from polyelectrolyte/surfactant mixtures are very similar throughout all investigated systems. A reduction of foam film stability is detected slightly below the nominal pI of the system and very stable foam films were found in the concentration regime above the pI. However, the surface characterization of the air/water interface revealed that this phenomenon is not due to a charge reversal at the interface. The shape of the stability curve is qualitatively the same for different surface coverage and is dependent on the polyelectrolyte/surfactant concentration ratio. Below the pI, hydrophobic polyelectrolyte/surfactant complexes adsorb at the surface, but due to the low amount of unbound surfactant molecules in the system, the foam films are not very stable. Furthermore, above the pI most of the surface active complexes are released from the interface and only more or less a pure surfactant layer is left at the surface. Nevertheless, in this concentration regime, the most stable foam films are found. The qualitative influence of the polyelectrolyte hydrophobicity on foam film stability was only minor, at least in the case of 100% charged polyelectrolytes. The addition of more hydrophobic polyelectrolytes only affects the absolute film stability, but not the shape of the stability curve.

5.1.2.3 Polyelectrolyte/Nonionic Surfactant Systems

Experiments show that mixing a positively charged polyelectrolyte like PDADMAC with nonionic surfactants like $C_{12}G_2$ has only a minor effect on the surface tension. This indicates that no surface active complexes are formed in the mixture. The neat water/air interface is assumed to be negatively charged due to OH^- adsorption. The addition of a neutral surfactant leads to a reduction of the surface charge; however, the surface has a negative potential over a large concentration range which could enhance the adsorption of the polyelectrolyte [64].

Foam films of these mixtures show an NBF transition at rather low disjoining pressures (approx. 800 Pa, see Figure 5.7). The addition of a cationic polyelectrolyte, however, screens the negative charge of the film surfaces thus decreasing the electrostatic repulsion Π_{EL} and leading to the formation of an NBF. Compared to the NBFs obtained by the addition of salt or by the increase of surfactant concentration in the case of nonionic surfactants, these NBFs have a rather low stability. This could be due to the fact that the surface coverage is less dense or that the fluctuating polymer chains disrupt the ordering in the film, which leads to film rupture. In mixtures with an anionic polymer like PSS, no NBF transition can be observed before the film rupture at 6000 Pa. This is an indication for a stronger electrostatic repulsion within the film in the case of $C_{12}G_2$/PSS compared to $C_{12}G_2$/PDADMAC.

5.1.2.4 Stratification Phenomena

In the semidilute polyelectrolyte concentration regime, another interesting phenomenon is observed. Above C', a transient polyelectrolyte network is formed both in the bulk solution and in the film core which leads to a discontinuous film thinning [78]. Such a stepwise thinning of the foam film is called a stratification process (see Chapter 2) and occurs due to an oscillation of the disjoining pressure in the film [79–82]. The origin of these steps in film thinning is assumed to be that layer by layer of the polyelectrolyte network is pressed out of the film when the applied pressure is increased. Since it is not possible to go back to another branch of the oscillation of the disjoining pressure, this is an irreversible process. The steps in film thickness follow a power law that scales with $\Delta h \sim c^{-1/2}$ for linear polyelectrolytes and $\Delta h \sim c^{-1/3}$ for branched polymers, for example, poly(ethylene imine) (PEI) [83] (c – monomer concentration). For linear polymers, the step corresponds to the mesh size ξ of the polymer network that is formed in the bulk solution [58,84,85].

The degree of charge of the polyelectrolyte and the salt concentration strongly affect the stratification behavior. In the case of charged polyelectrolytes, the step size remains constant down to the Manning threshold [86]. In films with polyelectrolytes with a lower degree of charge, Δh

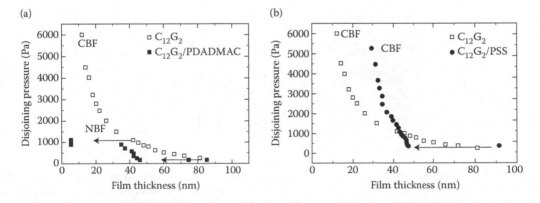

(a)   (b)

FIGURE 5.7 Π-h_w isotherm for foam films obtained from pure $C_{12}G_2$ and $C_{12}G_2$/PDADMAC (a) and pure $C_{12}G_2$ and $C_{12}G_2$/PSS (b) solutions with surfactant concentrations of 0.01 mM and polyelectrolyte concentrations of 3 mM. (Kristen N, von Klitzing R. Effect of polyelectrolyte/surfactant combinations on the stability of foam films. Soft Matter 2010; 6: 849. Reproduced by permission of The Royal Society of Chemistry [43].)

increases and all steps take place simultaneously at a very low disjoining pressure [79,87,88]. Nonionic polymers induce no stratification at all. Salt addition on the other hand reduces the disjoining pressure at which the stratification is induced. Above a threshold concentration, the steps in film thickness are completely suppressed.

The occurrence of the network of polymer chains has been widely discussed in literature. Many studies address the question of whether a real network is formed in confinement or if there rather is a layering of the polymers with a pronounced alignment parallel to the surface [84,89,90]. Experiments with fluorescent-labeled PAA (poly-acrylic acid) [91,92] conclude that there is indeed a layer wise arrangement of the polymer chains in which the distance between the layers corresponds to the random network of the same polymer in the respective bulk phase. The rigidity of the polymer backbone also plays an important role in the stratification process. The force oscillation of flexible polyelectrolytes can be easily observed as long as the velocity of the two approaching film surfaces is not too fast [90]. For rigid polymers, the stratification can only take place when the viscosity of the solution is large enough so that the network has time to equilibrate to the applied pressure. Otherwise no stepwise thinning can be observed.

The choice of the surfactant has no detectable influence on the structuring of the polyelectrolyte chains within the film core. This means that a certain polyelectrolyte concentration leads to a fixed period of force oscillation independent of the surfactant type [64,93]. Furthermore, the interaction between the polyelectrolytes and surfactant molecules can be neglected with respect to their effect on structural forces. For example, foam films containing the polycation PDADMAC in combination with either nonionic $C_{12}G_2$ or positively charged $C_{16}TAB$ readily show stratification [64]. The study concludes that the surfactant does not influence the step size, therefore, it does not influence the structuring of the polyelectrolytes within the film. Furthermore, the charge and the elasticity of the interfaces have no effect on the structural forces. The force oscillations in films of polyelectrolyte solutions were not only measured in foam films, but also in wetting films [94] and between two solid interfaces with atomic force microscopy (AFM) [89,95]. The period of the pressure oscillation remains constant in all cases.

The velocity of the stratification on the other hand is affected by the interactions between polyelectrolyte and surfactant. In the beginning of the stratification process, small black dots appear in the film. These small discontinuities in the film thickness spread over the whole film. The velocity of the domain growth, however, depends on certain boundary conditions of the surface, see Figure 5.8. In foam films with a polyelectrolyte that is linked to the surface, the growth of the domains is much slower than in the case of a depleted interface [97,98]. The driving force of this domain growth is the difference in film tension $\Delta\gamma_f$ between the thicker and the thinner film areas.

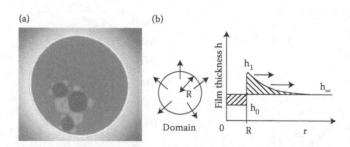

FIGURE 5.8 (a) Foam film during stratification process from the top. (Picture taken from Kristen N. PhD thesis: Interactions in thin liquid films: Oppositely charged polyelectrolyte/surfactant mixtures. 2010 Technical University, Berlin [76].) (b) Sketch of a growing domain of the radius R; the film thickness equals h_0 inside the domain and h_∞ at infinity; the film tension difference between the inside and the outside results in a rim with height h_1; material transport is marked by black arrows. (Reprinted with permission from Heinig P, Beltran C, Langevin D. Domain growth dynamics and local viscosity in stratifying foam films. *Phys Rev E* 2006; 73: 051607. Copyright 2006 American Chemical Society [96].)

The film tension in the inner part of the domain is smaller than that of the thicker film [96] and as the system favors the lower energy state of the thin film area, these domains are enlarged.

5.1.3 PROTEIN FOAM FILMS

Proteins, by their chemical structure, are zwitterionic polyelectrolytes which in aqueous solutions exhibit various structural conformations such as a random coil, for example, β-casein or a globule, for example, β-lactoglobulin (BLG), serum albumin, and so on. Proteins in aqueous solutions carry a certain electrical net charge which is determined by the solution pH as at the isoelectric point (pI), the molecules carry no/negligible net charge. Generally, proteins adopt tertiary structure (folding) which could be affected and changed when the molecules adsorb at interfaces (unfolding). Therefore, the behavior of proteins at interfaces and in thin liquid films appears to be an even more complicated issue than that of polymers. Proteins, in contrast to usual surfactants, adsorb virtually irreversibly at the water/air interface and produce viscoelastic layers [5,22,99,100], thus reducing the mobility of the surfaces of foam films. After formation, the films exhibit dimpling [22,23,25,29] and drain slowly (sometimes several hours) to equilibrium thickness [22].

It is interesting to note that pioneering experiments on protein foam films were performed more than a century ago by Joseph Plateau who classified a great variety of solutions, including albumin solution, according to their foamability and the persistence of single film caps on solution. In § 258 in his outstanding monograph *"Statique experimentale et theoryque des liquides soumis aux seules forces moleculaires"* published in 1873 [101], Plateau concluded that the persistence of film caps resting on solution is related to the surface tension and the *viscosity of the surface* (see also Chapter 2). Caps on albumin solutions lived sometimes hours and days as the drainage was quite slow.

5.1.3.1 Electrostatic Stabilization of Protein Foam Films

After the experiments of Plateau with large films and film caps on solution, Musselwhite and Kitchener [9] conducted the first systematic study on protein foam films using a measuring cell of a Scheludko type [1,12,16]. They investigated foam films from solutions of bovine serum albumin (BSA) and γ-globulin. The film thickness measurements have detected effects of pH and electrolyte concentration. The limiting real film thickness h of rupture of unstable black foam films from BSA at high C_{el} was found to be 3.5 nm. The authors posed some questions, such as the relation between the film thickness and the geometrical dimensions of the protein molecules in the bulk and at the surface of the solution.

BSA foam films have been further investigated by means of the Scheludko-Exerowa tube cell [1,12,16, see also Chapter 3]. Many reported results on foam films from protein solutions reveal that film thickness and stability strongly depend on both the protein concentration and the pH/electrolyte solvent conditions [9,10,22,23,28,39,41,47,50]. For the case of BSA, stable films with equivalent thickness h_w of 70–80 nm can be obtained at pH 5.8 (p$I \approx 4.7$) and a concentration as low as $C_{BSA} = 1$ nM [28]. Further increase in C_{BSA} leads to small variations (several nm) in film thickness. It is observed that within the approximate concentration range of 0.6–6 μM, the foam films can persist in two equilibrium states—as a thick film or as a black film with $h_w \approx 10$ nm. Adsorption of charged protein molecules gives rise to effective electrostatic stabilization Π_{EL} of the protein films. This is elegantly demonstrated by the dependence of the film stability on pH for films at protein concentrations lower than the respective critical concentration for the formation of stable black films (C_{BF}), but higher than the minimum concentration of black spot formation (C_{bl}) [28,30]. In the vicinity of pI (negligible Π_{EL}), the films rupture with the formation of black spots, while at pH $\neq$ pI, electrostatically stabilized thick films are formed (see Figure 5.9).

A similar trend in the h-pH dependence was observed for BLG foam films [30]. Figure 5.9 shows the h_w versus pH dependence for foam films obtained from buffered solutions of 10^{-5} M BLG for two buffer concentrations $C_{buff} = 1$ and 10 mM. On one hand, approaching the isoelectric point pI leads to a sizable reduction in the double layer potential $|\varphi_0|$ and, respectively, in the film thickness (slopes

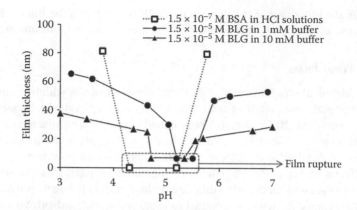

FIGURE 5.9 h_w vs pH dependence for foam films from 10^{-5} M β-lactoglobulin and 1.5×10^{-7} M BSA solutions. (Plots adapted from the data set reported in references Basheva ES, Kralchevsky PA, Christov NC, Danov KD, Stoyanov SD, Blijdenstein TBJ, Kim H-J, Pelan EG, Lips, A. *Langmuir* 2011; 27: 4481; Zawala J, Todorov R, Olszewska A, Exerowa D, Malysa K. *Adsorption* 2010; 16: 423.)

in Figure 5.9), because the adsorbing protein molecules carry much less or negligible electrical net charge [22,23]. On another hand, increasing C_{buff} (pH $\neq$ pI) leads to reduction of Π_{EL} in the films and, respectively, in film thickness (shift of the slopes in Figure 5.9), due to the screening effect of high ionic strength [9,22,23,30,39,42]. This fact is further confirmed by measurements of the disjoining pressure isotherm Π-h for foam films stabilized by different proteins [25,27,29,30,40,102]. Under the conditions of relatively low capillary pressures (<500 Pa) and high electrolyte concentrations or/and pH near the isoelectric point, the protein foam films practically do not reach homogenous thickness and persist in exhibiting random island-like patterns [26,29,30,103].

Figure 5.10 shows the Π-h curves for foam films from 10^{-4} M BLG solutions with pH 5 (nearby pI) and pH 7 (negative net charge) [30]. For pH 7, the films were studied at different ionic strengths. For the two lower buffer concentrations, the Π-h_w curves start from large thicknesses in analogy to the results in Figure 5.9. At the lowest $C_{buff} = 3$ mM, h_w gradually decreases with increasing Π due to the compression of the interacting diffuse double layers at film surfaces and the film ruptures at $h_w = 27$ nm (5 kPa), which thickness corresponds to those typical for the common black films in Figure 5.9. Under the conditions of low ionic strength and high $|\phi_0|$ potentials, the Π_{EL} barrier in the

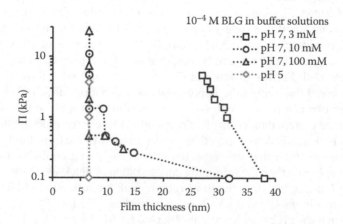

FIGURE 5.10 Π-h_w isotherm for foam films from 10^{-4} M β-lactoglobulin solutions with different pH and buffer concentrations, $C_{buff} = 3$, 10, and 100 mM, lines are guide to the eye. (Plots adapted from the data set reported in Gochev G et al. *Colloids Surfaces A* 2014; 460; 272–279 [30].)

disjoining pressure isotherm is high enough to prevent transition to a NBF and the CBF ruptures [29,30,102]. Such behavior is observed for foam films from low molecular weight surfactants as well [1, see also Chapter 3]. At $C_{buff} = 10$ mM, Π_{EL} is more strongly suppressed and h_w gradually decreases down to about 10 nm with increasing the pressure up to several hundred Pa. At around 1 kPa, transition to a lower thickness $h_w \approx 6.5$ nm occurs and this thickness is virtually independent of further pressure increase. The electrostatic origin of the stabilization of BLG foam films before the transition to the thinnest black film is most clearly demonstrated by the reduction of the Π_{EL}-barrier in the films. Further increase of C_{buff} up to 100 mM causes nearly a twofold reduction of the Π_{EL}-barrier compared to the case of films at $C_{buff} = 10$ mM.

5.1.3.2 Protein Black Foam Films

In Chapter 3, the formation, classification, and stability of black foam films (CBF and NBF) were discussed in detail in terms of interaction forces, that is, the film disjoining pressure isotherm Π-h. In this section, we discuss the properties and stability of protein black films. Formation of stable microscopic protein black films was reported for the first time by Yampolskaya et al. for BSA foam films [10]. Such equilibrium films are joined to the transition zone of the meniscus at comparatively low contact angles of 1°–2°, which indicates that these are CBFs [22,50]. At pH near the isoelectric point, the critical concentration for formation of a stable black film C_{BF} for BSA foam films was found to be 1.5 μM according to Reference 22 whereas an about one order of magnitude higher value is expected according to Reference 28. Various thicknesses were reported for BSA black films: 3.5–10.5 nm [9], 12–14 nm [22], 13.2–15.8 nm [23], 10 nm [28], 12–27 nm [40], and 8.6–17 nm [29]. The disjoining pressure isotherm for such CBFs shows a gradual decrease in the real film thickness down to $h = 11.5$ nm or 8.6 nm depending on pH [29]. In the latter work, the authors concluded that these thicknesses correspond, respectively, to three layer and bilayer film structures.

The formation of black film (spots) at protein concentrations below C_{BF} can be regarded as an instability issue [28,30], whereas at protein concentrations above C_{BF}, the black film formation plays a stabilizing role [28,30]. In Figure 5.9, the unstable BSA foam films rupture soon after the formation of black spots, while in the case of BLG, the black spots persist for a longer time (1–2 min, $h_w \approx 6.5$ nm) and can occupy most of the film area. This fact should be related rather to the higher protein concentration $C_{BLG} = 10^{-5}$ M used in comparison with that for BSA (1.5 × 10⁻⁷ M) than to the nature of the globular protein. For a higher BLG concentration of 10^{-4} M at pH 5 (p$I \approx 5.1$), the black foam film has virtually the same equivalent thickness but is much more stable (Figure 5.10). The same figure shows the experimental disjoining pressure isotherm Π-h_w for BLG foam films at various pH/electrolyte solvent conditions. For pH 5, black films with $h_w \approx 6.5$ nm are readily obtained at the lowest capillary pressure measured and further pressure increase does not lead to any change in the equivalent film thickness [30].

The observed independency of the thickness of the thinnest black films on the increasing disjoining pressure in Figure 5.10 suggests that such a protein black film can be regarded as a protein bilayer (protein NBF) where short range surface forces operate in accordance with the NBF from surfactants [1,12]. The formation of a self-assembled bilayer structure and estimation of its adhesive energy were reported for black foam films (∼6 nm) stabilized by the protein hydrophobin [26,27]. Accordingly, one could assume that the NBF from BLG ($h \approx 6.0$ nm, after correction of the equivalent film thickness $h_w \approx 6.5$ nm [30]) has a similar bilayer structure [30]. However, x-ray reflectivity measurements of vertical planar black films in a frame drown from BLG solutions at pH 5.2–5.3 and different protein concentrations [47–49] showed qualitatively and quantitatively rather different results. The thickness of the black films was found to vary with the BLG concentration and h-values of 11.3 nm at $C_{BLG} \approx 1.36 \times 10^{-5}$ M, 13.0 nm at $C_{BLG} \approx 2.72 \times 10^{-5}$ M, and 15.2 nm or even 20.0 nm at $C_{BLG} \approx 5.44 \times 10^{-5}$ M were reported. In all these cases, the analysis of the reflectivity data of the BLG black films has yielded a three slab film structure that consists of two identical outer layers (each ∼1.3 nm) and a film core with varying thickness that is obviously dependent of the protein concentration. The reasons behind the differences between the results obtained in Reference 30 and those reported in References 47–49 are not clear.

5.1.3.3 Influence of Surface Active Species on Protein Foam Films

Addition of surface active components to protein solutions can affect, sometimes remarkably, the properties and stability of the respective protein-alone foam films [24,31,41,49,104,105]. Foam films obtained from protein solutions where another protein [23,104] or a polymer [49] is added exhibit more complex behaviors than the protein-alone films which, however, depend on the mixing ratio. Films from solutions of hydrophobin and another protein, such as the random-coil β-Casein or the globular BLG and ovalbumin, or the nonionic low molecular weight surfactant Tween20 differ by their thickness and stability from the hydrophobin alone films [104]. The black films from the mixtures have different thicknesses depending on the type and size of the confined species between the film's surfaces covered with hydrophobin. However, the disjoining pressure isotherm for mixed films shows that final formation of a hydrophobin bilayer film occurs in all other cases except for that for β-Casein. The latter finding is explained with the strong incorporation of the β-Casein coils into the hydrophobin monolayers at film surfaces so that they cannot be expelled from the film. It was shown in a different study [40], that addition of Pluronics L62 and F68 to BSA solutions leads to reduction of the stability of the BSA alone foam films providing evidence of the antifoaming effect of these polymers over the protein system.

Formation of protein/surfactant complexes and their adsorption at film surfaces as well as replacement of such complexes and protein molecules from the water/air interface by surfactant molecules [105–108] strongly influence the thickness and stability of the foam films from mixed protein/surfactant solutions [36,38,39]. At high surfactant/protein ratios, replacement of the protein (or complexes) from the film surfaces by the surfactant molecules occurs as demonstrated for foam films from β-Casein/C_{12}DMPO [39,106], β-Casein/Tween20 [42], Lysozyme/C_{12}DMPO and Lysozyme/SDS [38], and BLG/Tween20 [31,105] mixed solutions. Recently, a new modification of the capillary cell of Scheludko and Exerowa was developed making possible exchanging the liquid in the meniscus which supports the film [24]. This allows applying experimental protocols with sequential introduction of new components in the system in order to gain deeper knowledge about the mechanisms of film stabilization. Such experiments with equilibrium BSA black films showed that subsequent exchange with protein free solution does not change film behavior, which reveals that protein does not (or negligibly) desorb from the film surfaces. On the contrary, after introduction of a solution (Tween20 or SDS) with enough high surfactant concentration in the system, the film behavior becomes identical to that of the surfactant alone films which confirms that the surfactant molecules do replace the protein molecules from the film surfaces.

REFERENCES

1. Exerowa, D., Kruglyakov, P.M. *Foam and Foam Films.* Elsevier; Amsterdam, 1998.
2. Fleer, J., Cohen-Stuart, M., Leermakers, F. Effect of Polymers on the Interaction between Colloidal Particles. In: Lyklema, J. (Ed.) *Fundamentals of Colloid and Interface Science V; Soft Colloids.* Elsevier; 2005. Ch. 1.
3. Stubenrauch, C., Rippner Blomqvist, B. Foam films, foams and surface rheology of non-ionic surfactants: Amphiphilic block copolymers compared with low molecular Weight Surfactants. In: Tadros, Th.F. (Ed.), *Colloid Stability: The Role of Surface Forces - Part I; Colloids and Interface Science Series*, Vol. 1. WILEY-VCH: Weinheim; 2007. Ch. 11.
4. Gochev, G. Thin liquid films stabilized by polymers and polymer/surfactant mixtures. *Curr Opin Colloid Interface Sci* 2015;20:115.
5. Miller, R., Liggieri, L. (Ed.) Interfacial rheology. *Progress in Colloid and Interface Science Series*, Vol. 1. Brill, Leiden, 2009.
6. Narsimhan, G. Characterization of interfacial rheology of protein-stabilized air–liquid interfaces. *Food Eng Rev* 2016; 8: 367.

7. Noskov, B.A., Bykov, A.G. Dilational surface rheology of polymer solutions. *Russian Chem Rev* 2015; 84: 634.
8. Mysels, K.J., Shinoda, C., Frankel, S. *Soap Films: Studies on their Thinning.* Pergamon; printed by The University Press, Glasgow, 1959.
9. Musselwhite, P.R., Kitchener, J.A. The limiting thickness of protein films. *J Colloid Interface Sci* 1967; 24: 80.
10. Yampolskaya, G.P., Rangelova, N.I., Bobrova, L.E., Platikanov, D., Izmailova, V.N. Obtaining black foam protein films. *Biophysics* 1977;22:975.
11. Lyklema, J., van Vliet, T. Polymer-stabilized free liquid films. *Faraday Discuss Chem Soc* 1978;65:25.
12. Scheludko, A. Thin liquid films. *Adv Colloid Interf Sci* 1967; 1: 391.
13. Sedev, R., Exerowa, D. DLVO and non-DLVO forces in foam films from amphiphilic block copolymers. *Adv Colloid Interface Sci* 1999; 83: 111.
14. Sedev, R. PEO-brush at the liquid/gas interface. *Colloids Surfaces A* 1999; 156: 65.
15. Rippner, B., Boschkova, K., Claesson, P.M., Arnebrant, T. Interfacial films of poly(ethylene oxide)-poly(butylene oxide) block copolymers characterized by disjoining pressure measurements, in situ ellipsometry, and surface tension measurements. *Langmuir* 2002; 18: 5213.
16. Exerowa, D., Platikanov, D. Thin liquid films from aqueous solutions of non-ionic polymeric surfactants. *Adv Colloid Interface Sci* 2009; 74: 147–148.
17. Millet, F., Benattar, J.J., Perrin, P. Vertical free-standing films of amphiphilic associating polyelectrolites. *Phys Rev E* 1999; 60: 2045.
18. Millet, F., Benattar, J.J., Perrin, P. Structures of free-standing vertical thin films of hydrophobically modified poly(sodium acrylate)s. *Macromolecules* 2001; 34: 7076.
19. Guenoun, P., Schalchli, A., Sentenac, D., Mays, J.W., Benattar, J.J. Free-standing black films of polymers: A model of charged brushes in interaction. *Phys Rev Lett* 1995; 74: 3628.
20. Gochev, G., Petkova, H., Kolarov, T., Khristov, K., Levecke, B., Tadros, T., Exerowa, D. Effect of the degree of grafting in hydrophobically modified inulin polymeric surfactants on the steric forces in foam and oil-in-water emulsion films. *Colloids Surfaces A* 2011; 391: 101.
21. Exerowa, D., Kolarov, T., Pigov, I., Levecke, B. Tadros ThF. *Langmuir* 2006; 22: 5013.
22. Yampolskaya, G.P., Platikanov, D. Proteins at fluid interfaces: Adsorption layers and thin liquid films. *Adv Colloid Interface Sci* 2006; 159: 128–130.
23. Clark, D.C., Coke, M., Mackie, A.R., Pinder, A.C., Wilson, D.R. Molecular diffusion and thickness measurements of protein-stabilized thin liquid films. *J Colloid Interface Sci* 1990; 138: 207.
24. Wierenga, P., Basheva, E.S., Denkov, N.D. Modified capillary cell for foam film studies allowing exchange of the film-forming liquid. *Langmuir* 2009; 25: 6035.
25. Wierenga, P., van Norél, L., Basheva, E.S. Reconsidering the importance of interfacial properties in foam stability. *Colloids Surfaces A* 2009; 344: 72.
26. Basheva, E.S., Kralchevsky, P.A., Christov, N.C., Danov, K.D., Stoyanov, S.D., Blijdenstein, T.BJ., Kim, H-J., Pelan, E.G., Lips, A. Unique properties of bubbles and foam films stabilized by HFBII hydrophobin. *Langmuir* 2011; 27: 4481.
27. Basheva, E.S., Kralchevsky, P.A., Danov, K.D., Stoyanov, S.D., Blijdenstein, T.B.J., Kim, H-J., Pelan, E.G., Lips, A. Self-assembled bilayers from the protein HFBII hydrophobin: Nature of the adhesion energy. *Langmuir* 2011; 27: 2382.
28. Zawala, J., Todorov, R., Olszewska, A., Exerowa, D., Malysa, K. Influence of pH of the BSA solutions on velocity of the rising bubbles and stability of the thin liquid films and foams. *Adsorption* 2010; 16: 423.
29. Cascão Pereira, L.G., Johansson, C., Radke, C.J., Blanch, H.W. Surface forces and drainage kinetics of protein-stabilized aqueous films. *Langmuir* 2003; 19: 7503.
30. Gochev, G., Retzlaff, I., Exerowa, D., Miller, R. Electrostatic stabilization of foam films from β-lactoglobulin solutions. *Colloids Surfaces A* 2014; 460; 272–279.
31. Coke, M., Wilde, P.J., Russell, E.J., Clark, D.C. The influence of surface composition and molecular diffusion on the stability of foams formed from protein/surfactant mixtures. *J Colloid Interface Sci* 1990; 138: 489.
32. Braunschweig, B., Schulze-Zachau, F., Nagel, E., Engelhardt, K., Stoyanov, S., Gochev, G., Khristov, K., Mileva, E., Exerowa, D., Miller, R. and Peukert, W. Specific effects of Ca^{2+} ions and molecular structure of ß-lactoglobulin interfacial layers that drive macroscopic foam stability. *Soft Matter* 2016; 12: 5995–6004.
33. Wilde, P.J., Nino, M.R.R., Clark, D.C., Rodríguez Patino, J.M. Molecular diffusion and drainage of thin liquid films stabilized by bovine serum albumin-tween 20 mixtures in aqueous solutions of ethanol and sucrose. *Langmuir* 1997; 13: 7151.

34. Husband, F., Wilde, P. The effects of caseinate submicelles and lecithin on the thin film drainage and behavior of commercial caseinate. *J Colloid Interface Sci* 1998; 205: 316.

35. Kharlov, A.E., Filatova, L.Yu., Zadymova, N.M., Yampolskaya, G.P. Black foam films stabilized with the mixtures of bovine serum albumin and nonionic surfactant tween 80. *Colloid Journal* 2007; 69: 117.

36. Angarska, Zh.K., Elenskyib, A.A., Yampolskayab, G.P., Tachev, K.D. Foam films from mixed solutions of bovine serum albumin and n-dodecyl-β-maltoside. *Colloids Surfaces A* 2011; 382: 102.

37. Gerasimova, A.Ts., Angarska, Zh.K., Tachev, K.D., Yampolskaya, G.P. Drainage and critical thickness of foam films from mixed solutions of bovine serum albumin and n-dodecyl-β-D-maltoside. *Colloids Surfaces A* 2013; 438: 4.

38. Alahverdjieva, V.S., Khristov, Khr., Exerowa, D., Miller, R. Correlation between adsorption isotherms, thin liquid films and foam properties of protein/surfactant mixtures: Lysozyme/C10DMPO and lysozyme/SDS. *Colloids Surfaces A* 2008; 323: 132.

39. Kotsmar, Cs., Arabadzhieva, D., Khristov, Khr., Mileva, E., Grigoriev, D.O., Miller, R., Exerowa, D. Adsorption layer and foam film properties of mixed solutions containing β-casein and C12DMPO. *Food Hydrocolloids* 2009; 23: 1169.

40. Sedev, R., Németh, Zs., Ivanova, R., Exerowa, D. Surface force measurement in foam films from mixtures of protein and polymeric surfactants. *Colloids Surfaces A* 1999; 149: 141.

41. Németh, Zs., Sedev, R., Ivanova, R., Kolarov, T., Exerowa, D. Thinning of microscopic foam films formed from a mixture of bovine serum albumin and Pluronic L62. *Colloids Surfaces A* 1999; 149: 179.

42. Maldonado-Valderrama, J., Langevin, D. On the difference between foams stabilized by surfactants and whole casein or β-casein. Comparison of foams, foam films, and liquid surfaces studies. *J Phys Chem B* 2008; 112: 3989.

43. Kristen, N., von Klitzing, R. Effect of polyelectrolyte/surfactant combinations on the stability of foam films. *Soft Matter* 2010; 6: 849.

44. Üzüm, K., Kristen, N., von Klitzing, R. Polyelectrolytes in thin liquid films. *Curr Opin Colloid Interface Sci* 2010; 15: 303.

45. Petkova, R., Tcholakova, S., Denkov, N.D. Foaming and foam stability for mixed polymer–surfactant solutions: Effects of surfactant type and polymer charge. *Langmuir* 2012; 28: 4996.

46. Fauser, H., Von Klitzing, R., Campbell, R.A. Surface adsorption of oppositely charged C14TAB-PAMPS mixtures at the air/water interface and the impact on foam film stability. *J Phys Chem* 2015; 119: 348.

47. Petkova, V., Sultanem, C., Nedyalkov, M., Benattar, J.J., Leser, M.E., Schmitt, C. Structure of a freestanding film of β-lactoglobulin. *Langmuir* 2003; 19: 6942.

48. Kolodziejczyk, E., Petkova, V., Benattar, J-J., Leser, M.E., Michel, M. Effect of fluorescent labeling of β-lactoglobulin on film and interfacial properties in relation to confocal fluorescence microscopy. *Colloids Surfaces A* 2006; 279: 159–166.

49. Liz, C.C.C., Petkova, V., Benattar, J.J., Michel, M., Leser, M.E., Miller, R. X-ray reflectivity studies of liquid films stabilized by mixed β-lactoglobulin-Acacia gum systems. *Colloids Surfaces A* 2006; 282–283: 109–117.

50. Nedyalkov, M., Sultanem, C., Benattar, J.J. Contact angles of protein black foam films under dynamic and equilibrium conditions. *Cent Eur J Chem* 2007; 5: 748.

51. Alexander, S. Adsorption of chain molecules with a polar head a scaling description. *J Phys France* 1977; 38: 983.

52. de Gennes, P.G. Conformations of polymers attached to an interface. *Macromolecules* 1980; 13: 1069.

53. de Gennes, P.G. Polymers at an interface; a simplified view. *Adv Colloid Interface Sci* 1987; 27: 189.

54. Sedev, R., Kolarov, T., Exerowa, D. Surface forces in foam films from ABA block copolymer: A dynamic method study. *Colloid Polymer Sci* 1995; 273: 906.

55. Krasowska, M., Hristova, E., Khristov, Khr, Malysa, M., Exerowa, D. Isoelectric state and stability of foam films, bubbles and foams from PEO–PPO–PEO triblock copolymer (P85). *Colloid Polym Sci* 2006; 284: 475.

56. Gotchev, G., Kolarov, T., Khristov, Khr, Exerowa, D. Electrostatic and steric interactions in oil-in-water emulsion films from Pluronic surfactants. *Adv Colloid Interface Sci* 2011; 168: 79.

57. Marinova, K., Alargova, R., Denkov, N., Velev, O., Petsev, D., Ivanov, I., Borwankar, R. Charging of oil-water interfaces due to spontaneous sdsorption of hydroxyl ions. *Langmuir* 1996; 12: 2045.

58. Stubenrauch, C., von Klitzing, R. Disjoining pressure in thin liquid foam and emulsion films—New concepts and perspectives. *J Phys: Condens Matter* 2003; 15: R1197.

59. Beattie, J.K., Djerdjev, A.M., Warr, G.G. The surface of neat water is basic. *Faraday Discuss* 2008; 141: 31.

60. Manev, E.D., Pugh, R.J. Diffuse layer electrostatic potential and stability of thin aqueous films containing a nonionic surfactant. *Langmuir* 1992; 7: 2253.

61. Tadros, T. Polymeric surfactants in disperse systems. *Adv Colloid Interface Sci* 2009; 281: 147–148.
62. Exerowa, D., Gotchev, G., Kolarov, T., Khristov, Khr., Levecke, B., Tadros, Th.F. Comparison of oil-in-water emulsion films produced using ABA or ABn copolymers. *Colloids Surfaces A* 2009; 335: 50.
63. Khristov, Khr., Petkova, H., Alexandrova, L., Nedyalkov, M., Platikanov, D., Exerowa, D., Beetge, J. Foam, emulsion and wetting films stabilized by polyoxyalkylated diethylenetriamine (DETA) polymeric surfactants. *Adv Colloid Interface Sci* 2011; 168: 105.
64. Kolaric, B., Jaeger, W., Hedicke, G., von Klitzing, R. Tuning of foam film thickness by different (poly) electrolyte/surfactant combinations. *J Phys Chem B* 2003; 107: 8152.
65. Ritacco, H., Kurlat, D., Langevin, D. Properties of aqueous solutions of polyelectrolytes and surfactants of opposite charge: Surface tension, surface rheology, and electrical birefringence studies. *J Phys Chem B* 2003; 107: 9146.
66. Bergeron, V., Claesson, P. Structural forces reflecting polyelectrolyte organization from bulk solutions and within surface complexes. *Adv Colloid Interface Sci* 2002; 96: 1.
67. Taylor, D.J.F., Thomas, R.K., Penfold, J. The adsorption of oppositely charged polyelectrolyte/surfactant mixtures: Neutron reflection from dodecyl trimethylammonium bromide and sodium poly(styrene sulfonate) at the air/water interface. *Langmuir* 2002; 18: 4748.
68. Asnacios, A., von Klitzing, R., Langevin, D. Mixed monolayers of polyelectrolytes and surfactants at the air-water interface. *Colloids Surfaces A* 2000; 167: 189.
69. Langevin, D. Polyelectrolyte and surfactant mixed solutions. Behavior at surfaces and in thin films. *Adv Colloid Interface Sci* 2001; 467: 89–90.
70. Asnacios, A., Langevin, D., Argillier, J.F. Complexation of cationic surfactant and anionic polymer at the sir/water interface. *Macromolecules* 1996; 29: 7412.
71. Asnacios, A., Langevin, D., Argillier, J.F. Mixed monolayers of cationic surfactants and anionic polymers at the air-water interface: Surface tension and ellipsometry studies. *Eur Phys J B* 1998; 5: 905.
72. Stubenrauch, C., Albouy, P.A., von Klitzing, R., Langevin, D. Polymer/surfactant complexes at the water/air interface: A surface tension and X-ray reflectivity study. *Langmuir* 2000; 16: 3206.
73. Bhattacharyya, A., Monroy, F., Langevin, D., Argillier, J.F. Surface rheology and foam stability of mixed surfactant/polyelectrolyte solutions. *Langmuir* 2000; 16: 8727.
74. Kristen, N., Simulescu, V., Vüllings, A., Laschewsky, A., Miller, R., von Klitzing, R. No charge reversal at foam films surfaces after addition of oppositely charged polyelectrolytes?. *J Phys Chem B* 2009; 133: 7986.
75. Kristen-Hochrein, N.., Andre, L.., Reinhard, M.., Regine von, K.. Stability of foam films of oppositely charged polyelectrolyte/surfactant mixtures: Effect of isoelectric point. *J Phys Chem B* 2011; 115: 14475.
76. Kristen, N. PhD thesis: Interactions in thin liquid films: Oppositely charged polyelectrolyte/surfactant mixtures. 2010 Technical University, Berlin.
77. Kristen, N., Vüllings, A., Laschewsky, A., Miller, R., Von Klitzing, R. Films from oppositely charged polyelectrolytes/surfactant mixtures: Effect of Surfactant and polyelectrolyte hydrophobicity. *Langmuir* 2010; 26: 9321.
78. Bergeron, V., Langevin, D., Asnacios, A. Thin-film forces in foam films containing anionic polyelectrolyte and charged surfactants. *Langmuir* 1996; 12: 1550.
79. Asnacios, A., Espert, A., Colin, A., Langevin, D. Structural forces in thin films made from polyelectrolyte solutions. *Phys Rev Lett* 1997; 78: 4974.
80. von Klitzing, R., Espert, A., Asnacios, A., Hellweg, T., Colin, A., Langevin, D. Forces in foam films containing polyelectrolyte and surfactant. *Colloids Surfaces A* 1999; 149: 131.
81. von Klitzing, R., Espert, A., Colin, A., Langevin, D. Comparison of different polymer-like structures in the confined geometry of foam films. *Colloids Surfaces A* 2001; 176: 109.
82. Toca-Herrera, J.L., von Klitzing, R. Fluorescence spectroscopy on polyelectrolyte free standing films. *Macromolecules* 2002; 35: 2861.
83. von Klitzing, R., Kolaric, B. Influence of polycation architecture on the oscillation pressure in liquid free-standing films. *Progr Colloid Polym Sci* 2003; 122: 122.
84. Theodoly, O., Tan, J.S., Ober, R., Williams, C.E., Bergeron, V. Oscillatory forces from polyelectrolyte solutions confined in thin liquid films. *Langmuir* 2001; 17: 4910.
85. Qu, D., Pedersen, J.S., Garnier, S., Laschewsky, A., Möhwald, H., von Klitzing, R. Effect of polymer charge and geometrical confinement on ion distribution and the structuring in semidilute polyelectrolyte solutions: Comparison between AFM and SAXS. *Macromolecules* 2006; 39: 7364.
86. Qu, D., Brotons, G., Bosio, V., Fery, A., Salditt, T., Langevin, D., von Klitzing, R. Interactions across liquid thin films. *Coloids Surfaces A* 2007; 303: 97.

87. Kolaric, B., Jaeger, W., von Klitzing, R. Mesoscopic ordering of polyelectrolyte chains in foam films: Role of electrostatic forces. *J Phys Chem B* 2000; 104: 5096.

88. von Klitzing, R., Kolaric, B., Jaeger, W., Brandt, A. Structuring of poly(DADMAC) chains in aqueous media: A comparison between bulk and free-standing film measurements. *Phys Chem Chem Phys* 2002; 4: 1907.

89. Milling, A., Kendall, K. Depletion, adsorption, and structuring of sodium poly(acrylate) at the water/silica interface. 1. An atomic force microscopy force study. *Langmuir* 2000; 16: 5106.

90. Kleinschmidt, F., Stubenrauch, C., Deacotte, J., von Klitzing, R., Langevin, D. Stratification of foam films containing polyelectrolytes. Influence of the polymer backbone's rigidity. *J Phys Chem B* 2009; 113: 3972.

91. Anghel, D.F., Toca-Herrera, J.L., Winnik, F.M., Rettig, W., von Klitzing, R. Steady-state fluorescence investigation of pyrene-labeled poly(acrylic acid)s in aqueous solution and in the presence of sodium dodecyl sulphate. *Langmuir* 2002; 18: 5600.

92. Rapoport, D.H., Anghel, D.F., Hedicke, G., Möhwald, H., von Klitzing, R. Spatial distribution of polyelectrolytes in thin free-standing aqueous films resolved with fluorescence spectroscopy. *J Phys Chem C* 2007; 111: 5726.

93. von Klitzing, R., Espert, A., Colin, A., Langevin, D. Structure of foam films containing additionally polyelectrolytes. In Sadoc, J., Rivier, N. (Eds.) Foams, emulsions and cellular materials. *Kluwer Academic Publishers*, Dordrecht, 1999; 73.

94. Letocart, P., Radoev, B., Schulze, H., Tsekov, R. Experiments on surface waves in thin wetting films. *Colloids Surfaces A* 1999; 149: 151.

95. Qu, D., Baigl, D., Williams, C., Möhwald, H., Fery, A. Dependence of structural forces in polyelectrolyte solutions on charge density: A combined AFM/SAXS study. *Macromolecules* 2003; 36: 6878.

96. Heinig, P., Beltran, C., Langevin, D. Domain growth dynamics and local viscosity in stratifying foam films. *Phys Rev E* 2006; 73: 051607.

97. Beltran, C., Guillot, S., Langevin, D. Stratification phenomena in thin liquid films containing polyelectrolytes and stabilized by ionic surfactants. *Macromolecules* 2003; 36: 8506.

98. Beltran, C., Langevin, D. Stratification kinetics of polyelectrolyte solutions confined in thin films. *Phys Rev Lett* 2005; 94: 217803.

99. Fainerman, V.B., Miller, R. Equilibrium and dynamic characteristics of protein adsorption layers at gas-liquid interfaces: Theoretical and experimental data. *Colloid Journal* 2005;67:393.

100. Benjamins, J., Lucassen-Reynders, E.H. Interfacial rheology of adsorbed protein layers. In: Miller, R., Liggieri, L. (Eds.), *Interfacial Rheology*. Progress in Colloid and Interface Science Series Vol. 1. Brill, 2009. Ch 7.

101. Plateau, J.A.F. Statique experimentale et theoryque des liquides soumis aux seules forces moleculaires. Gautier-Villars, Trubner et cie, F. Clemm, 2 Vols; 1873.

102. Dimitrova, T.D., Leal-Calderon, F., Gurkov, T.D., Campbell, B. Disjoining pressure vs thickness isotherms of thin emulsion films stabilized by proteins. *Langmuir* 2001; 17: 8069.

103. Rullier, B., Axelos, M.A.V., Langevin, D., Novales, B. β-Lactoglobulin aggregates in foam films: Effect of the concentration and size of the protein aggregates. *J Colloid Interface Sci* 2010; 343: 330.

104. Danov, K.D., Kralchevski, P.A., Rdulova, G.M., Basheva, E.S., Stoyanov, S.D., Pelan, E.G. Shear rheology of mixed protein adsorption layers vs their structure studied by surface force measurements. *Adv Colloid Interface Sci* 2015; 222: 148–161.

105. Wilde, P., Mackie, A., Husband, F., Gunning, P., Morris, V. Proteins and emulsifiers at liquid interfaces. *Adv Colloid Interface Sci* 2004; 63: 108–109.

106. Mileva, E. Impact of adsorption layers on thin liquid films. *Curr Opin Colloid Interface Sci* 2010; 15: 315.

107. Dan, A., Gochev, G., Krägel, J., Aksenenko, E.V., Fainerman, V.B., Miller, R. Interfacial rheology of mixed layers of food proteins and surfactants. *Curr Opin Colloid Interface Sci* 2013; 18: 302.

108. Dan, A., Gochev, G., Miller, R. Tensiometry and dilational rheology of mixed b-lactoglobulin/ionic surfactant adsorption layers at water/air and water/hexane interfaces. *J Colloid Interface Sci* 2015; 449: 383.

6 Biomedical Foam Films

Dotchi Exerowa and Roumen Todorov

CONTENTS

Theoretical and experimental results discussed in the previous sections of this chapter indicate that foam films are an excellent model with which to study molecular interactions between two surfaces. Foam films are an element of the dispersive system of foam, defining foam properties. The latter was dealt with in Chapter 12. Particular attention is paid to foam film applications for biomedical purposes and this chapter will be dedicated to that. As will be shown, the foam film is a tool to study the surface structure and stability of lung alveoli.

Surfactants present at the alveolar surface are referred to as *lung* or *pulmonary surfactants* in literature, have been studied extensively, and continue to draw active interest. Figure 6.1a shows a schematic of an alveolar surface covered by a phospholipid monolayer in contact with the hypophase, as stipulated by the widely adopted hypothesis of alveolar surface structure [1].

Figure 6.2 depicts a diagram of pulmonary surfactant (PS) composition [2], the main components being phosphatidylcholine (PC), phosphatidylglycerol (PG) and surfactant associated proteins (SP) SP-A, SP-B, SP-C and SP-D. Almost half the content of surfactant PC is composed of dipalmitoylphosphatidylcholine (DPPC), which is the main surface-active species in PS.

Issues associated with pulmonary surfactants refer to surfactant insufficiency in preterm neonates developing a respiratory distress syndrome (RDS) that gives rise to physiological pathologies and even death [3].

A special method was introduced to study alveolar surface and stability, and a new diagnostic method was proposed for fetal lung maturity assessment [4,5]. The model used by the method is the black foam film model, including bilayer films. Furthermore, the notion of alveolar surface structure was expanded by suggesting a "monolayer-bilayer" structure [6].

Compared to the largely employed "monolayer" model, the black foam film model accounts not only for lateral interactions between first neighbor molecules in the adsorption layer, but also normal interactions in a bidimensional ordered system such as the bilayer black film, that is, the Newton black film (NBF) (see Chapter 3).

In view of the purpose of black foam films from pulmonary surfactants, it is widely accepted to name them biomedical foam films. RDS treatment with animal-derived lung surfactant extracts, namely therapeutic pulmonary surfactants or therapeutic surfactant preparations (TSP), is largely applied [2]. Such preparations have been proven beneficial and medications such as Curosurf, Survanta or Alveofact are widely used in preterm infants. Some new synthetic products have recently been introduced although their usefulness has yet to be clinically confirmed.

Results obtained about biomedical foam films by means of black foam films will be presented herein with a focus on the alveolar surface structure and stability.

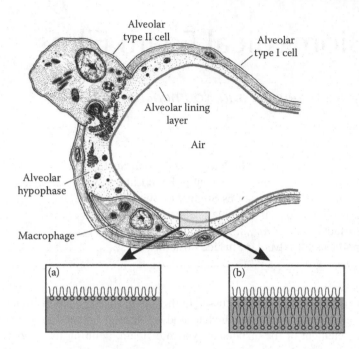

FIGURE 6.1 Schematic presentation of lung alveolus and alveolar lining layer: (a) monolayer model of pulmonary surfactant; (b) recently proposed three-dimensional model of surface-associated structures. (Adapted from Hawgood, S., Clements, J.A. *J Clin Invest* 1990; 86: 1[1].)

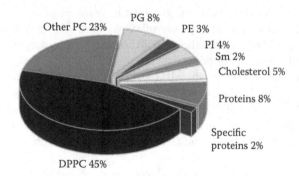

FIGURE 6.2 Pulmonary surfactant composition: DPPC—dipalmitoylphosphatidylcholine; PC—phosphatidylcholines; PG—phosphatidylglycerol; PI—phosphatidylinositol; PE—phosphatidylethanolamine; Sm—sphingomyelin.

6.1 BLACK FOAM FILMS AS A TOOL TO STUDYING THE ALVEOLAR SURFACE

This section will consider black foam films as a model to study alveolar stability and structure. A major advantage is the use of microscopic foam films studied under the conditions of lung alveolus, namely at a film radius of ~100 µm and capillary pressure between 0.03 and 3.0 kPa. This pressure range is close to the one in the alveolus during inhaling and exhaling.

Microscopic foam films have been studied by means of the microinterferometric technique (see Chapter 3) that allows introducing new parameters and measuring new dependences characterizing alveolar surface structure and stability. TSP black foam film dependences to be considered are as follows: film thickness h versus electrolyte concentration; probability W for black film formation versus pulmonary surfactant concentration; disjoining pressure versus film thickness $\Pi(h)$ (see Chapter 3).

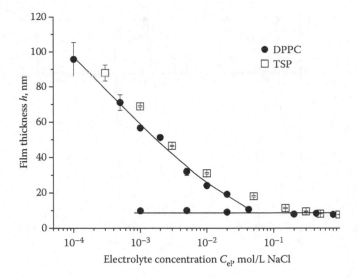

FIGURE 6.3 Film thickness of DPPC and TSP microscopic foam films as a function of electrolyte (NaCl) concentration.

Figure 6.3 depicts the dependence of film thickness h versus electrolyte concentration (C_{el}; NaCl) for DPPC and TSP. Film thickness decreases with increasing C_{el} as per DLVO theory (see Section 3.2). At $C_{el} = 10^{-1}$ mol/L, film thickness no longer changes and films become Newton films (8 nm) [7]. Therefore, at physiological electrolyte concentration (NaCl) of 0.15 mol/L, DPPC films are Newton black films (NBF).

The probability W for black foam film formation is proven to be a very suitable parameter to describe alveolar stability. The relation W versus PS concentration is given by Equation 3.47, Chapter 3. The probability W is determined after the entire film is taken up by black spots that emerge in a jump-like manner during the critical stability status of the film. The black film lifetime varies from very unstable to infinitely stable black films (see Equation 3.46, Chapter 3). Figure 6.4 shows a principle $W(C)$ dependence. As is seen, the dependence is steep, thus making it possible to introduce two new characteristics of black film stability: minimum concentration C_c of black film formation and concentration C_t of a 100% black film formation.

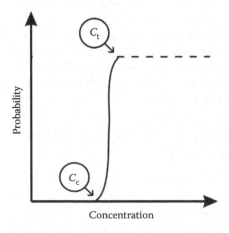

FIGURE 6.4 Probability for black film formation versus surfactant concentration $W(C)$. C_c—minimum concentration required for black film formation, C_t—concentration rendering 100% black films.

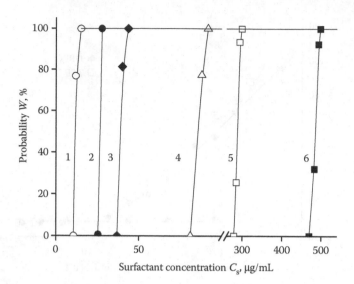

FIGURE 6.5 Dependence of probability W for black foam film formation on surfactant concentration: (1) DPPC in amniotic fluid; (2) PG; (3) Egg PC; (4) Lecithin; (5) PI; (6) Sphingomyelin (Sm).

Figure 6.5 presents $W(C)$ dependences for various phospholipids [8–10]. As is seen, the curves are very steep, and both C_c and C_t may be determined very precisely.

It is important to note that critical concentrations are very sensitive to the nature of the phospholipids and their structure, to the composition of the film-forming solution (presence of ions and surface-active agents), and to the temperature and phase state of the phospholipids [9–11]. At higher temperature, the $W(C)$ curve shifts to the right, that is, C_c and C_t are higher. A similar $W(C)$ curve shift is also observed for various lipid phases of phospholipid within the solution bulk [12]. These results indicate the method's capabilities regarding the study of the layer covering the alveolar air-water interface and are associated with the role played by the structure of PS components constituting such a layer.

Another important characteristic is the disjoining pressure Π, considered in Chapter 3. As is seen, Π is determined by molecular interactions and defines the thermodynamic state of the film. $\Pi(h)$ dependences, that is, disjoining pressure isotherms, may be measured directly with microinterferometric equipment, the so called pressure balance technique.

Figure 6.6 depicts $\Pi(h)$ isotherms of DPPC solutions in the presence of NaCl [7,10]. The symbols indicate experimentally obtained values while solid lines are calculated with the DLVO theory at constant charge σ and constant potential φ, respectively. As is seen, the experimental points fit the curve calculated at constant charge well and are in good agreement with the DLVO theory at $\Pi = \Pi_{el} + \Pi_{vw}$.

It is worth noting that the electrolyte concentration is relatively low in this case. At higher electrolyte concentrations, both films from the phospholipid fraction of PS and from PS show a deviation from the DLVO theory and a bilayer film is obtained [7,13]. Further $\Pi(h)$ isotherms for various TSP will be presented in Section 6.3.

6.2 BLACK FOAM FILM METHOD FOR ASSESSMENT OF FETAL LUNG MATURITY

As mentioned earlier, surfactant insufficiency in preterm neonates leads to respiratory distress syndrome (RDS), often resulting in death. For that reason, it is expedient to devise a method for the assessment of fetal lung maturity [5,14,15]. The black foam film method (BFF) has been proven suitable for this purpose, offering a number of advantages compared to other methods. Some rather distinguishable features of the method proposed refer to its high preciseness (90%), rapidness (about 20 min), and small quantities of amniotic fluid samples required [5,8,10,14].

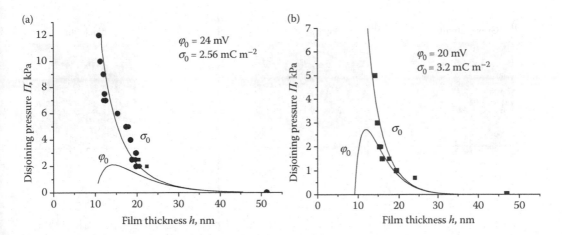

FIGURE 6.6 $\Pi(h)$ isotherm of films formed from DPPC solutions of ionic strength 2×10^{-3} mol/L (a) and 5×10^{-3} mol/L (b); symbols represent experimental data; lines—theoretical calculations at the limiting conditions of a constant potential (φ) and respective constant charge density (σ).

Considering the notions in previous sections, the black foam film is a tool to study alveolar structure and stability. Figure 6.5 (curve 1) presents the probability for black foam film formation depending on the DPPC concentration, the latter contained in amniotic fluid (AF). This phospholipid holds the highest AF percentage content and defines AF surface properties. As indicated by the figure, the curve is very steep and may be characterized by an initial concentration C_c of black film formation, and C_t, the concentration of 100% black film formation. C_c describes the condition at which a film is/ is not formed, and as such is a very sensitive parameter. C_t is also a sensitive parameter for a stable black film. Thus, both C_c and C_t are easily measurable parameters that may be used to characterize black film stability. Surfactant insufficiency, that is, immature lung, has been studied in view of a lack of black foam film formation. Amniotic fluid is used to establish a threshold dilution versus gestation weeks. The results obtained have been compared with clinical data so as to set "the threshold dilution that distinguishes mature from immature samples" [5,8,10,14]. Clinical application involves an AF sample that is diluted and then compared to "the threshold dilution" in order to establish sample maturity, that is, surfactant sufficiency/insufficiency. Thus, a rapid method to assess fetal lung maturity and to detect RDS was established in Reference 5. A schematic of the black foam film method for assessment of fetal lung maturity is presented in Figure 6.7. The threshold concentration distinguishes between surfactant insufficiency/surfactant sufficiency, that is, lung immaturity/lung maturity. Photos of the corresponding microscopic foam films are also given in this figure.

Timely RDS diagnosis allows for applying an adequate treatment of surfactant insufficiency in early gestation weeks for reducing the risk of preterm infants with RDS. The method proposed has been further expanded to include blood and meconium contaminated AF samples, as well as postnatal RDS diagnostics [10,14].

6.3 THERAPEUTIC SURFACTANT PREPARATIONS STUDIED BY BLACK FOAM FILMS

Thin liquid film studies, in particular foam films stabilized by lipids (phospholipids) and PS components, have made it possible to apply the black foam film method to another application area [11,14], that of assessing surface characteristics of exogenous surfactant preparations used in clinical practice for RDS treatment [15–19]. Commercially available therapeutic surfactant preparations (TSP) may be of synthetic or natural origin. Synthetic TSPs contain DPPC, the principal PS lipid component, and some other surfactants. A TSP of natural origin contains a hydrophobic fraction of natural PS obtained from animals—either from lung lavage or lung tissue [2]. Clinical data and model study results of TSP

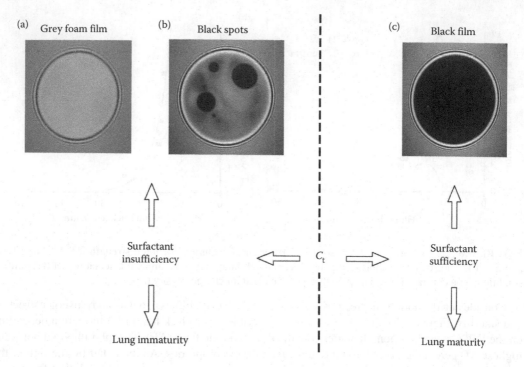

FIGURE 6.7 A schema of the black foam film method for assessment of fetal lung maturity; (a) unstable thin liquid film; (b) black spots in thin liquid film; (c) black foam film.

surface properties have proven the advantages of natural to synthetic preparations. It is believed that the hydrophobic surfactant proteins SP-B and SP-C, specific for PS, could account for the beneficial therapeutic effect observed. Currently, only preparations of natural origin are used for RDS treatment. Clinical trials of a new generation of synthetic TSPs is under way, with such preparations containing synthetic or recombinant analogues of SP-B and SP-C, along with DPPC and PG [20–22]. While improving existing preparations as well as introducing some new ones, the issue of establishing an adequate *in vitro* method to study preparation interfacial properties still remains open [15,23,24]. The first foam film method implementations to TSPs date back to their initial clinical application worldwide [16]. The obtained results indicate the method's applicability to such types of studies. As mentioned above, black foam film forms at the air/water interface, which is analogous to the interface at the alveolus and wherein such preparations are expected to realize the therapeutic effect designated. Additionally, films are formed and studied under conditions that are the closest to physiological conditions [16,19]. An important advantage to the BFF method is the possibility to account for the surface forces that define the molecular interactions both laterally and normally. The types and action range of surface forces are of crucial importance to the structure and stability of alveolar lining layers [7,13,19,25].

An update of the Scheludko-Exerowa microinterferometric method designed to form and study thin liquid films allows visually monitoring and documenting the BFF formation process. Film surface morphology can be registered directly along with the presence of molecular aggregates, structures, and so on. Video files can then be subjected to video processing in order to obtain data about: film drainage rate, time for black spot formation, black spot expansion rate, and so on. [16,26]. Photographs showing the formation of black foam film via black spots in a thicker foam film, taken under a microscope, are presented in Figure 3.5 of Chapter 3. The photos in series (a–d) show a smooth, regular drainage of the thicker film (a), formation of black spots (b), and black film (c). On the contrary, the photos in series (1–4) show irregular film drainage (1) and impeded formation of black spots (2,3) and black film (4) due to complicated rheological properties of the aqueous surfactant solution used. Depending on the TSP type, it is possible to observe films with a regular drainage, such as films from Curosurf and Infasurf or films with a pronounced irregular drainage, such as films from Survanta [16,17,19,26]. Survanta films drain slowly with dimples, and black spots are irregular in shape and enlarge very slowly. Even after 120 min, they fail to encompass the entire film area while being surrounded by dimple zones. The nonhomogenous drainage of Survanta films can be explained by the presence of rigid nanostructures formed by PS components and DPPC, tripalmitoylglycerol, and phosphatic acid added to the preparation [27]. Visual observation of the BFF formation process provides information about bulk micro- and nanostructure stability depending on TSP composition and the role of these substances in adsorption layer formation at the solution/air interface [18]. Modern perceptions stipulate that apart from participating in the formation of a dense layer covering alveolar interfaces, such structures form during the breathing process [24,25]. We may conclude that the more homogenous the film drainage is, the more unstable such structures are, thus making their disintegration faster in both the bulk and the alveolar surface during the process of breathing.

As already pointed out above, studies of the probability W for black foam film formation versus concentration make it possible to determine C_t and C_c parameters associated with film stability. $W(C)$ curves of TSP foam films are steep, just like those for phospholipids and PS [16,17,26]. In contrast to that, some preparations show a time-dependence regarding film formation, that is, a certain waiting time. Increasing time renders a shift of $W(C)$ curves towards smaller TSP concentrations, respectively, towards smaller C_t and C_c values. For clinically applied preparations, such a trend is observed for Curosurf and Infasurf. As for Survanta and Alveofact, such influence is not detected. Any increased waiting time has no impact on C_c for both preparations while for Alveofact, C_t decreases, as observed for Curosurf and Infasurf. The effect of waiting time on the critical concentration is probably due to the increase in adsorption layer density with increasing time by means of either an additional adsorption of surfactant molecules or slow spreading of some molecular nanostructures at the interface. The effect of waiting time on C_c of TSP is important for clinical practice because it allows estimating the treatment dose, and the time to register the therapeutic effect of the preparation.

The impact of waiting time prior to film formation is of importance for $W(C)$ and has to be kept in mind when different exogenous surfactants are compared with each other. Figure 6.8 depicts $W(C)$ dependences for five clinically applied TSPs at 30 min waiting time and facilitates preparation assessment regarding the capability of black film formation at one and the same conditions.

It is seen that for two of the preparations, Curosurf and Infasurf, the $W(C)$ curves are positioned at much lower concentrations than the others. C_c values for Survanta and Alveofact are significantly

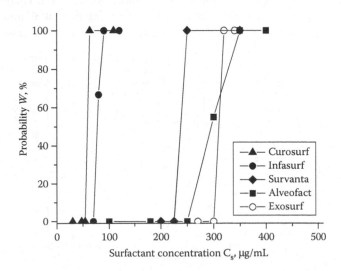

FIGURE 6.8 Dependence of probability W for black foam film formation on surfactant concentration for five therapeutic surfactants.

higher while the curve for Exosurf (a synthetic protein-free preparation) is shifted toward even higher concentrations. Lower C_c values for animal-derived TSPs as compared to that of Exosurf indicate that the role played by the other PS components contributes to better interfacial properties exhibited by surfactants of natural origin. In view of the probability for BFF formation depending on the TSP concentration, we may arrange the studied preparations in the following sequence: Curosurf > Infasurf > Survanta > Alveofact > Exosurf. This order gives a clear indication that with respect to BFF formation, preparation type and composition play an important role. It is well known that natural TSPs have a different composition regarding both the main DPPC and the other PS components: charged phospholipids, neutral lipids, surfactant proteins SP-B and SP-C, and so on. [2]. These differences are attributed not only to the source material either being lung lavage (Infasurf, Alveofact) or tissue (Curosurf, Survanta), but also to the extraction and purification procedures employed. The different phospholipid content can result in a significant difference of adsorption and hence different behavior of the surface layer formed. Neutral lipids (mainly cholesterol) and other additives are likely to play an important role in altering the lateral molecular interactions as well as the surface rheological properties of different preparations [2,11,16,18,28,29]. The BFF method provides options to assess the effect exerted by TSP components on the preparation's surface properties. Furthermore, the ability of bulk structures (incorporating the PS components) to disintegrate when subjected to film capillary pressure is also accounted for. The stability of such structure also depends on the ability of lipid molecules to self-assemble [11,29]. It has been shown that phospholipid lamellar structures form BFF at much lower bulk concentrations as compared to inverted hexagonal and bicontinuous cubic nonlamellar liquid crystalline phases [11,12]. On the other hand, a surface-associated surfactant reservoir is formed at the alveolar surface during exhaling and lipid molecules contained in it may arrange in diverse self-assemblies. The therapeutic effect of preparations depends directly on the ability of molecules in such structures to reacquire their position at the air/water interface during inhaling.

Usually, surfactant adsorption at the solution/air interface takes place during an experimentally controlled time (t) which may be insufficient for the concentration of adsorbed surfactant to reach saturation. For that reason, the adsorption time may have a considerable influence on the probability W for BFF observation and, accordingly, on the critical surfactant concentration C_c. A comparison of C_c values in $W(C)$ dependences with kinetic surface tension curves $\gamma(t)$ for Curosurf are presented in Figure 6.9. C_c decreases with increasing time from 108 µg/mL at a waiting time of 0 min, to 30 µg/mL after 60 min.

Changes in the $C_c(t)$ dependence become less pronounced after 30 min. Therefore, when C_c is the parameter selected for practical purposes, this particular waiting time of 30 min is the suitable one. Data for $\gamma(t)$ presented in Figure 6.9 at Curosurf concentrations within the range of the given $W(C)$

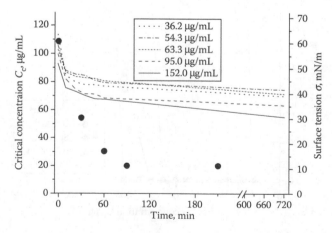

FIGURE 6.9 Dependences: (1) of critical surfactant concentration for black foam film formation on waiting time (black circles); (2) of solution's surface tension on time (lines).

dependences are obtained from measurements using PAT (drop shape profile analysis tensiometry) method. The kinetic curves follow the usual course of γ as a function of time. An equilibrium value γ_{eq} is reached after a certain time which increases with decreasing Curosurf concentration. Within the concentration range studied, the γ_{eq} values do not differ substantially, which gives us grounds to conclude that they are less sensitive than C_c.

So far, however, the effect of t on W and C_c has not been considered theoretically. This effect has been explored and comparison of the obtained expression for $C_c(t)$ with corresponding experimental data for foam films obtained from aqueous solutions of the therapeutic surfactant Infasurf has been reported in Reference 30. A good fit between the theoretical dependence and experimental data has been established. This allows carrying out a quantitative analysis of the $W(C_c, t)$ dependence and establishing the effect of t on C_c below which NBF in foam film practically cannot be observed. Thus, C_c proves to be a precise parameter characterizing bilayer formation and stability.

BFF as a model system opens new horizons to get an insight of molecular interaction forces, operating in TSP foam films and to establish a range of long- and short-range molecular interaction forces and the transitions between them. Molecular interaction forces in foam films have been assessed by means of direct measurements of film thickness versus electrolyte concentration and disjoining pressure/film thickness isotherms ($\Pi(h)$) [7,13]. The dependences of equilibrium foam film thickness on electrolyte concentration $h(C_{el})$ for films from Curosurf, Infasurf, and Alveofact have been obtained [31]. Due to the "rheologic" behavior of films, the thickness of films from Survanta has not been measured. For the TSPs studied, the results follow the general trend of a monotonous decrease in h with increasing C_{el} (Figure 6.3). In the case of Curosurf, the black foam film formation starts at $C_{el} = 0.05$ mol/L NaCl. Further increase in C_{el} leads to a decrease in h down to 7.8 nm at 1 mol/L NaCl. At a physiologically relevant electrolyte concentration, the black film thickness is 12 nm. The curves for Infasurf and Alveofact are similar but are located at larger thicknesses. Films from Infasurf reach $h = 11$ nm at $C_{el} = 0.5$ mol/L NaCl and the film thickness remains constant with further increase in C_{el}. The thickness of films obtained at physiological C_{el} (0.15 mol/L NaCl) amounts to 14 nm. At all C_{el} values, Alveofact films remain thicker than the corresponding Curosurf films. At physiological electrolyte concentration, Alveofact films are thicker ($h = 17.2$ nm) than Infasurf and Curosurf films. There is not a sharp transition observed in either of the $h(C_{el})$ for the three preparations, in contrast to DPPC films (Figure 6.3). This demonstrates the action of non-DLVO surface forces on stabilization of TSP films. In the case of Alveofact, it can be assumed that the acting forces are of steric origin due to the higher amount of surfactant proteins.

At physiological electrolyte concentration (0.15 mol/L NaCl), TSP films are common black but do not reach the DPPC bilayer film thickness even when pressure is applied (in the range of breathing pressure in the alveoli). The experimentally obtained $\Pi(h)$ isotherms for foam films from Curosurf and Infasurf at physiological $C_{el} = 0.15$ mol/L NaCl are shown in Figure 6.10. For Infasurf, an increase in Π causes a decrease of the film thickness down to 13 nm at 400 μg/mL and 11 nm at 100 μg/mL. The respective isotherms for Curosurf films are located in the range of smaller thicknesses. Out of the studied $\Pi(h)$, only those at Curosurf concentration of 100 μg/mL attain the thickness of DPPC black foam films at a pressure in the breathing pressure range. At higher concentrations, Curosurf films remain thicker even at higher pressure values. Thicker films are an indication that some additional non-DLVO forces are acting. For some of the Curosurf isotherms, the action of hydration (structural) surface forces has been established [32]. Hence, additional studies of films stabilized by TSPs can further contribute to the elucidation of the origin, role, and range of action of DLVO and non-DLVO surface forces in these films.

BFF parameters (critical concentrations, thickness, disjoining pressure, etc.) for the studied TSPs support the notion that BFFs are a realistic structural analogue of surface films existing *in vivo* in the lung. These parameters may be used as a standard in the process of developing a new generation of therapeutic surfactant preparations. The BFF method provides a unique visual record of foam film formation and stability, and clearly defines differences relative to both the nature and concentration of the TSPs.

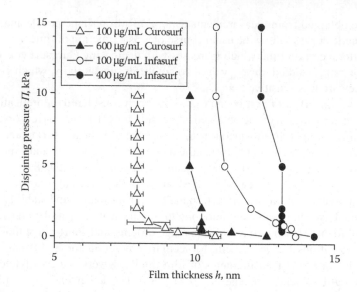

FIGURE 6.10 Disjoining pressure versus film thickness for foam films formed from Curosurf (triangles) and Infasurf (circles) at physiological electrolyte concentration (0.15 mol/L NaCl).

6.4 INFLUENCE OF INHIBITORS ON THERAPEUTIC SURFACTANT PREPARATIONS

Pulmonary surfactant (PS) inactivation caused by some factors (plasma proteins, lyso- and unsaturated phospholipids, free fatty acids, etc.) has become quite a topical issue in recent years [24,33]. The BFF method allows studying such factors' impacts on the interfacial activity of TSPs [26,31,34–36].

In inflammatory lung diseases, hydrolysis of surfactant phospholipids by phospholipase generates lysophospholipids. Such degradation cannot only deplete active surfactant lipids, but also releases products such as lysophosphatidylcholines (lysoPC) and free fatty acids that are severe biophysical inhibitors of surfactant's activity [24,37]. If such compounds are present in a mixture with TSP, their intrinsic surface-active behavior can reduce the adsorption of surfactant constituents into the interface or alter the properties of film adsorption layers.

Figure 6.11 shows the $W(C)$ dependence of lysoPC after 30 min waiting time [35]. The curve is shifted toward much lower concentrations compared to that of pure Infasurf (see Figure 6.8). The values of C_c and C_t are, respectively, 2 and 8 µg/mL. In order to study the inhibitory effect of lysoPC, it was added to an Infasurf solution with a constant surfactant concentration (equal to C_t), at which there is a 100% probability for a BFF to be formed. Addition of even 2 µg/mL lysoPC causes destabilization of Infasurf films ($W = 0$).

Such a destabilizing effect is observed within the range between 2 and 7 µg/mL lysoPC ($W < 100$). It is worth noting that these concentration values are higher than the normal endogenic surfactant and TSPs content [2]. The formation of stable BFF at lysoPC concentrations ≥ 8 µg/mL is probably due to the formation of mixed adsorption layers or to the predominant lysolipid adsorption. These results indicate that the BFF method can be successfully applied to study lysolipid inhibitory effects which diminishes TSP surface activity due to degradation of phospholipids contained in the sample occurring during long-term storage.

Some other low molecular weight surfactants may also be the reason for an inactivation of preparations, the former being produced by lung infection microorganisms, for example, rhamnolipids [38]. Figure 6.12 presents the $W(C)$ dependence for films from TSP and from a mixture of TSP and a rhamnolipid [34]. In this case, the inhibitor (rhamnolipid) concentration is maintained constant and equal to C_c. The rhamnolipid (RhL) present in the mixture destabilizes BFF and the $W(C)$

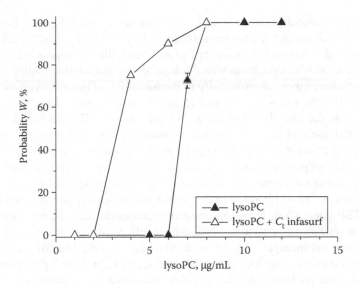

FIGURE 6.11 Dependence of the probability W for black foam film formation from lysoPC (filled triangles) and its mixtures with Infasurf (open triangles) on concentration of lysoPC at physiological electrolyte concentration (0.15 mol/L NaCl).

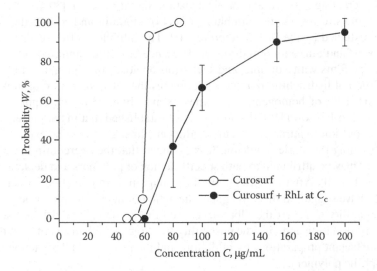

FIGURE 6.12 Probability W of formation of black foam films formed from TSP and mixtures of TSP and RhL ($C_c = 4.5$ μg/mL), $C_{el} = 0.15$ mol/L NaCl.

curve of the mixture is shifted toward higher TSP concentrations compared to the curve of the original preparation. The rhamnolipid molecules probably adsorb predominantly and obstruct the participation of TSP components in the film adsorption layer formation. As a way to overcome the rhamnolipid inhibitory effect, the TSP concentration has to be increased up to 200 μg/mL. This result may be used in clinical practice to treat lung infection-caused respiratory disorders with TSP.

Additional information about inhibitor's activity on TSP interfacial properties may be obtained from the dependence of foam film thickness on electrolyte concentration. At low electrolyte concentrations (up to 0.1 mol/L NaCl), the curves of the preparation and the rhamnolipid mixture are very close, in contrast to the pure rhamnolipid [34]. At the physiological concentration, film thicknesses for all three solutions studied coincide almost perfectly. Raising the NaCl concentration results in a curve for the mixture closely approaching the curve of the pure rhamnolipid. At concentrations above

0.8 mol/L NaCl, the rhamnolipid film thickness remains invariable at 5.5 nm. The films from TSP retain their thickness of about 8 nm at concentrations above 0.5 mol/L NaCl. Films of the mixture (TSP and rhamnolipid) continue to thin and at 1 mol/L NaCl, their thickness is closer to that of pure rhamnolipid. This result is in agreement with the assumption that the rhamnolipid is predominantly adsorbed and participates more pronouncedly in the formation of black film adsorption layers.

Though black film thicknesses at $C_{el} = 0.15$ mol/L NaCl is very close, the role of inhibitors is reflected clearly in the directly measured $\Pi(h)$ isotherms [34]. The curves of TSP alone and of mixtures with a rhamnolipid are very close at pressures within the breathing pressure range and reach a thickness of about 8 nm. Rhamnolipid ion effects on molecular interactions in the film become evident at higher pressure values. At pressures higher than 10 kPa, the curve for the mixture is very close to the curve for the pure rhamnolipid and reaches its film thickness. Another difference is the higher stability of films obtained from the mixture which rupture at much higher pressures that films from TSP alone. Thus, $\Pi(h)$ measurements demonstrate that the presence of rhamnolipids in a pulmonary surfactant suspension has a significant effect on surface forces acting in the films. The incorporation of rhamnolipid ions in the adsorbed layers at the film/air interfaces changes the surface electric parameters of the films, their thickness, and stability at higher pressures.

Leakage of plasma proteins into the alveolar space due to an impaired alveolar-capillary barrier is an early event in the pathogenesis of RDS in adults [24]. The mechanism by which PS is inactivated by albumin, fibrinogen, and hemoglobin is not yet fully understood [39]. However, there is evidence that once adsorbed, the protein film excludes the surfactant lipids from entering the interface by creating a steric and/or electrostatic energy barrier [39,40]. The capability of hydrophilic polymers to reverse the inhibitory effect of albumin and restore the surface activity of TSP has been studied by the BFF method [26,41]. The inhibitory effect is demonstrated by the change in Curosurf and Survanta film drainage. Presence of serum albumin leads to the formation of strong rheologic films with a dimple, and such films do not drain completely to form black foam films. The addition of hydrophilic polymers suppresses the effect on film morphology, and in the case of Curosurf, BFF of homogenous thicknesses are obtained again. Applying the dynamic method of Scheludko-Exerowa [10], the authors have established that in the presence of a polymer, a change from repulsion to attraction occurs in thicker films. This may be explained by depletion attraction overcoming the steric repulsion. It is assumed that the overcoming of albumin caused inhibition of TSP may be attributed to both specific action of polymers and depletion attraction via osmotic pressure [24,40]. After an estimation of the magnitude of depletion attraction, the authors in Reference 26 have suggested that either specific action or depletion attraction is responsible for the polymer capability to revert the albumin caused inhibition of TSP. Black foam film studies provide unique information and give the possibility for direct visualization of both the inactivation of exogenous surfactant preparations by inhibitors and the recovery of TSP action by the addition of some hydrophilic polymers.

Situated at one of the largest body/environment borders, the pulmonary surfactant happens to be the first barrier encountered by any agent attempting to enter the body through the respiratory tract [42]. This is why the pulmonary surfactant is responsible for the primary defense functions of the lung against harmful external agents. The damage from anesthetics and the high level of oxygen to PS can be connected with the direct chemical attack on surfactant components and with the intense processes of peroxidation of phospholipids, as well as with an effect on the biosynthesis and secretion of PS components. *In vitro* studies on black foam films obtained from lung lavages of treated rats showed that volatile anesthetics (halothane and penthrane) increase the threshold concentration (C_t) due to alterations in the surfactant composition [36]. Apart from anesthetics, some other medications pass through the alveolar-capillary barrier, for instance, corticosteroids which are active ingredients of asthma inhalation mixtures. These substances also affect surfactant interfacial properties and their actions may also be studied by the BFF method.

Compared to other models currently in use, foam films provide new opportunities for studying the properties and functions of the physiologically important alveolar surface layer.

6.5 SYNTHETIC SURFACTANTS

Although animal-derived TSPs are currently the state of the art, they have both practical and theoretical disadvantages [20]. These surfactant preparations are expensive to produce and supplies are limited. Thus, there is a need to develop synthetic surfactants which can be produced in large quantities at a reasonable cost. Currently, a new generation of synthetic pulmonary surfactants (SPS) is being developed that may replace animal-derived surfactants used in the treatment of the respiratory distress syndrome. They have the potential to increase the efficacy and cost effectiveness of this therapy. SPSs contain synthetic lipids (mainly DPPC and PG) and synthetic or recombinant analogues of the surfactant proteins SP-B and SP-C [43]. A new synthetic surfactant preparation (SPS) has been compared to Curosurf [21] and Survanta [22] with respect to treatment efficiency. It is the first time that a protein B and C analogue containing SPS has shown significant improvement of endogenous surfactant in inactivation in preterm lambs as compared to surfactant preparations of animal origin.

Data collected with the BFF method for natural PS presented in Section 6.3 can be used for a comparison with synthetic surfactants containing a mixture of DPPC, palmitoyloleoylphosphoglycerol (POPG) and synthetic analogues of SP-B and SP-C. It is worth noting that the process of foam film formation reveals some differences between SPS and the animal-derived surfactants Curosurf and Infasurf, such as significantly slower expansion of black spots, irregular (noncircular) shape of black spots formed, and occasionally existence of dimples in black films formed by synthetic surfactants. The drainage of films from insoluble surfactants is typically nonhomogenous due to the rigid structures formed in the bulk which do not spread easily at the solution/air interface. Similar behavior has been observed for Survanta, a DPPC modified surfactant formulation of natural origin [16,26]. In contrast to Survanta, SPS films drained completely to black foam films, thus enabling the determination of both $W(C)$ and $\Pi(h)$ dependences.

The $W(C)$ curves for SPS are similar to the ones shown in Figure 6.8. They are positioned at higher concentrations though closer to the curves of Curosur and Infasurf. The C_c values are very low as compared to the synthetic protein-free Exosurf. SPS films also exhibit an influence of waiting time prior to film formation, similar to Curosur and Infasurf. A prolongation of the waiting time resulted in a shift of $W(C)$ toward lower concentrations. In contrast to therapeutic surfactant formulations of natural origin where C_c decreases with increased waiting time, the SPS showed no significant decrease in C_c values after 30 min. These results indicate that up to 30 min, SPS black foam films may form at significantly higher concentrations as compared to those established for some formulations of natural origin (Curosurf and Infasurf). This may be attributed to the stability of bulk structures formed by formulation components in the course of production. Bulk structures from SPS incorporate two phospholipids (DPPC and POPG), and surfactant protein analogues that render them more rigid. Hence, compared to Curosurf, such structures would require more time to disintegrate and released molecules to participate in the adsorption layer formation at the solution/air interface. After 30 min of waiting time, almost all structures are destroyed and molecules are released to participate in the adsorption layers of the film. The result shows that irrespective of the longer time required by synthetic surfactants, the composition is very well selected so that the remaining components could assist adsorption of DPPC molecules.

The extension of waiting time also results in a decrease of the C_t values, however, the differences in values for 30 and 60 min are larger than those for C_c. Most probably, the effect of time on critical concentrations is due to an increase in the adsorption layers' density which occurs via both additional surfactant adsorption and spreading of molecular nanostructures at the solution/air interface.

The thicknesses of SPS films measured at physiological electrolyte concentrations are higher than those for Curosurf films, but very close to those of Infasurf films. The disjoining pressure isotherms $\Pi(h)$ indicate that there is a dependence on SPS concentration. At the lowest concentration studied (200 μg/mL), the isotherms fall within the thickness range of 15 and 14 nm. Within the pressure range corresponding to the breathing process, films a thicker than films of natural surfactant preparations, as shown in Figure 6.11. It is worth noting that for a pressure increase to slightly above 3 kPa, all films rupture. The increase of the SPS concentration leads to nonhomogenous film drainage and to thicknesses similar to those of natural preparations. As a whole, the concentration increase renders increased film stability, that is, film rupture occurs at pressures much higher than 4 kPa. Within the pressure range corresponding to breathing pressures, the film thickness varies between 14 and 12 nm. Raising the pressure above 4 kPa gives films with a thickness of 11 nm only and the film thickness does not alter up to the pressure of film rupture. In general, similar to animal-derived TSP, the films from SPS are stable at the exhale pressure. Compared to Curosurf films, they are less stable at higher pressure. It seems that the observed inhomogeneities, resulting in irreproducibility of the obtained film thicknesses, necessitate further statistical experiments on synthetic surfactant foam films.

The first results of SPS studies by means of the BFF method show that the method enables *in vitro* assessments of the surface properties of preparation and allows comparison with natural TSPs. The parameters obtained could be used to guide a production of new SPS. However, further experiments are needed in order to determine the precise composition for an optimal synthetic surfactant, and the BFF method could be a useful tool in such endeavors.

6.6   A NEW HYPOTHESIS FOR ALVEOLAR SURFACE STRUCTURE

Black foam films are not only a tool to explore the alveolar surface, but also offer an *in vitro* model enabling studies of the alveolar structure and stability. We consider the model as being close to the *in vivo* alveolar structure in view of the correlation of C_c and C_t with the RDS in newborn infants, that is, the absence of black film corresponds to surfactant insufficiency and, respectively, to disease condition. And, vice versa, the presence of black films corresponds to infant's lung maturity. As indicated, clinical results have proven this statement for preterm neonates [5,8].

At inhaling and exhaling pressures, black films show a thickness of 12–8 nm, that is, they thin down to bilayer black films. This gives reason to assume, that in an ordered bi-dimensional system first neighbor molecules interact within a short-range. The hydrophobic tails of phospholipid molecules separate the air and the hydrophobic tails of nanostructure formations in the alveolar hypophase (Figure 6.1b). "Head-to-head" contact points between the alveolar monolayer and a network of underlining bilayers should be realized during vesicle binding to the alveolar monolayers at junctions of the lipid monolayer with underlining membranes (of alveolar cells and macrophages, etc.), as well as in the multilaminated areas of the three-dimensional surface-associated structures of the alveolar lining layer.

Under different conditions, for instance, solution content, stratified films may also be obtained [4,11]. This means that with film thinning, bilayers may "slip" into a film dimple, that is, we may assume that there is a multilayer lamellar structure. Figure 6.13 shows a $\Pi(h)$ isotherm of PS stratified films, containing 150 µg/mL DPPC, in the presence of 47% ethanol, 7×10^{-2} mol/L NaCl, film radius 200 µm, and $t° = 25°C$. Ethanol probably plays a role in the formation of DPPC multilayer films.

Foam films from Curosurf in the presence of 0.15 mol/L NaCl show that some other nanostructure types may be obtained at the alveolar surface. Figure 6.14 depicts $\Pi(h)$ isotherms of Curosurf. Two types of curves are observed: a direct transition to bilayer films, and curves with a pronounced small plateau with kinks. In the majority of cases, these curves start with an exponential section and end with a vertical section, that is, the thickness is not altered with rising capillary pressure. Similar curves are obtained for a pressure range of 0.03–3 kPa, which corresponds to the inhaling and exhaling pressures.

The vertical section is clearly seen at higher pressures (note the interrupted scale on the ordinate). The reproducibility of the curves is not high because the foam film formation occurs in the presence of various nanostructures (such as micelles, liposomes, bilayers, and multilayers) [25,27] in the bulk solution which are in contact with the film. From that perspective, the kinks reflect a specific nanostructure while reaching a constant film thickness (vertical line in $\Pi(h)$ curves).

The thicknesses corresponding to the vertical lines are close to the size of different nanostructures in the bulk volume. The vertical line is moving toward the right to thicker black foam films. It is worth noting that the thickness of these structures is predominantly 9.5 nm, while for phospholipid bilayer films it is 8 nm. Figure 6.15 shows a schematic of the probable structures. The hydrophobic

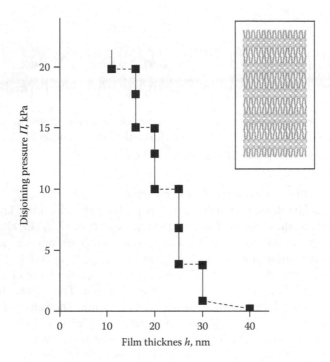

FIGURE 6.13 *In vitro* multilamellar structures of PS; $\Pi(h)$ dependence for foam films of the total phospholipid fraction of rat pulmonary lavages. Inset—the stratified structure of the film to be considered as multilayer structure consisting of several lamellae (bilayers).

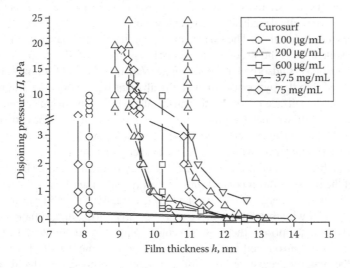

FIGURE 6.14 Disjoining pressure versus film thickness isotherms for films from aqueous solutions of Curosurf in the presence of 0.15 mol/L NaCl. O—100 μg/mL; Δ—200 μg/mL; □—600 μg/mL; ∇—37,5 mg/mL; ◇—75 mg/mL.

SP-B lies with its hydrophobic portion toward the bilayer chains, while the hydrophilic portion is directed toward the phospholipid heads.

Interaction forces acting at the onset of the isotherms, that is, where film thickness changes with pressure, is worth considering (Figure 6.14). The above mentioned forces correspond to electrostatic, Π_{el}, steric, Π_{ster}, and structural, Π_{str}, positive (repulsion) components of disjoining pressure counteracted by the van der Waals negative (attraction) component, Π_{vdW}. Not all components are

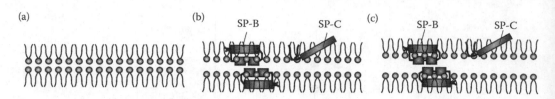

FIGURE 6.15 Schematic presentation of a nanostructure model in Curosurf black foam films of different thicknesses: (a) phospholipid bilayer; (b) thinnest black foam film, and (c) thicker black foam film. Molecule cartoons refer to: ⱱ-phospholipid; ⱳ-SP-B and ⱱ-SP-C.

operating in the case under consideration. In the presence of 0.15 M NaCl, Π_{el} is suppressed [13]. Furthermore, Π_{ster} would hardly matter in the case of phospholipids' adsorption layers having a head size of 0.4–0.7 nm [44]. On the other hand, the macroscopic theory of Π_{vdW} [45] allows its calculation to a sufficient degree of accuracy as indicated in Reference 46. Then any difference between the experimentally measured disjoining pressure, Π_{exp}, and the Π_{vdW} term could be attributed to Π_{str}. Since the latter changes exponentially with thickness [47], an exponential fit of Π_{exp}—Π_{vdW} with a characteristic decay length as a free parameter has been explored. The exponential dependence of Π_{exp}—Π_{vdW} runs well with a decay length between 1.1 and 2.9 nm. The values of the characteristic decay lengths observed as well as the scope of action presume the existence of secondary hydration (structural) forces [47]. Therefore, the interaction forces acting at the onset of the $\Pi(h)$ curve are structural forces, while short-range interactions operate in the region of the observed nanostructures.

Model studies of black foam films from alveolar surfactant via a direct disjoining (capillary) pressure determination indicate the presence of structures (11–8 nm) at alveolar surfaces, including down to attaining a bilayer. These structures provide new insight into the lung surfactant system and are important in various lung pathologies associated with respiratory disorders, respectively, surfactant insufficiency.

On the grounds of a theoretical analysis, a different model of surfactant layers on the alveolar surface was proposed in References 6 and 48. The surface energy of the alveolar surfactant layer is determined in the scope of a modification of the structural model of Larsson et al. [48] according to which this layer is built up of lipid monolayers adsorbed at the hypophase/air interface and supported by a network of lipid bilayers immersed into the hypophase, that is, the alveolar liquid. Equations were derived for the dependence of the specific surface energy of the surfactant layer on the distance between the bilayers constituting the layer. It is shown that at equilibrium, this energy can have values comparable to or less than the 1 mJ/m² needed for normal functioning of the alveolus during the respiration cycle. The specific surface energy of the surfactant layer with a monolayer-bilayer structure can have such low values only if the layer is of optimal thickness and if the specific line energy of the monolayer-bilayer contact lines is negative and that of the bilayer-bilayer contact lines is positive. It is found that in a dynamic regime the change in the specific surface energy of the alveolar surfactant layer with a bilayer-monolayer structure is in qualitative agreement with that determined experimentally during lung inflation and deflation [6].

In the structural model proposed, the principle role is played by the bilayer. The black foam film model, including bilayers, introduced in Reference 6, is in agreement with this theoretical model. We presume that the structure registered by $\Pi(h)$ isotherms and present at the alveolar surface are bilayer structures with surfactant proteins SP-B and SP-C incorporated into them.

Let us hope that the suggested models of alveolar structure and stability, described above in detail, are bringing us closer to the *in vivo* situation, and will enable alveolar respiratory disorders studies.

REFERENCES

1. Hawgood, S., Clements, J.A. Pulmonary surfactant and its apoproteins. *J Clin Invest* 1990; 86: 1.
2. Blanco, O., Pérez-Gil, J. Biochemical and pharmacological differences between preparations of exogenous natural surfactant used to treat respiratory distress syndrome: Role of the different components in an efficient pulmonary surfactant. *Eur J Pharmacol* 2007; 568: 1.

3. Kramer, B.W. The respiratory distress syndrome (RDS) in preterm infants: Physiology, prophylaxis and new therapeutic approaches. *Intensive Med* 2007; 44: 403.

4. Exerowa, D., Lalchev, Z. Bilayer and multilayer foam films: Model for study of the alveolar surface and stability. *Langmuir* 1986; 2: 668.

5. Exerowa, D., Lalchev, Z., Marinov, B., Ognianov, K. Method for assessment of fetal lung maturity. *Langmuir* 1986; 2: 664.

6. Kashchiev, D., Exerowa, D. Structure and surface energy of the surfactant layer on the alveolar surface. *Eur Biophys J* 2001; 30: 34.

7. Todorov, R., Cohen, R., Exerowa, D. Surface forces in foam films from DPPC and lung surfactant phospholipid fraction. *Colloids Surf A* 2007; 310: 32.

8. Exerowa, D., Lalchev, Z., Kashchiev, D. Stability of foam lipid bilayers of amniotic fluid. *Colloids Surf* 1984; 10: 113.

9. Exerowa, D. Chain-melting phase transition and short-range molecular interactions in phospholipid foam bilayers. *Adv Colloid Interface Sci* 2002; 96: 75.

10. Exerowa, D., Kruglyakov, P.M. Foam and foam films. In: Möbius, D., Miller, R. (Eds.), *Studies in Interface Science*, Vol. 5. Elsevier: Amsterdam, 1998.

11. Lalchev, Z. Phospholipid foam films: Types, properties and applications. In: Tadros, Th. (Ed.), *Colloids and Interface Science Series*, Vol. 1. Wiley-VCH: Weinheim, 2007.

12. Jordanova, A., Lalchev, Z., Tenchov, B. Formation of monolayers and bilayer foam films from lamellar, inverted hexagonal and cubic lipid phases. *Eur Biophys J* 2003; 31: 626.

13. Cohen, R., Todorov, R., Alexandrov, S.v., Christova, Y., Lalchev, Z., Exerowa, D. Direct measurement of molecular interaction forces in foam films from lung surfactant fraction. *Colloid Polym Sci* 2006; 284: 546.

14. Lalchev, Z., Christova, E. *Alveolar Surfactant and Neonatal Respiratory Distress Syndrome: Physiological Aspects and Therapy*. Sofia University Press "St. Kliment Ohridski": Sofia, 2010.

15. Cordova, M., Mautone, A.J., Scarpelli, E.M. Rapid in vitro tests of surfactant film formation: Advantages of the Exerowa black film method. *Pediatr Pulmonol* 1996; 21: 373.

16. Scarpelli, E., Mautone, A., Lalchev, Z., Exerowa, D. Surfactant liquid and black foam films formation and stability in vitro and correlative conditions in vivo. *Colloids Surf B* 1997; 8: 133.

17. Lalchev, Z., Georgiev, G., Jordanova, A., Todorov, R., Christova, E., Vassilieff, C. A comparative study of exogenous surfactant preparations and tracheal aspirate: Interfacial tensiometry and properties of foam films. *Colloids Surf B* 2004; 33: 227.

18. Antonova, N., Todorov, R., Exerowa, D. Rheological behavior and parameters of the in vitro model of lung surfactant systems: The role of the main phospholipid component. *Biorheology* 2003; 40: 531.

19. Lalchev, Z., Todorov, R., Exerowa, D. Thin liquid films as a model to study surfactant layers on the alveolar surface. *Curr Opin Colloid Interface Sci* 2008; 13: 183.

20. Jordan, B.K., Donn, S.M. Lucinactant for the prevention of respiratory distress syndrome in premature infants. *Expert Rev Clin Pharmacol* 2013; 6: 115.

21. Seehase, M., Collins, J.J.P., Kuypers, E., Jellema, R.K., Ophelders, D.R.M.G., Ospina, O.L., Perez-Gil, J. et al. New surfactant with SP-B and C analogs gives survival benefit after inactivation in preterm lambs. *PLoS ONE* 2012; 7: e47631.

22. Sato, A., Ikegami, M. SP-B and SP-C containing new synthetic surfactant for treatment of extremely immature lamb lung. *PLoS ONE* 2012; 7: e39392.

23. Wustneck, R., Perez-Gil, J., Wustneck, N., Cruz, A., Fainerman, V.B., Pison, U. Interfacial properties of pulmonary surfactant layers. *Adv Colloid Interface Sci* 2005; 117: 33.

24. Zuo, Y., Veldhuizen, R.A.W., Neumann, A.W., Petersen, N.O., Possmayer, F. Current perspectives in pulmonary surfactant—Inhibition, enhancement and evaluation. *Biochim Biophys Acta* 2008; 1778: 1947.

25. Pérez-Gil, J. Structure of pulmonary surfactant membranes and films: The role of proteins and lipid–protein interactions. *Biochim Biophys Acta* 2008; 1778: 1676.

26. Georgiev, G., Vassilieff, C., Jordanova, A., Tsanova, A., Lalchev, Z. Foam film study of albumin inhibited lung surfactant preparations: Effect of added hydrophilic polymers. *Soft Matter* 2012; 8: 12072.

27. Bernard, W., Mottaghian, J., Gebert, A., Rau, G., van der Hard, H., Poets, C. Commercial versus native surfactants: Surface activity, molecular components, and the effect of calcium. *Am J Respir Crit Care Med* 2000; 162: 1524.

28. Lalchev, Z., Todorov, R., Christova, V., Wilde, P., Mackie, A., Clark, D. Molecular mobility in the monolayers of foam films stabilized by porcine lung surfactant. *Biophys J* 1996; 71: 2591.

29. Casals, C., Cañadas, O. Role of lipid ordered/disordered phase coexistence in pulmonary surfactant function. *Biochim Biophys Acta* 2012; 1818: 2550.

30. Kashchiev, D., Exerowa, D. Effect of surfactant adsorption time on the observation of Newton black film in foam film. *J Colloid Interface Sci* 2009; 330: 404.
31. Exerowa, D., Todorov, R., Platikanov, D. Interfacial properties of therapeutical lung surfactants studied by thin liquid films. In: Ohshima, H., Makino, K. (Eds.), *Colloid and Interface Science in Pharmaceutical Research and Development*. Elsevier: Amsterdam, 2014.
32. Exerowa, D., Todorov, R., Cohen, R., Kolarov, T. Nanostructures at the lung surface (submitted).
33. Zasadzinski, J.A., Stenger, P.C., Shieh, I., Dhar, P. Overcoming rapid inactivation of lung surfactant: Analogies between competitive adsorption and colloid stability. *Biochim Biophys Acta* 2010; 1798: 801.
34. Cohen, R., Vladimirov, G., Todorov, R., Exerowa, D. Effect of rhamnolipids on pulmonary surfactant foam films. *Langmuir* 2010; 26: 9423.
35. Alexandrov, S., Todorov, R., Jordanova, A., Lalchev, Z., Exerowa, D. Inactivation of pulmonary surfactant by lysophosphatidylcholine. *Biotechnol Biotechnol Eq* 2009; 23: 684.
36. Lalchev, Z., Christova, Y., Todorov, R., Alexandrov, V., Stoichev, P., Petkov, R. Alterations of biochemical and physicochemical quantities of pulmonary lavages attending halothane and pentrane treatment in rats. *ACP Appl Cardiopul Pathophys* 1992; 4: 315.
37. Hite, R.D., Seeds, M.C., Jacinto, R.B., Grier, B.L., Waite, B.M., Bass, D.A. Lysophospholipid and fatty acid inhibition of pulmonary surfactant: Non-enzymatic models of phospholipase A2 surfactant hydrolysis. *Biochim Biophys Acta* 2005; 1720: 14.
38. Zulianello, L., Canard, C., Köhler, T., Caille, D., Lacroix, J.S., Meda, P. Rhamnolipids are virulence factors that promote early infiltration of primary human airway epithelia by *Pseudomonas aeruginosa*. *Infect Immun* 2006; 74: 3134.
39. Taeusch, H.W., de la Serna, J.B., Perez-Gil, J., Alonso, C., Zasadzinski, J.A. Inactivation of pulmonary surfactant due to serum-inhibited adsorption and reversal by hydrophilic polymers: Experimental. *Biophys J* 2005; 89: 1769.
40. Fernsler, J.G., Zasadzinski, J.A. Competitive adsorption: A physical model for lung surfactant inactivation. *Langmuir* 2009; 25: 8131.
41. Georgiev, G.A.s., Kutsarova, E., Jordanova, A., Tsanova, A., Vassilieff, C.S., Lalchev, Z. Tuning of surface properties of thin lipid-protein films by hydrophilic non-surface active polymers. *Biotechnol Biotechnol Eq* 2009; 23: 547.
42. Stephanova, E., Valtcheva-Sarker, R., Topouzova-Hristova, T., Lalchev, Z. Influence of volatile anaesthetics on lung cells and lung surfactant. *Biotechnol Biotechnol Eq* 2007; 21: 393.
43. Curstedt, T., Calkovska, A., Johansson, J. New generation synthetic surfactants. *Neonatology* 2013; 103: 327.
44. Seelig, J., Seelig, A. Lipid conformation in model membranes and biological membranes. *Q Rev Biophys* 1980; 13: 19.
45. Dzyaloshiuskii, I.E., Lifshitz, E.M., Pitaevskii, L.P. The general theory of van der Waals forces. *Adv Phys* 1961; 10: 165.
46. Donners, W.A.B., Rijubout, J.B., Vrij, A. Calculation of van der Waals forces in thin liquid films using Lifshitz' theory. *J Colloid Interface Sci* 1977; 60: 540.
47. Parsegian, Y.A., Zemb, T. Hydration forces: Observation, explanations, questions. *Curr Opin Colloid Interface Sci* 2011; 16: 618.
48. Larsson, M., Larsson, K., Andersson, S., Kakhar, J., Nylander, T., Ninham, B., Wollmer, P. The alveolar surface structure: Transformation from a liposome-like dispersion into a tetragonal CLP bilayer phase. *J Dispers Sci Technol* 1999; 20: 1.

7 Foam Films Stabilised by Particles

Gareth Morris and Jan Cilliers

CONTENTS

7.1 INTRODUCTION

The ability of particles to stabilize thin liquid films in emulsions was reviewed by Pickering [1], building on earlier work by Ramsden [2]. Many of the same fundamental processes occur whether the system is a water-in-oil film (emulsion) or a water-in-air film (foam). Particles attached to thin foam films can both stabilize or destabilize the film hence, to understand the conditions under which stability is promoted, consideration must also be given to those conditions under which stability is actively reduced. Research into particle stabilized foams has, therefore, progressed hand-in-hand with research into the use of particles as antifoaming agents.

In a two phase foam, liquid is drawn from the film into the Plateau borders under the action of capillary pressure, which arises from liquid draining out of the foam. As more liquid drains from the foam, capillary pressure rises, causing the films to thin until they fail. The capillary pressure at which this occurs at is known as the critical capillary pressure, P_{crit}. In the case of a particle stabilized froth (three phase foam), the particles attached to the foam films increase P_{crit}. As a higher capillary pressure is required to cause film failure, the thin liquid films have a longer lifetime, producing a more stable froth. The value of P_{crit} for a particle stabilized thin liquid film is determined by a complex relationship between the particle hydrophobicity, shape and packing density, and pattern on the film.

7.2 THE PARTICLE-FILM SYSTEM

The interaction between the particles and the thin liquid film is complex and challenging to study both experimentally and theoretically. A useful method for investigating the fundamental behavior of the system is to reduce it to two dimensions (2D) and examine a single circular particle bridging both sides of the thin liquid film, as shown in Figure 7.1.

7.2.1 THE 2D PARTICLE-FILM SYSTEM

The contact angle (θ), particle radius (R_P), separation distance (S_{PP}), radius of curvature of the liquid-vapor (LV) interface (R_L), and curvature of the film are used to describe the 2D system, with θ defined as the angle that forms between the LV interface and the solid-liquid (SL) interface at the point where the solid, liquid, and gas phases meet (Figure 7.1). This is called the three point contact (TPC) labelled in Figure 7.1. The Young-Laplace law (Equation 7.1) relates the curvature of the LV interface ($r_{1,2}$) and surface tension (γ) to the difference in pressure (ΔP) between the two phases. In three dimensions (3D), r_1 and r_2 are the orthogonal radii of curvature of the film, however, in 2D, r_2 tends to infinity and thus Equation 7.1 simplifies to Equation 7.2 (with $r = R_L$). In the case of a particle stabilized thin liquid film, ΔP is the capillary pressure drawing liquid out of the film via the Plateau borders.

$$\Delta P = \gamma \left(\frac{1}{r_1} + \frac{1}{r_2} \right) \tag{7.1}$$

$$\Delta P = \frac{\gamma}{r} \tag{7.2}$$

In the top set of images in Figure 7.2, it can be seen that at a capillary pressure of 0 ($P = 0$), the LV interfaces are flat and adopt a position on the particle surface that ensures θ is maintained at the TPC. As the contact angle increases to 90°, with $P = 0$, the liquid film thickness (2 h) decreases until at 90°, $h = 0$, and the film fails. This film failure at 90° is known as bridging-dewetting, first proposed by Garrett [3], and is widely accepted as the mechanism through which particles with large contact angles ($\theta > 90°$) destroy films. In this mechanism, an antifoaming particle bridges the film, but instead of preventing liquid from draining out of the film by holding the TPCs apart, its high contact angle causes them to be drawn together and the film fails once they meet the particle surface. A discussion of particles acting as antifoamers is beyond the scope of this chapter but has been reviewed recently by Garrett [4].

If the contact angle remains below 90°, the film will remain stable for a range of values for $P > 0$. The bottom row of images in Figure 7.2 show that as P increases, so does the curvature ($1/R_L$) of the LV interface until $h = 0$ and film failure occurs.

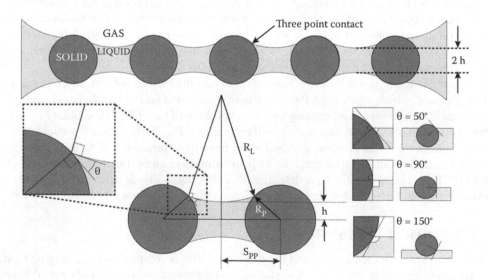

FIGURE 7.1 Showing the structure of a particle stabilized film, the geometric dimensions involved, and the effect of contact angle on the position of the LV interface.

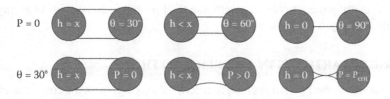

FIGURE 7.2 The effect of contact angle and increased capillary pressure (P) on the film thickness (h).

7.2.2 PARTICLES AS SURFACTANTS

Both surfactants and particles stabilize foams, but, arguably, they do so through different mechanisms and on different scales. They do, however, also exhibit similarities, as discussed by Binks [5]. It was proposed that just as a surfactant's properties can be described in terms of the hydrophile-lipophile balance (HLB), those of spherical particles can be described in terms of contact angle and size. The HLB is an important parameter used to characterize the relative efficiency of a surfactant's hydrophobic and hydrophilic parts, whereas the hydrophobicity of a particle affects its readiness to adsorb at an interface. Once adsorbed, a particle is held strongly at the two phase interface, whereas surfactant molecules are adsorbing and desorbing dynamically from the interface over very short time scales.

Binks [5] also reported that as nano- or microparticles are not dissolved in the liquid phase, there are no solubility based phenomena such as the formation of micelles, as is found with surfactants. It is, however, possible for the particles to form loose aggregates in the film and become trapped between the opposite interfaces. Here they hold the opposite sides apart and prolong film lifetime [6,7].

Binks [5] used Equation 7.3 to calculate the energy required to remove a particle from an air-water or oil-water interface, $\gamma_{\alpha\beta}$ is the surface tension of the interface the particle is attached at and E is the energy.

$$E = \pi R_P^2 \gamma_{\alpha\beta} (1 \pm \cos\theta)^2 \tag{7.3}$$

The plus or minus signs represent the energy required to remove the particle from the interface into one of the fluid phases. E is at a maximum when $\theta = 90°$ and decreases rapidly either side of this. E also decreases with the square of R_P and particles of a size comparable to a surfactant molecule (5 nm) are very easily detached, suggesting that there is a lower limit to the size of a particle that can effectively stabilize a film.

7.2.3 2D MODELS OF PARTICLES IN THIN LIQUID FILMS

In the case of the 2D model described in Figure 7.1, as the capillary pressure increases and the liquid drains from the film, the LV interfaces distort around the particles to maintain the contact angle at the particle's surface. As the LV interfaces thin between the particles, they maintain the radius of curvature defined by the Young-Laplace law. Eventually, the capillary pressure will reach P_{crit}, the opposite sides of the film will touch, and the film fails (Figure 7.2). Both Denkov et al. [8] and Ali et al. [9] investigated the 2D case but used different approaches.

Denkov et al. [8] approached the problem from the perspective of Pickering emulsions; the theory is applicable to particle stabilized foam films as well. Here the spherical particles are assumed to occupy a circular cell whose area corresponds to the packing density of the particles on the film. The distortion of the meniscus surrounding the particle is assumed to be axisymmetric and centered on the particle, so that the 2D case can be evaluated.

Ali et al. [9] also used a similar approach to investigate the 2D case, albeit approaching the problem from a more geometric perspective. They concluded, much like Denkov et al. [8], that the problem of particle stabilization can be treated elegantly by supposing the particles are evenly distributed throughout the film. In reality, small perturbations will cause unbalanced forces to act on each particle, causing them to clump together in the film and open up areas of empty film surface that reduce film stability.

7.3 SPHERICAL PARTICLES IN THIN LIQUID FILMS

The analytical model derived by Ali et al. [9] returns a value of P_{crit} from the contact angle and particle spacing (Equation 7.4), where S_{pp} is the distance from the center of the particle to the center of the film (Figure 7.1). These 2D results are plotted in Figure 7.3 along with results from 3D numerical simulations.

$$P_{crit} = \frac{2\gamma\cos\theta}{S_{pp}^2 - R_p^2} \tag{7.4}$$

From Equation 7.4, P_{crit} tends to infinity for closely packed particles when $S_{pp} = R_p$. The densest possible packing of spheres on a film is close packed hexagonal, when the spaces correspond to $S_{pp} = 1.155$. However, even using $S_{pp} = 1.155$ as the minimum value for the separation in Equation 7.4, the 2D case is still unable to take into account the complex topology that the LV interface adopts in 3D. It is clear that this requires a 3D approach to fully understand the film stability.

The 2D analysis shows that circular particles can both stabilize or destroy films. The shortcoming of the 2D analysis has been touched upon; particles are rarely smooth, uniform spheres and do not distribute evenly in a film. In reality, the LV interface surrounding the particles has a complex topology that affects the film stability. With access to more powerful computing power and numerical modelling tools, it has become possible to expand the 2D models into 3D.

7.3.1 3D MODELS OF PARTICLES IN THIN LIQUID FILMS

Morris et al. [10] investigated the effect of particle packing in 3D using the *Surface Evolver* [11] program. Here, regular hexagonal and square packing arrangements were simulated. Both models used idealized uniform spheres, akin to the 2D models, and were able to determine the topology of the film surrounding the particles.

Particle packing density on the film (A_{PP}) is calculated using Equation 7.5 (where n_p is the number of particles and A_C is the area of the periodic cell). Although both packing arrangements yield similar

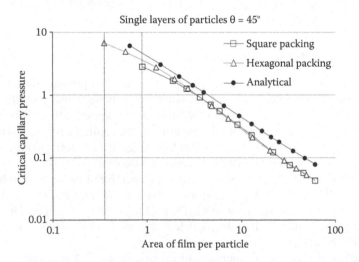

FIGURE 7.3 Showing a comparison of analytical and numerical models of P_{crit} for single layer of particles stabilizing the liquid film. The analytical model shown for single layers is from [9], the two vertical lines represent the minimum packing area for hexagonal (solid) and square (dashed) packing.

film stabilities for a given θ and A_{PP}, hexagonal packing can achieve a higher packing density and so the greatest film stability. Comparing 2D and 3D predictions of P_{crit} (Figure 7.3), it can be seen that the 2D model overpredicts.

$$A_{pp} = \frac{A_c - n_p R_p^2 \pi}{n_p} \tag{7.5}$$

When a thin liquid film is heavily loaded with ideal spherical particles, they can pack with square and hexagonal regularity [12,13]. For partial loading, the capillary forces will force the particles to draw together into irregular agglomerates [14]. This requires a statistical analysis of repeated model results from random particle distributions in the film.

Morris et al. [15] performed repeated (more than a thousand times) simulations of periodic cells containing up to 20 particles randomly placed in a thin liquid film. The simulations used the same packing densities but different packing arrangements to allow statistical analysis of the film stability. They produced a relationship between packing density, particle contact angle, and P_{crit} (Equation 7.6); K is a fitted constant (2.31) and A_{pp} the area of empty film per particle (defined in Equation 7.5).

$$P_{crit} = \frac{K\gamma \cos\theta}{A_{pp}} \tag{7.6}$$

7.3.2 DOUBLE LAYERS OF PARTICLES IN THIN LIQUID FILMS

Thus far, the analysis has considered only single layers of particles in the film. However, for particle stabilized foams that are heavily laden with particles, it is not uncommon to see double layers of particles in the film.

The stability and energy considerations here are slightly different [6,13,16–18]. Under certain conditions, a double layer of particles can stabilize a film, even when the particles have a contact angle greater than 90°, up to a theoretical limit of 129°. The case of double (or more) layers of particles stabilizing a thin liquid film has similarities with work done on porous flow and the capillary bridging of a pore. For example, Hilden and Trumble [19] and Cox et al. [20] both used the Surface Evolver to investigate the complex shape of the LV interface as it travels through a pore.

A similar approach was used by Morris et al. [21] to investigate the effect of contact angle and particle packing on the stability of double layers in 3D, which built on previous studies [13,18] with the different predictions for P_{crit} compared in Figure 7.4. The equation from [13] used to predict P_{crit} for a close packed double layer of particles is given in Equation 7.7; where γ is the surface tension at the oil-water interface and α_{max} is a geometric parameter $\alpha_{max} = \theta + \arccos(\sqrt{3}\sin\theta/2)$.

$$P_{crit} = \frac{2\gamma\sqrt{1 - \dfrac{3\sin^2\theta}{4}}}{R_p\left(\dfrac{2}{\sqrt{3}} - \sin\alpha_{max}\right)} \tag{7.7}$$

Morris et al. assume a static, uniformly packed, double layer of spherical particles [21] and identify three possible film failure modes; *particle bridging, capillary pressure driven failure*, and *film inversion*. In all three cases, as the capillary pressure increases, so too does the curvature of the LV interfaces. This draws the film surface toward the particles and interface on the opposite side of the film. The first case, *particle bridging* (Figure 7.5, top), occurs when the particles have a contact angle greater than 90°. Once the curvature of the LV interface brings it into contact with particles in the opposite layer, it bridges them and the film fails immediately as the particles cause bridging-dewetting of the film. The second mode, *capillary pressure driven failure* (Figure 7.5, middle), is

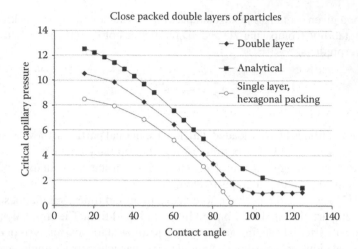

FIGURE 7.4 A comparison of analytical and numerical models of P_{crit} for single and double layers of particles stabilizing the liquid film. The analytical results shown for a single layer is from [9], that shown for double layers from [13].

analogous to the standard film failure mode and occurs when $\theta < 90°$; the LV interface touches the particle in the opposite layer and attaches to them. As the pressure continues to rise, the curvature of the LV interfaces eventually brings them into contact with each other and failure occurs. Finally, *film inversion* (Figure 7.5, bottom) is analogous to the bridging of a pore. In this case, the particles are closely packed together and the LV interfaces cannot sustain the curvature required to bridge the opposite side of the film. Instead, once the maximum sustainable capillary pressure is reached, the

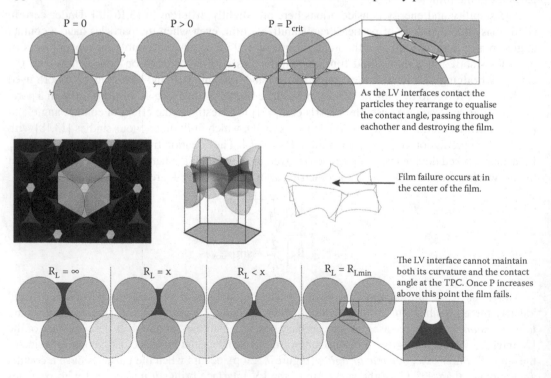

FIGURE 7.5 The three film failure mechanisms for a double layer of particles; *particle bridging* (top), *capillary driven failure* (middle), and *film inversion* (bottom).

LV interface inverts itself, bridging the film and causing immediate failure. Films stabilized by double layer systems are much more computationally intensive to model than single layers. As such, both dynamic and nonspherical particle systems have yet to be investigated theoretically in great detail.

7.4 NONSPHERICAL PARTICLES IN THIN LIQUID FILMS

7.4.1 ORTHORHOMBIC PARTICLES

While foams stabilized solely by spherical particles are found occasionally in industry and practice, many processes contain irregular particles with asperities and nonuniform shapes. One such industrial process is froth flotation, used in the mining industry to concentrate low grade ores. A key part of this process involves a froth stabilized by small mineral particles in the size range of a few hundred microns. The effect that sharp edges on particles in a thin liquid film have on the behavior of the particle and stability of the film is still poorly understood. However, Dippenaar [22] published groundbreaking work that captured the bridging-dewetting process predicted in [3]. A high speed camera was used to capture the moment when a small, orthorhombic galena particle bridged both sides of a thin liquid film and caused it to fail. Up to that point, the understanding of thin liquid film rupture by a particle was either theoretical or based on bulk experiments on particle stabilized foams. Dippenaar's work showed that it is not only the hydrophobicity or shape of the particle, but also the orientation that it adopts within the film that defines the extent to which the particle stabilizes or ruptures the film. Dippenaar also observed the apparent bridging-dewetting behavior of a particle with $\theta < 90°$.

There have been numerous studies that investigated the stable orientations of nonspherical particles in a film, but relatively few assessing how orientation affects film stability. Dippenaar used orthorhombic galena particles ($\theta = 80° \pm 8$) to identify two orientations that occurred with roughly equal probability; *flat* and *diagonal*. When the particles were placed in a thin liquid film, those that adopted the diagonal orientation often caused rupture of the film as soon as they bridged both sides. Morris et al. [23,24] created a numerical model of the particle film interaction and conducted a thorough investigation of the effects of particle aspect ratio and contact angle. They identified four energetically stable orientations dependent on shape and contact angle; *horizontal, vertical, rotated*, and *diagonal* (Figure 7.6). For a cubic particle, *horizontal* and *vertical* orientations are equivalent to the *flat* orientation. The *flat* orientation can be described as one in which two of the orthorhombic particle's faces are parallel to the interface, while the *rotated* orientation has three of the orthorhombic faces bridging the interface with their face all normal at an angle of 55° to the Z-axis. The *diagonal* orientation described by Dippenaar becomes stable for particles with higher aspect ratios and a more oblong shape, in this case, two of the facet normals are at 45° to the Z-axis and one at 90°.

From their simulation results, [23] found that a cubic particle with a contact angle of $\sim 65°$ has two stable orientations (*flat* and *rotated*), one of which (*flat*) has a P_{crit} up to four times greater than the other (*rotated*). This pattern of orientation behavior, while less prominent, is present for contact angles up to 80°, at which point the flat orientation becomes energetically unstable. Below a contact angle of 65°, the rotated orientation is not energetically stable. The behavior is also seen across a range of particle aspect ratios [24], with a third orientation *diagonal* (Figure 7.6) becoming energetically stable for particles with a more elongated shape, replacing the *rotated* orientation at high contact angles. Morgan et al.

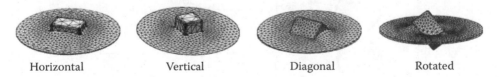

| Horizontal | Vertical | Diagonal | Rotated |

FIGURE 7.6 The four stable types of orientation for an orthorhombic particle in a thin liquid film. For a cubic particle, the orientations of *horizontal* and *vertical* become the same and collapse to one *flat* orientation.

[25] investigated the stable orientations of cubic hematite particles at an oil-water interface. Using the gel trapping technique, they managed to image the particle orientations at the interface with a scanning electron microscope (SEM), recording instances of both the *flat* and *rotated* orientations.

On a larger experimental scale, de Folter et al. [26] used cubic and peanut shaped hematite particles to make ultra stable Pickering emulsions that lasted more than a year. This was attributed to the packing density achieved with cubic particles that could approach up 90%. It was reported that the particles adopted orientations with one face parallel to the interface, equivalent to a *flat* orientation.

The interplay between shape, contact angle, particle orientation, and film stability is complex. For example, a uniform orthorhombic (or superellipsoidal) particle can cause either almost immediate film rupture [22] or an ultrastable foam [26]. It is expected that particles with surface heterogeneity or more irregular shapes will exhibit even greater complexity of their behavior which cannot be simulated easily at a level that will provide useful insight.

7.4.2 PARTICLES WITH SURFACE PATTERNING

The work of San Miguel and Behrens [27] has shown that particle surface heterogeneity has a significant effect on film (and foam) stability. They coated 0.9 μm spherical silica with 50 nm particles, which were then partially dissolved to tune the surface roughness and contact angle hysteresis. When these particles were used to create an emulsion, the P_{crit} (P_C^{Max}) passed through a peak as the surface roughness increased (Figure 7.7). As these particles were very regular in underlying shape and surface alteration, it is a valuable step toward quantifying the effect of surface roughness on film stability.

Janus particles have a surface divided into two hemispherical zones, often, but not always, along one plane of symmetry. The two zones have different surface energies, which affects the way the TPC line passes over them, as well as their preferred orientation in the film. Janus particles offer a much greater amount of orientation tuneability than uniform particles and create much more complicated deformations of the LV interface, resulting in particle interactions that are difficult to predict. Much research has recently investigated Janus particles, and while much concentrated on particle behavior at an interface rather than a thin liquid film, useful similarities can still be found. The majority of the literature focuses upon manufactured or idealized Janus particles, which tend to have their surface heterogeneity split equally into two hemispheres [28].

However, it is important to note that industrial applications do not always have control over the shape or surface heterogeneity of the particles they process. For example, froth flotation regularly deals with irregular heterogeneous particles. The comminution process used to liberate valuable (hydrophobic) mineral grains from their undesired (hydrophilic) host rock often generates particles that are made up of both materials. These particles often have an irregular surface distribution of minerals. The study of the interaction of these particles with the thin liquid films they become attached to during flotation is still in its infancy. However, Gautam and Jameson [29] have studied the

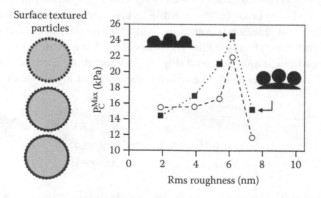

FIGURE 7.7 Showing the surface patterning of particles of varying roughness [27] and their effect on P_{crit}.

shape of the meniscus when patches of hydrophobic surface are surrounded by a hydrophilic surface matrix. The modelling approach they used has also resulted in a basic method for investigating the force of attachment of complex heterogeneous particles to a bubble film.

7.5   MULTIPLE PARTICLE INTERACTION

The methods and models discussed so far have treated the particle distribution on the film as fixed in its initial setting. However, in reality, the capillary forces acting on the particles cause them to draw together into aggregates [8,9]. Modelling large numbers of particles in a thin liquid film is computationally intensive, but small scale analysis of the interaction between particles is possible. Experimentally, the behavior of particles in a thinning film has been investigated by [13]. They found that particles initially packed in a double layer will 'unzip' to a single layer as capillary pressure increases, providing space is available for the increase in surface coverage of the film.

In the multipole method proposed in [30], the meniscus surrounding the particle is expanded as a Fourier series so that the attraction between spherical particles can be calculated. It was expanded in [31] to calculate the interaction of particles with the same charge at an interface. Lewandowski et al. [32] found that the ellipsoidal quadrupole could be used to describe the meniscus surrounding a cylindrical particle lying with its axis of symmetry parallel to the LV interface. The description of the meniscus was not accurate close to the particle but matched beyond a few radii. The model results were verified experimentally and used to investigate the self alignment of the particles, leading the way for future modelling and fine tuning of self organizing systems.

7.6   CONCLUSIONS

Recently, there have been many interesting developments in particle and thin liquid film research, with an explosion of work applied to heterogeneous and nonspherical particles. There has also been a concerted effort to expand longstanding analyses of the 2D system into 3D. However, it is clear that there is still much research to be done.

In spite of great increases in the complexity of numerical simulations, the scale still remains relatively small. Gaining new insight into the particle stabilized thin liquid film system will require new methods to simulate the particle-film interaction as well as an increase in the scale of the models. Once it is possible to simulate hundreds or thousands of particles on a thin liquid film while simultaneously capturing the shape of the interface, it will also be possible to investigate the most complex and dynamic characteristics of the system. These include the self-ordering of particles and the effect of shape and surface heterogeneity on particle interaction and film stability.

Experimentally, there have been new approaches to capture the shape of the film around particles, and new ways of manufacturing, measuring, and altering their surface properties. High speed photography has become much more widely accessible, allowing researchers to capture images of the film right up to the moment it fails. This provides immense insight into the mechanics of the film failure process which occurs over a very short time frame. As image resolution and film speeds continue to increase, it will be possible to observe in greater detail the film failure process.

Further computational and experimental advances will help to bridge the gap in understanding between the discrete interactions between a few particles and the complex self-organizing behavior of the thousands of particles on a film. With this knowledge will come the development of particle stabilized emulsions and foams as the functionalized and tunable materials of the future.

REFERENCES

1. Pickering, S.U. *J Chem Soc* 1907; 91: 2001–2021.
2. Ramsden, W. *Proc Royal Soc* 1903; 72: 156.
3. Garrett, P. *J Colloid Interface Sci* 1979; 69: 107–121.

4. Garrett, P. *Curr Opin Colloid and Interface Sci* 2015; 20: 81–91.
5. Binks, B.P. *Curr Opin in Colloid and Interface Sci* 2002; 7: 21.
6. Kaptay, G. *Colloids Surfaces A Physicochem. Eng Asp* 2006; 282–283: 387–401.
7. Dickinson, E. *Curr Opin Colloid Interface Sci* 2010; 15: 40–49.
8. Denkov, N., Ivanov, I., Kralchevsky, P., Wasan, D.T. *J Colloid Interface Sci* 1992; 150(2): 589–593.
9. Ali, S.A., Gauglitz, P.A., Rossen, W.R. *Ind Eng Chem Res* 2000; 39: 2742–2745.
10. Morris, G., Pursell, M.R., Neethling, S.J., Cilliers, J.J. *J Colloid Interface Sci* 2008; 327: 138–144.
11. Brakke, K. *Experimental Mathematics* 1992; 1: 141–165.
12. Bournival, G., Ata, S. *Miner Eng* 2010; 23: 111–116.
13. Horozov, T.S., Aveyard, R., Clint, J.H., Neumann, B. *Langmuir* 2005; 21: 2330–2341.
14. Horozov, T.S., Aveyard, R., Clint, J.H., Binks, B.P. *Langmuir* 2003; 19: 2822–2829.
15. Morris, G.D.M., Neethling, S.J., Cilliers, J.J. *Langmuir* 2011.
16. Nushtaeva, A.V., Kruglyakov, P.M. *Coll Journ* 2003; 65: 374–382.
17. Kruglyakov, P.M., Nushtayeva, A.V., Vilkova, N.G. *J Colloid Interface Sci* 2004; 276: 465–474.
18. Kaptay, G. *Colloids Surfaces A Physicochem Eng Asp* 2003; 230: 67–80.
19. Hilden, J., Trumble, K. *J Colloid Interface Sci* 2003; 267: 463–474.
20. Cox, A.J., Neethling, S., Rossen, W.R., Schleifenbaum, W., Schmidt-Wellenburg, P., Cilliers, J.J. *Colloids Surf A: Physicochem Eng* 2004; 245: 143–151.
21. Morris, G.D.M., Neethling, S.J., Cilliers, J.J. *Colloids Surfaces A Physicochem Eng Asp* 2014; 443: 44–51.
22. Dippenaar, A. *Int J Miner Process* 1982; 9: 1–14.
23. Morris, G., Neethling, S.J., Cilliers, J.J. *Miner Eng* 2010; 23: 979–984.
24. Morris, G., Neethling, S.J., Cilliers, J.J. *J Colloid Interface Sci* 2011; 361: 370–380.
25. Morgan, A.R., Ballard, N., Rochford, L.A., Nurumbetov, G., Skelhon, T.S., Bon, S.A.F. *Soft Matter* 2013; 9: 487.
26. de Folter, J.W.J., Hutter, E.M., Castillo, S.I.R., Klop, K.E., Philipse, A.P., Kegel, W.K. *Langmuir* 2014; 30: 955–964.
27. San-Miguel, A., Behrens, S.H. *Langmuir* 2012; 28: 12038–12043.
28. Kumar, A., Park, B.J., Tu, F., Lee, D. *Soft Matter* 2013; 9: 6604.
29. Gautam, A., Jameson, G.J. *Miner Eng* 2012; 36–38: 291–299.
30. Danov, K.D., Kralchevsky, P.A. *Adv Colloid Interface Sci* 2010; 154: 91–103.
31. Danov, K.D., Kralchevsky, P.A. *J Colloid Interface Sci* 2010; 345: 505–514.
32. Lewandowski, E.P., Cavallaro, M., Botto, L., Bernate, J.C., Garbin, V., Stebe, K.J. *Langmuir* 2010; 26: 15142–15154.

Section III

Foams

8 Historical Perspectives on Foams

M. Vignes-Adler

CONTENTS

Research on foams is a long story which started at the beginning of the 19th century. Although a foam is a simple set of bubbles separated by thin films albeit with very special properties, it is very unique in science as having attracted scientists from many different disciplines, including mathematicians, physicists, chemists, physico/biochemists, chemical/food engineers, rheologists, metallurgists, theoreticians, and/or experimentalists. Over the course of centuries, all contributed to the topic with their own intellectual tools toward their own scopes and at their own scales and achieved progress almost independently until rather recent years. More than 200 years after it started, foam is a topic which is still expanding rapidly.

In this chapter, the different steps leading to the present knowledge on liquid foams are presented.

8.1   THE 19TH CENTURY: THE GREAT PRECURSORS, LAPLACE, PLATEAU, KELVIN

In 1805, by establishing the celebrated law that relates the inner pressure to the outer one of an isolated spherical bubble, Laplace introduced the fundamental concept of surface tension γ that is the major physical property of a surface between 2 fluid phases, gas/liquid or liquid 1/liquid 2 [1]

$$P_i - P_e = 2\gamma/R \qquad\qquad (8.1)$$

Later in 1873, Plateau investigated the mechanical equilibrium of a cluster of 4 polyhedral bubbles and he established the so called Plateau's rules [2]. Three films seen as singular surfaces intersect along a Plateau border (PB) with angles equal to 60°, whereas 4 Plateau borders intersect at a node with an angle equal to the tetrahedral angle $\cos^{-1}(-1/3)$ as described in Chapter 2 of this book.

Kelvin was curious about everything and he wanted to know what space-filling arrangement of cells of equal volume has a minimum surface area? In 1887, he conjectured that foams made of tetrakaidecahedral cells consisting of 6 planar quadrilateral faces and 8 nonplanar hexagonal ones with a zero mean curvature packed in a bcc structure satisfy Plateau's laws of mechanical equilibrium and divide the space with minimum partitional area. It was the first study on foam morphology [3]. More than one century later, the analysis of foam structures, called the Kelvin problem, will again be very actively worked on by mathematical physicists [4].

It is remarkable that these early studies constitute the theoretical basis of modern studies of foam structures. Laplace, Plateau, and Kelvin can be seen as the authentic founders of the field. However, in their time, their results were considered as curiosities!

8.2   THE 20TH CENTURY BEFORE WWII

During the 19th century, studies on foams remained sporadic. Things radically changed at the turn of the 20th century with the development of modern industry. The industrial revolution started at the end of the 19th century. Very soon, it required valuable minerals and metals in large quantity and high quality. Froth flotation purposely appeared as an appropriate process to separate mineral particles from gangue based on their surface properties [6]. It is generally accepted that three British chemists who specialized in metallurgy, Sulman, Picard, and Ballot, are the originators of the froth flotation process for concentrating ores preliminary to the extraction of metal (see Figure 8.1).

Starting in 1926, Gaudin from the Montana School of Mines was the first to perform fundamental research on flotation chemistry. He proposed the selective flotation of particles and he successfully tackled the problem of nonflotation of colloidal sulfide mineral particles due to their being unable to come in contact with gas bubbles because of their fine size and state of dispersion [7]. Along with mechanized mining, froth flotation allowed the economic recovery of valuable materials from much lower grade ores than before.

8.2.1   FOAMING ABILITY OF A SOLUTION

To measure the foaming ability (foaminess) of a solution, several apparatuses were designed. There are essentially two types: (i) Ross–Miles test: a foaming solution of defined volume is dropped into the same solution from a well-defined height and the obtained foam height is recorded [8]; (ii) bubbling method: a gas flow is injected into the foaming solution placed in a cylindrical container through a sintered glass plate or from a capillary at a given flow rate for a given time; then, the gas flow is stopped and the foam collapses [9]. The volume of produced foam is measured as a function of time and the destabilization time once the gas flow is stopped. As discussed by S. Ross, the first method is static whereas the second one is dynamic because foam volume results from the balance between the generated bubbles and the collapsed ones as long as the gas flow is maintained [10]. Bikerman pointed out the necessity to characterize foaminess as a property independent of the

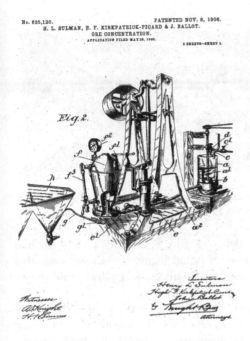

No. 835,120.   PATENTED NOV. 6, 1906.
    H. L. SULMAN, H. F. KIRKPATRICK-PICARD & J. BALLOT.
                ORE CONCENTRATION.
            APPLICATION FILED MAY 29, 1905.   2 SHEETS—SHEET 2.

Fig. 2.

The following is an example of the application of this invention to the concentration of a particular ore. An ore containing ferruginous blende, galena, and gangue consisting of quartz, rhodonite, and garnet is finely powdered and mixed with water containing a fraction of one per cent. or up to one per cent. of a mineral acid or acid salt, conveniently sulfuric acid or mine or other waters containing ferric sulfate. To this is added a very small proportion of oleic acid, (say from 0.02 per cent to 0.5 per cent on the weight of ore.) The mixture is warmed, say, to 30° to 40° centigrade and is briskly agitated in a cone mixer or the like, as in the processes previously cited, for about two and one-half to ten minutes, until the oleic acid has been brought into efficient contact with all the mineral particles in the pulp.

When agitation is stopped, a large proportion of the mineral present rises to the surface in the form of a froth or scum which has derived its power of flotation mainly from the inclusion of air-bubbles introduced into the mass by the agitation, such bubbles or air-films adhering only to the mineral particles which are coated with oleic acid. The

FIGURE 8.1 The 1905 patent: "This invention relates to improvements in the process for the concentration of ores, the object being to separate metalliferous matter from gangue by means of oils, fatty acids, or other substances which have a preferential affinity for such metalliferous matter over gangue." (http://www.freepatentsonline.com/0835120.pdf [5]).

apparatus; therefore, he defined three global foam parameters, foam maximum expansion, foaminess, foam lifetime, that are still in use today in industry [9]. The Ross-Miles test and the bubbling method are currently performed in academic laboratories and industrial sites.

8.2.2 FOAM VOLUME CONTROL: ANTIFOAMING AND DEFOAMING AGENTS, FOAM INHIBITORS

In some processes, formation of foam has to be either totally avoided or at least strongly inhibited. During World War II, aircraft engines were intensively used to fight in Europe and East Asia. It was crucial to avoid foam formation in the lubricants. In their contribution to the war effort, a group of chemists at Stanford University investigated this problem to provide recommendations to oil suppliers for formulating oils that avoided foam formation [11]. Challenging issues were the tricky composition of lubricant oils (base oil with many additives), the high temperature of the engine, and the low atmospheric pressure at the flight altitude that favored heteronucleation and the growth of bubbles. No model existed that clearly described the underlying physics, and nobody could decide which characteristics a molecule should have to be active as a foam killer or inhibitor. They tested many additives and their conclusions were largely empirical: "Heterogeneity of a bubble film is apparently the most effective cause of breakage. As a result, mixtures of two pure chemicals, an alkyl succinyl sodium sulfonate and glycerine, form an especially effective antifoaming agent. Since foam is always a result of adsorption or rearrangement of the surface of an otherwise nonfoaming liquid, suggestions are developed as to how an antifoaming agent can upset this arrangement or distribution" (from [11]).

8.2.3 FOAM FREE-DRAINAGE

Freshly made foams are wet with spherical bubbles and thick films and the excess of liquid drains out because of gravity and capillary forces. They evolve either toward a metastable equilibrium

(a)   (b)   (c)

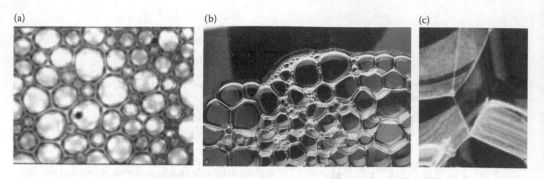

FIGURE 8.2 (a) Wet foam (courtesy of R. Höhler), (b) dry foam and (c) foam elements: film, Plateau border, and node (@Krüss GmbH Surface Science Instrumentation, www.kruss.de)

state where the foam is dry with thin films or they collapse (Figure 8.2)*. This phenomenon is called foam gravity syneresis. The metastability/instability of liquid foams is a serious issue that was simultaneously and independently investigated by chemical engineers at a global scale and by physicochemists at a local scale, respectively.

First global studies on free-drainage of foam were published in the US in the 1940's. Miles et al. [12] investigated the drainage of foams of relatively equal sized bubbles and uniform liquid fraction without concomitant foam breakdown by measuring the foam electrical conductivity that is linearly related to the foam liquid content. Using solutions of Sodium Dodecyl Sulfate (SDS) pure or with the addition of dodecanol and/or glycerol, they could show the respective influence of the higher shear dynamic bulk viscosity, the surface viscosity, and the smaller bubble size in slowing down the liquid flow through the foam. They warned against using foam drainage rate data to infer foam stability, because many different factors can influence the drainage rates.

8.2.4 THIN LIQUID FILM STABILITY

Film stability plays a decisive role in foam stability. Studies on the stability of a single liquid film sandwiched between two colloidal particles were independently performed by Deryaguin and Landau in Russia [14], and Verwey and Overbeek in the Netherlands [15] during the war years. They understood the role of the so called long-range surface forces, van der Waals attraction, electrostatic repulsion, and other short-range intermolecular forces in stabilizing the film. All these intermolecular forces give rise to the disjoining pressure that balances the capillary drainage and prevents film rupture and colloids' aggregation. This is the celebrated DLVO theory. The understanding of the action of these intermolecular forces has been a major breakthrough in the science of colloids.

8.3 POST-WWII RESEARCH ON FOAMS

After WWII, scientists went back to their academic laboratories and started again performing fundamental studies in their own scientific fields.

What is the foam state of art at the mid-20th century? The Laplace and Plateau laws are established; Kelvin introduced the space filling concept. The DLVO theory exists but is not yet applied to foam films[†]. Ore flotation is a major industrial process that still requires improvement for a more efficient recovery of solid particles. Antifoams and defoamers have been empirically developed. Foam free drainage has been investigated at a global scale. Intensive research is going to develop in this framework.

---

* Semantic point: physicochemists/physicists do not use same word for the same foam element. bubble/cell, film/face, Plateau border/edge, node/vertex. In this chapter, we use both.
† See Chapter 2 of this book.

In 1946, the botanist Matzke made an avant-gardist study [16]. By means of a syringe, he produced equal sized bubbles from a soap solution that he manually placed one by one in a cylindrical dish covered with a glass plate. He obtained disordered foam with equal sized polyhedral bubbles that he examined with a binocular. Matzke classified the bubbles according to their number m of n-edges faces. He did not observe any tetrakaidecahedral bubbles, but he observed that in the central part of the foam 99.8% of the faces are quadrilateral, pentagonal or hexagonal, with two-thirds being pentagonal, which is quite normal since the regular pentagon angle is 108°, a value that is very close to Plateau angle value of 109.47°. Sixty years after Kelvin, Matzke experimentally and precisely analyzed the morphology of dry and disordered foams.

At the same time, the crystallographers Bragg and his post-doc Nye made a raft of multiple layers of monodisperse bubbles, which formed a three-dimensional crystalline foam (Figure 8.3a) [17]. His friend Smith, who was a metallurgist, pointed out the analogy between the Bragg foam crystal and polycrystalline solids [18]. He considered "soap froth to evolve by a similar mechanism to that which governs grain growth in metals and sintered materials."

Again, these studies, considered as curiosities in their time, were forgotten for many years. Both the foam bubble morphologies and the crystalline arrangements of monodisperse foams were to be actively revisited at the very end of the 20th century.

After 1946, necessary improvements of the efficiency of industrial processes stimulated a great number of fundamental studies performed at the local scale.

8.3.1 FROTH/ORE FLOTATION

Ore flotation has two aims: (i) selective separation of particles of different mineral natures, (ii) aggregation of particles of the same nature to concentrate them. Hydrophobic particles tend to attach

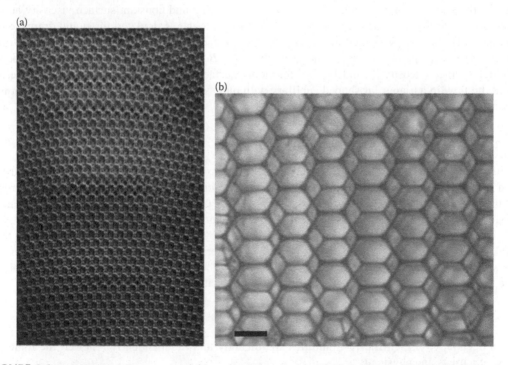

FIGURE 8.3 (a) Nye and Bragg crystal foam showing a stacking fault terminated by partial dislocations. Bubble diameter 700 μm [17]; (b) well-ordered monodisperse foam in a bcc arrangement. The scale bar represents 200 μm; Copyright (2008) American Chemical Society. (Reprinted with permission from Höhler, R. et al. *Langmuir* 2008; 24: 418–425.)

to bubbles and rise to the top of the flotation cell in the form of froth, whereas hydrophilic particles stay in the liquid phase and drain back into the tails. The most elementary model of froth flotation can be described as the capture of a specific particle by a rising bubble. For capture to occur between a particle and a bubble, they must undergo an encounter; this collision as a first step is dominated by hydrodynamics. Additionally, this encounter should be close and bring the particle and the bubble within the range of attractive forces, so that the intervening liquid film between the bubble and the particle drains and finally ruptures [19]. This second step is dominated by physicochemistry. This basic problem has been theoretically and experimentally studied over years for many materials: organic particles, coals, ink particles, hydrophilic or hydrophobic particles, any shaped, small, large, fiber-like, aggregates, mixtures of chemically different particles, and so on. The capture and selection are made efficient by the control of interfacial chemistry of appropriate surface active agents. Contributors were mostly from Australia [20], Russia [21], the United States [22], England [23], and Bulgaria [24] to name a very few. Many additional references can be found in [19]. In this context, the flotation cell technology was also much improved over the years, as an example, we cite the use of colloidal gas aphrons to capture very small particles [25] and the celebrated Jameson cell [26–29].

8.3.2 FOAM FRACTIONATION

Foam fractionation is the foaming off of dissolved material from a solution via adsorption of the solutes at the bubble surfaces. In a pioneering work, Leonard and Lemlich modeled the process by investigating a liquid flow in stationary or moving foams for steady state conditions [30]. They considered that the liquid is solely flowing in the Plateau borders' network and not in the films, and they introduced the surface viscosity to take into account the resistance to the flow caused by the adsorbed surfactants. They numerically solved the pertinent form of the Navier-Stokes equation in a PB-like capillary of noncircular cross section with finite and constant surface viscosity at its boundaries. Lemlich must be credited for being the first one who calculated the liquid drainage in a realistic PB and took into account the influence of surface viscosity by introducing the surface mobility M that is the inverse Boussinesq number $M = Bo^{-1} = \mu R_{Pb}/\mu_s$. $M = 0$ for rigid surfaces with high surface viscosity μ_s, and $M \gg 1$ for mobile surfaces with low μ_s. Later, Kumar and Desai solved the flow in a triangular shaped capillary with more sophisticated boundary conditions at its surface [31].

8.3.3 FOAM COARSENING/DYNAMICS/RIPENING*

Freshly made foams drain out of the excess liquid because of gravity and because of capillarity when the films are thin. Then as it coarsens, the smaller bubbles shrink and the larger ones grow because of gas transfer due to different capillary pressures. Coarsening is an important consequence of foam aging. Princen and Mason performed an elementary experiment on foam coarsening [32]. They investigated the permeability of foam films to gases by measuring and correctly analyzing the shrinkage of a single bubble attached below a gas-water interface. They conclude that the gas transfer simply follows a Fickian law, and together with Overbeek, they "strongly suggest that diffusion takes place through aqueous pores between the surfactant molecules" [33].

8.4 THE 1980s: FOAM RENEWAL

In this decade, two scientific communities significantly and independently renewed the topic—physicochemists and physicists.

---

* Semantic point: physicists use "coarsening" or "dynamics," whereas physicochemists use "ripening."

8.4.1 Physicochemists' Input

Formerly investigated topics were revisited, in particular those dealing with the measurement of foam liquid fraction and the foaminess control. In the 1940's, Miles et al. [12] and later Clark [13] used electrical conductivity to measure the average liquid fraction in foam. Lemlich made a further step [34]. He developed a theory for the electrical conductivity of polyhedral foam of sufficiently low density as to effectively conduct only through a random lattice of very narrow PB and not through foam films. If σ is the ratio of the electrical conductivity of the foam to the electrical conductivity of the liquid, and φ_l the foam liquid fraction, then Lemlich's relation is: $\sigma = 1/3\varphi_l$, an expression which is valid for very dry foams as $\varphi_l \rightarrow 0$. Lemlich's major contribution was the use of a realistic geometrical model for dry foam.

The inhibition of excessive foam volume and the absence of any foam have always been major industrial issues. Foam volume control requires the addition of antifoamers in the liquid before use, whereas foam breaking required defoamers to be sprinkled on the surface of an existing foam [37]. A typical antifoamer or defoamer consists of oil droplets, hydrophobic solid particles or a mixture of both. The knowledge on antifoams reported in [11] was mainly empirical. However, a two-step mechanism was recognized for the action of antifoams: an entering step when the oil droplet enters the film surface and forms a lens that spreads on the film surface, which then thins the film until its rupture. Another mechanism assumes that the film drains until the droplet emerges into its air-water surfaces; then the droplet bridges the film to form a nonuniform region extending through it, making a hole in the film and causing its rupture.

There was a need for simple criteria for the efficiency of antifoamers (defoamers). Three coefficients are involved in these processes, the entry coefficient $E = \gamma_{WA} + \gamma_{WO} - \gamma_{OA}$, the Harkins spreading coefficient $S = \gamma_{WA} - \gamma_{WO} - \gamma_{OA}$, and the bridging coefficient $B = \gamma^2_{WA} + \gamma^2_{WO} - \gamma^2_{OA}$, where W, A, and O denote the water, air, and oil phases, respectively.

In 1950, S. Ross proposed $E > 0$ for the oil droplet entering the film surfaces and $S > 0$ for spreading [35]. In 1980, Garrett considered that the oil lens spreading is not the only possible cause of rupture, which can also be caused by bridging the two film surfaces and proposed $B > 0$ as a criterion for film rupture [36]. In 2004, Denkov achieved a major contribution by investigating antifoamers made of oil and particles mixtures. He distinguished the fast antifoamers that rupture the foam films in seconds and destroy the whole foam in less than a minute, and the slow antifoams that do it in several minutes and tens of minutes (even several hours), respectively [38]. Fast antifoams rupture the films by a bridging-stretching or a bridging-dewetting mechanism. The oily globules of a slow antifoamer are unable to enter the film surfaces and are expelled in the Plateau border; only after being compressed do they enter the solution surface and destroy the adjacent foam films. In spite of obvious progress made by very talented teams [39], the antifoaming mechanisms are not yet fully understood.

8.4.2 Physicists' Input

Starting in 1983, physicists, many of them being former metallurgists or theoretical physicists, were attracted to foams in the wake of Smith [18].

Weaire and coworkers developed a novel approach by numerically simulating two-dimensional foams [40]. An example of numerical dry 2D foams satisfying the Plateau rules obtained with 2D-FROTH software is displayed in Figure 8.4a [41]. In their words, bubbles are cells separated by edges (zero thickness films), characterized by a surface tension equal to 1. They ignored surfactants and their physicochemical properties. Foams were generated from different initial networks (Voronoï [42] or Potts [43]), and concepts of mathematical physics valid for random cellular patterns were used to characterize the ordered or disordered structure and the structural rearrangements observed in the evolving numerical foams (Figure 8.4b) [44]. To some extent, they have conceptually revolutionized the topic.

(a)   (b)

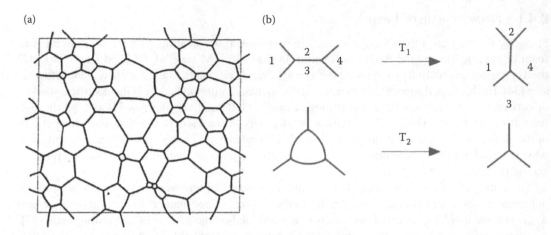

FIGURE 8.4   (a) Two-dimensional Weaire and Kermode numerical foam obtained with 2D-FROTH software [40]; (b) topological rearrangements: T_1 is observed when the edge between 2 adjacent cells decreases so much that it forms a thermodynamically unstable 4-edges node, that relaxes by swapping the cell edges. T_2 is observed as a bubble shrinks and eventually disappears without any film rupture in the foam. Similar topological rearrangements are observed in 3D-foams; (Reproduced by kind permission of C. Monnereau (1998) PhD thesis. Université de Marne-la-Vallée.)

Two-dimensional foam evolves in time for two reasons (i) coarsening, (ii) topological rearrangements also called structural changes (Figure 8.4b). Actually, in 1952, the mathematician von Neumann had demonstrated that the average area growth of an n-edges cell is a linear function of its number n of edges:

$$\frac{dA_n}{dt} = \frac{2\pi}{3}\gamma\kappa(n-6) \tag{8.2}$$

where γ is the liquid surface tension and κ is the permeability constant of the film [45]. This means that an n-edges cell shrinks if $n < 6$, it grows if $n > 6$, and remains constant if $n = 6$, the average number of edges being equal to 6. Glazier et al. [46] investigated the dynamics of a 2D soap foam made of a single layer of bubbles sandwiched between two flat parallel glass plates that they placed on the glass of a photocopier. The foam appeared as a polygonal-like cells pattern. They verified that von Neumann's law is valid on the average as long as there is no film rupture between two bubbles. They also analyzed their structural changes, that is, the topological rearrangements T_1 and T_2.

Like 2D foams, 3D foams evolve in time by gravity and capillary drainages, coarsening, and topological rearrangements. These phenomena may be coupled or not. They are observed whatever the nature and properties of the materials that compose the liquid phase.

In 1992, Brakke developed the Surface-Evolver software aimed at minimizing capillary surfaces [47], which made possible the numerical generation of dry 3D foams satisfying Plateau's laws [48]. Surface-Evolver is still considered as the best tool for representing evolving capillary surfaces [56].

As pointed out by Princen [49], foams have very peculiar rheological properties such as a high viscosity relative to their constituent bulk phases, a yield stress, and a shear thinning behavior. He investigated the elastic-plastic behavior for simple shearing flow of perfectly ordered fluid-fluid honeycombs that range in structure from a hexagonal film network to a close packed circle as the liquid fraction increases. He theoretically derived expressions for certain rheological properties, such as the stress-strain relationship, yield stress, and shear modulus of monodispersive ordered 2D foams of infinitely long cylindrical bubbles. The considered variables are the volume fraction of the dispersed phase, the bubble radius, the interfacial tension, the thickness of the films separating adjacent bubbles, and the films' associated contact angles. He made experiments to test the models by means of a specially designed rheometer [50]. Many groups investigated the rheology of ordered 2D foams, such as Armstrong et al. at MIT [51] and Kraynik et al. [52] at Sandia National Lab. Using 2D-FROTH software [40], Weaire and Fu calculated stress-strain curves for a number of numerical samples of two-dimensional, reasonably disordered foams under an extensional shear [53].

Three-dimensional wet foams with very small bubbles are white and opaque. In 1991, Durian et al. used light scattering techniques to noninvasively probe the foam time evolution, coarsening, and structural changes [54]. They employed both static and dynamic light scattering techniques and they exploited the strong multiple scattering characteristic of foams by approximating the transport of light as a diffusive process. Thus, the average bubble size can be determined directly from the total transmitted light through the foam, while the rate of rearrangement events can be determined from the temporal fluctuations of the scattered intensity. They observed a scaling behavior with the average bubble diameter growing in time as t^z where $z = 0.45 \pm 0.05$ [55] and the rate of these rearrangement events also exhibits a scaling behavior in time as t^y, where $y = 2.0 \pm 0.2$. This scaling behavior necessarily implies that the bubble size distribution is statistically self-similar and occurs in the absence of film ruptures.

8.5 FOAM GOLDEN AGE

In 1994, the EC network "Physics of foams" was created with five European teams which were formerly metallurgists (D. Weaire and M. Forte), a mathematical physicist (N. Rivier), an experimental physicochemist (D. Langevin), and an experimental physicist (J. Earnshaw). In this framework, Weaire invited the foam community to meet in Renvyle, a small village on the western Irish coast. It was the first conference in the world of this type: physicists, mathematicians, numerical analyst physicists, physicochemists, rheologists, and industrial researchers could share their common interest in foam science. Actually, physicists and physicochemists in the broad sense mostly ignored each other in Renvyle. However, within a few years, physicochemists discovered the world of numerical foams, random cell patterns, structural rearrangements, and the Weaire-Phelan foam that more efficiently partitions space than Kelvin's tetrakaidecahedral foam [57]. Physicists discovered that amphiphilic molecules named surfactants with very special surface properties were necessary to stabilize the foam films. Two worlds met and started working together in a remarkable and long-lasting fruitful collaboration regularly reporting their results at the biennial Eufoam Conference series.

Since then, foam research has known a fantastic development performed by many teams with different scientific backgrounds from many countries boosted by theoretical and technological progress, and many more uses of foams in industry and everyday life. One can distinguish four main directions.

8.5.1 FOAM FORCED DRAINAGE

In 1993, Weaire et al. generated a monodispersive ordered foam in a long vertical tube by blowing N_2 bubbles in a local commercial detergent solution at a constant flow rate [58]. Once the foam had drained to a state close to its equilibrium, further solution was introduced from the top at a steady rate Q, rewetting the dry foam*. They observed that the boundary to the dry foam remained very sharp, and that its velocity v and the gas fraction Φ follow a power law as

$$v \sim Q^{1/2} \quad \text{and} \quad (1 - \Phi) \sim Q^{1/2} \tag{8.3}$$

---

* Miles et al. mentioned the possibility of rewetting foam when it has fully drained to black films without breakdown [12].

They called it forced drainage by opposition to free drainage as wet foam is spontaneously draining [59]. Neglecting any contribution of liquid flow in films to the drainage, assuming that the flow in the PB's is of a Poiseuille type and using an orientational averaging procedure to take into account the random network of the PB's, they derived the so-called "foam drainage equation." It is a nonlinear partial differential equation for the foam density as a function of time and vertical position in the foam column [60].

Stone et al. reproduced the forced drainage experiment by using their local commercial detergent solution and they found a different exponent [61]

$$v \sim Q^{1/3} \tag{8.4}$$

The injected liquid flows through the PB's network. The flow is a Poiseuille one if the surface is rigid (very high surface viscosity and zero velocity at the PB surface) so that the viscous dissipation occurs in the PB or it is a plug flow if the surface is mobile (low surface viscosity) whereas the viscous dissipation occurs in the nodes. In 1999, using a well-defined surfactant solution of SDS with/without dodecanol*, with the very insoluble dodecanol rigidifying the bubble surfaces, Langevin et al. confirmed the above scaling laws and explained these two behaviors in terms of a transition between a node-dominated and a PB-dominated viscous dissipation, for which theory predicts, respectively, $\alpha = 1/3$ and $\alpha = 1/2$ [62].

Many more groups, physicists and physicochemists, worked at the global scale on forced or free foam drainage theoretically [63–66], experimentally by visualization of the flow streams [67,68], and at the local scale in a single Plateau border, theoretically [69] and experimentally [70], for different surface rheology, rigid, mobile, viscoelastic, for Newtonian or nonNewtonian liquids, for any kinds of surfactants, ionic, nonionic, zwitterionic, mixtures of surfactants, for proteins, polymers, and mixtures made of surfactants and polymers, and so on.

More complex aspects of foam drainage were also investigated, for example, the role of the films and of the Marangoni stresses generated by the liquid flow in Plateau borders [71]. The liquid is located in both the films and PB: although they are connected, they are not influencing the foam in the same way. In the context of foam forced drainage, related questions were the following. Is the liquid only flowing in the PB's network or are films swelling because of drainage [67]? Is there any influence of the topological rearrangements on the global foam drainage [72,73]? What is the influence of drainage on foam stability? Carrier and Colin pointed out a coupling between drainage and coalescence [73]. They suggest that the rearrangements of the bubbles during the drainage of the foam induce an increase of the area of the bubbles, which temporarily decreases the amount of adsorbed surfactant by unit area and weakens the interfaces.

8.5.2 Structure and Dynamics of Three-Dimensional Foams

The structure, disorder, and dynamics were topics of interest. The purpose was to investigate the 3D analog of von Neumann's law and to analyze their disordered structure. First studies were numerically performed by Glazier [74]. Then new experimental devices made possible the probing of detailed internal foam structure. Gonatas et al. used nuclear magnetic resonance imaging to visualize the bubbles inside the foam [75]. Monnereau and Vignes-Adler used optical tomography associated to a foam reconstruction algorithm to reconstruct the real structure of slightly disordered dry foam with rigid films. Using Surface-Evolver, they could determine the time dependency of the average volume of the f-faces internal bubbles $\langle V_f(t) \rangle$ as

$$\langle V_f(t) \rangle^{-1/3} \frac{d\langle V_f(t) \rangle}{dt} = k(f - f_0)$$

---

* Same composition was used by Miles et al. in 1945 [12].

where k is a diffusion coefficient and f_0 is the average number of faces per bubble [76] and characterize the disorder in the foams. Hilgenfeldt et al. obtained a similar relation for numerical foams [77].

Foam density variations were detected by electrical resistance tomography in [78,79]. Gopal and Durian used diffusing wave spectroscopy to investigate bubble dynamics [80], while Cohen-Addad and Höhler developed multispeckle diffusing wave spectroscopy [81] to investigate the structural changes in relaxing foams. Later, Lambert et al. used X-ray tomography [82], whereas the very powerful Fast X-ray tomography at the European Synchrotron Radiation Facility offered the possibility to examine the local changes in the structure of flowing foam allowing a better fundamental understanding of foam rheology and the validation of rheological models [83].

8.5.3 Rheology of 3D Foams

Despite being constituted only of fluids, aqueous foams exhibit either solid-like or liquid-like mechanical behavior, depending on the applied stress and their liquid content. Under a small shear, foam behaves as an elastic solid, its elasticity arising from the surface tension of the films. For sufficiently large deformations, topological changes are induced that are not immediately reversible if the deformation is reduced; the foam becomes progressively plastic. Above a certain yield stress, the foam starts flowing [84]. At the macroscopic scale, foam behaves as a shear-thinning fluid. The relation between shear stress and strain rate is described by the Herschel-Bulkley phenomenological law [85]. Many efforts have since been devoted to foam rheology [86].

8.5.4 Foams in Industry

Numerous uses of foams in industries fostered research. The use of foams to enhance oil recovery is a celebrated example [87,88] as well as lightweight metallic foams [89] and building materials [90,91]. Recently, food foams were developed to produce aerated foods with gas bubbles fully or partially replacing dispersed fat particles that can be found in emulsion based foods [92].

8.6 NEW FOAMS

In the 21st century, research focuses on "new" foams, more precisely monodisperse foams, particles stabilized foams, and responsive foams.

8.6.1 Monodisperse Foams

In 2001, Gañán-Calvo and Gordillo produced monodisperse foams by capillary flow focusing, from surfactant free mixtures of water-ethanol and water-glycerol [93]. The foams are wet, crystal-like ordered with equal sized spherical bubbles of diameter ranging from 5 to 120 μm. As long as the topological order of these crystalline foams is maintained, they are very stable in spite of the absence of surfactants, and do not coarsen because there is no pressure drop between 2 adjacent bubbles. Later, Lorenceau et al. were able to produce calibrated foams at a high flow rate and to observe their crystalline order [94], which was very systematically investigated by van der Net et al. [95]. Since these pioneering works, monodisperse foams have been studied by many groups in many configurations [96] and now they serve as the precursor for solid porous media with micrometric, regularly packed pores. A good review can be found in [97].

8.6.2 Particles Stabilized Foams

As early as 1903, Ramsden from Pembroke College in Oxford reported that bubbles armored by very small solid particles possess a very strong stability [98]. This result was forgotten for 100 years and has only recently been rediscovered [99]. In 2004, Alargova et al. described the first reported experiment on superstable wet foams solely stabilized by solid microparticles, more precisely by polymer microrods [100]. The physical reason for the improved efficiency of particles over surfactants in stabilizing foams is their attachment energy that can be up to several thousand kT per particle, where k is the Boltzmann constant and T the absolute temperature, whereas it is only a few kT per surfactant molecule. Hence, particle adsorption at the surface can reasonably be considered as irreversible. This requires a contact angle close to 90° so that the particle is not preferentially wetted either by the gas or by the liquid [101]. Much effort has been devoted to superstable foams stabilized by micro/nano particles. A few examples are listed below [102–105].

8.6.3 STIMULI-RESPONSIVE FOAMS

In 2006, the first stimuli-responsive foams were obtained when Middelberg designed pH sensitive peptide surfactants that can be reversibly switched between a mechanically strong cohesive film state and a mobile detergent state in response to changes in the solution conditions [106]. In the wake of Middelberg, many groups worked on stimuli responsive foams, and there are now foams that can reversibly change their stability under pH, thermal, magnetic triggering, and by UV illumination [107–111]. The list is definitely not limitative.

8.7 CONCLUSION

It is impressive to see how much knowledge about foams has progressed since the beginning of the last century. Let us summarize the major milestones.

The 1905 patent for ore flotation is the beginning of the industrial history of foams.

The 1940's DLVO theory for film stability is the first consideration of long-range intermolecular forces.

In 1946, the first investigation of foam morphology was made by Matzke.

In 1968, first calculations of the flow in a realistic Plateau border as a function of the interfacial mobility were proposed by Lemlich.

In 1983, first investigations of foam rheology by Princen were made, as well as first numerical foams and analysis of the foam internal structure by physicists were proposed. 1991 is the year of the first nondestructive probing of the foam internal structure by light scattering techniques.

In 1994, the seminal Renvyle conference was held and solely dedicated to foams where physicists, physicochemists, modelers, mathematicians, and food scientists met for the first time. It was a major event, the beginning of pluridisciplinary studies of foams.

Since 2001, we have a number of new foams which have renewed the field.

REFERENCES

1. Laplace, P.S. *Traité de Mécanique Céleste*. A Paris : de l'imprimerie De Crapelet. 1805.
2. Plateau, J.A.F. Statique expérimentale et theoryque des liquides soumis aux seules forces moléculaires. *Gauthier-Villars Trubner et cie F Clemm* 2: 1873.
3. Thomson, W. (Lord Kelvin) On the division of space with minimum partitional area. *Phil Mag* 1887; 24: 151–503.
4. Weaire, D. (Ed.) *The Kelvin Problem Foam Structures of Minimal Surface Area*. Taylor and Francis Ltd: London, 1996.
5. Sulman, H.L., Picard, H.F.K., Ballot, J. British Patent 7803 April 12; duplicated as US Patent 835 120, May 29, 1905.
6. Fuerstenau, M.C., Jameson, G., Yoon, R.H. (Eds.) *Froth Flotation: A Century of Innovation*. Society of Mining, Metallurgy, and Exploration Inc Publ, 2007.
7. Gaudin, A.M. *Flotation Mechanism, A Discussion of the Functions of Flotation Reagents, AIME Technical Publications 4*. American Institute of Mining and Metallurgical Engineers: New York, 1927.
8. Ross, J., Miles, G.D. An apparatus for comparison of foaming properties of soaps and detergents. *Oil & Soap* 1941; 18(5): 99–102.

9. Bikerman, J.J. The unit of foaminess Trans. *Faraday Soc* 1938; 34: 634–638.
10. Ross, S. Current methods of measuring foam. *Industrial and Engineering Chemistry* 1943; 15(5): 329–334.
11. McBain, J.W., Ross, S., Brady, A.P., Robinson, J.V., Abrams, I.M., Thorburn, M.R.C., Lindquist, C.G. *Foaming of aircraft-engine oils as a problem in colloid chemistry*, NACA AAR No 4105, War time report, Sept. 1944.
12. Miles, G.D., Shedlovsky, L., Ross, J. Foam drainage. *J Phys Chem* 1945; 49(2): 93–107.
13. Clark, N.O. The electrical conductivity of foam. *Trans Faraday Soc* 1948; 44: 13–15.
14. Derjaguin, B.V., Landau, L.D. Theory of the stability of strongly charged lyophobic sols and of the adhesion of strongly charged particles in solutions of electrolytes. *Acta Physicochim USSR* 1941; 14: 633.
15. Verwey, E.J.W., Overbeek, J.Th.G. *Theory of the Stability of Lyophobic Colloids*. Elsevier, 1948.
16. Matzke, E.B. The three-dimensional shape of bubbles in foam—An analysis of the role of surface forces in three-dimensional cell shape determination. *Am J Bot* 1946; 33: 58–80.
17. Bragg, L., Nye, J.F. A dynamical model of a crystal structure. *Proc Roy Soc London Series A* 1947; 190: 474–481.
18. Smith, C.S. On blowing bubbles for Bragg's dynamic crystal model. *J Applied Phys* 1949; 20(6): 631.
19. Ralston, J., Fornasiero, D., Grano, S. Pulp and solution chemistry, 227–58 in [6].
20. Sutherland, K.L. Physical chemistry of flotation. XI: Kinetics of the flotation process. *J Phys Chem* 1948; 52(2): 394–425.
21. Deryaguin, B.V., Dukhin, S.S. Theory of flotation of small and medium-sized particles. *Trans Inst Min Metall* 1960–1961; 70: 221–246.
22. Nutt, C.W., Froth flotation: The adhesion of solid particles to flat interfaces and bubbles. *Chem Eng Sci* 1960; 12(2): 133–141.
23. Kitchener, J.A. The froth flotation process: past, present and future - in brief. In: Ives, K.J. (Ed.) *The Scientific Basis of Flotation, Proc Nato Advanced Study Inst* Cambridge, U.K., jul. 5-16, 1982.
24. Scheludko, A., Tschaljowska, S.I., Fabrikant, A. Contact between a gas bubble and a solid surface and froth flotation. *Spec Discuss Faraday Soc* 1970; 1: 112–117.
25. Halsey, G.S., Yoon, R.H., Sebba, F. Cleaning of fine coal by flotation using colloidal gas aphrons. *Proc Technical Program - International Powder and Bulk Solids Handling and Processing* 1982; 67–75.
26. Jameson, G.J., Nam, S., Young, M.M. Physical factors affecting recovery rates in flotation. *Mineral Sci Eng* 1977; 9(3): 103–118.
27. Clayton, R., Jameson, G.J., Manlapigi, E.V. The development and application of the Jameson cell. *Minerals Engineering* 1991; 4(7-a11): 925–933.
28. Trahar, W.J., Warren, L.J. The flotability of very fine particles --a review. *Int J Mineral Processing* 1976; 3: 103–131.
29. Ahmed, N., Jameson, G.J. The effect of bubble size on the rate of flotation of fine particles. *Int J Mineral Processing* 1985; 14: 195–215.
30. Leonard, R.A., Lemlich, R. A study of interstitial liquid flow in foam. *Part I. Theoretical model and application to foam fractionation AIChE J* 1965; 11: 18–25.
31. Desai, D., Kumar, R. Flow through a Plateau border of a cellular foam. *Chem Eng Sci* 1982; 37(9): 1361–1370.
32. Princen, H.M., Mason, S.G. Permeability of soap films to gases. *J Colloid Sci* 1965; 20: 353–375.
33. Princen, H.M., Overbeek, Th.G., Mason, S.G. Permeability of soap films to gases II. A simple mechanism of monolayer permeability. *J Colloid Interface Sci* 1967; 24: 125–130.
34. Lemlich, R. A theory for the limiting conductivity of polyhedral foam at low density. *J Colloid Interface Sci* 1978; 64(1): 107–110.
35. Ross, S. coworkers, Inhibition of foaming. *J Phys Colloid Chem* 1950; 54(3): 429–36, *J Phys Chem* 1953; 57(7): 684–686, 1954; 58(3): 247–250, 1956; 60(9): 1255–1258, 1957; 61(10): 1261–1265, 1958; 62(10): 1260–1264.
36. Garrett, P.R. Preliminary considerations concerning the stability of a liquid heterogeneity in a plane parallel liquid film. *J Colloid Interface Sci* 1980; 76(2): 587–590.
37. Pugh, R.J. Foaming, foams films, antifoaming and defoaming. *Adv Colloid Interface Sci* 1996; 64: 67–142.
38. Denkov, N.D. Mechanisms of foam destruction by oil-based antifoams. *Langmuir* 2004; 20: 9463–9505.
39. Denkov, N.D., Marinova, K.G., Tcholakova, S.S. Mechanistic understanding of the modes of action of foam control agents. *Adv Colloid Interface Sci* 2014; 206: 57–67.

40. Weaire, D., Kermode, J.P. The evolution of the structure of a two-dimensional soap froth. *Phil Mag B* 1983; 47(3): L29–L31; Computer simulation of a two-dimensional soap froth - I. method and motivation. *Phil Mag B* 1983; 48(3): 245–259; - II. Analysis of results. *Phil Mag B* 1984; 50(3): 379–395.

41. Kermode, J.P., Weaire, D. 2D-FROTH: A program for the investigation of 2-dimensional froths. *Computer Physics Communications* 1990; 60(1): 75–109.

42. Weaire, D., Kermode, J.P., Wejchert, J. On the distribution of cell areas in a Voronoi network. *Phil Mag B* 1986; 53(5): 101–105.

43. Glazier, J.A., Anderson, M.P., Grest, G.S. Coarsening in the two-dimensional soap froth and the large-Q Potts model: A detailed comparison. *Phil Mag B* 1990; 62(6): 615–647.

44. Rivier, N. Structure of random cellular networks and their evolution. *Physica D: Nonlinear Phenomena* 1986; 23(1-3): 129–137; Order and disorder in packings and froths in Disorder and Granular Media Eds Bideau D. and Hansen A. 1993; 3: 55-102, Elsevier.

45. von Neumann, J. *Discussion – Shape of Metal Grains in "Metal Interfaces"*. American Society for Metals, (Publ.): Cleveland, 1952, p. 108.

46. Glazier, J.A., Gross, S.P., Stavans, J. Dynamics of two-dimensional soap froths. *Phys Rev A* 1987; 36(1): 306–312.

47. Brakke, K.A. The Surface Evolver. *Exp Math* 1992; 1: 141–165.

48. Phelan, R., Weaire, D., Brakke, K. Computation of equilibrium foam structures using the Surface Evolver. *Exp Math* 1995; 4(3): 181–192.

49. Princen, H.M., coworkers. Rheology of foams and highly concentrated emulsions: I. Elastic properties and yield stress of a cylindrical model system. *J Colloid Interface Sci* 1983; 91(1): 160–175, 1985; 105(1): 150–171, 1986; 112(2): 427–437.

50. Princen, H.M. A novel design to eliminate end effects in a concentric cylinder viscometer. *J Rheology* 1986; 30(2): 271–283.

51. Armstrong, R.C., coworkers. Rheology of foams. *J. Non-Newtonian Fluid Dynamics* 1986; 22: 1–22, 1987; 25(1): 61–92, 1988; 22: 69–92.

52. Kraynik, A.M. Foam flows. *Ann Rev Fluid Mech* 1988; 20: 327–357.

53. Weaire, D., Fu, T.L. The mechanical behavior of foams and emulsions. *J Rheology* 1988; 32(3): 271–283.

54. Durian, D.J., Weitz, D.A., Pine, D.J. Multiple light-scattering probes of foam structure and dynamics. *Science* 1991; 252: 686–688.

55. Durian, D.J., Weitz, D.A., Pine, D.J. Scaling behavior in shaving cream. *Phys Rev A* 1991; 44: R7902(R).

56. Cox, S.J. A viscous froth model for dry foams in the Surface Evolver. *Colloids Surfaces A: Physicochemical and Engineering Aspects* 2005; 263: 81–89.

57. Weaire, D., Phelan, R. A counter-example to Kelvin's conjecture on minimal surfaces. *Phil Mag Lett* 1994; 69(2): 107–110.

58. Weaire, D., Pittet, N., Hutzler, S., Pandal, D. Steady-state drainage of an aqueous foam. *Phys Rev Lett* 1993; 71(16): 2670–2673.

59. Bhakta, A., Ruckenstein, E. Decay of standing foams: Drainage, coalescence and collapse. *Adv Colloid Interface Sci* 1997; 70(1-3): 1–123.

60. Verbist, G., Weaire, D., Kraynik, A.M. The foam drainage equation. *J Phys Condensed Matter* 1996; 8(21): 3715–3731.

61. Koehler, S.A., Hilgenfeldt, S., Stone, H.A. Liquid flow through aqueous foams: The node-dominated foam drainage equation. *Phys Rev Lett* 1999; 82(21): 4232–4235.

62. Durand, M., Martinoty, G., Langevin, D. Liquid flow through aqueous foams: From the Plateau border-dominated regime to the node-dominated regime. *Phys Rev E* 1999; 60: R6307–R6308.

63. Goldfarb, I.I., Kann, K.B., Schreiber, I.R. Liquid flow in foams. *Fluid Dynamics* 1988; 23(2): 244–249.

64. Koehler, S., Hilgenfeldt, S., Stone, H.A. Generalized view of foam drainage: Experiment and theory. *Langmuir* 2000; 16(15): 6327–6341.

65. Neethling, S.J., Lee, H.T., Cilliers, J.J. A foam drainage equation generalized for all liquid contents. *J Physics Condensed Matt* 2002; 14: 331–342.

66. Durand, M., Langevin, D. Physicochemical approach to the theory of foam drainage. *Eur Physical J E* 2002; 7(1): 35–44.

67. Carrier, V., Destouesse, S., Colin, A. Foam drainage: A film contribution? *Phys Rev E* 2002; 65(6): 061404–061409.

68. Koehler, S.A., Hilgenfeldt, S., Weeks, E.R., Stone, H.A. Drainage of single Plateau borders: Direct observation of rigid and mobile interfaces. *Phys Rev E* 2002; 66(4): 040601–040604.

69. Nguyen, A.V. Liquid Drainage in Single Plateau Borders of Foam. *J Colloid Interface Sci* 2002; 249(1): 194–199.

70. Pitois, O., Fritz, C., Vignes-Adler, M. Liquid drainage through aqueous foam: Study of the flow on the bubble scale. *J Colloid Interface Sci* 2005; 282: 458–465.

71. Pitois, O., Louvet, N., Rouyer, F. Recirculation model for liquid flow in foam channel. *Eur Physical J E* 2009; 30(1): 27–35.

72. Saint-Jalmes, A., Langevin, D. Time evolution of aqueous foams: drainage and coarsening. *J Phys Condensed Matt* 2002; 14(40): 9397–9412.

73. Carrier, V., Colin, A. Coalescence in draining foams. *Langmuir* 2003; 19(11): 4535–4538.

74. Glazier, J.A. Grain growth in three dimensions depends on grain topology. *Phys Rev Lett* 1993; 70: 2170–2173.

75. Gonatas, C.P., Leigh, J.S., Yodh, A.G., Glazier, J.A., Prause, B. Magnetic resonance images inside a foam. *Phys Rev Lett* 1995; 75(3): 573–576.

76. Monnereau, C., Vignes-Adler, M. Dynamics of real three-dimensional foams. *Phys Rev Lett* 1998; 80(23): 5228–5231.

77. Hilgenfeldt, S., Kraynik, A.M., Koehler, S.A., Stone, H.A. An accurate von Neumann's law for three-dimensional foams. *Phys Rev Lett* 2001; 86(12): 2685–2688.

78. Phelan, R., Weaire, D., Peters, E.A.J.F., Verbist, G. The conductivity of a foam. *J Phys Condensed Matt* 1996; 8(34): L475–L482.

79. Wang, M., Cilliers, J.J. Detecting non-uniform foam density using electrical resistance tomography. *Chem Eng Sci* 1999; 54(5): 707–712.

80. Gopal, A.D., Durian, D.J. Shear-induced "melting" of an aqueous foam. *J Colloid Interface Sci* 1999; 213(1): 169–178.

81. Cohen-Addad, S., Höhler, R. Bubble dynamics relaxation in aqueous foam probed by multispeckle diffusing-wave spectroscopy. *Phys Rev Lett* 2001; 86(20): 4700–4703.

82. Lambert, I., Cantat, I., Delannay, R., Renault, A., Graner, F., Glazier, J.A., Veretennikov, I., Cloetens, P. Extraction of relevant physical parameters from 3D images of foams obtained by X-ray tomography. *Colloids Surfaces A: Physicochem Eng Aspects* 2005; 263: 295–330.

83. Davies, I.T., Cox, S.J., Lambert, J. Reconstruction of tomographic images of dry aqueous foams. *Colloids and Surfaces A: Physicochemical and Engineering Aspects* 2013; 438: 33–40.

84. Rouyer, F., Cohen-Addad, S., Vignes-Adler, M., Höhler, R. Dynamics of yielding observed in a three-dimensional aqueous dry foam. *Phys Rev E* 2003; 67: 021405.

85. Höhler, R., Cohen-Addad, S. Rheology of liquid foam. *J Physics Condensed Matter* 2005; 17(41): R1041–R1069.

86. Cohen-Addad, S., Höhler, R., Pitois, O. Flow in foams and flowing foams. *Ann Rev Fluid Mechanics* 2013; 45: 241–267.

87. Bergeron, V., Fagan, M.E., Radke, C.J. Generalized entering coefficients: A criterion for foam stability against oil in porous media. *Langmuir* 1993; 9(7): 1704–1713.

88. Rossen, W.R. Foams in enhanced oil recovery. In: Prud'homme, R.K., Khan, S. *Foams: Theory Measurement and Applications*. Marcel Dekker, New York City, 1996.

89. Banhart, J. Manufacture, characterisation and application of cellular metals and metal foams. *Progress in Materials Science* 2001; 46: 559–632.

90. Gutiérrez-González, S., Gadea, J., Rodríguez, A., Junco, C., Calderón, V. Lightweight plaster materials with enhanced thermal properties made with polyurethane foam wastes. *Construction Building Materials* 2012; 28(1): 653–658.

91. Akthar, F.K., Evans, J.R.G. High porosity (> 90%) cementitious foams. *Cement and Concrete Research* 2010; 40(2): 352–358.

92. Murray, B.S., Ettelaie, R. Foam stability: Proteins and nanoparticles. *Curr Opinion Colloid Interface Sci* 2004; 9(5): 314–320.

93. Gañán-Calvo, A.M., Gordillo, J.M. Perfectly monodisperse microbubbling by capillary flow focusing. *Phys Rev Lett* 2001; 87: 274501.

94. Lorenceau, E., Sang, Y.Y.C., Höhler, R., Cohen-Addadm, S. A high rate flow-focussing foam generator. *Phys Fluids* 2006; 18: 097103.

95. van der Net, A., Delaney, G.W., Drenckhan, W., Weaire Hutzler, S. Crystalline arrangements of microbubbles in monodisperse foams. *Colloids Surfaces A: Physicochem. Eng Aspects* 2007; 309: 117–124.

96. Höhler, R., Sang, Y.Y.C., Lorenceau, E., Cohen-Addad, S. Osmotic pressure and structures of monodisperse ordered foam. *Langmuir* 2008; 24: 418–425.

97. Drenckhan, W., Langevin, D. Monodisperse foams in one to three dimensions. *Curr Opinion Colloid Interface Sci* 2010; 15(5): 341–348.

98. Ramsden, W. Separation of solids in the surface-layer of solutions and 'suspensions' (observation on surface-membranes, bubbles, emulsions, and mechanical coagulation). – preliminary account. *Proc Roy Soc London* 1903; 72: 156–164.

99. Du, Z., Bilbao-Montoya, M.P., Binks, B.P., Dickinson, E., Ettelaie, R., Murray, B.S. Outstanding stability of particle-stabilized bubbles. *Langmuir* 2003; 19: 3106–32108.

100. Alargova, R.G., Warhadpande, D.S., Paunov, V.N., Velev, O.D. Foam superstabilization by polymer microrods. *Langmuir* 2004; 20: 10371–10374.

101. Dickinson, E., Ettelaie, R., Kostakis, T., Murray, B.S. Factors controlling the formation and stability of air bubbles stabilized by partially hydrophobic silica nanoparticles. *Langmuir* 2004; 20(20): 8517–8525.

102. Binks, B.P., Horozov, T.S. Aqueous foams stabilized solely by silica nanoparticles. *Angew Chem Int Ed* 2005; 44: 3722–3725.

103. Hunter, T.N., Pugh, R.J., Franks, G.V., Jameson, G.J. The role of particles in stabilizing foams and emulsions. *Adv Colloid Interface Sci* 2008; 137: 57–58.

104. Gonzenbach, U.T., Studart, A.R., Tervoort, E., Gauckler, L.J. Ultrastable particle-stabilized foams. *Angew Chem Int Ed* 2006; 45: 3526–3530.

105. Kunieda, H., Shrestha, L.K., Acharya, D.P., Kato, H., Takase, Y., Gutiérrez, J.M. Super-stable nonaqueous foams in diglycerol fatty acid esters—non polar oil systems. *J dispersion Sci Tech* 2007; 28: 133–142.

106. Malcom, A.S., Dexter, A.F., Middelberg, A.P.J. Foaming properties of a peptide designed to form stimuli-responsive interfacial films. *Soft Matter* 2006; 2: 1957–1966.

107. Salonen, A., Langevin, D., Perrin, P. Light and temperature bi-responsive foams. *Soft Matter* 2010; 6: 5308–5311.

108. Fameau, A.-L., Saint-Jalmes, A., Cousin, F., Houinsou-Houssou, B., Novales, B., Navailles, L., Nallet, F., Gaillard, C., Boué, F., Douliez, J.-P. Smart foams: Switching reversibly between ultrastable and unstable foams. *Angew Chem Int Ed* 2011; 50: 8264–8269; erratum Angew Chem Int Ed 2011;50:8414.

109. Chevallier, E., Monteux, C., Lequeux, F., Tribet, C. Photofoams: remote control of foam destabilization by exposure to light using azobenzene surfactant. *Langmuir* 2012; 28: 2308–23012.

110. Lam, S., Blanco, E., Smoukov, S.K., Velikov, K.P., Velev, O.D. Magnetically responsive Pickering foams. *J Am Chem Soc* 2011; 133: 13856–13859.

111. Carl, A., von Klitzing, R. Smart foams: New perspective towards responsive composite materials. *Angew Chem Int Ed* 2011; 50: 11290–11292.

RECENT BOOKS

112. Weaire, D., Hutzler, S. *The Physics of Foams*. Cambridge Univ. Press: London, 2000.

113. Stevenson, P. (Ed.) *Foam Engineering: Fundamentals and Applications*. Wiley, 2012.

114. Cantat, I., Cohen-Addad, S., Elias, F., Graner, F., Höhler, R., Pitois, O., Rouyer, F., Saint-Jalmes, A. *Foams: Structure and Dynamics*. CPI Group (UK) Ltd: Croydon, 2013.

9 Fundamentals of Foam Formation

Wiebke Drenckhan, Aouatef Testouri, and Arnaud Saint-Jalmes

CONTENTS

The formation of foams takes place when closely packed gas bubbles are generated within a foaming solution. Numerous techniques have been developed in the past to obtain control over the different structural parameters of foams, such as the gas fraction Φ (=gas volume/foam volume) or the bubble size distribution [1]. Depending on the technique, bubble sizes may range from micrometers to centimeters, bubbles can be monodisperse or polydisperse, and gas fractions can vary over the entire range. Unfortunately, to most users' despair, each individual technique typically covers only a relatively small range of these parameters. Moreover, the choice of a foaming technique is not only guided by the properties of the obtained foam, but also by its production rate. Hence, any academic or industrial foaming application needs to start with a wise choice of the appropriate foaming technique(s).

The common task in all foaming techniques is the generation of bubbles within a liquid. This requires the creation of a gas/liquid interface of surface tension γ which, in turn, implies an energy input of $U = 4\gamma\pi r_B^2$ for a bubble of radius r_B. For typical values of surface tension and bubble radius, this energy is several orders of magnitude larger than thermal energies (kT). This means that bubble formation is not a spontaneous process and requires a large energy input into a liquid when creating foam. Different foaming techniques put this energy into the liquid in different ways, using physical, chemical or even biological means. Physical means include mechanical action, such as gas sparging, whipping, shaking or phase transitions like boiling, cavitation or effervescence. In chemical techniques, bubbles are formed either by a gas-releasing chemical or electrochemical reaction (electrolysis), whereas most common biological approaches rely on gas-generating species such as yeast. In this chapter, we concentrate on physical foaming techniques.

There are one-step or two-step processes for forming a foam. In one-step processes, the generated bubbles immediately form the final foam at a well defined gas fraction. However, most foaming techniques are based on two-step processes. In one scenario, isolated bubbles are generated in a liquid and subsequently compacted to create the final foam. Alternatively, a coarse foam with large bubbles is produced, which is then broken up into smaller bubbles to produce the final foam.

This chapter is dedicated to the description of the fundamental mechanisms which lead to the initial bubble formation. These may be grouped conceptually into two categories. The first category requires a topological change in order to create a bubble. This topological change can occur in many different ways, such as the bubble detachment from a nozzle or the break-up of a large bubble into smaller ones. These mechanisms differ from each other in the way the gas and the liquid phase flow with respect to each other, as is discussed in Section 9.3. Even if the initial mechanism of creating a pocket-like gas/liquid interface may be very different, the final break up mechanism leading to the topological change is the same: the gas/liquid interface has to be deformed into a slender filament which is physically unstable and breaks to make the topological transition. The physics of this instability is described in more detail in Section 9.2.

The second category does not require any topological change as a freely floating spherical bubble is directly created within the liquid by a phase transition or by a chemical reaction. For this type of bubble creation, various physical mechanisms are discussed in more detail in Section 9.4. Some bubble generation mechanisms may even belong to both classes. For example, a phase transition can be favored by the presence of walls or impurities from which the bubbles need to detach via a topological change.

In both categories, a number of different stresses are involved in the process of bubble creation. We briefly review these in Section 9.1. and discuss how they may be grouped into nondimensional numbers for a coherent description of the different bubbling processes.

Let us start, however, with some comments on the physicochemical aspects of foaming. A foaming solution has to contain stabilizing agents as will be shown in subsequent Chapters of this book sections of this chapter (see Chapters 13–15). The properties of any obtained foam depend sensitively on the extent of bubble coalescence occurring during the foaming process. In many foaming techniques, the properties of final foam are then given by a subtle equilibrium between the generation and coalescence of bubbles. Moreover, the stabilizing agents may have characteristic times of adsorption, which have to be compared with those of bubble generation and the input of energy during the foaming process. For example, if the bubbling process is faster than the equilibration time of the foaming agent at the bubble surface, the surface tension will be higher than the equilibrium surface tension [2]. Moreover, the foaming agents provide viscoelastic properties to the bubble surfaces [3], adding important interfacial stresses to the bubble formation process. In the following, to simplify our discussion, we shall neglect these additional effects. We will assume that bubbles are indefinitely stable and their interfaces have a constant surface tension value γ which is reached immediately after the creation of each bubble.

The formulation of the foaming solution can also significantly modify the bulk rheology of the liquid phase. This could lead to a non-Newtonian behavior of the liquid and consequently to different conditions for the bubble formation [4]. For the sake of clarity, we will limit our description to the foaming of simple, that is, Newtonian, fluids.

9.1 THE PHYSICS OF INTERFACIAL DEFORMATION

The process of bubble formation involves the creation and deformation of a gas/liquid interface. Even if the bubble is formed in a nonmechanical way (such as by chemical reactions), the overall process involves several mechanical stresses (= force per area). A very important stress is caused by the surface tension γ, leading to a normal stress σ_γ. This stress leads to a pressure drop Δp across the curved bubble surface of mean curvature κ, captured by the Young–Laplace Equation [5]

$$\sigma_\gamma = \Delta p = 2\gamma\kappa. \tag{9.1}$$

For later purposes, κ is often captured by $1/L$, where L is a characteristic radius of curvature of the interface. It is often chosen to be the bubble radius R_B.

Hydrostatic stresses arise from graviational effects. They can be approximated by

$$\sigma_g = \rho g L, \tag{9.2}$$

where ρ is the liquid density, g the graviational acceleration and L a characteristic length scale.

This force plays a major role in the detachment of bubbles and their compaction in the foam. The surface tension and gravitational stresses are "static" in nature and do not require any fluid motion. However, in a dynamic situation, "dynamic stresses" may also arise. One is the tangential viscous stress σ_V. It results from a viscous shear flow close to the interface and depends on the fluid's viscosity η_L and the respective deformation rate dU/dL of the fluid. It can be approximated as

$$\sigma_V = \eta_L \frac{U}{L}. \tag{9.3}$$

Here, U is the characteristic flow velocity and L the characteristic length of the liquid flow. In general, there are viscous contributions from both the gas phase and the liquid phase. Any viscous flow can also create a dynamic pressure, which applies additional normal stresses on the interface. The motion of the gas and liquid phases can also create inertial stresses σ_I in the normal direction to the bubble surfaces caused by the momentum of the flow, which can be approximated by

$$\sigma_I = \rho U^2. \tag{9.4}$$

As discussed above, the presence of stabilizing (foaming) agents can give rise to important interfacial stresses [6] which can influence the generation of bubbles [7]. Additional stresses may act in particular systems including those arising from acoustic, electric or magnetic forces. However, these will not be considered here.

Note that even if all the mentioned stresses act simultaneously during the process of bubble formation, some of them can often be neglected. For example, in "quasi-static" situations ("quasi-static regime"), the flow velocities are small and the system can be described by a sequence of static states. Hence, quasi-static processes do not depend on the flow velocities, meaning that viscous and inertial stresses can be neglected. In contrast, for dynamical conditions ("dynamic regime"), viscous and inertial stresses have to be taken into account. Since bubble formation is often dominated by one of the two stresses, one often differentiates between a "viscosity-dominated" and an "inertia-dominated" regime.

The analysis of the different stresses is highly complex. This complexity can be simplified by the use of nondimensional numbers which measure the relative importance of different stresses and which are associated with a theoretical framework that relies on "scaling arguments." The most relevant dimensionless numbers for the description of bubble formation processes are summarized in Table 9.1. In Figure 9.1, the interrelations between these different numbers are shown schematically. In this chapter, we use these dimensionless numbers for the classification of different regimes of foaming techniques.

The definition of the dimensionless numbers in Table 9.1 include the following parameters: surface tension γ, density ρ, viscosity η, gravitational acceleration g, characteristic velocity U, and a characteristic length L. U and L have to be chosen carefully to appropriately capture the flow conditions. As bubble formation is a two-phase flow problem, one also has to clarify whether the stress of one or the other of the two phases dominates or if both have to be considered.

A parameter frequently required in many theoretical approaches is the capillary length

$$l_c = \sqrt{\frac{\gamma}{\rho g}}, \tag{9.5}$$

which results from setting $Bo = 1$, that is, from saying that capillary and gravitational stresses have the same order of magnitude. Beyond this length, one, therefore, has to take into account gravity when dealing with surface tension effects.

TABLE 9.1
Dimensionless Numbers Used Here to Classify the Different Regimes of Bubble Formation

Name	Symbol	Definition	Ratio of Stress
Bond number	Bo	$\dfrac{\Delta\rho g L^2}{\gamma}$	$\dfrac{\text{Gravitational stress}}{\text{Interfacial stress}}$
Capillary number	Ca	$\dfrac{\eta U}{\gamma}$	$\dfrac{\text{Viscous stress}}{\text{Interfacial stress}}$
Froude number	Fr	$\dfrac{U}{\sqrt{gL}}$	$\sqrt{\dfrac{\text{Inertial stress}}{\text{Gravitational stress}}}$
Galilei number	Ga	$\dfrac{g\rho^2 L^3}{\eta^2}$	Gravity vs viscosity
Ohnesorge number	Oh	$\dfrac{\eta}{\sqrt{\rho\gamma L}}$	$\dfrac{\text{Viscous stress}}{\sqrt{\text{Interfacial} * \text{inertial}}}$
Reynolds number	Re	$\dfrac{\rho U L}{\eta}$	$\dfrac{\text{Inertial stress}}{\text{Viscous stress}}$
Weber number	We	$\dfrac{\rho U^2 L}{\gamma}$	$\dfrac{\text{Inertial stress}}{\text{Interfacial stress}}$

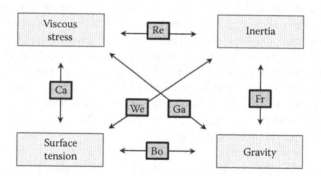

FIGURE 9.1 Interrelations between dimensionless numbers used to classify different bubble formation regimes.

Finally, the ratios of viscosity, density, and flow rate for the gas (index G) and liquid (index L)

$$\lambda_V = \frac{\eta_G}{\eta_L}, \ \lambda_D = \frac{\rho_G}{\rho_L}, \ \lambda_Q = \frac{Q_G}{Q_L} \tag{9.6}$$

are required when describing bubble formation processes.

9.2 RUPTURING OF GAS LIGAMENTS

In this section, we briefly discuss the physics associated with the topological change required for the detachment of a bubble (Figure 9.2). The different situations have one element in common: a sphere-like gas pocket is deformed into an elongated cylinder, which then becomes physically unstable and breaks (i.e., it undergoes a topological transition). If this cylinder is short compared to its width, it

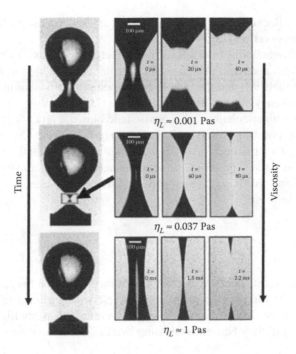

$\eta_L \approx 0.001$ Pas

$\eta_L \approx 0.037$ Pas

$\eta_L \approx 1$ Pas

FIGURE 9.2 Bubble pinch-off in different mixtures of water and glycerol with different liquid viscosity η_L. (Adapted from Burton, J.C. et al. *Phys Rev Lett* 2005; 94(18): 184502 [8].)

breaks in one point and generates one bubble (Figure 9.2), while when it is very long, it can break up into many bubbles.

The pioneering work for understanding the break-up of a cylindrical ligament into bubbles was done by Savart [9] and Plateau [10]. Rayleigh was then the first to derive a physical model [11] known as the "Rayleigh–Plateau instability." Note that a ligament does not necessarily break into bubbles of equal size. What is even more important is the geometrical confinement which has dramatic effects on the ligament's curvatures and therefore on its stability. If a ligament is confined in one direction (e.g., in a flat channel), it remains stable until the confinement is released. This situation is used to control bubble formation, for example, in microfluidic devices [12]. A more detailed analysis of the instability of fluid ligaments was presented in Reference 1.

9.3 BUBBLE GENERATION WITH TOPOLOGICAL CHANGES

Various mechanisms exist which lead to the detachment of a bubble from a gas pocket. Among them, the gravity-driven detachment is the most common one (Section 9.3.1). At small length scales (small *Bo*), however, gravity-based mechanisms are not very efficient. Therefore, in many cases, a co-flow of gas and liquid is used in order to take advantage of viscous or inertial forces for the bubble break-up process. Such a co-flow can be set up either in unconfined or confined conditions, the latter being much more efficient to induce dynamic stresses. We mainly discuss here bubble blowing in confined "co-flow" and "cross-flow" (Section 9.3.2), bubble break up under shear (Section 9.3.3), and bubble formation by gas entrainment (Section 9.3.4). In what follows, the various flow problems are discussed in terms of flow rates Q and flow velocities U.

9.3.1 BUBBLE FORMATION IN A STATIONARY LIQUID

The slow bubbling into a stationary liquid can be described rather easily, while the dynamic case is a highly complex problem due to the intricate coupling between the dynamic stresses and the bubble shape. Moreover, the influence of the dynamic contact angle between the liquid interface and the orifice is not well understood yet [13].

Let us discuss the slow bubbling (quasi-static regime) at constant gas flow rate Q_G into a stationary foaming solution through a vertical orifice with a circular cross section of radius R_0. In this case, Ca $\ll$ 1 and We $\ll$ 1 (Table 9.1) and the bubble generation process is dominated by surface tension and gravity forces (buoyancy), that is, Bo > 1.

Photographs along with simulations of a quasi-static blowing process at constant flow rate are shown in Figure 9.3 for a wetting (a) and non-wetting (b) nozzle orifice. During the blowing process, the bubble goes through well-defined pressure states which are schematically shown in Figure 9.4. First, in the "nucleation stage," the pressure increases until reaching a maximum when the bubble has a hemispherical shape, that is, when the bubble radius $R_B = R_0$. Hence, according to the Young–Laplace Equation 9.1, we have

$$\Delta P_{max} = \frac{2\gamma}{R_O} \tag{9.7}$$

This pressure is important, as in any bubble formation application the applied pressure must be larger than this value. As the blowing pressure is inversely proportional to the orifice radius, the required pressures for very small values of R_0 could become unfeasibly high. A certain reduction of the maximum pressure may be obtained using orifices with more complex geometrical cross sections [14].

If the applied pressure is higher than the maximum pressure, the bubble grows beyond the hemispherical shape (leading to a pressure decrease) and detaches via an instability mechanism (discussed in Section 9.2) when the surface tension force $F_\gamma = 2\pi\gamma R_0$ (keeping the bubble at the

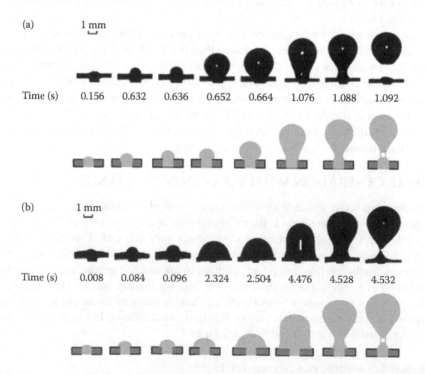

FIGURE 9.3 Evolution of bubble formation as a function of time for a wetting (a) and a nonwetting (b) orifice; top–experiments, bottom—simulations. (Adapted from Gnyloskurenko, S.W. et al. *Colloids Surf A* 2003; 218: 73–87 [13].)

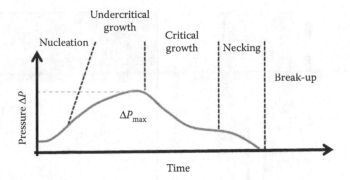

FIGURE 9.4 Schematic presentation of the pressure evolution during the bubble generation process.

orifice) is of the order of the buoyancy force $F_G = 4/3\pi\rho g R_B^3$. The obtained bubble size can, therefore, be approximated by $R_B \sim R_0 Bo^{-1/3}$, after using $R_0 = L$ in the definition of Bo.

Beyond the quasi-static bubbling conditions (Figure 9.5), we have to consider different dynamic forces acting in the bubbling process (see Figure 9.6a). The viscous drag of the liquid acts on the bubble and slows down the bubble detachment. Thus, larger bubbles are formed. The same is true for the inertial forces of the liquid, while on the contrary, the inertial forces of the gas support the bubble separation from the orifice.

It was found in Reference 15 that for negligible surface tension and inertial effects, the bubble radius can be approximated by

$$R_B \sim \left(\frac{\eta_L Q_G}{\Delta\rho g}\right)^{1/4} \sim R_0 \left(\frac{Ca}{Bo}\right)^{1/4}, \tag{9.8}$$

	Quasi-Static	Dynamic	
		Viscous-dominated	Inertia-dominated
	Low Ca Low Re Low We	High Ca Low Re Low We	High Ca High Re High We
Bubbling from nozzle			
Co-flow			
Cross-flow			

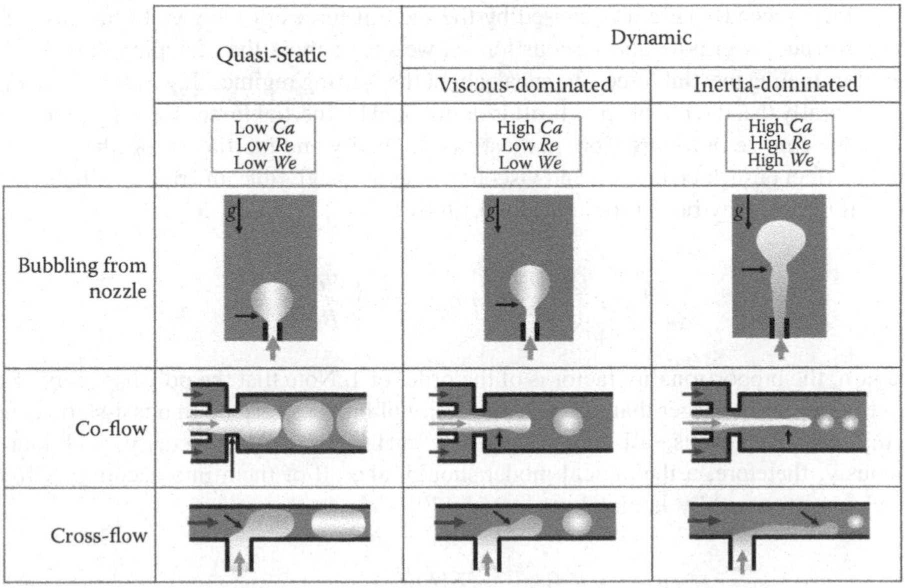

FIGURE 9.5 Different bubbling conditions depending on the importance of the static and dynamic forces and on the confinement. They gray arrows indicate the motion of the fluid, while the black arrow indicates the location of bubble pinch-off (leading to the topological transition). (According to Drenckhan, W., Saint-Jalmes, A. *Adv Colloid Interface Sci* 2015; 222: 228–259 [1].)

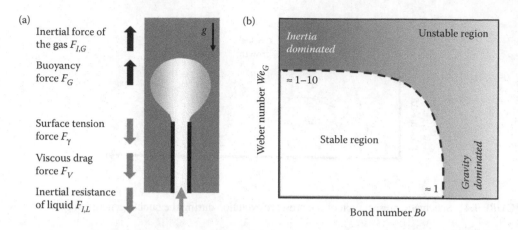

FIGURE 9.6 (a) Forces acting on a bubble during its creation. (b) Diagram of bubble stability; bubbles detach from the orifice when either B_o or We_G of the gas are sufficiently high. (Inspired from Drenckhan, W., Saint-Jalmes, A. *Adv Colloid Interface Sci* 2015; 222: 228–259 [1].)

with a proportionality constant of the order of 1. Note, in contrast to the quasi-static regime, the bubble size depends now on the liquid viscosity η_L and the gas flow rate Q_G.

Upon further increase of Q_G, inertial forces start to play a significant role (Weber number of the gas $We_G > 1$). For such large We_G, the formed bubble becomes increasingly elongated since inertial forces begin to outweigh the restoring surface tension forces. The point where the gas thread pinches off moves away from the nozzle (middle column in Figure 9.5), leading ultimately to the formation of a gas jet (right column in Figure 9.5). This jet eventually breaks up into bubbles and widens away from the orifice as a consequence of viscous friction decelerating the gas. When We_G is sufficiently high, the bubbles detach entirely due to the dynamic bubble shape while gravity is negligible [16]. Thus, two extreme scenarios exist: one where the bubble pinch-off is driven entirely by gravity ($Bo > 1$) and the second where it is caused by the inertial forces of the gas. In the case of bubble formation governed by gravity and viscous forces, we speak about the "dripping regime," while in the case of dominating inertial forces we speak about the "jetting regime." It was shown theoretically and experimentally that the bubble pinch-off is dominated by inertial forces for We_G larger than 1–10 [16–18]. As one can see in Figure 9.6b, in most cases, gravity and inertia act together.

When the effect of surface tension and viscous forces are negligible, the radius of bubbles formed in the inertial regime may be approximated [16,19] by

$$R_B \sim \frac{Q_G^{2/5}}{g^{1/5}} \sim R_0 Fr_G^{2/5} = R_0 \left(\frac{We_G}{Bo}\right)^{1/5}. \tag{9.9}$$

Here again, the proportionality factor is of the order of 1. Note that the polydispersity of bubbles formed in this regime is larger than the polydispersity of bubbles formed in quasi-static conditions.

In many practical cases, all forces (gravity, surface tension, viscosity, and inertia) act simultaneously: therefore, a theoretical model should take all of them into account. A frequently used model was proposed by Jamialahmadi et al. [20]

$$R_B = R_0 \left[\frac{5.0}{Bo^{1.08}} + \frac{9.261 Fr_G^{0.36}}{Ga_G^{0.39}} + 2.14 Fr_G^{0.51}\right]^{1/3} \tag{9.10}$$

The Froude number Fr is an important parameter in this relationship and represents the ratio of inertial and gravitational stresses (Figure 9.1). It can also be defined by $Fr = \sqrt{We / Bo}$.

9.3.2 Bubble Formation in a Flowing Liquid

A large number of bubbling techniques makes use of a flowing liquid as sketched in the examples shown in Figure 9.5. This creates additional viscous and inertial stresses which can play an important role in bubble formation. There are techniques which involve "co-flows" and "cross-flows" of the gas and the liquid flow may arise in confined or unconfined conditions. Figure 9.5 shows examples of confined co- and cross-flow. Smith was one of the first to use a confining geometry in order to blow small bubbles [21]. Later, at the end of the 20th century, microfluidic techniques started to integrate similar approaches [22–24].

In the *quasi-static regime*, the gas and liquid flow rates are small and dynamic forces are negligible. Thus, any unconfined flow gives the same results as bubble blowing into a stationary liquid. When gravity is absent, there is no mechanism for detachment and bubbles cannot be generated. This is very different when bubbles are formed under confined conditions. Then the liquid flow leads to a filling of the channel or the constriction (Figure 9.5), which pinches off the bubble and the slow flow takes it away [24]. At a given constant gas flow rate Q_G and liquid flow rate Q_L, a simple relationship can be derived [25]

$$V_B \sim R_B^3 \sim V_C \frac{Q_G}{Q_L} \tag{9.11}$$

where V_B is the obtained bubble volume and V_C the characteristic volume of the channel/constriction. The prefactor depends on the experimental geometry. This regime is called the "squeezing regime" and allows to form highly monodisperse foams.

The simplest case in the *dynamic regime* arises when the viscous drag exerted by the flowing liquid on the bubble becomes larger than the capillary force attaching the bubble to the orifice, while inertial stresses remain negligible (middle column of Figure 9.5). Here, the viscous Stokes force ($F_V = 6\pi R_B \eta_L U_L$) plays the role of gravity. The balance of the viscous and capillary forces ($2\pi\gamma R_O$) allows to predict the size of the resulting bubble

$$R_B \sim \frac{\gamma}{\eta_l U} R_O = Ca_L^{-1} R_O \tag{9.12}$$

We can see that the size of the bubble is inversely proportional to the capillary number Ca_L of the liquid and proportional to the radius R_O of the orifice. In confined geometries, one observes a well-defined transition from the quasi-static "squeezing regime" to the viscosity-dominated "dripping regime" at a critical Ca_L [26].

At even higher gas or liquid velocities, inertial forces start to play a role (right column of Figure 9.5), which moves the pinch-off point away from the injection point, forming a jet in the "jetting regime." Since the obtained bubble size is of the order of the jet diameter, the formation of very thin jets (high liquid velocity) can be used for the generation of very small bubbles, even down to micrometers in diameter [27]. More details about the dynamic regime of co-flow were discussed in Reference 1.

9.3.3 Breakup of Bubbles Under Shear

To create bubbles for building up foam, we can also consider breaking up larger into smaller bubbles by applying stresses via an external shear flow. Although the concept is rather simple, the physics behind it is highly complex. To describe the situation, we consider a bubble of viscosity η_G dispersed in a second fluid of viscosity η_L and exposed to a shear flow, as shown in Figure 9.7a. At low shear, the bubble becomes elongated depending on the capillary number Ca of the imposed shear flow [28–30]. With increasing Ca, the shape of the bubble becomes increasingly elongated, which is captured by

(a)   (b)
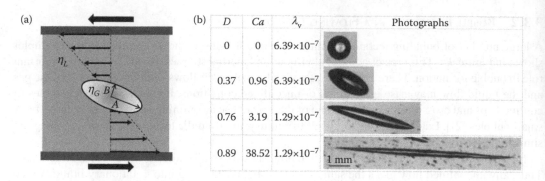

D	Ca	λ_v	Photographs
0	0	6.39×10^{-7}	
0.37	0.96	6.39×10^{-7}	
0.76	3.19	1.29×10^{-7}	
0.89	38.52	1.29×10^{-7}	1 mm

FIGURE 9.7 (a) Deformation of a bubble in a shear flow. (b) Photographs of equilibrium bubble shapes in a shear field for different Ca values and viscosity ratios λ_v. (Adapted from Muller-Fischer, N. et al. *Exp Fluids* 2008; 45: 917–926 [30].)

the aspect ratio $D = (A - B)/(A + B)$. Photographs of bubble shapes obtained for different Ca and ratios of viscosity $\lambda_v = \eta_G/\eta_L$ are shown in Figure 9.7b.

When the capillary number reaches a critical value Ca^*, the bubble assumes a critical deformation D^* beyond which the bubble cannot keep a steady shape and hence elongates indefinitely before breaking into smaller bubbles by the Rayleigh–Plateau instability discussed in Section 9.2. When the viscosity of the inner fluid is much lower than that of the outer fluid (for an air bubble in water we have $\lambda_v \sim 0.01$), the bubble can resist flows of rather high capillary numbers of $Ca^* \approx 3$ before breaking up.

In addition to the Rayleigh–Plateau instability, there are two more important mechanisms which can lead to the formation of smaller bubbles, even well below the critical capillary number Ca^*. In the first case, destabilization of the large bubble arises when the shear is switched off. That is, the elongated bubble shape (as shown in Figure 9.7b), which is stable under shear, becomes unstable as it retracts to its spherical shape after the shear is stopped. This process is called "end-pinching" and can break the bubbles into two or more smaller bubbles depending on the initial deformation and the viscosity ratio. For more details and references, see Reference 1. In the second case, the pointed corners of highly deformed bubbles become unstable and eject a fine gas jet which destabilizes into very small bubbles (order of a few microns). The presence of surfactants seems to play an important role in this instability which is called "tip streaming" [31–34].

9.3.4 GAS ENTRAINMENT AT FREE SURFACES

Another important mechanism is related to the entrainment of gas (most frequently air) at the surface of a flowing liquid. This commonly arises upon sudden changes of the cross-section of a flow or of the speed between two fluid zones [35] encountered, for example, in breaking waves, hydraulic jumps or water falls. A well known example in foam-related work is the "plunging jet" which generates bubbles upon the impact of a liquid jet on a foaming solution [36], also used in the standardized Ross–Miles test (see Chapter 10). The key process parameter is the impact velocity U of the jet, and the corresponding capillary number $Ca = \eta_L U/\gamma$. Using simple model systems, one can show that entrainment occurs only above a critical capillary number Ca_c of the jet (cf. Figure 9.8). Surprisingly, the rate of gas entrainment depends only on the viscosity of the gas (and not of the fluid) [36]. This is because the gas film decouples the liquid jet from the rest of the fluid (Figure 9.8).

The physics of plunging jets for low viscosity liquids like water is characterized by low capillary numbers (Ca < 1) and high Reynolds numbers ($Re > 100$). Under these conditions, inertia-driven instabilities can appear at the gas/liquid interface, and therefore modify the surface of the impacting jet. One of these is the Kelvin–Helmholtz instability [37]. Beyond a critical velocity difference

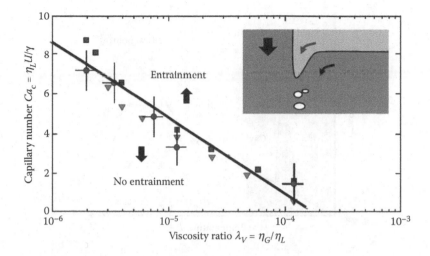

FIGURE 9.8 Dependence of the critical capillary number Cac on the viscosity ratio λ_V. (Adapted from Kiger, K.T., Duncan, J.H. *Annu Rev Fluid Mech* 2012; 44: 563–596 [36].)

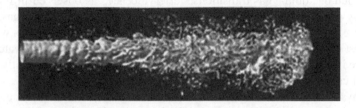

FIGURE 9.9 Kelvin–Helmholtz instability on a liquid jet. (Taken from Shinjo, J., Umemura, A. *Int J Multiph Flow* 2010; 36: 513–532 [38].)

between the liquid and the surrounding gas, interfacial fluctuations become unstable and are amplified by the flow. These fluctuations interfere significantly with the entrainment process. At sufficiently high Weber numbers $We = ReCa$, the Kelvin–Helmholtz instability can even lead to another mechanism of air entrainment on the jet itself. An example is shown in Figure 9.9. Here, the destabilization of the interface is strongly amplified and leads to complex flow structures, that is, to the formation of smaller ligaments, which can again lead to the kind of capillary instabilities discussed in Section 9.2. In the absence of stabilizing agents, this fluid break up process generates a spray of small drops and is therefore called "atomization" or "fragmentation." The addition of surfactants can turn this into an efficient foaming mechanism.

9.4 BUBBLE GENERATION WITHOUT TOPOLOGICAL CHANGES: PHASE TRANSITIONS

A liquid→gas phase transition is another option to create bubbles, this time without the need to undergo a topological change since the created bubbles do not need to be detached from a larger gas pocket. Two different types of phase transitions are generally considered [39–43]. In the first group, a pure liquid is evaporated locally, leading to a vapor bubble (Figure 9.10). This can be induced by a sudden decrease in pressure (called cavitation) or by raising the temperature (called boiling). Cavitation appears in rapid flows of liquids which can lead to strong, localized pressure drops. Well-known examples include constrictions in pipes or rapid propeller motion. The formation of bubbles by boiling is a well known phenomenon which we can observe in the kitchen, for example, when heating up milk.

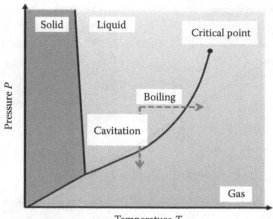

FIGURE 9.10 Liquid→gas transition of a pure liquid via pressure decrease ("cavitation") or temperature increase ("boiling").

The second group of liquid→gas transitions, "effervescence," is observed in liquids which are (super)saturated by a dissolved gas at a given pressure. Since the solubility of a gas decreases with decreasing pressure, a pressure drop leads to the formation of bubbles by the gas which comes out of solution. In most cases, the gas is carbon dioxide due to its high solubility in water and due to its natural occurrence in fermentation processes. Effervescence is, therefore, frequently observed when opening bottles of carbonated drinks like champagne, beer or cola [42,43].

The general physical mechanism behind these types of bubble generation processes is to locally drive the liquid into a supercritical state so that nucleation leads to the creation of small bubbles which then grow. This is required to overcome a critical nucleation barrier ΔG_C which arises from the energy ($\sim \gamma R^2$) required to create the gas/liquid interface. This barrier leads to a critical bubble radius R_c below which spontaneously formed bubbles disappear, while bubbles larger than the critical size grow. The nucleation rate n of bubbles can be estimated by [39–41]

$$n \sim \exp\left(-\frac{\Delta G_C}{kT}\right). \tag{9.13}$$

The nucleation barrier is greatly reduced or even removed in the presence of impurities or preformed gas cavities which may be contained in the liquid or on the solid surface in contact with the liquid. These serve as nucleation sites and are well known to soda, champagne or beer drinkers since they lead to the formation of the observed bubble trains which rise elegantly to the surface of the drink [42,43]. In these cases, the created bubbles have to detach from an object, leading again to a topological transition as discussed in Section 9.3.

Studies of the different bubble nucleation (and growth) processes in a liquid have a long history with many open questions [4]. More details on the formation of bubbles via phase transitions are given in Reference 1.

ACKNOWLEDGMENTS

The authors thank Bill Rossen, Emmanuelle Rio, Dominique Langevin, Nikolai Denkov, Cosima Stubenrauch and Jean-Eric Poirier for very useful comments on this manuscript. W. Drenckhan acknowledges funding from an ERC Starting Grant (agreement 307280-POMCAPS) and an "Attractivity Chair" of the IDEX UNISTRA (managed by the French National Research Agency as part of the "Investments for the future" program).

REFERENCES

1. Drenckhan, W., Saint-Jalmes, A. The science of foaming. *Adv Colloid Interface Sci* 2015; 222: 228–259.
2. Hsu, S.-H., Lee, W.-H., Yang, Y.-M., Chang, C.-H., Maa, J.-R. Bubble formation at an orifice in surfactant solutions under constant-flow conditions. *Ind Eng Chem Res* 2000; 39: 1473–1479.
3. Miller, R., Liggieri, L. Interfacial rheology. In: Miller, R. (Ed.) *Progress in Colloid and Interface Science.* CRC Press: Boca Raton, 2009.
4. Li, H.Z., Mouline, Y., Midoux, N. Modelling the bubble formation dynamics in non-Newtonian fluids. *Chem Eng Sci* 2002; 57(3): 339–346.
5. Hoorfar, M., Neumann, A.W. Recent progress in axisymmetric drop shape analysis (ADSA). *Adv Colloid Interface Sci* 2006; 121: 25–49.
6. Langevin, D. Influence of interfacial rheology on foam and emulsion properties. *Adv Colloid Interface Sci* 2000; 88(1–2): 209–222.
7. Golemanov, K., Tcholakova, S., Denkov, N.D., Ananthapadmanabhan, K.P., Lips, A. Breakup of bubbles and drops in steadily sheared foams and concentrated emulsions. *Physical Review E* 2008; 78(5): 051405.
8. Burton, J.C., Waldrep, R., Taborek, P. Scaling and instabilities in bubble pinch-off. *Phys Rev Lett* 2005; 94(18): 184502.
9. Savart, F. Constitution of a liquid jet flowing through a circular orifice in a thin diaphragm. *Ann Chim Phys* 1833; 53: 337–386.
10. Plateau, J.F. *Acad Sci Bruxelles Mem* 1843; 16: 3.
11. Rayleigh, L. On the instability of jets. *Proc London Math Soc* 1878; 10: 4–13.
12. Dollet, B., van Hoeve, W., Raven, J.P., Marmottant, P., Versluis, M. Role of the channel geometry on the bubble pinch-off in flow-focusing devices. *Phys Rev Lett* 2008; 100(3): 034504.
13. Gnyloskurenko, S.W., Byakova, A.V., Raychenko, O.I., Nakamura, T. Influence of wetting conditions on bubble formation at orifice in an inviscid liquid. Transformation of bubble shape and size. *Colloids Surf A* 2003; 218: 73–87.
14. McGuinness, P., Drenckhan, W., Weaire, D. The optimal tap: Three-dimensional nozzle design. *J Phys D-Appl Phys* 2005; 38(18): 3382–3386.
15. Davidson, J.F., Schueler, B.O.G. Bubble formation at an orifice in a viscous liquid. *Chem Eng Comm* 1960; 38: 335.
16. Pamerin, O., Rath, H.-J. Influence of buoyancy on bubble formation at submerged orifices. *Chem Eng Sci* 1995; 50(19): 3009–3024.
17. Kulkarni, A., Joshi, J. Bubble formation and bubble rise velocity in gas-liquid systems: A review. *Ind Eng Chem Res* 2005; 44(16): 5873–5931.
18. Zhao, Y.F., Irons, G.A. The breakup of bubbles into jets during submerged gas injection. *Metall Trans B* 1990; 21(6): 997–1003.
19. Kumar, R., Kulorr, N.R. The formation of bubbles and drops. *Adv Chem Eng* 1970; 8: 228–368.
20. Jamialahmadi, M., Zehtaban, M.R., Muller-Steinhagen, H., Sarrafi, A., Smith, J.M. Study of bubble formation under constant flow conditions. *Chem Eng Res Des* 2001; 79(A5): 523–532.
21. Smith, C.S. On blowing bubbles for Bragg's dynamic crystal model. *J Appl Phys* 1949; 20(6): 631–631.
22. Christopher, G.F., Anna, S.L. Microfluidic methods for generating continuous droplet streams. *J Phys D-Appl Phys* 2007; 40: R319–R336.
23. Marmottant, P., Raven, J.P. Microfluidics with foams. *Soft Mat* 2009; 5(18): 3385–3388.
24. Garstecki, P., Ganan-Calvo, A.M., Whitesides, G.M. Formation of bubbles and droplets in microfluidic systems. *Bull Pol Acad Sci Tech Sci* 2005; 53(4): 361.
25. Lorenceau, E., Sang, Y.Y.C., Höhler, R., Cohen-Addad, S. A high rate flow-focusing foam generator. *Phys Fluids* 2006; 18(9): 097103.
26. De Menech, M., Garstecki, P., Jousse, F., Stone, H.A. Transition from squeezing to dripping in a microfluidic T-shaped junction. *J Fluid Mech* 2008; 595(1): 141–161.
27. Castro-Hernandez, E., van Hoeve, W., Lohse, D., Gordillo, J.M. Microbubble generation in a co-flow device operated in a new regime. *Lab Chip* 2011; 11(12): 2023–2029.
28. Acrivos, A. The breakup of small drops and bubbles in shear flows. *Ann NY Acad Sci* 1983; 404: 1–11.
29. Stone, H.A. Dynamics of drop deformation and breakup in viscous fluids. *Annu Rev Fluid Mech* 1994; 26: 65–102.
30. Muller-Fischer, N., Tobler, P., Dressler, M., Fischer, P., Windhab, E.J. Single bubble deformation and breakup in simple shear flow. *Exp Fluids* 2008; 45: 917–926.
31. Eggers, J. Non-linear dynamics and breakup of free-surface flows. *Rev Mod Phys* 1997; 69: 865–929.

32. Anna, S.L., Mayer, H.C. Microscale tipstreaming in a microfluidic flow focusing device. *Phys Fluids* 2006; 18(12): 121512.

33. Baret, J.-C. Surfactants in droplet-based microfluidics. *Lab Chip* 2012; 12(3): 422–433.

34. Souidi, K., Mardaru, A., Roudet, M., Marcati, A., DellaValle, D., Djelveh, G. Effect of impellers configuration on the gas dispersion in high-viscosity fluid using narrow annular gap unit. Part 1: Experimental approach. *Chem Eng Sci* 2012; 74: 287–295.

35. Chanson, H. Air Bubble Entrainment in Free Surface Turbulent Shear Flows. Academic Press, 1995.

36. Kiger, K.T., Duncan, J.H. Air-entrainment mechanisms in plunging jets and breaking waves. *Annu Rev Fluid Mech* 2012; 44: 563–596.

37. Charu, F., de Forcrand-Millard, P. *Hydrodynamic Instabilities. Cambridge Texts in Applied Mathematics.* Cambridge University Press: Cambridge, 2011.

38. Shinjo, J., Umemura, A. Simulation of liquid jet primary breakup: Dynamics of ligament and droplet formation. *Int J Multiph Flow* 2010; 36: 513–532.

39. Brennen, C.E. *Cavitation and Bubble Dynamics.* Oxford University Press: Oxford, 1995.

40. Jones, S.F., Evans, G.M., Galvin, K.P. Bubble nucleation from gas cavities: A review. *Adv Colloid Interface Sci* 1999; 80(1): 27–50.

41. Lugli, F., Zerbetto, F. An introduction to bubble dynamics. *Phys Chem Chem Phys* 2007; 9(20): 2447–2456.

42. Liger-Belair, G., Polidori, G., Jeandet, P. Recent advances in the science of champagne bubbles. *Chem Soc Rev* 2008; 37(11): 2490–2511.

43. Vignes-Adler, M. The fizzling foam of champagne. *Angew Chem-Int Ed* 2013; 52(1): 187–190.

10 Foam Formation Techniques

Wiebke Drenckhan, Aouatef Testouri,
and Arnaud Saint-Jalmes

CONTENTS

A brief presentation of some of the most frequently applied foaming techniques is the subject of this chapter. These techniques are based on the fundamental mechanisms of bubble generation presented in Chapter 9. Note, however, that most of the existing techniques rely on a combination of different mechanisms. In addition, at the scale of a foaming generator, we cannot separate the formation of an isolated bubble from its environment where thousands of other bubbles are simultaneously created. These bubbles may also coalesce or exchange gas with each other (Chapter 11), hence creating bubble size distributions which are not only determined by the generating mechanism, but also by foam aging effects. Thus, although the main intrinsic mechanisms of foam generation are identified, the properties of the resulting foams cannot be easily predicted from the knowledge presented in the previous section on single bubbles (Chapter 9). As a consequence of this complexity, only few foam generators have been investigated in depth and their understanding remains a major challenge.

For this reason, only a rather qualitative description is provided here. To include these techniques into a general framework, our strategy is to group them according to the underlying physical mechanisms presented in Chapter 9. Each time, we start with the simplest designs and move toward more sophisticated ones. As a guide to the reader, the key features of the most frequently used methods are summarized in Table 10.1. We would like to remind the reader that we concentrate on the description of physical foaming techniques. A wealth of chemical and biological foaming techniques exists but is not covered here.

10.1 MECHANICAL METHODS FOR FOAM FORMATION

10.1.1 BUBBLE FORMATION IN A STATIONARY LIQUID

Bubbling a gas into a stationary liquid (Chapter 9.3.1) is an important industrial process, encountered in bubble column reactors [2] or in flotation tanks [3]. In many applications, the bubbling process is ensured by sparging a gas through many parallel orifices like in a perforated plate (Figure 10.1a) or a porous plate (Figure 10.1b), the first having uniform and the second nonuniform orifice sizes and arrangements. The most frequently used spargers are made of sintered glass particles. They have various advantages, such as being hydrophilic, easy to clean, and being available with various pore sizes. In a polydisperse sparger, the gas will preferentially go through the largest pores due to the

199

TABLE 10.1
Summary of the Most Frequently used Foaming Methods

Principle	Technique	Bubble Size (mm)	Poly-Dispersity (%)	Initial Gas Fraction φ	Formation Rate (L/min)
Bubbling into a stationary liquid	Individual orifice	0.1–10	2–40	<0.60	<1
(Active gas phase, passive fluid phase)	Many orifices/ Perforated plate	0.1–10	2–40	<0.60	<5 (per cm²)
	Porous disc	0.1–10	10–50	<0.70	<5 (per cm²)
Co-flow of gas and liquid	Individual orifice	0.01–1	2–30	<0.99	<0.1
(Active gas phase, active fluid phase)	Porous media	0.01–1	>30	<0.90	<1
	Static mixer	0.01–1	>30	<0.95	<15
	Straight tube	0.01–1	>30	<0.97	<10
Air entrainment and break-up under shear	Ross-Miles (plunging jet)	0.1–5	10–40	<0.60	<1
(Passive gas phase, active fluid phase)	Venturi hose	0.01–0.1	<30	<0.90	<5
	Propeller	>1	<30	<0.97	<1000
	Kitchen blender	0.01–1	<40	<0.97	<1
	Rotor–stator-mixer	0.01–0.1	<30	<0.90	<10
	Foaming by shaking	0.1–10	<70	<0.80	>1
Phase transition	Nucleation	0.001–0.01	10–40	<0.60	<1
	Aerosol cans	>0.01	>30	<0.99	<1
	Cavitation	≈0.001	>20	<0.60	<0.001
Electrochemical	Electrolytic cell	0.001–0.01	10–30	<0.60	<0.001 (per cm²)

Source: According to Drenckhan, W., Saint-Jalmes, A. *Adv Colloid Interface Sci* 2015; 222: 228–259 [1].

required pressure conditions (Section 9.3.1). With increasing gas pressure (gas flow rate), the number of active pores increases and the average size of active pores decreases [4–6]. As a rule of thumb, the bubble size is roughly proportional to the mean pore size of a sparger.

The characteristic size of bubbles generated by sparging are of the order of tens of micrometers up to one centimeter. Foams formed in such a way are reasonably monodisperse, since the spargers tend to have a characteristic pore size and since the pressure conditions in the sparger tend to select a narrow range of pores (generally the largest). Using a porous sparger, it is possible to create about one liter of foam per minute and per cm² of pore surface. Since the bubbles are created within a liquid

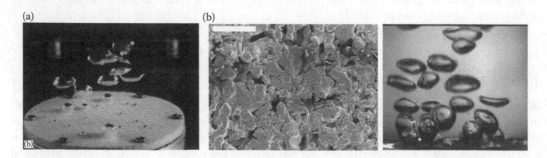

FIGURE 10.1 Examples of sparging through many orifices contained in (a) a regularly perforated plate (From Loimer, T. et al. *Chem Eng Sci* 2004; 59: 809–18 [7].) or in (b) a porous material (From Kazakis, N.A. et al. *Chem Eng J* 2008; 137(2): 265–281 [8].)

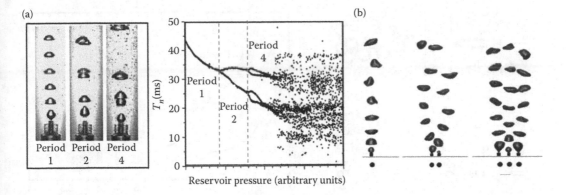

FIGURE 10.2 (a) Range of bubbling patterns (periods) at increasing injection pressures (gas flow rate). (According to Loimer, T., Machu, G., Schaflinger, U. *Chem Eng Sci* 2004; 59: 809–18 [5].) (b) Self-organizing structure of bubble departures due to the hydrodynamic interaction of closely spaced orifices (From Mosdorf, R., Wyszkowski, T. *Int J Heat Mass Transf* 2013; 61: 277–286 [10].)

pool, the resulting foams have generally a high liquid content close to the surface of the sparger. However, due to the foam rising in the experimental vessel, liquid drains by gravity and foams of much lower liquid contents are formed away from the sparger.

It was found that, although the fundamental bubble formation process at each orifice can be described similarly to that of an isolated orifice (see Chapter 9.3.1), there are significant interactions between the many simultaneously formed bubbles which must be taken into account [9]. These interactions arise from a pressure coupling through the reservoir, from hydrodynamic interactions through the fluid [10,11] or from direct contact between bubbles. To understand these different interactions and their interplay, researchers have simplified the configuration and considered devices consisting of only a few orifices with well-defined size and spacing (Figure 10.2). Moreover, complex bubbling regimes arise even at one single orifice. For example, while Chapter 9.3.1 was dealing only with simple, periodic regimes of bubble formation, highly nonlinear bubbling modes can be observed at an orifice under certain flow conditions. As shown in Figure 10.2a, a bifurcation of the bubbling period T may set in at intermediate gas flow rates (gas pressures). For some values of the gas flow rate, a completely chaotic bubble formation can even be observed [12–15]. The existence of several bubbling periods can result in bi- or tridisperse or completely polydisperse bubble size distributions. The observed regimes depend on various parameters, such as the liquid properties, the gas flow rate, and the orifice type. It seems that such complex bubble formation behavior is mainly caused by the dynamics of the gas/liquid meniscus inside the orifices [16,17].

In many geometries of porous or perforated plates, the number of active orifices depends on the gas flow rate. For several applications, it is required to work with low flow rates to obtain foams with low polydispersities. In [18], it was reported that regular bubbling can also be found at sufficiently high gas flow rates, such that bubbling takes place in the jetting regime (see Chapter 9.3.1). A way to avoid pressure coupling of several orifices and hence polydisperse bubbling is to use long slender orifices [19,20].

10.1.2 Simultaneous Injection of Gas and Liquid

The techniques for creating bubbles discussed in Chapter 10.1.1 can be improved by generating a flow of the foaming solution, creating a so called "co-flow" of the gas and the liquid (Figure 10.3 and Chapter 9.3.2). This ensures that the created bubbles are actively detached and carried away from the orifice. It provides at the same time a better control of the bubble sizes and of the liquid content of the foam. As for the bubbling through a single pore inside a stationary fluid, a single co-flow set-up can

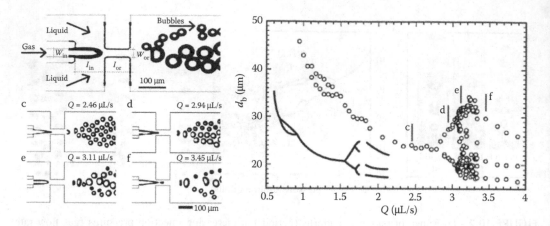

FIGURE 10.3 Dependence of bubble diameter d_b as a function of the liquid flow rate Q, at constant gas pressure obtained after simultaneously injecting gas and liquid through a narrow orifice (Adapted from Garstecki, P. et al. *Phys Rev Lett* 2005; 94(23): 234502 [21].)

already display a complex behavior [21]. Figure 10.3 shows an example, evidencing period doubling and chaotic bubbling through a constriction as the liquid flow rate Q is increased.

Although it is possible to create more than 10^6 bubbles/min with co-flow devices, the amount of resulting foam is small. Often, single co-flow devices are run in parallel to increase the throughput. This leads to a highly complex foaming behavior if the devices are not properly decoupled [22].

The co-flow of gas and liquid through an interconnected random network (like a porous material) can be an efficient route to foam generation (Figure 10.4). This kind of foam formation is actually actively used and studied for enhanced oil recovery [23–25]. Depending on the flow rates of the fluids, different regimes can be found. At low gas/liquid flow rates, both fluids follow separate paths through the porous material, so that only a small pressure drop across the porous material is observed (case A in Figure 10.4a). When the flow rates are increased, the pressure drop also increases leading to a coarse ("weak") foam (case B in in Figure 10.4a). Above a critical pressure gradient ∇P^* in the porous medium, the pressure drop jumps to a much higher value and the foaming becomes much more efficient [26] (case C in Figure 10.4a). Different foaming mechanisms

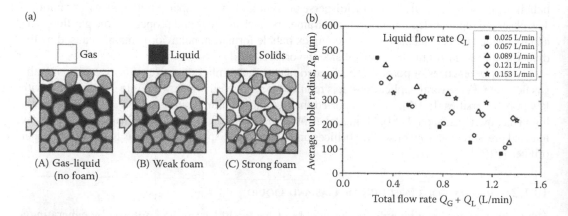

FIGURE 10.4 (a) Flow scenarios arising from a gas/liquid co-flow in porous media. (According to Lee, S., Kam, S.I. In: Enhanced Oil Recovery Field Case Studies. J. Sheng (Ed.) 2013, Gulf Professional Publishing [23].). (b) Average bubble radius as a function of the flow rate of a co-flow of gas and liquid through a porous material (closely packed 2 mm glass beads). (According to Guillermic, R.-M. et al. *J Rheol* 2013; 57: 333 [28].)

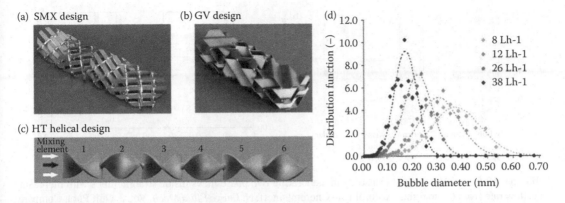

(a) SMX design   (b) GV design   (d)

(c) HT helical design

FIGURE 10.5 High speed co-flow of gas and liquid through purpose-designed static mixers efficiently generates foams. (a–c) Examples of static mixer geometries. (d) Typical bubble size distributions obtained with a static mixer at a fixed gas fraction of $\Phi = 0.92$ and at four different gas flow rates. (According to Talansier, E. et al. *AIChE J* 2013; 59: 132 -145 [30].)

can be identified, including "Lamella division" and "Leave-behind" or "Snap-off", as discussed in more detail in [27].

The critical pressure gradient ∇P^* has been shown to be related to the characteristic radius R_P of the pore throats R_P as $\nabla P^* \sim R_P^{-2}$ [26], that is, very high pressure gradients are needed to form foam in the presence of narrow pore throats. An efficient foam production is given above this critical pressure gradient and the corresponding critical flow rates. The size of the bubbles is roughly of the order of the pore sizes. Both the bubble size and the gas fraction of the foam can be controlled via the total and relative flow rates of the gas and the liquid. The average bubble radius R_B is approximately proportional to the inverse of the total flow rate Q [29], that is, $R_B \sim 1/Q = 1/(Q_G + Q_L)$. Figure 10.4b shows an example of this rule for co-flow through a pack of spherical glass beads of 2 mm diameter [28]. It is important to note that real porous media have microscopic pore sizes, therefore, the required pressures for foam formation are very high. As a consequence, bubble coalescence plays a vital role in fixing the bubble size distribution of the obtained foam. An important result is that in a wide range of pressure gradients, foam formation is rather unstable with oscillating properties. Experts relate this to the existence of complicated feedback mechanisms in the porous media [29]. In order to reliably generate a homogeneous foam, an optimized range of flow rates should, therefore, be selected.

For higher foaming rates, special "porous materials" have been designed (Figure 10.5a–c) into which the liquid and gas can be co-injected at velocities of up to several meters per second. Devices known as 'static mixers' create a strong shear flow of gas and liquid through a given network of pores, which results in enhanced bubble break up by shear (see Chapter 9.3.3). Such mixers are typically used in a laminar regime and characteristic bubbles sizes of the obtained foams are of the order of 100 μm (Figure 10.5d) with a wide range of gas fractions. When used at high flow rates, that is, in a turbulent regime, smaller bubbles and lower gas fractions are obtained. A guideline for the use of static mixers can be found in [30].

There are numerous types of static mixers, including fine grids or tubes filled with compact steel wool (or with other types of fiber materials [31]). These systems have the advantage of ensuring a high permeability, while maintaining a small characteristic length leading to the generation of small bubbles. Such mixers can be mounted at the entrance of more complex foam generators, as in the case of firefighting foam generators [31]. Foams made in such a way give access to a broad range of bubble sizes and liquid fractions.

Note that even a simple straight tube geometry may be sufficient: at sufficiently high flow rates, a co-injection of liquid and gas into a tube can lead to complex two phase flows (see Figure 10.6a),

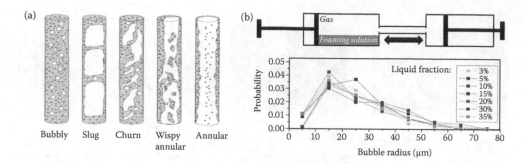

Bubbly   Slug   Churn   Wispy   Annular
                        annular

FIGURE 10.6 (a) Various patterns of foam generating two phase flows inside straight tubes with increasing gas flow rate from left to right. (According to Cheremisinoff, N. *Encyc Fluid Mech*, Vol. 3. Gulf Publ. Company, 1986, 1536 [32].) (b) Foam generation via pushing a mixture of gas and liquid many times back and forth through a tube connecting two syringes. The resulting bubble size distributions are roughly independent of the liquid fraction (from 3% to 35%). (According to Gaillard, T. et al. *Int J Multiphase Flow* [34].)

eventually creating foams [32]. The physical elements of such a procedure are very complex and include air entrainments and bubble break up. Based on this principle, researchers have found that very efficient foam generators can be built which provide foams with well-defined bubble sizes and gas fractions [33,34]. Small amounts of foams with a highly accurate control over bubble size and liquid fraction can be produced using the so called "Double-Syringe technique" (Figure 10.6b [34]) which uses two interconnected syringes to push repeatedly well-defined volumes of liquid and gas through the channel connecting the two syringes. The two fluids are very efficiently mixed, leading to foams with small bubbles of about 10 μm, whose size distributions are independent of the liquid fraction, as can be seen in Figure 10.6b.

10.1.3 Entrainment of Gas at Free Surfaces

At free liquid surfaces, gas entrainment (Chapter 9.3.4) often arises at high fluid velocities and at sudden changes of the flow velocity. This phenomenon can be observed in nature, for example in cascades, in breaking waves on the sea or at hydraulic jumps [35]. A number of lab-scale instruments have been set up to simplify the complexity of the processes [35].

One of the most studied examples of air entrainment is that of breaking waves. Even though a clear description of this highly complex process has not yet emerged from the studies [35], the main time sequences of the process are beginning to be disentangled and are summarized in Figure 10.7.

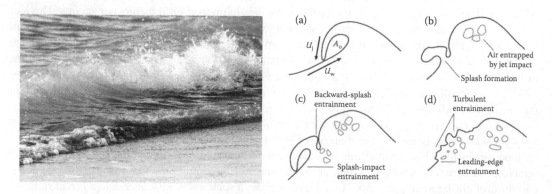

FIGURE 10.7 Left: photograph of a breaking wave with air entrainment. Right a-d: Different steps in air entrainment of a breaking wave. (According to Kiger, K.T., Duncan, J.H. *Ann Rev Fluid Mech* 2012; 44: 563–596 [35].)

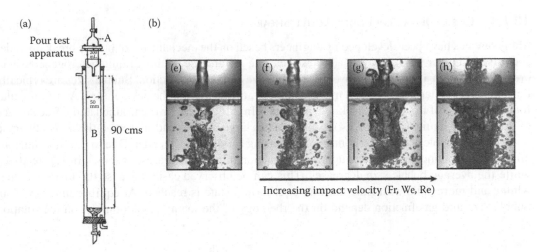

FIGURE 10.8 (a) Foam generation by a plunging jet, proposed in 1941 and known as the Ross–Miles test. (According to Ross, J., Miles, G.D. 1941; 18(5): 99–102 [37].) (b) Images of plunging jets and the accompanying bubble generation with increasing impact velocity (from (e) to (h)). Note that scale bars are 10 mm. (From Kiger, K.T., Duncan, J.H. *Ann Rev Fluid Mech* 2012; 44: 563–596 [35].)

The first step of air entrainment occurs when a plunging wave hits its front face (Figure 10.7a). This is the most frequently studied step of air entrainment, as it is similar to the physics of a plunging jet (Chapter 9.3.4). Thereafter, several post-jet impact entrainment modes appear. They are mainly connected with the splash of the first impact, as is schematically shown in Figure 10.7b,c. Finally, some turbulent entrainment occurs over all subsequent splashes and at the leading edge of the turbulent region (Figure 10.7d). Note that although plunging waves are rather efficient in entraining air into water, this has not (yet) been used for the design of a foaming device. The same is true for gas entrainments at hydraulic jumps [36].

Air entrainment for foam generation is being used in one of the most famous foaming techniques, the Ross–Miles test [37], which creates foam by a plunging jet. This technique reproduces household conditions of filling bathtubs or sinks. Figure 10.8a shows the historic set-up published in 1941 [37]. This foaming test refers to a norm and is registered under the number ASTMD1173-53 (2001). Although many other automated test methods exist, this historical one is still in practice. The protocol of the test is comprised of the following steps: 50 ml of the test solution is placed at the bottom of a cylinder, while a pipette with a funnel (A in Figure 10.8a) containing 200 mL of the same solution is mounted at the top of B at a fixed height of 90 cm. The amount of 200 mL solution is allowed to fall into the cylinder. After the 200 mL solution is completely flown out of pipette A, the height of the produced foam in B is recorded with time. Note that initially, the falling liquid impacts not only on the liquid layer, but also on the vessel bottom (this also changes once the first foam layers are formed in B). Bubbles can also be destroyed by the falling liquid. Hence, the amount of generated foam is a complex result of foam production and destruction. Foams formed in this way are commonly polydisperse. The average size of the bubbles is between some a hundred μm and up to a few mm, and the foam formation rate is rather low and of the order of one liter per minute. Initially, the liquid content of the foams is rather high and decreases due to gravity-driven drainage.

More efficient foam generation techniques have been developed based on controlled air entrainment. Efficient air entrainment happens at the outlet of nozzles with specifically designed cross sections (Venturi-type nozzles), where the flowing liquid creates a partial vacuum entraining the surrounding air. This type of arrangement yields low and intermediate gas fractions, and air has to be additionally injected to increase the foam gas fraction. For more information see Reference 1.

10.1.4 Bubble Break Up under Active Shear

Many devices have been developed by engineers based on the mechanisms of bubble break up under shear (see Chapter 9.3.3). These machines shear the gas and the foaming liquid heterogeneously to create foams with a good control over the bubble size and the gas fraction. Bubble sizes are typically between a few micrometers and a mm in diameter, depending on the device used. A very popular foaming device of this kind is the kitchen blender. In this device, air is entrained at the free surface of the blended liquid (Chapter 9.3.4), which creates large bubbles and subsequently breaks them up into smaller bubbles during the continuous shearing action of the blender. Due to the simultaneous air entrainment and bubble break up, the gas fraction initially increases, as shown in Figure 10.9a, while the average bubble size decreases. This can be observed easily because the foam becomes whiter and more solid with time until an equilibrium state is reached. At equilibrium, the mean bubble size and gas fraction depend on the rheology of the foaming solution and on the rotation

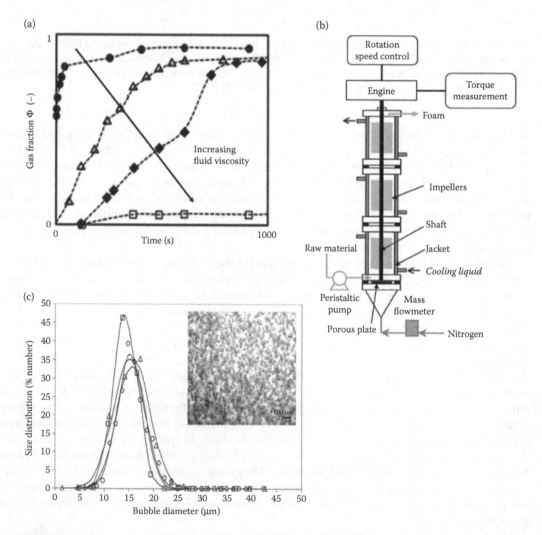

FIGURE 10.9 (a) Gas fraction Φ as a function of time for different foaming mixtures containing particles, foam generated by a kitchen blender (Data taken from Lesov, I., Tcholakova, S., Denkov, N. *J Colloid Interface Sci* 2014; 426: 9–21 [38].) (b) Common design of a "Narrow Annular Gap Unit"; (According to Thakur, R.K., Vial, C., Djelveh, G. *J Food Eng* 2003; 60: 9–20 [39].) (c) Examples of bubble size distributions obtained from a Narrow Annular Gap Unit; (From Souidi, K. et al. *Chemical Engineering Science* 2012; 74: 287–295.)

velocity of the blender. As a general rule, smaller bubbles and lower gas fractions are obtained for liquids of higher viscosity. At long times and high speeds of beating, the size of bubbles can be as small as some tens of micrometers. For extreme shearing conditions and special formulations of foaming solutions, even bubbles of submicron size can be obtained [38]. For low viscosity liquids, the initial foaming rate can be rather high (almost 1 L/min), however, the subsequent bubble break up can take much longer. The main disadvantage of kitchen blenders for industrial applications is the fact that it is a batch process.

In order to produce foams via shearing action in a continuous manner, many innovative blenders have been designed. The so called rotor–stator machines represent a large group of such improved blenders, which also replace the passive air entrainment by an active gas injection. They consist of sets of narrowly spaced "stators" and quickly rotating "rotors" through which the gas (air) and the foaming solution have to pass. Many papers are dedicated to the optimization of such blenders [41,42]. As a criterion, the ratio between the mechanical driving force and the inertial force (Newton number Ne) is often used in these processes. It turns out that the input of mechanical energy is the decisive parameter that controls the quality of the produced foam, which is a function of the rotor speed, the viscosity of the fluid, and the blending time [42–45]. The most famous rotor/stator blender is the "Ultra-Turrax®" which allows producing foam with very small bubbles (as small as a few µm).

Another class of devices is based on the "Narrow Annular Gap Unit" (Figure 10.9b) [40], which, similar to the rotor-stator devices, contains a number of propellers which rotate in a jacket through which a gas/liquid premix is driven. The high shear rates in the narrow gap between the propellers and the shaft surface lead to efficient foaming with characteristic bubble sizes of some tens of micrometers, as shown in Figure 10.9c.

In addition to the design of any kind of mixers, one also has to consider effects of bubble "cooperativity," as mentioned before in the introduction. As soon as the bubbles in the foam are closely packed, they cannot be considered as individual objects—as we did in Chapter 9. As an example, the role of the gas fraction on bubble break up has been studied experimentally in classical shear rheometers (see for example [46,47]). A critical capillary number for the break up is observed well below the one obtained for a single bubble. This is explained by a so called "microstructure-induced" instability induced by pinch-off processes of elongated bubbles by their neighbors [46].

Probably the simplest method of foaming is shaking a solution in a partially filled container. This can be made in a controlled way, that is, at a certain frequency and amplitude over a certain time. The foamability of the studied solution can then be determined from the amount of generated foam and the size distribution of the generated bubbles can be determined as a function of the experimental parameters. This is called the "Bartsch test." Note that the foam generation by shaking combines almost all different foaming mechanisms discussed in Chapter 9 and a quantitative description of such a complex phenomenon is out of reach at the moment. Nevertheless, a range of different effects could be studied separately in much simpler experiments, as it was done by Caps et al. [48] using rotating flat cells (Hele-Shaw cells) (Figure 10.10).

10.2   FOAMING THROUGH PHASE TRANSITIONS AND BY ELECTROCHEMISTRY

The generation of bubbles through bubble *nucleation* and growth in a supersaturated liquid is another way of making foam and is well known to us when dealing with beverages: (more or less stable) foams form naturally in fizzy drinks, beer or champagne. The supersaturated gas is typically CO_2 because of its high solubility in water and because it arises naturally in the fermentation process used for the production of the alcoholic fizzy drinks. It has been shown that most of the formed bubbles in supersaturated liquids are generated by heterogeneous nucleation at the wall surface or around impurities (see Chapter 9.4 and [49,50]). It can be easily observed that continuous streams of bubbles

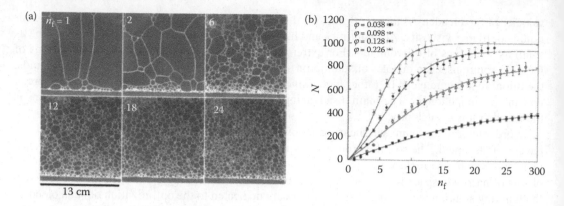

FIGURE 10.10 (a) Images of progressive foam formation after flipping n_f times a flat cell containing a well-defined fraction φ of liquid. (b) Evolution of number of bubbles N created in the cell as a function of flipping. (From Caps, H. et al. *Appld Phys Lett* 2007; 90: 214101-1–214101-3 [49].)

rise from distinct nucleation sites, as shown in Figure 10.11a. Therefore, the supersaturation in such beverages can be comparatively low. For champagne, it is typically fixed at pressures between 5 and 10 bar and the release is about 5 liter of CO_2 [50].

The foaming of large quantities of highly viscous liquids is more difficult and standard extruder screws are often used as shown in Figure 10.11b. This method also allows the foaming of thermosensitive polymers. These polymers are mixed with gas at high pressures and temperatures. A fast foaming, in parallel to solidification, can be reached when the mixture is released to ambient pressure and temperature.

Boiling as a controlled method of foaming is not often used. We mostly remember it as a unwelcome surprise in the kitchen when soup or milk are left on the stove for too long (Figure 10.11c). One already existing application is related to shaving gels which foam when they are put on the skin. This effect is generally generated by substances like isopentane, which is liquid at room temperature and turns into its gaseous state at body temperature. Moreover, the foaming is enhanced by the subsequent shearing of the gel on the skin, which breaks its microstructure and initiates the required nucleation process. A similar foaming system is found in aerosol cans [51], like those used to produce shaving foams. Here, the phase transition is driven by a pressure change rather than a

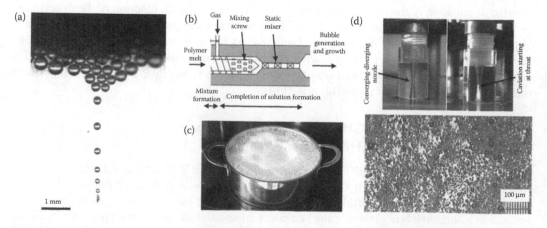

FIGURE 10.11 (a) Bubbles rising from a nucleation site in a beer bottle. (b) Extruder screw for polymer foaming. (c) Foam formation by boiling in the kitchen. (d) Hydrodynamic cavitation via rapid flow through a constriction and example of obtained foams. (From Raut, J.S. et al. *Soft Matter* 2012; 8(17): 4562–4566 [54].)

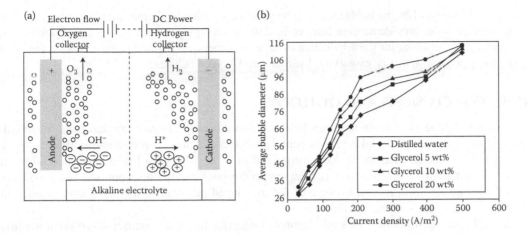

FIGURE 10.12 (a) Generation of bubbles and foam via electrolysis: general concept of the generation of oxygen and hydrogen bubbles at the anode and cathode, respectively. (b) Variation of mean bubble size via the current density in an electroflotation cell. (According to Mansour, L.B., Chalbi, S., Kesentini, I. *Indian J Chem Technol* 2007; 14(3): 253–257 [56].)

temperature change. The can contains two immiscible liquids which create an emulsion under the pressure conditions in the can (several bars). The dispersed phase is a liquid of low vapor pressure (most commonly butane/propane mixtures) which remains liquid inside the can. Once the liquid is released to ambient pressure, the dispersed liquid changes into a gas and the entire emulsion droplet turns into a gas bubble. The final bubble size is, therefore, determined by the initial droplet volume, which is commonly controlled by the spray head and improved by shaking the can before using it. Bubble sizes in these applications are typically around 10–100 μm.

Techniques based on *cavitation* are increasingly used (see also Chapter 9.4). This is well suited when bubbles in the μm-range are required. Initially, sonication was used for bubble generation, however, its efficiency is rather low. This is why researchers are increasingly using hydrodynamic cavitation [52,53]. In these devices, a rapid flow of the foaming liquid is created in such a manner that a sudden, strong pressure drop occurs which leads to the generation of cavitation bubbles. The most popular scenario consists of rapid flow through a constriction (Figure 10.11d), leading to a pressure drop just after the constriction and to the formation of tiny bubbles in this zone. These quickly formed bubbles move with the flow and have to be stabilized against coalescence by adding stabilizers [53].

Electrolysis is an electrochemical way for creating gas bubbles [54]. When a voltage is applied to two electrodes immersed in an aqueous solution (Figure 10.12a), the water molecules decay ($2H_2O \leftrightarrow 2H^+ + 2OH^-$), and the formed protons H^+ combine with electrons from the cathode to produce hydrogen gas (H_2). At the anode, the OH^- ions liberate electrons and form water and oxygen according to the following reaction scheme

Cathode:	$4H^+ + 4e^- \leftrightarrow 2H_2$
Anode:	$4OH^- \leftrightarrow 2H_2O + O_2 + 4e^-$.

Thus, hydrogen bubbles are formed at the cathode and oxygen bubbles at the anode. Both types of bubbles adhere at the electrode surfaces and are released by gravity or by an external liquid flow. The bubbles are generally formed at small defects of the electrode surface and attach only weakly so that their size is rather small (Figure 10.12b) [55]. As the number of active sites at an electrode surface can be rather large, this method can produce large amounts of foam with very small bubbles. Due to the small bubble size, the main range of application is electroflotation as its efficiency increases as the bubble size is reduced. This method is also used in waste water treatment or mineral flotation. As one can see in Figure 10.12b, the bubble size can be varied conveniently via the current density. Purpose-designed electrodes provide an even finer control to obtain highly monodisperse microbubbles [56]. An important constraint for the application of an electrochemical method is the restricted choice of gases to be produced, that is, essentially hydrogen and oxygen bubbles.

10.3   CONCLUSIONS AND OUTLOOK

Although the process of foaming has a long-standing history, real progress has been made only recently in trying to understand the fundamental processes which are involved in the different bubble generation processes (Chapter 9). As summarized in Table 10.1, each foaming method allows to produce foams with characteristic gas fractions, corresponding size distributions, and production rate. To create foams over a wide range of the mentioned parameters, a combination of different foaming techniques is often required.

A wealth of publications appeared recently to describe the main features of different foaming methodologies. However, at present, only the simplest methods can be properly explained, such as the bubbling from individual orifices and the breakup of bubbles in a shear field. But even in these very simple cases, quite a number of important questions remain open. An important reason for the slow progress in understanding foam formation techniques is not only the fact that they often combine different physical effects (Chapter 9), but that surfactant dynamics (Chapter 1) and foam aging effects (Chapter 11) tend to play a non-negligible role in fixing the final bubble size distributions. Moreover, it is inherently difficult to rapidly and correctly determine the bubble size distributions with currently available techniques. Fast X-ray tomography may play an important role in this exercise in the future.

Various challenges need to be tackled in the future in order to progress in our understanding of foam formation. One needs to move from the description of single bubbles to a much more complex consideration of the various types of bubble interactions arising in real foams, including foam aging effects. A second challenge will be to understand the influence of the complex mechanical properties of the gas/liquid interfaces, which arise from the presence of the stabilizing agents. Last but not least, we need to tackle the description of the foaming of complex, non-Newtonian liquids whose flow properties depend on the flow conditions. Such liquids include polymer solutions, polymer melts, emulsions or particle suspensions which are heavily used in foaming applications.

In summary, the intensive work of engineers over the last decades has produced many reliable foaming methods. The challenge of today and in the future is to establish a better description and prediction of the different foaming processes. This would help us not only to optimize existing foaming techniques, but also to design new foaming methods in order to produce tailored foams for a wide range of foam applications.

ACKNOWLEDGMENTS

The authors would like to thank W.R. Rossen, E. Rio, D. Langevin, N. Denkov, C. Stubenrauch and J.-E. Poirier for very useful comments on this manuscript. W. Drenckhan acknowledges funding from an ERC Starting Grant (agreement 307280-POMCAPS) and an "Attractivity Chair" of the IDEX UNISTRA (managed by the French National Research Agency as part of the 'Investments for the future' program).

REFERENCES

1. Drenckhan, W., Saint-Jalmes, A. The science of foaming. *Adv Colloid Interface Sci* 2015; 222: 228–259.
2. Kantarci, N., Borak, F., Ulgen, K.O. Bubble column reactors. *Process Biochemistry* 2005; 40: 2263–2283.
3. Lu, S., Pugh, R., Forssberg, E. Interfacial separation of particles. In: *"Studies in Interface Science"*, Vol. 20. Elsevier Sci., Amsterdam, NL, 2005, ISBN 9780444516060.

4. Kazakis, N.A., Mouza, A.A., Paras, S.V., Experimental study of bubble formation at metal porous spargers: effect of liquid properties and sparger characteristics on the initial bubble size distribution. *Chem Eng J* 2008; 137(2): 265–281.
5. Loimer, T., Machu, G., Schaflinger, U., Inviscid bubble formation on porous plates and sieve plates. *Chem Eng Sci* 2004; 59: 809–18.
6. Zahradnik, J, Kastanek, F. Gas holdup in uniformly aerated bubble column reactors. *Chem Eng Commun* 1979; 3: 413–29.
7. Loimer, T., Machu, G., Schaflinger, U. Inviscid bubble formation on porous plates and sieve plates. *Chem Eng Sci* 2004; 59: 809–18.
8. Kazakis, N.A., Mouza, A.A., Paras, S.V. Experimental study of bubble formation at metal porous spargers: effect of liquid properties and sparger characteristics on the initial bubble size distribution. *Chem Eng J* 2008; 137(2): 265–281.
9. Kulkarni, A., Joshi, J. Bubble formation and bubble rise velocity in gas-liquid systems: A review. *Industrial & Engineering Chemical Research* 2005; 44(16): 5873–5931.
10. Mosdorf, R, Wyszkowski, T. Self-organising structure of bubble departures. *Int J Heat Mass Transf* 2013; 61: 277–286.
11. Pereira, P.A.C., Colli, E., Sartorelli, J.C. Synchronization of two bubble trains in a viscous fluid: Experiment and numerical simulation. *Phys Rev E* 2013; 87: 022917.
12. Tufaile, A., Sartorelli, J.C. Chaotic behavior in bubble formation dynamics. *Physica A: Statistical Mechanics and its Applications*, 2000; 275(3–4): 336–346.
13. Mosdorf, R., Shoji, M. Chaos in bubbling — nonlinear analysis and modelling. *Chemical Eng Sci* 2003; 58(17): 3837–4086.
14. Pereira, F.A.C., Colli, E., Sartorelli, J.C. Period adding cascades: Experiment and modeling in air bubbling. *Chaos* 2012; 22: 013135.
15. Tritton, D.J., Egdell, C. Chaotic bubbling. *Phys Fluids A* 1993; 5(2): 503.
16. Ruzicka, MC, Bunganic, R, Drahos, J. Meniscus dynamics in bubble formation. Part I: Experiment. *Chem Eng Res Des* 2009;87:1349–56.
17. Stanovsky, P, Ruzicka, MC, Martins, A, Teixeira, JA. Meniscus dynamics in bubble formation: A parametric study. *Chem Eng Sci* 2011;66:3258–67.
18. Ruzicka, M.C., Drahos, J., Zahradnik, J., Thomas, N.H. Structure of gas pressure signal at two-orifce bubbling from a common plenum. *Chem Eng Sci* 2000; 55: 421–429.
19. Loimer, T., Machu, G., Schaflinger, U. Inviscid bubble formation on porous plates and sieve plates. *Chemical Eng Sci* 2004; 59: 809–818.
20. Xie, S., Tan, R.B.H. Bubble formation at multiple orifces—bubbling synchronicity and frequency. *Chemical Eng Sci* 2003; 58: 4639–4647.
21. Garstecki, P., Fuerstman, M.J., Whitesides, G.M. Nonlinear dynamics of a flow-focusing bubble generator: An inverted dripping faucet. *Phys Rev Lett* 2005; 94(23): 234502.
22. Hashimoto, M., Shevkoplyas, S.S., Zasonska, B., Szymborski, T., Garstecki, P., Whitesides, G.M. Formation of bubbles and droplets in parallel, coupled flow-focusing geometries. *Small* 2008; 4(10): 1795–1805.
23. Lee, S., Kam, S.I. Enhanced oil recovery by using CO_2 foams: Fundamentals and field applications. In: *Enhanced Oil Recovery Field Case Studies*. J. Sheng (Ed.) 2013, pp. 23–61. Gulf Professional Publishing.
24. Rossen, W.R. Foams in enhanced oil recovery. In: Prud'homme, R.K. (Ed.) *Foams: Theory, Measurements, and Applications*. Marcel Dekker, Inc., New-York, USA, 1996.
25. Schramm, L.L., Isaacs, E.E. Foams in enhancing petroleum recovery. In: Stevenson, P. (Ed.), *Foam Engineering: Fundamentals and Applications*. Wiley-Blackwell, London, GB, 2012.
26. Gauglitz, P.A., Friedmann, F., Kam, S.I., Rossen, W.R. Foam generation in homogeneous porous media. *Chem Eng Sci* 2002; 57(19): 4037–4052.
27. Rossen, W.R. A critical review of Roof snap-off as a mechanism of steady-state foam generation in homogeneous porous media. *Colloids Surf A Physicochem Eng Asp* 2003; 225(1–3): 1–24.
28. Guillermic, R.-M., Volland, S., Faure, S., Imbert, B., Drenckhan, W. Shaping complex fluids—How foams stand up for themselves. *J Rheol* 2013; 57: 333.
29. Friedmann, F., Jensen, J.A. Some parameters influencing the formation and propagation of foams in porous media. 1996.
30. Talansier, E., Dellavalle, D., Loisel, C., Desrumaux, A., Legrand, J. Elaboration of controlled structure foams with the SMX static Mixer. *AIChE J* 2013; 59: 132–145.
31. Gardiner, B.S., Dlugogorski, B.Z., Jameson, G.J. Rheology of fire-fighting foams. *Fire Saf J* 1998; 31(1): 61–75.

32. Cheremisinoff, N. Encyclopedia of fluid mechanics: Gas-liquid flows. *Encyc Fluid Mech*, Vol. 3. Gulf Publ. Company, Houston, TX, USA, 1986, 1536.

33. Saint-Jalmes, A., Vera, M.U., Durian, D.J. Uniform foam production by turbulent mixing: New results on free drainage vs. liquid content. *Eur Phys J* 1999; 12(1): 67–73.

34. Gaillard, T., Roché, M., Honorez, C., Jumeau, M., Balan, A., Jedrzejczyk, C., Drenckhan, W. Controlled foam generation using cyclic diphasic flows through a constriction. *Int J Multiphase Flow* 2017; 96: 173–187.

35. Kiger, K.T., Duncan, J.H. Air-entrainment mechanisms in plunging jets and breaking waves. *Ann Rev Fluid Mech* 2012; 44: 563–596.

36. Chanson, H. *Air Bubble Entrainment in Free Surface Turbulent Shear Flows*. Academic Press, London, GB, 1996.

37. Ross, J., Miles, G.D. An apparatus for comparison of foaming properties of soaps and detergents oil and soap. *J Am Oil Chem Soc* 1941; 18(5): 99–102.

38. Lesov, I., Tcholakova, S., Denkov, N. Factors controlling the formation and stability of foams used as precursors of porous materials. *J Colloid Interface Sci* 2014; 426: 9–21.

39. Thakur, R.K., Vial, C., Djelveh, G. Influence of operating conditions and impeller design on the continuous manufacturing of food foams. *J Food Eng* 2003; 60: 9–20.

40. Souidi, K., Mardaru, A., Roudet, M., Marcati, A., DellaValle, D., Djelveh, G. Effect of impellers configuration on the gas dispersion in high-viscosity fluid using narrow annular gap unit. Part1:Experimental approach. *Chem Eng Sci* 2012; 74: 287–295.

41. Kroezen, A., Wassink, J.G. Foam generation in rotor-stator mixers. *JSDC* 1986; 102: 397–403.

42. Muller-Fischer, N., Suppiger, D., Windhab, E.J. Impact of static pressure and volumetric energy input on the microstructure of food foam whipped in a rotor–stator device. *J Food Eng* 2007; 80: 306–316.

43. Hanselmann, W., Windhab, E. Flow characteristics and modelling of foam generation in a continuous rotor/stator mnixer. *J Food Eng* 1999; 38: 393–405.

44. Garrett, P.R. Recent developments in the understanding of foam generation and stability. *Chem Eng Sci* 1993; 48(2): 367–392.

45. Mary, G., Mezdour, S., Delaplace, G., Lauhon, R., Cuvelier, G., Ducept, F. Modelling of the continuous foaming operation by dimensional analysis. *Chem Eng Res Des* 2013; 91(12): 2579–2586.

46. Golemanov, K., Tcholakova, S., Denkov, N.D., Ananthapadmanabhan, K.P., Lips, A. Breakup of bubbles and drops in steadily sheared foams and concentrated emulsions. *Phys Rev E* 2008; 78(5): 051405.

47. Tcholakova, S., Lesov, I., Golemanov, K., Denkov, N.D., Judat, S., Engel, R., Danner, T. Efficient emulsification of viscous oils at high drop volume fraction. *Langmuir* 2011; 27: 14783–14796.

48. Caps, H., Vandewalle, N., Broze, G., G. Zocchi. Foamability and structure analysis of foams in Hele-Shaw cell. *Appld Phys Lett* 2007; 90: 214101-1–214101-3.

49. Vignes-Adler, M. The fizzling foam of champagne. *Angew Chem Int Ed* 2013; 52(1): 187–190.

50. Liger-Belair, G., Polidori, G., Jeandet, P. Recent advances in the science of champagne bubbles. *Chem Soc Rev* 2008; 37(11): 2490–2511.

51. Arzhavitina, A., Steckel, H. Foams for pharmaceutical and cosmetic application. *Int J Pharm* 2010; 394(1–2): 1–17.

52. Gogate, P.R., Pandit, A.B. Hydrodynamic cavitation reactors: A state of the art review. *Rev Chem Eng* 2001; 17(1): 1–85.

53. Raut, J.S., Stoyanov, S.D., Duggal, C., Pelan, E.G., Arnaudov, L.N., Naik, V.M. Hydrodynamic cavitation: a bottom-up approach to liquid aeration. *Soft Matter* 2012; 8(17): 4562–4566.

54. Fernandez, D., Maurer, P., Martine, M., Coey, J.M.D., Möbius, M.E. Bubble formation at a gas-evolving microelectrode. *Langmuir* 2014; 30: 13065–13074.

55. Mansour, L.B., Chalbi, S., Kesentini, I. Experimental study of hydrodynamic and bubble size distributions in electroflotation process. *Indian J Chem Technol* 2007; 14(3): 253–257.

56. Hammadi, Z., Morin, R., Olives, J. Field nano-localization of gabubble production from water electrolysis. *Appl Phys Lett* 2013; 103: 223106.

11 Foam Stabilization Mechanisms

D. Langevin

CONTENTS

11.1 INTRODUCTION

Foams are dispersions of air bubbles in a liquid, frequently stabilized by surfactant molecules (Figure 11.1), but polymers, proteins or particles can also be used. The stabilizer's role is to slow down the different mechanisms of foam aging: drainage, coalescence, and coarsening. Liquid foams *drain* rapidly under the influence of gravity until the liquid volume fraction ϕ reaches values smaller than a few percent. They evolve slowly afterward due to gas transfer between bubbles (*coarsening*) and

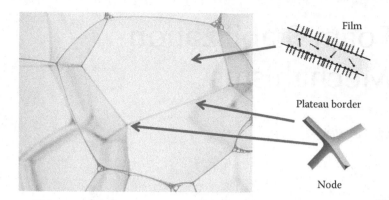

FIGURE 11.1 Aqueous foam stabilized by surfactants with polyhedral bubbles. Liquid film between bubbles (top right) covered by surfactant monolayers and Plateau border junction (bottom right). Surfactant molecules are represented by a circle (polar head) in contact with water, and a hydrophobic chain in contact with air. The surfactant is also solubilized in water and present in the bulk liquid.

rupture of the films separating the bubbles (*coalescence*), until they fully disappear, typically a few hours later.

Industrial applications necessitate larger liquid (or solid) volume fractions (frequently around 50%). To stabilize such wet foams, the continuous phase needs to be solidified. The solid continuous phase of the foam is melted to allow the production of the foam and resolidified before drainage takes place. The continuous phase may also be liquid and contain polymerization precursors allowing to gel or solidify the foam immediately after production. The specificities of the stabilization of these solid-like foams have been discussed in Section V. Applications and will not be addressed in this chapter.

Aside from the three fundamental mechanisms which control foam stability, drainage, coarsening, and coalescence, liquid evaporation may also play an important role. When left to open air, foams can be destroyed because the films thin more rapidly and rupture. Compact surface monolayers can be used to reduce evaporation rates [1]. Although few studies have been devoted to the issue in foams, a recent paper reports on the substantial role of humidity on foam stability [2].

In the following, we will concentrate on the three main stabilization mechanisms. In Chapter 8, foam production was discussed independently of the surface-active agents used. We will begin complementing this description by discussing the limitations introduced by these agents. We will provide then a brief overview of the three main mechanisms involved in the destabilization of foams. Examples will be given, including foams stabilized by surfactants, proteins, and particles.

11.2 FOAMING

11.2.1 SURFACE COVERAGE

In the case of stable foams, the amount of foam produced by a given device is limited by the amount of surface-active agent available to cover the bubble surfaces and to prevent coalescence. This problem has been investigated in the case of emulsions and for *good* surfactants, that is, able to efficiently protect the drops against coalescence. When there is not enough surfactant to cover all the drops, the drop radius is larger than expected from the production device, and is given by:

$$r = 3\phi\Gamma_\infty/[(1 - \phi)C] \tag{11.1}$$

where C is the bulk concentration of the surface-active agents (weight per unit volume), Γ_∞ their saturation surface concentration (mass per unit area at full coverage), and ϕ the continuous phase volume fraction. One sees in Equation 11.1 that when C increases, r decreases until a minimum is reached as determined by the foaming procedure. If C continues to increase, the radius remains constant [3].

The calculation is less simple for foams: even during formation, the bubbles may not be spherical due to rapid drainage. Let us consider a very dry foam whose bubbles have the shape of a tetrakaidecahedron (Kelvin cell). The bubble volume is $V = \delta_V L^3$, where L is the length of the Plateau border (PB) with $\delta_V \sim 11.3$; the bubble area is $A = \delta_A L^2$ with $\delta_A \sim 26.8$. One also has L $\sim 0.72r$, where r is the radius of a sphere having the same volume as the bubble. This radius is then given by an expression very similar to Equation 11.1, with a factor 3.29 instead of 3.

Studies of monolayers have shown that the saturation surface concentration Γ_∞ is of the order of 1 mg/m^2 for surfactants and 2–3 mg/m^2 for proteins. For particles, Γ_∞ depends on the particle size and nature: for instance, Γ_∞ is about 50 mg/m^2 for the 100 nm radius silica particles for which an example of studies will be shown later. The concentration Γ_∞ generally corresponds to the point where the surface pressure Π, shows a sharp increase when increasing Γ. In the case of surfactants, this occurs at a bulk concentration of the order of 1/10^e of the critical micellar concentration (CMC), and is associated with the onset of formation of Newton black films (see Chapter 3). In the case of the silica particles, the maximum surface concentration of 50 mg/m^2 corresponds to an incomplete surface coverage by these particles, which form rather a rigid percolated network at the surface [4]. This peculiar behavior was also observed in particle stabilized emulsions [5].

The required bulk concentration depends on the method used: this amount is less for low energy mixers than for high energy ones, because high energy mixing produces small bubbles, hence high surface area, with the amount of required surface-active agents being larger.

Taking the example of foams made by turbulent mixing, the bubble radii are about 50 μm for surfactants, a little less, about 35 μm for proteins and for particles. To obtain foams with liquid volume fractions of $\phi = 0.2$ during the mixing process, the minimum concentrations are 2 x 10^{-2} wt% for surfactants, 10^{-1} wt% for proteins, and around 1 wt% for particles. Despite particles being very good stabilizers, they have to be added in a significant amount to produce stable foams. When the amount of surface-active agents is less than this minimum value, there is a significant amount of coalescence, the foam volume decreases rapidly to a limit value corresponding to the saturation surface concentration and remains afterwards constant. This mechanism is called "limited coalescence" [6].

Coalescence during foam formation is also more important with surfactants that are not as good stabilizers. In some cases, the equilibrium between the number of bubbles that are generated and those that disappear cannot be established, the bubble sizes are larger than given by Equation 11.1, and the amount of foam created is small.

Some foams are made using emulsions instead of simple surfactant solutions. In this case, the surfactant also needs to cover the emulsion drops, and the required amount of surfactant is larger. It was shown that once both the coverage of the bubbles and the drops reach the saturation coverage, the foam becomes stable [7].

11.2.2 Mixing Energy

In general, foams can be produced by low energy procedures such as bubbling, hand shaking or microfluidic techniques. Foams stabilized by particles are a notable exception, but these foams are difficult to produce. The particles generally bear electric charges and they create large adsorption barriers, much larger than ionic surfactants [8,9]. Getting the particles onto the interfaces is difficult, and these foams cannot generally be produced by bubbling or by microfluidic techniques. *In situ* hydrophobization methods have been devised to surmount this difficulty [10,11].

In order to produce particle-stabilized foams, procedures involving large energy inputs are required: energetic shaking or turbulent mixing, thus allowing to overcome the adsorption barriers described above. Similar difficulties were mentioned by Golemanov et al. in the case of emulsions [12]. Tcholakova et al. discussed the problem later on by comparing the repulsive forces between a particle and a drop with the hydrodynamic forces pushing the particle toward the drop surface [8]. The results were adapted to air bubbles covered by the 100 nm silica particles mentioned earlier [9] and are summarized afterward. The force between a particle and the bubble surface can be written as:

$$F \approx 2\pi a \int_h^\infty \pi(h)dh \qquad (11.2)$$

where a is the particle radius, $\pi(h)$ is the disjoining pressure of the aqueous film formed between the particle and the bubble surface, and h is the distance between them. It was assumed that $\pi(h)$ can be written as in the DLVO theory with the sum of an electrostatic contribution and of a van der Waals contribution. The force F* required to overcome the barrier is in the range 4×10^{-13} to 4×10^{-14} N for the concentrations used in the dispersions. The adsorption barriers also exist for surfactants, but they are much larger for particles since their value is proportional to the particle size (see Equation 11.2).

Tcholakova et al. state that the hydrodynamic force F_h in laminar flow is:

$$F_h \sim \eta a^2 \dot{\gamma}, \qquad (11.3)$$

with η being the liquid viscosity and $\dot{\gamma}$ the shear rate. In an inertial turbulent flow,

$$F_h \sim \rho a^2 \mathfrak{I}^{2/3} r^{2/3} \qquad (11.4)$$

where ρ is the fluid density, $\mathfrak{I}$ is the rate of energy dissipation per unit mass, and r is the bubble radius. With the turbulent mixer used, $\mathfrak{I} \sim 10^5$ J/kg/s, and $F_h \sim 10^{-10}$ N with r ~ 35 μm. The force F* opposing the adsorption of the particles can, therefore, be overcome. It can even be overcome if vigorous hand shaking is used ($\mathfrak{I} \sim 10^3$ J/kg/s and $F_h \sim 10^{-12}$ N), as observed [13,14]. With other devices such as gas bubbling (at most ten bubbles of millimeter radius produced per second) or in microfluidic devices (gas and liquid flow rates below 100 mL/h, bubble sizes of the order of 100 μm), foam could not be produced with these particles. Indeed, the shear rates are below $\dot{\gamma} \sim 10^2$ s^{-1} and $F_h \sim 10^{-15}$ N for a particle of radius equal to 100 nm, that is, well below F*.

11.2.3 ADSORPTION KINETICS

For surfactant concentrations of the order of weight %, the adsorption times calculated for a diffusion-controlled process are much shorter than the characteristic times for foaming. In practical conditions of foam formation, there is in addition an important convective transport, which makes adsorption times still smaller.

In the case of proteins, the adsorption kinetics is slower than predicted by diffusion mechanisms. The minimum protein concentration required for foaming strongly depends on the protein: the amount of foam is larger with the flexible proteins that adsorb faster than with the globular ones [15]. Slowly adsorbing proteins such as lysozyme and ovoalbumin produce hardly any foam. In the case of emulsions produced by membrane emulsification, it has been shown that because of slow protein adsorption, the final emulsion drop size depends on the timescale at which the droplets are formed [16]. The same is likely to be true for foams.

Foaming is difficult with dilute protein solutions. The foam is produced within seconds, at which time no significant protein adsorption has occurred, and with *poor* surfactants, the foam partially collapses during the process. In practical applications, large protein concentrations are used to ensure good foaming. Alternatively, mixed solutions of proteins and surfactants can be used: the surfactant adsorbs quickly and prevents the foam from collapsing, then the protein adsorbs later and enhances the long term stability [17].

11.2.4 SURFACE RHEOLOGY

The surface rheology also plays a role in foaming, although limited. In one of the few existing studies addressing this issue, the threshold for rupture of an oil drop covered by surfactant in a well defined shear flow was studied. It was shown that the threshold follows approximately linearly the variation of the surface compression elasticity calculated at a frequency equal to the shear rate [18]. The energy used during the emulsification process is then smaller if the surface elasticity is smaller. However, if the elasticity is decreased, coalescence can occur and a compromise needs to be found. The behavior of oil drops covered with β-casein is similar, but when β-lactoglobulin is used, the threshold is even smaller than for pure liquids. This has been explained by the presence of a rigid surface layer that limits the droplet deformation and changes the rupture mechanism [19]. Although no detailed studies of this kind exist for foams, the general trends are likely to be the same.

11.3 COARSENING

Coarsening involves the transport of gas between bubbles of different sizes, leading to the growth of the average bubble radius r with time t: $r \sim t^{1/2}$ [20,21]. Coarsening is sometimes called *bubble disproportionation*. Its origin is similar to the phenomenon of Ostwald ripening in dilute dispersions, where the gas diffuses from the smaller to the larger bubbles due to a difference in Laplace pressure. In the latter situation however, $r \sim t^{1/3}$. The law $r \sim t^{1/2}$ arises from the fact that in foams, the gas mainly diffuses trough the thin films between bubbles for which the diffusion path is the smallest.

11.3.1 OSTWALD RIPENING

Ostwald ripening was extensively studied in emulsions [22]. The pressure in emulsion drops is increased by the capillary pressure which is larger in smaller drops, hence the liquid diffuses from the small drops to the large ones across the continuous phase; this results in an increase of drop radius with time. Lifshitz and Slyozov showed that when the drop concentration is very small, the drop radius increases as the cubic root of time t: $r \sim t^{1/3}$, with a rate $\Omega = dr^3/dt$ given by:

$$\Omega = \frac{8\gamma D_m S V_m}{9\,RT} \tag{11.5}$$

where γ is the interfacial tension between dispersed and continuous phases, D_m is the diffusion coefficient of molecules of the dispersed phase into the continuous one, V_m is their molar volume, S is the solubility of the dispersed phase in the continuous phase, R is the gas constant, and T is the absolute temperature.

Although equation 11.5 accounts well for the influence of the oil-water solubility S, the measured rate is usually higher than predicted by this equation [23–25]. This could of course occur from the fact that Equation 11.5 is valid only for very dilute systems, and corrections for drop volume fraction have been proposed [26]. However, these corrections cannot account for the rate differences observed with different surfactants [27].

Meinders and van Vliet addressed the issue with numerical simulations. They took into account the compression elastic modulus E for the layer of surface active species covering the oil-water interfaces. They showed that Ostwald ripening in emulsions can be slowed down by increasing the elastic modulus [28]. When the compression elastic modulus E reaches the value $E = \gamma/2$, their simulations show that Ostwald ripening stops as predicted earlier by Gibbs [29]. The Gibbs argument holds for a single bubble and makes use of the derivative of the Laplace pressure P with respect to the bubble radius r. For an isolated bubble covered by layers with a compression elastic modulus E, one has :

$$\frac{dP}{dr} = \frac{d\left(\dfrac{2\gamma}{r}\right)}{dr} = -\frac{2\gamma}{r^2} + \frac{2}{r}\frac{d\gamma}{dr} = \frac{2}{r^2}(2E - \gamma) \tag{11.6}$$

since $E = A\,(d\gamma/dA) = (r/2)(d\gamma/dr)$ with the area of the bubble $A = 4\pi r^2$. If $E < \gamma/2$, the pressure inside the bubble increases when its radius decreases, which leads to a self-accelerated dissolution of the bubble and to its complete disappearance. If, on the contrary, $E > \gamma/2$, the pressure inside the bubble decreases upon decrease of the bubble radius. The dissolution of the bubble, therefore, slows downs and eventually stops when the Laplace pressure approaches zero. In this case, the bubble will distort and adopt faceted shapes: this has been observed both experimentally and in simulations on bubbles covered by particle monolayers [30].

Note that in their simulations, Meinders and van Vliet assume a constant surface elastic modulus E, which might not be the case in practice as the surface layers become increasingly compressed (or expanded). The experiments of Reference 27 evidenced a correlation between ripening rates and the elastic modulus E.

11.3.2 COARSENING OF FOAMS

In a foam structure, the gas mainly diffuses trough the thin films between bubbles for which the diffusion path is the smallest. As a consequence, the bubble growth law is different from that of classical Ostwald ripening: $r \sim t^{1/2}$, with r being the equivalent bubble radius (radius of the sphere having the same volume as the bubbles). Here, the growth rate depends on the number of faces of the bubbles rather than on their size; bubbles with small numbers of faces shrink, bubbles with large numbers of faces grow. The problem is quite difficult to tackle, but an order of magnitude of the coarsening rate $\Omega_f = dr^2/dt$ can be estimated as $\Omega_f \sim D_{eff}\,f(\phi)$, where $f(\phi)$ is the fraction of total area A of the bubble covered by thin films and D_{eff} is an effective diffusion coefficient: $D_{eff} \sim \gamma D_m S V_m/h$, h being the film thickness [31]. This particular process of bubble growth is called *coarsening*.

The growth rate of an individual bubble of volume V in a foam can be written as: $dV/dt = V^{1/3}\,\mathcal{G}$, $\mathcal{G}$ depending both on the shape of the bubble and on the physicochemical properties of the liquid, gas, and surfactant through the effective gas diffusion coefficient D_{eff}:

$$\mathcal{G} = -D_{eff} \int_S \frac{\mathcal{H}\,dS}{V^{1/3}} \tag{11.7}$$

where $\mathcal{H}$ is the mean curvature of a surface element dS of the bubble [32,33]. Bubbles with a small number of faces shrink, while those with a large number of faces grow; bubbles with around 12 faces remain stationary [34]. Averaging over suitable distributions of bubble geometry [31], it follows that $dL^2/dt = 2\,D_{eff}\,f(\phi)$, with:

$$D_{eff} = \frac{4\delta_A}{3\delta_V\beta}\frac{D_m He \gamma V_m}{h} \tag{11.8}$$

where L is the average bubble edge (PB) length, δ_A, δ_B, and β are geometrical factors, and D_m, He, γ, V_m, and h are the gas diffusion coefficient, Henry constant (expressed in mole m^{-3} Pa^{-1}), surface tension, molar gas volume, and film thickness, as defined before. Equation 11.8 is the analog of Equation 11.5 for a dispersed gas phase and diffusion through films between bubbles.

For the Kelvin tetradecahedra, the bubble volume is $V = \delta_V L^3$, with $\delta_V \sim 11.3$ and the bubble area $A = \delta_A L^2$ with $\delta_A \sim 27$. The effective mean curvature is $\mathcal{H} \approx 1/(\beta L)$ with $\beta \sim 10$ for dry foams. The function $f(\phi)$ is the fraction of the area of the bubble covered by thin films; $f(\phi)$ is close to one for very dry foams and decreases when ϕ increases such as: $f(\phi) \approx (1 - 1.52\, \phi^{1/2})^2$ [35].

It follows that $\Omega_f = dr^2/dt$ is equal to $2\, D_{eff}\, f(\phi)$ multiplied by $\delta_A/(4\pi)$, hence:

$$\Omega_f = \frac{2\delta_A^2}{3\pi\delta_V\beta} \frac{D_m He \gamma V_m}{h} f(\phi) \tag{11.9}$$

The quantity $D_m He/h$ is the film permeability K_0 when He is expressed as the gas volume fraction in water. It is sometimes assumed that the surfactant molecular layers at the film surface slow down gas transfer. The permeability is then:

$$K = \frac{D_m He}{h + 2D_m/K_s} \tag{11.10}$$

where K_s is the gas permeability across a surfactant monolayer. For pure water films in air, $He = 0.018$ and $D_m = 2 \times 10^{-9}$ m^2/s [36], hence for a film 10 nm thick: $K_0 \sim 3.6$ mm/s, comparable but somewhat higher to the permeabilities measured for such thin films $K \sim 1$ mm/s [37].

11.3.3 INFLUENCE OF SURFACE LAYERS' ELASTICITY

The influence of the layers of surface-active substances at the surface of bubbles has not been investigated much to date. It has been shown that protein stabilized foams coarsen more slowly than surfactant stabilized foams, but this was attributed to the larger film thickness in the case of proteins [38]. Recently, differences in coarsening rates were observed with different surfactants, but were attributed to differences in permeability of the thin liquid films to gases due to the presence of surfactant monolayers covering the film surfaces [39].

The presence of the monolayer can have another, potentially more important influence on the coarsening process due to its mechanical properties. The numerical simulations by Meinders and van Vliet discussed in §3.1 showed that Ostwald ripening in emulsions can be slowed down and even stopped entirely by increasing the compression elastic modulus E of the monolayer [28].

Coarsening of foams is a much more complex issue: foams are assemblies of close-packed bubbles and coarsening depends on the number of faces of the bubbles rather than on their size [34] (but both parameters are strongly correlated). While the coarsening behavior of an individual bubble is reasonably straightforward to analyze, the collective coarsening process of bubbles within a foam is much more complex. We will see later, however, that coarsening can be suppressed when the Gibbs criteria $E > \gamma/2$ is fulfilled.

11.3.4 INFLUENCE OF GAS

Let us also mention that the nature of the gas used for foaming plays a crucial role through the parameter D_{eff}: gases soluble in water such as CO_2 give less stable foams than less soluble ones such as N_2, because the CO_2 transport across water films is faster. The stability of CO_2 foams can be improved by adding small amounts of nitrogen: since the gas composition in each bubble cannot change (otherwise the chemical potential would vary locally), the gas diffusion process is slowed down [40]. An even stronger effect can be obtained by adding small amounts of less water soluble gases, such as C_2F_6 [41].

As in the case of evaporation, the diffusion of gas molecules through thin films is affected by the presence of surfactant monolayers. However, it is not yet clear if the contribution of the monolayer is always significant [22].

11.3.5 SPECIFIC EXAMPLES

11.3.5.1 Surfactant Foams

Some values for the rate Ω_f can be found in the literature and are reported in Table 11.1.

The coarsening rate measured in Reference 31 for dry SDS foams is in good agreement with the calculated value using Equation 11.9 and a film thickness h~35 nm. Since the initial bubble size is large, the disjoining pressure is small and the films are likely thick enough for the contribution of monolayer permeability (Equation 11.10) to be negligible [37].

The coarsening rates measured with wetter SDS foams are smaller. The foams were held in rotating containers, hence the liquid volume fraction could remain large. However, the ratio of the coarsening rate values measured for wet and dry SDS foams is much larger than the ratio of the volume fraction corrections: f(0.15) = 0.17 while f(0.01) = 0.72.

It has been remarked that the Ostwald ripening of emulsions depends on surfactant concentration. This has been attributed to the presence of surfactant micelles that can act as surfactant carriers, although a quantitative explanation is far from established [44]. In Reference 31, the films could contain micelles [45], whereas in Reference 41 they are thinner: this might contribute to the large difference in coarsening rates. Unfortunately, no information about the surfactant concentration in the aqueous phase of the foams is given in the measurements reported.

The difference between coarsening rates observed for wet foams and different gases in Reference 41 can be accounted for by the different solubility of these gases in water. The values measured for a slightly wetter foam made using air [42] is comparable to that of Reference 41 for nitrogen, for which water solubility is similar.

Other experiments have been made with air and mixtures of ionic and nonionic surfactants known to form compact surface layers [39, 43]. Larger Ω_f values were found, of the order of 10^{-10} m^2/s, but the films could have been thinner. Apparent fast coarsening could also arise from coalescence events as the polydispersity increases with time in the experiments of Reference 39.

Foams made with other surfactant mixtures were shown recently to be very stable [46]. They were made with two surfactants having opposite charges that associate in the solutions in the form of vesicles [47]. These surfactant mixtures are sometimes called *catanionic*. The mixed monolayer which is formed by the rupture of vesicles at the air-water interface is extremely rigid (large elastic moduli E). The coarsening of these foams is much slower than that of standard surfactant foams. This could be due either to the very closely packed monolayers which act as gas barriers, to the high elastic modulus E of the monolayer which counteracts coarsening or to the presence of closely packed vesicles between the bubbles.

Foams made with other types of catanionic mixed solutions, hydroxyl stearic fatty acid and ethanol amine, were also found to be very stable [48]. At the difference of the myristic acid-cetyltrimethylammonium chloride (CTACl) mixtures, they contain long tubes which form spontaneously and reversibly and block Plateau borders. The sizes of the foam films were found to be much smaller than for standard surfactant foam films, with a very thick meniscus full of tubes.

TABLE 11.1

Coarsening Rates for Foams Made with the Different Surfactants with Different Initial Radius r_0, Different Liquid Fractions, and Different Gases

Ω_f (μm^2/sec)	C$_2$F$_6$, Dry	C$_2$F$_6$, Wet (ϕ = 0.15)	N$_2$, Wet (ϕ = 0.15)	Air, Wet (ϕ = 0.20)	Air, Dry
SDS	11[31] (r_0 ~200 μm)	0.095[41] (r_0 ~20 μm)	14[41] (r_0 ~50 μm)	12[42] (r_0 ~140 μm)	
Mixtures					130(39) (r_0 ~200 μm)
					120(43) (r_0 ~70 μm)

According to Equation 11.9, the fractional area of the films is small and the coarsening time large. It is also probable that the surface layers are rigid, slowing down further coarsening.

In practice, the condition $E = \gamma/2$ can never be reached with soluble surfactants: coarsening is a slow process, and surfactants can desorb and adsorb easily, so the resistance to compression (and expansion) of the layer at the bubble surface vanishes in the long time limit. Catanionics form ion pairs at the surface and have large desorption energies, hence the resistance to compression does not vanish.

11.3.5.2 Protein Foams

Protein-stabilized bubbles injected below the air-water surface also shrink and disappear [49]. This is at first sight a contradiction with the Gibbs criteria, because protein layers can have very large compression moduli, well above $\gamma/2$, and the layers exchange little with bulk (proteins are frequently irreversibly adsorbed). However, these layers can slowly collapse upon increasing compaction, forming multilayers and offering no resistance to bubble shrinkage.

Currently known protein foams are only very stable if the foaming liquid is gelified, with the notable exception of foams made with hydrophobins. These proteins form solid-like layers at the surface of water which have high elastic moduli E and do not collapse upon compression [50]. When standard proteins are used and when the foaming solution is fluid, the foam is nevertheless more stable than standard surfactant foams. Even when the protein concentration is small, protein foam films are thick and irregular [38,51,52] (Figure 11.2 left). Apparently, protein aggregates are trapped in the film and stop the film thinning process. They form at the film surfaces since no aggregates are seen in the bulk solution. When the protein concentration is large enough, the films are gel-like. The coarsening of foams made from protein solutions is somewhat slower than that of surfactant foams (Figure 11.2 right). This feature was attributed to the larger film thicknesses according to Equation 11.9 [38].

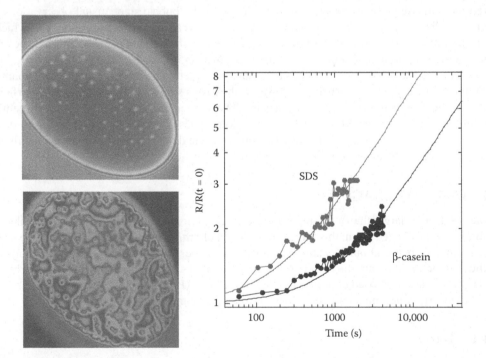

FIGURE 11.2 Optical microscopy pictures of foam films made with SDS (top left) and β-casein solutions (bottom left) showing that thickness is heterogeneous when proteins are used. Right: coarsening of a β-casein foam compared to an SDS foam; the lines are fits with a squared root of time variation. (Data from Saint-Jalmes, A. et al. *Colloids Surf A, Physicochem Eng Aspects* 2005; 263(1–3): 219–25 [38].)

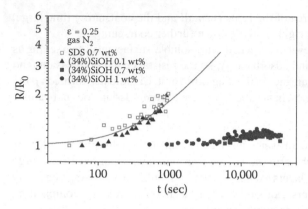
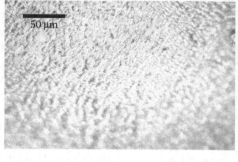

FIGURE 11.3 Coarsening of foams made by turbulent mixing using SDS and particles. Left, bubble radius as measured using multiple light scattering (DWS). The line is a fit with r $\sim t^{1/2}$. Right: Optical microscopy picture of a bubble in a foam stabilized with particles. The bubble surface is corrugated, suggesting a resistance to shrinkage. (After Cervantes Martinez, A. et al. *Soft Matter* 2008; 4(7): 1531–5 [53].)

11.3.5.3 Particle Foams

Coarsening arrest requires both high surface elastic modulus and resistance to collapse. Particle layers are irreversibly adsorbed, they resist collapse and buckle (Figure 11.3), possibly the reason why hydrophobin foams and particle foams do not coarsen [50,53].

X-ray tomography of particle foams made by shaking revealed that some bubbles shrink, but that other bubbles become larger. This contradiction with the results shown in Figure 11.3 could be due to the fact that the bubbles could not be fully covered or to the fact that the difusing wave spectroscopy (DWS) method used only probes the averaged value of the bubble size. Visual observations after several months of foams made by turbulent mixing also revealed the presence of bubbles which are sometimes broken, but seemingly without having influenced their neighbors [53]. This may be related to the fracture of particle monolayers seen upon expansion [54].

In order to facilitate foam production, surfactants are sometimes added and, in this case, foams can be obtained using hydrophilic particles. However, the foams stabilized by mixtures with surfactants are less stable than foams stabilized by particles. The surface of the growing bubbles is seemingly never sufficiently well covered and further expansion of these bubbles can never be arrested. They coalesce with other large bubbles, leading to their complete disappearance. A foam with only small bubbles remains at the end of the aging process [55].

11.4 FOAM DRAINAGE

Bubbles with sizes larger than a few microns rise quickly due to gravity and the liquid is collected at the bottom of the created foam: this is the phenomenon of *drainage*. When the liquid volume fraction of the foam falls below about 36%, the bubbles are no longer spherical, they distort into polyhedral and the flattened regions are the liquid films. The liquid flows through the interstitial spaces between bubbles, which are composed of thin films, *Plateau borders* (PBs) made of connections of three films and junctions or *nodes* made of connections of four PBs (Figure 11.1).

11.4.1 Theory

The problem of foam drainage is quite complex in view of the structure of the network in which the liquid flows [20]. When the foam is dry enough, the contribution of the films can nevertheless be neglected. The simplest result is obtained in the case of *forced drainage*: the liquid is added at the top of a dried foam. A drainage front then forms and travel downward at a constant velocity V. When the surface mobility is low, the contribution of the PBs is dominant; when the surface mobility is large, the contribution of the nodes is dominant [56]. The theoretical velocity is:

$$v = K_{drain} \, \rho g L^2 \phi^\alpha / \eta \qquad (11.11)$$

where ρ is the liquid density, η is its viscosity, g is the gravity acceleration, K_{drain} is a foam permeability ($K_{drain} \sim 1/150$), and α is an exponent depending on the surface mobility: $\alpha = 1$ for rigid surfaces, ½ for mobile surfaces [41] (Figure 11.4). With L $\sim$100 µm and $\phi^\alpha = 0.1$, one finds v $\sim 10^{-4}$ m/s and a time of order of 2 minutes for the drainage across 1 cm of foam. The order of magnitude is in agreement with observations.

The Stokes velocity for a single bubble is $v_{st} = 2r^2 \rho \, g/(9\eta)$. Since L $\sim$0.72r, one sees that the velocity of drainage of a foam is much smaller than v_{st}: the numerical factor is about 100 times smaller and the term ϕ^α is also small.

The notion of surface mobility can be related to surface rheology: the elastic moduli and viscosities will be small for a mobile surface and large for a rigid one. The compression moduli and viscosities for the surfactant solutions used to make foams are small because these solutions are concentrated and exchanges between surface and bulk are rapid. Because the surface shear modulus is zero, one is left with the surface shear viscosity η_s. One should note that the notion of surface mobility is not intrinsic; indeed, a surface may behave as mobile even if the surface viscosity is large when the bubble radius is large as well or when the bulk viscosity is large. What matters is the relative contributions of surface and bulk viscosities, easily estimated with the Boussinesq number B = η_s/(ηr) [57]. When B > 1, the surface viscosity dominates and the surface behaves as rigid; when B < 1, the bulk viscosity dominates and the surface behaves as mobile. The transition between the two regimes has been observed [58].

Free drainage is more difficult to handle. Numerical simulations showed the drained volume varies first linearly with time, then exponentially at longer times, the characteristic time being the ratio of foam height to the velocity v given by Equation 11.11 [21].

After drainage is completed, an equilibrium vertical liquid fraction profile is established (Figure 11.5). In these conditions, gravity is compensated by the osmotic pressure in the foam [59].

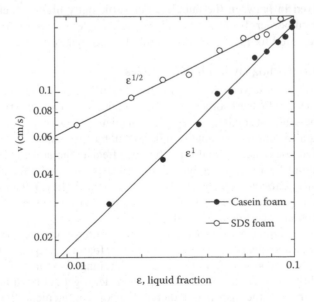

FIGURE 11.4 Typical forced drainage curves for foams made from an SDS solution and from a casein solution. (After Saint-Jalmes, A. *Soft Matter* 2006; 2(10): 836–49 [41].)

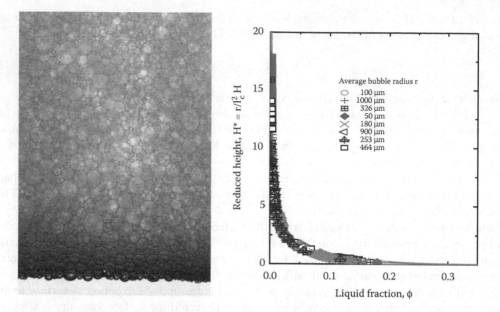

FIGURE 11.5 Draining foam (left): the top of the foam is *dry* and composed of polyhedral bubbles. The bottom of the foam, which is in contact with the liquid, is *wet* and contains spherical bubbles. Right: vertical profile of the equilibrium volume fraction $\phi(z)$ calculated (line) and measured (points) for a foam with different mean bubble radius r; H* is the reduced height H r/l_c^2, being the capillary length; (Adapted from Maestro, A. et al. *Soft Matter* 2013; 9(8): 2531–40 [59].)

11.4.2 Specific Examples

11.4.2.1 Surfactant Foams

Standard surfactant foams drain in a few minutes, somewhat more slowly if the bubble size is small (v ~r^2, according to Equation 11.11). Catanionic foams drain much more slowly [60]. As liquid drains out between the bubbles, an increasing trapping and compaction of the vesicles and vesicle aggregates is observed in between the bubbles. For sufficiently high concentrations, a complete blocking of the Plateau borders by vesicles can lead to a complete arrest of drainage [60]. A similar behavior has been reported for foams containing tubular aggregates [61].

11.4.2.2 Foams Containing Hydrophilic Particles

Situations similar to those observed with vesicles are seen in foams containing hydrophilic particles: the particles can block the Plateau borders. A first example was given by Friberg who noticed that foams made in the presence of small particles with lamellar structures (produced when the foaming liquid is water in equilibrium with a lamellar phase) can be very stable. When the liquid contains many particles, it becomes viscous, and one obvious effect is the slowing down of the drainage process. As pointed out by Friberg, a more subtle effect is due to the balance of the interfacial tensions: if the particle cannot enter the surface of films and is larger than the film thickness, it will be trapped in the Plateau borders, as he observed [62]. Similar effects were reported later in individual Plateau border experiments [63,64].

Another interesting example of particles arresting the drainage has been observed with solutions where a colloidal clay, Laponite, is dispersed in SDS before foaming. Laponite forms gels in aqueous solutions containing SDS, and a nonclassical arrest of drainage is seen in foams made with these dispersions [65]. While the foam drains, the laponite particles get confined in the Plateau borders, and since the yield stress of the dispersions increases upon confinement, the interstitial fluid gels and drainage is arrested after a time t_j. As coarsening continues, the bubble size increases, the size of the Plateau borders increases as well, and drainage then starts again at a time t_u (Figure 11.6).

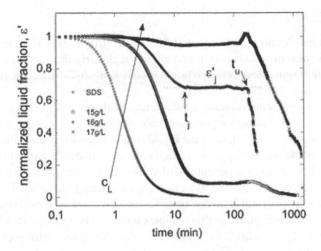

FIGURE 11.6 Time evolution of the liquid fraction for SDS foams and various concentration of laponite C_L. The times t_j and t_u are indicated on the curve at $C_L = 16g/L$. (After Guillermic, R.M. et al. *Soft Matter* 2009; 5(24): 4975–82 [65].)

11.4.2.3 Foamed Emulsions or "Foamulsions"

At small oil volume fractions, the foams made from emulsions drain and coarsen like classical foams made from SDS only. At larger oil volume fractions, very different features are observed. When $\phi_{oil} > 0.63$ (random close packing of spheres), the emulsion droplets are densely packed [66], and the emulsion becomes viscoelastic, with a finite shear modulus and yield stress. Microscope images of such a foam ($\phi_{oil} = 70\%$) are shown in Figure 11.7. One can see that droplets are actually confined and crowded between bubbles, which stay anomalously far away from each other.

The presence of such a dense assembly of droplets trapped and jammed in between the bubbles has several effects. The local viscosity increases, slowing down both film thinning and Plateau borders' shrinking (slower drainage) as in the case of particles. In addition, for initial bubble diameters of the order 100 μm, hydrodynamic stresses in the Plateau borders become comparable to the yield stress of the emulsion (of the order of a few Pa [67,68]). Drainage can, therefore, not only be slowed down, but even be arrested if the yield stress of the emulsion becomes higher than the local hydrodynamic stresses as seen before for the laponite-SDS foam Figure 11.6) and confirmed by Goyon et al [69].

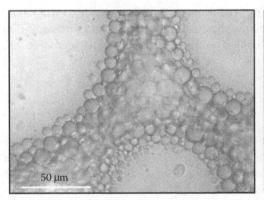
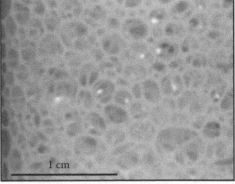

FIGURE 11.7 Optical microscopy photograph of a foam made from a rapeseed oil emulsion with $\phi_{oil} = 70\%$ immediately after preparation (left) and 6 hours after preparation where the Plateau borders are still very thick (right). (After Salonen, A. et al. *Soft Matter* 2012; 8(48): 12135 [7].)

11.5   COALESCENCE

The stability of foams is controlled by the stability of the thin films which separate bubbles. Foam films of standard surfactant solutions thin and either break early or reach an equilibrium thickness controlled by the interaction forces between liquid surfaces (h $\sim$5–20 nm). The rupture of equilibrium films is a stochastic process (see Chapter 3).

So far, very little is understood about the main mechanisms of film rupture, and still less about coalescence in foams. Some authors report that coalescence occurs once the bubbles have reached a critical size [70] as in emulsions [71], once the liquid fraction has reached a critical value [72] or when the applied pressure or the capillary pressure reach a critical value [73]. Even if these mechanisms are very different, it is difficult to experimentally discriminate between them since capillary pressure, liquid fraction, and bubble size are linked as in the case of emulsions [74]. The different behaviors observed in the literature might also be due to different flow conditions and accordingly to different coalescence processes. More details on this complex issue can be found in Reference 75. Very stable foams frequently have thick gelified or solid foam films, for which film rupture mechanisms are different. They will not be discussed here.

De Gennes claimed that in most practical cases, coalescence in foams were due to antifoam-like agents [76]. Whether a particle acts as antifoam or not depends in particular on the interfacial tensions between the three phases (air, water, particles). One generally introduces various coefficients (such as the entry coefficient E_n and the bridging coefficient B_r) to describe the particle antifoam potential [77,78]. These coefficients E_n and B_r are defined as follows:

$$E_n = \gamma_{aw} + \gamma_{pw} - \gamma_{ap}$$
$$B_r = \gamma_{aw}^2 + \gamma_{pw}^2 - \gamma_{ap}^2 \qquad (11.12)$$

where γ are the different surface tension, and p, a, and w standing, respectively, for the particle, the air, and the aqueous phases. The entry coefficient E_n is linked to the potential of the dispersed particles to penetrate into the air-water interface. It should be positive for the particle to act as antifoam. The bridging coefficient B_r is linked to the ability of the particles to bridge the foam films. The actual antifoam action also depends on an energy barrier for entering the gas-water interface: a potential antifoam which cannot enter the air/water interface does not work as well as a less strong antifoam which can easily enter the interface [79]. Particles that cannot enter the surface of films will be trapped in the Plateau borders and can act as foam stabilizers [80]. More details can be found in Chapter 5.

In some cases, the antifoam can be produced *in situ*. For instance, when foams made with nonionic surfactants are heated above the cloud point, the aqueous phase starts to separate into surfactant-rich and surfactant-poor phases, with the surfactant-rich phase nucleating in the form of microdroplets, which act as a defoamer [81].

Let us first discuss the case of coalescence at rest.

11.5.1   COALESCENCE AT REST

In foams and concentrated emulsions (ϕ below 30%–40%), bubbles and drops are in contact. For emulsions with small drop size, sedimentation or creaming are virtually absent and two regimes can be identified: at short times, Ostwald ripening dominates, with a drop size varying as $t^{1/3}$, until the drops reach a critical radius r* after which coalescence starts (Figure 11.8) [71].

In the experiments of Figure 11.8, the drops were initially rather monodisperse, but the polydispersity p was larger than predicted for the asymptotic drop size distribution in Ostwald ripening ($p_{or} \sim$20%). Hence, p decreases with time in this regime until it reaches p_{or}. In the coalescence regime, p increases rapidly. A separation between Ostwald ripening and coalescence regimes has been also observed in

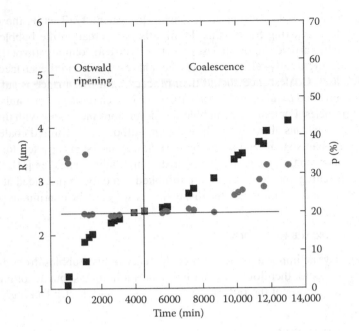

FIGURE 11.8 Drop radius versus time for octane in water emulsions stabilized by tetradecyltrimethylammonium bromide (TTAB) (black squares) and drop polydispersity (red circles). (Adapted from Georgieva, D. et al. *Langmuir* 2009; 25(10): 5565–73 [27].)

foams [70]. The existence of a rather well defined drop/bubble size threshold is consistent with the fact that coalescence time increases with drop area.

In extremely dry foams, another coalescence mechanism has been identified: bubble reorganization events require a minimum amount of water and if the liquid fraction is less than a critical value (which is r-dependent), the foam collapses [82]. Similarly, coalescence in draining foams [72] and in foams submitted to a constant gas flow rate [83] has been shown to occur above a very small liquid volume fraction. However, in draining foams, the critical liquid fraction has been observed to be independent of bubble size [72], a feature not expected in the model of Reference 82. It is, therefore, not clear that this model could account for all the reported coalescence studies.

As in the case of small φ, a coalescence event can induce coalescence of neighboring bubbles/drops because of the motion induced by the first event. It has been reported [84] that increasing liquid viscosity suppresses this effect in 2D bubbles rafts, as expected. A simple model based on kinetic energy transfer can account for the results [84]. Another possible origin of the cascade of coalescence events has been proposed that is stochastic in nature: if the probability of rupture is proportional to area, coalescence makes the bubbles/drop area larger, which accounts for a large coalescence probability [85]. Dynamical scaling associated with the distribution of cell edge lengths L has been evidenced [86], as well as a peculiar dependence on liquid volume fraction (the averaged L increases with time for wet foams and decreases for dry foams), a feature remaining unexplained so far. Avalanche phenomena are also commonly observed in 3D foams [87].

11.5.2 LIMITED COALESCENCE

We have mentioned that particle-stabilized foams are ultrastable only if the particle concentration in the initial dispersion is large enough. It was reported, using multiple light scattering techniques (see Figure 11.8), that the stability of foams produced by turbulent mixing with silica particle concentrations below 0.7 wt.% was limited, and comparable to that of a foam made with a standard surfactant such as SDS [9].

However, when the particle concentration reaches the value of 0.7 wt.%, the stability becomes remarkable, with foams lasting for months. If initially after creation the bubble surfaces are not sufficiently covered by particles, upon coalescence, the surface to volume ratio of the created bubbles decreases, hence the eventually released particles could re-adsorb and the surface concentration of the particles increases. Coalescence should then proceed until the surface is sufficiently covered. This phenomenon is called *limited coalescence* [6] and has been observed in emulsions stabilized by the same type of particles. For foams, the bubble size never stops increasing with time, even when the concentration reaches values above the stability limit. Although the limited coalescence has never been observed with foams stabilized solely by particles, it has been reported for foams stabilized by mixtures of a short chain amine and small hydrophilic silica particles [11]. Ultrastable foams for which both coarsening and coalescence are inhibited can only be produced at high particle and amine concentrations or those for which the dispersions also gel in the continuous phase of the foam.

11.5.3   ROLE OF SURFACE RHEOLOGY

Surface elasticity plays an important role in the coalescence of two bubbles/drops and in film rupture (Chapter 3). Its role should, therefore, be equally important in foams, but as for bubbles, drops, and films, a complete description of the influence of surface elasticity is still lacking.

11.5.4   FORCED COALESCENCE

As for bubbles, drops, and films, coalescence in foams is very different in dynamical conditions and in particular during their generation.

11.5.4.1   Coalescence under Compression

Coalescence may occur under an applied osmotic compression [71]. This phenomenon is well known in emulsions: when ϕ is decreased below the limit volume fraction ϕ_c, a "catastrophic" phase inversion is observed [88], giving rise to unstable inverse and multiple emulsions. Note that ϕ_c varies with experimental conditions (mixing protocol, surfactant concentration, stirring velocity). Coalescence is observed in emulsions above a certain critical pressure, applied either by centrifugation or osmotic compression, which increases the dispersed volume fraction [8]. A nonstochastic process described for films in §2.2.2 has been proposed to account for the critical pressure: coalescence occurs when the capillary pressure becomes larger than the DLVO disjoining pressure maximum [8]. Such a mechanism should not operate in the presence of short range forces (i.e., with well covered interfaces) or with nonionic surfactants. Furthermore, being deterministic, it cannot explain the stochastic character of film rupture in foams [89].

11.5.4.2   Coalescence under Flow

Recent experiments showed that, at the difference of equilibrium films, a surfactant stabilized vertical film entrained on a frame ruptures after a rather well defined time that depends on the pulling velocity. The film thickness at rupture is about 10 nm [90]. The rupture appears to involve hydrodynamic instabilities likely similar to those occurring for draining films which did not yet reach equilibrium [91], called by Mysels "marginal regeneration" [92]. This is likely why many coalescence events take place during foam formation when the foam films are rapidly stretched, the lifetime of these films being much shorter than that of the equilibrium films.

Let us also mention that large shear stresses may also induce coalescence in surfactant foams. Smaller stresses are, however, likely required with protein and particle foams. It has been reported that small shears can easily break the films between drops in protein-stabilized emulsions: concentrated emulsions stabilized by proteins, although very stable at rest, destabilize under moderate shear [93].

11.6 CONCLUSIONS

Foam stability involves either slow coarsening, slow drainage or slow coalescence. In many cases, these phenomena are coupled. This coupling does not always enhance the foam stability, as for instance in the SDS-laponite foams, where coarsening is responsible for the restart of drainage. Coarsening is slowed down when the films between bubbles are thick (or small) or when low water soluble gases are used, and it can be arrested when the surface compression modulus E is large enough. Drainage can be slowed down when the surfactant layers are rigid or when particles accumulate in the Plateau borders; it can be arrested only if the liquid of the foam either gels or solidifies with a yield stress sufficiently high. The parameters governing coalescence are not yet fully elucidated. Coalescence seems, however, slowed down when the surface layers have larger elastic moduli E or when the films are very thick and/or rigid. One drawback systematically encountered with very stable foams is that they are difficult to produce.

Although there have been significant advances in the qualitative understanding of the stabilizing mechanisms of foams, they will be less obvious to model in order to obtain quantitative predictions. Drainage is perhaps the easiest process to model, although it is not always simple to estimate the surface mobility. The role of film permeability and of the surface elasticity in coarsening remains less clear.

Coalescence is by far the most complex problem. The state of the models is more advanced in the case of pure fluids, where the coalescence of two isolated bubbles is now well understood. Once the surfaces are covered by dense surfactant layers, surfactant motion gives rise to surface tension gradients, and the nature of the problem changes totally. It is not yet clear if there is still a critical size or rather a critical volume fraction or a critical capillary pressure. When polymers, proteins or particles are used as stabilizers, the foams are usually more stable, at least at rest, and new coalescence schemes appear. Many different coalescence phenomena have been identified, stochastic and nonstochastic, including heterogeneous ones (e.g., antifoams). The state of the modeling of these processes is very preliminary due to their complexity. The same holds for the experimental side. Coalescence of a population of bubbles with cooperative effects such as avalanches are for the time being mostly not understood.

A full comprehension which would allow for the complete control over the aging of foams, is still eluding us. Further investigations are clearly desired, combining experiments at different scales: surface layers, films, Plateau borders, bubbles, and foam itself. Indeed, the mechanisms evidenced so far can act at any of these scales and partial measurements can miss important features.

REFERENCES

1. Barnes, G.T. The effects of monolayers on the evaporation of liquids. *Adv. Colloid Interface Sci* 1986; 25(2): 89–200.
2. Li, X.L., Karakashev, S.I., Evans, G.M., Stevenson, P. Effect of environmental humidity on static foam stability. *Langmuir* 2012; 28(9): 4060–8.
3. Taisne, L., Walstra, P., Cabane, B. Transfer of oil between emulsion droplets. *J. Colloid Interface Sci* 1996; 184(2): 378–90.
4. Stocco, A., Drenckhan, W., Rio, E., Langevin, D., Binks, B.P. Particle-stabilised foams: An interfacial study. *Soft Matter.* 2009; 5(11): 2215–22.
5. Vignati, E., Piazza, R., Lockhart, T.P. Pickering emulsions: Interfacial tension, colloidal layer morphology, and trapped-particle motion. *Langmuir* 2003; 19(17): 6650–6.
6. Arditty, S., Whitby, C.P., Binks, B.P., Schmitt, V., Leal-Calderon, F. Some general features of limited coalescence in solid-stabilized emulsions (vol 11, pg 273, 2003). *Eur Phys J E* 2003; 11(3):273–281.
7. Salonen, A., Lhermerout, R., Rio, E., Langevin, D., Saint-Jalmes, A. Dual gas and oil dispersions in water: Production and stability of foamulsion (vol 8, pg 699, 2012). *Soft Matter* 2012; 8(48): 699–706.
8. Tcholakova, S., Denkov, N.D., Lips, A. Comparison of solid particles, globular proteins and surfactants as emulsifiers. *Phys Chem Chem Phys* 2008; 10(12): 1608–27.

9. Stocco, A., Rio, E., Binks, B.P., Langevin, D. Aqueous foams stabilized solely by particles. *Soft Matter* 2011; 7(4): 1260–7.

10. Park, J.I., Nie, Z., Kumachev, A., Abdelrahman, A.I., Binks, B.P., Stone, H.A., Kumacheva, E. A Microfluidic Approach to Chemically Driven Assembly of Colloidal Particles at Gas–Liquid Interfaces. *Angew Chem Int Ed* 2009; 48(29): 5300–4.

11. Arriaga, L.R., Drenckhan, W., Salonen, A., Rodrigues, J.A., Iniguez-Palomares, R., Rio, E. et al. On the long-term stability of foams stabilised by mixtures of nano-particles and oppositely charged short chain surfactants. *Soft Matter* 2012; 8(43): 11085–97.

12. Golemanov, K., Tcholakova, S., Kralchevsky, P.A., Ananthapadmanabhan, K.P., Lips, A. Latex-particle-stabilized emulsions of anti-Bancroft type. *Langmuir* 2006; 22(11): 4968–77.

13. Binks, B.P., Horozov, T.S. Aqueous foams stabilized solely by silica nanoparticles. *Angew Chem Int Ed* 2005; 44(24): 3722–5.

14. Stocco, A., Garcia-Moreno, F., Manke, I., Banhart, J., Langevin, D. Particle-stabilised foams: Structure and aging. *Soft Matter* 2011; 7(2): 631–637.

15. Martin, A.H., Grolle, K., Bos, M.A., Stuart, M.A., van Vliet, T. Network forming properties of various proteins adsorbed at the air/water interface in relation to foam stability. *J Colloid Interface Sci* 2002; 254(1): 175–83.

16. Schubert, H., (ed.) *Proceedings of 1st European Symposium Process Technology in Pharmaceutical and Nutricional Sciences* 1998.

17. Bos, M.A., van Vliet, T. Interfacial rheological properties of adsorbed protein layers and surfactants: A review. *Adv Colloid Interface Sci* 2001; 91(3): 437–71.

18. Janssen, J.J.M., Boon, A., Agterof, W.G.M. Influence of dynamic interfacial properties on droplet breakup in simple shear flow, *Aiche J* 1994; 40(12): 1929–39.

19. Williams, A., Janssen, J.J.M., Prins, A. Behaviour of droplets in simple shear flow in the presence of a protein emulsifier. *Colloids Surf A, Physicochem Eng Aspects* 1997; 125(2–3): 189–200.

20. Weaire, D., Hutzler, S. *The Physics of Foams*. Clarendon Press: Oxford, 1999.

21. Cantat, I., Cohen-Addad, S., Elias, F., Graner, F., Hohler, R., Pitois, O. et al. *Foams - Structure and Dynamics*. Oxford University Press, 2013. 265 p.

22. Taylor, P. Ostwald ripening in emulsions. *Adv Colloid Interface Sci* 1998; 75(2): 107–63.

23. Schmitt, V., Cattelet, C., Leal-Calderon, F. Coarsening of alkane-in-water emulsions stabilized by nonionic poly(oxyethylene) surfactants: The role of molecular permeation and coalescence. *Langmuir* 2004; 20(1): 46–52.

24. Schmitt, V., Leal-Calderon, F. Measurement of the coalescence frequency in surfactant-stabilized concentrated emulsions. *Europhysics Letters* 2004; 67(4): 662–8.

25. Arditty, S., Schmitt, V., Giermanska-Kahn, J., Leal-Calderon, F. Materials based on solid-stabilized emulsions. *J Colloid Interface Sci* 2004; 275(2): 659–64.

26. Baldan, A. Review Progress in Ostwald ripening theories and their applications to nickel-base superalloys - Part I: Ostwald ripening theories. *J Mater Sci* 2002; 37(11): 2171–202.

27. Georgieva, D., Schmitt, V., Leal-Calderon, F., Langevin, D. On the Possible Role of Surface Elasticity in Emulsion Stability. *Langmuir* 2009; 25(10): 5565–73.

28. Meinders, M.B.J., van Vliet, T. The role of interfacial rheological properties on Ostwald ripening in emulsions. *Advances* 2004; 108: 119–26.

29. Gibbs, J.W. *The Collected Works*. Green and co.: Longmans, 1928.

30. Abkarian, M., Subramaniam, A.B., Kim, S-H., Larsen, R.J., Yang, S-M., Stone, H.A. Dissolution Arrest and Stability of Particle-Covered Bubbles. *Phys Rev Lett* 2007; 99(18): 188301.

31. Hilgenfeldt, S., Koehler, S.A., Stone, H.A. Dynamics of coarsening foams: Accelerated and self-limiting drainage. *Phys Rev Lett* 2001; 86(20): 4704.

32. Mullins, W.W. The statistical self-similarity hypothesis in grain-growth and particle coarsening. *J Appl Phys* 1986; 59(4): 1341–9.

33. Glazier, J.A. Grain-growth in 3 dimensions depends on grain topology. *Phys Rev Lett* 1993; 70(14): 2170–3.

34. Lambert, J., Cantat, I., Delannay, R., Mokso, R., Cloetens, P., Glazier, J.A. et al. Experimental growth law for bubbles in a moderately "Wet" 3D liquid foam. *Phys Rev Lett* 2007; 99(5): 058304.

35. Princen, H.M. Osmotic-pressure of foams and highly concentrated emulsions. 1. Theoretical considerations. *Langmuir* 1986; 2(4): 519–24.

36. Cussler, E.L. *Diffusion: Mass Transfer in Fluid Systems*. 2nd ed. Cambridge University Press: New York, 1997.

37. Farajzadeh, R., Krastev, R., Zitha, P.L.J. Foam film permeability: Theory and experiment. *Adv Colloid Interface Sci* 2008; 137(1): 27–44.
38. Saint-Jalmes, A., Peugeot, M.L., Ferraz, H., Langevin, D. Differences between protein and surfactant foams: Microscopic properties, stability and coarsening. *Colloids Surf A, Physicochem Eng Aspects* 2005; 263(1–3): 219–25.
39. Tcholakova, S., Mitrinova, Z., Golemanov, K., Denkov, N.D., Vethamuthu, M., Ananthapadmanabhan, K.P. Control of Ostwald Ripening by Using Surfactants with High Surface Modulus. *Langmuir* 2011; 27(24): 14807–19.
40. Weaire, D., Pageron, V. Frustrated froth - evolution of foam inhibited by an insoluble gaseous component. *Philosophical Magazine Letters* 1990; 62(6): 417–21.
41. Saint-Jalmes, A. Physical chemistry in foam drainage and coarsening. *Soft Matter* 2006; 2(10): 836–49.
42. Isert, N., Maret, G., Aegerter, C.M. Coarsening dynamics of three-dimensional levitated foams: From wet to dry. *Eur Phys J E* 2013; 36(10): 1–6.
43. Magrabi, S.A., Dlugogorski, B.Z., Jameson, G.J. Bubble size distribution and coarsening of aqueous foams. *Chem Eng Sci* 1999; 54(18): 4007–22.
44. Ariyaprakai, S., Dungan, S.R. Influence of surfactant structure on the contribution of micelles to Ostwald ripening in oil-in-water emulsions. *J Colloid Interface Sci* 2010; 343(1): 102–8.
45. Bergeron, V., Radke, C.J. Equilibrium measurements of oscillatory disjoining pressures in aqueous foam films. *Langmuir* 1992; 8(12): 3020–6.
46. Varade, D., Carriere, D., Arriaga, L.R., Fameau, A.-L, Rio, E., Langevin, D., Drenckhan, W. On the origin of the stability of foams made from catanionic surfactant mixtures. *Soft Matter* 2011; 7(14): 6557–6570.
47. Michina, Y., Carriere, D., Mariet, C., Moskura, M., Berthault, P., Belloni, L. et al. Ripening of Catanionic Aggregates upon Dialysis. *Langmuir* 2009; 25(2): 698–706.
48. Fameau, A-L., Saint-Jalmes, A., Cousin, F., Houssou, B.H., Novales, B., Navailles, L. et al. Smart foams: Switching reversibly between ultrastable and unstable foams. *Angew Chem Int Ed* 2011; 50(36): 8264–9.
49. Dickinson, E., Ettelaie, R., Murray, B.S., Du, Z.P. Kinetics of disproportionation of air bubbles beneath a planar air-water interface stabilized by food proteins. *J. Colloid Interface Sci* 2002; 252(1): 202–13.
50. Blijdenstein, T.B.J., de Groot, P.W.N., Stoyanov, S.D. On the link between foam coarsening and surface rheology: Why hydrophobins are so different. *Soft Matter* 2010; 6(8): 1799–808.
51. Wilde, P.J., Nino, M.R.R., Clark, D.C., Patino, J.M.R. Molecular diffusion and drainage of thin liquid films stabilized by bovine serum albumin Tween 20 mixtures in aqueous solutions of ethanol and sucrose. *Langmuir* 1997; 13(26): 7151–7.
52. Senee, J., Robillard, B., Vignes-Adler, M. Films and foams of Champagne wines. *Food Hydrocolloids* 1999; 13(1): 15–26.
53. Cervantes Martinez, A., Rio, E., Delon, G., Saint-Jalmes, A., Langevin, D., Binks, B.P. On the origin of the remarkable stability of aqueous foams stabilised by nanoparticles: Link with microscopic surface properties. *Soft Matter* 2008; 4(7): 1531–5.
54. Safouane, M., Langevin, D., Binks, B.P. Effect of particle hydrophobicity on the properties of silica particle layers at the air-water interface. *Langmuir* 2007; 23: 11546–53.
55. Salonen, A., Gay, C., Maestro, A., Drenckhan, W., Rio, E. Arresting bubble coarsening: A two-bubble experiment to investigate grain growth in the presence of surface elasticity. *Europhysics Lett* 2016; 116(4): 46005.
56. Durand, M., Martinoty, G., Langevin, D. Liquid flow through aqueous foams: From the plateau border-dominated regime to the node-dominated regime. *Phys Rev E* 1999; 60(6): R6307–R8.
57. Langevin, D. Rheology of adsorbed surfactant monolayers at fluid surfaces. In: Davis, S.H., Moin, P. (Eds.), *Annual Review of Fluid Mechanics*, Vol. 462014. 2014; 47–65.
58. Safouane, M., Saint-Jalmes, A., Bergeron, V., Langevin, D. Viscosity effects in foam drainage: Newtonian and non-Newtonian foaming fluids. *Eur Phys J E* 2006; 19(2): 195–202.
59. Maestro, A., Drenckhan, W., Rio, E., Hohler, R. Liquid dispersions under gravity: Volume fraction profile and osmotic pressure. *Soft Matter* 2013; 9(8): 2531–40.
60. Varade, D., Carriere, D., Arriaga, L.R., Fameau, A.L., Rio, E., Langevin, D. et al. On the origin of the stability of foams made from catanionic surfactant mixtures. *Soft Matter* 2011; 7(14): 6557–70.
61. Fameau, A.L. Assemblages d'acides gras: du volume aux interfaces. *PhD thesis*, Université de Nantes; 2011.
62. Friberg, S., Saito, H. *Foam stability and association of surfactants*. In: Akers, RJ. (Ed.) Foams. Academic Press, 1976; 31–4.
63. Carn, F., Colin, A., Pitois, O., Vignes-Adler, M.l., Backov, R.n. Foam drainage in the presence of nanoparticle/surfactant mixtures. *Langmuir* 2009; 25(14): 7847–56.

64. Guignot, S., Faure, S., Vignes-Adler, M., Pitois, O. Liquid and particles retention in foamed suspensions. *Chem Eng Sci* 2010; 65(8): 2579–85.

65. Guillermic, R.M., Salonen, A., Emile, J., Saint-Jalmes, A. Surfactant foams doped with laponite: Unusual behaviors induced by aging and confinement. *Soft Matter* 2009; 5(24): 4975–82.

66. Mason, T.G., Bibette, J., Weitz, D.A. Yielding and flow of monodisperse emulsions. *J Colloid Interface Sci* 1996; 179(2): 439–48.

67. Pal, R. Effect of droplet size on the rheology of emulsions. *Aiche Journal* 1996; 42(11): 3181–90.

68. Pal, R. Yield stress and viscoelastic properties of high internal phase ratio emulsions. *Colloid Polym Sci* 1999; 277(6): 583–8.

69. Goyon, J., Bertrand, F., Pitois, O., Ovarlez, G. Shear Induced Drainage in Foamy Yield-Stress Fluids. *Phys Rev Lett* 2010; 104(12).

70. Georgieva, D., Cagna, A., Langevin, D. Link between surface elasticity and foam stability. *Soft Matter* 2009; 5(10): 2063–71.

71. Bibette, J., Leal Calderon, F., Schmitt, V., Poulin, P. *Emulsion Science: Basic Principles*. second ed: Springer; 2007.

72. Carrier, V., Colin, A. Coalescence in draining foams. *Langmuir* 2003; 19(11): 4535–8.

73. Khristov, K., Exerowa, D., Minkov, G. Critical capillary pressure for destruction of single foam films and foam: Effect of foam film size. *Colloids Surf A Physicochem Eng Aspects* 2002; 210(2–3): 159–66.

74. Bibette, J., Morse, D.C., Witten, T.A., Weitz, D.A. Stability-criteria for emulsions. *Phys Rev Lett* 1992; 69(16): 2439–42.

75. Langevin, D., Rio, E. Coalescence in foams and emulsions In: Somasundaran P., (ed.) *Encyclopedia of Surface and Colloid and Science*, second edition ed. Taylor and Francis: New York, 2012. p. 1–15.

76. de Gennes, P.G. Some remarks on coalescence in emulsions or foams. *Chem Eng Sci* 2001; 56(19): 5449–50.

77. Garrett, P. (ed.) *Defoaming, Theory and Industrial Applications*. Dekker, M., 1993.

78. Denkov, N.D. Mechanisms of foam destruction by oil-based antifoams. *Langmuir* 2004; 20(22): 9463–505.

79. Denkov, N.D., Marinova, K.G., Christova, C., Hadjiiski, A., Cooper, P. Mechanisms of action of mixed solid-liquid antifoams: Exhaustion and reactivation. *Langmuir* 2000; 16(6): 2515–28.

80. Cohen-Addad, S., Krzan, M., Hohler, R., Herzhaft, B. Rigidity percolation in particle-laden foams. *Phys Rev Lett* 2007; 99(16).

81. BonfillonColin, A., Langevin, D. Why do ethoxylated nonionic surfactants not foam at high temperature? *Langmuir* 1997; 13(4): 599–601.

82. Biance, A.L., Delbos, A., Pitois, O. How Topological Rearrangements and Liquid Fraction Control Liquid Foam Stability. *Phys Rev Lett* 2011; 106(6): 068301.

83. Hutzler, S., Losch, D., Carey, E., Weaire, D., Hloucha, M., Stubenrauch, C. Evaluation of a steady-state test of foam stability. *Philos Mag* 2011; 91(4): 537–52.

84. Ritacco, H., Kiefer, F., Langevin, D. Lifetime of bubble rafts: Cooperativity and avalanches. *Phys Rev Lett* 2007; 98(24).

85. Hasmy, A., Paredes, R., Sonneville-Aubrun, O., Cabane, B., Botet, R. Dynamical transition in a model for dry foams. *Phys Rev Lett* 1999; 82(16): 3368.

86. Burnett, G.D., Chae, J.J., Tam, W.Y., Dealmeida, R.M.C., Tabor, M. Structure and dynamics of breaking foams. *Phys Rev E* 1995; 51(6): 5788–96.

87. Vandewalle, N., Lentz, J.F., Dorbolo, S., Brisbois, F. Avalanches of popping bubbles in collapsing foams. *Phys Rev Lett* 2001; 86(1): 179.

88. Salager, J.L., Forgiarini, A., Marquez, L., Pena, A., Pizzino, A., Rodriguez, M.P. et al. Using emulsion inversion in industrial processes. *Adv Colloid Interface Sci* 2004; 108: 259–72.

89. Tobin, S.T., Meagher, A.J., Bulfin, B., Mobius, M., Hutzler, S. A public study of the lifetime distribution of soap films. *Am J Phys* 2011; 79(8): 819–24.

90. Saulnier, L., Champougny, L., Bastien, G., Restagno, F., Langevin, D., Rio, E. A study of generation and rupture of soap films. *Soft Matter* 2014; 10(16): 2899–906.

91. Joye, J.L., Hirasaki, G.J., Miller, C.A. Numerical simulation of instability causing asymmetric drainage in foam films. *J Colloid Interface Sci* 1996; 177(2): 542–52.

92. Mysels, K., Shinoda, K., Frankel, S. *Soap Films*. Pergamon Press, 1959.

93. van Aken, G.A., van Vliet, T. Flow-induced coalescence in protein-stabilized highly concentrated emulsions: Role of shear-resisting connections between the droplets. *Langmuir* 2002; 18(20): 7364–70.

12 Role of Foam Films in Foam Stability

Khristo Khristov

CONTENTS

As it is pointed out in Chapter 3, foam properties correlate directly with the properties of the foam films, which are an essential element of any foam system. This statement is generally acknowledged by the scientific community and much work has been done on both microscopic foam films and foams to quantify this correlation, most of which are summarized in [1–7]. Four types of foam films are distinguished: common thin film (CTF), common black film (CBF), Newton black film (NBF), and bilayer film (BF). CTF and CBF are stabilized by long range van der Waals and electrostatic DLVO forces, but CTF drain continuously until reaching their equilibrium thickness while CBF reach an equilibrium thickness via black spot formation [8–11]. NBF and BF are stabilized predominately by steric non-DLVO forces, which are short range for low molecular and long range for polymeric surfactants (see Chapter 6 and [3,7]). The thermodynamic and hydrodynamic properties exhibited by these foam films differ [2,3,5,7]. It is reasonable to expect that the respective foams consisting of CTFs, CBFs, NBFs or BFs will also express different properties [3,8,9,12–15]. Elucidation of the role of the foam film type on foam stability is not a trivial task. The reason is that foam is a three-dimensional system of interrelated films and Plateau borders and the two principal processes in foams (foam drainage and film rupture) are related in a very specific way. For example, liquid drainage causes an increase in capillary pressure (capillary pressure is the difference between the gas and liquid phase pressures). As capillary pressure rises, the work required to rupture the film decreases. At a significantly high capillary pressure, this work may become so small that mechanical disturbances or even thermal fluctuations may rupture the film. Film rupture leads to liberation of excess liquid which delays both drainage rate and reaching hydrostatic equilibrium. The other issue is the fact that studies of microscopic foam films are performed at well defined capillary pressures P_c whereas in foams, P_c is a function of the foam column height. A way out of this dilemma is to examine films and foams at the same P_c. A suitable method for foam film studies is the *Thin Liquid Film—Pressure Balance Technique* (TlF-PBT), see Chapter 4. The corresponding method for foams is the *Foam Pressure Drop Technique* (FPDT) where the capillary pressure in the foam can be regulated thus making foam study conditions very close to TLF-PBT' regarding capillary pressure, film size, thickness, and so on. [3,9,16]. FPDT allows a characterization of aqueous foams in a reasonably short time even if the foam is very stable. A combination of these two methods is a way to find quantitative correlations between single foam films and foams. A schematic of both techniques is given in Figure 12.1.

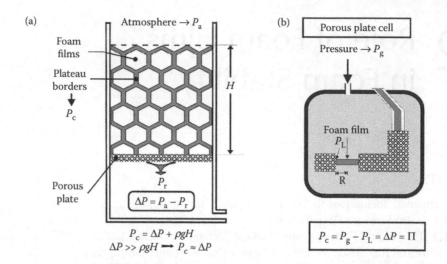

FIGURE 12.1 Schematic of foam pressure drop technique (a) and thin liquid film-pressure balance technique (b). Reprinted from *Studies in Interface Science*, Vol. 5, Exerowa, D., Kruglyakov, P.M. Foam and foam films, Copyright 1998, with permission from Elsevier. Redrawn from Khristov, Khr. et al. *Rev Sci Inst* 2004; 75: 4797 [3,16].)

12.1 EXPERIMENTAL TECHNIQUES FOR FOAM AND FOAM FILM STUDIES

12.1.1 FOAM PRESSURE DROP TECHNIQUE

In a gravitational field, capillary pressure is represented by $P_c = \rho g H$, where H is foam column height; ρ is foaming solution density, and g is the acceleration due to gravity (9.81 ms^{-2}). According to this formula, P_c in the foam is a function of foam column height, therefore, all foam properties (film thickness, Plateau border radius, foam stability, etc.) would change along with the height of the foam column. However, if a reduced pressure (P_r) is created under the porous plate that is in contact with the foam, a pressure difference $\Delta P = P_a - P_r$ will appear, where P_a is atmospheric pressure (Figure 12.1a). This ΔP should not exceed the capillary pressure in the plate pores. Such a condition ensures semi-permeability of the plate, that is, only the foam liquid phase can pass through the pores but not the gas. Due to the pressure difference, a pressure gradient will originate in foam Plateau borders thus causing an accelerated liquid drainage and an increase in capillary pressure. When a hydrostatic equilibrium is reached, the capillary pressure in the borders will be $P_c = \Delta P + \rho g H$. If $\Delta P \gg \rho g H$, then $P_c \cong \Delta P$ and the pressure in the foam liquid phase will practically be equal along the whole foam column height. In that case, the above mentioned foam properties can be considered equal along the whole foam column. FPDT allows measuring foam lifetime τ_p at different ΔP as well as the critical pressure for fast foam destruction—"avalanche" like $P_{cr.foam}$ [3,9,12,16]. Measuring τ_p as a function of ΔP, one obtains characteristic curves for foams consisting of CTFs, CBFs, NBFs, and BFs, respectively. This is considered to be the main advantage of FPDT, thus making it unique and superior to all other foam characterization techniques [3,9,17].

12.1.2 THIN LIQUID FILM-PRESSURE BALANCE TECHNIQUE

A foam film in the porous plate measuring cell of Exerowa and Scheludko is formed in a hole drilled in a porous plate (Figure 12.1b). TLF-PBT enables a direct measurement of disjoining pressure (Π) versus film thickness (h) isotherms and gives an opportunity to follow film behavior in a large pressure range. Films of various radii can be investigated using porous plates of various hole diameters in the film holder. It is seen (Figure 12.1a and b) that if the film radius in the porous plate and the film radius in the foam, that is, foam dispersity, are properly chosen, the described techniques for the study of both single foam films and foams involve very close conditions as far as the capillary pressure, film size, thickness, Plateau border, radius, and so on are concerned. The critical pressure of single foam film rupture $P_{cr.film}$ has been determined by a smooth increase in P_a in the measuring cell up to film rupture. TLF-PBT has been described in detail in Chapter 4.

12.2   EXPERIMENTAL RESULTS ON THE ROLE OF FOAM FILMS IN FOAM STABILITY

Historically, the first attempt to find a correlation between foam film type and the respective foams stabilized by sodium dodecyl sulfate (SDS) was done by Khristov et al. [18], studying foam film $\Pi(h)$ isotherms and $\tau_p(\Delta P)$ dependences of foams.

As is described in Chapter 3, the disjoining pressure dependence on film thickness is called *disjoining pressure isotherm* Π (h) and is believed to be an important thermodynamic foam film characteristic. Isotherms of different foam film types are shown in Figure 12.2. Curve 1 is the $\Pi(h)$ isotherm for CTF. It is seen that film thickness decreases with the increase in Π and films rupture within a capillary pressure range marked with arrows on curve 1. The $\Pi(h)$ isotherm (curve 2) of CBF follows the same trend, but the change in film thickness is smaller and films rupture in a narrow range of capillary pressure. The $\Pi(h)$ isotherm of NBF follows a different course (curve 3). There, film thickness is not altered by pressure increase and films rupture at an almost constant capillary pressure (marked with an arrow on curve 3). These results are a clear indication that the type of surface forces acting in different foam film types governs their properties and hence, the properties of the respective foams as indicated below.

The effect of foam film type on foam stability has also been studied with FPDT. Figure 12.3 depicts $\tau_p(\Delta P)$ dependence for foams obtained from aqueous SDS solutions at various electrolyte (NaCl) concentrations, that is, the respective foams are built up of different types of foam films. In all experiments, the surfactant concentration used provides for maximum adsorption layer saturation. All three curves have different courses, corresponding to different film types: CTF (curve 1), CBF (curve 2), and NBF (curve 3). Increasing ΔP leads to a decrease in τ_p of CTF and CBF foams from hours in the gravitational field to a few minutes at ΔP above 10^4 Pa. The course of $\tau_p(\Delta P)$

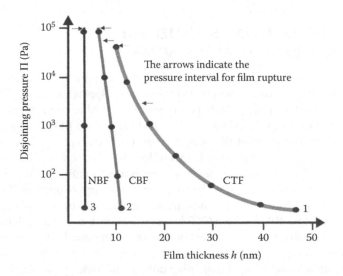

FIGURE 12.2 Isotherms of disjoining pressure for microscopic foam films from 2×10^{-3} mol/L SDS solutions; Curve 1: CTF (10^{-3} mol/L NaCl), curve 2: CBF (10^{-1} mol/L NaCl) and curve 3: NBF (4×10^{-1} mol/L NaCl). (Reprinted from *Studies in Interface Science*, Vol. 5, Exerowa, D., Kruglyakov, P.M. Foam and foam films, Copyright 1998, with permission from Elsevier [3].)

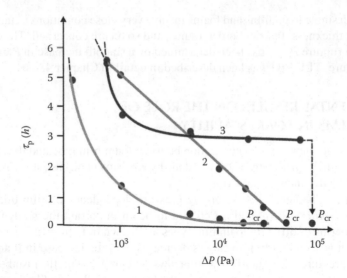

FIGURE 12.3 Foam lifetime τ_p versus pressure drop ΔP. Curve 1: CTF (10^{-3} M NaCl), curve 2: CBF (10^{-1} M NaCl) and curve 3: NBF (4×10^{-1} M NaCl). (Reprinted from *Studies in Interface Science*, Vol. 5, Exerowa, D., Kruglyakov, P.M. Foam and foam films, Copyright 1998, with permission from Elsevier.)

dependence for NBF foam is different; after an initial decrease in τ_p to 10^3 Pa, a further increase in ΔP of about two orders of magnitude does not lead to change in τ_p and results in a curved plateau. Figure 12.3 makes it obvious that the foam film type affects foam stability and the course of $\tau_p(\Delta P)$ dependences, as well as the value of the critical pressure for foam destruction. Analyzing $\tau_p(\Delta P)$ and $\Pi(h)$ dependences gives ground to conclude that the type of foam films (CTF, CBF, and NBF) in a foam are decisive with respect to the properties of single foam films and the respective foams.

A number of experimental investigations have followed the above studies in order to establish a relation between foam film type/foam stabilized by low molecular, natural, and synthetic polymeric surfactants. Some examples are described below.

12.3 FOAM FILMS AND FOAMS STABILIZED BY SYNTHETIC AND NATURAL SURFACTANTS

This paragraph looks at the correlation between foam film type and respective foam stabilized by F108 and P85 (polyethylene oxide–polypropylene oxide–polyethylene oxide (PEO–PPO–PEO)) ABA block copolymer surfactants [8]. Only two types of foam films are formed from solutions of nonionic polymeric surfactants: CTFs stabilized by DLVO forces and BFs stabilized by steric forces. Figure 12.4 presents $\Pi(h)$ isotherms of BFs (curve 1) and CTFs (curve2).

As is seen from Figure 12.4, both $\Pi(h)$ isotherms follow the same course: film thickness decreases with pressure increase, but the dependence for BFs is much steeper (curve 1, Figure 12.4). The course of $\Pi(h)$ isotherm of CTF (curve 2) indicates that with a pressure range of up to 10^4 Pa, the electrostatic component of disjoining pressure plays a significant role in film thickness [1,7–10,12]. This is clearly expressed at pressures below 10^3 Pa where CTF is much thicker ($h \approx 110$ nm) than BF ($h \approx 45$ nm). In the case of BFs, the electrostatic component of disjoining pressure is suppressed and these films are stabilized by steric forces only. In the range of pressures higher than 10^4 Pa, both isotherms come closer and the steric component of disjoining pressure becomes predominant. As is described in Chapter 3 and by other authors [1,6,12], it is reasonable to consider that in the case of CTFs, the disjoining pressure should be regarded as consisting of three components: $\Pi = \Pi_{vw} + \Pi_{el} + \Pi_{st}$. The steric component additionally stabilizes CTFs and both types of foam films rupture at almost equal pressure ($P_{cr.film} = 4$–7×10^4 Pa for CTF and around 10^5 Pa for BF).

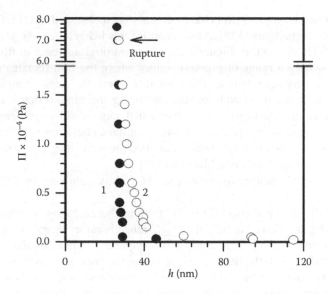

FIGURE 12.4 Isotherms of disjoining pressure for microscopic foam films from 10^{-5} mol/L F108 solutions. Curve 1: BF ($C_{el} = 10^{-1}$ mol/L) and curve 2: CTF ($C_{el} = 10^{-4}$ mol/L) NaCl. (According to Khristov, Khr. et al. *Colloids and Surfaces A* 2001; 186: 93 [8].)

Figure 12.5 presents $\Delta P(\tau_p)$ dependences for foams from F108 solutions at the same surfactant and electrolyte concentrations as in Figure 12.4.

Curve 1 depicts $\tau_p(\Delta P)$ dependence for BF foam. It runs similar to the one of SDS foam with NBFs (see curve 3, Figure 12.3). After an initial decrease in τ_p down to 5×10^3 Pa, rising further ΔP to 10^5 Pa does not lead to a significant change in τ_p, reaching a plateau in the curve. Similar behavior exhibits the $\tau_p(\Delta P)$ dependence for CTF foams. Here again a plateau is established in the

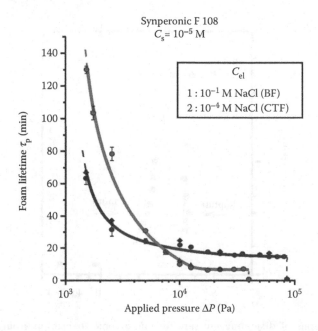

FIGURE 12.5 Foam lifetime τ_p versus pressure drop ΔP. Curve 1: bilayer film (BF) and curve 2: common thin film (CTF). (According to Khristov, Khr. et al. *Colloids and Surfaces A* 2001; 186: 93 [8].)

10^4 to 4×10^4 Pa pressure range. Such behavior has not been observed in CTF foams stabilized by low molecular weight surfactants [3,18]. An account for this behavior can be given considering the $\Pi(h)$ isotherms for CTF. As seen in Figure 12.4, after an initial decrease in film thickness, rising pressure leads to reaching a range of pressure values where the film remains almost constant in thickness. This range corresponds to a film thickness where the steric component of disjoining pressure (Π_{st}) appears to act, thus additionally stabilizing the film so that it no longer alters its thickness. This means that if the film does not change in thickness with a pressure increase, a plateau in $\tau_p(\Delta P)$ dependence is observed [17,18]. The other similarity between SDS foam with NBFs and F108 foam with BFs is that with the increase in applied pressure, a $P_{cr.foam}$ is reached at which the foam destructs very quickly, "avalanche" like [8,17,18].

Figure 12.6 depicts $\Pi(h)$ isotherms for single foam films stabilized by P85 block copolymer surfactant [8].

Curve 1 refers to BL films and curve 2 to CTF. Both isotherms follow a course similar to the one observed for F108 (Figure 12.4), but here the $P_{cr.film}$ range is rather large and much lower in value for both film types (curve 1 and 2, Figure 12.6). It is important to indicate that Π_{st} starts to act in a pressure range coinciding with the pressure range where the critical pressure of single film rupture is reached (marked by arrows in Figure 12.6). The lower arrows point to the lowest pressure of film rupture while the upper arrows point to the highest pressure value, that is, they cover the pressure range in which all films rupture. Each isotherm is based on at least 30 films studied [8].

Figure 12.7 presents $\tau_p(\Delta P)$ dependences for foams from P85 solutions. Curve 1 refers to BLF foam (C_{el} 10^{-1} mol/L NaCl), curve 2—for CTF foam ($C_{el} = 5 \times 10^{-4}$ mol/L NaCl). Both curves follow a similar course, namely, foam lifetime monotonously decreases with increase in applied pressure. A short plateau is reached at $\Delta P = 5 \times 10^3$ Pa for BLF and at $\Delta P = 7 \times 10^3$ Pa for CTF foam. The presence of the plateau in both curves (especially for CTF foam) can be understood by considering the $\Pi(h)$ isotherms for single foam films obtained from the same solutions (Figure 12.6). As is indicated above, Π_{st} starts to act in a pressure range coinciding with the pressure range where the critical pressure of single film rupture is reached. There are two factors involved: the presence of Π_{st} additionally stabilizing the film (as in the case of F108

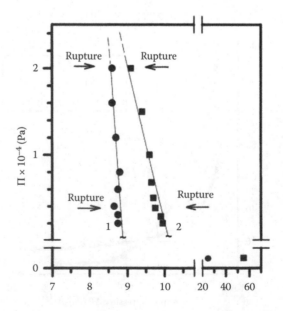

FIGURE 12.6 Isotherms of disjoining pressure for microscopic foam films from 7×10^{-5} mol/L P85 solutions. Curve 1: BF ($C_{el} = 10^{-1}$ mol/L) and curve 2: CTF ($C_{el} = 10^{-4}$ mol/L) NaCl. (According to Khristov, Khr. et al. *Colloids and Surfaces A* 2001; 186: 93 [8].)

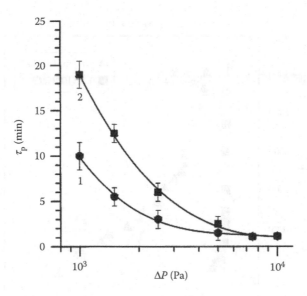

FIGURE 12.7   Foam lifetime τ_p versus pressure drop ΔP for foam from 7×10^{-5} mol/L P85 solutions. Curve 1: BF ($C_{el} = 10^{-1}$ mol/L) and curve 2: CTF ($C_{el} = 10^{-4}$ mol/L) NaCl. (According to Khristov, Khr. et al. *Colloids and Surfaces A* 2001; 186: 93 [8].)

foam with CTF, curve 2, Figure 12.5) and the critical pressure range for film rupture is reached accounting for no further alteration of τ_p with ΔP increase. The predominance of either factor in the CTF foam plateau of $\tau_p(\Delta P)$ dependence is hard to distinguish. Most probably, both factors exercise a mutual effect.

Correlations between the $\Pi(h)$ isotherms of single foam films and $\tau_p(\Delta P)$ dependences for foams stabilized by the natural surfactant INUTEC SP1 are described below. INUTEC SP1 is a graft polymer consisting of an inulin backbone of linear poly-fructose chains (loops) hydrophobically modified by random grafting with dodecyl chains [19,20]. The surfactant concentration used ensures maximum absorption layer saturation in the foam film and foam experiments.

Figure 12.8 shows the $\Pi(h)$ isotherms of BLFs (curve 1) and CTFs (curve 2). Increasing the pressure leads to BLF thinning of from 15 nm at 50 Pa to 9 nm at 1500 Pa. Further increase in pressure up to 2750 Pa (their critical pressure of rupture) does not change film thickness. For CTF, the change in film thickness with pressure increase is significantly larger, from 62 nm at 50 Pa to 14 nm at 3000 kPa. In this case, the critical pressure of film rupture is about 2500–3000 Pa.

Figure 12.9 presents $\tau_p(\Delta P)$ dependences for foam from 10^{-4} mol/L INUTEC SP1 solution. The surfactant concentration used ensures maximum absorption layer saturation. Both $\tau_p(\Delta P)$ dependences monotonically decrease with ΔP increase, until the critical pressure $P_{cr.foam}$ is reached above which foams collapse (avalanche-like) as indicated by arrows in the figure. It is seen that $P_{cr.\,foam}$ for CTF foams is lower than $P_{cr.foam}$ for BF foams, but in both cases, $P_{cr.film}$ and $P_{cr.foam}$ are close in value.

In summary, combining the use of the *Foam Pressure Drop Technique* and the *Thin Liquid Film-Pressure Balance Technique,* it is possible to establish quantitative correlations between the properties of isolated microscopic foam films and macroscopic foams. The examples presented clearly demonstrate the role that the type of the corresponding foam films play in the properties and stability of respective foams. Such comparisons are made possible because both techniques permit conducting studies of single foam films and foams under identical conditions. Furthermore, both techniques provide for a step-wise increase in capillary pressure which can be controlled and maintained constant during measurements. The difference is the parameter measured: lifetime $\tau_p(\Delta P)$ for foams and film thickness $\Pi(h)$ for foam films, that is, two situations are compared that

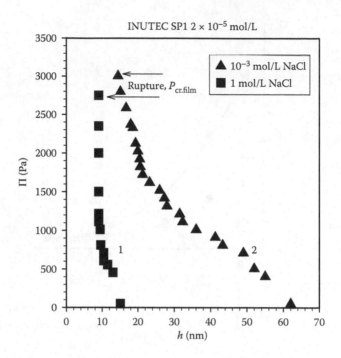

FIGURE 12.8 $\Pi(h)$ isotherms of microscopic foam films from 2×10^{-5} mol/L INUTEC SP1 solutions. Curve 1: BF ($C_{el} = 1$ mol/L) and curve 2: CTF ($C_{el} = 10^{-3}$ mol/L) NaCl. (According to Khristov, Khr. et al. *Comptes Rendus de L'Academie Bulgare Des Sciences* 2013; 66: 369. Gochev, G. et al. *Colloids Surfaces A* 2011; 391: 101 [19,20].)

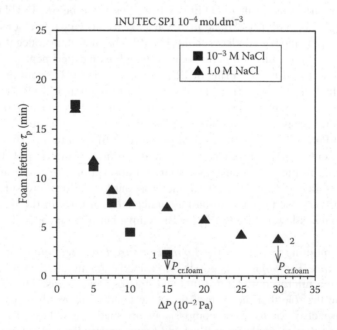

FIGURE 12.9 $\tau_p(\Delta P)$ dependences for foams stabilized by INUTEC SP1, 10^{-4} mol/L. Curve 1: BF foam ($C_{el} = 1$ mol/L) and curve 2: CTF foam ($C_{el} = 10^{-3}$ mol/L) NaCl. (According to Khristov, Khr. et al. *Comptes Rendus de L'Academie Bulgare Des Sciences* 2013; 66: 369. Gochev, G. et al. *Colloids Surfaces A* 2011; 391: 101 [19,20].)

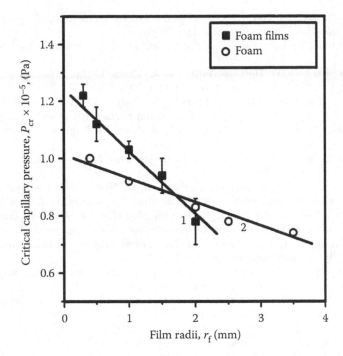

FIGURE 12.10 Critical capillary pressures for foam film rupture and foam destruction. Curve 1: foam films and curve 2: foams. (According to Khristov, Khr. et al. *Colloids Surfaces A* 2002; 210: 159 [21].)

seem interrelated at first glance. Foam lifetime is a measure for the stability whereas foam film thickness is a measure for the acting surface forces. Regardless of that, however, the correlation established is that both the lifetime τ_p and the thickness h decrease with increase in applied capillary pressure. In other words, the thinner the film, the more sensitive is the foam to external disturbances leading to a higher probability of foam film rupture. An obvious further step here would be to try and correlate the lifetime of foams and isolated films at constant capillary pressure. However, the lifetime of the latter can be orders of magnitude higher than that of the corresponding foam and can, therefore, rarely be measured in reasonable times.

Khristov et al. [21] have proposed a quantitative comparison of the critical capillary pressure of film rupture $P_{cr.film}$ and of foam destruction $P_{cr.foam}$ as measures for both film and foam stability. It has been demonstrated that $P_{cr.foam}$ and $P_{cr.film}$ can only be compared quantitatively if the size of films in the foam equals that of a single foam film [21]. As shown in Figure 12.10, the larger the film size, the lower the stability of foam and film, respectively. By a proper choice of single film size and film size in the foam (foam dispersity), it is possible to distinguish the stability of very stable films and foams [21]. For the time being, there are no other methods available to characterize stable films and foams. Furthermore, the parameter "critical capillary pressure' has been used for evaluating the stability of foam films and foams from mixed surfactant solutions [22–27], emulsion (oil/water/oil and water/oil/water) films, and emulsions [20,28–30], as well as the efficiency of a foam displacement process in enhanced oil recovery [31]. This is important from both fundamental and applied points of view since foams and emulsions enjoy a vast practical application.

REFERENCES

1. Pugh, R.J. Foaming, foam films, antifoaming and defoaming. *Adv. Colloid Interface Sci* 1996; 64: 67.
2. Ivanov, I.B., (Ed.), *Thin Liquid Films. Fundamentals and Applications.* Marcel Dekker; New York and Basel, 1988.

3. Exerowa, D., Kruglyakov, P.M. Foam and foam films. In: Möbius, D., Miller, R. (Eds.), *Studies in Interface Science*. Vol. 5. Elsevier: Amsterdam, 1998.

4. Mittal, K.L., Kumar, P., (Eds.), *Emulsions, Foams, and Thin Films*. Marcel Dekker; New York and Basel, 2000.

5. Platikanov, D., Exerowa, D. Thin liquid films. In: Lyklema, J., (Ed.), *Fundamentals of Interface and Colloid Science. Soft Colloids*. Elsevier: Amsterdam, 2005. Ch. 6.

6. Stubenrauch, C., Blomqvist, B.R. Colloid stability. In: Tadros, T.F. (Eds.), *The Role of Surface Forces; Part I*, Vol. 1, Chapter 11, Foam Films, Foams and Surface Rheology of Non-Ionic Surfactants: Amphiphilic Block Copolymers Compared with Low Molecular Weight Surfactants. Wiley-VCH Verlag GmbH & Co: Weinheim, 2007, p. 263.

7. Exerowa, D., Platikanov, D. Thin liquid films from aqueous solutions of non-ionic polymeric surfactants. *Adv Colloid Interface Sci* 2009; 47: 74.

8. Khristov, Khr., Jachimska, B., Malysa, K., Exerowa, D. "Static" and steady-state foams from ABA triblock copolymers: Influence of the type of foam films. *Colloids and Surfaces A* 2001; 186: 93.

9. Stubenrauch, C., Makievski, A.V., Khr, K., Exerowa, D., Miller, R. A new experimental technique to measure the drainage and lifetime of foams. *Tenside, Surfactants, Detergents* 2003; 40: 196–7.

10. Cohen, R., Ozdemir, G., Exerowa, D. Free thin liquid films (foam films) from rhamnolipids: Type of the film and stability. *Colloids Surfaces B* 2003; 29: 197.

11. Coons, J.E., Halley, P.J., McGlashan, S.A., Tran-Cong, T. Scaling laws for the critical rupture thickness of common thin films. *Colloids and Surfaces A* 2005; 263: 258.

12. Kruglyakov, P.M., Exerova, D., Khristov, Khr. New Possibilities for foam investigation: Creating a pressure difference in the foam liquid phase. *Adv. Colloid Interface Sci* 1992; 40: 257.

13. Waltermo, Å, Claesson, P.M., Simonsson, S., Manev, E., Johansson, I., Bergeron, V. Foam and thin-liquid-film studies of alkyl glucoside systems. *Langmuir* 1996; 12: 5271.

14. Stubenrauch, C. On foam stability and disjoining pressure isotherms. *Tenside, Surfactants, Detergents* 2001; 38: 350.

15. Stubenrauch, C., Khristov, Khr. Foams and foam films stabilized by C_nTAB: Influence of the chain length and of impurities. *J Colloid Interface Sci*, 2005; 286: 710.

16. Khristov, Khr., Exerowa, D., Christov, L., Makievski, A.V., Miller, R. Foam analyzer: An instrument based on the foam pressure drop technique. *Rev Sci Inst* 2004; 75: 4797.

17. Khristov, Khr., Exerowa, D., Malysa, K. Surfactant foaming: New concepts and perspectives on the basis of model studies. In: Zitha, P., Banhart, J., Verbist, G., (Eds.), *Foams, Emulsions and Their Applications*, Verlag Metall Innovation Technologie: Bremen, 2000, p. 21.

18. Khristov, Khr., Exerowa, D., Kruglyakov, P.M. Influence of the type of foam films and the type of surfactant on foam stability. *Colloid & Polymer Sci* 1983; 261: 265.

19. Khristov, Khr., Gochev, G., Petkova, H., Levecke, B., Tadros, T., Exerowa, D. Foams Stabilized by Hydrophobically Modified Inulin Polymeric Surfactant. *Comptes Rendus de L'Academie Bulgare Des Sciences* 2013; 66: 369.

20. Gochev, G., Petkova, H., Kolarov, T., Khristov, Khr., Levecke, B., Tadros, T., Exerowa, D. Effect of the degree of grafting in hydrophobically modified inulin polymeric surfactants on the steric forces in foam and oil-in-water emulsion films. *Colloids Surfaces A* 2011; 391: 101.

21. Khristov, Khr., Exerowa, D., Minkov, G. Critical capillary pressure for destruction of single foam films and foam: Effect of foam film size. *Colloids Surfaces A* 2002; 210: 159.

22. Schlarmann, J., Stubenrauch, C. Stabilization of Foam Films with Non-Ionic Surfactants: Alkyl Polyglycol Ethers Compared with Alkyl Polyglucosides. *Tenside Surfactants Detergents* 2003; 40: 190.

23. Alahverdjieva, V.S., Khristov, Khr., Exerowa, D., Miller, R. Correlation between adsorption isotherms, thin liquid films and foam properties of protein/surfactant mixtures: Lysozyme/C_{10}DMPO and lysozyme/ SDS. *Colloids Surfaces A* 2008; 323: 132.

24. Wang, L., Yoon, R.-H. Effects of surface forces and film elasticity on foam stability. *International Journal of Mineral Processing* 2008; 85: 101.

25. Kotsmar, C., Arabadjieva, D., Khristov, Khr., Mileva, E., Grigoriev, D.O., Miller, R., Exerowa, D. Adsorption Layer and Foam Film Properties of Mixed Solutions Containing Beta-Casein and C12DMPO. *Food Hydrocolloids* 2009; 23: 1169.

26. Carey, E., Stubenrauch, C. Foaming properties of mixtures of a non-ionic (C_{12}DMPO) and an ionic surfactant (C_{12}TAB). *J Colloid Interface Sci* 2010; 346: 414.

27. Andersson, G., Carey, E., Stubenrauch, C. Disjoining pressure study of formamide foam films stabilized by surfactants. *Langmuir* 2010; 26: 7752.

28. Taylor, S.D., Czarnecki, J., Masliyah, J. Disjoining pressure isotherms of water-in-bitumen emulsion films. *J Colloid Interface Sci* 2002; 252: 149.
29. Khristov, Khr., Petkova, H., Alexandrova, L., Nedyalkov, M., Platikanov, D., Exerowa, D., Beetge, J. Foam, emulsion and wetting films stabilized by polyoxyalkylated diethylenetriamine (DETA) polymeric surfactants. *Adv Colloid Interface Sci* 2011; 168: 105.
30. Khristov, Khr., Czarnecki, J. Emulsion films stabilized by natural and polymeric surfactants. *Current Opinion Colloid Interface Sci* 2010; 15: 324.
31. Farajzadeh, R., Andrianov, A., Krastev, R., Hirasaki, G.J., Rossen, W.R. Foam–oil interaction in porous media: Implications for foam assisted enhanced oil recovery. *Adv Colloid Interface Sci* 2012; 183–84: 1.

13 Surfactant-Stabilized Foams

Arnaud Saint-Jalmes

CONTENTS

13.1 INTRODUCTION AND SCOPE

In this chapter, we focus on foams made from solutions of *surfactants*. The main goal of this chapter is to determine and present which specificities—at all the different length scales of a foam (interfaces, films, bubbles, and macroscopically)—are associated with the use of such molecules as foam stabilizers.

We first define which molecules are considered in this chapter, that is, the *low molecular mass surfactants*. We recall some of their properties in bulk and at interfaces and present some examples of classical surfactants. We also introduce what is often referred to as cosurfactants. Then, we focus on the interfaces and thin film scales, in relation with the foaming ability of such solutions. We discuss the specificities of the foam aging in terms of drainage and coarsening, introducing concepts of interfacial mobility. Finally, we present some of the main features of the rheology of foams made with such low molecular mass surfactants.

We consider the following two cases. The first one corresponds to the simplest solution of a single soluble surfactant. The important point here is that, in this situation, the surface tension can be considered as constant in all cases: equilibrium is recovered so fast that γ is constant. In other words, there is no interfacial viscoelasticity. In that respect, we will define here a "reference foam" to which more complex formulations can be compared. This case will be linked to what is often referred to as the regimes of mobile interfaces in the literature.

Second and still using these low molecular mass molecules, we add a first level of complexity and consider the cases of mixtures of surfactants and cosurfactants. These mixtures of surfactants of different solubilities are discussed because they are the simplest system providing some interfacial viscoelasticity. We discuss here what this interfacial viscoelasticity can induce at the macroscopic scale. Here also, the case of mixtures of surfactants will give another reference state to which other formulations could be compared: this second state corresponds to a system with interfacial viscoelasticity, but without complex behavior in bulk and within the thin films separating the bubbles.

13.2   SURFACTANTS: DEFINITIONS, EXAMPLES, AND ORGANIZATION IN BULK AND AT INTERFACES

The solubility in water of a molecule depends on its chemical structure. This is linked to how much this added molecule disturbs the local organization of the water molecules. In bulk, water molecules interact through hydrogen bonds, linking hydrogen atoms to the oxygen atoms of neighbor molecules. The presence of an external molecule, like nonpolar ones such as hydrocarbons, distorts the bond network. The amplitude of the distortion dictates the solubility of this nonpolar compound. This interaction is known as the hydrophobic effect [1–3]. Summing the distortion due to each single molecule, one comes to a finite amount of possible disturbance, implying that all chemicals have a critical concentration for solubility, describing the maximal amount which can be solubilized in water (before phase separation). In that spirit, an alkane chain has an almost infinite hydrophobicity. However, the simple substitution of one of its end methyl group hydrogen atoms with a polar moiety confers hydrophilicity to this molecule, counteracting the hydrophobic part. This combination then becomes an *amphiphile* molecule, having two parts of opposite solubility [4–5]. Gaining such an amphiphilic nature has a strong impact: the solubility, even once a tiny hydrophilic part has been added, can be orders of magnitude greater than the solubility limit of the pure hydrocarbon chain [4–5].

Here we focus on amphiphilic molecules usually called *surfactants*. The first point is that it concerns "small" molecules, that is, having low molecular mass (typically ranging from 200 g/mol to 800 g/mol). This separates these molecules from proteins, polymers or colloids. Secondly, these molecules have generally well separated hydrophilic and hydrophobic parts (Figure 13.1). The hydrophobic part is often based on an aliphatic chain, with the number of carbons typically between 10 to 20, and the hydrophilic head is often based on an ionic moiety. We give below some examples of common surfactants found in cosmetics, detergents, and food industries.

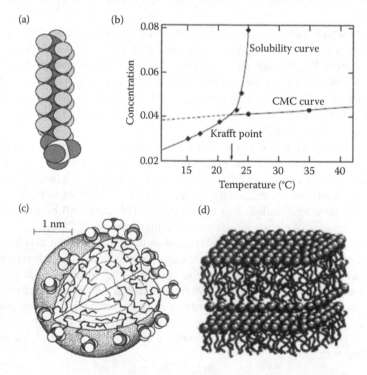

FIGURE 13.1   (a) Drawing of a typical low molecular mass surfactant; (b) solubility curve as a function of temperature: above T_k, micelles can exist and solubility drastically increases [5]; (c) drawing of a micelle; (d) drawing of a portion of a lamellar phase.

Back to the solubility of such surfactants, it first depends on how temperature compares with a critical temperature, called the Krafft temperature T_K [4–6]. For $T < T_K$, the aqueous mixture reaches saturation in monomers, and surfactant molecules phase separate to form crystals. But for $T > T_K$, the solubility increases by several orders of magnitude (Figure 13.1b). The origin of this effect is linked to the ability of amphiphiles to form finite-size aggregates, called *micelles* (Figure 13.1c). The bulk concentration above which the number of micelles increases sharply is called the *critical micelle concentration* (*cmc*). At higher concentrations, more complex and denser phases are obtained, like lamellar phases or hexagonal phases (Figure 13.1d). Note also that, besides the chemical structure of the surfactant, there are external parameters controlling the *cmc*, like the temperature T or the addition of salt (when dealing with ionic surfactants) [4–6].

In parallel to the formation of micelles in bulk, amphiphiles can also adsorb to free interfaces to satisfy their ambivalent structure (Figure 13.2). The adsorption tends to reduce the surface tension. From a static point of view, the bulk concentration, surface concentration, and surface tension are interconnected: the occurrence of micelles in bulk above the *cmc* corresponds to a saturation of the interfacial concentration and surface tension (Figure 13.2). All these features can be compiled within various theoretical frameworks, such as the widely used one from Gibbs [4,6,8].

Thus, the impact of the hydrophobic effect is clear in this equilibrium picture: the more hydrophobic the molecule, the lower its *cmc* (Figure 13.2b). In that spirit, the *cmc* is a crucial parameter as it describes the hydrophilic/hydrophobic balance of the surfactant. In parallel, the

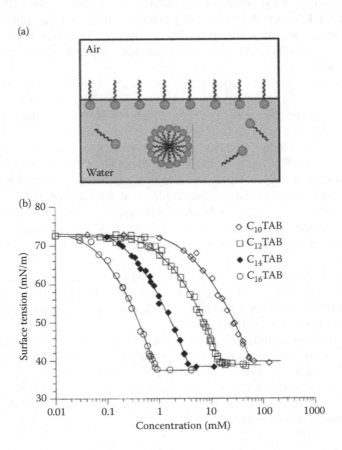

FIGURE 13.2   (a) Surfactants at the air water interface, in equilibrium with monomers and micelles in bulk; (b) at rest, surface tension versus concentration curve for 4 C_nTAB molecules. *Note*: The longer the alkyl chain, the lower the *cmc*, though the surface tension above the *cmc* remains almost independent of the amphiphilicity. (From Bergeron, V. *J Phys Condens Matt* 1999; 11: R215 [20].)

hydrophilic-lipophilic balance (HLB) scale can be used also to describe the amphiphilicity of the molecules, but it is more generally used in the emulsion sciences [5].

On the other hand, surfactants are also at the origin of a dynamical effect known as the Marangoni effect [9–10]. Heterogeneities in the distribution of surfactant molecules at the interface between two fluids trigger a Marangoni flow, that is, the bulk flow of both phases due to a gradient of their interfacial tension. Fluid flows induced by gradients of interfacial tension have received a great deal of attention and it is striking to encounter them in very different situations which might apparently share no common features. Indeed, Marangoni flows are crucial to the understanding of transport phenomena in lipid nanotubes, the dynamics of thin liquid films in foams and emulsions, the capillary locomotion of insects, surfactant replacement therapy for neonates suffering from respiratory distress syndrome, coating and printing processes or the control of water evaporation (references to these examples can be found in [11]). In that spirit, it was recently shown that the *cmc* is also a relevant scale for characterizing these dynamical flows induced by surfactant gradients, in particular when dealing with spreading of a local excess [11]).

If one wants to classify surfactants, a usual sorting relies on the ionic nature of these molecules; one generally separates the anionic, cationic, nonionic, and amphoteric (which charge depends on the pH) molecules. We present below some examples of these surfactant families which are currently used. This list illustrates in particular the large chemical differences which exist between surfactants, and the variety of compounds used.

Anionic surfactants: Alkyl sulfates (like sodium dodecylsulfate, or SDS, with the formula $C_{12}H_{25}OSO^{-3}Na^{+}$), alkylethersulfates and alkylbenzenesulfates are the standard ones. There are also alpha-olefin sulfonates and sulfoccinates (like sodium bis(2-ethylhexyl) sulfosuccinate, or AOT: $(C_8H_{17}-O-COCH_2)_2SO^{-3}Na^{+}$). Anionic surfactants are frequently found in detergent products, washing liquids, and so on.

Cationic surfactants: Cationic surfactants are often chosen for their corrosive and bactericidal effects. These include alkyl trimethyl ammonium bromides (C_nTAB: $C_nH2_{n+1}N(CH3)_3^{+}$ Br^-) and alkyl trimethyl ammonium chlorides (like DTAC: $C_{12}H_{25}N(CH3)^{+}$ Cl^-) and cetylpyridinium chloride (CPCl: $C_{16}H_{33}N(CH2)_5^{+}$ Cl^-).

Nonionic surfactants: These are present in shampoos and other personal cleaning products. These include polyoxyethylene alkyl ethers (C_nEO_m), alkyl glucosides C_iG_j, the *spans* and *tweens* families (used for emulsions), the family of Triton X ($C14H_{22}O(C_2H_4O)n$), and polyglycerol ethers. In the food industry, one also finds the diglycerides of fatty acids, the sodium stearoyl lactylate, and the lécithines.

Amphoteric surfactants: These are molecules having a positive or negative charge depending on the pH of the solution: for instance, sodium carboxylates, RCO_2Na (like natural soaps, in which R represents a carbon chain), betaines (like cocoylamidopropylbetaine), and tetradecyldimethylamineoxide ($C_{14}H_{29}NO(CH_3)_2$).

Cosurfactants: as discussed below, these are added to primary surfactant solutions to provide some interfacial viscoelasticity. They are usually solubilized within the surfactant micelles. For instance, one can find simple fatty acids or fatty alcohols as common cosurfactants, but one can also find some amphoteric species like the betaines.

There are obviously many other structures for surfactants, like amino acid based molecules [12] or those named *gemini*. Gemini surfactants are basically two single-chain surfactants covalently linked at the level of their hydrophilic heads. This creates a molecule having a large and anisotropic head and two chains [13–14]; this can be further complicated by designing trimers and higher order oligomers [15]. Finally, it is also important to note that the counter ion of ionic species also play a role, with nontrivial specific ion effects [16].

13.3 SPECIFICITIES OF FLUID INTERFACES COVERED BY SURFACTANTS

As pointed out before, the molecules usually referred to as *surfactants* correspond both to "small" molecules, and those having a strong amphiphilic nature, that is, with two well identified and separated parts. Therefore, despite their high solubility (optimized by the ability to form micelles), they still have a strong affinity for the interface.

A first aspect of surfactant interfacial adsorption is its dynamics. As discussed in previous chapters on interfacial adsorption, the rate of adsorption of a chemical species depends strongly on its bulk concentration c_b. In the framework of an adsorption controlled by diffusion, we recall here that the typical time scale for adsorption is $t_{diff} = \Gamma^2/D_v c_b^2$, where Γ is the interfacial concentration (always of the order of mg/m^2) and D_v is the diffusion coefficient of the surfactants in the bulk liquid. It is, therefore, very sensitive to both the diffusion of the surfactants in bulk and to their bulk concentration. Low molecular mass surfactants have diffusion coefficients of the order of $D_v = 5 \times 10^{-10}$ m^2 s^{-1}: with such high values, the time t_{diff} can be decreased well below 1 s as soon as concentrations are of the order of the g/L. For instance, with $c = 1$ g $\cdot$ L^{-1}, we find $t_{diff} = 0.002$ s. This dynamic can be further increased in real situations of foaming thanks to the gas and fluid mixing processes. Thus, such surfactants are indeed very efficient to rapidly cover new interfaces.

Once adsorbed, it is then important to determine whether or not these layers provide some 2D viscoelasticity (see previous chapters). Let us start with solutions made with a single surfactant, such as an SDS solution for instance. In that case, as soon as the concentration is not too low (cmc/10), it turns out that it is difficult to create surface tension gradients: exchanges with the bulk are always faster than the deformation rates. Consequently, the surface concentration is always instantaneously equilibrated, and the surface tension can then be considered as a constant. With the available experimental tools, either in compression/dilatation or in shear, the viscoelasticity of such surfactant covered interfaces remains within the limits of resolution.

However, some interfacial viscoelasticity can be obtained by making mixtures of different surfactants and/or with cosurfactants (generally less soluble, thus less efficient to equilibrate gradients). Generally, there is almost no effect in terms of response to shear: even mixtures of surfactants and insoluble cosurfactants (like dodecanol) still provide low values of shear viscosities, and never some have elastic responses under shear [17]. Indeed, the existence of elastic shear moduli rely on the existence of interaction within the plane of the interface; expect a crowding effect when cosurfactants are added (in agreement with lower surface tensions), there are no in-plane interactions between these adsorbed molecules. By contrast, the viscoelasticity in compression/dilatation can be significantly increased with cosurfactants (as here, it is the exchange with the bulk which is more crucial). Therefore, the addition of molecules of low solubility (solubilized only thanks to the presence of the surfactant micelles)—like fatty alcohols or fatty acids—can easily provide elastic moduli of a few tens of mN/m [18].

13.4 FOAMING AND SPECIFICITIES OF THIN FILMS STABILIZED BY SURFACTANTS

The ability of a given solution to foam is often called "foamability" or "foaminess." This property can be investigated and measured by various techniques and devices (see previous chapters and [19]), providing estimations of how efficient a solution is to incorporate gas.

In fact, the foamability relies on two features: first, it depends strongly on the capacity to rapidly and homogeneously cover the freshly made interfaces. A rapid and homogenous adsorption is granted by the low molecular mass of the surfactants and their diffusion coefficient (see previous section).

Second, a good foamability also relies on the subsequent ability to create stable thin films between bubbles. These only exist if repulsive forces are created: the covered interfaces must induce disjoining pressures, resisting the capillary suction of the Plateau borders.

Low molecular weight surfactant solutions also provide such a requirement: in particular, the ionic surfactants are at the origins of very efficient electrostatic repulsions which can stabilize the thin films (see other chapters of the book). By comparison, the nonionic surfactants generally create steric effects to stabilize the thin films; this is often less efficient than the electrostatic effects, and consequently the foamability is lower (in practice, the shampoos for kids, using nonionic surfactants, actually do not foam well).

(a)

(b)

(c)

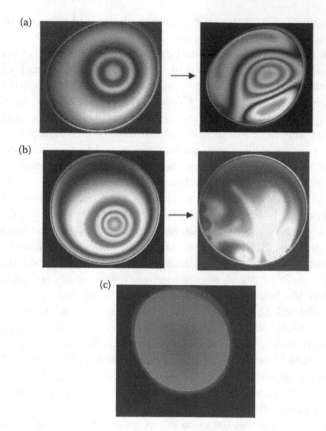

FIGURE 13.3 Pictures obtained by the thin film balance apparatus: (a) and (b) two examples of the drainage of a surfactant film, with a dimple escaping asymmetrically into the surrounding meniscus; (c) the final equilibrium black film, with a thickness of 40 nm.

For ionic surfactants, the thin film balance apparatus is a useful tool to test the stability of a single thin film and to understand its origins of stability [20–21]. It is also helpful to visualize the film uniformity and texture. For simple surfactant solutions, one always gets flat and homogeneous stable films as shown in Figure 13.3; such a behavior is in agreement with a stabilization mechanism based on electrostatic repulsion [20–21]. Moreover, as a consequence of this origin of repulsive forces, the thickness is indeed small: such surfactant solutions typically give films with thickness in the range of tens of nm ("black films", Figure 13.3). When considering that the diameter of such films can be up to millimeters, it is striking that such fragile films can exist. Dynamically speaking, these films usually drain within seconds, with dimples escaping the film in an asymmetric way (Figure 13.3) [22]. Adding cosurfactants, one still gets the same final state, but the interfacial viscoelasticity generally induces a slower drainage with a much longer lifetime of the central dimple [22–23].

In summary, the main point here is that low molecular mass surfactant solutions have generally optimal foaming properties. This is one of the reasons for their wide use in formulations for a large pallet of applications. In other words, all methods of foaming [19] can be used with such surfactant solutions. This is indeed a significant difference with more complex formulations [24], for which foamability is generally technique-dependent. Obviously, this foaming depends on the concentration: being above the *cmc* is not necessary and often a concentration of about the *cmc*/10 is sufficient to obtain a good foaming. In parallel, one must also be aware that, in dynamical conditions, some depletion effects might occur [25] once there is no longer enough surfactant inside the initial bulk.

13.5 AGING OF SURFACTANT-STABILIZED FOAMS

Once formed, the foams can be characterized by two physical parameters: the bubble radius R, and the foam liquid fraction ϕ_l [26–28]. This liquid fraction of the foam, ϕ_l, is defined as the volume of liquid divided by the volume of foam (respectively, ϕ_g is the gas fraction and $\phi_l + \phi_g = 1$). Low liquid fractions imply a shrunk network with long and slender Plateau borders (PBs) and films covering a large part of the bubble area. Conversely, wet foams have a small area of films between bubbles, as swollen PBs and nodes widely cover the bubbles. From the bubble point of view, the liquid fraction describes how packed the bubbles are. Together with the size of the bubble, the liquid fraction is a key parameter of a foam, as most of the foam properties depend on it [26–28]. Moreover, this liquid fraction and the bubble size are not constant with time. Foams are out-of-equilibrium materials which drain and coarsen, resulting in a decrease of liquid fraction and increase of bubble size with time. Therefore, bubbles get bigger and more and more packed with time, and the liquid network gets emptied. We discuss below in detail these aging mechanisms.

13.5.1 DRAINAGE OF SURFACTANT-STABILIZED FOAMS

Drainage is a major destabilizing effect for aqueous foams (Figure 13.4a): it is the irreversible flow of liquid through the continuous liquid network as a consequence of gravity. As already explained in previous chapters, the foam liquid is confined into the network of Plateau borders (PBs), which are connected by four at nodes (or vertices), while some liquid is also trapped in the flat films formed between two bubble faces. In the framework of drainage models and in agreement with experiments, this fluid contained in the films can be neglected, and foam drainage is thus generally considered as a problem of flow through Plateau borders connected by nodes/vertices (Figure 13.4b) [29–31].

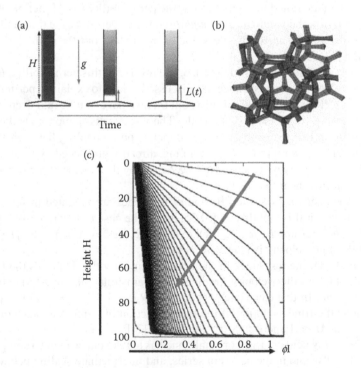

FIGURE 13.4   (a) Drainage at the macroscopic view; (b) the network of Plateau borders connected by nodes, in which the fluid is driven by gravity and capillarity; (c) free-drainage simulations of the normalized liquid fraction profiles as a function of time, for a foam column of height H; at $t = 0$, the liquid fraction is 1 at all vertical positions and the final equilibrium profile is eventually reached.

In this skeleton of PBs, the fluid primarily flows downward by gravity. But there is also some liquid transport due to capillary effects which are related to liquid fraction gradients. Such gradients of ϕ_l imply pressure gradients in the liquid. In a wet part of the foam, the PB radius is bigger than in a dryer part: due to the Laplace-Young law, the capillary pressure in the liquid in the wet part must be higher than that in the dry part. Thus, a capillary flow is induced, bringing liquid from high regions to the low ones. In that sense, one can see that capillary effects tend to smooth out liquid fraction gradients. The resulting steady-state flow in a foam is obtained by balancing these gravity and capillary effects, together with some viscous dissipation occurring within the fluid network.

In a situation of *free-drainage*(as described above, where a foam is left to drain), gravity and capillary effects are opposed (Figure 13.4c). There is another situation, called *forced-drainage*, where one injects the surfactant solution on top of a dry foam, and where gravity and capillarity are in the same direction (the top is wetter) [29–31]. This leads to a well defined front propagation, whose velocity can be easily measured (by optical or electrical means, for instance [30–32]). This latter approach has been widely used to compare models and experiments.

Note also that there is an equilibrium profile at the end of drainage. This equilibrium state is obtained as some liquid is always kept inside the foam sitting on its drained liquid (as in Figure 13.4a). This equilibrium profile originates from the balance of gravitational and capillary effects; the latter sucking liquid from the underneath pool. In practice, this finally corresponds to the creation of a wet foam layer at the bottom of a foam whose thickness is $\xi = \gamma/\rho g R$ [28,29].

As drainage corresponds to the flow within an interconnected network, models have been based on the formalism developed for porous media. These models use Darcy's law, which relate the liquid velocity to the driving pressure gradient G (including both gravitational forces ρg, and capillary pressure gradients) via a permeability k and the fluid viscosity μ; it is $G = \mu v/k$ [31].

A first major difference with solid porous media is that the pore size (i.e., the PB section) is dynamically coupled to the liquid fraction ϕ_l: thus, the permeability k is a function of liquid content, $k(\phi_l)$. Once $k(\phi_l)$ is known, the *foam drainage equation*, which describes the time and space variation of $\phi_l(r, t)$, is then derived by injecting the velocity obtained from the Darcy's law into a usual continuity equation [29–31,33].

Therefore, the key point is to determine $k(\phi_l)$, or the inverse of this permeability, $R_e(\phi_l)$ which can be understood as a hydrodynamic resistance. As discussed below, to explain experimental results and to determine the correct form of k or R_e, it turns out that we have to take into account the details of the flow at the scales of a single PB and of a node. The viscous dissipation, thus the hydrodynamic resistance of the PBs and of the nodes, actually depends on a coupling between the flow in bulk and at the interface of these structures. This coupling, already introduced by Lemlich in the 1960's [34], is quantified by a mobility parameter, $M = \mu R/\mu_s$ (being the inverse of the Boussinesq number) where μ_s is the interfacial shear viscosity.

Assuming that the resistances of the PBs and of the nodes are mounted in series, one can have two regimes depending on this mobility M [31,35,36]. In the case of low mobility, the resistance in the PBs dominates (*channel-dominated*regime). In that case, $k(\phi_l) \sim R_e(\phi_l)^{-1} \sim \phi_l R^2$, which implies, for instance, that the fluid velocity is $v \sim \rho g R^2 \phi_l/\mu$.

For a high mobility M, the resistance in the nodes eventually dominates as the type of flow has changed within the PBs so that velocity gradients are vanishing (Figure 13.5). This is called the *node-dominated* regime. In that case, one gets: $k(\phi_l) \sim R_e(\phi_l)^{-1} \sim \phi_l^{1/2} R^2$ and $v \sim \rho g R^2 \phi_l^{1/2}/\mu$. As well, if one monitors the time evolution of the liquid fraction in a free drainage configuration, one gets that $\phi_l(t) \sim t^{-1}$ for $M \ll 1$ and $\phi_l(t) \sim t^{-2}$ for $M \gg 1$ [31].

In the cases where M is neither large nor small, one has to take into account a complete model with the resistances in the PBs and in the nodes in series, and intermediate scaling behaviors are found.

Back to foams made of a single surfactant solution, we already pointed out that the interfacial shear viscosity is extremely small and close to the limit of resolution of the methods; $\mu_s \sim 10^{-8}$ kg/s [17]. Consequently, one generally obtains a drainage corresponding to the limit of high mobility for such single surfactant stabilized foams (even if one must keep in mind that M also depends on the

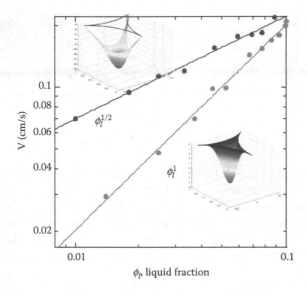

FIGURE 13.5 Experimental results—fluid velocity as a function of the foam liquid fraction—evidencing the existence of different drainage regimes, tuned by the interfacial mobility. At low mobility (high shear viscosity, for instance), one gets $v \sim \phi_l$, while it is $v \sim \phi_l^{1/2}$ for high mobility. This latter regime is usually found for low molecular mass surfactant as the surface shear viscosity is almost zero. The two insets represent simulations of the flow profile inside a Plateau border associated with each regime: strong gradients and high resistances in the border for low interfacial mobility and fewer gradients in bulk and interfacial shear for the high mobility regime. (From Drenckhan, W. et al. *Phys of Fluids* 2007; 19: 102101 [17].)

bubble diameter and the bulk viscosity). Then, when mixing surfactants and/or adding cosurfactants, it is indeed possible to increase slightly, but sufficiently, the interfacial shear viscosity, so that M varies enough to shift from a high mobility to a low mobility regime.

These different regimes are illustrated in Figure 13.5, showing forced-drainage data. For an SDS foam, the regime of high mobility is observed with $v \sim \phi_l^{1/2}$, while the addition of dodecanol or fatty acids viscosify the interfaces (μ_s can be increased by about a factor 10 [17]) to get the other limit with $v \sim \phi_l$. In this graph, we also show the computed flow profile in the PBs in these two limits [17]. Strong gradients and dissipation correspond to the *channel-dominated* regime (and inversely).

Finally, we want to point out that these drainage studies can be used to indirectly measure these interfacial shear viscosities [17], which are particularly difficult to measure for interfaces covered by soluble surfactants.

In summary, the macroscopic foam drainage features are controlled by the interfacial properties via the coupling parameter M which includes the interfacial shear viscosity. When playing with low molecular mass surfactants, a situation of high interfacial mobility is the most probable (though it also depends on the bubble size). Adding other surfactants and cosurfactants can nevertheless be sufficient to induce a shift to the low mobility regime [36].

13.5.2 COARSENING OF SURFACTANT-STABILIZED FOAMS

In parallel to drainage, an aqueous foam also evolves with time by gas diffusion through the thin films from bubbles to bubbles. Diffusion is due to pressure differences between bubbles, which can be evidenced by looking at the curvature of the bubble faces. For well separated droplets or grains, this process is known as *Ostwald ripening*, and is called *coarsening* for cellular materials like foams [29–31]. Concretely, coarsening tends to increase the volume of certain bubbles at the expense of others. In average, the net result is that the mean bubble diameter grows with time (Figure 13.6) [37].

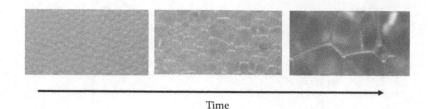

Time

FIGURE 13.6   Pictures of a portion of foam as a function of time, illustrating foam coarsening.

A major feature of coarsening is that it is a self-similar process (obtained after some initial transient). The statistical distributions of topological and dimensionless geometrical quantities remain invariant in time while the characteristic scales continue to increase. Completing initial light scattering studies [38], high speed x-ray tomography has definitively shown that the self-similar growth regime exists. Simply speaking, it was found that—after a rapid transient regime—the successive tomographic images look alike, except that the average bubble size increases with the square root of time [39].

To go further, let us recall some results from models developed from the pioneering works of Von Neumann, Mullins (for 3D grains), and Glazier (for 3D foams) [40–46]. The gas volume flow rate through a bubble face of area S is $dV/dt \sim S\,\Delta P$ with ΔP the Laplace pressure difference being equal to $4\gamma H$ and H is the mean face curvature. The proportionality coefficient between dV/dt and SH is an effective diffusion coefficient D_{effV} (in cm²/s). For dry systems, consistent with a self-similar regime, one has for the mean values that $V^{-1/3}\partial V/\partial t$, $R\partial R/\partial t$, and $\partial S/\partial t$ are constant.

More precisely, $V^{-1/3}\,\partial V/\partial t$ is equal to $C.D_{effV}$, with C being a dimensionless function describing the topology of each bubble [33]. On average, simulations showed that $H.R \sim 0.1$, which implies that $C \sim 1$. Shifting from $V^{1/3}$ to R, one gets $R\partial R/\partial t = C.D_{effR}$, and $D_{effR} = D_{effV}/3\ \delta^{2/3}$ (with δ as $V = \delta\,R^3$) [33].

It is then straightforward to get a prediction for the time evolution of the bubble radius (consistent with the square root behavior of the scaling regime at long times):

$$R^2(t)/R_0^2 = 1 + (t - t_0)/t_c. \tag{13.1}$$

with R_0 the bubble radius at t_0, and:

$$t_c = R_0^2/2C\,D_{effR} \quad \text{and} \quad D_{effR} = D_f[(\gamma v_m He)/h]\cdot(4/3\delta^{2/3})\cdot f(\phi_l). \tag{13.2}$$

Here, γ is the surface tension, D_f is the diffusivity of the gas inside the thin liquid film (considered equal to the one in bulk). As D_{effR} and D_f have the same unit, the rest of Equation 13.2 is dimensionless. It includes v_m, the ideal gas molar volume, He the gas Henry constant, reflecting the solubility of the gas, and h the film thickness. Equation 13.2 also includes a dimensionless function $f(\phi_l)$ representing the normalized effective bubble area for gas diffusion. Considering only gas diffusion through the thin films covering a bubble, and no gas transfer through the area covered by the thick Plateau borders, one gets $f(\phi_l) \sim (1-1.52\ \phi_l^{1/2})^2$ [33]. Slightly different forms are proposed in the literature for this function $f(\phi_l)$, and some issues are still open [47].

The experimental results for foams made of a single surfactant solution, without interfacial viscoelasticity, are consistent with these theoretical considerations. This especially implies that the adsorbed molecules only play a role via surface tension. Such a behavior also shows the central role of the physical parameters—bubble size and liquid fraction—which are indeed much more relevant to tune the coarsening dynamics than the chemical recipe of the solution. To illustrate this point, we show in Figure 13.7 some data for foams made with an SDS solution and for different liquid fractions. There is a clear dependence with ϕ_l, in agreement with the function $f(\phi_l)$ given previously.

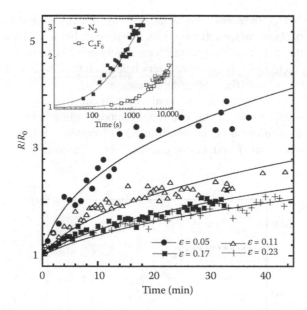

FIGURE 13.7 Normalized bubble radius as a function of time for different constant liquid fractions. The liquid fraction is kept constant thanks to a rotating device (clinostat) to avoid foam drainage. The evolution with the liquid fraction follows the model described in the text. In the insert, data evidencing the role of the gas properties.

Here, a constant liquid fraction has been obtained thanks to a rotating device preventing drainage. We also show results for two gases of different Henry constant: coarsening can actually be tuned by the choice of the gas (or by using gas mixtures [36]).

When interfaces are becoming viscoelastic (with mixtures of surfactant and co-surfactants), results tend to show that there might be some decrease in the coarsening dynamics [18,48]. However, this is not as strong as that found with proteins and particles, which are able to fully stop the coarsening process [49–50]. In fact, an effect on coarsening is expected if the interfacial dilatational modulus E remains always high when compared to the surface tension (Gibbs criterion) [7,51]. This is feasible for irreversibly adsorbed molecules at interfaces (like solid particles) [49]. However, for surfactant systems, it is not clear that at the low frequencies f corresponding to the coarsening rate ($f \sim 10^{-3}$ Hz), interfaces still have some elasticity. Therefore, there are still open questions on the coarsening dynamics of complex surfactant systems, and for finding out what could be the origins of a possible slower coarsening rate.

Finally, it is interesting to compare the dynamics of drainage and coarsening. Drainage is slowed down when making small bubbles (as it was shown previously, the fluid velocity scales with R²). On the opposite side, Equation 13.2 shows that making small bubbles provides a faster coarsening process. So, if one wants to optimize drainage by the bubble size, this will be done at the expense of coarsening (and vice versa). As a consequence, there are in fact two opposite limits: a situation with drainage without coarsening for large bubbles, and a situation with strong coarsening without drainage for small bubbles (in that limit, one even gets a drainage driven by the coarsening process [33]). In all cases, as both aging mechanisms depend on the liquid fraction and the bubble size and are simultaneously changing with time, drainage and coarsening are highly coupled [33,36].

13.6   SPECIFICITIES OF THE RHEOLOGY OF SURFACTANT-STABILIZED FOAMS

Depending on the relative proportions of gas and liquid, an aqueous foam has various mechanical behaviors, ranging from the ones of an elastic solid to those of a viscous liquid. Foam rheology is a very active field, as illustrated in recent reviews [52,53]: first, for fundamental reasons as foams are often considered as a model system for soft materials, and second, because of the numerous industrial applications where foams are used in dynamic conditions and under flow. The development of multiple light scattering techniques has also been very helpful on these issues for linking microscopic mechanisms at the bubble scale to the macroscopic features [54].

A first family of foam rheological studies deals with linear viscoelasticity and yielding. This corresponds to experiments made in the oscillatory modes, allowing to measure viscoelastic moduli—G' and G''—as a function of the amplitude and frequency (Figure 13.8a), but also as a function of the bubble size, liquid fraction, and chemical formulation [55].

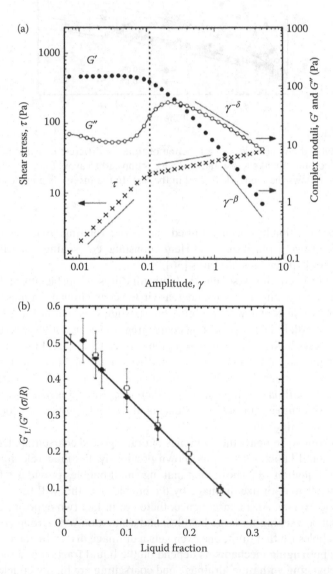

FIGURE 13.8   (a) Typical oscillatory data obtained on a surfactant foam: here, for an amplitude-sweep test, the elastic modulus G', the viscous modulus G'', and the stress σ are monitored. The foam yielding is well evidenced by the change of slope of σ, corresponding to the beginning of the decrease of G'. One can infer from such data the yield stress and strain and the values of the elastic moduli in the linear regime, at low strain. (b) evolution of the elastic modulus as a function of the liquid fraction for two different surfactant systems: once normalized by the Laplace pressure, all data collapse on a master curve, only depending on the liquid fraction.

For solutions of a single surfactant, some universal features have been reported once the rheological quantities are normalized by the Laplace pressure $P_c = \gamma/R$, with γ the gas-liquid surface tension and R a characteristic bubble dimension [52,53,55]. After normalization, the elastic moduli G' and the yield stress σ_y are finally only a function of the liquid fraction (Figure 13.8b). Both of them vanish as the liquid fraction approaches 0.36, the limit where the bubbles are no longer packed [52,55]. The origin of the elastic behavior of foams is indeed linked to the deformation of the bubbles away from a spherical shape. The effect of adding cosurfactant to induce interfacial viscoelasticity turns out to have almost no effect for such data: surface tension generally dominates the elastic contribution.

Understanding the viscous moduli and its dependence on the foam properties is more complex. The dissipation mechanism remains the tricky part to understand: energy dissipation in foam may have very different local origins and their identification, as a function of the characteristic time scale and of the foam properties, is a long lasting fundamental issue. For simple shear, it has been shown that dissipation is directly coupled to the foam coarsening at very low frequency (10^{-2} Hz) [56] and arises from viscous dissipation between bubbles for frequencies around the Hz [57]. At higher frequencies, the origin of how the foam loss modulus varies with the square root of the frequency is not yet fully understood [56].

Beside oscillatory experiments, steady shear measurements are also performed: in such experiments, the total stress σ_t is measured as a function of a shear rate $\dot{\gamma}$. Playing with the rheometer cell wall roughness, one can study how the foam is sheared when using rough surfaces, but also how the foam slips (when using smooth surfaces). This difference is illustrated in Figure 13.9 [58], and both types of experiments have received a lot of interest in the last decades.

With rough surfaces and no slip, the foams are steadily sheared, and measurements reveal that surfactant foams have a typical Herschel-Bulkley behavior, with a yield stress σ_y and a viscous stress σ_v scaling with $\dot{\gamma}$ [59–61]. First, note that the features of the yielding observed in such experiments are consistent with the yielding studied in the oscillatory mode [58]. Regarding the viscous stress ($\sigma_v = \sigma_t - \sigma_y$), it is found to scale as $\sigma_v \sim \dot{\gamma}^\alpha$, with $\alpha \sim 1/2$ for foams made of simple surfactant solutions. In parallel to this scaling regime, there is a strong dependence of this stress on the liquid fraction, as shown in Figure 13.9b.

For surfactant mixtures providing dilational viscoelasticity, a different scaling behavior is found; the exponent α is reduced toward 0.2, as shown in [61]. These different scalings have been summarized within a global framework, where again an interfacial mobility is introduced [61]. By

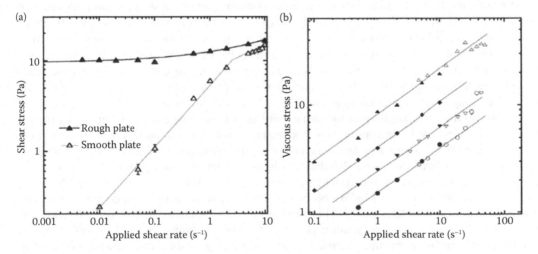

FIGURE 13.9 (a) Steady-shear curves (stress versus shear rate) for a surfactant-stabilized foam: effect of the slip at walls controlled by the roughness of the device surface. This allows to study either a pure shear flow or only the slip of the foam. For rough surfaces, data can be fitted by a Herschel-Bulkley behavior, with a yield stress. (b) viscous stress (total stress—yield stress) for different liquid fractions.

adding cosurfactants, one can shift from a low-mobility regime (corresponding to simple surfactant recipes) to a high-mobility one. Such a change of the dynamic properties of the interfaces creates clear differences at the macroscopic scale.

Concerning the slip of foams on solid surfaces and the impact on the flow properties, there are still some questions not completely solved. Many recent works have focused on these slip issues, implementing the seminal works of Bretherton in the early 1960s, and trying to deepen our understanding of the hydrodynamics within the liquid layers separating the bubbles and the walls [62–64]. The existing models predict that—once again - different slip regimes can be obtained depending on an interfacial "mobility" (controlled by the surfactants adsorbed at the interfaces) and on the liquid volume fraction. Different power laws between the slip velocity and the stress characterize these regimes. Experimentally, these different regimes have been observed [58,62–64], and when simple surfactant solutions are used, one gets the behavior associated with mobile interfaces, whereas by using surfactant mixtures and due to the occurrence of bubble surface viscoelasticity, a high mobility regime is found.

13.7 REMARKS AND CONCLUSIONS

Despite such a huge variety of chemical structures, surfactants have many common features which make them a key element in foam and emulsion formulations: they are "small" molecules, they are soluble and easy to disperse in water, but they are also able to adsorb rapidly on interfaces. As a consequence, surfactant solutions foam well and independently of the foaming techniques. In other words, surfactant-based foams are the simplest to produce and the simplest to understand. Recently, a standard protocol has even been proposed to study such foams [65]. As well, their behaviors (with time or in terms of rheology) are generally consistent with existing models.

But—when using a single surfactant solution—the obtained foams may also be the less "interesting," especially in terms of long-lasting stability. The absence of interfacial viscoelasticity prevents having further stabilizing mechanisms, obtained when interfaces are rigid and have low mobilities. Indeed, the single surfactant foams usually correspond to the limiting case of highly mobile interfaces, and this has a direct impact on the way the foams evolve in time and flow. Remember first that these *interfacial mobilities* are not only an intrinsic interfacial parameter, but also include structural and geometrical quantities (the liquid fraction and the bubble size). Second, experiments reveal that some simple mixtures of surfactants and cosurfactants can produce a shift toward a regime of low mobility (especially providing longer stability). Therefore, interfacial properties can actually tune the macroscopic behavior: in many cases, low or high interfacial mobility implies very different macroscopic features. These extreme limits can already be obtained by using only combinations of low molecular mass surfactants, without the need for more complex chemicals. Nevertheless, more complex molecules—proteins, polymers, particles—will provide other foam regimes, as discussed in the following chapters, and are impossible to get with only surfactants.

Regarding the time evolution of the surfactant foams by drainage and coarsening, as shown in this chapter, one must keep in mind how important the physical parameters of the foams (bubble size and liquid fraction) are. For surfactant solutions, the aging can be widely tuned and controlled by these physical quantities. Note also that we discussed here the drainage and coarsening, though the final stage of the foam life is driven by bubble coalescence (when a film between bubbles breaks). Regarding the mechanisms behind bubble coalescence, this remains the most complex issue, and is not much understood yet; elements can be found in a recently published review [66]. We also want to point out that, as noted previously, drainage and coarsening are highly coupled: as well, coarsening and drainage are also important ingredients in foam rheology, especially because they modify R and ϕ_l, and second, because coarsening is a source of relaxation of low frequencies [56]. Therefore, a global approach is always necessary on these issues.

Finally, we want to emphasize that there is still some room for research on foams stabilized by surfactants; recent new directions deal with responsive molecules [67,68], taking advantage of the fact that the foam macroscopic behavior can be very sensitive to tiny changes at the scales of the interfaces and films. In that sense, foams or emulsions can be seen as interesting templates for designing smart and stimuli-responsive materials.

REFERENCES

1. Tanford, C. *Science* 1978; 200: 1012.
2. Tanford, C. *The Hydrophobic Effect: Formation of Micelles and Biological Membranes*, 2nd Edition. Wiley-Interscience: New York, 1980.
3. Chandler, D. *Nature* 2005; 437; 640.
4. Rosen, M.J. *Surfactants and Interfacial Phenomena*, 3rd Edition. Wiley-Interscience: Hoboken NJ, USA, 2004.
5. Holmberg, K., Jonsson, B., Kronberg, B., Lindman, B. *Surfactants and Polymers in Aqueous Solution*, 2nd Edition. Wiley editions, Copyright 2002 John Wiley & Sons, Ltd.
6. Israelachvili, J. *Intermolecular and Surface Forces*, 3rd edition. Academic Press: New York, 2011.
7. Gibbs, J.W. *The Scientific Papers of J. Willard Gibbs*. Oxbow Press: Woodridge, 1933.
8. Chang, C.H., Franses, E.I. *Colloids and Surfaces A* 1995; 100: 1.
9. Marangoni, C. *Il Nuovo Cimento Series 2* 1871; 5–6: 239.
10. Levich, V.G., Krylov, V.S. *Ann Rev Fluid Mech* 1969; 1: 293.
11. Roché, M., Li, Z., Griffiths, I.M., LeRoux, S., Cantat, I., Saint-Jalmes, A., Stone, H.A. *Phys Rev Lett* 2014; 112: 208302.
12. Bordes, R., Holmberg, K. *Adv Colloid Interface Sci* 2015; 222: 79–91.
13. Espert, A., von Klitzing, R., Poulin, P., Colin, A., Zana, R., Langevin, D. *Langmuir* 1998; 14: 4251.
14. Pinazo, A., Perez, L., Infante, M.R., Franses, E. *Colloids and Surface A*, 2001; 189: 225.
15. Salonen, A., In, M., Emile, J., Saint-Jalmes, A. *Soft Matter* 2010; 6: 2271.
16. Schelero, N., von Klitzing, R. *Curr Opin Colloid Interface Sci* 2015; 20: 124.
17. Drenckhan, W. et al. *Phys of Fluids* 2007; 19: 102101.
18. Golemanov, K., Denkov, N.D., Tcholakova, S., Vethamuthu, M., Lips, A. *Langmuir* 2008; 24: 9956.
19. Drenckhan, W., Saint-Jalmes, A. *Adv Coll Int Science* 2015; 222: 228–259.
20. Bergeron, V. *J Phys Condens Matt* 1999; 11: R215.
21. Stubenrauch, C., von Klitzing, R.J. *Phys Condens Matt* 2003; 15: R1197.
22. Singh, G., Hirasaki, G., Miller, C. *J Coll Int Sci* 1996; 184: 92.
23. Stoyanov, S., Denkov, N. *Langmuir* 2001; 17: 1150.
24. Tcholakova, S., Denkov, N.D., Lips, A. *Phys Chem Chem Phys*, 2008; 10: 1608.
25. Boos, J., Drenckhan, W., Stubenrauch, C. *Langmuir* 2012; 28: 9303.
26. Prud'homme, R.K., Khan, S.A. *Foams: Theory, Measurements, and Applications*. Marcel Dekker Inc.: New York, 1997.
27. Saint-Jalmes, A., Durian, D.J., Weitz, D. *Kirk-Othmer Encyclopedia of Chemical Technology*, 5th Edition 8 2005, 8, p. 697, 5th Ed, John Wiley and Sons: New York, 2004.
28. Cantat, I., Cohen-Addad, S., Elias, F., Hohler, R., Graner, F., Pitos, O., Royer, F., Saint-Jalmes, A. *Foams: Structure and Dynamics*. Oxford University Press: Oxford, UK, 2013.
29. Verbist, G., Weaire, D., Kraynik, A. *J Phys: Condens Matt* 1996; 8: 3715.
30. Weaire, D., Hutzler, S., Verbist, G., Peters, E. *Adv Chem Phys* 1997; 102: 315.
31. Koehler, S.A., Hilgenfeldt, S., Stone, H.A. *Langmuir* 2000; 16: 6327.
32. Feitosa, K., Marze, S., Saint-Jalmes, A., Durian, D.J. *J Phys Condens Matter* 2005; 17: 6301.
33. Hilgenfeldt, S., Koehler, S.A., Stone, H.A. *Phys Rev Lett* 2001; 86: 4704.
34. Leonard, R., Lemlich, R. *A.I.Ch. E. Journal* 1965; 11: 18.
35. Stone, H.A., Koehler, S.A., Hilgenfeldt, S., Durand, M.J. *Phys Condens Matter* 2003; 15: S283.
36. Saint-Jalmes, A. *Soft Matter* 2006; 2: 836.
37. Jurine, S., Cox, S., Graner, F. *Coll Surf A* 2005; 263: 18.
38. Durian, D.J., Weitz, D.A., Pine, D.J. *Phys Rev A* 1991; 44: 7902.
39. Lambert, J., Mokso, R., Cantat, I., Cloetens, P., Glazier, J., Graner, F., Delanna, R. *Phys Rev Lett* 2010; 104: 248304.
40. von Neumann, J. *Metal Interfaces*. American Society for Metals, Cleveland, 1952. p 108.
41. Mullins, W.J. *Appl Phys* 1986; 59: 1341.
42. Glazier, J., Gross, S., Stavans, J. *Phys Rev A* 1987; 36: 306.
43. Mullins, W. *Acta Metall* 1989; 37: 2979.

44. Glazier, J., Stavans, J. *Phys Rev A* 1989; 40: 7398.
45. Stavans, J., Glazier, J. *Phys Rev Lett*, 1989; 62: 1318.
46. Glazier, J. *Phys Rev Lett* 1993; 70: 2170.
47. Feitosa, K., Durian, D.J. *Eur Phys J E* 2008; 26: 309.
48. Giorgeva, D., Cagna, A., Langevin, D. *Soft Matter* 2009; 5: 3063.
49. Horozov, T.S. *Curr Opin Colloid Interface Sci* 2008; 13: 134.
50. Blijdenstein, T.B.J., de Groot, P.W.N., Stoyanov, S.D. *Soft Matter* 2010; 6: 1799.
51. Kloek, W., von Vliet, T., Meinders, M.J. *Colloid Interface Sci* 2001; 237: 158.
52. Hohler, R., Cohen-Addad, S. *J Phys Condens Matter* 2005; 17: R1041.
53. Dollet, B., Raufaste, C., *C R Physique* 2014; 15: 731.
54. Höhler, R., Cohen-Addad, S., Durian, D.J. *Curr Opin Colloid Interface Sci* 2014; 19: 242.
55. Marze, S., Guillermic, R.M., Saint-Jalmes, A. *Soft Matter* 2009; 5: 1937.
56. Gopal, A.D., Durian, D.J. *Phys Rev Lett* 2003; 91: 188303–1880301.
57. Denkov, N.D., Tcholakova, S., Golemanov, K., Ananthapadmanabhan, K.P., Lips, A. *Phys Rev Lett* 2008; 100: 138301.
58. Marze, S., Langevin, D., saint-jalmes, A. *J Rheol* 2008; 52: 1091.
59. Tcholakova, S., Denkov, N.D., Golemanov, K., Ananthapadmanabhan, K.P., Lips, A. *Phys Rev E* 2008; 78: 011405.
60. Denkov, N.D., Tcholakova, S., Golemanov, K., Lips, A. *Phys Rev Lett* 2009; 103: 118302.
61. Denkov, N.D., Tcholakova, S., Golemanov, K., Ananthpadmanabhan, K.P., Lips, A. *Soft Matter* 2009; 5: 3389.
62. Denkov, N.D., Subramanian, V., Gurovich, D., Lips, A. *Colloids Surf A* 2005; 263: 129.
63. Denkov, N.D., Tcholakova, S., Golemanov, K., Subramanian, V., Lips, A. *Colloids Surf A*, 2006; 282: 329.
64. Cantat, I. *Phys Fluids* 2013; 25: 031303.
65. Boos, J., Drenckhan, W., Stubenrauch, C. Protocol for studying aqueous foams stabilised by surfactant mixtures. *J Surf and Detergents* 2013; 16(1): 1–12.
66. Langevin, D. *Curr Opin Colloid Interface Sci* 2015; 20: 92.
67. Fameau, A.L., Carl, A., Saint-Jalmes, A., von Klitzing, R. *Chem Phys Chem* 2015; 16: 66.
68. Zhang, L., Mikhailovskaya, A., Yazghur, P., Muller, F., Cousin, F., Langevin, D., Wang, N., Salonen, A. *Angew Chem Int Ed* 2015; 54: 1.

14 Protein-Stabilized Foams

Christophe Schmitt, Deniz Z. Gunes,
Cécile Gehin-Delval, and Martin E. Leser

CONTENTS

14.1 INTRODUCTION

Proteins represent a major element of naturally occurring or industrially manufactured foams. The most likely reason is the variety of amino acid compositions (owing to the combination of the 20 elementary amino acid building blocks), chain lengths and 3D structures that can be found in proteins from animal, plant, microbial or fungal sources [1]. These intrinsic properties confer to proteins various levels of hydrophobicity, net charge or structural flexibility that are essential for controlling their diffusion to the air/water interface and their interfacial spreading, as well as their subsequent reorganization in order to form more or less compact, thick, and viscoelastic layers [2–4]. The aim of this book chapter is not give an exhaustive review of all the developments related to protein foams as this was already the subject of a very comprehensive book that we invite the readers to refer to [5]. Here, we will focus on the latest findings and novel trends in the field of protein-stabilized foams since 1999 and we will mainly use examples of food-grade proteins that are readily commercially available and relatively well characterized. These ingredients are widely used by the food industry in order to design numerous aerated products including meringues, ice cream, cakes, bread, foamed desserts or coffee/cappuccino foams [6]. These food applications and the processing conditions required to produce them will be further described in Chapter 20 of this book.

It was described early on that the interfacial properties of proteins were mainly driven by the balance between the distribution of their hydrophobic and charged residues along the polypeptide chain [7]. Thus, the most hydrophobic proteins such as globular β-lactoglobulin or random coil β-casein exhibited the highest foaming capacity compared to the least hydrophobic ribonuclease [8,9]. Consequently, any physical or bio/chemical treatments leading to an increase of the protein hydrophobicity will, in general, improve the foaming properties compared to the native proteins [10]. The most common ways to increase protein hydrophobicity is to reduce the surface charge by adjusting the pH close its isoelectrical pH (IEP) or by screening its charges by addition of counterions contained in salts such as NaCl or $CaCl_2$ [11]. Additional ways to modify the hydrophobicity of proteins are to partially or fully denature those using specific physical treatments such as heat, pressure, shear, pulsed electrical field or ultrasounds in order to disrupt covalent or noncovalent interactions responsible for the native state structure [12]. This chapter is attempting to review the effect of such physical parameters and treatments on the foaming properties of a series of food-grade proteins that are mostly used in complex aerated food manufacture.

In the first section, we will mainly describe foaming properties of proteins having limited content of secondary structures and no covalent bounds, the so-called random coil proteins which behave closely to low molecular weight surfactants or block copolymers [1,13]. Most of the examples will concern specific casein fractions, caseinates, and micellar caseins. The next section will be devoted to globular proteins, with special emphasis on "traditionally used" whey, egg white, and soy—as a model of plant—proteins. A specific part of this section will review recent findings on hydrophobins (a peculiar class of small globular proteins) found in some filamentous fungi which do exhibit exceptional interfacial properties [14]. Then the foaming properties of protein aggregates obtained by heat treatment will be considered since this topic has attracted quite a lot of scientific interest in the last decade [15]. Finally, the foaming properties of electrostatic protein/polysaccharides complexes that offer a variety of combinations between the proteins listed above and food-grade charged polysaccharides will be discussed. It should be noticed here that the use of enzymatic hydrolysis to modulate the foaming properties of proteins will not be considered, this subject being too vast on its own [16–20]. As well, the use of so-called protein/polysaccharide conjugates (which are chemically of biochemically covalently bound species) in foams will not be part of this chapter as they have been reviewed recently [21]. Finally, we will not report on the foaming properties of mixtures of proteins and surfactants which have been discussed in excellent review papers [22,23].

Before concluding this introduction section, we would like to mention recent findings on the nature of the compounds responsible for the stability of the nest foam of the tungara tropical frog because it summarizes on its own all the sections to be covered in this chapter. A decade ago, Cooper and coworkers [24] reported the very high stability of the foam nests of the frog *Engystomops pustulosus* since these "naturally manufactured" foams were stable for more than 10 days in tropical conditions of relative humidity and temperature without visible signs of destabilization or microbial contamination (Figure 14.1).

Preliminary investigations of the interfacial properties of the foam compounds had revealed the presence of a set of at least six proteins, the so-called ranaspumins, which were forming a complex structure in the systems and were responsible for foam stability. Interestingly, the same research group recently conducted further biochemical characterization of the ranaspumins' structure. They concluded that this cocktail comprised a surfactant-like protein responsible for interfacial activity. This protein was interacting with lectin-like proteins to generate aggregates that were able thereafter

FIGURE 14.1   Picture of Engystomops pustulosus colonial nests. Individual nests are approximately 10 cm in diameter. (Reprinted from Fleming, R.I. et al. *Proc R Soc B* 2009;276:1787–1795 [25].)

to build complexes with polysaccharides in order to form rigid and stable interfacial layers, leading to very stable foams [25].

14.2 RANDOM COIL PROTEINS

Caseins are typical examples of random coil proteins which are characterized by a limited content of secondary structures and no covalent bonds so that they can be modeled as block copolymers [13]. These specific structural features confer to caseins interfacial behavior similar to some low molecular weight surfactants in terms of surface activity, but with additional interfacial stabilization properties owing to their large molecular weights of around 20,000 g · mol^{-1} and their chain flexibility [26]. Bovine milk caseins are mainly composed of four fractions: α_{s1}, α_{s2}, β, and κ-caseins which mainly differ by their charge and hydrophobicity distribution [27]. The most abundant casein in bovine milk is α_{s1}-casein ($\approx$45%), followed by β-casein ($\approx$32%), α_{s2}-casein ($\approx$12%), and κ-casein ($\approx$11%). Interestingly, in milk, these fractions organize into colloidal structures in the size range of about 100–300 nm, the so-called casein micelles which are maintained by electrostatic calcium phosphate bridges [28,29]. The micellar structure of casein micelles has revealed that the different fractions are organized from the most hydrophobic (α_{s1}, α_{s2}, and β) to the most hydrophilic (κ) caseins from the interior to the surface of the micelle, respectively [28,30]. The last possible form of caseins is the so-called caseinates which are obtained by acidic precipitation of the caseins by dissolution of calcium-phosphate bridges followed by neutralization by addition of sodium, calcium or potassium hydroxide. The most common caseinate is sodium which is composed by the main casein fractions organized in nanometer-scale aggregates [31]. Due to their different structure and size, micellar caseins, sodium caseinate, and the casein fractions do not behave similarly at air/water interface and in foams.

It has been shown that β-casein is the most surface active casein fraction at the air/water interface, the likely reason being its large C-terminal hydrophobic region and its high chain flexibility enabling fast diffusion to the interface and interfacial rearrangements [32]. Thus, β-casein was able to displace adsorbed α_{s1}-casein from an air/water interface [32]. It also strongly competes for interfacial coverage when mixed with globular proteins such as bovine serum albumin, 11S soy globulin or α-lactalbumin as evidenced by interfacial phase separation determined by fluorescent microscopy [33,34]. In a mixture of caseins originating from skimmed milk powder at 1.4 wt% and pH 6.8, the superior surface activity of caseins was shown by the calculation of an enrichment ratio of 2.8 for β-casein compared to about 2.3 for the other caseins [35] or 1.7 for β-lactoglobulin and α-lactalbumin, the major whey proteins. Addition of 2 mM CaCl$_2$ in the system reduced enrichment of caseins in the foam, the reason being the formation of non-native casein micelles that were less prone to diffuse and rearrange at interface compared to isolated casein fractions due to an increase in size and a reduction of flexibility [35]. β-casein was able to produce a larger foam volume at pH 6.7 and 0.14 wt% compared to both wheat gliadin or soy globulin [36]. Authors explained this difference by the highest adsorption rate of β-casein to the air/water interface, leading to a fast decrease of the surface tension and the creation of an interface necessary for air bubble production. The protein concentration and molecular changes of casein chains of β-casein and β-casein peptides were investigated using adsorption at Teflon interface at pH 6.7 and at an ionic strength of 20 mM [37]. The highest absorbed amount of 3 mg · m^{-2} for intact β-casein of its hydrophobic C-terminal peptide were correlated with good foam formation, but better foam stabilization compared to the more hydrophilic peptides. No specific influence of the secondary structure of this random coiled protein or peptides could be related to foam stability since α-helix content of 20%–25% could lead to either stable or unstable foams. This result was recently confirmed upon separating β-casein and lysozyme mixtures using a foaming column [38]. Investigation of the secondary structure changes of β-casein using circular dichroism (CD) spectroscopy and infrared reflection absorption spectroscopy (IRRAS) revealed no changes in the β-casein enriched in the foam compared to that present in solution, even in conditions of maximum enrichment of 10 times (pH 8.7, temperature 30°C, ionic strength 100 mM). Likely, random coiled proteins adsorb at the air/water interface exhibiting their most hydrophobic regions without any change in the secondary structure, but likely with changes in the interfacial conformation leading to high protein interfacial concentration, but weak viscoelastic networks [36]. The importance of the thickness of casein interfacial layers on the stability (bubble coarsening at constant liquid fraction of 15%) of foams was shown by comparing casein-based foam with SDS-based foams at pH 5.6 [26,39]. SDS adsorbed faster at the interface than caseins (mixture of fractions) and lead to smaller air bubbles at comparable concentration, however, casein-based foam coarsening was much slowed down at a comparable initial air bubble size of 120 μm. Investigation of the structure of casein thin films indicated that the stability was mostly due to some interfacial aggregation leading to thick heterogeneous films that were able to strongly reduce the rate of gas exchange and coalescence, possibly due to film elasticity. The thickness of casein films was shown to be able to reach around 80 nm at disjoining pressure of about 10 Pa and shrunk to 16 nm at 150 Pa [40]. This thickness was larger than surfactant film of Tween 20 (11 nm), leading to much stabler foams.

Foaming properties of sodium caseinate were investigated in detail and it was found that concentrations of about 0.5 wt% were leading to maximum foam capacity at a natural pH of 6.8 [42]. This concentration corresponded to the interfacial layer with protein and also to constant interfacial tension. Foam stability was highest at neutral pH while it was lowest close to the isoelectric pH (IEP) of caseins (4.6). Interestingly, pH influence on foam stability for sodium caseinate was very similar to that of β-casein, implying that this fraction was mostly governing interfacial and foaming properties as discussed earlier. Authors explained that close to IEP, casein chains were forming less thick interfacial layers due to reduced repulsions at the interface [43].

Having started with caseins fractions and caseinates, skimmed milk represents a more practically used complex mixture of caseins in micellar and soluble forms together with soluble globular whey proteins. Its foaming properties will of course depend mostly on similar parameters affecting the composition and aggregation state of its constituting proteins and heat treatment combined with pH have been shown to be the most important variables [44]. Most stable skimmed milk foams were obtained at 45°C for pH values slightly lower (6.4) than the natural pH of milk (6.8) and for a protein content of about 4 wt% [45]. Similar foaming improvement was described by the addition of calcium chloride into skimmed milk [41]. Such salt addition led to the increase of the micellar caseins content in milk and an interfacial stabilization of the air bubble by soluble casein/whey protein fractions, but also partially spread casein micelles as shown in Figure 14.2. Addition of too

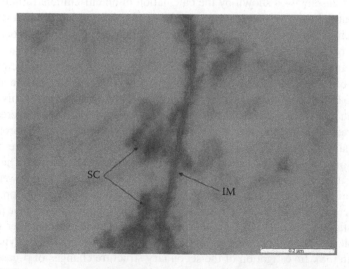

FIGURE 14.2 TEM micrograph of the spreading of casein micelles at the surface of collapsed interfacial membrane. IM = collapsed interfacial membrane, SC = spread casein micelles. Bar = 0.2 μm. (Reprinted with permission from Kamath, S. et al. *J Dairy Sci* 2011;94:2707–2718 [41].)

large amounts of calcium chelating agents (citrate, Ethylenediaminetetraacetic acid [EDTA]) or acidification of skimmed milk led to complete disruption of casein micelles and reduced its foaming properties because interfacial layers formed by caseins fractions were not thick enough to reduce drainage and air bubble coalescence [41,46]. Controlled dissociation of casein micelles by EDTA addition has been shown to be very effective for stabilization of air bubbles in a complex ice cream recipe and TEM observation revealed partial absorption of micellar caseins and soluble β-casein at the air/water and fat/water interfaces [47].

Besides random coiled proteins such as milk caseins, important protein ingredients used for production and stabilization of foams are globular proteins. In the next section, we will mainly focus on whey proteins, but will give relevant examples for soy and egg white proteins. To conclude this section, the very special case of the globular class II hydrophobins will be discussed and recent findings will be reviewed.

14.3 GLOBULAR PROTEINS

Globular proteins are characterized by high contents of secondary structure originating from the fact that the most hydrophobic regions of the peptidic chain associate to form so-called β-sheets while the most polar residues generally form helical structures [1]. In most cases, the structure of globular proteins is maintained by covalent disulfide bonds, the ability of the protein to unfold being inversely proportional to the number of disulfide bonds present [12]. As for random coil proteins, globular proteins interfacial properties are driven by their average hydrophobicity and are, therefore, strongly pH and salt dependent. However, partial or total unfolding of the protein due to heat treatment is generally an important additional parameter to trigger the interfacial and foaming properties of this class of proteins [11].

The most commonly used globular proteins in foams are whey proteins and egg white proteins, and to a lesser extent, soy proteins. Whey proteins are obtained by removal of the casein fraction from milk and are mainly composed by β-lactoglobulin (≈50%–60%), α-lactalbumin (≈15%–20%), and bovine serum albumin (5%–10%) which are mostly driving the interfacial properties of whey protein isolates or concentrates [48]. β-lactoglobulin is a dimeric protein (≈36,000 g · mol^{-1}) mostly containing β-sheet structures and its globular structure is maintained by two disulfide bonds and an additional free thiol group. Its isoelectrical pH is about 5.2 for the dimer [49]. On the contrary, α-lactalbumin is a small monomeric (14,200 g · mol^{-1}) calcium binding protein with an IEP of 4.3, mostly consisting of helical structures and containing four disulfide bounds responsible for its high structural stability [50]. Egg white proteins are mostly composed by ovalbumin (≈54%), ovotransferrin (≈13%), ovomucoïd (≈11%), ovomucin (≈3.5%), and lysozyme (≈3.5) [51]. As for whey proteins, ovalbumin is mostly responsible for the overall foaming properties of the mixture. It is a rather hydrophobic protein containing β-sheets, exhibiting a large molar mass (45,000 g · mol^{-1}), and an isoelectrical pH of 5.0. It contains one disulphide bond and four free thiol groups. Soy proteins are mainly composed of two major fractions, glycinin (so called 11S globulin) and β-conglycinin (so-called 7S globulin), both fractions being strongly involved in the foaming properties of the soy proteins [52]. Soy glycinin is a hexameric protein formed by association of an acidic and a basic trimer linked by a disulfide bond. It has a molar mass of about 300,000 g · mol^{-1} and mostly contains helical structures. β-conglycinin is a trimeric protein made from assembly of α′, α, and β subunits leading to a molar mass of 170,000 g · mol^{-1}.

14.3.1 WHEY PROTEINS

Recent investigation of the foaming properties of β-lactoglobulin as a function of pH revealed that the most stable foams were obtained near its isoelectrical pH of 5.2 [53]. This condition also corresponds to a minimum in surface tension, implying that the protein had strongly adsorbed at the interface and was able to form thick and densely packed layers. Such assumption was correlated with a multilayer organization of the interfacial film (8 nm of thickness determined by ellipsometry) and a low mobility of water molecules within the film due to strong protein attractive interactions. Interestingly, such multilayer structure of β-lactoglobulin had been already described for free-standing films at 0.1 wt% protein concentration for which a thickness of about 11 nm was determined by x-ray scattering at pH 5.2 (corresponding to 3 layers of proteins) and leading to a film lifetime of almost 2 hours [54]. β-lactoglobulin film structure formation at its IEP led to high dilatational viscoelasticity [53] but also high shear interfacial elasticity [55]. Viscoelasticity of β-lactoglobulin interfacial layers was proportional or even larger than the rheological properties of the foams in conditions where repulsive interfacial layers were formed, that is, below and above IEP [53], confirming the hypothesis that mostly bulk viscoelastic properties were driving bubble stability [56,57]. However, at the isoelectrical pH, foams exhibited their highest storage modulus and yield stress with a loss of proportionality with interfacial properties of the protein films. Authors explained this discrepancy by the complex attractive structure built in the interfacial layers via protein aggregation and reported that a similar effect was induced by divalent cations as compared to monovalent salts which had a similar effect on the pH of foam stability and rheology [58]. It is worth mentioning here that very similar results were reported for the effect of pH or ionic strength on rheology of whey protein isolate foams [43,59]. The stability of β-lactoglobulin foams against disproportionation at various pHs was followed from the variation of the total foam interfacial area and the mean bubble size [60]. The slowest coarsening rate was obtained close to IEP (Figure 14.3), confirming that the densely packed interfacial layers coupled with the high bulk viscosity were able to counterbalance gas diffusion due to Laplace pressure in the bubbles.

Croguennec et al. [61] slightly modified the structure of β-lactoglobulin by blocking its free sulfhydryl group while the remainder of the structure was unmodified. Interestingly, such modification increased the protein hydrophobicity and improved surface activity leading to good foamability while reducing the film viscoelacticity. Very similar improvement of interfacial and foaming properties has been observed upon blocking the thiol group of β-lactoglobulin using allyl isothiocyanate [62]. However, in this case, further modification of the protein led to detrimental effects on foaming properties due to too high hydrophobicity. Kim and coworkers [63] increased the surface hydrophobicity of β-lactoglobulin by submitting a 0.1 wt% solution to a controlled heat

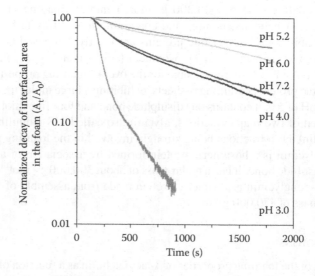

FIGURE 14.3 Evolution of the foam interfacial area determined by pressure measurements as a function of pH of 1 wt% BLG dispersion at 25°C. (Reprinted with permission from Rami-Shojaei, S. et al. *Image Vis Comput* 2009;27:609–622 [60].)

treatment at 80°C up to 30 min and showed an improved foam stability that could be explained by increased protein flexibility at the interface as shown by adsorption isotherms.

The interfacial properties of α-lactalbumin at the air/water interface were compared between the protein in its native state (pH 7.0) and a partially denatured state, the so-called molten globule, obtained at pH 2.0 [64]. The surface load of α-lactalbumin increased from 1.5 in the native state to 3.0 mg · m^{-2} in the molten globule state. This higher surface load could be due to the overall 15-fold increase of protein hydrophobicity. On the other hand, compared to the native state, the surface shear viscoelasticity was markedly reduced, indicating that partially denatured α-lactalbumin layers were less prone to build strongly interacting networks. The effect of calcium binding was also shown to impact strongly α-lactalbumin surface and foaming properties [65]. Removal of calcium by use of EDTA improved the foaming capacity of the protein, but also the foam stability which might be both related to some conformational changes of the protein as well as a reduction of its overall charge resulting in increased hydrophobicity.

Mixtures of whey proteins such as whey protein isolates (WPI) have been investigated for their interfacial and foaming properties and compared to those of their main protein constituents, α-lactalbumin and β-lactoglobulin. In general, most of the functional properties (foamability, foam stability, foam rheology) of WPI were close to those reported for β-lactoglobulin, however, some improvements that could be due to synergistic effects between proteins in mixtures or due to impact of ionic composition have been reported [59,66,67].

14.3.2 EGG WHITE PROTEINS

Interfacial structure formation of ovalbumin and its thermostable conformer S-ovalbumin at air/water interface revealed a maximum in shear elasticity around their IEP of 5.0 for diluted solutions around 0.01 wt% [68]. Investigation of the protein secondary structure at the interface using polarization modulation infrared reflection-absorption spectroscopy (PM-IRRAS) showed that protein reorganization in the film involved mostly β-sheet structures which are strongly involved in protein aggregation phenomena. This observation was in accordance with the increase of the interfacial concentration of ovalbumin and S-ovalbumin, leading to the formation of multilayers. Such specific aggregated interfacial structure has been confirmed by studies of air bubble disproportionation in 0.05 wt% ovalbumin dispersions at pH 7.0 leading to protein particles ghosts after complete dissolution of the bubble [69]. Interestingly enough, the interfacial shear viscoelastic properties of ovalbumin were close to those of β-lactoglobulin whereas the dilatational ones were 2–3 times higher, indicating much stronger viscoelastic films due to protein aggregation [69]. In an attempt to understand the influence of the complex mixture of egg white proteins on their interfacial properties compared to single proteins, ovalbumin/ovotransferrin/lysozyme mixture was foamed at pH 7.0 and changes in the protein structure, composition, and hydrophobicity were recorded [70]. Clearly, it was a synergistic effect on interfacial organization of the proteins being in a mixture compared to single proteins, especially for lysozyme that was fully denatured and involved in disulfide bonds formation. According to circular dichroism measurements, all three proteins lost significant contents in α-helical or β-sheets features at the expense of β-turns and random coil structures. In binary mixtures of ovalbumin and lysozyme at 1 wt% and pH 7.0, ellipsometry revealed that lysozyme was the most surface active species due to the small size of the protein (14,300 g · mol^{-1}), but that the overall interfacial and foaming (capacity, density, stability) properties of the mixture were controlled by ovalbumin [71]. The thickness of an interfacial layer of pure lysozyme solution at equilibrium surface tension was about 3–5 nm, indicating almost monolayer organization. Interestingly, PM-IRRAS measurements revealed little structural changes in the molecule which was mainly exhibiting a-helical and random coil conformations [72]. The effect of temperature pretreatment (20–90°C) of 0.5 wt% solutions of lysozyme and ovalbumin at pH 7.0 and 0.1 M NaCl was investigated with respect to foaming properties and foam quality (bubble size). It was clear that when the heating temperature was above the denaturation temperature of ovalbumin (72°C), the foaming capacity increased markedly, seemingly due to some structural changes of the protein and an increase of its hydrophobicity. A similar effect was reported for lysozyme, but at a higher temperature of 85°C [73]. Both heated proteins foams exhibited a strong reduction of the air bubble size compared to nonheated proteins as well as a reduction of the drainage rate in the foams, likely due to increased interfacial viscosity due to enhanced protein-protein interactions.

The microstructure (bubble size) and rheology (yield stress) of egg white protein whipped foams generated at 10 wt% and pH 7.0 were compared to that of whey protein isolate foam obtained in same conditions [74]. Egg white foams exhibited smaller air bubbles than WPI-based ones as well as a much higher yield stress. Additionally, air bubble coarsening in egg white foams was reduced compared to WPI due to increased interfacial elasticity and a higher liquid volume fraction in lamellae and Plateau borders [74]. The foam stability of egg white foams at pH 7.0 could be further improved by addition of sucrose, likely due to the increase of bulk viscosity [75], and mixtures of egg white proteins and whey protein isolate were also shown to exhibit a synergistic effect on foam stability and firmness depending on the mixing ratio and pH [76]. The best foam quality was obtained at pH 9.0 for a 50/50 mixture, which could be explained by attractive electrostatic interactions between the main protein fractions of egg white and WPI [76].

14.3.3 SOY PROTEINS

Interfacial properties of soy glycinin (11S) depended strongly on pH [77]. Interfacial adsorption was faster at pH 3.0 compared to 6.7 due to the dissociation of glycinin in its acidic and basic 3S subunits. The protein surface load also increased at acidic pH reaching $3 \ mg \cdot m^{-2}$ while it was $2 \ mg \cdot m^{-2}$ at neutral pH. This interfacial behavior of glycinin subunits had a direct impact on the dilatational elasticity of the layers that were much more elastic at pH 3.0. Interestingly, even if the initial shear interfacial viscoelasticity of the acid fractions was higher with a short adsorption time compared to neutral pH, both values were close after 20 hours of adsorption. A direct consequence was that no stable foams were obtained at pH 6.7, while stable foams were generated at pH 3.0 [77]. PM-IRRAS experiments were used to follow changes in the structure of soy glycinin at air/water interface and it was shown that β-sheets were mostly formed at pH 3.0, indicating protein aggregation and possibly explaining the high viscoelasticity described for these conditions [78,79]. The ability of glycinin to retard single air bubble disproportionation was the highest compared to all globular proteins discussed in previous sections, the likely reason again being a combination of high dilatational elasticity and shear viscosity [69]. Such air bubble stabilization properties were also reported in glycinin based foams at pH 3.0, where experimental air bubble size distribution was shown to correlate well with a model incorporating sufficiently high film dilatational elasticity as well as relaxation time [80]. Differences in surface activity and foaming properties of glycinin and β-conglycinin have been investigated as a function of pH and ionic strength [81]. Both proteins were shown to be less surface active at pH 5.0 which is close to their IEP, which might be due to protein over-aggregation in the bulk. However, β-conglycinin (7S) was more surface active than glycinin (11S) at pH 7.0, which could be explained by a faster diffusion to the interface due to its smaller molecular weight. In addition, 7S globulin exhibited much stabler foams than 11S at neutral pH in the presence of 0.5 M NaCl owing to the development of a significant surface viscoelasticity [81]. Addition of sucrose in these systems lead to increased foam stability most likely by increasing bulk viscosity since interfacial properties of the two proteins remained unchanged [82]. In an attempt to improve the surface and foaming properties of 11S globulin, Wagner and Guéguen [83] showed that acidic deamidation increased protein surface hydrophobicity and thereby its foam stabilizing properties at pH 7.6. Additionally, reduction of the disulfide bridges led to improved foaming properties of 11S globulin due to a reduction of its molecular weight and an increase it its hydrophobicity. Recently, it was evidenced that the foaming properties of 11S soy globulin could be improved by hydrophobic interactions with a steviol glycoside that was inducing conformational changes in the protein as revealed by fluorescence spectroscopy [84]. Interestingly, an optimum protein to steviol glycoside weight mixing ratio of 1:2.5–1:5 was described for increasing surface elasticity and foam stability. An excess of steviol glycoside was detrimental to foam stability due to direct adsorption of the glycoside at the interface and the formation of a weaker film. It is worth mentioning that the foaming properties of major whey proteins (α-lactalbumin and β-lactoglobulin) were compared to those of soy globulins at pH 8.0 in the presence of 0.01 M NaCl [85]. Both whey proteins were more surface active and exhibited high foamability, but β-conglycinin showed the closest data compared to milk proteins, which might imply that it could replace them in some applications. Further, a synergistic improvement on surface viscoelasticity and foam stability was described upon mixing β-conglycinin with β-lactoglobulin at pH 7.0, which could be explained by the formation of heteroprotein complexes [86].

14.3.4 HYDROPHOBINS

To conclude the section on globular proteins, it is important to mention the peculiar case of hydrophobins, a group of small globular proteins found in fungi, which exhibit surface properties and foam stabilization behavior close to those of rigid particles [87]. Hydrophobins (HFB) are small globular proteins (7,200 g · mol^{-1}) which are found in several species of filamentous fungi such as *Agaricus bisporus*, *Trichoderma reesei* or *Neurospora crassa*. Depending on the hydrophilic/ hydrophobic amino acid pattern along their sequence, hydrophobins were classified as class I or II [14]. Despite these differences in hydrophobicity pattern, class I and II hydrophobins share a conserved motive of 8 cystein residues forming 4 disulfide bonds which are responsible for the extremely high structural stability of the protein. In addition, about half of the total hydrophobic amino acid residues of the sequence are organized at the surface of the protein, forming the so-called "hydrophobic patch" on one side of the protein, as opposed to a hydrophilic patch on the other side. In this respect, hydrophobins exhibit overall structural similarities to low molecular weight surfactants.

Class II hydrophobins I and II from *T. reesei* were investigated for their interfacial properties and it was found that surface tensions as low as 25 mN · m^{-1} could be achieved around pH 5.0, that is, close to the IEP of the proteins [88]. Interestingly enough, interfacial shear elasticity moduli for 1–10 μM hydrophobin I and II dispersions upon full protein adsorption reached 1,000 mN · m^{-1}, which is 1–5 orders of magnitude higher than the best described globular proteins such as in whey or soy. Upon foaming, the obtained air bubbles were extremely stable against disproportionation for several hours as compared to β-casein of β-lactoglobulin stabilized bubbles that shrunk after several minutes [88]. Light microscopy images of the air bubbles revealed a very highly structured and rigid interface that was able to burst upon increasing pressure, but that kept its shape after bursting, indicating a very

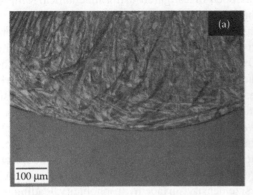
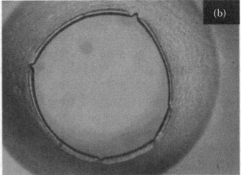

FIGURE 14.4 Optical images of bubbles and collapsing bubbles made using a 0.7 mM solution of HFBII. (a) Shows the surface of a collapsed bubble that has been forced to burst. (b) Shows the top surface of a burst bubble and the stabilizing film present. Scale bar is 100 μm. (Reprinted with permission from Cox, A. et al. *Langmuir* 2007;23:7995–8002. Copyright 2007 American Chemical Society [88].)

high rigidity (Figure 14.4). Class II hydrophobins II were used to produce aqueous foams at a protein concentration of 0.1 wt% in the presence of xanthan as a bulk viscosifying agent and air bubbles were shown to noticeably resist disproportionation for periods up to 4 months at room temperature. One specific foam showed an impressive stability over 28 months in chilled air, with a reduced loss of volume (10%) compared to fresh foam [89]. The ability for hydrophobins to form such stable foams was explained by the specific 2D structure formed by the protein at the interface, leading to a highly viscoelastic modulus over a wide range of deformation. This enables hydrophobin-stabilized air bubbles to easily resist high internal Laplace pressure [90]. Further investigation of the structure of interfacial layers of hydrophobins revealed a thickness of 12 nm, corresponding to 2 protein layers (S-bilayers) separated by a layer of water molecules. Upon compression, the interface was able to wrinkle, forming sandwiches of hydrophobin layers (up to tetramers) which enables these layers to reach a tension free layer and to overcome Laplace pressure effects, explaining the high stability of bubbles and foams [91–93]. Recently, studies have combined mixtures of hydrophobins together with milk proteins such as β-casein, β-lactoglobulin or caseinate. It was shown that depending on the weight mixing ratio, the shear viscosity of the mixed film increased while its elasticity remained constant. This led to an improved resistance of air bubbles to Laplace pressure [94,95].

14.4 PROTEIN AGGREGATES

As shown above in the case of elastic hydrophobins interfaces or viscous micellar casein multilayers, it is important to maintain a contribution from surface elasticity and/or from bulk viscoelasticity in order to reduce disproportionation rates in protein foams [96]. Such a combination of interfacial and bulk stabilization effects can also be achieved by the use of protein aggregates obtained in conditions where proteins are partially or fully denatured and hold together via noncovalent and/or covalent bonds. In this respect, protein aggregates can be viewed as a peculiar type of polymeric particles [97]. Protein aggregates can be obtained by several ways, but the most common one is to heat a dispersion of native proteins above their denaturation temperature in order to promote their three dimensional aggregation [15]. Depending on the balance between attractive/repulsive forces in the system (set by pH and ionic strength) as well as protein concentration, aggregates of various size, shape, charge, flexibility, and density can be obtained [98]. The most commonly used protein aggregates are made of β-lactoglobulin or whey protein isolate. In this case, it was shown that heating at very low pH (<2.5) led to fibrillar aggregates, while spherical aggregates were obtained close to IEP (4.0 < pH < 6.0) and self-similar fractal aggregates close to neutrality (pH > 6.5) [99]. Whey protein aggregates were generated at 10 wt% at pH 7.0 and the impact of native/ aggregated ratio on foamability and stability has been investigated using mechanical whipping [100]. Protein aggregation markedly increased bulk viscosity from 2 to 300 mPas, which in turn drastically reduced the ability of protein aggregates to lower surface tension due to strong hindering of diffusion/adsorption to the interface in a short time. Interestingly, an optimum of foamability/ foam stability was determined for a mixture of 80% protein aggregates/20% native protein. In these conditions, native proteins promoted high foamability due to rapid interfacial adsorption while aggregates reduced strong drainage through increased bulk viscosity [100]. Very comparable results were obtained on similar systems using membrane foaming apparatus, indicating that the foaming stabilization properties of protein aggregates were moderately affected by shear stress applied during the aeration phase [101]. Unterhaslberger and coworkers [102] reported a similar optimum ratio (20:80) between native and aggregated β-lactoglobulin at 1.0 wt% and pH 7.0, but controlled aggregation via ionic strength modulation. Interestingly enough, the stabilizing properties of the aggregates seems to be fairly controlled by their size/density/surface hydrophobicity, leading to an interfacial elasticity of 50 mN · m^{-1} and a marked reduction of air bubble disproportionation rate [102].

Influence of size, density and surface hydrophobicity of whey proteins was investigated upon heating 1 wt% whey protein isolate solution in the pH range 6.0–7.0 with or without NaCl in order to maximize the amount or aggregate formed and reduced the amount of nonaggregated proteins [104]. The most efficient aggregates regarding foam formation and stability were obtained at pH 6.8 in presence of 0.1 M NaCl. Thus, in these conditions, 90% of the proteins were in aggregated form, while 10% proteins kept similar properties to native whey proteins, that is, faster adsorption speed due to their globular and small size. The fractal aggregates formed at pH 6.8 were characterized by a low surface charge but a relatively high surface hydrophobicity. The resulting films were characterized by low gas permeability values and the foams by slow drainage and disproportionation rates [104]. Surface properties of fractal ($d_f = 2.0$) whey protein aggregates were further investigated by varying their size from 35 to 200 nm and mixing them with native proteins [105]. Interestingly, no stable foams were obtained for pure aggregate solutions at 0.1 wt%. The presence of at least 4% native proteins was necessary to reduce surface tension by rapid adsorption followed by subsequent stabilization by aggregates. The most volume stable foams were observed for the smallest protein aggregates while the less dry ones were those produced with the largest aggregates which could be explained by the presence of a disjoining pressure in the thin film created by protein 2D aggregation. This assumption was confirmed by the study of the structure of foam films which indicated the formation of a gelled network in the thin film, enabling the air bubble to withstand high pressures (up to 100 Pa) for several cycles as shown in Figure 14.5 for 35 nm aggregates [103,106]. Spherical and fractal whey protein aggregates were shown to be able to deform at the air/water interface using atomic force microscopy (AFM) imaging, leading to the formation of viscoelastic layers with dilatational elasticity in the order of 60 mN · m^{-1} [107,108]. Recently, the foaming properties of spherical whey protein microgels of 200 nm in diameter were investigated as a function of pH [109]. From the high protein concentration of 5 wt% needed to build a stable foam and the high foam stabilizing properties of these aggregates obtained at their IEP of 5.0, it was concluded that mostly bulk aggregation of these particles was responsible for the low drainage rate and air bubble stability, as reported for synthetic particles [97]. The presence of large insoluble whey protein aggregates was shown to be detrimental for air bubble stability if their size was larger than 100 μm and their concentration higher than 5% in the mix with native proteins. This effect was attributed to the breaking of the thin film between air bubbles due to the large size of the aggregates [102,110,111]. Recently, fibrillar micrometric aggregates obtained by submitting micellar casein/whey protein mixtures to high pressure-low temperature treatment were shown to efficiently stabilize foams, likely due to their high bulk viscosifying properties [112]. The foaming properties of 250 μm whey protein gel particles obtained after freeze-drying of a 10 wt% heat set gel at 10 wt% and pH 8.0 were shown to be very high, especially their liquid stability [113]. Confocal scanning laser microscopy revealed that whey gel particles were fully incorporated in

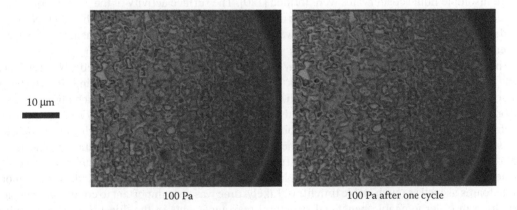

FIGURE 14.5 Photos taken from a foam film containing 96% of β-lactoglobulin aggregates (R_h = 35 nm) (bulk concentration 1 g/L) for a pressure cycle 100 Pa–0 Pa–100 Pa. Scale bar is 10 μm. (Reprinted with permission from Rullier, B. et al. *J. Colloid Interface Sci* 2009;336:750–755 [103].)

the thin film surrounding air bubbles and ensured stabilization via high bulk viscosity. It should be finally noticed that in complex systems such as ice cream, the use of aggregated whey protein obtained by controlled heat treatment was beneficial to air bubble stability upon storage due to more homogeneous air bubble size distribution [114].

14.5 PROTEIN-POLYSACCHARIDE COMPLEXES

In this last section, the foaming properties of electrostatic complexes between protein and polysaccharide will be discussed. Thus, as discussed before, the use of a single protein might not be sufficient to ensure long term stability of protein foams unless proper protein aggregates are used. A potential approach to nonaggregated proteins use could be to mx them with charged polysaccharides in order to combine the surface activity of the protein and the bulk properties of the polysaccharide [32,115]. The interaction of β-lactoglobulin with sulfated polysaccharide at pH 7.0 was shown to slightly increase the surface viscoelasticity of the film compared to protein alone [116]). However, because the profile of surface pressure relatively close to that of the pure protein, it was concluded that the interaction with the polysaccharide was occurring with an already adsorbed protein layer. This scenario was investigated in a mixture of β-lactoglobulin and pectin at various pHs (4.0–7.0) in order to promote complex formation, before or after protein adsorption [117]. In general, the surface shear modulus of sequentially adsorbed layers of protein and pectin was larger than for simultaneously adsorbing complexes. The likely explanation was that the pectin chains were reinforcing the already adsorbed protein layer via specific electrostatic interactions with the protein. The interfacial properties of already formed complexes were dependent on the mixing ratio, the most surface active complexes being obtained for charge neutrality of the complexes [117]. The structure of both types of protein/polysaccharide interfacial layers was elucidated by neutron reflectivity and time resolved fluorescence anisotropy [118]. Indeed, it was found that negatively charged complexes adsorbed much less at the interface compared to neutral ones, the adsorbed amounts being 4 and 14 mg $\cdot$ m^{-2}, respectively. Interestingly, the thickness of the adsorbed layers was also very different, neutral complexes being able to become much thinner (20 nm), but consequently more packed layers of negatively charged complexes (50 nm layers) for which there were chain repulsions explained the less dense packing. Improvement of foam stabilization has been described in mixtures of rapeseed globulin 2S and napin, together with pectins with various charge density [119]. If foamability was mainly due to free protein molecules, thick layers of 30 nm composed of complexes enables to maintain a high level of liquid within the foams and to prevent long term coarsening. A detailed study reported the impact of complex formation between β-lactoglobulin and acacia gum in conditions where insoluble liquid coacervates were formed [120]. The surface activity of the neutral complexes was equivalent to the protein alone, but if the dilatational elasticities were comparable (55 mN $\cdot$ m^{-1}), the dilatational viscosity of the complexes was almost double, with a strong frequency dependence indicating structural rearrangements.

The structure of the thin film stabilized by the β-lactoglobulin at pH 4.2 revealed a Newton black film, while for the complexes, a very thick film entrapping particles and forming an interfacial network was observed. The resistance of such bubbles to disproportionation was much higher than for the protein film (Figure 14.6) and consequently, the corresponding gas permeability coefficients were much lower. Additionally, for air bubble stability, the presence of protein/polysaccharide complexes was shown to reduce thin film drainage. The importance of time of the stabilization properties of the β-lactoglobulin/acacia gum system was further highlighted by comparing systems with freshly formed complexes and 24 hours after complex formation [121]. A fresh system led to much more stable foams as well as much stable thin films. Likely, time was very important to controlled complex formation in order to enable beneficial structural rearrangements in the thin films. This mobility of liquid/molecules was lost after 24 hours, leading to interfacial films that were not elastic enough and were too permeable to gas to prevent air bubble disproportionation. Such control of time on the foaming properties of complexes was successfully applied to complex food systems such as ice cream

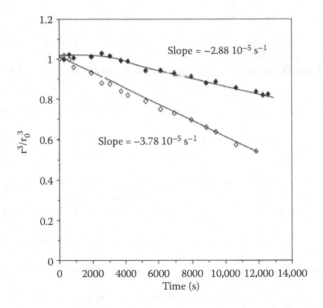

FIGURE 14.6 Time evolution of the normalized size of a single air bubble, r^3/r_0^3, stabilized by 0.1 wt% β-lactoglobulin or 0.1 wt% β-lactoglobulin/acacia gum complexes at pH = 4.2, ratio 2:1, and 25°C: (empty diamonds), β-lactoglobulin; (full diamonds) β-lactoglobulin/acacia gum complexes. (Reprinted with permission from Schmitt, C. et al. Interfacial and foam stabilization properties of β-lactoglobulin-acacia gum electrostatic complexes. In: Dickinson, E. (Ed.), *Food Colloids: Interactions, Microstructure and Processing.* Royal Society of Chemistry: Cambridge, 2005, pp. 284–300.. Copyright 2005 Royal Society of Chemistry [120].)

or acid dessert mousses for which the process of complex formation and foaming stages were well controlled on a pilot line [122].

14.6 CONCLUSIONS

From the nonexhaustive review of the developments on protein stabilized foams during the last 15 years, we can conclude that better understanding has been generated on the effects of structure and composition for the major food proteins, that is, caseins, whey, egg, and soy proteins. It should be noticed, however, that none of these sources is able to generate shelf-stable foams even if their foamability is high. A noticeable exception to that list of proteins is fungi hydrophobins. They exhibit very specific foam stabilizing properties owing to the specific particle-like structure formed at the air/water interface and leading to the most highly viscoelastic layers reported for proteins. The main limitation for the wide use of these proteins in food applications is their current availability and consumer acceptance. An interesting way to improve the stabilization properties of globular proteins is to induce their controlled aggregation in order to generate protein particles which are able to build thicker and more resistant viscoelastic networks at interfaces. One should, however, take care to combine these aggregates with a fraction of native proteins in order to maintain sufficient foamability. Complexes between proteins and charges polysaccharides could be another technical alternative to generate stable foams by combining the surface activity of proteins and the bulk stabilizing properties of the polysaccharide. Here, the main limitations reside in the conditions necessary to form the complexes which might not be compatible with all final applications. The latest developments in the screening of novel plant protein sources might help to identify new proteins with adequate structural and functional properties suitable for a larger range of applications. Nevertheless, coming back to the introductory remark on the natural foaming system developed by the tropical frog, it seems essential that (food) chemists and physicists consider efficient "cocktails" of native/aggregated/complexed/glycated mixtures of proteins to solve the technical challenge of designing and delivering stable foams.

REFERENCES

1. Schwenke, K.D. Proteins: Some principles of classification and structure. In: Möbius, D., Miller, R. (Eds.), *Proteins at Liquid Interfaces*. Elsevier: Amsterdam, 1998, pp. 1–50.
2. Wilde, P.J. Interfaces: Their role in foam and emulsion behaviour. *Curr Opin Colloid Interface Sci* 2000;5:176–181.
3. Damodaran, S. Protein stabilisation of emulsions and foams. *J Food Sci* 2005;70:54–66.
4. Wierenga, P.A., Gruppen, H. New views on foams from protein solutions. *Curr Opin Colloid Interface Sci* 2010;15:365–373.
5. Möbius, D., Miller, R. *Proteins at Liquid Interfaces*, Vol. 7. Elsevier: Amsterdam, 1998.
6. Campbell, G.M., Mougeot, E. Creation and characterisation of aerated food products. *Trends Food Sci Technol* 1999;10:283–296.
7. Damodaran, S. Functional properties. In: Nakai, S., Modler, H.W. (Eds.), *Food Proteins. Properties and Characterization*. VCH Publishers Inc.: New York, 1996, pp. 167–234.
8. Townsend, A., Nakai, S. Correlations between hydrophobicity and foaming capacity of proteins. *J Food Sci* 1983;48:588–594.
9. Siebert, K.J. Modeling protein functional properties from amino acid composition. *J Agric Food Chem* 2003;51:7792–7797.
10. Wierenga, P.A., van Norel, L., Basheva, E.S. Reconsidering the importance of interfacial properties in foam stability. *Colloids Surf A* 2009;344:72–78.
11. Foegeding, E.A., Luck, P.J., Davis, J.P. Factors determining the physical properties of protein foams. *Food Hydrocoll* 2006;20:284–292.
12. Kilara, A., Harwalkar, V.R. Denaturation. In: Nakai, S., Modler, H.W. (Eds.), *Food Proteins. Properties and Characterization*. VCH Publishers Inc.: New York, 1996, pp. 71–165.
13. Dickinson, E. Proteins at interfaces and in emulsions. Stability, rheology and interactions. *J Chem Soc Faraday Trans* 1998;94:1657–1669.
14. Linder, M.B. Hydrophobins: Proteins that self assemble at interfaces. *Curr Opin Colloid Interface Sci* 2009;14:356–363.
15. Mezzenga, R., Fischer, P. The self-assembly, aggregation and phase transitions of food protein systems in one, two and three dimensions. *Rep Prog Phys* 2013;76:046601. (43pp).
16. van der Ven, C., Gruppen, H., de Bont, D.B.A., Voragen, A.G.J. Correlations between biochemical characteristics and foam-forming and -stabilizing ability of whey and casein hydrolysates. *J Agric Food Chem* 2002;50:2938–2946.
17. Davis, J.P., Doucet, D., Foegeding, E.A. Foaming and interfacial properties of hydrolyzed β-lactoglobulin. *J Colloid Interface Sci* 2005;288:412–422.
18. Hiller, B., Lorenzen, P.C. Surface hydrophobicity of physiologically and enzymatically treated milk proteins in relation to techno-functional properties. *J Agric Food Chem* 2008;56:461–468.
19. Pizones Ruiz-Henestrosa, V., Carrera Sanchez, C., Pedroche, J.J., Millan, F., Rodriguez Patino, J.M. Improving the functional properties of soy glycinin by enzymatic treatment. Adsorption and foaming characteristics. *Food Hydrocoll* 2009;23:377–386.
20. Raikos, V. Enzymatic hydrolysis of milk proteins as a tool for modification of functional properties at interfaces of emulsions and foams—A review. *Curr Nutr Food Sci* 2014;10:134–140.
21. Dickinson, E. Hydrocolloids as emulsifiers and emulsion stabilizers. *Food Hydrocoll* 2009;23:1473–1482.
22. Bos, M.A., Van Vliet, T. Interfacial rheological properties of adsorbed protein layers and surfactants: A review. *Adv. Colloids Interface Sci* 2001;91:437–471.
23. Rodriguez Patino, J.M., Carrera Sanchez, C., Rodriguez Nino, M.R. Implications of interfacial characteristics of food foaming agents in foam formulations. *Adv Colloid Interface Sci* 2008;140:95–113.
24. Cooper, A., Kennedy, M.W., Fleming, R.I., Wilson, E.H., Videler, H., Wokosin, D.L., Su, T., Green, R.J., Lu, J.R. Adsorption of frog foam nest proteins at the air-water interface. *Biophys J* 2005;88:2114–2125.
25. Fleming, R.I., Mackenzie, C.D., Cooper, A., Kennedy, M.W. Foam nest components of the tungara frog: A cocktail of proteins conferring physical and biological resilience. *Proc R Soc B* 2009;276:1787–1795.
26. Saint-Jalmes, A., Peugeot, M.-L., Ferraz, H., Langevin, D. Differences between protein and surfactant foams: Microscopic properties, stability and coarsening. *Colloids Surf A* 2005;263:219–225.
27. Swaisgood, H.E. Chemistry of the caseins. In: Fox, P.F., (Ed.), *Advanced Dairy Chemistry—1. Proteins*. Blackie Academic & Professional: London, 1992, pp. 63–110.
28. Rollema, H.S. Casein association and micelle formation. In: Fox, P.F. (Ed.), *Advanced Dairy Chemistry—1: Proteins*. Blackie Academic & Professional: London, 1992, pp. 111–140.

29. Horne, D.S. Casein micelle structure: Models and mubbles. *Curr Opin Colloid Interface Sci* 2006;11:148–153.

30. Dalgleish, D.G. On the structural models of bovine casein micelles—Review and possible improvements. *Soft Matter* 2011;7:2265–2272.

31. HadjSadok, A., Pitkowski, A., Nicolai, T., Benyahia, L., Moulai Mostefa, N. Characterisation of sodium caseinate as a function of ionic strength, pH and temperature using static and dynamic light scattering. *Food Hydrocoll* 2008;22:1460–1466.

32. Dickinson, E. Mixed biopolymers at interfaces: Competitive adsorption and multilayer structures. *Food Hydrocoll* 2011;25:1966–1983.

33. Sengupta, T., Damodaran, S. Incompatibility and phase separation in a bovine serum albumin/β-casein/water ternary film at the air/water interface. *J Colloid Interface Sci* 2000;229:21–28.

34. Sengupta, T., Damodaran, S. Lateral phase separation in adsorbed binary protein films at the air-water interface. *J Agric Food Chem* 2001;49:3087–3091.

35. Zhang, Z., Dalgleish, D.G., Goff, H.D. Effect of pH and ionic strength on competitive protein adsorption to air/water interfaces in aqueous foams made with mixed proteins. *Colloids Surf. B: Biointerfaces* 2004;34:113–121.

36. Bos, M.A., Dunnewind, B., van Vliet, T. Foams and surface rheological properties of β-casein, gliadin and glycinin. *Colloids and Surfaces B: Biointerfaces* 2003;31:95–105.

37. Caessens, P.W.J.R., de Jongh, H.H.J., Norde, W., Gruppen, H. The adsorption-induced secondary structure of β-casein and of distinct parts of its sequence in relation to foam and emulsion properties. *Biochim Biophys Acta* 1999;1430:73–83.

38. Barackov, I., Mause, A., Kapoor, S., Winter, R., Schembecker, G., Burghoff, B. Investigation of structural changes of β-casein and lysozyme at the gas-liquid interface during foam fractionnation. *J Biotechnol* 2012;161:138–146.

39. Saint-Jalmes, A., Marze, S., Langevin, D. Coarsening and rheology of casein and surfactant foams. In: Dickinson, E. (Ed.), *Food Colloids: Interactions, Microstructure and Processing*. Royal Society of Chemistry: Cambridge, 2005, pp. 273–283.

40. Maldonado-Valderrama, J., Langevin, D. On the difference between foams stabilized by surfactants and whole casein or β-casein. Comparison of foams, foam films, and liquid surfaces studies. *J Phys Chem B* 2008;112:3989–3996.

41. Kamath, S., Webb, R.E., Deeth, H.C. The composition of interfacial material from skim milk foams. *J Dairy Sci* 2011;94:2707–2718.

42. Carrera Sanchez, C., Rodriguez Patino, J.M. Interfacial, foaming and emulsifying characteristics of sodium caseinate as influenced by protein concentration in solution. *Food Hydrocoll* 2005;19: 407–416.

43. Marinova, K.G., Basheva, E.S., Nenova, B., Temelska, M., Mirarefi, A.Y., Campbell, B., Ivanov, I.B. Physico-chemical factors controlling the foamability and foam stability of milk proteins: Sodium caseinate and whey protein concentrates. *Food Hydrocoll* 2009;23:1864–1876.

44. Huppertz, T. Foaming properties of milk: A review of the influence of composition and processing. *Int J Dairy Technol* 2010;63:477–488.

45. Borcherding, K., Lorenzen, P.C.H.R., Hoffmann, W. Effect of protein content, casein-whey protein ratio and pH value on the foaming of skimmed milk. *Int J Dairy Technol* 2009;62:161–169.

46. Nogueira Silva, N., Piot, M., Fernandes de Carvalho, A., Violleau, F., Fameau, A.-L., Gaucheron, F. pH-induced demineralization of casein micelles modifies their physico-chemical and foaming properties. *Food Hydrocoll* 2013;32:322–330.

47. Zhang, Z., Goff, H.D. Protein distribution at air interfaces in dairy foams and ice cream as affected by casein dissociation and emulsifiers. *Int Dairy J* 2004;14:647–657.

48. Kinsella, J.E. Milk proteins: physicochemical and functional properties. *Crit Rev Food Sci Nutr* 1984;21:197–262.

49. Hambling, S.G., McAlpine, A.S., Sawyer, L. β-lactoglobulin. In: Fox, P.F. (Ed.), *Advanced Dairy Chemistry*. Elsevier Science Publishers Ltd: London, 1992, pp. 141–190.

50. Brew, K., Grobler, J.A. α-lactalbumin. In: Fox, P.F. (Ed.), *Advanced Dairy Chemistry*. Elsevier Science Publishers Ltd.: London, 1992, pp. 191–229.

51. Anton, M., Nau, F., Lechevallier, V. Egg proteins. In: Phillips, G.O., Williams, P.A. (Eds.), *Hanbook of Hydrocolloids*, Second Edition. Woodhead Publishing Ltd.: Oxford, 2009, pp. 359–382.

52. Gonzalez-Perez, S., Arellano, J.B. Vegetable protein isolates. In: Phillips, G.O., Williams, P.A. (Eds.), *Handbook of Hydrocolloids*. Second edition. Woodhead Publishing Ltd.: Oxford, 2009, pp. 383–419.

53. Engelhardt, K., Lexis, M., Gochev, G., Konnerth, C., Miller, R., Willenbacher, N., Peukert, W., Braunschweig, B. pH effects on the molecular structure of β-lactoglobulin modified air-water interfaces and its impact on foam rheology. *Langmuir* 2013;29:11646–11655.

54. Petkova, V., Sultanem, C., Nedyalkov, M., Benattar, J.-J., Leser, M.E., Schmitt, C. Structure of a freestanding film of β-lactoglobulin. *Langmuir* 2003;19:6942–6949.

55. Xu, R., Dickinson, E., Murray, B.S. Morphological changes in adsorbed protein films at the air-water interface subjected to large area variations, as observed by Brewster angle microscopy. *Langmuir* 2007;23:5005–5013.

56. Dickinson, E., Ettelaie, R., Murray, B.S., Du, Z. Kinetics of disproportionation of air bubbles beneath a planar air-water interface stabilized by food proteins. *J Colloid Interface Sci* 2002;252:202–213.

57. Sharma, V., Jaishankar, A., Wang, Y.-C., McKinley, G.H. Rheology of globular proteins: Apparent yield stress, high shear rate viscosity and interfacial viscoelasticity of bovine serum albumin solutions. *Soft Matter* 2011;7:5150–5160.

58. Lexis, M., Willenbacher, N. Relating foam and interfacial rheological properties of β-lactoglobulin solutions. *Soft Matter* 2014;10:9626–9636.

59. Davis, J.P., Foegeding, E.A., Hansen, F.K. Electrostatic effects on the yield stress of whey protein isolate foams. *Colloids Surf. B: Biointerfaces* 2004;34:13–23.

60. Rami-Shojaei, S., Vachier, C., Schmitt, C. Automatic analysis of 2D foam sequences. Application to the characterization of aqueous proteins foams stability. *Image Vis Comput* 2009;27:609–622.

61. Croguennec, T., Renault, A., Bouhallab, S., Pezennec, S. Interfacial and foaming properties of sulfydryl-modified bovine β-lactoglobulin. *J Colloid Interface Sci* 2006;302:32–39.

62. Rade-Kukic, K., Rawel, H., Schmitt, C. Formation of conjugates between β-lactoglobulin and allyl isothiocyanate: Effect on protein heat aggregation, foaming and emulsifying properties. *Food Hydrocoll* 2011;25:694–706.

63. Kim, D.A., Cornec, M., Narsimhan, G. Effect of thermal treatment on interfacial properties of β-lactoglobulin. *J Colloid Interface Sci* 2005;285:100–109.

64. Cornec, M., Kim, D.A., Narsimhan, G. Adsorption dynamics and interfacial properties of α-lactalbumin in native and molten globule state conformation at air-water interface. *Food Hydrocoll* 2001;15:303–313.

65. Ibanoglu, E., Ibanoglu, S. Foaming behaviour of EDTA-treated α-lactalbumin. *Food Chem* 1999;66:477–481.

66. Luck, P.J., Foegeding, E.A. The role of copper in protein foams. *Food Biophys* 2008;3:255–260.

67. Luck, P.J., Bray, N., Foegeding, E.A. Factors determining yield stress and overrun of whey protein foams. *J Food Sci* 2001;67:1677–1681.

68. Renault, A., Pezennec, S., Gauthier, F., Vié, V., Desbat, B. Surface rheological properties of native and S-ovalbumin are correlated with the development of an intermolecular β-sheet network at the air-water interface. *Langmuir* 2002;18:6887–6895.

69. Murray, B.S., Dickinson, E., Du, Z., Ettelaie, R., Kostakis, T., Vallet, J. Disproportionation kinetics of air bubbles stabilized by food proteins and nanoparticles. In: Dickinson, E. (Ed.), *Food Colloids: Interactions, Microstructure and Processing*. Royal Societry of Chemistry: Cambridge, 2005, pp. 259–272.

70. Lechevalier, V., Croguennec, T., Pezennec, S., Guérin-Dubiard, C., Pasco, M., Nau, F. Evidence for synergy in the denaturation at the air-water interface of ovalbumin, ovotransferrin and lysozyme in ternary mixture. *Food Chem* 2005;92:79–87.

71. Le Floch-Fouéré, C., Pezennec, S., Lechevalier, V., Beaufils, S., Desbat, B., Pézolet, M., Renault, A. Synergy between ovalbumin and lysozyme leads to non-additive interfacial and foaming properties mixtures. *Food Hydrocoll* 2009;23:352–365.

72. Alahverdjieva, V.S., Grigoriev, D.O., Ferri, J.K., Fainerman, V.B., Aksenenko, E.V., Leser, M.E., Michel, M., Miller, R. Adsorption behavior of hen egg-white lysozyme at the air/water interface. *Colloids Surf A* 2008;323:167–174.

73. Hagolle, N., Relkin, P., Popineau, Y., Bertrand, D. Study of the stability of egg white protein-based foams: effect of heating protein solution. *J Sci Food Agric* 2000;80:1245–1252.

74. Yang, X., Foegeding, E.A. The stability and physical properties of egg white and whey protein foams explained based on microstructure and interfacial properties. *Food Hydrocoll* 2011;25:1687–1701.

75. Davis, J.P., Foegeding, E.A. Comparisons of the foaming and interfacial properties of whey protein isolate and egg white proteins. *Colloids Surf. B* 2007;54:200–210.

76. Kuropatwa, M., Tolkach, A., Kulozik, U. Impact of pH on the interactions between whey and egg white proteins as assessed by the foamability of their mixtures. *Food Hydrocoll* 2009;23:2174–2181.

77. Martin, A.H., Bos, M.A., van Vliet, T. Interfacial rheological properties and conformational aspects of soy glycinin at the air/water interface. *Food Hydrocoll* 2002;16:63–71.
78. Martin, A.H., Meinders, M.B.J., Bos, M.A., Cohen Stuart, M.A., van Vliet, T. Conformational aspects of proteins at the air/water interface studied by infrared reflection-absorption spectroscopy. *Langmuir* 2003;19:2922–2928.
79. Martin, A.H., Cohen-Stuart, M.A., Bos, M.A., van Vliet, T. Correlation between mechanical behavior of protein films at the air/water interface and the intrinsic stability of protein molecules. *Langmuir* 2005;21:4083–4089.
80. Meinders, M.B.J., Bos, M.A., Lichtendonk, W.J., van Vliet, T. Effect of stress relaxation in soy glycinin films on bubble dissolution and foam stability. In: Dickinson, E., Miller, R. (Eds.), *Food Colloids. Biopolymers and Materials*. Royal Society of Chemistry: Cambridge, 2003, pp. 156–164.
81. Pizones Ruiz-Henestrosa, V., Carrera Sanchez, C., Yust Escobar, M.d.M., Pedroche Jimenez, J.J., Millan Rodriguez, F., Rodriguez Patino, J.M. Interfacial and foaming characteristics of soy globulins as a function of pH and ionic strength. *Colloids Surf A: Physicochem Eng Aspects* 2007;309:202–215.
82. Pizones Ruiz-Henestrosa, V., Carrera Sanchez, C., Rodriguez Patino, J.M. Effect of sucrose on the functional properties of soy globulins: Adsorption and foam characteristics. *J Agric Food Chem* 2008;56:2512–2521.
83. Wagner, J.R., Guéguen, J. Surface functional properties of native, acid-treated, and reduced soy glycinin. 1. Foaming properties. *J Agric Food Chem* 1999;47:2173–2180.
84. Wan, Z.-L., Wang, L.-Y., Wang, J.-M., Yuan, Y., Yang, X.-Q. Synergistic foaming and surface properties of a weakly interacting mixture of soy glycinin and biosurfactant stevioside. *J Agric Food Chem* 2014;62:6834–6843.
85. Medrano, A., Abirached, C., Araujo, A.C., Panizzolo, L.A., Moyna, P., Anon, M.C. Correlation of average hydrophobicity, water/air interface rheological properties and foaming properties of proteins. *Food Sci Technol Int* 2012;18:187–193.
86. Pizones Ruis-Henestrosa, V.M., Martinez, M.J., Carrera Sanchez, C., Rodriguez Patino, J.M., Pilosof, A.M.R. Mixed soy globulins and β-lactoglobulin system behaviour in aqueous solutions and at the air-water interface. *Food Hydrocoll* 2014;35:106–114.
87. Murray, B.S. Stabilization of bubbles and foams. *Curr Opin Colloid Interface Sci* 2007;12:232–241.
88. Cox, A., Cagnol, F., Russell, A.B., Izzard, M.J. Surface properties of class II hydrophobins from Trichoderma reesi and influence on bubble stability. *Langmuir* 2007;23:7995–8002.
89. Cox, A.R., Aldred, D.L., Russell, A.B. Exceptional stability of food foams using class II hydrophobin HFBII. *Food Hydrocoll* 2009;23:366–376.
90. Blijdenstein, T.B.J., de Groot, P.W.N., Stoyanov, S.D. On the link between foam coarsening and surface rheology: Why hydrophobins are so different. *Soft Matter* 2010;6:1799–1808.
91. Basheva, E.S., Kralchevsky, P.A., Danov, K.D., Stoyanov, S.D., Blijdenstein, T.B.J., Pelan, E.G., Lips, A. Self-assembled bilayers from the protein HFBII hydrophobin: Nature of the adhesion energy. *Langmuir* 2011;27:4481–4488.
92. Basheva, E.S., Kralchevsky, P.A., Christov, N.C., Danov, K.D., Stoyanov, S.D., Blijdenstein, T.B.J., Kim, H.-J., Pelan, E.G., Lips, A. Unique properties of bubbles and foam film stabilized by HFBII hydrophobin. *Langmuir* 2011;27:2382–2392.
93. Stanimirova, R., Gurkov, T.D., Kralchevsky, P.A., Balashev, K.T., Stoyanov, S.D., Pelan, E.G., Surface pressure and elasticity of hydrophobin HFBII layers on the air-water interface: Rheology versus structure detected by AFM imaging. *Langmuir* 2013;29:6053–6067.
94. Wang, Y., Bouillon, C., Cox, A., Dickinson, E., Durga, K., Murray, B.S., Xu, R. Interfacial study of class II hydrophobin and its mixtures with milk proteins: Relationship to bubble stability. *J Agric Food Chem* 2013;61:1554–1562.
95. Burke, J., Cox, A., Petkov, J., Murray, B.S. Interfacial rheology and stability of air bubbles stabilized by mixtures of hydrophobin and β-casein. *Food Hydrocoll* 2014;34:119–127.
96. Kloek, W., Van Vliet, T., Meinders, M. Effect of bulk and interfacial rheological properties on bubble dissolution. *J Colloid Interface Sci* 2001;237:158–166.
97. Dickinson, E. Food emulsions and foams: Stabilization by particles. *Curr Opin Colloid Interface Sci* 2010;15:40–49.
98. Saglam, D., Venema, P., van der Linden, E., de Vries, R. Design, properties, and applications of protein micro- and nanoparticles. *Curr Opin Colloid Interface Sci* 2014;19:428–437.
99. Nicolai, T., Britten, M., Schmitt, C. β-lactoglobulin and WPI aggregates: Formation, structure and applications. *Food Hydrocoll* 2011;25:1945–1962.

100. Davis, J.P., Foegeding, E.A. Foaming and interfacial properties of polymerized whey protein isolate. *J Food Sci* 2004;69:404–410.

101. Bals, A., Kulozik, U. Effect of pre-heating on the foaming properties of whey protein isolate using a membrane foaming apparatus. *Int Dairy J* 2003;13:903–908.

102. Unterhaslberger, G., Schmitt, C., Shojaei-Rami, S., Sanchez, C. β-lactoglobulin aggregates from heating with charged cosolutes: formation, characterization and foaming. In: Dickinson, E., Leser, M.E. (Eds.), *Food Colloids: Self-Assembly and Material Science*. Royal Society of Chemistry: Cambridge, 2007, pp. 175–192.

103. Rullier, B., Axelos, M.A.V., Langevin, D., Novales, B. β-lactoglobulin aggregates in foam films: Correlation between foam films and foaming properties. *J Colloid Interface Sci* 2009;336:750–755.

104. Schmitt, C., Bovay, C., Rouvet, M., Shojaei-Rami, S., Kolodziejczyk, E. Whey protein soluble aggregates from heating with NaCl: Physicochemical, interfacial, and foaming properties. *Langmuir* 2007;23:4155–4166.

105. Rullier, B., Novales, B., Axelos, M.A.V. Effect of protein aggregates on foaming properties of β-lactoglobulin. *Colloids Surf A* 2008;330:96–102.

106. Rullier, B., Axelos, M.A.V., Langevin, D., Novales, B. β-lactoglobulin aggregates in foam films: Effect of the concentration and size of the protein aggregates. *J Colloid Interface Sci* 2010;343:330–337.

107. Mahmoudi, N., Axelos, M.A.V., Riaublanc, A. Interfacial properties of fractal and spherical whey protein aggregates. *Soft Matter* 2011;7:7643–7654.

108. Mahmoudi, N., Gaillard, C., Boué, F., Axelos, M.A.V., Riaublanc, A. Self-similar assemblies of globular whey proteins at the air-water interface: Effect of the structure. *J Colloid Interface Sci* 2010;345:54–63.

109. Schmitt, C., Bovay, C., Rouvet, M. Bulk self-aggregation drives foam stabilization properties of whey protein microgels. *Food Hydrocoll* 2014;42:139–142.

110. Nicorescu, I., Loisel, C., Vial, C., Riaublanc, A., Djelveh, G., Cuvelier, G., Legrand, J. Combined effect of dynamic heat treatment and ionic strength on the properties of whey protein foams—Part II. *Food Res Int* 2008;41:980–988.

111. Nicorescu, I., Riaublanc, A., Loisel, C., Vial, C., Djelveh, G., Cuvelier, G., Legrand, J. Impact of protein self-assemblages on foam properties. *Food Res Int* 2009;42:1434–1445.

112. Baier, D., Schmitt, C., Knorr, D. Changes in functionality of whey protein and micellar casein after high pressure-low temperature treatments. *Food Hydrocoll* 2015;44:416–423.

113. Lazidis, A., Hancocks, R.D., Spyropoulos, F., Kreuss, M., Berrocal, R., Norton, I.T. Stabilization of foams by whey protein gel particles. In: Williams, P.A., Phillips, G.O. (Eds.), *Gums and Stabilisers for the Food Industry—17*. Royal Society of Chemistry: Cambridge, 2014, pp. 252–262.

114. Relkin, P., Sourdet, S., Smith, A.K., Goff, H.D., Cuvelier, G. Effect of whey protein aggregation on fat globule microstructure in whipped-frozen emulsions. *Food Hydrocoll* 2006;20:1050–1056.

115. Schmitt, C., Turgeon, S.L. Protein/polysaccharide complexes and coacervates in food systems. *Adv Colloid Interface Sci* 2011;167:63–70.

116. Baeza, R., Carrera Sanchez, C., Pilosof, A.M.R., Rodriguez Patino, J.M. Interactions of polysaccharides with β-lactoglobulin spread monolayers at the air-water interface. *Food Hydrocoll* 2004;18:959–966.

117. Ganzevles, R.A., Zinoviadou, K., van Vliet, T., Cohen Stuart, M.A., de Jongh, H.H.J. Modulating surface rheology by electrostatic protein/polysaccharide interactions. *Langmuir* 2006;22:10089–10096.

118. Ganzevles, R.A., Fokkink, R., van Vliet, T., Cohen Stuart, M.A., de Jongh, H.H.J. Structure of mixed β-lactoglobulin/pectin adsorbed layers at air/water interface; a spectroscopy study. *J Colloid Interface Sci* 2008;317:137–147.

119. Schmidt, I., Novales, B., Boué, F., Axelos, M.A.V. Foaming properties of protein/pectin electrostatic complexes and foam structure at nanoscale. *J Colloid Interface Sci* 2010;345:316–324.

120. Schmitt, C., Kolodziejczyk, E., Leser, M.E. Interfacial and foam stabilization properties of β-lactoglobulin-acacia gum electrostatic complexes. In: Dickinson, E. (Ed.), *Food Colloids: Interactions, Microstructure and Processing*. Royal Society of Chemistry: Cambridge, 2005, pp. 284–300.

121. Schmitt, C., Palma da Silva, T., Bovay, C., Rami-Shojaei, S., Frossard, P., Kolodziejczyk, E., Leser, M.E. Effect of time on the interfacial and foaming properties of β-lactoglobulin/acacia gum complexes and coacervates at pH 4.2. *Langmuir* 2005;21:7786–7795.

122. Schmitt, C., Kolodziejczyk, E. Protein-polysaccharide complexes: from basics to food applications. In: Williams, P.A., Phillips, G.O. (Eds.), *Gums and Stabilisers for the Food Industry 15*. Royal Society of Chemistry: Cambridge, 2010, pp. 211–221.

15 Foams Stabilized by Particles

Marcel Krzan, Agnieszka Kulawik-Pióro,
and Bożena Tyliszczak

CONTENTS

15.1 INTRODUCTION

Stabilization of aqueous foams or emulsions by various solid particles was noticed by researchers more than one hundred years ago [1,2]. However, during the last century, researchers mostly focused on emulsions. Therefore, the "stabilization of foams by particles" (also known as "particle laden foams") is a relatively new subject in this discipline of science. People are still gathering new data and try to find general rules describing the stability and rheology of "particle-laden foams". Hence, the influence of particle size, concentration, wettability, and many other factors influencing the foam parameters are studied now [3–9].

Flotation technology faced the problem of foam stabilized by particles more than a hundred years ago, but it was just considered to be a technological phenomenon. Froth bubbles carry the particles of the useful material outside the frothing reactor [3] (see also Chapter 21).

The situation changed at the end of the last century, when it was noticed that foam stabilized by particles could be used in various technological applications. It was connected to the foam stability established without using any surfactants [9]. The additional particle-laden foams parameters, like the noticeable yield stress (and flow of the whole foam fraction), also extended the numbers of applications. It is also important to note that particles help control the foam lifetime. In some foams, specially designed particles can be used, which initially stabilize them but later destroy them [10–13]. As an example, iron or iron oxide particles can be mentioned, which at the beginning can work as foam stabilizers, but later destroy the foam via the application of an external magnetic field. Carbon particles can also initially stabilize foam and later with ultraviolet radiation (UV) destroy it.

The problem of particle attachment to the interface is more difficult to understand than in the case of surfactants adsorption. Particles mostly exhibit neither surface activity nor foamability properties. As a result, the attachment of particles in the interfacial layer is not spontaneous like the adsorption of surfactant molecules [14] (see also Chapter 7).

In this chapter, we will discuss various problems connected with the influence of particles on the properties of foam. For this, measurements of surface tension or contact angle are not enough to understand foam stability and/or rheology. The aspects of drainage, coarsening, and coalescence mechanisms in the presence of particles with different sizes, shapes, charge, and materials have to be discussed, also taking into account solution conditions. Synergetic influences of the particles on foam rheology must be mentioned as well. It is completely clear that rules describing the evolution of classical foams (foams without particles) cannot be directly used for particle-stabilized foams. In each new experimental case, all synergetic interactions occurring between particles and foam films must be carefully studied and analyzed.

Before starting to present various methods to stabilize aqueous foams by particles, a few additional statements should be made. First, there is no generally accepted standard method for the generation of particle stabilized foams. Experience has clearly shown that with the same particle suspension, very stable or completely metastable foams can be formed, regardless of the applied foaming method. This is connected with different ways of gas dispersion, various sizes of the bubbles, or other unique and characteristic experimental parameters. Second, the methods of foam analysis (stability, rheology, etc.) are also not standardized. It should be also stressed that only in very few studies was the problem of foam investigated on all relevant levels, starting from the molecular adsorption of surfactants on the particles or liquid surfaces, through particle attachment and particle-particle mutual interactions before and after attachment in the foam film surface, up to the stability and rheology of a whole real foam.

15.2 FORMATION OF PARTICLE-LADEN FOAMS—CAPABILITIES AND LIMITATIONS

Typical hydrophobic (or partially hydrophobic) particles do not show surface active properties. Only partially hydrophobized nanoparticles under certain specific conditions could spontaneously attach to the interface and in this way diminish the surface tension of an aqueous suspension [14]. Moreover, solid particles in aqueous suspension are surrounded by an electrical double layer (EDL). This creates an additional energetic barrier hampering the coalescence and interactions with interfaces. In such a case, only strong external forces can lead to the formation of three phase contacts and the attachment of particles at the interface. Therefore, the generation of aqueous foams stabilized solely by solid particles requires strong mechanical methods, such as mechanical homogenization, shaking, turbulent mixing or whipping. As can be expected, such requirement significantly reduces the applicability of these foam systems.

Hydrophilic particles, which cannot create three phase contacts at interfaces, remain essentially in the bulk of the solution between the foam films, making the problems even more difficult. However, in such cases, the particles could have been hydrophobized before foam generation. This could be arranged permanently before introduction into the foaming solution or in the solution via the interactions with particular chemicals. Bubbling is a method for the generation of particle stabilized foams only in specific cases, mostly when the additional surfactants (foaming agents) are presented in the foaming solution. Additional methods will be described further below in this chapter.

15.2.1 EFFECT OF PARTICLE SIZE AND SHAPE

For "particle laden-foam" generation spherical nano- or micro-particles are most frequently used. Their degree of hydrophobization and size is crucial. Of course, surface heterogeneity and the density of the material of which the particle are made also have great importance. Stable thin foam films and consequently stable foam can be attained only under specific conditions, that is, when the particles are sufficiently strongly attached at the interface. A degree of hydrophobization given by a contact angle θ between 40° and 90° guarantees particles to attach at interfaces and does not destroy a foam film. The underlying details have been already described in the previous Chapter 7. Recently, modern experimental methodologies have been designed allowing us to exactly measure the contact angles describing the real position of single attached particles at interfaces, such as ellipsometry, Brewster angle microscopy and Scanning Electron Cryo-Microscopy (cryo-SEM) [15–17].

The description of heterogeneous particles or those of irregular shapes is much more complicated. It is generally accepted that the size of particles and the structure of their surface have a strong influence on the particle-interface interactions [8,18–23]; however, relevant available literature data are rather limited. A foam can be stabilized by various particles, with different sizes and shapes (spherical, rod-like, flat or disk-like shapes), and made of various materials.

Rod-like particles were used for the stabilization of foams by Alargova et al. [8]. They synthetized rod-like hydrophobic microparticles (1 μm diameter and a few tens μm length) from epoxy-type photoresist SU-8 (see Figure 15.1). As the result, they obtained super stable foams (over 20 days of stability) just by shaking. The high stability was strictly connected to the thick and "hairy" structure built by the rod-like particles in the foam films. SEM images and observations of single foam films proved that the whole surface of the film was densely covered with intertwined rods and that the foam film was thicker than at least two opposing layers of rods (ca. 2 μm). It was proposed that dense layers of particles were formed, completely impenetrable for gas or liquid. Consequently, all aging processes in the foam were inhibited. It has also been shown that even small amounts of elongated particles can act as foam stabilizers, while for the same number of spherical particles, it is impossible to obtain stable foams. Some authors tried to explain this fact in terms of the larger value of free

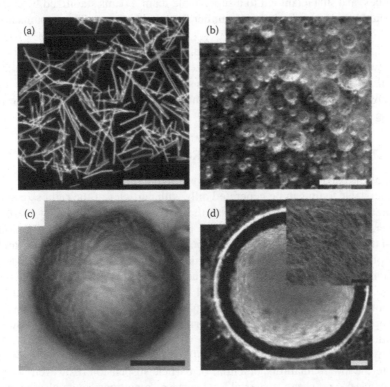

FIGURE 15.1 (a) Microscopic photos of SU-8 microrods; (b) Foam stabilized by rods; (c) Photo of single bubble covered by rods; (d) Dense surface layer of SU-8 rods (scale bar—50 μm for images a and c; 200 μm for images b and d). (Taken from Alargova, R.G. et al. *Langmuir* 2004; 20: 10371–10374 [8].)

energy for a spontaneous desorption of elongated particles as compared with spherical ones. Strong capillary forces between neighboring elongated particles in the interfacial layer were also considered [21]. The stabilization effects of rod-like (or disk-like/cone-like) particles are also connected with the lower three phase contact angle as compared with those for spherical particles [8,21]. However, foam stability is mostly explained just by the structure of the rigid shell of intertwining rods in the foam film, which provides a mechanical rigidity to the foam [8,23–25].

Recently, Sham and Notley presented data about foams stabilized by elongated exfoliated graphene particles [26]. The authors proved that it is possible to obtain strong stabilization of foam even for low amounts of particles (below 0.1% w/w). This was possible due to the specific thin and elongated shape and low density of the used graphene particles. The foam aging analysis proved that disproportionation was complete inhibited. The authors claimed that the high stability was achieved by a complete coverage of the foam films by the graphene plates. What is also interesting to note is the fact that the graphene particles can attach spontaneously at the foam film surfaces.

It must be pointed out here that particles with irregular shapes do not guarantee a high foam stability. The performance of various prolonged (rod-like) particles in foam stabilization was discussed by Wege et al. [24]. It was clearly shown that in some specific experimental conditions, foams formed with rod-like particles with highest foamability are not permanently stable.

15.2.2 OTHER SOLID AND PROLONGED NANOMATERIALS AS ADDITIVES IN AQUEOUS FOAMS

Only recently, Kramer et al. [27] presented another approach for the generation of stable particle-laden foams. They used multicomponent mixtures containing not only nanoparticles and surfactant, but also carbon nanotubes. As particles, they used two types of fumed silica suspensions and as surfactant, they used cocamidopropyl betaine. The mutual interaction between the nanoparticles, carbon nanotubes, and surfactant led to a superstable foam. Foams stabilized by carbon nanotubes show narrow bubble size distribution and smaller foam bubble sizes. These foams also present low density and porosity materials.

Another methodology, developed by Al-Qararah et al. [28–31], was based on rigid tubes combined with cellulose (wood) fibers. This type of stabilizer system leads to synergetic interactions between the fibers and the foam film interfaces, forming a special fiber network with unique big micropores. The authors have shown that it is possible to control the material properties of the fiber networks via the foaming protocol.

15.3 EFFECT OF THE PARTICLES' HYDROPHOBICITY

The influence of the particles' hydrophobicity on thin foam film stability was already analyzed in Chapter 7. Superstable foam can be generated only with particles of the right size and right degree of hydrophobization. This is connected with the irreversible attachment of the particles to the interfacial layer. As it was proved by Binks et al. [4,9], the energy of particle detachment can be as high as several thousand kT. This means the particles' attachment energy is a thousand times higher than the adsorption energy of a surfactant molecule [4]. Surfactant molecules can desorb relatively easily from the interface because their energy of adsorption is comparable with the thermal energy.

As was already mentioned above, a special degree of hydrophobization ($40° < \theta < 90°$) is necessary to obtain stable thin foam films [32] (see Figure 15.2). This means that the attachment of super hydrophobic particles at foam film surfaces would immediately destroy the foam film ($\theta > 90°$), while hydrophilic particles ($\theta < 40°$) will never attach there [4,21,32,33]. However, if the degree of hydrophobization is in the range of "partial hydrophobization," stable foams can be created, despite the low foamability potential of particle suspension. However, additional factors, such as the particles' size, shape, material, and surface smoothness must also be considered. The proper selection of all the mentioned parameters allows generating stable foams without using any surfactants.

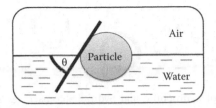

FIGURE 15.2   Schematic description of a particle attachment at the liquid interface.

The range of necessary hydrophobization ($40° < \theta < 90°$) cannot be taken as a strict law. As discussed in Chapter 7, in special cases even particles with a degree of hydrophobization of 126° can be used to form stable foams.

15.4   CHANGING OF PARTICLE HYDROPHOBICITY

Some degree of hydrophobicity of particles is necessary to generate stable foams. This can be established by permanent or "*in situ*" methods. A permanent hydrophobization is always made before the particles are introduced into the foaming solution. Therefore, we count these particles as "originally" hydrophobic. Of course, various particles and different methods of permanent hydrophobization lead to particles with different degrees of hydrophobization. The most frequently applied method of silanization is using the vapor phase of dichlorodimethylsilane as presented in Figure 15.3.

Reversible ("*in situ*") hydrophobization means that the particle suspensions contain chemicals interacting with their surfaces. As the result, the surface of the hydrophilic particle is hydrophobized. Various mixtures with efficient foam properties can be formulated, which will be presented further in Section 15.4.2.

15.4.1   HYDROPHILIC PARTICLES

Highly hydrophilic particles do not attach on foam film surfaces but remain dispersed in the bulk of the foam films. However, partially hydrophilic particles can improve the stability of a foam. Depending on the size, shape, and concentration of partly hydrophilic particles, they can aggregate and accumulate in the Plateaus borders and slow down the liquid drainage [34,35]. Such particles can also form a network structure (some kind of a weak gel). As the result, the foam stability can be improved. Faure et al. [36] used hydrophilic particles (sizes ca. 5.6 μm), which at a certain concentration diminish the liquid flow during drainage due to viscosity effects. When the foam

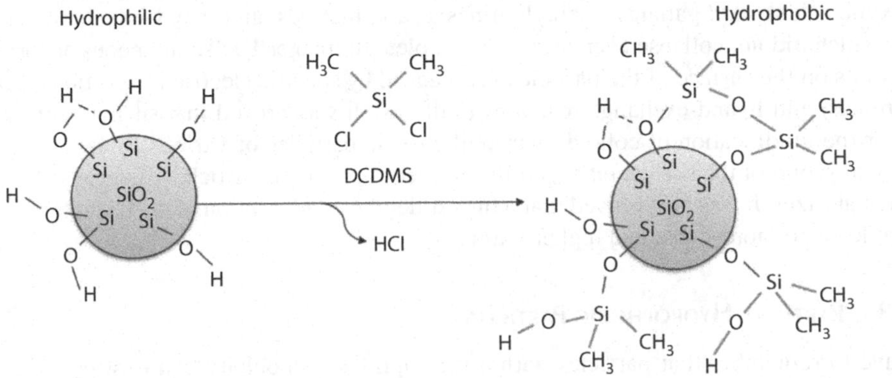

FIGURE 15.3   Schematic method of the permanent hydrophobization of various hydrophilic particles.

FIGURE 15.4 Arrangement of hydrophilic and hydrophobic particles in the interfacial layer (black—hydrophilic surface, gray—hydrophobic).

dries, the particle agglomerates and stay as plugs in the foam film leading to elastic properties of the foam [34–36].

Figure 15.4 schematically presents the interaction of hydrophilic and hydrophobic particles with an interfacial layer. Completely hydrophilic particles do not attach to the interface but stay completely immersed in the bulk of the aqueous solution. Also, fully hydrophobic particles do not attach to the interface. Figure 15.4 shows the situation for hydrophilic (which means less than 40%) and hydrophobic (more than 90% of surface hydrophobization) particles. However, this scheme does not count for the influence of the particles' density and weight. Wege et al. [24] have shown that the weight of the particles cannot be neglected. Thus, we can consider a "hydrophobic" heavy particle immersed into the aqueous liquid as if it is hydrophilic, and conversely, a mainly hydrophilic but very light particle appears as if it is "hydrophobic".

15.4.2 HYDROPHILIC PARTICLES: INTERACTIONS WITH SURFACTANTS AND OTHER CHEMICALS

In this chapter, we discuss the situation of hydrophilic particles which gain some hydrophobic properties only in the solution right before foam generation [4,37–42]. It was confirmed that such "*in situ*" hydrophobization can improve the efficiency of foam formation and foam stability of particle suspensions. However, we also have to consider that the added surfactants cannot only act as hydrophobization agents, but also as a foaming agent. Such simultaneous effects must be carefully investigated case by case. Hydrophobization of particles can happen due to their interaction with various types of surfactants (ionic or nonionic), polyelectrolytes, proteins, biopolymers, or other compounds. All existing factors and synergetic interactions have to be simultaneously considered. Added an electrolyte can diminish the influence of the electric double layer. The presence of other chemicals can modify the particle suspension's bulk viscosity. The impact of gelling agents and cross-linking components must also be considered. The number of other compounds to be added is almost unlimited.

In a series of papers [37–40] and in Chapter 25, Gonzenbach et al. presented the partial hydrophobization of metalloid particles by various surfactants. In their work, they used short chain carboxylic acids, alkyl gallates or alkylamines as amphiphiles, and they were able to create stable foams which did not collapse even after the complete drainage. The spontaneous adsorption of the surfactants on the surface of the particles was caused by specific electric interactions (carboxylates and amines) and ligand-exchange reactions (gallates). It was proved that such an approach can be used for the modification of colloid hydrophilic oxide particles of various surface chemistry. The functional group of the surfactant was tailored according to the particles' surface to be selected as a foam stabilizer. It was also proved that a higher degree of adsorption of surfactants on the particle surface leads to more stable and higher foams.

15.4.3 PARTIALLY HYDROPHOBIC PARTICLES

We have to remember that particles with a too high hydrophobicity can destroy foam films due to a particle bridge-dewetting mechanism (see Chapter 18). It is generally accepted that particles with hydrophobicity characterized by a contact angle larger than 90° are not suitable to form stable foam. Therefore, the literature mainly presents systems in which partially hydrophobic particles ($\theta = 90°$) are mixed with various types of surfactants. It was proved that such mixtures provide much higher foamability than solutions of surfactants or particles alone. Note, however, that higher foamabilities do not entail higher foam stabilities or better viscoelastic properties. It can happen that foams generated from such composed mixtures have a shorter lifetime than those formed solely by dispersions of partial hydrophobic particles.

15.5 COMPLEX PARTICLES

So far, we have considered uniform particles having the same surface characteristics. Using such uniform particles minimalizes the number of factors influencing the foam parameters. Recently, new types of particles known as "complex particles" (or "surface active particles") have been discussed in the literature. Such particles can more easily be incorporated into interfacial layers [43,44].

15.5.1 JANUS PARTICLES

Among the "complex particles," Janus particles are most famous. They consist of two regions with different chemical or topological and hence surface properties [43] (see Figure 15.5). Such a type of particles can be synthesized, for example, from dendrimers or block copolymer micelles. From the point of view of foam stabilization, we can distinguish two kinds of Janus particles. First, the particles have two regions of different wettability. Such particles behave exactly like surfactant molecules adsorbing at an interface: the hydrophobic site is directed toward the air while the hydrophilic one is immersed into the aqueous solution. A second type of Janus particles has two sites of opposite (or at least different) charges. Such particles have a large dipole moment, which allows them to take an organized orientation in the electric field [44].

A good example of Janus particles was given by Reculusa et al. [43]. In this work, the authors synthetized hybrid asymmetric colloid particles of 50–150 nm diameter, containing organic (hydrophobic) and inorganic (hydrophilic) parts. The synthesis was performed by a seeded emulsion polymerization process. The obtained Janus particles consist of one silica particle and a single polystyrene nodule. Blanco et al. [45] also presented organic-inorganic patchy particles for foam and emulsion stabilization. In this case, the original hydrophilic silica particles were used as cores of the Janus particles. An additional heterogeneous nucleation of zein (abundant natural protein extracted from corn) leads to the formation of anisotropic composites of Janus particles. The authors found that the stabilizing effect was strictly related to the particles' surface coverage by a zein layer. A maximum foam stability was obtained with 40% of zein coverage, and any further increasing of the zein surface coverage drastically diminished the foam ability. Studies at different suspension pHs have shown that most stable foams were generated close to the isoelectric point (about pH 5) of the protein.

Some years ago, Brown et al. [22] presented disk-shaped Janus particles which were hydrophobic on their convex face and hydrophilic on their concave face. They proved that such particles could distort the solution interface increasing anisotropic forces between them. As result, they obtained ordered flocculated structures of particles in the interfacial layer.

FIGURE 15.5 Scheme of a Janus particle and its interactions with the interfacial layer (black—hydrophilic part, gray—hydrophobic part).

15.5.2 BIOPOLYMERS AS NATURAL PARTICLES

In real applications, various biopolymer particles are gaining more importance. This is especially evident in food applications, where colloidal particles used in fundamental research (such as colloidal silica particles) are not accepted. In this case, various types of aggregates, complexes, and coacervates of biopolymers (protein or/and polysaccharide) are used [44].

The effect of adsorption of polymers and proteins on the foam film and foam stability was already discussed in other chapters of this book. Here, we only want to discuss the situations where proteins or polymers aggregate and act as complex particles. In literature, we can find several types of particles used in the food industry, such as protein or polysaccharide coacervates [46,47], various protein-polysaccharide complexes [48,49], thermally modified protein aggregates [50,51], ethyl cellulose [52,53], wax crystals [52–54], or even drops of oils [52]. A review on foam stabilization by organic complex particles was published recently by Dickinson [55].

The interactions of biopolymer particles with foam film surfaces are more complex than is the case for the adsorption of simple particles. In this case, we also have to consider biopolymer structure variations going on under various solution conditions (pH, electrolyte, temperature, etc.). Synergetic interactions lead to additional opportunities, but also problems. The variations of pH in the suspension containing the protein particles can lead to structural changes. As a consequence, the adsorption rate of these protein particles and the suspension's interfacial shear viscosities can be simultaneously varied [52]. Both effects can influence the foam stability via coalescence and disproportionation processes.

15.6 INFLUENCE OF ADDED ELECTROLYTE AND CHANGES IN THE PH

The presence of an electrolyte in the particle suspension can cause combined effects on foam properties. It is generally accepted that the surface activity of ionic surfactants is enhanced by the presence of an inorganic electrolyte in the solution. This is connected to the influence on the electric double layer, formed during the adsorption of ionic surfactant molecules [56,57]. The EDL creates an energetic barrier which hampers further adsorption of ionic surfactant molecules. This in turn results in a lower surface activity of ionic surfactants when compared with their nonionic analogues having the same hydrocarbon chain length. The phenomenon has a strong impact on foam formation.

First, it influences the adsorption process at the interfaces of rising bubbles during foam formation, that is, it affects the bubbles' local and terminal velocities. The problem is complex and cannot be simply connected with the increased surface activity, as was recently discussed by Jarek et al. [58] and Krzan et al. [59–61]. Second, in foam, the compression of the electric double layer by an electrolyte can accelerate the thinning of the foam films during drainage and, as the result, reduce the foam stability and elasticity [62,63]. Third, the EDL compression enhances the adsorption of ionic surfactants and in this way enhances the foam properties [64]. Fourth, some past papers claim that the presence of a strong ionic strength (high concentration of electrolyte) can increase the thickness of the foam films due to the charge at the film surfaces. As the result, film drainage can be reduced which improves foam stability [65].

In conclusion, the addition of an electrolyte can improve or reduce the elasticity and stability of foam. There is no general rule for the positive or negative effects of an electrolyte. Therefore, the foaming properties of each solution composition containing an electrolyte must be studied case by case and the impact of each component (and influence of its concentrations) must be considered separately [66,67].

The problem is a lot more complicated in the case of multiple interactions occurring between particle, electrolyte, and ionic or nonionic surfactants. In this case, each of the mentioned factors has to be considered. Note, particles dispersed in aqueous solutions generate their own electrical double layer. As the result, some repulsion forces exist between particles and bubbles in each foaming formulation. Electrolytes, due to their compression of the EDL, can support the adhesion of particles to foam films [68–70]. The thinning of the EDL can also facilitate the aggregation of additional particles which could lead to a destabilization and sedimentation in the suspension [58–61]. However, the destabilization of a particle suspension does not always entail the diminishing of foaming properties. As we have shown in our recent research (Jarek et al. [71]), precipitated particles can be introduced into foam via a froth flotation process. The bubbles, used for foam formation, can act simultaneously as a carrier for previously precipitated particles. Hence, all particles can become incorporated into the foam films and Plateau borders (where they act as big plugs), leading to very stable, but inflexible foam structures.

15.7 RHEOLOGY OF PARTICLE LADEN FOAMS

Details on the rheology of foam are described in Chapter 17 of this book. Here, the rheology of particle-stabilized foams is very briefly discussed. It is a difficult subject due to the complexity of particle-laden foam structures and the nature of the different components in foam (liquid, gas, surfactants or/and biosurfactants, other surface active polymers or biopolymers, particles, etc.). This complexity leads to some additional problems and unexpected effects not observed in the rheology of classical foams (without particles).

Note, "rheology" is not only a method of measuring the viscosity and elasticity. For foams, we consider the "rheology" as an additional external force. Therefore, it can be counted as another dynamic "foam aging" process, in addition to drainage, coarsening or coalescence. Like other aging processes, the rheology depends on other processes and even a slow creep deformation below the yield stress can influence the foam structure and affect the stability of foam [72–84]. The viscoelastic response of particle-stabilized foams is different from that of classic foams [85,86], which typically show just a viscous behavior without an elastic modulus or yield stress. The presence of yield stress in particle-laden foams is directly connected with the irreversible attachment of the particles at the foam film surfaces. The particles are already irreversibly attached and immobilized in some places of the foam films and engage in mutual interaction to generate some rigid structure. Now, external stresses can either shift the whole foam to a new position (we can discuss the foam's yield stress and flow behavior) or destroy the particles' mutual interactions and contacts with the foam films leading to a collapse of the foam.

In recent papers, Guillermic et al. studied the rheology of foams stabilized by nanoparticles of laponite [85], while Cohen-Addad et al. [86] used spherical glass and carbon microparticles and irregular talc microparticles. Similar work was performed by Erasov et al. [87] with large silica and bentonite nanoparticles. However, in almost all studies, the structure of foam and its rheological properties are not related to the adsorption of the surfactant or biosurfactant on a molecular level.

In 2007, we presented the rheological analysis of model foams stabilized by particles added to the Gillette shaving cream [86]. Glass, carbon, and talc microparticles of various sizes (20–130 μm) were used in a series of experiments. The particles were introduced into the shaving cream foam by mechanically whipping the blend over 1 minute at 8000 rpm. It was found that the addition of particles enhances the elasticity and viscosity of the Gillette foam by one order of magnitude, independent of the kind of particles (see Figure 15.6). The variations of the viscoelasticity with the solid volume fraction was in qualitative agreement with that predicted by an effective medium rigidity particle concentration model. It was proved that the network of central force (uniform and equal) springs relevant for classical foams was transformed into another structure consisting of some very strong springs generated by rigid bonds between particles. It was also found that there is a maximum concentration of particles above which the viscoelastic properties are no longer improved. This maximum concentration increases with the particles' size. It was concluded that the studied particle-laden Gillette foams behaved like a superelastic material.

In 2009, Guillermic et al. [85] presented a similar analysis, where sodium n-dodecyl sulfate (SDS) foams were stabilized by clay laponite nanoparticles (colloidal disks with a diameter of ca. 30 nm and a height of ca. 1 nm). The particles' size was about half that of the foam films' thickness. The foams were generated by stirring at 1000 rpm over 10 min. Exactly as described in [86], the viscoelastic properties were enhanced by the particles. Some characteristic maximum concentration of particles was also found, above which their enhancing effect on the viscoelasticity was lost.

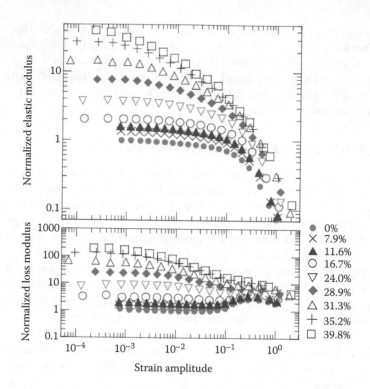

FIGURE 15.6 Evolution of the normalized elastic and loss moduli with the applied strain amplitude for Gillette foam laden with glass beads (D = 41 μm) at increasing solid volume fraction. (Redrawn from Cohen-Addad, S. et al. *Phys Rev Lett* 2007; 99: 168001–168004 [86].)

In 2015, Erasov et al. [87] tried to stabilize foams generated from mixtures containing acylamidopropyl betaine and xanthum gum with added silica or bentonite clay particles (in the size range 250–450 nm). The foams were obtained by using a kitchen mixer at 12,000 rpm over an unknown time. The studies were dedicated just to the effective viscosity of the foam, and it was found that the addition of silica or bentonite clay particles increases the foam effective viscosity, while no influence on the yield stress or drainage rate was observed.

Özarmut and Steeb [88] determined the effective yield stress of model Gillette shaving cream foams additionally stabilized by glass beads as particles (Silibeads Glassbeads Type S with particle sizes between 0.25–4.40 mm). The results confirmed that some threshold concentration of particles exists, above which rheological experiments could not be performed. The authors also proved that there is a direct relation between the particle concentration and the foam yield stress.

Zhang et al. [89] proved recently that foam, generated by stirring SDS solutions, can be additionally stabilized by starch particles via reducing their drainage. However, the presence of starch particles simultaneously diminished the mixtures' foamability. The rheological viscoelastic properties were enhanced by the particle concentration, exactly as in the other mentioned papers.

In conclusion, we can clearly state that the presence of particles improves the rheological properties of all studied aqueous foams. However, the number of published investigations is still too low to find any general rules.

15.8 SOLIDIFICATION OF FOAMS STABILIZED BY PARTICLES

Solid foams possess low density and very high specific area. Therefore, they are useful in different applications, such as thermal and sound insulations. There are various ways of foam solidification. First, there are weak or strong gelifications of the foam structure. A weak gelification can be connected with interactions in the bulk of the foam film liquid due to the interactions between hydrophilic particles. As we described further above, if the concentration of such particles exceeds some necessary threshold, a weak gelification can decrease the drainage rate of the foam.

Strong gelification in foams can be obtained using special additional gelling (crosslinking) agents. In such cases, the time of liquid solidification is crucial for all processes during foam formation. Therefore, the final mixing of the gelling components should be done just before foam formation. An example of such a foam was recently presented by Song et al. [90]. They generated (by stirring at 300 rpm) foam from suspensions containing poly(vinyl alcohol) and cellulose nanocrystals with the addition of formaldehyde as a crosslinker. The final formation (and gelation) of the foam was performed in an oven at 80°C.

Gelled particle-laden foam can be later transformed into some kind of solid monolith by using heating at a high temperature. Exactly this methodology was presented in a recent series of papers by Zabiegaj et al. [91–93]. The authors proved that dense wet foams containing various carbon particles (hydrophilic, hydrophobic, activated carbon of different structure and properties, etc.) can be transformed into solid monolithic structures of high porosity. In their studies, they used different surfactants as foaming agents and various gelling agents. The foams were generated by strong mixing at 8000–15,000 rpm and an almost identical gelling procedure in the oven at 80°C, as described in [90]. The final stable monoliths developed in this way were further solidified in an additional thermal procedure in a special heater at 900°C or 1300°C. Tseng and Hsu [94] applied a procedure similar to that described in [91–93] and prepared macroporous ZrO_2/Ti composite foams. However, the initial wet foam was formed by bubbling rather than mixing.

15.9   "SMART" OR "INTELLIGENT" PARTICLE-STABILIZED FOAMS

The term of "smart" or "intelligent" foams can be understood in different ways, that is, first when foams are sensitive to physical factors such as temperature, UV radiation, magnetic field, presence of chemicals, pH, redox, or CO_2/N_2. This allows us to control not only the formation, but also the lifetime of foams.

The attachment of particles to foam film surfaces leads to the formation of a more rigid structure, hence particles improve the foam stability. Additional mutual interactions between the attached particles can, under certain conditions, inhibit the aging processes in foam. The yield stress shown by particle laden foams allows using them in different technological processes, which is impossible with classical foams. However, from the technological point of view, foam stability can be a positive or negative feature. Stable liquid foam can be used in continuous industrial processes, supposed it collapses finally immediately. Example of such applications you can find below.

Fameau et al. [12,95] proved that the multilayer multitubular structure of the adsorbed ethanolamine salt of 12-hydroxy stearic acid changes immediately into micelles upon heating, which leads to a rapid foam destabilization and collapse of the whole foam. More examples of such foam responsive systems were presented by Salonen et al. [11].

Similar effects can be obtained with carbon particles. Such particles can interfere with UV light [11]. Light of proper frequency and intensity warm carbon particles and above a certain threshold value, the foams collapse. Focusing light on particular regions let the foam collapse locally instead of the whole foam. Magnetic particles can also be used and via an external magnetic field, the foam can be destabilized at any moment [96–99].

Chemicals can be also used as stabilization—destabilization factors. Zhu et al. [100] has shown that surfactants can act as stabilizer-destabilizer. The addition of a cationic surfactant to hydrophilic silica particles leads to a partial hydrophobization and stabilization of foam, which can be "neutralized" by adding an identical amount of an anionic surfactant so that the silica particles can no longer stabilize the foam.

15.10 POSSIBLE DIRECTIONS OF FUTURE RESEARCH

The formation of modern innovative particle-laden foams requires the synthesis of new kinds of particles with various shapes, structures, and/or properties, such as Janus and other complex or "intelligent" particles which provide new features. The application, for example, of the rigid carbon nanotubes and elastic nanowires could lead to the development of completely new kinds of foam structures. These foams can be completely rigid (nanotubes—no elasticity) or elastic in some controlled way, that is, when using flexible wires.

One can also imagine various other applications of "smart foams," such as those stabilized by chitosan particles which have UV screening properties. In this way, we would get UV protection cream which at the same time, is safe for humans (hypo-allergic). In a similar way, other smart particles (appropriate for certain applications) can be introduced into foam.

In the present situation, we must understand that there is still a large number of open questions. For example, foams mostly are formed by sparging and the created bubbles rise and get in contact with particles in the suspension. Such processes are really complex and require separate studies, where the hydrodynamic interactions (hydrodynamic fields created by the particle and bubble, the wake structures in the proximity of the bubble and particle, etc.) will be studied simultaneously with the adsorption processes. The recently published studies where the bubble hits various flat surfaces [101–103] can be only considered just as a rough approximation for the real interactions. Some more direct methods for investigating elementary processes in foam formation and stabilization should be developed in the near future.

ACKNOWLEDGMENTS

Financial support from National Science Centre of Poland (grants no. 2016/21/B/ST8/02107 and 2011/01/B/ST8/03717) is gratefully acknowledged. Part of this work has been also supported by the research and/or staff mobility actions: COST CA15124, COST MP1106, COST CM1101, European Union Erasmus+ program (project numbers: 2014-1-PL01-KA103-000225 & 2016-1-PL01-KA103-000225), the Polish-Italian Bilateral Research cooperation project titled "Biocompatible particle-stabilized foams and emulsions as carriers for healing agents" between ICMATE-CNR and ICSC-PAS and by the Polish-Bulgarian Bilateral Research cooperation project titled "Biocompatible particle-stabilized foams and emulsions for biomedical applications" between IPC BAS and ICSC-PAS.

The authors would also like to thank Dr. Ewelina Jarek for helpful suggestions and discussion.

REFERENCES

1. Pickering, S.U. *J Chem Soc* 1907; 91: 2001–2021.
2. Ramsden, W. *Proc R Soc London* 1903; 72: 156–164.
3. Nguyen, A.V., Schulze, H.J. *Colloidal Science of Flotation*. Marcel Dekker: New York, 2004.
4. Binks, B.P. *Curr Opin Colloid Int* 2002; 7: 21–41.
5. Gao, T. *Metall Mater Trans A* 2002; 33: 3285–3292.
6. Du, Z.P., Bilbao-Montoya, M.P., Binks, B.P., Dickinson, E., Ettelaie, R., Murray, B.S. *Langmuir* 2003; 19: 3106–3108.
7. Dickinson, E., Ettelaie, R., Kostakis, T., Murray, B.S. *Langmuir* 2004; 20: 8517–8525.
8. Alargova, R.G., Warhadpande, D.S., Paunov, V.N., Velev, O.D. *Langmuir* 2004; 20: 10371–10374.
9. Binks, B.P., Horozov, T.S. *Angew Chem Int Ed* 2005; 44: 3722–3725.
10. Fameau, A.L., Lam, S., Velev, O.D. *Chem Sci* 2013; 4: 3874–3881.
11. Salonen, A., Langevin, D., Perrin, P. *Soft Matter* 2010; 6: 5308–5311.
12. Fameau, A.L., Saint-Jalmes, A., Cousin, F., Houssou, B.H., Novales, B., Navailles, L., Nallet, F. et al. *Chem Int Ed* 2011; 50: 8264–8269.
13. Kim, S., Barraza, H., Velev, O.D. *J Mater Chem* 2009; 19: 7043–7049.
14. Ravera, F., Santini, E., Loglio, G., Liggieri, L. *J Phys Chem B* 2006; 110: 19543–19551.
15. Azzam, R.M.A., Bazhara, N.M. *Ellipsometry and Polarized light*. Elsevier: Amsterdam, 1977.

16. Khlebtsov, B.N., Khanadeev, V.A., Khlebtsov, N.G. *Langmuir* 2008; 24: 8964–8970.
17. Jin, H., Zhou, W., Cao, J., Stoyanov, S.D., Blijdenstein, T.B.J., de Groot, P.W., Arnaudov, L.N., Pelan, E.G. *Soft Matter* 2012; 8: 2194–2205.
18. Stamou, D., Duschl, C., Johannsmann, D. *Phys Rev E Stat Phys Plasmas Fluids Relat Interdiscip Topics* 2000; 62: 5263–5272.
19. Lucassen, J. *J Colloids Surf* 1992; 65: 131–137.
20. Kralchevsky, P.A., Denkov, N.D., Danov, K.D. *Langmuir* 2001; 17: 7694–7705.
21. Hunter, T.N., Pugh, R.J., Franks, G.V., Jameson, G.J. *Adv Colloid Interface Sci* 2008; 137: 57–81.
22. Brown, A.B., Smith, C.G., Rennie, A.R. *Phys Rev E Stat Phys Plasmas Fluids Relat Interdiscip Topics* 2000; 62: 951–960.
23. Alargova, R.G., Paunov, V.N., Velev, O.D. *Langmuir* 2006; 22: 765–774.
24. Wege, H.A., Kim, S., Vesselin, P.N., Zhong, Q., Velev, O.D. *Langmuir* 2008; 24: 9245–9253.
25. Dickinson, E., *Food Hydrocolloids* 2012; 28: 224–241.
26. Sham, A.Y.M., Notley, S.M. *J Colloid Interf Sci* 2016; 469: 196–204.
27. Krämer, C., Kowald, T.L., Butters, V., Trettin, R.H.F. *J Mater Sci* 2016; 51: 3715–3723.
28. Al-Qararah, A.M., Hjelt, T., Koponen, A., Harlin, A., Ketoja, J.A. *Colloids and Surfaces A* 2013; 436: 1130–1139.
29. Al-Qararah, A.M., Hjelt, T., Koponen, A., Harlin, A., Ketoja, J.A. *Colloids and Surfaces A* 2015; 467: 97–106.
30. Al-Qararah, A.M., Hjelt, T., Koponen, A., Harlin, A., Ketoja, J.A., Kiiskinen, H., Koponen, A. *J Timonen Colloids and Surfaces A* 2015; 482: 544–553.
31. Al-Qararah, M. Aqueous foam as the carrier phase in the deposition of fibre networks, *Academic Dissertation for the Degree of Doctor of Philosophy*, Jyväskylä: Finland, November 2015.
32. Johansson, G., Pugh, R.J. *Int J Miner Process* 1992; 34: 1–21.
33. Stocco, A., Drenckhan, W., Rio, E., Langevin, D., Binks, B.P. *Soft Matter* 2009; 5: 2215–2222.
34. Kruglyakov, P.M., Taube, P.R. *Colloid J* 1972; 34: 194–196.
35. Pugh, R.J. *Adv Colloid Interface Sci* 1996; 64: 671–742.
36. Faure, S., Volland, S., Crouzet, Q., Boutevin, G., Loubat, C. *Colloids and Surfaces A* 2011; 382: 139–144.
37. Gonzenbach, U.T., Studart, A.R., Tervoort, E., Gauckler, L.J. *Angew Chem Int Ed* 2006; 45: 3526–3530.
38. Gonzenbach, U.T., Studart, A.R., Tervoort, E., Gauckler, L.J. *Langumir* 2007; 23: 1025–1032.
39. Studart, A.R., Gonzenbach, U.T., Tervoort, E., Gauckler, L.J. *J Am Ceram Soc* 2006; 89: 1771–1789.
40. Gonzenbach, U.T., Studart, A.R., Tervoort, E., Gauckler, L.J. *Langmuir* 2006; 22: 10983–10988.
41. Binks, B.P., Lumsdon, S.O. *Langmuir* 2000; 16: 8622–8631.
42. Sun, X., Chen, Y., Zhao, J. *RSC Adv* 2016; 6: 38913–38918.
43. Reculusa, S., Poncet-Legrand, C., Perro, A., Duguet, E., Bourgeat-Lami, E., Mingotaud, C., Ravaine, S. *Chem Mater* 2005; 17: 3338–3344.
44. Murray, B.S., Durga, K., Yusoff, A., Stoyanov, S.D. *Food Hydrocolloids* 2011; 25: 627–638.
45. Blanco, E., Smoukov, S.K., Velev, O.D., Velikov, K.P. *Faraday Discuss* 2016; 191: 73–88.
46. Moschakis, T., Murray, B.S., Biliaderis, C.G. *Food Hydrocolloids* 2010; 24: 8–17.
47. Schmitt, C., da Silva, T.P., Bovay, C., Rami-Shojaei, S., Frossard, P., Kolodziejczyk, E. *Langmuir* 2005; 21: 7786–7795.
48. Turgeon, S.L., Schmitt, C., Sanchez, C. *Curr Opin Colloid In* 2007; 12: 166–178.
49. Schmitt, C., Aberkane, L., Sanchez, C., Proteinepolysaccharide complexes and coacervates. In: Phillips, G.O., Williams, P.A. (Eds.), *Handbook of Hydrocolloids*, Woodhead: Cambridge, 2009, pp. 420–476.
50. Unterhaslberger, G., Schmitt, C., Shojaei-Rami, S., Sanchez, C., Lactoglobulin, C., Aggregates from heating with charged cosolutes: formation, characterization and foaming. In: Dickinson, E., Leser, M.E. (Eds.), *Food Colloids: Self-Assembly and Material Science*. Royal Society of Chemistry: Cambridge, UK, 2007, pp. 177–194.
51. Paunov, V.N., Cayre, O.J., Noble, P.F., Stoyanov, S.D., Velikov, K.P., Golding, M. *J Colloid Interf Sci* 2007; 312: 381–389.
52. Campbell, A.L., Stoyanov, S.D., Paunov, V.N. *Soft Matter* 2009; 5: 1019–1023.
53. Campbell, A.L., Stoyanov, S.D., Paunov, V.N. *Chem Phys Chem* 2009; 10: 2599–2602.
54. Campbell, A.L., Holt, B.L., Stoyanov, S.D., Paunov, V.N. *J Mater Chem* 2008; 18: 4074–4078.
55. Dickinson, E. *Curr Opin Colloid In* 2010; 15: 40–49.
56. Warszynski, P., Lunkenheimer, K., Cichocki, K. *Langmuir* 2002; 18: 2506–2514.
57. Para, G., Jarek, E., Warszynski, P. *Adv Coll Int Sci* 2006; 122: 39–55.
58. Jarek, E., Warszynski, P., Krzan, M. *Colloid Surf A* 2016; 505: 171–176.
59. Krzan, M., Jarek, E., Warszynski, P., Rogalska, E. *Colloid Surf B* 2015; 128: 261–267.

60. Krzan, M., Malysa, K. *Physicochem Probl Miner Process* 2012; 48: 49–62.
61. Krzan, M., Malysa, K. *Physicochem Probl Miner Process* 2009; 43: 43–58.
62. Lai, K.Y., Dixit, N. Additives for foams. In Prud'homme, R.K., Khan, S.A. (Eds.), *Foams: Theory, Measurements, and Applications*, Vol. 5. Marcel Dekker: New York, 1996, pp. 315–338.
63. Sarma, D.S.H.S.R., Khilar, K.C. *J Colloid Interface Sci* 1990; 137: 300–303.
64. Chattopadhyay, A.K., Ghalcha, L., Oh, S.G., Shah, D.O. *J Phys Chem* 1992; 96: 6509–6513.
65. Ingram, B.T. *J Chem Soc Faraday Trans* 1972; 1: 2230–2238.
66. Staszak, K., Wieczorek, D., Michocka, K. *J Surfactants Deterg* 2015; 18: 321–328.
67. Powale, S.R., Bhagwat, S.S. *J Disper Sci Technol* 2006; 27: 1181–1186.
68. Lu, S., Pugh, R.P., Forssberg, E. *Interfacial Separation of Particles*, 1st Edition. Elsevier: Amsterdam, 2005.
69. Kostakis, T., Ettelaie, R., Murray, S. *Langmuir* 2006; 22: 1273–1280.
70. Carale, T.R., Pham, Q.T., Blankschtein, D. *Langmuir* 1994; 10: 109–121.
71. Jarek, E., Petkova, H., Santini, E., Szyk-Warszynska, L., Ravera, F., Ligierri, L., Mileva, E., Warszynski, P., Krzan, M. Influence of chitosan on surface and properties of lauryol ethyl arginate solutions containing silica nanoparticles, unpublished paper.
72. Weaire, D., Hutzler, S. *The Physics of Foams*. Clarendon Press: Oxford, 1999.
73. Prud'homme, R.K., Khan, S. *Foams: Theory, Measurements and Applications*. Marcel Dekker: New York, 1996.
74. Princen, H.M., Kiss, A.D. *J Colloid Interf Sci* 1986; 112: 427–437.
75. Princen, H.M., Kiss, A.D. *J Colloid Interf Sci* 1989; 128: 176–187.
76. Weaire, D., Hutzler, S., Drenckhan, W., Saugey, A., Cox, S.J. *Progr Colloid Polym Sci* 2006; 133: 100–105.
77. Princen, H.M. *J Colloid Interface Sci* 1983; 91: 160–175.
78. Khan, S.A., Armstrong, R.C. *J Non-Newton Fluid* 1986; 22: 1–22.
79. Kraynik, A.M. *Annu Rev Fluid Mech* 1988; 20: 325–357.
80. Höhler, R., Cohen-Addad, S. *J Phys Condens Matter* 2005; 17: R1041–R1069.
81. Weaire, D. *Curr Opin Colloid Interface Sci* 2008; 13: 171–176.
82. Cohen-Addad, S., Höhler, R., Khidas, Y. *Phys Rev Lett* 2004; 93: 028302–028304.
83. Marze, S.P.L., Saint-Jalmes, A., Langevin, D., *Colloid Surf A* 2005; 263: 121–128.
84. Krzan, M. *Tech Trans Chem* 2013; 1-Ch: 9–27.
85. Guillermic, R.M., Salonen, A., Emile, J., Saint-Jalmes, A., *Soft Matter* 2009; 5: 4975–4982.
86. Cohen-Addad, S., Krzan, M., Höhler, R., Herzhaft, B. *Phys Rev Lett* 2007; 99: 168001–168004.
87. Erasov, V.S., Pletnev, M.Y., Pokidko, B.V. *Colloid J* 2015; 77: 614–621.
88. Özarmut, AÖ, Steeb, H. *J Phys: Conf Series* 2015; 602: 012031.
89. Zhang, Y., Chang, Z., Luo, W., Gu, S., Li, W., An, J. *Chinese J Chem Eng* 2015; 23: 276–280.
90. Song, T., Tanpichai, S., Oksman, K. *Cellulose* 2016; 23: 1925–1938.
91. Zabiegaj, D., Santini, E., Guzmán, E., Ferrari, M., Liggieri, L., Buscaglia, V., Buscaglia, M.T., Battilana, G., Ravera, F. *Colloid Surf A* 2013; 438: 132–140.
92. Zabiegaj, D., Santini, E., Ferrari, M., Liggieri, L., Ravera, F. *Colloid Surf A* 2015; 473: 24–31.
93. Zabiegaj, D., Buscaglia, M.T., Giuranno, D., Liggieri, L., Ravera, F. *Micropor Mesopor Mat* 2017; 239: 45–53.
94. Tseng, W.J., Hsu, K.T. *Adv Powder Technol* 2016; 27: 839–844.
95. Fameau, A.L., Houinsou-Houssou, B., Novales, B., Navailles, L., Nallet, F., Douliez, J.P. *J Colloid Interface Sci* 2010; 341: 38–47.
96. Hutzler, S., Weaire, D., Elias, F., Janiaud, E., Philos, E. *Mag Lett* 2002; 82: 297–301.
97. Drenckhan, W., Elias, F., Hutzler, S., Weaire, D., Janiaud, E., Bacri, J. *J Appl Phys* 2003; 93: 10078–10083.
98. Lam, S., Blanco, E., Smoukov, S.K., Velikov, K.P., Velev, O.D. *J Am Chem Soc* 2011; 133: 13856–13859.
99. Blanco, E., Lam, S., Smoukov, S.K., Velikov, K.P., Saad, A., Khan, A., Velev, O.D. *Langmuir* 2013; 29: 10019–10027.
100. Zhu, Y., Pei, X., Jiang, J., Cui, Z., Binks, B.P. *Langmuir* 2015; 31: 12937–12943.
101. Krasowska, M., Krzan, M., Malysa, K. *Physicochem Probl Miner Process* 2003; 37: 37–50.
102. Krasowska, M., Krzan, M., Malysa, K. Frother Inducement of the Bubble Attachment to Hydrophobic Solid Surface, Proceedings of the 5th UBC-McGill Bi-Annual International Symposium of Fundamentals of Mineral Processing, August 22–25, Canadian Institute of Mining, Metallurgy and Petroleum paper, 2004, pp. 121–135. DOI: 10.13140/RG.2.1.2856.7766.
103. Malysa, K., Krasowska, M., Krzan, M. *Adv Coll Interface Sci* 2005; 114–115: 205–225.

Section IV

Structure of Foams and Antifoaming

16 The Structure of Liquid Foams

Stefan Hutzler and Wiebke Drenckhan

CONTENTS

16.1 INTRODUCTION

Understanding foam structure is a key to understanding many foam properties. The opacity of a foam, for example, is a consequence of light being scattered from the thin films separating the bubbles or from the Plateau borders where the films meet. The drying out of a foam under gravity is governed mainly by the flow of liquid through the Plateau border network. This network also governs how the electrical conductivity of a foam varies with its liquid volume fraction and how foam flows when sheared (foam rheology, Chapter 7). The existence of a yield stress and the property of shear thinning require an understanding of the structural changes due to bubble rearrangements.

This chapter, which is based on our recent review [1], introduces a number of key descriptors of foam structure, such as liquid volume fraction, bubble mono- or polydispersity or order and disorder. The aim is to introduce the reader to the large range of different types of foam structures that are found in nature or may be produced in the laboratory.

After a brief overview of foam structures (Section 16.2), we will in Section 16.3 justify several approximations that simplify their analysis. Section 16.4 is dedicated to the analysis of dry foams with liquid volume fraction $\varphi < 0.05$, consisting mainly of bubbles with almost polyhedral shapes. Section 16.5 describes experimental studies of wet foams and Section 16.6 concerns the nature of bubble interactions in such foams. Section 16.7 is a brief summary of experimental, analytical, and computational methods for understanding foam structure. A concluding Section 16.8 completes this chapter.

16.2 OVERVIEW OF DIFFERENT FOAM STRUCTURES

The left photograph in Figure 16.1 shows a typical polydisperse, disordered liquid foam under gravity, sitting on top of a liquid pool. The shape of the bubbles in a foam depends on the local liquid fraction φ (the ratio of the liquid volume to foam volume). At the bottom of the foam column, that is, at the foam-liquid interface, the bubbles are nearly spherical. The liquid fraction corresponding to this *wet limit* is around 36%, that is, $\varphi_c = 0.36$, for foams of low polydispersity. This is the value corresponding to the packing fraction $1 - \varphi_c = 0.64$ of a random close packing of equal volume hard spheres (see also Section 16.6).

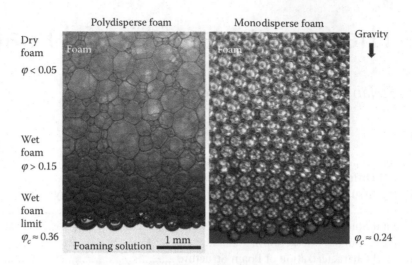

FIGURE 16.1 Photographs of a polydisperse (left) and a monodisperse (right) foam floating on top of the foaming solution. These foams were produced with ordinary dishwashing solution ("Fairy Liquid").

Further up in the foam column, the bubbles are more and more deformed and take on near polyhedral shapes, with thin curved films between them. For liquid fractions $\varphi < 0.05$, we call the foams *dry* (c.f. Section 16.4). Their topology is described by Plateau's rules (Section 16.4). Foams of liquid fraction exceeding about 0.15 are often called *wet* foams, but this value is chosen simply as lying about halfway between the wet and dry limits (see [2]).

Taking this definition of a wet foam, we can estimate its height on top of a liquid pool as $l_c^2/2R$, where R is the mean bubble radius and l_c is the capillary length, given by $l_c = \sqrt{\gamma/\Delta\rho g}$ (γ is the surface tension, g is the gravitation constant, and $\Delta\rho$ is the gas/liquid density difference). The number of bubble layers in this wet foam is called the *Princen number* Pri [3], and it is given by

$$\text{Pri} = \left(\frac{l_c}{2R}\right)^2. \tag{16.1}$$

For a typical foaming solution ($\gamma \approx \gamma_{\text{water}}/2 = 0.036$ N/m), resulting in $l_c \approx 1.6$ mm, the average radius of the bubbles should thus be smaller than 0.25 mm in order to form more than about 10 layers of wet foam.

The polydisperse foam of Figure 16.1 is typical for examples found in nature or many industrial applications. In fundamental research, there is often a preference for studying *monodisperse foams*, consisting of bubbles with a polydispersity of less than 5%. The right photograph in Figure 16.1 shows an example of such a monodisperse foam, again sitting on top of the foaming solution. Such foams have a tendency to order when confined into tubes of width of only a few bubble diameters. They can also order spontaneously in bulk when produced from bubbles that are smaller than the capillary length, resulting in a crystalline wet foam. For more details, see Section 16.5 and [4–7].

In Figure 16.2, we present an overview of different foam structures, emphasizing the role of bubble polydispersity and liquid fraction. Here, polydispersity p_σ is defined via a normalized standard deviation of the bubble radii $R_i = ((3/4\pi)V_i)^{1/3}$ from that of spheres of equivalent volumes V_i,

$$p_\sigma = \sqrt{\frac{<R^2>}{<R>^2} - 1}. \tag{16.2}$$

The polydispersity of foams is generally smaller than 50%, with bubble size distributions depending on the method of foam production [8] (see Chapter 10) or on the age of the foam, since interbubble

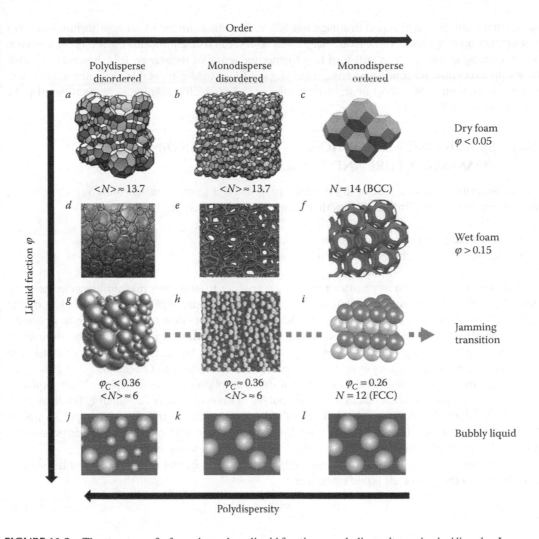

FIGURE 16.2 The structure of a foam depends on liquid fraction φ, polydispersity, and order/disorder. Images: a, b, e, g, thanks to A. M. Kraynik, see also (Kraynik, A.M. et al. *Phys Rev Lett* 2004; 93: 208301–208304 [9]; Kraynik, A.M. *Adv Engineering Materials* 2006; 8: 900–906 [10]); c and f by S. Cox, see also (Cox, S. et al. In: Scheffler, P.C.a.M. (Ed.) *Cellular Ceramics: Structure, Manufacturing, Properties and Applications*, Wiley, 2005 [11]); h (Meagher, A.J. et al. Slow crystallisation of a monodisperse foam stabilised against coarsening. *Soft Matter* 2015; 11: 4710–4716. Reproduced by permission of The Royal Society of Chemistry [12]), i (Reprinted with permission from Heitkam, S. et al. Packing spheres tightly: Influence of mechanical stability on close-packed sphere structures. *Phys Rev Lett* 2012; 108: 148302. Copyright 2012 by the American Physical Society.)

gas diffusion will lead to foam coarsening. Film coalescence may also alter the distribution. Foams are called monodisperse if their polydispersity is less than about 5%. It will be shown in Section 16.4 that in some cases, the Sauter mean radius $R_{32} = <R^3>/<R^2>$ is used rather than the average radius $<R>$, so that we can define the polydispersity parameter p_{32} as [9]

$$p_{32} = \frac{R_{32}}{<R^3>^{1/3}} - 1 = \frac{<V>^{2/3}}{<V^{2/3}>} - 1. \tag{16.3}$$

All of this chapter is dedicated to the structure of a foam *in equilibrium*, which is determined by the minimization of surface energy for fixed bubble volumes. In the presence of gravity, this refers to a foam in which liquid drainage has led to the establishment of an equilibrium profile of liquid fraction (see Figure 16.1). Note that a foam is never in true equilibrium, since processes such as coarsening or coalescence will lead to a further reduction of its energy. It is, however, possible to set up experimental conditions (e.g., using a gas of low solubility, stable surfactants, sufficient environmental humidity, short observation times, etc.) for which a foam may be considered to be *effectively* in equilibrium.

16.3 SIMPLIFYING ASSUMPTIONS FOR UNDERSTANDING FOAM STRUCTURE AND ENERGY

Before describing the structure of dry (Section 16.4) and wet foams (Section 16.5) in greater detail, we will lay out the key simplifications that we make for this analysis. These are

- the surface tension γ is assumed to be constant;
- the bubbles are assumed to be incompressible.

Treating surface tension as constant restricts our analysis to structures in equilibrium in the sense defined above. This means a restriction to times scales which are *sufficiently long* to exceed time scales in which dynamic effects, such as the Marangoni effect (which may play a role in structural relaxation after bubble rearrangements) can be neglected. The time scales must also be *sufficiently short*, so that foam aging effects (coalescence, coarsening) can be neglected. For the discussion of the physicochemistry of foams in relation to foam structure we refer to Chapter 4.

The assumption of incompressibility may be justified as follows. From the Laplace-Young equation we can estimate pressure differences between a bubble of radius R and its surrounding liquid as $2\gamma/R$, where γ is the interfacial tension. For bubbles of radius 0.1 mm and a typical value for γ of 30 mN/m, the resulting pressure difference amounts to about 10^3 Pa. Compared to the atmospheric pressure of 10^5 Pa, this is small enough to consider such bubbles in a foam as incompressible.

The energy of a foam is then simply its total surface energy E, that is, the product of the surface tension γ with the sum of all liquid interfaces S,

$$E = \gamma S. \tag{16.4}$$

For a foam consisting of bubbles with average radius of 0.1 mm, the surface energy is, therefore, of the order of 10^{-8} J per bubble, that is, 10^{13} times larger than the classical thermal energy of $1\ kT \sim 10^{-21}$ J per bubble. In addition, the change in potential energy when vertically displacing a bubble by its radius is proportional to $\Delta\rho g R^4$. This is about 10^9 times larger than kT. Hence, we can conclude that the thermal energy is negligible for the process of packing bubbles.

When comparing (surface) energy of different foam structures, it is convenient to introduce the dimensionless energy $\hat{E}$, defined as

$$\hat{E} = \frac{<S>}{<V>^{2/3}}, \tag{16.5}$$

where $<S>$ and $<V>$ are mean bubble surface area and volume, respectively. For a monodisperse foam consisting of spherical bubbles, this results in $\hat{E} = 3^{2/3}(4\pi)^{1/3} \approx 4.836$. A hypothetic cubic bubble would have $\hat{E} = 6$.

For finite values of liquid fraction, it is often more convenient to introduce the *relative surface excess* $\varepsilon(\varphi)$ as

$$\varepsilon(\varphi) = \frac{<S(\varphi)> - <S_0>}{<S_0>}, \tag{16.6}$$

where $<S(\varphi)>$ is the average total surface area of the foam at liquid fraction φ and $<S_0>$ is the average surface area of the foam, if all its bubbles are treated as spheres. Note that various other related nondimensional quantities are used in the literature.

The energy landscape of a foam is very complex and the bubbles are generally trapped in local energy minima. In the absence of thermal fluctuations, topological changes which would be required to exit such minima, do not occur spontaneously. However, structural changes leading to a decrease of total energy occur as a foam coarsens due to interbubble gas diffusion [2]. They may also be obtained via mechanical shearing of the foam [14] or following the injection of foaming solution into a foam (*forced drainage*). A general consequence of this is that the overall foam structure is strongly history dependent.

16.4   DRY FOAMS

The structure of *dry foams*, that is, foams of liquid fraction of less than $\varphi \approx 0.05$ (Figure 16.2), is well described by Plateau's laws [2,15,16]:

1. Three foam films meet symmetrically under angles of 120° in channels, called Plateau borders.
2. Four such Plateau borders meet symmetrically in a node under tetrahedral angles of $arccos\ (-1/3) \approx 109.47°$ (Maraldi angle).

Plateau's laws are a consequence of the minimization of surface area. They dictate the local topology and geometry of a dry foam. The laws were stated by the Belgian scientist Joseph Antoine Ferdinand Plateau in the 19th century, based on observations of soap films formed in metal wire frames [15].

Most of the liquid in dry foams is contained in the Plateau borders and their nodes. Films of foams in equilibrium are only some tens of nanometers thick and in structural considerations are, therefore, often approximated as having infinitesimal thickness.

The Young-Laplace equation describes the shape of a liquid film between two bubbles having a pressure difference of ΔP as

$$\Delta P = 4\gamma\kappa, \tag{16.7}$$

where κ is the mean curvature of the film surface, given by $\kappa = 1/2((1/r_1)+(1/r_2))$ with r_1 and r_2 as the two principal radii of curvature. The neighboring bubbles in a foam adjust position and shape so as to reach an equilibrium configuration, consistent with Plateau's rules and given bubble volumes.

Computer simulations showed that the pressure in a bubble is correlated with its number of neighbors, which, in turn is correlated with its volume [2,16]. This is important for understanding gas diffusion between neighboring bubbles, called coarsening, disproportionation or Oswald ripening [17].

Equipped with Plateau's rules, in 1887, Lord Kelvin set out to find the equilibrium structure of a monodisperse space-filling foam [18]. His considerations of how space can be divided into equal-size cells with a minimum total interfacial area led him to the shape of a truncated octahedron as the optimal structure (see Figure 16.3 left), consisting of bubbles with $N = 14$ neighbors and arranged in a *bcc* (body-centered cubic) structure [19]. To fulfill Plateau's rules, Kelvin had to introduce a slight curvature into the eight hexagonal faces; the six square faces are flat.

It took more than a century until Weaire and Phelan [20] were able to surpass Kelvin's result by finding a foam structure with 0.3% less interfacial area (see Figure 16.3 right). This so called Weaire-Phelan structure consists of eight equal-volume bubbles, of two different types, and an average number of neighbors of $<N> = 13.5$ [20].

The Weaire-Phelan structure ($\hat{E} = 5.288$) has a lower energy than the Kelvin structure ($\hat{E} = 5.306$). However, due to its complexity, it does not readily form in nature, although it was produced in the laboratory [21].

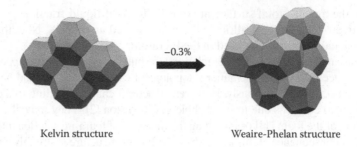

Kelvin structure   Weaire-Phelan structure

FIGURE 16.3 Surface Evolver simulations of the Kelvin (left) and Weaire-Phelan foam structure, simulations performed by Simon Cox.

Table 16.1 summarizes some of the properties of various dry foam structures. For comparison, the table also includes values for a cubic tiling (which disobeys Plateau's rules) and a hypothetical bubble, with 13.4 faces, whose scaled energy of $\hat{E} = 5.1$ serves as a lower bound [22].

Statistical data about disordered dry foams is mainly based on extensive computer simulations [2,9,10,16,24] using the Surface Evolver software [26]. The left of Figure 16.4 shows the distribution of the number of neighbors of bubbles in foams with different polydispersities. The average number of neighbors of bubbles in a disordered, monodisperse foam is $<N> \approx 13.7$. This is close to the value determined experimentally by Matzke [24] (full circles in the left of Figure 16.4) and to the values for both Kelvin and Weaire-Phelan structures. In polydisperse foams the average number of neighbors decreases with increasing polydispersity.

The right of Figure 16.4 shows that the average number of neighbors of a bubble $<N>$ is correlated with its volume: the larger a bubble is, the more neighbors it tends to have. The simulations also show that with increasing polydispersity, $\hat{E}$ decreases (Figure 16.5).

Furthermore, it was found that each individual bubble in a foam adjusts its shape to obtain a well defined dimensionless energy $\hat{E}^* \approx 5.33 \pm 0.03$ [9], independent of its size [9,27]. This value corresponds to a surface excess of $\varepsilon^*(\varphi = 0) = 0.100 \pm 0.008$ and may be used to calculate how the scaled energy $\hat{E}$ of the entire foam depends on its polydispersity [9] via the relation

$$\hat{E} = \frac{\hat{E}^*}{1 + p_{32}}. \tag{16.8}$$

The line which corresponds to this equation is shown along with the simulation data in Figure 16.5, evidencing excellent agreement.

TABLE 16.1

Summary of Scaled Energy $\hat{E}$ (Equation 16.5), the Relative Surface Excess ε (Equation 16.6), and the Average Number of Neighbors $<N>$ for a Dry Foam ($\varphi = 0$)

Type of Structure	Scaled Energy $\hat{E}$	Relative Surface Excess ε ($\varphi = 0$)	Number of Neighbors N (or $\langle N \rangle$)	Year + Ref
Cubic tiling	6	0.241	6	
Ideal bubble (not space filling)	5.1	0.055	13.4	1992 [23]
Kelvin	5.306	0.097	14	1887 [19]
Weaire-Phelan	5.288	0.093	13.5	1994 [20]
Random monodisperse foam	5.330 ± 0.006	0.102 ± 0.001	13.7	1946/2003 [24,25]
Random polydisperse foam (for $p_{32} < 0.5$) (Equation 16.3)	$3.6 < \hat{E} < 5.33$	0.100 ± 0.008	$11.4 < \langle N \rangle < 13.7$	2004 [9]

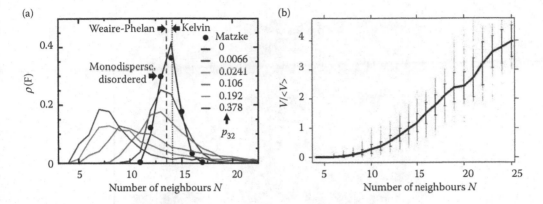

FIGURE 16.4 Left: Probability distribution of neighbors N of bubbles in foams of different polydispersities p_{32}, (●)—experimental data by Matzke [24]; (Reprinted with permission from Kraynik, et al. Structure of random foam. *Phys Rev Lett* 2004; 93: 208301–208304. Copyright 2004 by the American Physical Society [9].) Right: Relationship between the volume of a bubble and its number of neighbors; (Reprinted from *Colloids Surf A*, 263, Jurine, S. et al., Dry three-dimensional bubbles: Growth-rate, scaling state and correlations, 18–26, Copyright 2005, with permission from Elsevier [17].)

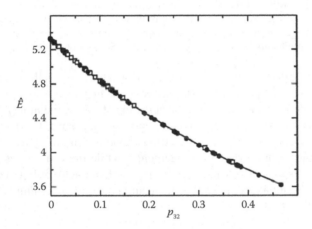

FIGURE 16.5 Scaled energy density $\widehat{E}$ as a function of polydispersity p_{32} for a bidispersive foam (open symbols) and a polydisperse foam (filled symbols); solid line (Equation 16.8) (Reprinted with permission from Kraynik, et al. Structure of random foam. *Phys Rev Lett* 2004; 93: 208301–208304. Copyright 2004 by the American Physical Society [9].)

16.5 EXPERIMENTAL STUDIES OF WET FOAMS

The structure of wet foams is no longer governed by Plateau's laws (see Section 16.4) and 8-fold vertices are possible [28]. Microgravity experiments on soap films spanned by a cubic wire frame showed their stability for a corresponding liquid fraction as low as $\varphi \approx 0.02$ [29]. In earth-bound experiments with ordered bulk foams, Höhler et al. [5] observed a coexistence of a *bcc* structure (with 4-fold vertices) with an *fcc* (face-centred cubic) structure (with 8-fold vertices) at about $\varphi = 0.07$. Simulations with the Surface Evolver agree with this finding. With further increase of the liquid fraction, deviations from the predicted angles at Plateau borders and vertices are also observed, caused by the increasing effective line tension. For more details, see [30].

In order to analyze wet foams experimentally, it is necessary to overcome the drying out of a foam due to gravitationally driven drainage. Several techniques are available for this. The use of bubbles

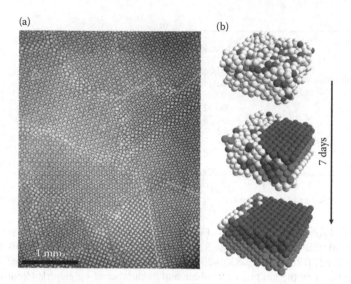

FIGURE 16.6 (a) Photo of the surface of crystalline monodisperse foam, showing different grain orientations, boundaries, and stacking defaults; (van der Net, A. et al. The crystal structure of bubbles in the wet foam limit. *Soft Matter* 2006; 2: 129–134. Reproduced by permission of The Royal Society of Chemistry [4].) (b) X-ray tomography data showing the crystallization of bulk monodisperse foam over the course of seven days. (Meagher, A.J. et al. Slow crystallisation of a monodisperse foam stabilised against coarsening. *Soft Matter* 2015; 11: 4710–4716. Reproduced by permission of The Royal Society of Chemistry [12].)

much smaller than the capillary length, resulting in many bubble layers in the wet regime (large Princen number, Equation 16.1), is one possibility. Alternatively, one may replenish the draining liquid via the addition of a constant flow of surfactant solution at the top of a foam column (*forced drainage* [2]). A third option is provided by experimenting in a microgravity environment [31].

Surprisingly, monodisperse wet foams consisting of bubbles smaller than a few hundred microns crystallize spontaneously into ordered arrangements. While the ordering is initially confined to regions where the foam is in contact with a confining vessel, or the foam-liquid or foam-gas interface [7,32], as in Figure 16.6a, it has also been seen to occur in the bulk of a sample and to extend in time, as shown in Figure 16.6b [12].

In the ordered regions, it was found that the bubbles arrange in both *fcc* (face centered cubic) and *hcp* (hexagonal close packing), that is, the arrangements associated with the packing of (hard) spheres, with a slight preference for the fcc arrangement. This preference was also found in computer simulations, where the bubbles are treated as soft spheres which aggregate, driven by buoyancy [13]; it was interpreted as a consequence of mechanical stability.

There are yet many open questions regarding the spontaneous ordering in monodisperse wet foams. These include:

- Does coarsening or drainage play a role in restructuring the foam?
- Can mechanical perturbations lead to an annealing of the foam?
- Is the spontaneous ordering a consequence of the specific form of the interaction potential between contacting bubbles, which deviates from that of spheres which interact via Hooke or Hertz contact forces (see Section 16.4)?

16.6 UNDERSTANDING BUBBLE INTERACTION

Rheological measurements and simulations have shown that shear modulus, yield stress, and yield strain decrease with liquid fraction [16,33]. They vanish at the critical liquid fraction close to $\varphi_c \approx 0.36$ ("rigidity loss transition"). In granular materials, this value corresponds to the random packing of spheres ("Bernal packing") and is also referred to as "jamming fraction" [34,35]. The value of φ_c decreases with increasing polydispersity, but remains close to 0.36 for the modest polydispersities generated by most foaming techniques [8] (Chapter 8). At the jamming transition, the average contact number of bubbles is six, i.e. the value that is obtained from constraint arguments for the packing of frictionless hard spheres [34,35].

Wet foams are often modeled as packings of (overlapping) soft spheres with a pairwise harmonic potential between spheres in contact [34,36]. Computer simulations have shown that this results in an increase ΔZ of the average contact number per bubble of the form $\Delta Z \sim (\Delta \varphi)^{1/2}$, where $\Delta \varphi = \varphi_c - \varphi$ [34].

Although being both computationally and conceptually attractive, the soft sphere model with its fundamental assumption of pairwise-additive interaction energies is only an approximation of bubble interaction. Computer simulations using the Surface Evolver [37,38], as well as analytical studies [39], have shown that the fact that bubbles adjust their shapes upon contact introduces many body coupling between all contacts, together with a logarithmic softening of the interaction. Further studies are required to investigate if and how this affects the scaling of ΔZ with distance $\Delta \varphi$ from the jamming point. Recent computer simulations of two-dimensional foams in which bubble shapes are accurately represented have suggested a linear scaling [40,41].

The Z-cone model presents a novel analytical approach which captures both the logarithmic softening and the dependence of the interaction potential on the number of contacts of a bubble. In the model, a bubble with Z neighbors is decomposed into Z equivalent pieces, followed by their approximation as *circular* cones. Both liquid and gas are treated as incompressible and the minimal surface area of the cap of each cone is computed analytically under the constraint of volume conservation [42,43]. This allows for the computation of the excess energy for periodic foam structures with Z neighboring bubbles as a function of liquid fraction, as

$$\varepsilon(\varphi) \sim -\frac{Z}{18(1-\varphi_c)^2} \frac{(\varphi_c - \varphi)^2}{ln(\varphi_c - \varphi)}, \quad (16.9)$$

in good agreement with simulations obtained with the Surface Evolver in the wet limit [37,42].

While $\varepsilon(\varphi)$ cannot be directly measured experimentally, it can be inferred from measurements of the so called *osmotic pressure* ($\Pi(\phi)$) in a foam, defined as

$$\Pi = -\gamma \left(\frac{\partial S}{\partial V_f} \right)_{V_{gas}=const.}, \quad (16.10)$$

where S is the total surface area of all bubbles in the foam of volume V_f, and V_{gas} is the constant gas volume [44–46]. In the case of experimental data for wet ordered foams, the deduced variation for $\varepsilon(\varphi)$ appears to be consistent with the expression obtained from the cone model (Equation 16.9). For details, we refer to our recent review article [1].

16.7 EXPERIMENTAL CHARACTERIZATION OF FOAM STRUCTURE

As we have shown above, foam structure depends crucially on the (local) liquid fraction φ. A number of experimental methods exist for its determination. One of the most elegant methods for measuring the total amount of liquid uses a U-tube filled with foam on one side and the foaming solution on the other side. As the amount of liquid on both sides is equal, the average liquid fraction of the foam can be estimated.

Liquid fraction *profiles* can be obtained via measurements of local electrical conductivity σ, using an array of electrodes in contact with the foam. This technique is based on the well established semi-empirical relationship $\sigma/\sigma_0 = 2\varphi(1 + 12\varphi)/(6 + 29\varphi - 9\varphi^2)$ [47], which expands on the simple

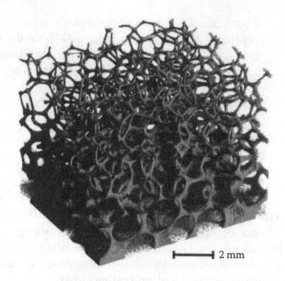

├────────┤ 2 mm

FIGURE 16.7 Tomographic 3D reconstruction of a dry aqueous foam made from ordinary detergent solution and nitrogen bubbles with an average diameter of 1.5 mm; (Meagher, A.J. et al. Slow crystallisation of a monodisperse foam stabilised against coarsening. *Soft Matter* 2015; 11: 4710–4716. Reproduced by permission of The Royal Society of Chemistry [12].)

linear relationship $\sigma/\sigma_0 = \varphi/3$ derived for dry foams by Lemlich [48] (σ_0 is the conductivity of the bulk solution). Various commercial devices are available for such measurements.

Recently, a new technique has been proposed whereby bulk liquid fraction is estimated from optical measurements of the surface liquid fraction of bubbles in contact with a solid confinement [49]. This makes use of semi-empirical expressions of foam excess energy that were referred to in Section 16.6.

The bubble size distribution in a foam may be determined experimentally by placing foam samples between two parallel plates of known separation. The resulting quasi two-dimensional foam can then be analyzed using standard imaging techniques. A rough noninvasive estimation of the average bubble size can also be obtained from photographs of surface bubbles. As in the measurement of the surface liquid fraction, there is the possibility of a bias due to the possibility of size segregation between surface and bulk [50]. A further noninvasive technique is diffusive light scattering combined with a separate measurement of the average liquid fraction [51].

Detailed structural data is hard to obtain. Photography is appropriate only for a few outer layers of foam, but fails as a method for a quantitative analysis of foam structure. The same is true for confocal microscopy. Recently, tomographic techniques, such as X-ray tomography, have proven very successful in determining the Plateau border network and bubble volumes (see Figures 16.6b and 16.7) [12,52,53]. The detection of foam films would benefit from higher energy synchrotron radiation.

16.8   CONCLUSIONS

We have shown that the structure of a foam depends crucially on the method for its generation and on the value of local liquid fraction. The foaming method also controls the degree of polydispersity and introduces the possibility of obtaining ordered, crystalline foams. Bubble shapes range from almost spherical for wet foams to polyhedral for dry foams.

The structure of dry foams is well understood (see Section 16.4) due to the large body of both experimental data and numerical simulations using the Surface Evolver software. This software, developed and maintained by Ken Brakke [26], is presently the most successful tool for simulating foam structure. Example simulations were already shown in Figures 16.2 and 16.3, while Figures 16.4 and 16.5 display Surface Evolver data.

Simulations of disordered wet foams, however, remain challenging, mainly due to the problems associated with detection and execution of topological changes. The Potts model [54,55] provides an alternative Monte Carlo type approach. The recently developed multiscale model of Sethian and Saye [56,57] could also be a candidate for exploring wet foams. Future progress in understanding wet foams will benefit from both advances in experimental imaging, together with new theoretical initiatives, see for example [38].

ACKNOWLEDGMENTS

The work was financially supported by the European Research Council (307280-POMCAPS), the Science Foundation Ireland (13/IA/1926), and by COST Actions MP1106 and MP1305. This work has been published within the IdEx Unistra framework and has benefited from funding from the state, managed by the French National Research Agency as part of the 'Investments for the future' program We would like to thank R. Miller for many suggestions concerning the manuscript.

REFERENCES

1. Drenckhan, W., Hutzler, S. Structure and energy of liquid foams. *Adv Colloid Interface Sci* 2015; 224: 1–16.
2. Weaire, D., Hutzler, S. *The Physics of Foams*. Clarendon Press: Oxford, 1999.
3. Weaire, D., Langlois, V., Saadatfar, M., Hutzler, S. Foam as a granular matter. In: Aste, T., Di Matteo, T., (Eds.), *Granular and Complex Materials*. World Scientific: Singapore, 2007, pp. 1–26.
4. van der Net, A., Drenckhan, W., Weaire, D., Hutzler, S. The crystal structure of bubbles in the wet foam limit. *Soft Matter* 2006; 2: 129–134.
5. Höhler, R., Sang, Y.Y.C., Lorenceau, E., Cohen-Addad, S. Osmotic pressure and structures of monodisperse ordered foam. *Langmuir* 2007; 24: 418–425.
6. Drenckhan, W., Langevin, D. Monodisperse foams in one to three dimensions. *Current Opinion Colloid Interface Sci* 2010; 15: 341–358.
7. Meagher, A.J., Mukherjee, M., Weaire, D., Hutzler, S., Banhart, J., Garcia-Moreno, F. Analysis of the internal structure of monodisperse liquid foams by X-ray tomography. *Soft Matter* 2011; 7: 9881–9885.
8. Drenckhan, W., Saint-Jalmes, A. The science of foaming. *Adv Colloid Interface Sci* 2015; 222: 228–259.
9. Kraynik, A.M., Reinelt, D.A., van Swol, F. Structure of random foam. *Phys Rev Lett* 2004; 93: 208301–208304.
10. Kraynik, A.M. The structure of random foam. *Adv Engineering Materials* 2006; 8: 900–906.
11. Cox, S., Weaire, D., Brakke, K. Liquid foams - Precursors for Solid Foams. In: Scheffler, P.C.a.M. (Ed.) *Cellular Ceramics: Structure, Manufacturing, Properties and Applications*, Wiley: Weinheim, 2005.
12. Meagher, A.J., Whyte, D., Banhart, J., Weaire, D., Hutzler, S., Garcia-Moreno, F. Slow crystallisation of a monodisperse foam stabilised against coarsening. *Soft Matter* 2015; 11: 4710–4716.
13. Heitkam, S., Drenckhan, W., Fröhlich, J. Packing spheres tightly: Influence of mechanical stability on close-packed sphere structures. *Phys Rev Lett* 2012; 108: 148302.
14. Durand, M., Kraynik, A.M., van Swol, F., Kafer, J. Quilliet, C., Cox, S., Talebi, S.A., Graner, F. Statistical mechanics of two-dimensional shuffled foams: Geometry-topology correlation in small or large disorder limits. *Phys Rev E* 2014; 89: 062309.
15. Plateau, J.A.F. *Statique Expérimentale et Théorique des Liquides soumis aux seules Forces Moléculaires*. 2, Gauthier-Villars: Paris, 1873.
16. Cantat, I., Cohen-Addad, S., Elias, F., Graner, F., Höhler, R., Pitois, O., Rouyer, F., Saint-Jalmes, A. *Foams - Structure and Dynamics*, Cox, S. (Ed.) Oxford University Press: Oxford, UK, 2013, p. 300.
17. Jurine, S., Cox, S., Graner, F. Dry three-dimensional bubbles: Growth-rate, scaling state and correlations. *Colloids Surf A* 2005; 263: 18–26.
18. Thomson, W., LXIII. On the division of space with minimum partitional area (Reprinted). *Phil Mag Lett* 2008; 88: A503–A514.
19. Weaire, D. *The Kelvin Problem*. CRC Press: London, 1997.
20. Weaire, D., Phelan, R. A counterexample to Kelvin's conjecture on minimal surfaces. *Phil Mag Lett* 1994; 69: 107–110.
21. Gabbrielli, R., Meagher, A.J., Weaire, D., Brakke, K.A., Hutzler, S. An experimental realization of the Weaire–Phelan structure in monodisperse liquid foam. *Phil Mag Lett* 2012; 92: 1–6.

22. Sullivan, J.M. The geometry of bubbles and foams. In: Rivier, N. Sadoc, J.F. (Eds.), *Foams and Emulsions*, Kluwer: Dordrecht, 1999, p. 403.
23. Kusner, R. The number of faces in a minimal foam. *Proc R Soc Lond Series A: Math Phys Eng Sci* 1992; 439: 683–686.
24. Kraynik, A.M., Reinelt, D.A. van Swol, F. Structure of random monodisperse foam. *Phys Rev E* 2003; 67: 031403.
25. Matzke, E.B. The three-dimensional shape of bubbles in foam-an analysis of the role of surface forces in three-dimensional cell shape determination. *Am J Botany* 1946; 33: 58–80.
26. Brakke, K. The surface evolver. *Exp Math* 1992; 1: 141–165.
27. Hilgenfeldt, S., Kraynik, A.M., Reinelt, D.A., Sullivan, J.M. The structure of foam cells: Isotropic plateau polyhedra. *Europhys Lett* 2004; 67: 484–90.
28. Brakke, K. Instability of the wet cube cone soap film. *Colloids Surf A* 2005; 263: 4–10.
29. Barrett, D.G.T., Kelly, S. Daly, E.J. Dolan, M.J. Drenckhan, W. Weaire, D. Hutzler, S. Taking Plateau into microgravity: The formation of an eightfold vertex in a system of soap films. *Microgravity Sci Technol* 2008; 20: 17–22.
30. Weaire, D., Vaz, M.F. Teixeira, P.I.C. Fortes, M.A. Instabilities in liquid foams. *Soft Matter* 2007; 3: 47–57.
31. Langevin, D. Vignes-Adler, M. Microgravity studies of aqueous wet foams. *Eur Phys J E*, 2014; 37:16.
32. van der Net, A., Delaney, G.W., Drenckhan, W., Weaire, D., Hutzler, S. Crystalline arrangements of microbubbles in monodisperse foams. *Colloids Surf A* 2007; 309: 117–124.
33. Cohen-Addad, S., Höhler, R. Pitois, O. Flow in foams and flowing foams. *Annu Rev Fluid Mech* 2013; 45: 241–267.
34. van Hecke, M. Jamming of soft particles: Geometry, mechanics, scaling and isostaticity. *J Phys Condensed Matter* 2010; 22.
35. Lespiat, R., Cohen-Addad, S. Höhler, R. Jamming and flow of random close packed spherical bubbles: An analogy with granular materials. *Phys Rev Lett* 2011; 106: 148302.
36. Durian, D.J. Foam mechanics at the bubble scale. *PRL* 1995; 75: 4780–4783.
37. Lacasse, M-D., Grest, G.S., Levine, D. Deformation of small compressed droplets. *Phys Rev E* 1996; 54: 5436.
38. Hoehler, R. Cohen-Addad, S. Many-body interactions in soft jammed materials. *Soft Matter* 2017; 13: 1371–1383.
39. Morse, D.C. Witten, T.A. Droplet elasticity for weakly compressed emulsions. *EPL* 1993; 22: 549–555.
40. Winkelmann, J., Dunne, F.F., Langlois, V.J., Möbius, M.E., Weaire, D., Hutzler, S. 2D foams above the jamming transition: Deformation matters, *Colloids and Surfaces A: Physicochemical and Engineering Aspects* 2017; 534: 52–57.
41. Dunne, F.F., Bolton, F., Weaire, D., Hutzler, S. Statistics and topological changes in 2d foams from the dry to the wet limit. *Philoso Mag* 2017; 97: 1768–781.
42. Hutzler, S., Murtagh, R., Whyte, D., Tobin, ST., Weaire, D. Z-cone model for the energy of an ordered foam. *Soft Matter* 2014; 10: 7103–7108.
43. Whyte, D., Murtagh, R., Weaire, D., Hutzler, S. Applications and extensions of the Z-cone model for the energy of a foam. *Colloids Surfaces A, Physicochem Eng Aspects* 2015; 473: 115–122.
44. Princen, H.M. The structure, mechanics, and rheology of concentrated emulsions and fluid foams. In: Sjoblom, J. (ed.) *Encyclopedic Handbook of Emulsion Technology*. Marcel Dekker: New York, Basel, 2000, p. 243.
45. Princen, H.M., Kiss, A.D. Osmotic-pressure of foams and highly concentrated emulsions. 2. Determination from the variation in volume fraction with height in an equilibrated column. *Langmuir* 1987; 3: 36–41.
46. Princen, H.M. Osmotic pressure of foams and highly concentrated emulsions. I. Theoretical considerations. *Langmuir* 1986; 2: 519–524.
47. Feitosa, K., Marze, S., Saint-Jalmes, A., Durian, D.J. Electrical conductivity of dispersions: From dry foams to dilute suspensions. *J Phys Condens Matter* 2005; 17: 6301–6305.
48. Lemlich, R. A theory for the limiting conductivity of polyhedral foam at low density. *J Colloid Interface Sci* 1978; 64: 107–110.
49. Forel, E., Rio, E., Schneider, M., Beguin, S., Weaire, D., Hutzler, S., Drenckhan, W. The surface tells it all: Relationship between volume and surface fraction of liquid dispersions. *Soft Matter* 2016; 12: 8025–8029.
50. Wang, Y.J. Neethling, S.J. The relationship between the surface and internal structure of dry foam. *Colloids Surf A* 2009; 339: 73–81.
51. Vera, M.U., Saint-Jalmes, A. Durian, D.J. Scattering optics of foam. *Applied Optics* 2001; 40: 4210–4214.

52. Lambert, J., Cantat, I., Delannay, R., Mokso, R., Cloetens, P., Glazier, J.A., Graner, F. Experimental Growth Law for Bubbles in a Moderately "Wet" 3D Liquid Foam. *Phys Rev Lett* 2007; 99: 058304.

53. Mader, K., Mokso, R., Raufaste, C., Dollet, B., Santucci, S., Lambert, J., Stampanoni, M. Quantitative 3D characterization of cellular materials: Segmentation and morphology of foam. *Colloids Surf A* 2012; 415: 230–238.

54. Jiang, Y., Glazier, J.A. Extended large-Q Potts model simulation of foam drainage. *Phil Mag Lett* 1996; 74: 119–128.

55. Thomas, G.L., de Almeida, R.M.C., Graner, F. Coarsening of three-dimensional grains in crystals, or bubbles in dry foams, tends towards a universal, statistically scale-invariant regime. *Phys Rev E (Stat Nonlin Soft Matter Phys)* 2006; 74: 021407–021418.

56. Saye, R.I., Sethian, J.A. Multiscale modeling of membrane rearrangement, drainage, and rupture in evolving foams. *Science* 2013; 340: 720–724.

57. Saye, R.I., Sethian, J.A. Multiscale modelling of evolving foams. *J Comput Phys* 2016; 315: 273–301.

17 Foam Rheology

Norbert Willenbacher and Meike Lexis

CONTENTS

17.1 INTRODUCTION

Liquid foams are concentrated gas dispersions in a liquid with a packing fraction higher than 2/3 where the dispersed gas bubbles are no longer spherical. These jammed systems possess peculiar mechanical properties. The deformation under low stresses is mainly elastic due to capillary effects in the inclined foam films [1]. Above a critical stress, called the yield stress, the bubbles are forced to move past each other and the foam flows like a liquid. This special mechanical behavior gives rise to various industrial applications. Foams are used as drilling fluids in oil production or as firefighting agents. They are also used in everyday products, giving a special taste and feel to food products, enhancing the application properties of cosmetics or providing better cleaning properties for detergents. In all these applications, rheological properties need to be adapted to meet the according product requirements.

In this chapter, we will first give an overview of the basics of rheology and rheometry. We consider peculiarities in foam rheological measurements like wall slip and shear banding. Then we intensively discuss the rheological quantities yield stress and elasticity of foams in terms of their determination as well as the physical parameters determining their rheological properties. Finally, we focus on the correlation between microscopic interfacial phenomena and macroscopic foam behavior.

17.2 BASICS OF RHEOLOGY

Rheology describes the flow and deformation behavior of a material exposed to external mechanical stresses. In principal, this response can be viscous or elastic. For purely viscous materials, for example, water or honey, all of the imposed energy is dissipated. Deformation energy applied to purely elastic materials, for example, rubber, is completely stored. Most of the existing materials possess both properties, they behave viscoelastically.

First, we will consider the two limiting cases of ideal viscous and ideal elastic behaviors. With the help of a parallel plate model (Figure 17.1), the basic shear rheological parameters can be defined. A homogeneous medium is placed between the two plates with area A and distance y. The lower plate is fixed while the upper plate is moved with a constant velocity v. Provided that the material sticks to the plates, it is sheared to the distance x. The flow is assumed to be stationary and laminar.

The shear stress σ is defined as the force F acting per area:

$$\sigma = \frac{F}{A} \tag{17.1}$$

The deformation γ_s resulting from the applied shear stress is:

$$\gamma_s = \frac{x}{y} \tag{17.2}$$

The slope of the velocity versus position curve is the velocity gradient $\dot{\gamma}_s$:

$$\dot{\gamma}_s = \frac{dv}{dy} \tag{17.3}$$

An imposed stress σ accordingly results in a velocity gradient $\dot{\gamma}_s$. In the simplest case, $\dot{\gamma}_s$ is constant within the gap and proportional to σ. The proportionality constant η is called shear viscosity.

$$\sigma = \eta\dot{\gamma}_s \tag{17.4}$$

Materials that exhibit this linear relationship, for example, water, honey or glycerol, are called Newtonian fluids. For many materials, the viscosity is a function of applied shear rate and/or shearing time (Figure 17.2). They are called non-Newtonian fluids. If the viscosity decreases with increasing shear rate, the material is termed to exhibit shear thinning behavior. If the viscosity increases with shear rate, this is called shear thickening or dilatant behavior. The material is called rheopectic or thixotropic if the viscosity reversibly increases or decreases with time, respectively. Another class of materials, called yield stress fluids, essentially behaves like an elastic solid under low stresses and flows like a liquid when a certain critical stress, the yield stress σ_y, is exceeded. Depending on the flow behavior at stresses $\sigma > \sigma_y$ the material is called Bingham fluid or Herschel-Bulkley fluid if the response is Newtonian or shear thinning, respectively. A wide range of materials such as concentrated suspensions, emulsions, foams, pastes, and composites show yield stress behavior. This will be discussed in more detail in Chapter 17.5.

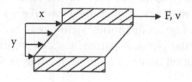

FIGURE 17.1 Parallel plate model.

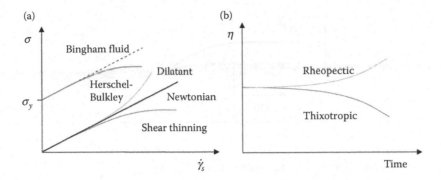

FIGURE 17.2 Different flow behavior of materials in dependence of (a) shear rate and (b) time.

For ideal elastic material, the deformation behavior can be described by Hooke's law:

$$\sigma = G_S \gamma_S \tag{17.5}$$

where G_S is the shear modulus and γ_S the deformation or strain. After relief of the strain, the material relaxes to the initial state without any remaining deformation. This behavior is independent of shear stress and duration of shear load.

Viscoelastic materials possess both viscous and elastic properties. At constant strain, shear stress shows a time dependent decrease and finally relaxes to zero for a viscoelastic fluid or to a finite value for viscoelastic solids. The range in which the resulting stress is proportional to the applied strain is called the linear viscoelastic regime (LVE). Strains that exceed this regime lead to a change of the microstructure and to a decrease of the apparent shear modulus. The most commonly used method for characterization of viscoelastic properties are oscillatory shear measurements. A sinusoidal deformation with small amplitude γ_0 and defined angular frequency Ω is applied to the material.

$$\gamma_S = \gamma_0 \cdot \sin(\Omega t) \tag{17.6}$$

The derivative of the deformation with respect to time gives the shear rate:

$$\dot{\gamma} = \Omega \cdot \gamma_0 \cdot \cos(\Omega t) \tag{17.7}$$

The response of the shear stress is characterized by the same sinusoidal shape as the input signal with a phase shift δ that is specific to each material:

$$\sigma = \sigma_0 \cdot \sin(\Omega t + \delta) \tag{17.8}$$

The phase angle δ lies between $0°$ (ideal elastic solids) and $90°$ (ideal viscous liquids).

The reduced elastic stress contribution which is in phase with the applied strain is termed storage modulus:

$$G' = \frac{\sigma_0}{\gamma_0} \cdot \cos(\delta) \tag{17.9}$$

whereas

$$G'' = \frac{\sigma_0}{\gamma_0} \cdot \sin(\delta) \tag{17.10}$$

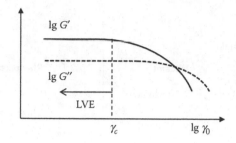

FIGURE 17.3 Oscillatory deformation amplitude sweep of G' and G''.

characterizes the viscous response which is in phase with the applied strain rate. The storage modulus is a measure of the deformation energy stored in the material, while the loss modulus in turn stands for the dissipated energy. Both quantities are used to define the complex shear modulus G^*:

$$G^* = G' + iG'' \qquad (17.11)$$

The ratio of loss and storage modulus gives the tangent of the phase angle and is called the loss factor:

$$\tan(\delta) = \frac{G'}{G''} \qquad (17.12)$$

Usually, the first step in characterizing viscoelastic material properties via oscillatory shear measurements is performing an amplitude sweep, where the deformation amplitude is varied at constant frequency. In this manner, the linear viscoelastic regime, where G' and G'' are constant, can be determined (Figure 17.3).

With knowledge of the critical deformation γ_c, a frequency sweep within the linear viscoelastic regime can be performed at a constant deformation amplitude $\gamma_0 < \gamma_c$. This experiment gives information about the time dependent deformation behavior. At low frequencies, the microstructure of a material has time to relax the applied stress and the viscous character dominates ($G'' > G'$). At high frequencies, materials appear more rigid since stress cannot relax fast enough and hence, the elastic properties dominate ($G' > G''$). Foams usually exhibit a broad frequency range where the elastic response dominates ($G' \gg G''$) and a terminal flow regime with $G'' > G'$ is hardly detectable.

17.3 ROTATIONAL RHEOMETRY

Rotational rheometers are widely used for rheological characterization of all kinds of fluids. A manifold of instruments with different specifications is commercially available. These rheometers are either used in a controlled stress or controlled strain/strain rate mode. This means either a torque is applied to the investigated sample and the corresponding angular deflection or speed is measured or vice versa. The torque is related to the stress acting on the fluid via the geometrical specifications of the sample fixture and angular deflection or speed are related to strain and strain rate, respectively. Today, most commercial rheometers are stress-controlled devices. Advanced instruments include a sophisticated control loop such that they can also be used in a controlled strain mode. Basically, rheological measurements are performed in two different modes. The material is either sheared continuously or oscillatorily.

Continuous or steady shear measurements are often used to measure the viscosity in dependence of the shear rate, shear strain, shear stress or time. Oscillatory measurements are used to determine the viscoelasticity of materials and the time dependent deformation behavior. There are several fixture geometries used for these measurements briefly described in the next section.

17.3.1 CONCENTRIC CYLINDER/VANE GEOMETRY

In the concentric cylinder measuring system (Figure 17.4a), a shear gap is generated by placing a bob into a cup, both with the same rotational axis. If the shear gap is small, a uniform shear rate can be assumed. Therefore, the ratio between radius of the inner bob R_B and outer cylinder R_C should be $0.97 < R_B/R_C < 1$ [2]. There are two possibilities to carry out the measurement. Either the bob rotates (Searle method) or the outer cylinder rotates (Couette method). In both cases, the torque acting on the inner bob is measured.

For dispersive systems like suspensions, emulsions or foams, the required shear gap is given by the particle, droplet or bubble size and hence, the restriction to a narrow shear gap cannot always be met. As a rule of thumb, the system can be treated as a continuum when the ratio of shear gap to dispersed phase size is at least ten. When adapting the measuring system by increasing the gap width, the inaccuracy arising due to varying shear rate across the gap needs to be taken into account.

Consider a sample placed between two concentric cylinders. A rotation of angular frequency Ω causes a shear rate $\dot{\gamma}_S$ at the inner wall that is given by:

$$\dot{\gamma} = 2\Omega \frac{R_C^2}{R_C^2 - R_B^2} \tag{17.13}$$

By measuring the torque M on the bob, the shear stress σ can be calculated as follows:

$$\sigma = \frac{M}{2\pi R_B^2 L} \tag{17.14}$$

where L is the immersion length of the bob.

The measured torque has to be corrected for contributions to the fluid deformation at the flat or conically shaped bottom end of the bob. The concentric cylinder system is especially suitable for low viscosity fluids, as they cannot flow out of the gap and a large area of shear is provided.

17.3.2 4-BLADED VANE GEOMETRY

Often, the bob is replaced by a vane (Figure 17.4b) especially for yield stress determination [3–5]. It is tolerant of large particles, droplets or bubbles and most importantly, wall slip is avoided. The

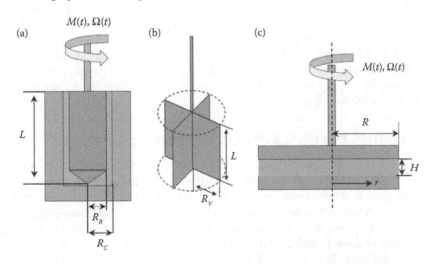

FIGURE 17.4 Illustration of (a) concentric cylinder (b) vane measuring system (c) plate/plate.

onset of vane rotation clearly indicates a structural change within the sample. Furthermore, sample damage during loading is reduced due to the thin blade-like profile [6]. The 4-bladed vane is most often used but 2-, 6- or 8-bladed vanes also exist.

For a 4-bladed vane properly filled to the upper edge of the blades, the following equation relates the shear stress acting on the fluid at the rim of the vane to the measured torque [5]:

$$\sigma = M \frac{1}{4\pi R_V^3} \left(\frac{L}{2R_V} + \frac{1}{6} \right)^{-1} \quad \text{for } \frac{L}{R_V} > 4 \tag{17.15}$$

17.3.3 PARALLEL PLATE

In the parallel plate measuring system (Figure 17.4c,d), the sample is placed between two plates with a gap width H. One of these plates is moving with angular velocity Ω while the other one is fixed. The tangential velocity depends on the radius r and the gap height h:

$$v(r,h) = \Omega r \frac{h}{H} \tag{17.16}$$

The shear rate is given by:

$$\dot{\gamma}_S = \frac{\Omega r}{H} \tag{17.17}$$

Hence, the shear rate depends on the radial position and it increases from zero in the center to its maximum value at the outer edge.

The shear stress at the rim is related to the torque:

$$\sigma = \frac{3M}{2\pi R^3} \left(1 + \frac{1}{3} \frac{d\ln(M)}{d\ln(\dot{\gamma}_S)} \right) \tag{17.18}$$

The adjustable gap height makes the parallel plate geometry very useful, especially for the measurement of dispersive systems like foams where large bubble radii r_{bubble} require correspondingly large shear gaps ($2r_{bubble} < H/10$) thus permitting evaluation of experimental data within the framework of continuum mechanics.

Another frequently used geometry is the cone and plate system. An advantage of this measuring system is that the shear rate within the gap is constant. However, the cone and plate geometries are not suitable for fluids including large dispersive objects because the continuum approximation is not valid in the vicinity of the gap center.

17.4 PECULIARITIES IN FOAM RHEOLOGICAL MEASUREMENTS

Foams exhibit a complex rheological behavior in various respects very similar to that of highly concentrated emulsions. Accurate determination of the rheological quantities like yield stress or viscosity is often disturbed by phenomena like wall slip, shear localization or shear banding. In general, foams are even less stable than emulsions and foam structure may change during the course of a rheological measurement. Drainage, coalescence, and Ostwald ripening are responsible for the thinning and rupture of foam lamellae. Accordingly, average bubble size as well as the gas volume fraction increase with time. Beyond that, flow-induced bubble coalescence and structural changes may occur depending on foam stability and have to be considered carefully.

17.4.1 WALL SLIP

Wall slip occurs due to inhomogeneous fluid properties at a boundary wall that causes a regime with a high shear rate gradient called a slipping layer [7]. As a result, the deformation in a rotational rheometer will not be affine anymore and in pipes plug flow will occur [8]. The situation for flow in parallel plates with slip is shown in Figure 17.5. Wall slip velocities v_{slip} depend on the type of wall material [9], but also on interfacial rheology [10]. Denkov et al. measured the foam wall friction of dry foams ($\phi = 0.9$) and found different friction laws for foams characterized by different surface elastic moduli.

Wall slip can be detected, for example, by comparing measurements of different geometries or by executing a measurement series using different gap heights. As the width of the slip layer is independent of gap width, its relative contribution to the deformation and flow within the gap decreases and the apparent viscosity increases with increasing gap height.

In Figure 17.6, apparent flow curves of a commercial shaving foam measured with different plate materials at different gap heights (3–6 mm) are shown. Sandblasted (almost smooth) plates, a serrated plate, and plates covered with sandpaper (grain size around 420 μm) have been used. Obviously, wall slip occurs for the first two measuring systems as the viscosity increases with

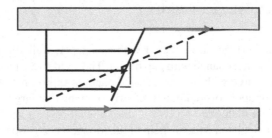

FIGURE 17.5 Parallel plate velocity field. The figure shows the velocity field of any particular radius r. The wall slip velocity v_{slip} is the same at each wall. Also shown are the actual shear rate in the fluid $\dot{\gamma}_S$ determined from the gradient of the linear velocity field and the apparent shear rate $\dot{\gamma}_{app}$ calculated from the relative velocity of the plates without knowledge of the slip layer. (Picture redrawn from Yoshimura, A. *J Rheol* 1988; 32: 53 [34].)

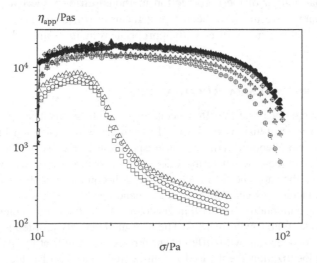

FIGURE 17.6 Apparent viscosity of a commercial shaving foam (Balea Men, Mann & Schröder GmbH) versus shear stress measured with a plate/plate set-up using different plate materials and gap heights. Sandblasted plates (open symbols), upper plate serrated/lower plate sandpaper (crossed symbols), sandpaper attached to both plates (closed symbols). Gap height: (■) 3 mm, (●) 4 mm, (▲) 5 mm, (◆) 6 mm.

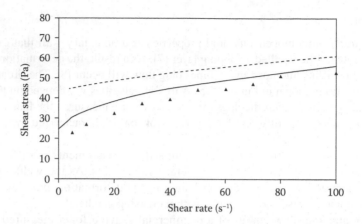

FIGURE 17.7   Shear stress versus shear rate for a foam containing 2% of surfactant dissolved in an aqueous solution of 2.5 g/L of Xanthan, 1 g/L of Aquapac regular (anionic cellulose), and 0.5 g/L of NaCl adjusted to pH 9: (▲) experimental data obtained with grooved plate surfaces ($\dot{\gamma}_S = 5 - 120 \ \mathrm{s}^{-1}$), (---) data computed with Mooney hypothesis, (–) data computed with Oldroyd–Jastrzebski hypothesis. (Data taken from Alderman, N.J. et al. *J Nonnewton Fluid Mech* 1991; 39: 291–310 [13].)

gap height. When the plate surfaces are covered with sandpaper, wall slip can be excluded as the corresponding flow curves do not change with gap height. The yield stress (explained in Chapter 17.5) extracted from the data measured with the sandblasted plates is approximately 4–5 times lower than the one measured with sandpaper covered plates, clearly showing that large measurement errors can arise when using wrong plate materials.

True shear rates can be calculated with the help of the Mooney-correction, which assumes that the slip velocity v_{slip} only depends on the shear stress at the wall ($v_{\mathrm{slip}} \sim \sigma$) [11], or the Oldroyd-Jastrzebski method, which assumes that v_{slip} depends not only on the wall stress, but also on the pipe diameter ($v_{\mathrm{slip}} \sim \sigma/H$) [12]. The different methods have been compared to measurements with grooved plate surfaces where wall slip could be excluded for foams made from surfactant mixtures [13]. Figure 17.7 demonstrates that neither of these correction methods retrieves the true flow curve for this surfactant foam and similar results have been reported earlier [14,15]. Hence, it is recommended to avoid or minimize wall slip using fixtures with appropriate roughness and selecting a large enough gap width [13,16].

17.4.2   SHEAR BANDING AND SHEAR LOCALIZATION

Shear banding and shear localization often occur in yield stress materials like granular matter, concentrated emulsions, suspensions or foams. In both cases, undeformed and sheared regions coexist [17], but these two phenomena differ with respect to the velocity profile in the shear gap.

Shear banding refers to the case where the velocity changes drastically within a narrow region at a critical position inside the gap. The shear rate exhibits a discontinuity at this position. The origin of shear banding is not understood so far. However, shear banding of foams has been reported upon shear start-up [18] and during continuous shear [19] in dry foams. The latter experiments were carried out in a parallel plate rheometer with rough surfaces. The measured velocity profile inside the gap is shown in Figure 17.8. Clearly, two regimes with different shear rates evolve indicating shear banding behavior.

In the case of shear localization, the sheared and unsheared regions evolve due to an inhomogeneous stress distribution across the shear gap as is found in Couette flows. But the transition from the sheared to the unsheared region is supposed to be continuous and the shear rate decreases gradually to zero. Consider a yield stress material in a wide gap Couette rheometer with the shear stress decreasing as $\sigma \sim (1/R^2)$ where R is the radial coordinate.

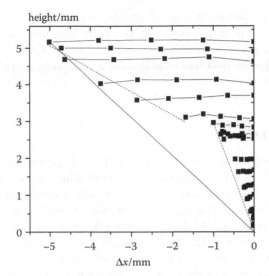

FIGURE 17.8 Velocity profile of a surfactant foam in a plate-plate rheometer at a gap height of 5 mm and constant shear rate $\dot{\gamma} = 0.2$ s^{-1}. (Adapted from Lexis, M., Willenbacher, N. *Colloids Surf A Physicochem Eng Asp* 2014; 459: 177–185 [19].)

At stresses below the yield stress, that is, at positions at which $\sigma < \sigma_y$, the material will not flow. With increasing applied stress, the material in the vicinity of the rotating bob will start to move because in this region the yield stress is already exceeded, but due to the inhomogeneous stress distribution, the applied stress close to the outer cylinder wall is still below the yield stress and the material will not flow there. The sheared region will become larger with increasing applied stress until eventually the yield stress is exceeded in the whole gap [17,18]. Plug flow of pasty materials is a well known example for shear localization [20].

Ovarlez et al. [21] observed a continuous decrease of $\dot{\gamma}_S(r)$ finally approaching zero within the experimental resolution at a distance $r = r_{\text{yield}}$ corresponding to the position at which $\sigma = \sigma_y$ (Figure 17.9) for surfactant foams with either rigid or mobile interfaces steadily flowing in a wide gap

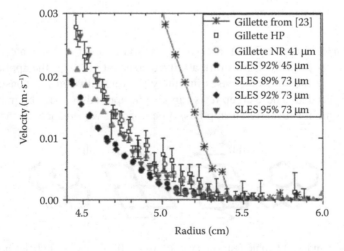

FIGURE 17.9 Velocity profiles of foams obtained in a Couette geometry ($R_B = 4.1$ cm, $R_C = 6$ cm, and $L = 11$ cm) using magnetic resonance imaging (MRI) [21,22]. Two commercial foams (Gilette HP, NR) and four surfactant foams made from a sodium lauryl ether sulfate (SLES) solution but differing in gas volume fraction and bubble size as indicated in the legend.

geometry much larger than the bubble size. Obviously, these foams behave as simple yield stress fluids without shear banding behavior. Note that in Figure 17.9, there are also data plotted from Rodts et al. [22] where shear banding behavior was found for the same commercial shaving foam as used by Ovarlez et al. The authors speculate that the foams used by Rodts et al. contained small impurities like solid particles which can lead to thixotropic effects and induce shear banding. This phenomenon is well known for densely packed emulsions [23].

17.5 THE YIELD STRESS

Foams, especially in the dry limit $\phi \to 1$, are densely packed, jammed systems [24]. They do not flow and deform elastically under stresses below a critical value called yield stress. If this critical value is exceeded, the microstructure, consisting of many interacting bubbles, is forced to rearrange [25] and the foam begins to flow. The yield stress is often defined as the minimum stress that is needed to make a material flow [4,26]. However, the existence of a true yield stress in such soft matter is controversial. Barnes et al. [27,28] for example claim that all fluids that flow under high shear stresses would do so under low stresses. The viscosity would always possess a finite value and it is just a question of measuring technique to determine it in the low stress regime. Other studies [16,19] argue that the yield stress is an "engineering reality" and many fluids exhibit drastic orders of magnitude drop in viscosity within a narrow range of applied stresses. Even if it may not be a true material constant, it is a useful measurable quantity characterizing processing and application properties of foam systems [29]. But it should be kept in mind that the absolute value of a measured yield stress not only depends on the material but may be strongly influenced by the type and time scale of sample load and deformation.

In the case of foams, the yield stress σ_y is considered as the point where the bubbles start sliding past each other. The bubble rearrangement scheme as proposed in a pioneering work of Princen [30] is shown in Figure 17.10. Stresses $\sigma < \sigma_y$ deform the bubbles but do not induce a structure change. When the stability limit is exceeded ($\sigma \geq \sigma_y$) the bubbles reorient as depicted in Figure 17.10d.

In real foams, bubble rearrangements are additionally induced by destabilization processes like disproportionation, drainage, and coarsening.

There are several methods to determine a yield stress value. The most common ones are explained in the following.

17.5.1 CREEP TEST

In a creep experiment (Figure 17.11a), a certain stress is applied to the material and the strain response is observed over time. If the strain attains a constant value γ_e, the applied stress is lower than the yield stress. If the strain increases infinitely to eventually reach a constant shear rate $\dot{\gamma}_e$, the yield stress is exceeded. These creep tests can also be applied in series by increasing the stress continuously. The time for this stress ramp has to be selected with respect to the relevant process

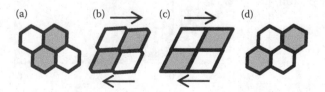

FIGURE 17.10 Shear induced bubble rearrangement: (a) $\sigma = 0$, (b) $\sigma < \sigma_y$, bubbles get deformed but no structure change is induced. After release of the stress, the bubbles would go back to their initial position, (c) $\sigma \approx \sigma_y$: hexagonal structure turns into a tetragonal structure, stability limit is reached, (d) $\sigma > \sigma_y$: bubbles retain their hexagonal structure and reorient.

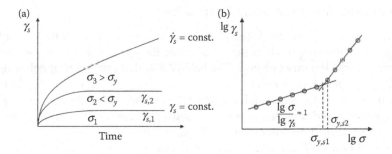

FIGURE 17.11   Determination of yield stress by (a) creep tests, (b) continuously increasing stress.

and material. The resulting strain curve $\gamma_s(\sigma)$ (Figure 17.11b) can be divided into two regions with different slopes. In the first part, the slope is close to one and deformation is only small. In the second part, the slope drastically increases indicating the onset of flow. Either the intersection of the tangents of the two regions $\sigma_{y,S1}$ or the first point that belongs to the second region $\sigma_{y,S2}$ can be selected to characterize the yield stress.

In real foams, the transition between the two regions is not always as sharp as depicted in Figure 17.11b, instead there is a transition zone in between. Figure 17.12 shows a deformation versus shear stress curve for a surfactant and a protein foam, respectively. Video recordings have been used to determine the point where the bubbles started sliding past each other. In Figure 17.12, this stress value has been marked with a circle. For the surfactant foam (Figure 17.12b), the transition zone is narrow so that $\sigma_{y,S1} \approx \sigma_{y,S2}$. In contrast, the transition zone for the whey protein isolate (WPI) foam (Figure 17.12b) spans over a relatively wide range of shear stresses so that $\sigma_{y,S1} \approx 2\sigma_{y,S2}$. The inserts in Figure 17.12 showing a magnification of the transition zone indicate that $\sigma_{y,S2}$ seems to be closer to the point where bubbles start to slide past each other.

Another protocol for yield stress determination in rotational rheometry is to apply a very slow, constant flow rate and to observe the stress response. The stress increases to a maximum (the yield stress) before it reaches a steady state value [31]. The inclined plane method is another approach to determine the yield stress [32]. Here, a foam is placed on a plane which can be inclined to different angles. From the angle where flow is visually observed, the yield stress can be calculated.

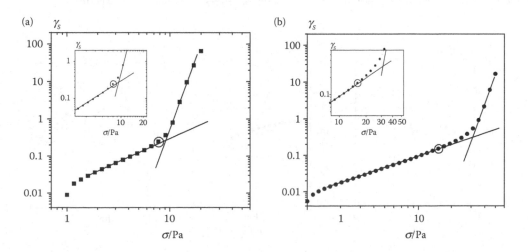

FIGURE 17.12   Deformation versus shear stress for (a) surfactant foam (2% Triton X-100, 0.2% sodium dodecyl sulfate (SDS) in water), (b) 1% whey protein isolate (WPI) foam. The circles mark the yield point that is visually observable from video recordings of the shear gap and the inserts show a magnification of the transition zone.

17.5.2 OSCILLATORY AMPLITUDE SWEEP EXPERIMENTS

An alternative way to access the yield stress can be found in oscillatory amplitude sweep measurements. The stress amplitude is increased while the frequency is kept constant. The end of the LVE is coupled with the onset of flow. The behavior well above and well below the yield stress can be described by power laws corresponding to straight lines in a logarithmic plot as shown schematically in Figure 17.13. Some authors define the intersection of these lines as yield stress $\sigma_{y,i}$ [7,27,32]. For emulsions and some foams, this is a robust and reproducible method. But not all foams show these two well defined deformation regimes (see Figure 17.14).

The crossover of G' and G'' can also be defined as yield stress $\sigma_{y,c}$ because this is the point where the viscous properties start to dominate over the elastic ones. For foams, it has been reported in several studies that yielding occurs before the crossover point is reached and $\sigma_{y,c}$ does, therefore, not seem to be an appropriate value [7,25,32].

The stress amplitude at which a certain deviation from the plateau value of G' occurs may also be defined as yield stress $\sigma_{y,o}$. However, it is not possible to specify a unique deviation

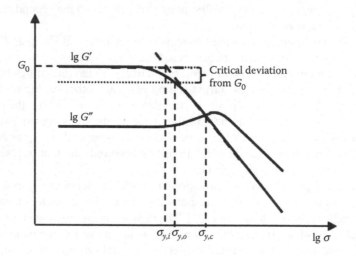

FIGURE 17.13 Yield stress determination from oscillatory stress amplitude sweep experiments.

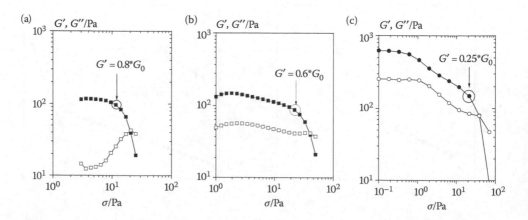

FIGURE 17.14 Storage (closed symbols) and loss moduli (open symbols) versus shear stress amplitude at $f = 1$ Hz for (a) surfactant foam (2% TX-100, 0.2% SDS), (b) 5% casein foam and (c) 1% whey protein isolate foam. The circles mark the yield point that is visually observable from video recordings of the shear gap.

criterion that applies for a broad variety of foams. Hence, it is recommended to visualize the bubbles inside the shear gap during the measurement to find an appropriate method for yield stress determination corresponding to the onset of bubbles sliding past each other in a given foam system.

17.5.3 Comparison of Yield Stresses Determined from Rotational and Oscillatory Measurements

Comparison of foam yield stresses determined via different methods is not well discussed in literature since a broad data basis is still lacking. However, Rouyer et al. [32] gathered data from different sources where foam and emulsion yield stresses have been measured by steady (inclined plane method or measurement of the stress response under low shear rate) and oscillatory ($\sigma_{y,i}$) shear experiments as shown in Figure 17.15a. The yield stresses are normalized by G_0 and as both quantities depend on (γ/r_{32}) effects of bubble size and polydispersity can thus be excluded. All data collapse on a master curve with the exception of the steady shear experiments of the emulsions carried out by Mason et al. [27]. This deviation was attributed to shear banding phenomena. Shear banding or foam fracture may occur at different shear stresses due to the different kind of deformation that is applied to the foam in oscillatory and steady shear thus leading to different apparent yield stress values. This comparison does not comprise a large variety of foams and, therefore, it is not yet clear whether the observed agreement is generally valid.

In Figure 17.15b, yield stress values $\sigma_{y,c}$ determined from oscillatory shear measurements are plotted versus $\sigma_{y,S2}$ determined from steady shear measurements for different protein and surfactant foams. The yield stress determined from the crossover of G' and G'' is always approximately 1.5 times higher than $\sigma_{y,S2}$. This correlation is useful since this crossover can be determined in a straightforward and highly reproducible manner. But it has to be kept in mind that large deviations

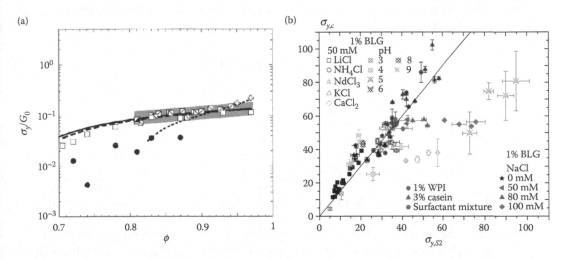

FIGURE 17.15 (a) Yield stress normalized by elastic shear modulus vs. volume fraction. Comparison of data from Princen and Kiss [15,16] (dotted line), Saint-Jalmes and Durian (oscillatory) [7] (continuous black line), Khan et al. [31] (⊙), Mason et al. [27]: oscillatory (□) and steady shear (●) experiments. Rouyer et al. [32]: Gillette foam for oscillatory (◁) and inclined plane (◢) measurements, tetradecyltrimethylammonium bromide (TTAB) foam (oscillatory) (▱). The dashed line corresponds to the curve: $\sigma_y/G_0 = 0.39(\phi - \phi_x)/\phi$. (b) Yield stresses determined from oscillatory shear measurements $\sigma_{y,c}$ versus yield stresses determined from rotational shear measurements σ_{S1} 1% ß-lactoglobulin (BLG) foams containing different amounts of NaCl and 50 mM of different kind of salt (more information in Reference 33), 1% whey protein isolate (WPI) foams, 3% casein foams, and foams made from a surfactant mixture containing 2% TX-100 and 0.2% SDS (more information in Reference 34). The continuous straight line corresponds to $\sigma_{y,c} = 1.5\sigma_{y,S1}$.

are found for β-lactoglobulin foams containing 50 mM of the divalent salt $CaCl_2$, 100 mM NaCl, and at pH 5. Under these conditions, protein aggregates occur more frequently and substantially different microstructure within the lamellae could lead to differences in the deformation response to oscillatory and steady shear as already mentioned above.

17.5.4 PREDICTION OF YIELD STRESS

Various studies have confirmed that the yield stress of so called liquid foams including a low viscosity of the continuous phase depend on the average bubble size, the surface tension, the gas volume fraction, and the liquid viscosity [34–39] as

$$\sigma_y = k \cdot \left(\frac{\gamma}{r_{32}} \right) \cdot \left(\frac{\eta_L}{\eta_W} \right)^{0.3} \cdot (\phi - \phi_c)^2 \tag{17.19}$$

where γ is the surface tension, r_{32} the Sauter mean radius, ϕ the gas volume fraction, and ϕ_c represents the maximum packing fraction of the bubbles before they start to deform into nonspherical shapes. This is the point where the system becomes jammed so that small stresses result in an elastic deformation and a minimum stress is needed to move the bubbles past each other. Usually, ϕ_c has been an estimated value [40,41], but can also be calculated from the measured bubble size distribution, as proposed by [19]. They used an empirical model equation established by Sudduth et al. [42]. The equation is based on a large number of experimental data for suspensions of non-Brownian particles. This restriction to undeformable spheres is not a serious constraint here since even the gas bubbles in foams at such low gas volume fractions are essentially spherical. The maximum packing fraction ϕ_c is calculated from the size distribution of the suspended particles. Assuming n-modal discrete distribution results in Equation 17.20, where $\phi_{c,mono}$ is the maximum packing fraction of a monodispersive suspension ($\phi_{c,mono} = 0.63$) and r_x is the x-th moment of the particle size distribution.

$$\phi_c = \phi_n - (\phi_n - \phi_{c,mono}) \exp\left[0.271 * \left(1 - \frac{r_5}{r_1} \right) \right] \tag{17.20}$$

with $\phi_n = 1 - (1 - \phi_{c,mono})^n$ and $r_x = \left(\sum_{i=1}^n N_i r_i^x \right) / \left(\sum_{i=1}^n N_i r_i^{x-1} \right)$.

The prediction of the yield stress also includes an empirically determined factor for the (weak) contribution of the liquid viscosity, where η_L is the continuous phase viscosity and η_w the water viscosity at the same conditions. This phenomenological extension of the model equation proposed by Lexis et al. [19,43] has been derived from measurements on foams made from casein, whey protein isolate, and a mixture of synthetic surfactants (Figure 17.16a). The solvent viscosity was varied using different water/glycerol mixtures and sugar solutions.

Furthermore, the equation includes a numerical prefactor k that varies depending on the kind of adsorbed amphiphile molecule at the interface (Figure 17.16b) In the literature, k values between 0.5 and 30 are found [11,19,27,33]. In Reference 33, it was shown that these k values are directly correlated with interfacial viscoelastic properties which will be discussed in Chapter 17.7 in more detail.

17.6 FLOW BEHAVIOR OF FOAMS UNDER STEADY SHEAR

In Figure 17.17a, flow curves of surfactant foams are shown. All curves possess an almost constant apparent viscosity η_{app} at low stresses $\sigma < \sigma_y$ followed by a drastic decrease of η_{app} in a narrow range of shear stresses around $\sigma \approx \sigma_y$. For $\sigma > \sigma_y$, the foams behave as shear thinning liquids. The apparent

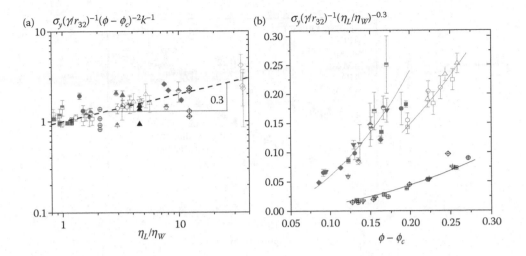

FIGURE 17.16 (a) Apparent yield stress σ_y normalized by Laplace pressure (γ/r_{32}), $(\phi - \phi_c)$ and k versus viscosity ratio (η_Λ/η_W), (b) Apparent yield stress τ_y normalized by Laplace pressure (γ/r_{32}) and solution viscosity ratio (η_Λ/η_W) vs. $\phi - \phi_c$ for foams made from different proteins and a surfactant mixture, vertically halved symbols: 1% WPI dissolved in various water glucose mixtures, closed symbols: 1% WPI, semiclosed symbols: 0.1% WPI, open symbols: casein and crossed symbols: surfactant mixture dissolved in various water/glycerol mixtures (glycerol or glucose content (■) 0%, (●) 20%, (▼) 30%, (▲) 40%, (◆) 60%).

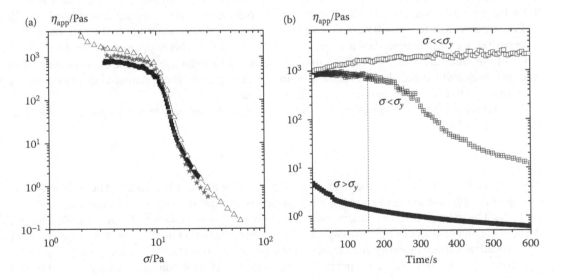

FIGURE 17.17 (a) Apparent viscosity versus shear stress for a foam made from a surfactant mixture (2% TX 100, 0.2% SDS) dissolved in water under different measurement conditions at a constant measurement time of 60 s: (■) shear stress continuously increasing from 3 to 25 Pa, (★) stepwise increase of shear stress from 3 to 30 Pa (2 s per data point), (△) stepwise increase of shear stress from 1 to 60 Pa (2 s per data point), (b) creep tests for the same surfactant foam with $\sigma \ll \sigma_y$ at 1 Pa (open symbols), $\sigma < \sigma_y$ at 8 Pa (crossed symbols) and $\sigma > \sigma_y$ at 20 Pa (closed symbols), (b) (Data taken from Lexis, M., Willenbacher, N. *Colloids Surf A Physicochem Eng Asp* 2014; 459: 177–185 [19].) All measurements were carried out with a plate/plate rheometer.

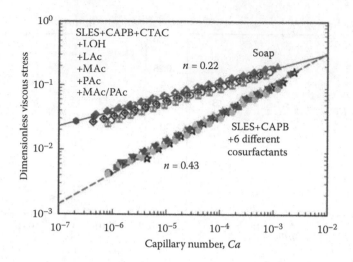

FIGURE 17.18 Dimensionless viscous stress $\sigma_v(\dot{\gamma}_S)/(\gamma/r_{32})$ versus capillary number $Ca = \eta\dot{\gamma}_S r_{32}\gamma$ for foams stabilized by sodium lauryl ether sulfate (SLES) + cocamidopropyl betaine (CAPB) mixture (0.33 + 0.17 wt%), without and with different cosurfactants added (0.02 wt%), which differ in their headgroups and for foams made from soap solution (mixture of potassium salts of fatty acids, pH = 10.2). The lines represent power law behavior with indices n as given in the figure. (Data taken from Denkov, N.D. et al. *A Lips, Soft Matter* 2009; 5: 3389 [1].)

viscosity below the yield stress is a result of bubble coalescence and rearrangement resulting in a motion of the upper plate [35,40]. For a given foam system, this quantity also depends on measuring parameters like initial and final stress of the measurement or the measurement time per data point. In Figure 17.17b, it can be seen that at stresses far below the yield stress, ($\sigma \ll \sigma_y$) of η_{app} increases over time. The reason for this is that the coalescence rate decreases with time due to decreasing number of separating lamellae [19].

For stresses close to but still below the yield stress ($\sigma > \sigma_y$), the foam starts to flow after a certain time period, for the example shown in Figure 17.17b after 160 s. This happens because the absolute value of the yield stress ($\sigma_y \sim 1/r_{32}$) of a foam decreases with foam age because the average bubble size increases and the distribution broadens with time. For $\sigma > \sigma_y$, the apparent viscosity decreases over time due to (shear induced) coalescence of the bubbles.

Beyond the yield stress, foams flow as shear thinning fluids well described by a phenomenological Herschel-Bulkley law:

$$\sigma = \sigma_y + \sigma_v(\dot{\gamma}_S) = \sigma_y + k_v\dot{\gamma}_S^n \tag{17.21}$$

where σ_y is the yield stress, k_v the foam consistency, and n a power law index. The term $\sigma_v(\dot{\gamma}_S)$ is the rate dependent fraction of the total stress which is estimated as $\sigma(\dot{\gamma}_S) - \sigma_y$ and scales as γ/r_{32} [1,10,16]. The index $n < 1$ is a characteristic of the shear-thinning behavior of foams and depends on the specific mechanism of viscous dissipation during flow. Denkov et al. [1,10,38] found the exponent n to depend on surface mobility and viscoelasticity. For different surfactant foams ($0.88 \leq \phi \leq 0.95$), they found $n \approx 1/2$ in the case of mobile interfaces (low surface modulus E^* of a few mN/m) and $n \approx 1/4$ in the case of rigid interfaces ($E^* > 60$ mN/m) as depicted in Figure 17.18.

17.7 STORAGE AND LOSS MODULUS

17.7.1 PLATEAU MODULUS

When small stresses below the yield stress are applied to foams, the response is linearly viscoelastic and the storage modulus G' which represents the elastic behavior is usually found to be much higher than the viscous modulus G''. The elasticity arises from the interfacial energy density ($\approx\gamma/R$) [36].

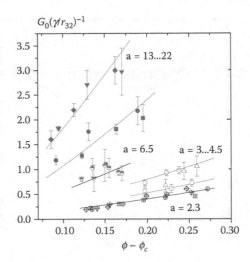

$G_0(\gamma/r_{32})^{-1}$

a = 13...22

a = 6.5

a = 3...4.5

a = 2.3

$\phi - \phi_c$

FIGURE 17.19 Plateau moduli normalized by Laplace pressure versus $\phi - \phi_c$ for foams made from 1% WPI (closed symbols), 0.1% WPI (semi-closed symbols), casein (open symbols), and surfactant mixture (crossed symbols) dissolved in various water/glycerol mixtures (glycerol content (■) 0%, (●) 20%, (▼) 30%, (▲) 40%, (◆) 60%).

At low frequencies between 0.1 and 10 Hz, the storage modulus is usually frequency independent, thus it is often denoted as plateau modulus G_0 and can be predicted as

$$G_0 = a \cdot \left(\frac{\gamma}{r_{32}}\right) \cdot \phi(\phi - \phi_c) \tag{17.22}$$

where r_{32} is the average Sauter radius, γ the surface tension, ϕ the gas volume fraction, and ϕ_c the critical gas volume fraction (see Chapter 17.5.4). The prefactor a varies for different foaming systems (Figure 17.19). Values between 0.5 and 30 have been reported, depending on interfacial rheological properties [11,19,33,44]. This will be discussed in more detail in Chapter 17.8.

17.7.2 VARIATION OF FREQUENCY

The frequency dependence of the complex modulus G^* for jammed systems is described by the following scaling law:

$$G^*(f) = G_0\left(1 + \sqrt{\frac{if}{f_c}}\right) + 2i\pi\eta_\infty f \tag{17.23}$$

where G_0 is the plateau modulus (Equation 17.22), f_c the characteristic relaxation frequency, and η_∞ the fluid viscosity in the limit of high frequency. This prediction has been confirmed experimentally in various investigations on foam systems [36,45,46]. The characteristic relaxation frequency is assumed to be proportional to the ratio of the dilational modulus E' and an effective interfacial viscosity including the surface viscosity E''/f_c, the solution viscosity, and the lamellar thickness as well as the bubble diameter [45]. Different scaling laws relating f_c to the foam modulus G_0 are predicted for rigid and mobile interfaces. This was confirmed experimentally for two different types of surfactant foams with $E' = 67$ mN/m and $E' \leq 20$ mN/m as shown in Figure 17.20.

For foams with even higher interfacial rigidity ($E^* \geq 100$ mN/m), it was shown that G^* cannot be described by Equation 17.23 in the whole frequency range and for each bubble size, above 10 Hz $G'' \sim f^\Delta$ with $\Delta < 1/2$ was found. This deviation becomes more pronounced with increasing interfacial rigidity and increasing bubble size [36].

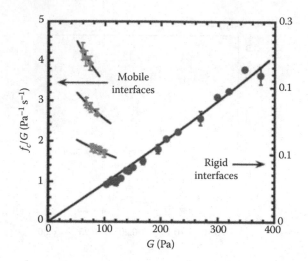

FIGURE 17.20 Scaled characteristic frequency f_c/G_0 of the collective bubble relaxation mode versus the elastic modulus G_0: (●) Gilette shaving foam, (◆) SLES (sodium lauryl ether sulfate) 40% glycerol, (▲) SLES 50% glycerol and (▼) SLES 60% glycerol foams. (Data taken from Krishan, K. et al. *Phys Rev E* 2010; 82 [45].)

More insight into relaxation mechanisms and how they are affected by bubble size and continuous phase viscosity is gained from the loss factor G''/G'. The curves for foams with moderate interfacial rigidity ($100 < E^* < 130$ mN/m) were found to collapse on a master curve when the frequency was rescaled by a factor $\tilde{\Omega}(d,\eta)$ (Figure 17.21a,b).

$$\tilde{\Omega}(d,\eta) = \Omega(d)H(\eta) = \left(\frac{d}{d_{\text{ref}}}\right)^2 \frac{\eta}{\eta_{\text{ref}}} \tag{17.24}$$

For foams with $E^* > 130$ mN/m, there is more than one characteristic frequency of viscoelastic response and no master curve could be found so far.

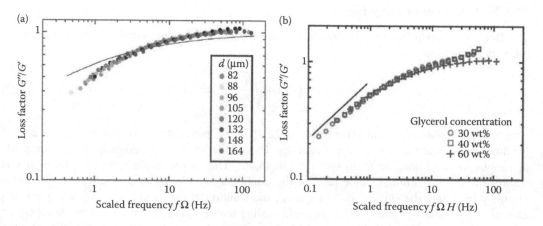

FIGURE 17.21 Loss factor of SLS-CAPB-LOH (sodium lauryl-dioxyethylene sulfate—cocoamidopropyl betaine—lauryl alcohol) foams versus scaled frequency (a) for different bubble sizes ($d_{\text{ref}} = 132$ μm, $E^* = 132$ mN/m, $\eta = 10.5$ mPas). The continuous line represents Equation 17.21 with the best fitted parameters $f_c = 0.2$ s^{-1}, $\eta_\infty = 0$, (b) for different continuous phase viscosities ($\eta = 2.5$–10.5 mPas, $E^* = 88$–132 mN/m, $d_{\text{ref}} = 133$ μm and $\eta_{\text{ref}} = 10.5$ mPas). d varies between 80 and 160 μm. The straight line has a slope = 1/2 and is a guide to the eye. (Data taken from Costa, S. et al. *Soft Mat* 2012; 9: 1100 [36].)

17.7.3 VARIATION OF SHEAR STRESS AMPLITUDE

Depending on the type of adsorbed molecule at the interface, G' and G'' can show different behaviors in dependence of applied shear stress amplitude as shown in Figure 17.14, exemplary for different aqueous foaming systems containing either 1% whey protein isolate (WPI), 3% casein or a surfactant mixture (0.2% SDS, 2% Triton X-100). More details about these foams can be found in Reference 19. All systems show a linear regime with $G' > G''$ at low stress amplitudes and a flow regime with $G'' > G'$ at high stress amplitudes. For the casein system, there is a sharp transition between both regimes, but for the WPI and the surfactant foams a third regime can be distinguished between linear viscoelastic and flow regimes. For the former, the moduli decrease simultaneously and for the latter G' decreases while G'' increases before crossing. The 1% WPI foams exhibit high storage modulus values at very low stress amplitudes. In Reference 19, it was concluded that the whey proteins build a network across the lamella that causes such high moduli. Intermediate stress amplitudes probably destroy this network without moving the bubbles past each other and hence, G' decreases but is still higher than G''. The increase in G'' for the surfactant foams can be explained as follows. As the stress amplitude is applied, some of the foam films get stretched while others are being compressed leading to regions with lower and regions with higher surfactant concentrations. In order to equilibrate this imbalance, a Marangoni flow from the compressed regions to the stretched ones is induced [47]. This is a dissipative process that becomes stronger with higher stress amplitudes and, therefore, leads to an increase in G''.

17.8 RELATING INTERFACIAL AND FOAM ELASTICITY

Foam rheological properties are affected by interfacial rheology since shearing a foam induces stretching and compression of the lamellae and hence, the surfactant layer at the air liquid interface. In Chapter 17.6 and 17.7.2, it was already shown that viscosity and moduli of foams with rigid and mobile interfaces exhibit distinctly different behaviors. Here, we focus on direct correlations between interfacial and bulk foam rheological properties. Besson et al. [48] investigated the dynamic response to sinusoidal variations of the distance between the bubble centers of two adjacent bubbles connected by a single lamella (Figure 17.22a). In the linear viscoelastic regime, the dimensionless complex angular modulus $A^*(\Omega) = A' + iA''$ can be deduced. The quantity $A^*(\Omega)$ is further assumed to scale with the ratio of dilational surface modulus to surface tension ($A^* \sim E^*/\gamma$) and based on the model

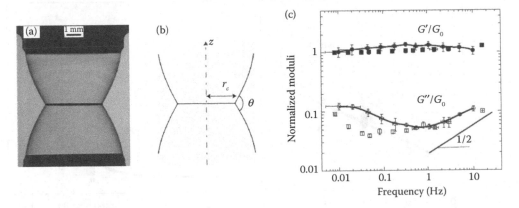

FIGURE 17.22 (a) Image of two contacting bubbles, (b) adhesion profile obtained after fitting the shape of the bubbles using the Young-Laplace equation. The contact radius r_c and the contact angle Θ are determined from the intersection of the reconstructed profiles, (c) measured ($\bullet$ G', $\circ$ G'') and predicted by Equation 17.25 ($\blacksquare$ G', $\square$ G') normalized elastic and loss moduli versus frequency with $G_0 = 206$ Pa, $\tau = 246$ s and $\alpha = 0.12$ is based on experimental angular moduli. The lines are guides to the eye. (Data taken from Besson, S. et al. *Phys Rev Lett* 2008; 101 [48].)

of Princen [49], a relationship between the complex foam modulus $G^*(\Omega)$ and the complex angular modulus $A^*(\Omega)$ measured for a single lamella is proposed:

$$\frac{G^*}{G_0} = \frac{i\omega\tau}{1+i\omega\tau}(1+\alpha A^*) \tag{17.25}$$

where τ is a characteristic time that is needed to equilibrate surface tension gradients between adjacent interfaces via coupled surface and bulk diffusive transport and α is a geometrical constant equal to $\sqrt{3}$ for a 2D hexagonal dry film network.

In Figure 17.22c, G'/G_0 and G''/G_0 data for a surfactant foam are compared to values plotted together with the data predicted by Equation 17.25 based on the two bubble response experiments. The good agreement between the data demonstrates that the fast relaxation processes observed in foams are determined by surfactant transport within the liquid films.

A direct empirical correlation between the yield stress and the interfacial dilational modulus E' of whey protein foams made at different pH, concentration, and valency of added salt has been proposed by Davis et al. [50]. However, they do not take into account the effect of bubble size (distribution) and gas volume fraction on σ_y, although pH and ionic strength are known to substantially affect the absolute value of this quantity. A systematic investigation of the influence of interfacial layer properties on the yield stress and the plateau modulus of different protein foams (mainly ß-lactoglobulin) was done by Lexis et al. [19,33] who found that foam rheology is tightly related to surface rheological properties of the corresponding protein solutions. The interfacial rheology was varied by adding different amounts and kinds of salt or changing the pH. Interfacial elastic moduli in shear and dilation were found to directly correlate with the normalized bulk foam plateau modulus G_0 (equal to prefactor a according to Equation 17.22) as depicted in Figure 17.23. Exceptions were found for foams where protein aggregation and structure or network formation across foam lamellae are supposed to be decisive for the bulk foam elastic modulus, for example, around the isoelectric point, at high ionic strength or for foams made from 1% whey protein isolate solutions.

Reduced yield stresses (equal to prefactor k in Equation 17.19) of the same ß-lactoglobulin (BLG) foams as in Figure 17.23 are plotted in Figure 17.24a and b versus the critical stress $\sigma_{c,\text{surface}}$ or the

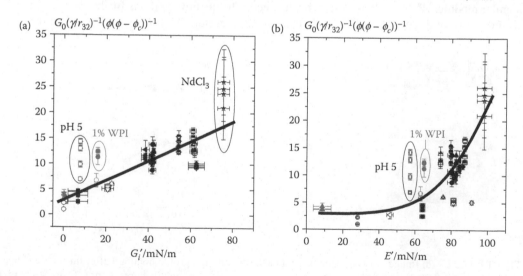

FIGURE 17.23 Storage moduli normalized by Laplace pressure (γ/r_{32}) and $\phi(\phi - \phi_c)$ versus (a) surface shear elastic modulus G_i' and (b) surface dilational eslastic modulus E'; (○) pH 3, (△) pH 4, (□) pH 5, (△) pH 6, (◇) pH 8, (▽) pH 9, pH 6.8 and NaCl: (■) 0 mM, (◆) 10 mM, (◄) 50 mM, (●) 80 mM, (▲) 100 mM, 50 mM (▥) KCl, (▲) NH_4Cl, (◉) LiCl, (◆) $CaCl_2$, (★) $NdCl_3$ (○) 0.1% WPI, (●) 1% WPI, (▲) 3% casein.

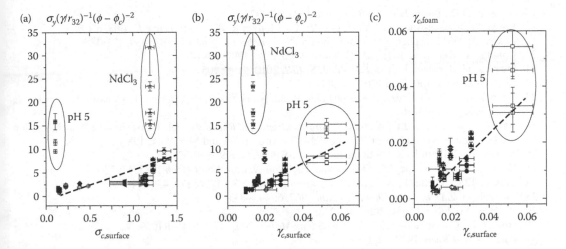

FIGURE 17.24 Yield stresses normalized by Laplace pressure and $(\phi - \phi_c)^2$ versus (a) critical shear deformation of the surface $\sigma_{c,\text{surface}}$, (b) critical shear deformation of the surface $\gamma_{c,\text{surface}}$ and (c) critical deformation of the foams $\gamma_{c,\text{foam}}$ versus critical deformation of the surface $\gamma_{c,\text{surface}}$, (O) pH 3, (△) pH 4, (□) pH 5, (▲) pH 6, (◇) pH 8, (▽) pH 9, pH 6.8 and NaCl: (■) 0 mM, (◆) 10 mM, (◀) 50 mM, (●) 80 mM, (▲) 100 mM, 50 mM (■) KCl, (▲) NH₄Cl, (●) LiCl, (◆) CaCl₂, (★) NdCl₃.

critical deformation $\gamma_{c,\text{surface}}$. These latter two quantities were determined from interfacial shear rheology and characterize the onset of a nonlinear response of the corresponding protein solution. The clear correlations between the parameter k and $\sigma_{c,\text{surface}}$ or $\gamma_{c,\text{surface}}$ show that k is determined by surface rheological features as well. Exceptions are again found for the foams made at pH 5 and those made from the BLG solutions including 50 mM NdCl₃. For these foams, structure or network formation across foam lamellae are supposed to dominate foam yielding as discussed above. Also, a clear correlation between the critical deformation $\gamma_{c,\text{foam}}$ characterizing the onset of nonlinear response during oscillatory shear of the foams and the critical deformation $\gamma_{c,\text{surface}}$ obtained in oscillatory surface shear experiments was found (Figure 17.24c) except for the foams at pH 5. Hence, as long as no structure or network formation inside the foam lamellae occurs, bulk foam and interfacial rheological properties are directly correlated.

REFERENCES

1. Denkov, N.D., Tcholakova, S., Golemanov, K., Ananthpadmanabhan, K.P. *A Lips, Soft Matter* 2009; 5: 3389.
2. Carozza, S. Rheological characterisation of gels and foams for food, Nr. 101, VDI-Verl., Düsseldorf 2001.
3. Cheng, D.C.-H. *Rheol Acta* 1986; 25(5): 542–554.
4. Scott, G.D., Kilgour, D.M. *Brit J Appl Phys* 1969; 2: 863–866.
5. Herzhaft, B. *J Colloid Interface Sci* 2002; 247: 412–423.
6. Bird, R.B., Gance, D., Yarusso, B.J. *Rev Chem Eng* 1982; 1: 1–70.
7. Saint-Jalmes, A., Durian, D.J. *J Rheology* 1999; 43: 1411–1422.
8. Barnes, H.A. *J Non-Newtonian Fluid Mech* 1999; 81(1–2): 133–178.
9. Herzhaft, B. *Oil & Gas Science and Technology—Rev IFP* 1999; 54: 587–596.
10. Denkov, N.D., Subramanian, V., Gurovich, D. *A. Lips, Colloids Surf A: Physicochem Eng Aspects* 2005; 263: 129–145.
11. Marze, S., Saint-Jalmes, A., Langevin, D. *Colloids Surf A: Physicochem Eng Aspects* 2005; 263: 121–128.
12. Schurz, J. *Rheol Acta* 1990; 170–171.
13. Alderman, N.J., Meeten, G.H., Sherwood, J.D. *J Nonnewton Fluid Mech* 1991; 39(3): 291–310.
14. Heller, J.P., Kuntamukkula, M.S. *Ind Eng Chem Res* 1987; 318–325.
15. Princen, H.M. *J Colloid Interface Science* 1986; 112(2): 427–437.
16. Princen, H.M., Kiss, A.D. *J Colloid Interface Science* 1989; 128: 176–187.

17. Nguyen, Q.D., Boger, D.V. *Annu Rev Fluid Mech* 1992; 47–88.
18. Rouyer, F., Cohen-Addad, S., Vignes-Adler, M., Höhler, R. *Physical Review E* 2003; 67(2).
19. Lexis, M., Willenbacher, N. *Colloids Surf A Physicochem Eng Asp E* 2014; 459: 177–185.
20. Bhakta, A., Ruckenstein, E. *Adv Colloid Interface Sci* 1997; 70: 1–124.
21. Ovarlez, G., Krishan, K., Cohen-Addad, S. *EPL* 2010; 91: 68005.
22. Liu, A.J., Nagel, S.R. *Nature* 1998; 396: 21–22.
23. Ragouilliaux, A., Ovarlez, G., Shahidzadeh-Bonn, N., Herzhaft, B., Palermo, T., Coussot, P. *Phys Rev E* 2007; 76(5): 051408.
24. Barnes, H.A., Hutton, J.F., Walters, K. *An Introduction to Rheology*, Vol. 3. Elsevier and Distributors for the U.S. and Canada. Elsevier Science Pub. Co: Amsterdam and New York, 1989.
25. Marze, S., Guillermic, R.M., Saint-Jalmes, A. *Soft Mat* 2009; 5: 1937.
26. Keentok, M., Milthorpe, J.F., O'Donovan, E. *J Non-Newton Fluid Mech* 1985; 17: 23–35.
27. Mason, T.G., Bibette, J., Weitz, D.A. *J Colloid Interface Sci* 1996; 179: 439–448.
28. Jastrzebski, Z.D. *Ind Eng Chem Fundam* 1967; 6: 445–453.
29. Mooney, M. *J Rheol* 1931; 2(2): 210–222.
30. Höhler, R., Cohen-Addad, S. *J Phys Condens Matter* 2005; 17: R1041–R1069.
31. Khan, S.A. *J Rheol* 1988; 32: 69.
32. Rouyer, F., Cohen-Addad, S., Höhler, R. *Colloids Surf A: Physicochem Eng Aspects* 2005; 263: 111–116.
33. Lexis, M., Willenbacher, N. *Soft Mat* 2014.
34. Yoshimura, A. *J Rheol* 1988; 32(1): 53–67.
35. Cohen-Addad, S., Höhler, R., Khidas, Y. *Phys Rev Lett* 2004; 93(2): 028302.
36. Costa, S., Höhler, R., Cohen-Addad, S. *Soft Mat* 2012; 9: 1100.
37. Coussot, P., Raynaud, J., Bertrand, F., Moucheront, P., Guilbaud, J., Huynh, H., Jarny, S., Lesueur, D. *Phys Rev Lett* 2002; 88(21): 218301.
38. Denkov, N., Tcholakova, S., Golemanov, K., Ananthapadmanabhan, K. A Lips, *Phys Rev Lett* 2008; 100(13): 138301.
39. Da Cruz, F., Chevoir, F., Bonn, D., Coussot, P. *Phys Rev E* 2002; 66: 051305.
40. Vincent-Bonnieu, S., Höhler, R., Cohen-Addad, S. *Europhys Lett (EPL)* 2006; 74: 533–539.
41. Hartnett, J.P. *J Rheol* 1989; 33: 671.
42. Sudduth, R.D. *J Appl Polym Sci* 1993; 48: 37–55.
43. Lexis, M., Willenbacher, N. *Chem Ing Tech* 2013; 85: 1317–1323.
44. Mason, T.G., Bibette, J., Weitz, D.A. *Phys Rev Lett* 1995; 75: 2051–2054.
45. Krishan, K., Helal, A., Höhler, R., Cohen-Addad, S. *Phys Rev E* 2010; 82(1): 011405.
46. Cohen-Addad, S., Hoballah, H., Höhler, R. *Phys Rev E* 1998; 57: 6897–6901.
47. Buzza, D.M.A., Lu, C.-Y.D., Cates, M.E. *J Phys II France* 1995; 5: 37–52.
48. Besson, S., Debrégeas, G., Cohen-Addad, S., Höhler, R. *Phys Rev Lett* 2008; 101(21): 214504.
49. Princen, H.M. *J Colloid Interface Sci* 1983; 91(1): 160–175.
50. Davis, J.P., Foegeding, E.A., Hansen, F.K. *Colloids Surf B: Biointerfaces* 2004; 34: 13–23.

18 Mechanisms of Antifoam Action

P. R. Garrett

CONTENTS

18.1 INTRODUCTION

Unwanted foam can represent a serious problem in many contexts. These include industrial processes and product formulations [1–3]. Examples of the former are gas-crude oil separation during oil production, desulfurization of natural gas, jet dying of textiles, radioactive waste treatment, the kraft pulp process, and fermentation using oxygen bubble columns. Formation of foam during the use of certain products can also be unacceptable. Formulation of such products usually involves the inclusion of antifoams. Examples include waterborne latex paints and detergents for machine washing of textiles and crockery. By contrast, products such as shampoos and hand dishwashing liquids must be formulated to produce copious amounts of foam despite the ever-present antifoam effects of triglyceride soils [4–7]. Defoaming also has medical applications, which include the use of antifoams to treat gastrointestinal gas in order to eliminate any foam obscuring the view of colonoscopy cameras and filters for removal of foam bubbles from the aspirated blood derived from surgery in order to permit recirculation [1].

These foam problems are for the most part addressed by resorting to the use of antifoams. Such antifoams are usually liquid based and are effective when dispersed as finely divided drops throughout the foaming liquid. In the case of aqueous foaming liquids, the antifoams usually consist of mixtures of hydrophobic oils and particles. In some cases, however, either hydrophobic particles or oils alone find application. In the case of nonaqueous foaming liquids, antifoams are usually simple liquids. Antifoam liquids, for both aqueous and nonaqueous use, are characterized by lower surface tensions than those of the foaming liquid. They should also be essentially insoluble in that medium. In both contexts, the mode of action usually appears to concern formation of unstable configurations where antifoam entities bridge across liquid films or Plateau borders [8–11].

Despite these practical generalizations, there would appear to be no agreed upon definition of the term "antifoam." However, the generalizations suggest we follow Denkov [8,11] and define an antifoam as a substance which can be dispersed as a separate phase (or phases) in a liquid subject to gas entrainment, so that foam films are destabilized to produce a defoaming effect. Removal of that separate phase should eliminate the effect. The latter can be achieved, in principle at least, by filtration or centrifugation. However, we should stress that defoaming effects can also be caused by the addition of substances to a solution of a surface active solute subject to gas entrainment which fall outside this definition. For example, addition of Ca^{2+} ions to an aqueous solution of a sodium alkylbenzene sulfonate usually produces a marked reduction in foamability [12]. This is due to the precipitation of essentially nonlabile, hydrophilic, mesophase particles which reduces the overall effectiveness with which surfactant can be transported to air-water surfaces, thereby diminishing the stability of foam films formed during rapid aeration. However, removal of the precipitate does not restore the foamability. Ca^{2+} is not, therefore, an antifoam. Similar foam behavior is reported when aqueous solutions of certain zwitterionic surfactants are mixed with solutions of double chain sodium alkyl benzene sulfonate solutions [13]. In this case, precipitation of a mixed mesophase produces lower foamability than that of solutions of either component surfactant alone. Which surfactant in such systems is the "antifoam"? Another example of confusion over the definition of the term "antifoam" is illustrated by Jha et al. [14] who describe 2-ethylhexanol, tributyl phosphate, and tetrabutylammonium chloride as antifoams. Essentially, these materials dissolve in aqueous micellar sodium dodecyl sulfate (SDS) solutions to produce changes in the micellar relaxation time, τ_2, which correlate with foamability. Since transport of SDS molecules to the air-water surface must involve breakdown of micelles, then changes in τ_2 will mean changes in the rate of transport of surfactant to air-water surfaces, and therefore, in stability of foam films during aeration. Jha et al. [14] in fact show that increasing the concentration of these "antifoams" can either increase or decrease foamability. Their effect on SDS foam behavior clearly indicates that they lie outside the definition of an antifoam given here. Another example of defoaming involving a *homogeneous* phase is the marked effect of certain propoxy/ethoxy compounds on the foam behavior of certain proteins. This is exemplified by the destabilizing effect of EO-PO-EO triblock ("pluronic") copolymers on the foam behavior of bovine serum albumin (BSA) solutions described by Nemeth et al. [15]. In this case, the protein forms stable foam with rigid, extremely slow-draining films where marginal regeneration is suppressed, presumably as a consequence of the viscoelastic shear or dilatational behavior of the adsorbed monolayer. However, addition of certain more surface active pluronic polymers appears to displace (at least partially) the protein to produce rapidly draining mobile unstable films. The available evidence appears to suggest that the relative stability of the BSA foam is due largely to slow drainage rather than to any significant difference in disjoining pressures with respect to the fast draining pluronic polymer contaminated foam films [15,16]. Similar observations are reported by Clark et al. [17] for the effect of Tween20 on the foam of α-lactalbumin. Marinova et al. [18] also report defoaming of sodium caseinate by pluronic polymers in homogeneous solution.

Although all these examples describe significant defoaming phenomena, they each involve processes specific to their context. They, therefore, do not afford the generality expected of antifoams as defined here where, for example, polydimethylsiloxane oils form the basis of antifoam action in most aqueous and nonaqueous contexts [1,2]. This generality appears to extend to the mode of action of antifoams, which invariably appears to involve formation of unstable bridging configurations in liquid films regardless of the context. In the case of antifoam action in aqueous liquids, for example, the films are air-water-air foam films, solid-water-air films where solid particles are involved and oil-water-air, so called pseudoemulsion films, where liquid antifoams are involved. It has been shown in particular that the stability of pseudoemulsion films plays a key role in determining antifoam action by oils in aqueous media [8,10,11,19].

This chapter is a summary of only the main aspects of present understanding of the mode of action of antifoams in both aqueous and nonaqueous contexts. A recent detailed comprehensive review of defoaming, including antifoams and their applications, is to be found elsewhere [20].

18.2   ANTIFOAM EFFECTS WITH HYDROPHOBIC PARTICLES IN AQUEOUS FOAM FILMS

Hydrophobic particles are known to act as antifoams in aqueous solutions of surfactants. These effects are dependent upon the contact angle and the geometry of the particles. The simplest cases concern spherical particles rods where the air-water contact angle, θ_{AW}, measured through the aqueous phase is $>90°$. A mechanism where the particle bridges both air-water surfaces in a foam film to produce a capillary pressure which enhances foam film drainage leads to the two three phase contact lines becoming coincident upon the particle (or rod). The particle surface is thereby eliminated from the foam film, leaving a hole which, if large enough, will expand to cause foam film rupture (see e.g., [10]). This simple mechanism has been verified by experiment using spherical particles and rods of known contact angle including direct observation using cinematography [21–23].

Particles with contact angles $<90°$ are also known to yield antifoam effects. Such particles are, however, invariably nonspherical [21,22,24]. Perhaps the most striking illustration of the effect of particle geometry concerns the comparison of the antifoam effectiveness of hydrophobed glass spheres with hydrophobed glass shards made by Frye and Berg [22] for solutions of a variety of surfactants. Results are presented in Figure 18.1 where it is obvious that, despite the similarity of equilibrium contact angles, the presence of sharp edges in the shards has significantly enhanced antifoam effectiveness.

It should be noted, however, that the weak antifoam effects by spheres with contact angles $<90°$ shown in the figure probably concern either dynamic effects or contamination by shards of glass. High dynamic contact angles can arise as a consequence of relatively slow transport of surfactant to air-water surfaces during foam generation which fails to maintain surfactant adsorption as new surfaces are formed. A symptom of this effect is that the particles decrease the amount of foam formed during aeration, but have little effect on foam stability after foam generation ceases (see e.g., [24]).

Perhaps the first insight into the role of sharp edges and corners in determining the antifoam effectiveness of particles was due to Dippenaar [21,26]. He observed that hydrophobed orthorhombic galena crystals could function as antifoams for aqueous solutions of a "frother" – 1,1,3-triethoxy butane. A high speed cinematographic study of the effect of such particles on the stability of films of *distilled water* revealed insights into the mechanism of this phenomenon. Given the relevant contact angle, he noted that such particles adopt two different orientations at the air-water surface where that surface hinges on the edges of the crystal after the manner suggested by Mason and coworkers [27,28] and provided $45° < \theta_{AW} < 135°$ (where the measured angle was $80 \pm 8°$). Each of these orientations, if adopted in a foam film, are depicted in Figure 18.2. It is obvious from the figure that only the diagonal orientation can give rise to an antifoam effect. This of course introduces an explanation for antifoam effects with particles exhibiting corners and sharp edges where $\theta_{AW} < 90°$. By contrast, the horizontal orientation can only inhibit drainage of liquid from the foam film leading to enhanced stability.

The orientations shown in Figure 18.2a and b are, however, two-dimensional depictions and are only exclusive in the case of particles where the third dimension is infinite (aspect ratio $\gg 1$). In the case of real orthorhombic particles, the air-water surface must not only hinge on the edges as depicted in the figure, but must also intercept the faces perpendicular to those edges. At equilibrium, the capillary pressure at the particle must be zero if the air-water surface remote from the particle is planar. Therefore, the air-water surface must present a catenoid profile against those faces where the contact angle is everywhere satisfied as depicted in Figure 18.2c. These requirements can lead to other orientations in both free air-water surfaces and in the air-water surfaces of foam films in addition to those depicted in Figure 18.2a and b.

In Figure 18.3a, we depict the sequence of film frames obtained by Dippenaar [21] showing the rupture of an aqueous film by a hydrophobed orthorhombic galena particle (of contact angle $80 \pm 8°$). As the particle approaches the point of rupture, it is seen to adopt a twisted orientation, which differs

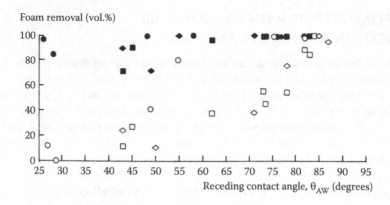

FIGURE 18.1 Antifoam effect of hydrophobed glass particles as a function of receding contact angles with various surfactant solutions [22]. Open symbols smooth spherical particles; filled symbols, ground glass particles. Foam generation by the Bartsch method [25]. Particle concentration 1 g dm^{-3}. (Reprinted from *J Colloid Interface Sci*, 130(1), Frye, G.C., Berg, J.C., Mechanisms for the synergistic antifoam action by hydrophobic solid particles in insoluble liquids, 54–59, Copyright 1989, with permission from Elsevier [61].)

in detail from the simple image presented in Figure 18.2b. Morris et al. [30] have repeated these experimental studies of Dippenaar [21] in combination with a complementary computer simulation of the process of film rupture. The latter utilized the surface energy minimization technique of Morris et al. [32,33] combined with the Surface Evolver software of Brakke [31]. The relevant Surface Evolver model of a particle bridging the two surfaces of a film formed in a capillary is tessellated and is depicted in Figure 18.3b. The curvature of the film is controlled by variation of the assumed gas volume. An example of a simulation of the orientation of a galena particle, with an aspect ratio of unity in such a film, is shown in Figure 18.3c where the twisted orientation clearly replicates that shown in the photographic frames in Figure 18.3a with the exception that the particle has moved away from the thinnest part of the film. However, Morris et al. [30] show that film rupture does not occur at the surface of the particle, as apparently shown in Figure 18.3a, but rather some distance into the foam film. The points of rupture in the simulation are depicted in Figure 18.3d. Observational evidence of the exact location of film rupture was not possible. However, if the simulation is run with the particle in the center of the film and with the same orientation, then film rupture occurs at the particle surface as shown by Dippenaar [21] and depicted in Figure 18.3a. We, therefore, have some ambiguity concerning the exact mode of antifoam action of particles with edges. However, these observations and simulations concern only pure water films in air where disjoining pressures may be neglected. That is not generally true in the case of typical aqueous surfactant solutions. Disjoining pressures may then be expected to inhibit foam film rupture sites from being located in films for the same reason that the foam is present at all. Unfortunately, simulation of disjoining pressures is not at present possible with the Surface Evolver software [31].

18.3   THE ROLE OF LIQUID DROPS BRIDGING AQUEOUS FOAM FILMS IN ANTIFOAM ACTION

There exists a general antifoam phenomenon concerning the destabilizing effect of a bridging liquid heterogeneity in a liquid foam film, which occurs provided certain criteria are satisfied. The liquid heterogeneity may, for example, be a hydrocarbon or a polydimethylsiloxane oil drop present in an aqueous foam film [8–10] or a drop of an alkoxylated-polydimethylsiloxane present in a crude oil foam film [34]. Other examples concern drops of the dispersed conjugate phase present in foam films of the continuous conjugate phase [10,35] in systems exhibiting partial miscibility. The well known antifoam

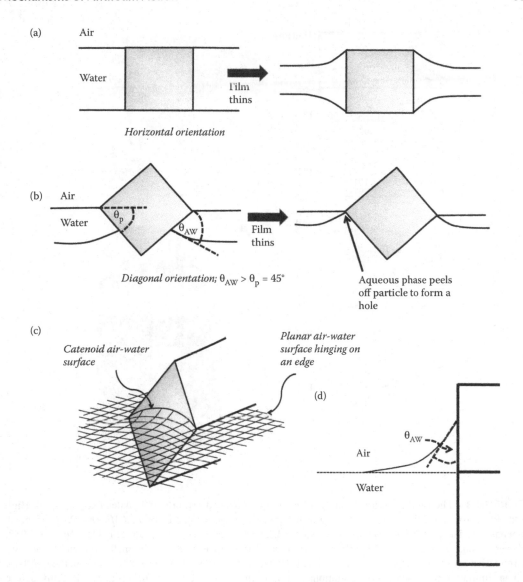

FIGURE 18.2 Orientation of an orthorhombic particle in a foam film. (a) and (b) Horizontal orientation stabilizes a draining foam film whereas a diagonal orientation can lead to film rupture (After reprinted from *Int J Mineral Processsing*, 9, Dippenaar, A. The destabilization of froth by solids. I. The mechanism of film rupture, 1–14, Copyright 1982, with permission from Elsevier [21].) (c) Mechanical equilibrium requires that the capillary pressure over the whole air-water surface be zero. The air-water surface contacting the end faces of such a particle must, therefore, form a catenoid profile. (d) Vertical slice through the center of an end face showing a concave air-water surface satisfying the relevant contact angle. The orthogonal radius of curvature must, therefore, be convex as shown in (c) (After Garrett, P.R. *Current Opinion in Colloid Interface Sci* 2015; 20: 81–91 [29].)

effect of drops of the cloud phase in aqueous solutions of ethoxylated alcohols typifies that type of behavior [36]. Despite this generality, we will describe the behavior of bridging liquid heterogeneities in foam films using the example of oil drops of low solubility dispersed in an aqueous medium. In this case, we characterize the system by an air-water surface tension, σ_{AW}, an oil-water surface tension, σ_{OW} and an oil-air surface tension, σ_{OA}. These surface tensions are supposed to be those present in foam films at any stage during their formation and are not necessarily equilibrium values.

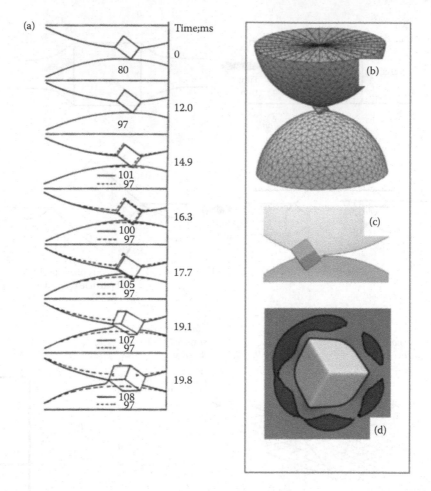

FIGURE 18.3 The rupture of air-water-air films by hydrophobed orthorhombic galena particles. (a). High speed cinematographic film frames showing film rupture after (Reprinted from *Int J Mineral Processsing*, 9, Dippenaar, A. The destabilization of froth by solids. I. The mechanism of film rupture, 1–14, Copyright 1982, with permission from Elsevier [21].) (b–d) Simulation of this process [30] by the method of surface energy minimization using the Surface Evolver software [31]. (b). Surface Evolver model of the particle in the capillary. (c). The simulation reveals twisted orientations similar to those observed in the film frames. (d). Film rupture is predicted by the simulation to occur away from the particle surface – the dark patches in the image indicate rupture zones. (Reprinted from Int J Mineral Processing, 131, Morris, G.D.M., Cilliers, J.J. Behaviour of a galena particle in a thin film, revisiting Dippenaar, 1–6, Copyright 2014, with permission from Elsevier [30].)

18.3.1 Emergence of a Drop into the Air-Water Surface

A fundamental requirement for a liquid heterogeneity to cause an antifoam effect is that it should emerge into the gas-foaming liquid surface. The overall free energy change accompanying that process is determined by the classical entry coefficient, E, which in the case of an oil drop dispersed in an aqueous medium is defined by

$$E = \sigma_{AW} + \sigma_{OW} - \sigma_{OA} \tag{18.1}$$

Essentially, E determines the free energy change accompanying the replacement of the air-water and oil-water surfaces of a thick oil-water-air duplex pseudoemulsion film by an air-oil surface. A duplex pseudoemulsion film is simply a film which is so thick that disjoining forces are absent and the oil-water and air-water surface tensions have their bulk phase values. At equilibrium, application of Antonow's rule [37] means that there are limits to the possible values of E so that

$$0 \leq E^e \leq \sigma_{OW}^e \qquad (18.2)$$

where the superscript denotes equilibrium properties.

Emergence of a drop into the air-water surface, and therefore antifoam action, requires that $E > 0$. However, this is a necessary condition but not a sufficient condition. This limitation derives from the nature of the asymmetric oil-water-air "pseudoemulsion" film, which separates the drop from emergence into the air-water surface. That film may be stabilized by an energy barrier, which must first be overcome if the drop is to emerge. Such a barrier could, for example, be generated by a combination of overlapping electrostatic double layers, van der Waals forces, and short range forces due to the presence of adsorbed anionic surfactant on both the air-water and oil-water surfaces of the pseudoemulsion film. Formation of an unstable oil bridge in a foam film is a critical aspect of antifoam action, which requires that an oil drop can emerge into the air-water surface. Presence of a metastable pseudoemulsion film could represent a significant barrier to the emergence of that oil drop into the air-water surface which could even potentially eliminate any antifoam action by the oil.

Bergeron et al. [38] have generalized the entry coefficient so that it also accounts for the formation of a metastable pseudoemulsion film rather than exposure of the oil phase to air with no intervening film. For this they use the concept of a film tension, $\sigma_{AWO}(h_{pf})$, where h_{pf} is the thickness of the pseudoemulsion film. This is defined as the work done in formation of a pseudoemulsion film of a given thickness against the disjoining pressure present in thinning the film from infinite thickness where the disjoining pressure is zero. We can then follow Bergeron et al. [38] and define a so called generalized entry coefficient, E_g, by replacing σ_{OA} with $\sigma_{AWO}(h_{pf})$ so that

$$E_g = \sigma_{AW} + \sigma_{OW} - \sigma_{AWO}(h_{pf}) \qquad (18.3)$$

Bergeron et al. [38] allow for the additional contribution due to the disjoining pressure in the case of a thin film by writing for the film tension, $\sigma_{AWO}(h_{pf})$ of a pseudoemulsion film of thickness h_{pf},

$$\sigma_{AWO}(h_{pf}) = \sigma_{AW} + \sigma_{OW} + \int_{\Pi_{AWO}(h_\infty)=0}^{\Pi_{AWO}(h_{pf})} h d\Pi_{AWO}(h_{pf}) \qquad (18.4)$$

where $\Pi_{AWO}(h_{pf})$ is the disjoining pressure in the pseudoemulsion film at thickness h_{pf}. In deriving the analogous relationship for a symmetrical foam film, the relevant Gibbs-Duhem equation is integrated assuming that the disjoining pressure is always equal to any capillary pressure due to any adjacent meniscus (see e.g., de Feijter et al. [39]. That assumption is also implicit in the derivation of Equation 18.4.

Combining Equations 18.3 and 18.4 means that

$$E_g = - \int_{\Pi_{AWO}(h_\infty)=0}^{\Pi_{AWO}(h_{pf})} h d\Pi_{AWO}(h). \qquad (18.5)$$

In the simple case where only attractive contributions are present (Π_{AWO} everywhere <0), then $E_g > 0$ and emergence occurs spontaneously. A metastable equilibrium at a given film thickness requires the application of a positive capillary pressure opposing film thinning. Conversely, where only repulsive contributions to the disjoining pressure are present (Π_{AWO} everywhere >0), then $E_g < 0$ and emergence cannot occur spontaneously so that both disjoining pressure and capillary

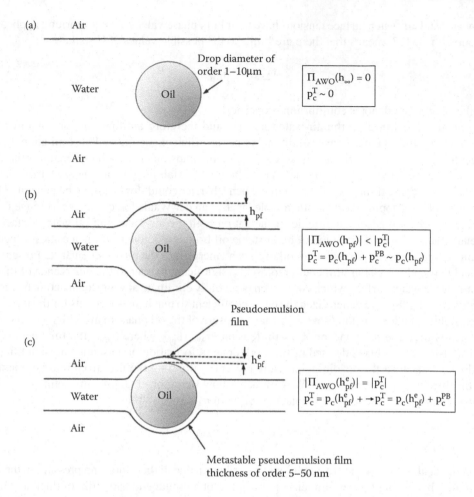

FIGURE 18.4 Simplified schematic of the evolution of the pressure in a pseudoemulsion film of a drop in a thinning foam film (not to scale). (a). Initially antifoam drop present in foam film where drop diameter $\ll$ foam film thickness, therefore both disjoining pressure, h_∞, and capillary pressure, p_c^T in the pseudoemulsion film are zero. (b). Drop becomes trapped in thinning foam film so that air-water surface in the vicinity becomes curved. Pseudoemulsion film continues to drain because capillary pressure exceeds disjoining pressure. (c). Pseudoemulsion film thins until disjoining pressure and capillary pressure equalize in a condition of mechanical equilibrium at thickness h_{pf}^e. Note, the symbols are defined in the text.

pressure are zero and duplex film behavior prevails. A sufficiently large drop of solution (at near-zero capillary pressure) should, therefore, spontaneously spread over the oil-air surface to form such a duplex film. A metastable equilibrium at a given *finite* pseudoemulsion film thickness requires the application of an opposing capillary pressure.

Now consider the actual situation for an antifoam drop present in a foam film as depicted in Figure 18.4. As the foam film drains, due to the capillary pressure in the Plateau border, the drop may become trapped in that film as illustrated in the figure. The drop will then be subject to a capillary pressure, p_c^T, tending to thin the pseudoemulsion film.

This capillary pressure is made up of two components – that due to Plateau border suction in the enclosing foam film, p_c^{PB}, and that due to curvature of the pseudoemulsion film, $p_c(h_{pf})$. Here the small dimensions of the drop relative to the Plateau border imply that in general $p_c(h_{pf}) \gg p_c^{PB}$ The pseudoemulsion film will then continue to thin until the modulus of the applied capillary pressure, p_c^T equals the disjoining pressure, $\Pi_{AWO}(h_{pf})$. A state of mechanical equilibrium at a film thickness h_{pf}^e will then prevail and the resulting pseudoemulsion film will cease to drain further. That equilibrium will be stable if the derivative of the disjoining pressure with respect to film thickness is negative; that is, $d\Pi_{AWO}(h_{pf})/dh < 0$. This follows because the disjoining pressure increases as the film thins so that thinning is resisted. By contrast, if that derivative is positive, then the disjoining pressure would decrease as the film thins leading to instability.

We may illustrate these arguments by considering a simple schematic disjoining pressure isotherm as shown in Figure 18.5a. The isotherm is typical of those likely to be formed in submicellar aqueous solutions of anionic surfactant at low ionic strengths. The surfactant is supposed to adsorb at both oil-water and air-water surfaces. The isotherm is seen to consist of inner and outer maxima with two minima. At low ionic strengths, the disjoining pressure isotherm for relatively large film thicknesses is likely to be dominated by long range electrostatic repulsion forces derived from overlapping electrostatic double layers which gives rise to the outer maximum. As the film thins, van der Waals attraction forces become increasingly important, eventually giving rise to the deep minimum in the isotherm. At extremely low film thicknesses (of <5 nm), short range hydrophobic attraction and steric/hydration repulsion forces are likely to dominate as the film is reduced to little more than the dimensions of a bimolecular leaflet [38]. There are no detailed observations or theoretical calculations of the disjoining pressure isotherm in a pseudoemulsion film at such low thicknesses. Therefore, somewhat speculatively, we depict, in Figure 18.5a, that region as a maximum representing steric/hydration repulsion followed by a minimum representing the attractive hydrophobic effect which it is assumed leads to a clean air-oil surface as the equilibrium state. A simple practical test of whether the latter represents the true equilibrium state for any system concerns tensiometry with a Wilhelmy plate. The air-oil surface tension should be probed with the plate. Once equilibrium is obtained, a drop of the relevant solution should be placed on the surface of the oil at a point remote from the plate

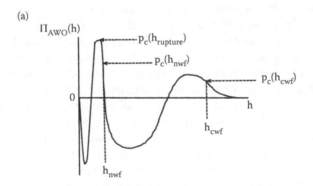

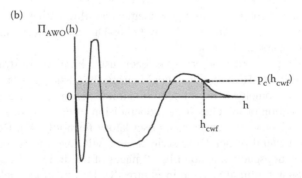

FIGURE 18.5 Schematic disjoining pressure isotherm for an oil-water-air pseudoemulsion film. (a). Effect of increasing capillary pressure to generate a common white film (cwf), a Newton white film (nwf), and eventually rupture. (b). Evaluation of the integral in Equations 18.4–18.5 (see text) to calculate the film tension using the argument of Bergeron et al. (From Bergeron, V., Fagan, M.E., Radke, C.J. Langmuir 1993; 9: 1704–1713 [38].)

[6]. If the clean air-oil surface represents the equilibrium state then the solution should not spread over the oil and there should be no change at all in the oil-air surface tension.

In Figure 18.5a, we illustrate the interaction of the capillary pressure with the disjoining pressure isotherm. At capillary pressures, $p_c(h_{cwf})$ and $p_c(h_{nwf})$, two different metastable films are formed. These are the common white film (cwf) and the Newton white films (nwf) of thicknesses h_{cwf} and h_{nwf} ,respectively. They are formed at points on the isotherm where, in both cases, the capillary pressure equals the disjoining pressure and $d\Pi_{AWO}(h)/dh < 0$. If sufficiently thin, such pseudoemulsion films are white in reflected light because interference between light reflected from the two surfaces of the film is constructive as distinct from the destructive interference of light reflected from the corresponding two surfaces of thin symmetrical foam films which are black in reflected light. These films may be seen, using an incident light microscope, as oil drops press up against the air-water surface. The thickness of the common white film (cwf) is of order 50 nm. The Newton white film (nwf) thickness is of order 5 nm [38]. The thickness region where a Newton white film prevails is depicted to range up to an inner maximum in the isotherm. Application of a capillary pressure, $p_c(h_{rupture})$, greater than the disjoining pressure at that maximum will reach film thicknesses where $d\Pi_{AWO}(h)/dh > 0$ and pseudoemulsion film rupture will then be inevitable. There is some evidence, however, that pseudoemulsion film rupture can occur at lower capillary pressures than this argument would suggest [8].

Bergeron et al. [38] apply Equation 18.5 to a similar disjoining pressure isotherm. Their integration is represented in Figure 18.5b in the case of the putative formation of common white film by the application of a capillary pressure, $p_c(h_{cwf})$, equal to the disjoining pressure, $\Pi_{AWO}(h_{cwf})$, and where the lower limit of the integral must be $\Pi_{AWO}(h_\infty) = p_c(h_\infty) = 0$. Since the disjoining pressure is then always positive, the generalized entry coefficient, defined by Equation 18.5, is negative and the common white film cannot form – only a duplex pseudoemulsion film with zero capillary pressure is permitted. However, it is possible to query this analysis. As we consider in Figure 18.4, the equilibrium between the capillary pressure and the disjoining pressure is achieved by applying a significant capillary pressure to a thick pseudoemulsion film at zero disjoining pressure until the film thins to the point where $\Pi_{AWO}(h_{cwf}) = p_c(h_{cwf})$. Equality between the capillary pressure and the disjoining pressure only occurs when the film thickness equals h_{cwf}. However, the derivation of Equations 18.4–18.5 assumes equality of the capillary pressure and the disjoining pressure at all disjoining pressures (see e.g., [39]). Here we should note that the pseudoemulsion film must thin to dimensions ≤ 0.1 μm before disjoining pressures become significant. However, antifoam drop radii are typically in the range of 0.5–5 μm. The radii of the curved pseudoemulsion films as film thicknesses approach 0.1 μm will, therefore, be little different from the radii of the drops. The capillary pressures will then become essentially constant at film thicknesses well before disjoining pressures are significant. The total pressure in the film would then be positive relative to ambient for thick films with low values of the disjoining pressure because $\Pi_{AWO}(h_{cwf}) < p_c(h_{cwf})$. We, therefore, propose that the condition $\Pi_{AWO}(h_{cwf}) = p_c(h_{cwf})$ is attainable and represents a metastable common white pseudoemulsion film. Perhaps we should better use the total pressure, $\Pi_{AWO}(h_{cwf}) - p_c(h_{cwf})$ in the relevant Gibbs-Duhem equation for the derivation of an expression for the film tension in this context?

A simple disjoining pressure isotherm has been used here to illustrate the concept of a generalized entry coefficient. However, we should stress that disjoining pressure isotherms may exhibit significantly more complex behavior. One well known example concerns micellar surfactant solutions where organization of micelles in pseudoemulsion films leads to so called stratification. The disjoining pressure isotherm then takes an oscillatory form [40–42]. Despite the complexity of such isotherms, the general principles described here will nevertheless prevail in determining the stability of the relevant pseudoemulsion films. Images of the drainage of a hexadecane-aq. SDS solution-air pseudoemulsion film are shown in Figure 18.6 by way of example where a drop of oil has been pressed up against an air-water surface by hydrostatic pressure [43]. Application of that pressure causes the film to drain asymmetrically until it becomes white in reflected light. The initial white film formed from this micellar solution is rapidly stratified, nucleating three successive white films of decreasing thickness. These strata consist of layers of micelles – each stratum containing one

5 s   10 s   20 s   30 s   147 s   180 s

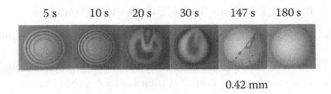

0.42 mm

FIGURE 18.6 Images of a draining hexadecane-aqSDS-air pseudoemulsion film at various times in a Scheludko cell. Sodium dodecyl sulfate concentration 0.03 M in aqueous solution. An interference pattern of Newton's rings is apparent as the film is formed. The irregular interference pattern at 20 s indicates asymmetric drainage. At 147 s two coexisting white films are apparent where the most stable is seen to prevail at 180s. (From Garrett, P.R., Wicks, S.P., Fowles, E. *Colloids Surfaces A Physicochem Eng Aspects* 2006; 282–283: 307–328 [43].)

less layer. Presumably, the applied pressure exceeded the maximum for each film until a metastable white pseudoemulsion film was formed. That film was stable for >1800 s. The antifoam effect of hexadecane drops in SDS solutions of that concentration was in consequence negligible (with no effect on foamability and where the resultant foam was stable for ≥600 s [43]).

Disjoining pressure isotherms where electrostatic double layer effects are essentially absent may be associated with nonaqueous fluids. Neat polydimethylsiloxanes, for example, often find applications as antifoams for hydrocarbons. High dielectric constants limit charge separation in such media so that significant electrostatic effects seem unlikely. Shearer and Akers [44,45] have, for example, made a study of the antifoam effect of polydimethylsiloxane oils on the foaming of lube oils, where the latter are presumably blends of, mainly, hydrocarbons. They have shown that, indeed, electrostatic effects are absent. In such circumstances it seems probable that disjoining pressures in the lube oil-polydimethylsiloxane-air pseudoemulsion films are dominated by attractive van der Waals forces and are, therefore, negative. Metastability of pseudoemulsion films will be absent and generalized entry coefficients positive, even at zero capillary pressure. Antifoam action in this system would not require any extra agency for destabilizing the relevant pseudoemulsion films. The results of Shearer and Akers [44,45] are consistent with that expectation.

Unfortunately, despite their theoretical importance, actual measurement of these disjoining pressure isotherms is extremely difficult (see e.g., [40]) and rarely performed. Study of the stability of pseudoemulsion films is, therefore, more conveniently provided by the Film Trapping Technique (FTT) [8]. This technique actually measures the critical capillary pressure for pseudoemulsion film rupture using an arrangement which is partially analogous to that depicted in Figure 18.4c. However, the drop is not confined in a symmetrical foam film, but rather in an aqueous film covering a hydrophilic glass surface. Increasing the applied pressure increases both the capillary pressure in an adjacent meniscus and the capillary pressure due to the curved air-water surface at the drop. Eventually, film rupture may be observed at a critical capillary pressure (a measure of the "critical entry barrier," p_c^{crit}). The arrangement permits direct observation of drop emergence so that the phenomenon can be correlated with p_c^{crit}. Denkov and coworkers have shown that the FTT is extremely useful in studying drop emergence and, therefore, understanding of antifoam action [8]. We give examples elsewhere.

The importance of this step of emergence into the gas-liquid surface of a foaming liquid has not always been realized. As we have emphasized here, antifoam action *requires* the rupture of pseudoemulsion films. This may be spontaneous in nonaqueous systems, but is usually not spontaneous in aqueous systems. Different rules are, therefore, required for the formulation of antifoams for these different contexts. In particular, as we shall see, antifoams for aqueous systems usually require the presence of hydrophobic particles whose sole function is rupture of pseudoemulsion films at low overall capillary pressures. The absence of such particles means that pseudoemulsion films can only be ruptured by the application of high capillary pressures as described here. In turn, that can often only be achieved if antifoam drops become trapped in the Plateau border nodes after the foam has drained. This gives rise to the "slow" antifoam effects described by Denkov et al. [8,46] which take many minutes to become apparent and are of limited practical application in defoaming.

18.3.2 Spreading Behavior of Oils on the Air-Water Surface

The relevance of the process of emergence of antifoam drops into the gas-foaming liquid surfaces has, until recently, been relatively neglected compared to that accorded to the surface behavior of the emerged drops. In the main this derives from early views [45,47,48] about the nature of antifoam action. Essentially it has often been proposed that such action *requires* that the antifoam spread over the surface of foam films. This behavior is supposed to result in the application of a shear force on the fluid in the foam film which causes catastrophic thinning and eventual film rupture [45,47,48]. That some oils function as effective antifoam ingredients for aqueous foams despite showing the absence of such spreading behavior (see e.g., [49]) has, however, rather undermined claims of generality for this theory of antifoam action.

There are essentially three types of spreading behavior expected of oils after emergence into the air-water surface. The first concerns spreading as a duplex film – the air-water surface is replaced by relatively thick layer of oil where the air-oil and oil-water surfaces have the same surface tensions as the relevant bulk phases. This behavior, where the air-water surface is completely wetted by bulk oil phase, is simply described by Brochard-Wyatt et al. [50] as *complete wetting*. Some polydimethylsiloxanes exhibit this type of behavior on aqueous surfactant solutions [51]. At the other extreme are oils which, after emergence into the air-water surface, do not at all contaminate that surface either by forming mixed monolayers or by forming any kind of film. Some hydrocarbon mineral oils can behave in this manner on the surfaces of anionic surfactant solution [49]. Such behavior is often described as *partial wetting* and is characterized by the formation of oil lenses covering part of the air-water surface. Finally, there is intermediate behavior. This is more complex. For example, an oil may spread over the air-water surface to form an unstable duplex liquid film which subsequently breaks up to form lenses in equilibrium with a thin liquid film which is so thin that disjoining forces are significant. This type of behavior is described by Brochard-Wyatt et al. [50] as *pseudo-partial wetting*. Denkov et al. [52] show for example that a particular polydimethylsiloxane oil shows pseudo-partial wetting if it is spread on an aqueous solution of sodium dioctyl sulphosuccinate.

We illustrate these three types of behavior in Figure 18.7. They can be defined quantitatively using the so called spreading coefficient.

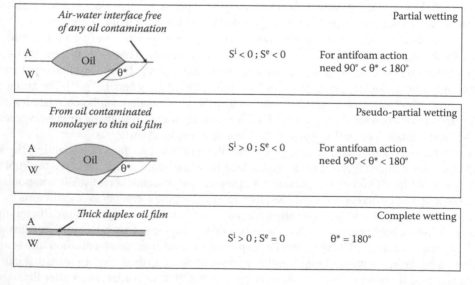

FIGURE 18.7 Schematic illustration of possible spreading behavior of a hydrophobic oil after emergence into the air-water (A/W) surface.

Here we distinguish between the initial spreading coefficient, S^i, and the equilibrium spreading coefficient, S^e. The former describes the behavior of the system before mutual saturation of the three relevant phases has occurred. We can, therefore, define these spreading coefficients as

$$S^i = \sigma^i_{AW} - \sigma^i_{OW} - \sigma^i_{OA} \qquad (18.6)$$

and

$$S^e = \sigma^e_{AW} - \sigma^e_{OW} - \sigma^e_{OA}. \qquad (18.7)$$

In the case of complete wetting where the oil spreads to form a duplex film, we must have $S^i > 0$ and $S^e = 0$. Such behavior implies that sufficient oil is present on the surface so that the spread film has the properties of the bulk phase.

In the case of partial wetting, we must have $S^i < 0$ and $S^e < 0$. In the case of pseudo-partial wetting, the emerged drop must first spread so that $S^i > 0$, but at equilibrium we must have stable oil lenses if sufficient oil is present so that $S^e < 0$.

For all these types of wetting behavior at the air-water surface, we must have, in general, at equilibrium

$$S^e \leq 0 \qquad (18.8)$$

which means that spontaneous spreading is a transient phenomenon. We should note also that the spreading coefficients are dependent upon both the air-water and oil-water surface tensions. This means that the spreading behavior is not only dependent upon the properties of the oil, but also on the effect of surfactant adsorption on those surface tensions.

18.3.3   FOAM FILM RUPTURE BY BRIDGING OIL DROPS

A requirement for antifoam action is that the oil must emerge into the air-water surface. As we have stated, in the case of aqueous foams, this usually requires the presence of hydrophobic particles to enable rupture of pseudoemulsion films. There is, however, no evidence that it also *requires* the specific type of the spreading behaviors shown in Figure 18.7. Indeed, we find that there are examples of oil-surfactant combinations where antifoam action is accompanied by each of the possible spreading phenomena. For example, polydimethylsiloxane oils can show either complete wetting or pseudo-partial wetting on the air-water surfaces of different surfactant solutions [53] whereas hydrocarbon mineral oils can sometimes show partial wetting [49]. Fraga et al. [34] even apparently show pseudo-partial wetting in the case of alkoxylated-polydimethylsiloxanes as antifoams for crude oil-air foams. Since this is a nonaqueous foaming system, electrostatic repulsion contributions to the disjoining pressures in pseudoemulsion films are likely to be absent and the presence of hydrophobic particles is not required.

We first consider the simple case of complete wetting where a duplex film forms as an oil drop emerges into the air-water surface. It has been shown that the mere presence of the film does not actually destabilize the foam. Thus, neither Ross and Nishioka [54] nor Trapeznikov and Chasovnikova [55] noticed any diminution of the stability of bubbles released under air-water surfaces over which a duplex polydimethylsiloxane film had been spread. However, over 60 years ago, Ross [56] proposed that unstable configurations, leading to foam film collapse, could occur if such oil drops bridge foam films. Figure 18.8 represents a schematic of the supposed process of film collapse by this mechanism.

Consider the case of a drop of equivalent spherical diameter slightly larger than the thickness of the aqueous phase present in a plane parallel foam film decorated with duplex films. Then, provided some agency is able to rupture the oil-water-oil emulsion films between the duplex films and the drop, a bridging configuration will form as shown in Figure 18.8b. The oil-air surface is planar even in the vicinity of the

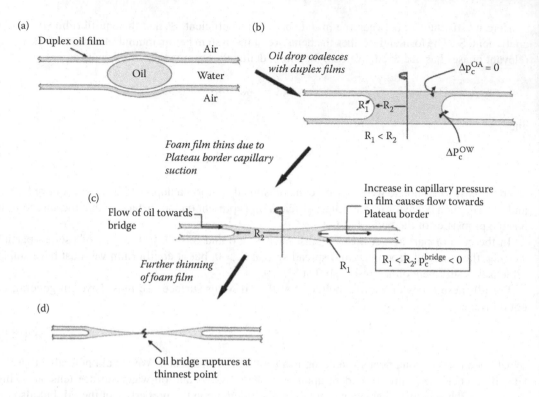

FIGURE 18.8 Schematic illustration (not to scale) of rupture of an aqueous foam film by bridging oil drop where the oil exhibits complete wetting (i.e., forming a duplex oil film over the foam film). (a). Large oil drop present in foam film decorated with duplex oil film. (b). The oil-water-oil emulsion film is ruptured by a suitable agency to form an oil bridge where the capillary pressure jumps across the oil-water, Δp_{OW}, and oil-air, $\Delta p_{OA} = 0$, surfaces are not equal so a condition of mechanical stability cannot be achieved. (c). The increasing capillary pressure jump at the oil-water surface, augmented by the thinning film, drives the interface away from the axis of revolution, enhancing stretching of the drop. (d). After further thinning, the two concave air-oil surfaces approach sufficiently closely to cause film rupture. (After Garrett, P.R. In: *The Science of Defoaming; Theory, Experiment and Applications*, Vol. 155. Taylor & Francis: Boca Raton, Surfactant Sci. Series, 2013, Chpt. 4, pp. 115–308.)

drop and, therefore, there is no capillary pressure at that surface. The oil-water surface of the bridging drop is on the other hand characterized by two radii of curvature, R_1 and R_2, which, although of different sign, are not in general equal. The curvature of the oil-water surface cannot, therefore, be a catenoid which means that a capillary pressure must be present at that interface which must be different from that at the oil-air surface. Thus, a condition of mechanical equilibrium cannot exist. As the foam film thins under the influence of the Plateau border suction, the axial radius, R_2, of the oil-water surface will increase and the convex radius, R_1 (shown in the figure in the plane of the paper) will decrease. This will produce an increase in the capillary pressure in the aqueous foam film adjacent to the drop.

The overall capillary pressure in the drop will also become progressively negative as the oil-air surfaces become concave, therefore, the drop becomes thinner at the rotational axis. The increasing capillary pressure in the foam film adjacent to the drop, as that film thins, will drive away the aqueous phase. This will necessarily cause the drop to stretch, thereby accentuating the tendency of the drop to become thinner at the rotational axis. Rupture as a result of van der Waals forces across the drop will eventually occur at that point leading to film rupture. This is of course the bridging stretching mechanism of Denkov and coworkers for which there is observational evidence [8,52].

This mechanism may be generalized to other types of wetting behavior by oils at the air-water surface. As we have seen with either partial or pseudo-partial wetting, oil lenses will be present on the surface of foam films provided again the rupture of the relevant pseudoemulsion films is

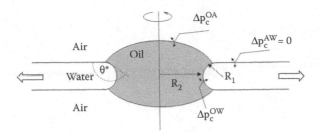

FIGURE 18.9 Configuration of a bridging drop in a plane parallel aqueous foam film with $\theta^* > 90°$. Mechanical equilibrium is not possible because $\Delta p_c^{OA} \neq \Delta p_c^{OW}$ for the jumps in capillary pressure at the oil-air and oil-water surfaces, respectively [57]. As the foam film thins the ratio R_2/R_1 increases and, therefore, Δp_c^{OW} increases, Δp_c^{OA} changes sign and decreases, Δp_c^{AW} becomes finite in the vicinity of the bridge. These processes may lead to film rupture in the center of a stretched bridge analogous to that shown in Figure 18.8 for the case of a duplex film forming oil.

arranged. If such lenses bridge a plane parallel foam film, then we obtain a configuration similar to that shown in Figure 18.9. Here the angle θ^*, and indeed the geometry of the lens is defined by Neumann's triangle of surface tensions.

Garrett [57] has shown that if the angle $\theta^* < 90°$, then mechanical equilibrium is possible, albeit for one unique foam film thickness. At other film thicknesses, this configuration can in principle lead to reduced rates of drainage from the foam film and thereby enhanced foam film stability [57]. However, if $\theta^* > 90°$ as shown in the figure, the capillary pressure at the oil-air surface cannot equal that at the oil-water surface and a condition of mechanical equilibrium cannot prevail [57]. Denkov [58] in a more detailed analysis has shown that relaxation of the requirement that the foam film be everywhere plane parallel gives rise to some conditions of stability for such configurations which become insignificant as the ratio of drop volume to foam film thickness increases. However, despite the clear identification of the condition for instability by a bridging drop in a foam film, we have no rigorous theoretical analysis of the processes which may derive from that circumstance. As we have stated, there is clear experimental evidence that the bridging-stretching mechanism actually occurs at least in the case of polydimethylsiloxane (PDMS) oils where $\theta^* \rightarrow 180°$ [8,52]. There is also a suggestion that other mechanisms such as the bridging-dewetting process may occur where $\theta^* \rightarrow 90°$ while we still have $\theta^* > 90°$ [8,9]. Here, we explore the possibility that there are limits to the range of values of θ^* which permit the bridging-stretching process to produce foam film rupture.

If $\theta^* > 90°$, then foam film thinning as a result of Plateau border suction will again give rise to an increasing imbalance of capillary pressure leading to film rupture by a bridging-stretching mechanism similar to that shown in Figure 18.8 for the case of an oil exhibiting complete wetting. However, the formation of a configuration of nascent film rupture, with a drop stretched to near zero thickness at the center of the rotational axis would appear to depend on the size of the angle θ^*. Consider the case shown in Figure 18.10a where θ^* is close to 180° and assume that the three relevant surfaces must satisfy Neumann's triangle of surface tensions. By analogy with the observation of bursting foam films [59,60], we would also expect the aqueous phase, displaced by the stretching drop, to accumulate as a swollen rim. Indeed, Denkov et al. [52] report observation of "fish-eyes" accompanying foam film rupture by drops of PDMS+ hydrophobed silica. Such fisheyes represent deformation of the air-water surface as the oil drop is stretched. This means that the air-water surface cannot be plane parallel adjacent to the oil bridge. The oil-water surface is, therefore, necessarily concave with respect to that bridge provided we have

$$\theta^* > 90° + \alpha \tag{18.9}$$

as shown in Figure 18.10a where α is the angle made by the tangent to the air-water surface, at the three phase contact line and the horizontal (defined as the bisector of the configuration). There is an

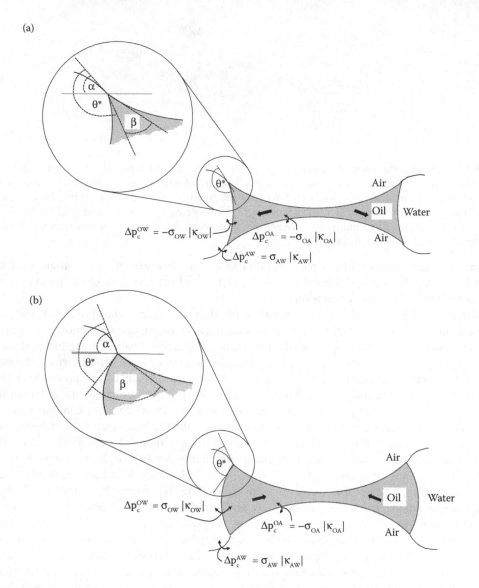

FIGURE 18.10 Schematic illustrating the effect of the magnitude of $\theta^\star$ on the configuration of a drop by the bridging-stretching process (approaching nascent film rupture). α is the angle made by the tangent to the air-water surface at the three phase contact line and the horizontal. β is the dihedral angle at the three phase contact line between the tangents to the oil-water and air-water surfaces. (a) $\theta^\star \to 180°$ where $\theta^\star > 90° + \alpha$; (b) $\theta^\star \to 90°$ and where $90° < \theta^\star < 90° + \alpha$.

additional requirement which must be satisfied if the oil-air surface is also to be concave with respect to the oil phase. The dihedral angle, β, between the tangents to the oil-water and air-oil surfaces at the three phase contact angle must satisfy the inequality

$$180° - \theta^* < \beta < 180° - \theta^* + \alpha. \qquad (18.10)$$

These requirements are illustrated by the inset in Figure 18.10a. The configuration has a high capillary pressure at the edge of the drop, therefore, no balance of the capillary pressure jumps at the three relevant surfaces can exist. Suppose then that all three relevant surfaces are of constant curvature (i.e., nodoid or unduloid segments) where $|\kappa_{AW}|$, $|\kappa_{OW}|$, and $|\kappa_{OA}|$ are the moduli of the curvatures of those surfaces in the vicinity of the bridging drop. It can be shown that provided

$$\sigma_{OW}\left|\kappa_{OW}\right| > \sigma_{AW}\left|\kappa_{AW}\right| + \sigma_{OA}\left|\kappa_{OA}\right|, \tag{18.11}$$

then the capillary pressure in the oil adjacent to the oil-water surface will be less (i.e., more negative) than that at the air-oil surface leading to a flow of oil from the latter which will tend to draw the two air-oil surfaces together. We note also that as the drop stretches, $\left|\kappa_{OW}\right|, \rightarrow 0$ and inevitably $\left|\kappa_{OW}\right| > \left|\kappa_{AW}\right|$ as the aqueous rim expands (which should compensate for $\sigma_{OW} < \sigma_{AW}$). All of this would aid film rupture by the bridging-stretching mechanism.

Even in the putative case where θ^* is close to 90°, as with the configuration shown in Figure 18.10a, the air-water surface cannot be plane parallel adjacent to the putative oil bridge if it is stretching. However, as shown in Figure 18.10b, there then exists the possibility that the oil-water surface must become convex with respect to the oil bridge as the foam film thins. As illustrated in Figure 18.10b, this requires that

$$90° < \theta^* < 90° + \alpha. \tag{18.12}$$

The condition given by Equation 18.10 must again also be satisfied if the oil-air surface is to be concave with respect to the oil phase. Again, no balance of capillary pressure jumps will be present. A large positive capillary pressure difference, Δp_c^{bridge}, will, however, exist within the oil bridge so that

$$\Delta p_c^{bridge} = \left(\sigma_{AW}\left|\kappa_{AW}\right| + \sigma_{OW}\left|\kappa_{OW}\right| + \sigma_{OA}\left|\kappa_{OA}\right|\right) > 0. \tag{18.13}$$

Such a capillary pressure difference will tend to cause oil to flow from the vicinity of the oil-water surface to the vicinity of the oil-air surface. In turn this would force apart the two oil-air surfaces in the bridge and inhibit the formation of a configuration like that depicted in Figure 18.10b. It is, therefore, possible that film rupture by the bridging-stretching process will not occur as $\theta^* \rightarrow 90°$. This suggests that perhaps the bridging-dewetting process of foam film rupture described by Denkov et al. [8,52] is likely in that case. This process is shown schematically in Figure 18.11 where the aqueous phase is depicted to peel off the bridging drop to form a hole in the foam film much as would occur with an inert hydrophobic particle provided we still have $\theta^* > 90°$.

Denkov et al. [9] have in fact claimed that mineral oil-based antifoams, exhibiting partial wetting behavior, have been observed to break foam films by the bridging-dewetting process. Details are yet to be published. Another possibility which has been suggested by Frye and Berg [61], also involves some drop stretching, but with both the oil-water and air-oil surfaces concave with respect to the oil phase. Here again the two three phase contact lines meet at a point so that the aqueous phase peels off the drop to form a hole. However, this process requires extremely high values of the dihedral angle (i.e., $\beta > 90°$), which are unlikely to be realized in practice.

We find then that the overall condition for antifoam action by a bridging oil drop, regardless of the details of the process, is given by

$$90° < \theta^* \leq 180° \tag{18.14}$$

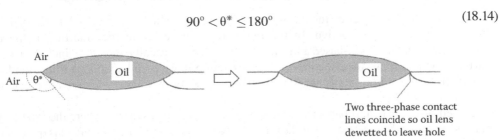

Two three-phase contact lines coincide so oil lens dewetted to leave hole

FIGURE 18.11 Schematic illustration of the bridging-dewetting mechanism involving elimination of the oil-water surface. (From Denkov, N.D., Cooper, P., Martin, J. Langmuir, 15, 8514–8529, 1999 [52].)

where the upper limit of the inequality, $\theta^* = 180°$, represents duplex film formation. Equation 18.14 is quite general and should, therefore, be equally applicable to nonaqueous systems (substituting the relevant oil-based surface tensions for the water-based tensions used here). It should again be stressed, however, that in the case of aqueous systems, where significant electrostatic repulsion contributions to disjoining pressures are likely, this condition usually only has predictive value if the classical entry coefficient is positive and if some agent for the rupture of metastable pseudoemulsion films is present.

The magnitude of θ^* is determined by Neumann's triangle of forces. This means that Equation 18.14 can be rewritten

$$0 < B \leq 2\sigma_{OW}\sigma_{AW} \tag{18.15}$$

where B is the so called bridging coefficient which is given by

$$B = \sigma_{AW}^2 + \sigma_{OW}^2 - \sigma_{OA}^2. \tag{18.16}$$

As with the spreading coefficient, B may have (nonequilibrium) initial and equilibrium values. However, the magnitude of B derives from Neumann's triangle so that a condition of mechanical equilibrium must exist between the three relevant surface tensions. Such a condition does not exist if for example we have $S^i > 0$. Therefore, calculations of B made for systems with $S^i > 0$ are not meaningful. This is not always realized (see e.g., [4,8,62–64].

There remain several theoretical challenges concerning these bridging mechanisms of antifoam action. Although there has been thorough theoretical study of the stability of bridging oil drops [57,58], there has, for example, been no simulation of the capillarity and hydrodynamics of the actual bridging-stretching process. Issues such as the role of the magnitude of the angle θ^*, the role of foam film drainage in the process, the effect of oil viscosity, and the ultimate fate of the oil drop after film rupture have either not been addressed at all or only superficially. Apart from the argument given here (Figure 18.10), there is no understanding of where bridging-dewetting should prevail, if at all, as a mechanism of foam film rupture over bridging-stretching or any other process. A recent study [65] of the bridging-stretching process using a "steered molecular dynamic" approach to "mimic" the forces on a foam film during bridging-stretching would appear to have no predictive value [29] in this context. Experimental studies of the bridging-stretching phenomenon are limited to the impressive studies by Denkov et al. [8,52]. There is, however, an absence of published experimental observations of the putative bridging-dewetting process. Perhaps nonaqueous systems would present fewer experimental challenges to studying these bridging processes where the complication of issues associated with emergence of the drop into the foaming liquid-gas surface is likely to be minimized.

18.4 MODE OF ACTION OF MIXED OIL-PARTICLE ANTIFOAMS

Antifoam action in aqueous media by hydrophobic oils which satisfy the criteria of Equations 18.14–18.15 is rarely effective. In the main, this derives from the stability of the relevant pseudoemulsion films. Oil-based antifoams for this context usually, therefore, include hydrophobic particles which, as we describe here, serve to rupture those pseudoemulsion films, rendering the oil effective by a bridging mechanism. Some of the key observations which led to this conclusion are summarized here.

Mixed oil-particle antifoams are characterized by a marked synergy where the mixed system is markedly more effective than either the oil alone or the particles alone. The effect is quite general for such systems with examples ranging from mixtures of hydrophobed silica with polydimethylsiloxanes to mixtures of hydrocarbon oils with organic particles such as alkylenedistearamides. An example of this type of behavior is shown in Figure 18.12 for mixtures of hydrophobed silica and liquid paraffin

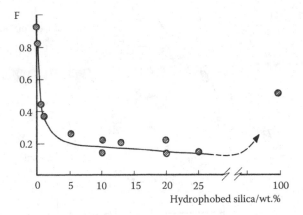

FIGURE 18.12 Illustrating oil-particle antifoam synergy with a plot of antifoam effectiveness against antifoam composition. F = volume of air in foam with antifoam/volume of air in foam without antifoam. Surfactant solution aqueous $\sim$0.7 mM sodium (C_{10}–C_{14}) alkylbenzene sulfonate. Antifoam liquid paraffin-hydrophobedsilica at concentration 1.2 g dm^{-3}. Measurements made by Bartsch method [25] at ambient temperature 22 $\pm$ 2°C. (From Garrett, P.R., Davis, J., Rendall, H.M. Colloids Surfaces A Physicochem Eng Aspects 1994; 85: 159–197 [49].)

in aqueous solutions of a sodium alkyl benzene sulfonate [49]. Here, F is the ratio of volume of air in the foam in the presence of antifoam/volume of air in the absence of antifoam.

The properties of this antifoam, as either a freely dispersed drop or as an oil lens, are presented in Figure 18.13. The particles are seen to adhere to the oil-water surface with a contact angle, θ_{OW}, measured through the aqueous phase >90° (measured advancing contact angles on compressed discs of the relevant hydrophobed silica were $\sim$140°). The oil exhibits partial wetting behavior on the surface of this surfactant solution with a positive classical entry coefficient, negative initial and equilibrium spreading coefficients, and where $\theta^* > 90°$ so that the bridging coefficient is positive under near equilibrium conditions. Antifoam effects shown in Figure 18.12 concerned the Bartsch method [25] with foam generated by shaking cylinders, so significant deviation of air-water surfaces from equilibrium would then be expected during that process [49].

However, destruction of single foam films drawn extremely slowly, using frames (at dilatational viscosities $\leq 10^{-3}$ s^{-1}), under such near equilibrium conditions also revealed significant antifoam effects as suggested by the positive equilibrium bridging coefficient [49]. We therefore find, with this system, negligible antifoam effects with the oil alone, some antifoam effect with the hydrophobed silica, and a marked synergy for the mixture.

Not only does the presence of the particles appear to potentiate the antifoam action of the liquid paraffin in this context, it also changes the emulsion behavior. Shaking equal volumes of this liquid paraffin and the relevant aqueous surfactant solution produces a stable oil-in-water emulsion. However shaking equal volumes of a mixture of liquid paraffin+10 wt.% hydrophobed silica with the same aqueous surfactant solution produces an extremely stable water-in-oil emulsion. This suggests not only that these hydrophobic particles stabilize such an emulsion against coalescence according to the well known Pickering (or Ramsden!) effects (see e.g., [66]), but they must also rupture oil-water-oil films. The latter was confirmed by timing the coalescence of drops of the oil released under a layer of liquid paraffin, with and without the hydrophobed silica, covering the aqueous surfactant solution [49]. Similar experiments [49] in which drops of oil with and without hydrophobed silica were released under the air-water surface of the surfactant solution suggested that the presence of the silica was causing rupture of the relevant pseudoemulsion film and, therefore, facilitating emergence of the oil into the air-water surface. The results of these observations are reproduced in Figure 18.14.

There have been many studies of the antifoam behavior of mixtures of polydimethylsiloxane (PDMS) oils and hydrophobed silica, most of which have been recently reviewed by Garrett [10].

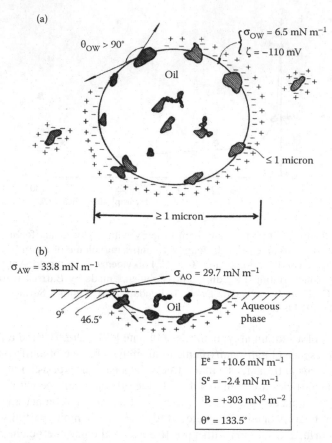

FIGURE 18.13 Summary of properties of entities formed upon dispersal of mixtures of liquid paraffin-hydrophobed silica (90/10 wt.%) in aqueous ~0.7 mM sodium (C_{10}–C_{14}) alkyl benzene sulfonate solution the antifoam effect of which is shown in Figure 18.11. (a). Dispersed drops. (b). Lenses formed at air-water surface. (From Garrett, P.R., Davis, J., Rendall, H.M. *Colloids Surfaces A Physicochem Eng Aspects* 1994; 85: 159–197 [49].)

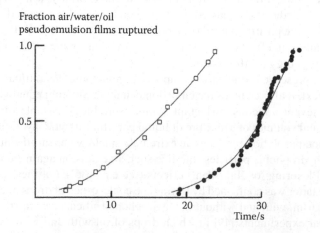

FIGURE 18.14 Effect of hydrophobed silica on the rupture of a pseudoemulsion film as revealed by the time of emergence of liquid paraffin oil drops into the air-water surface of a 0.7 mM aqueous solution of sodium (C_{10}–C_{14}) alkyl benzene sulfonate solution. □ Liquid paraffin-hydrophobed silica (90/10 wt.%). ● Liquid paraffin. (From Garrett, P.R., Davis, J., Rendall, H.M. *Colloids Surfaces A Physicochem Eng Aspects* 1994; 85: 159–197 [49].)

Arguably, the most thorough of these studies have been those by Denkov and coworkers [8,9]. Such oils usually exhibit either complete wetting or pseudo-partial wetting on aqueous surfactant solutions. Denkov et al. [52], for example, used a PDMS oil–hydrophobed silica mixture as a model antifoam for aqueous solutions of sodium dioctylsulphosuccinate. This oil + surfactant combination showed pseudo-partial wetting behavior (see Figure 18.7) with lenses of PDMS in equilibrium with a thin spread oil film. In contrast to the lenses of liquid paraffin depicted in Figure 18.13, the lenses of PDMS in this system had extremely low dihedral angles of only 0.4° which implies that, although $\theta^* < 180°$, it is close to 180°. Direct observation by video-microscopy of antifoam action in aqueous foam films prepared in a Scheludko cell [67] was made as part of this study. Rupture of the foam films was always preceded by the formation of at least one "fish eye" – a region in the foam film caused by a bridging drop surrounded by a deformed air-water surface. Film rupture was initiated by expansion of the fish eye in a process which appeared to take only a few milliseconds. These films were covered with a prespread thin film to ensure equilibrium with the oil and against which the oil would necessarily exhibit a zero spreading coefficient. Formation of the fish eye by spreading of the oil was not then possible, leaving bridging-stretching of drops as the only alternative. It seems, therefore, that Denkov et al. [52] had made convincing direct observations of the bridging-stretching process.

Denkov and coworkers [8] have explored the role of the hydrophobed silica in this context by using the "Film Trapping Technique" (the FTT). As we have seen, in this technique an antifoam drop is trapped in an aqueous film between the air-water and a hydrophilic glass surface [8]. Application of an overpressure increases the capillary pressure on the drop until it enters the air-water surface. The capillary pressure in an adjacent meniscus is taken as a measure of the "critical entry barrier," p_c^{crit}, to emergence of a drop into the air-water surface.

Measurements of p_c^{crit} have shown that the presence of hydrophobed silica does clearly reduce the barrier to emergence of drops of PDMS oil [68]. Examples of typical results are shown in Table 18.1 for aqueous solutions of sodium dioctylsulphosuccinate and Triton X100 (Octylphenol EO_{10}). Here we see, however, that p_c^{crit} is not only reduced by the presence of hydrophobed silica but is also influenced by the presence of the spread layer of oil which accompanies the wetting behavior of this oil on aqueous surfactant solution. In the absence of hydrophobed silica, the spread layer has little effect. However, in the presence of hydrophobed silica, the spread layer appears to significantly

TABLE 18.1

Effect of Hydrophobed Silica and Spread Layers on the Critical Capillary Pressures p_c^{crit} for PDMS Oil in Aqueous Micellar AOT[a] And Triton X100 (Octylphenol.EO_{10}) Solutions

		p_c^{crit}/Pa	
Antifoam	Spread layer?	AOT[a]	Triton X 100[b]
Neat PDMS oil[c]	no	28 ± 1	>200
	yes	19 ± 2	>200
PDMS/h.silica[d]	no	8 ± 1	30 ± 1
	yes	3 ± 2	5 ± 2

Source: Reprinted with permission from Denkov, N.D. et al. Role of spreading for the efficiency of mixed oil-solid antifoams. *Langmuir* 2002; 18: 5810–5817. Copyright 2002 American Chemical Society [68].

[a] Aqueous sodium dioctylsulphosuccinate solution at 10 mM.

[b] at 1 mM.

[c] of dynamic viscosity 1000 mPa s.

[d] 4.2 wt.% hydrophobed pyrogenic silica.

further reduce p_c^{crit}, thereby further facilitating the emergence of PDMS drops into the air-water surface [68]. We should also note that both Koczo et al. [69] and Bergeron et al. [51] also report that hydrophobed silica causes pseudoemulsion film rupture in the case of PDMS oils.

The mode of action of the hydrophobic particles in causing rupture of the pseudoemulsion film in the absence of spread oil layers would appear to have two principle steps [4,8,10]. The first step concerns emergence into the air-water surface of particles which adhere to the oil-water surfaces acting as asperities on antifoam drop surfaces. Such particles are usually irregular in shape and often as much as an order of magnitude smaller than the drops [8,10]. The second step concerns rupture of the oil-water-air pseudoemulsion film by the emerged particles.

The overall *force* exerted in pressing the particle against the air-water surface as the antifoam drop is squeezed by the Plateau border suction will be smaller than that exerted overall by the drop as a consequence of its small particle size relative to that of the latter. However, the capillary *pressure*, which is inversely proportional to the radius of the particle, will be higher in the more highly curved pseudoemulsion film formed between the particle and the air-water surface. The critical capillary pressure for rupture of the particle-water-air pseudoemulsion film may, therefore, be readily exceeded without being exceeded for rupture of the whole oil-water-air pseudoemulsion film [4]. However, this only allows the particle to emerge into the air-water surface. It must also rupture the otherwise metastable oil-water-air pseudoemulsion film if the entire drop is to emerge.

We have seen that hydrophobic particles can rupture symmetrical air-water-air foam films. Similar arguments can also be applied to the rupture of a pseudoemulsion film by a hydrophobic particle [8,10,70]. The process is shown in Figure 18.12 where we assume a spherical geometry, partial wetting so that thin oil films do not decorate the water-air surface, and that the particle is small so that the effect of gravity can be neglected. When the particle emerges into the air-water surface, it must adopt a bridging configuration which satisfies both contact angles as shown in the figure. The particle is depicted to have an oil-water contact angle, $\theta_{OW} > 90°$, and a finite air-water contact angle, $\theta_{AW} < 90°$, where both angles are measured through the aqueous phase. The geometry of the particle together with the air-water contact angle means that an unbalanced capillary pressure will develop in the vicinity of the particle. That will augment thinning of the pseudoemulsion film until the two three phase contact lines on the particle become coincident and the film is ruptured. The condition for this to occur is simply

$$\theta_{AW} > 180° - \theta_{OW}. \tag{18.17}$$

Here we note that if $\theta_{OW} > 90°$, then Equation 18.17 implies that a spherical particle with $\theta_{AW} < 90°$ will rupture the relevant pseudoemulsion film. However, for such a particle to rupture symmetrical air-water-air foam films [8,10,70], the contact angle condition is $\theta_{AW} > 90°$. Therefore, such a particle would be an effective promoter of the antifoam action of a suitable oil but be ineffective alone. This of course represents an explanation for the well known synergy in antifoam effectiveness of mixed oil-particle antifoams, at least for spherical particles.

Another consideration here concerns the extent to which the particles penetrate the oil-water-air pseudoemulsion film. If the particle does not protrude into the film to a distance at least equal to the thickness of the film, h_{pf}, then rupture of the film by the mechanism suggested by Figure 18.15 will not occur. It can be shown by simple geometry that for the particle to protrude a distance h_{pf} into the film we must have [71]

$$\cos\theta_{OW} = \frac{h_{pf}}{R_p} - 1s \tag{18.18}$$

where R_p is the radius of the particle. High values of θ_{OW} for a given particle size may mean the particle does not protrude far enough into the pseudoemulsion film to cause rupture until it thins as a result of increasing capillary pressure in the adjacent foam film.

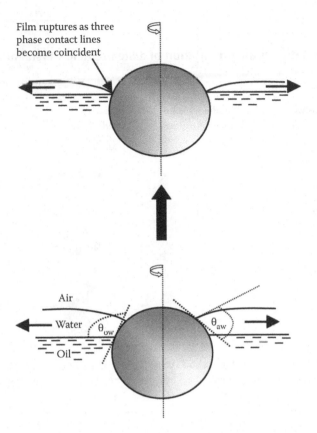

FIGURE 18.15   Rupture of air-water-oil pseudoemulsion film by a spherical particle with $90° < \theta_{OW} < 180°$ and $\theta_{AW} > 180° - \theta_{OW}$ so that $\theta_{AW} < 90°$. (From Garrett, P.R. In: *The Science of Defoaming; Theory, Experiment and Applications*, Vol. 155. Taylor & Francis: Boca Raton, Surfactant Sci. Series, 2013, Chpt. 4, pp. 115–308. Garrett, P.R. (Ed.). In: *Defoaming: Theory and Industrial Applications*, Vol. 45. Marcel Dekker: N.Y., Surfactant Sci. Series, 1993, Chpt. 1, 1–117. [10,70].)

Very few measurements of these contact angles for mixed oil-particle antifoams have been made. Garrett [10] lists most and finds all appear to satisfy the inequality Equation 18.17. For example, the oil-water contact angle, θ_{OW}, for the hydrophobed silica/liquid paraffin system, whose properties are summarized in Figure 18.13, is $143 \pm 4°$ and the corresponding air-water contact angle, θ_{AW} is $39 \pm 4°$ [49]. Both measurements were made using the sessile drop method on substrates prepared as compressed discs of particles, which methodology introduces some uncertainties! The silica particles were essentially fractal objects, rendering direct estimation of the contact angles on individual particles inaccessible to present experimental methodology.

In general, the relevant particles are not spherical but are usually of crystalline or amorphous geometry. Recently, Garrett et al. [7] in a study of the antifoam behavior of stearic acid/triolein mixtures in aqueous surfactant solution have shown that it is possible to calculate the orientation of the stearic acid crystals adhering to the triolein-waterinterface using the surface energy minimization technique of Morris et al. [32,33]. Stearic acid crystalizes from triolein as simple rhombic platelets (with six faces) and triolein exhibits partial wetting on the selected surfactant solution surface but with a negative bridging coefficient which becomes positive only under the nonequilibrium conditions prevailing during foam generation [7]. Using the approach of Morris et al. [32,33], the minimum required air-water contact angle for pseudoemulsion film rupture by a mechanism analogous to that depicted in Figure 18.15 may be inferred from the various orientations of each of the stearic acid crystal faces with respect to the oil-water surface. Results are presented in Table 18.2.

TABLE 18.2

Effect of the Aspect Ratio on the Orientation of a Stearic Acid Crystal at a Triolein-Water Surface

Aspect Ratilo AR	Probability of Orientation	Critical θ_{AW} for Pseudoemulsion Film Rupture/deg.	Images of Flat-Rotated Orientation	
			Side View	Top View
0.1	0.9	>0.0	AR = 0.1	AR = 0.1
0.4	0.9	>10.1	AR = 0.4	AR = 0.4
1.0	0.52	>25.2	AR = 1	AR = 1

Source: Garrett, P.R., Ran, L. Morris, G.D.M. *Colloids Surfaces A Physicochem Eng Aspects* 2017; 513: 28–40 [7].

Note: Calculated using the surface energy minimization method of Morris et al. [32,33] assuming a contact angle, θ_{OW}, of 120° measured through the aqueous phase.

The orientations were calculated with respect to the aspect ratio of the crystals (defined as the platelet diagonal/thickness). Extremely low air-water contact angles were inferred for pseudoemulsion film rupture by a particular crystal face. They are seen in the table to be significantly lower than those suggested by the inequality Equation 18.17. However, at low aspect ratios, the predicted orientation was flat in the surface with essentially no protrusion into the putative pseudoemulsion film so that antifoam effects cannot occur despite the requirement of only a near-zero contact angle. This suggests that an optimum antifoam effectiveness may exist at intermediate aspect ratios where significant protrusion is associated with minimal contact angles. These predicted air-water contact angles are so low that the stearic acid crystals characterized by such angles could not cause air-water-air foam film rupture. Therefore, again we predict antifoam synergy. Experimental observation of the antifoam behavior in this system is consistent with the expected synergy and low air-water contact angles with finite values only actually attainable under nonequilibrium conditions [5–7].

A further complication concerning the role of pseudoemulsion film rupture in antifoam action concerns the effect of a wetting film of oil decorating the air-water surface. As shown by Denkov et al. [68] and summarized in Table 18.1, the presence of such a film, in the case of mixtures of hydrophobed silica and PDMS oils, significantly reduces the critical capillary pressure for pseudoemulsion film rupture. Denkov et al. [68] offer an explanation for this phenomenon. Consider then the configuration formed if a spherical particle located at the oil-water surface of an oil drop, with $\theta_{OW} > 90°$, should bridge a pseudoemulsion film where the air-water surface is decorated by a spread film of oil. Such a configuration is depicted in Figure 18.16a in the case where the oil film is duplex.

As seen by Denkov et al. [68], the oil in the oil film will wet the surface of the particle to form an oil "collar" into which oil will be sucked by a negative capillary pressure. In the case of pseudo-partial wetting where the oil film is so thin that a significant disjoining pressure exists, then the capillary pressure in the collar should exceed that pressure in order to suck oil from the film. As the oil content of the collar increases, the meniscus will move toward the oil-water surface of the

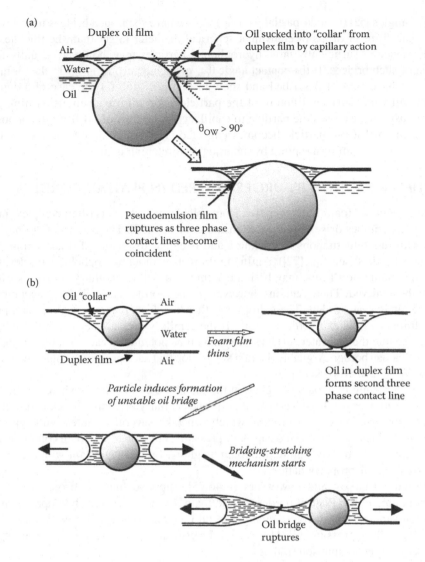

FIGURE 18.16 Schematic illustration showing potential roles of spherical particle with duplex film forming oil where $90° < \theta_{OW} < 180°$. (a). Particle facilitates rupture of air-water-oil pseudoemulsion film by forming an oil bridge. (b). In the case of a large particle, it will form a collar of oil as a result of capillary action and subsequently bridge the foam film. Foam film rupture may then occur by the bridging-stretching process. (From Garrett, P.R. In: *The Science of Defoaming; Theory, Experiment and Applications*, Vol. 155. Taylor & Francis: Boca Raton, Surfactant Sci. Series, 2013, Chpt. 4, pp. 115–308 [10].)

drop. An oil bridge will then form with an unbalanced capillary pressure in the pseudoemulsion film leading to film rupture. The process clearly differs significantly from that depicted in Figure 18.15. That it produces pseudoemulsion film rupture at lower overall critical capillary pressures presumably concerns differences in the relevant solid-water-air disjoining pressure isotherms for these different processes. Extending this argument to particles with more realistic geometries is of course desirable for which a preliminary attempt has already been reported [10]. Typical silica particles used in antifoam mixtures are, we remember, fractal objects!

Formation of unstable bridges in the case of oils which form spread oil films on the surfaces of aqueous surfactant solutions cannot involve the presence of oil lenses which effectively protrude into a foam film. Either lenses are absent (complete wetting duplex film) or lenses may have extremely small dihedral angles [52] (pseudo-partial wetting). As we have seen, such bridges must involve the presence in foam films of drops of equivalent spherical diameter larger than the thickness of the foam film. Presence of sufficiently large hydrophobed particles may, however, actually facilitate the formation of such bridges. If the contact angle θ_{OW} of such a particle is $>90°$, then it may form a "collar" as we have already described and show in Figure 18.16b for the case of a duplex film decorating the air-water surface. Rupture of the particle-water-oil pseudoemulsion film, possibly initiated by an asperity, exposes the particle to capillary driven flow of oil from the opposite side of the foam film so that the particle becomes engulfed. The bridge may then assume unstable proportions leading to foam film rupture by the bridging-stretching process.

18.5 ANTIFOAM ACTION BY DROPS TRAPPED IN PLATEAU BORDERS

The drainage of aqueous foam films prepared from soluble surfactants is often complex, involving hydrodynamic instabilities driven by surface tension gradients [72]. The pattern of drainage can be asymmetric with rapid fluctuations in thickness and irregular discharge of thick regions into the adjacent Plateau borders (see e.g., [72]). Antifoam droplets without hydrophobic particles, forming metastable pseudoemulsion films, may be readily removed before capillary pressures exceeding $p_c(h_{rupture})$ can be achieved. There remains, however, the possibility that such drops accumulate and become trapped in Plateau borders. These drops may then be subject to increasing capillary pressures as the foam drains and Plateau borders shrink. If the capillary pressures are high enough, so that the drops emerge into the air-water surfaces, they may then potentially cause rupture of foam films adjacent to the Plateau border. Arguably the first suggestion that such processes could occur in foams was made by Wasan and coworkers [73].

The capillary pressure in a foam is highest at the top of the foam column where it approximates the hydrostatic head once drainage has ceased [74]. The capillary pressure, therefore, progressively increases until drainage ceases, a process which can take several minutes with typical foam generation methodologies. Provided the capillary pressure at the top of the foam column is not then so high that it causes foam film rupture, the foam height will remain constant with the only decay caused by diffusional disproportionation.

Now, following Denkov and coworkers [8,46,75], suppose that oil drops with antifoam potential accumulate in the Plateau borders of the foam. In the absence of hydrophobic particles, pseudoemulsion rupture, and therefore emergence into the air-water surface, will require that the Plateau border capillary pressure, $p_c^{PB}(H)$, exceed the critical rupture capillary pressure $p_c(h_{rupture})$ in the oil-water-air pseudoemulsion so that

$$p_c^{PB}(H) > p_c(h_{rupture}) \tag{18.19}$$

where H is the height of the top of the foam column. Therefore, if the drop, upon emergence, is an effective antifoam, the foam height will decline until

$$p_c^{PB}(H_{res}^c) = \rho_W g H_{res}^c \tag{18.20}$$

where H_{res}^c is the residual height of the foam column. Basheva et al. [75] equate $p_c^{PB}(H_{res}^c)$ with the critical entry barrier, p_c^{crit} measured by the FTT, whilst acknowledging a lack of exact equivalence.

However, a limitation concerns the relative size of the Plateau border and antifoam drops. If these latter are too small then the air-water surface cannot be subject to distortion by the capillary pressure in the Plateau border so that emergence cannot occur as shown in Figure 18.4. Since Plateau borders are thickest at the bottom of the foam column, this will lead to residual foam heights, H_{res}^{size}, which are determined by drop diameter rather than by capillary pressure. Simple geometrical considerations lead to the requirement that for antifoam action we must have [8,46,75]

$$D \geq 0.31 \, \sigma_{AW} / \rho_W g H_{res}^{size} \tag{18.21}$$

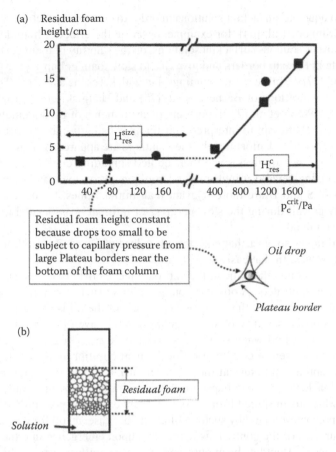

FIGURE 18.17 (a). Experimental observations of residual foam height (by Ross Miles method) as a function of the critical entry barrier, p_c^{crit} (measured by FTT on a variety of surfactant solutions). Antifoam was neat PDMS oil emulsified at 0.1 wt.%. The residual foam height, H_{res}^c, is seen to be proportional to p_c^{crit} at high foam heights where Plateau borders are thin (Equation 18.19) and independent of p_c^{crit} at low foam heights where Plateau borders are too thick for oil drops to bridge those borders (Equation 18.20). (b). Measurement of residual foam height. (After Hadjiiski, A.D. et al. In: Mittal K., Shah, D. (Eds.), *Role of Entry Barriers in Foam Destruction by Oil Drops in Adsorption and Aggregation of Surfactants in Solution*, Vol. 109. Marcel Dekker: NY, Surfactant Sci. Series., 2002, Chpt. 23, pp. 465–500 [46].)

where D is the drop diameter. A plot of residual foam height against p_c^{crit} by Hadjiiski et al. [46] is reproduced in Figure 18.17 for the foam behavior of a PDMS oil in a variety of aqueous surfactant solutions. This clearly reveals the two different regions where, respectively, the residual foam height is proportional to p_c^{crit} as implied by Equation 18.20 and independent of p_c^{crit} as implied by Equation 18.21. We should also note that attainment of these residual foam heights can take significant fractions of an hour as the capillary pressure in the foam column slowly increases. The phenomenon would, therefore, appear to have little value as an approach to foam control in most practical foam control contexts.

That antifoam action occurs in Plateau borders may, however, have relevance in the context of the use of foam floods for enhanced crude oil recovery. This represents a situation where aqueous foam films contact oil, trapped in rock, directly if capillary pressures are high enough to rupture pseudoemulsion films. Farajzadeh et al. [76] for example describe a "pinch-off" mechanism for foam film rupture. Here, a foam film is supposed to move over an oil drop trapped in a rock pore and separated from that film by an unstable pseudoemulsion film. Rupture of the pseudoemulsion film then means the oil dewetting into the Plateau border of the foam film leading in turn to rupture of the latter. This problem has stimulated interest in the effect of neat hydrocarbon oils on the stability of foams formed from aqueous surfactant solutions in order to develop surfactant systems which resist such film rupture. Simjoo et al. [64], for example, describe the effect of homologous hydrocarbon oils, dispersed in various aqueous surfactant solutions, on the stability of foam generated by sparging. The oils accumulate in Plateau borders and give rise to slow foam collapse over time periods >100 minutes. Wang et al. [77] describe similar findings, but with kerosene as the oil. This is all consistent of course with the observations of Basheva et al. [75] and Hadjiiski et al. [46]. However, neither Simjoo et al. [64] nor Wang et al. [77] report any measurements of pseudoemulsion film stability. Rather Simjoo et al. [64] attempt to interpret their findings using initial spreading coefficients and "initial bridging coefficients." Unfortunately, these authors are apparently unaware that their quoted values of the latter are meaningless since their quoted spreading coefficients are positive. Since film rupture is occurring several minutes after foam generation by sparging it seems likely that equilibrium values of S and B are more relevant than initial values. In that case, $S^e \leq 0$. Similar conditions probably prevail during the slow progress of intimately associated finely divided fluids during an actual foam flood.

The latter would rather suggest that equilibrium values of S and B would in any case be more relevant than initial values. In that case, $S^e \leq 0$.

All of this still leaves open the issue of the antifoam mechanism of drops emerging into the air-water surfaces of Plateau borders. A possible configuration of such a drop in cross section is shown in Figure 18.18. It is assumed that the bridging coefficient of the oil is positive so that $\theta^* > 90°$. A preliminary analysis of the stability of such a configuration, given by Garrett [10], suggests that, for the respective air-oil and oil-water surfaces, we must have $\Delta p_c^{AO} \neq \Delta p_c^{OW}$. This suggests that the configuration cannot represent a condition of mechanical equilibrium. It could then suggest that the relatively high capillary pressure at the oil-water surface could draw out oil bridges into the contiguous foam films leading to a collapse mechanism similar to the bridging-stretching process. In the case of oils which form spread films, however, there exists the possibility that the contiguous foam films are not precontaminated by such oil films. In that case, an oil film would spread from the oil drop *onto* the surfaces of the contiguous foam films upon emergence into the air-water surface. The resulting Marangoni effect has been observed to cause capillary waves which spread over the whole foam film [8]. Film rupture is attributed to the concomitant formation of locally thin regions in the foam films despite the overall average foam film thicknesses of ~1micron. That does not, however, preclude the possibility that the actual rupture event is associated with oil bridges being drawn out into the foam films as we describe here.

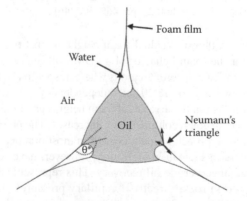

FIGURE 18.18 Schematic illustration of a cross section through a Plateau border containing an oil drop after pseudoemulsion films ruptured so that air-oil surfaces formed (E > 0). In this case, Neumann's triangle must be satisfied at the three phase contact lines where $\theta^* > 90°$ (i.e., bridging coefficient B > 0). (From Garrett, P.R. In: *The Science of Defoaming; Theory, Experiment and Applications*, Vol. 155. Taylor & Francis: Boca Raton, Surfactant Sci. Series, 2013, Chpt. 4, pp. 115–308 [10].)

18.6   ROLE OF DYNAMIC EFFECTS IN ANTIFOAM ACTION

Foam generation can involve the rapid formation of air-water surfaces. Stability of foam films under these dynamic conditions is dependent upon maintaining an adequate rate of transport of stabilizing surfactant to those freshly generated air-water surfaces so that adsorption levels are maintained. Otherwise levels of surfactant adsorption will be reduced, potentially leading to, for example, lower disjoining pressures, lower surface tension gradients, lower surface elasticities and viscosities, higher surface tensions, and therefore, higher Plateau border capillary pressures. All of these factors can lead to diminished foam film stability and diminished foamabilities under dynamic conditions. The rate of transport of surfactant to surfaces is in turn determined mainly by the state of dispersal of the surfactant. If the surfactant solution is mainly micellar, then the rate of micelle breakdown is key to the rate of adsorption because micelles generally do not adsorb and must breakdown first so that surfactant monomers can adsorb [78,79]. The rate of micelle breakdown is usually slower the lower the critical micelle concentration (CMC). Other states of dispersal for surfactant in solution include lamellar phase and crystalline particles. Such particles will be characterized not only by slow breakdown rates, but also low diffusion coefficients all factors which lead to slow transport and diminished foamability [12].

Another factor concerns differences in the rate at which air-water surfaces are formed with different methods of foam generation. For example, Patist et al. [78] report a study of the effect of dodecanol on the foamability of aqueous sodium dodecyl sulfate, SDS, solutions. Here, dodecanol increases equilibrium adsorption in SDS solutions but also slows the rate of micelle breakdown. The effect of dodecanol on the foamability of SDS solutions in consequence differs with different methodologies. Foam generation by the Bartsch method [25] of shaking cylinders produces more rapid air-water surface generation than foam generation by sparging. As a consequence, dodecanol has an adverse effect on foamability by cylinder shaking because of the adverse effect on micelle breakdown. However, that factor is of less importance during relatively gentle foam generation by sparging. Therefore, dodecanol increases foamability using that technique as a consequence of enhanced adsorption.

These dynamic factors can also affect the relative performance of antifoams especially in situations where the sign of the bridging coefficient can change from positive during foam generation to negative as air-water surface tensions decline to near equilibrium values after foam generation ceases. Extremely low air-oil surface tensions of $\sim$20 mN m^{-1} in the case of PDMS oils mean that with aqueous solutions of hydrocarbon chain surfactants such a transformation is unlikely. However, in the case of hydrocarbon and triglyceride-based oils it has been observed. For example, with a triglyceride like triolein with an ambient air-oil surface tension of $\sim$32 mN m^{-1}, the bridging coefficient is negative under near equilibrium conditions in some surfactant solutions so that antifoams based on that oil have little effect on foam stability [4–7]. However triolein-based antifoams can have significant effects on foam generation during which process only small dynamic effects are necessary to change the sign of the bridging coefficient [4–7].

A study of the effectiveness of a mineral oil-hydrophobed silica antifoam on the foam behavior of saline aqueous micellar solutions of a homologous series of sodium alkyl benzene sulfonate isomer blends is particularly illustrative of these dynamic effects [79]. Results of foam measurements are presented in Figure 18.19a. A pronounced maximum in initial foam height, measured immediately after foam generation ceased, is observed at an optimum chain length, both with and without the antifoam. This behavior correlates well with an equivalent minimum in measurements of dynamic air-water surface tensions as shown in Figure 18.19b. Surface activity as revealed by equilibrium surface tensions decreased monotonically with increasing chain length. However, surface tensions at low surface ages increased markedly with increasing chain lengths, which coincides with lower CMCs, and therefore, slower micelle breakdown rates. All of this behavior is found with other homologous series of anionic surfactants such as potassium carboxylate soaps and sodium fatty acid ester sulfonates [Garrett, P.R. Unpublished results].

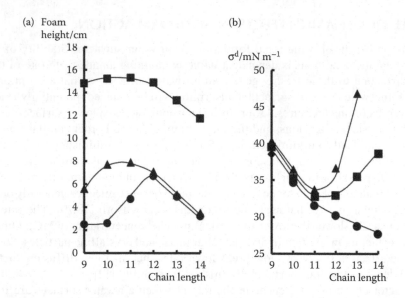

FIGURE 18.19 Foam and dynamic surface tensions for saline aqueous micellar solutions of homologous sodium alkyl benzene sulfonate blends (5×10^{-3} M surfactant in 0.038 M NaCl at 25°C). (a). Foam heights by Ross Miles method; □ in the absence of antifoam. (initial and after 960s). ▲ initial foam height in the presence of 1 g dm^{-3} hydrophobed silica/mineral oil antifoam; ● 960s after foam generation ceased. In the presence of antifoam. (b). Dynamic surface tensions as function of surface age, σ^d; ▲ 0.1s; ■ 0.5s; ● t = ∞. (After Garrett, P.R., Moore, P.R. *J Colloid Interface Sci* 1993; 159: 214–225 [79].)

In the case of the saline aqueous solutions of homologous sodium alkyl benzene sulfonates shown in Figure 18.19a, the foam initially generated in the absence of antifoam is stable for at least 960 s. The mineral oil-based antifoam is clearly effective for all homologues during foam generation as revealed by the initial foam heights measured immediately after cessation of foam generation. However, in the case of the higher homologues, the foam was stable for at least 960 s, despite the presence of antifoam and unstable in the case of the lower homologues. If we make the assumption that the hydrophobed silica is able to rupture all the relevant pseudoemulsion films regardless of circumstance, then the antifoam behavior should correlate with the sign of the relevant bridging coefficients. Calculation of the magnitude of those coefficients during the process of foam generation is problematic because of a lack of certainty concerning the prevailing surface tensions.

However, comparing the minimum in dynamic surface tensions with the maximum in foamability suggests a surface age of ~0.5 s for this method of foam generation. Using the dynamic surface tensions at that surface age together with the equilibrium oil-water interfacial tensions reveals that B is for the most part positive for solutions of these homologues wherever it could be meaningfully calculated. This clearly correlates with antifoam action during the process of foam generation for the relevant homologues shown in Figure 18.19b. However, we see from the same figure that antifoam action essentially ceases for the higher homologues after foam generation ceases. A measure of antifoam action in that circumstance is suggested by the use of the ratio F = volume of air in foam with antifoam/volume of air in foam without antifoam. Therefore, if we calculate the ratio R_F = F(t = 960 s)/F(t = 0 s) then we will find R_F = 1 if antifoam action has ceased up to 960 s after foam generation has stopped and find R_F < 1 if antifoam action has continued during that interval. Since R_F concerns a situation after foam generation has ceased where surface tensions tend towards equilibrium values then the relevant bridging coefficients can be calculated with greater certainty.

In Table 18.3 we compare R_F with B and $\theta^\star$ for the effect of this hydrophobed silica-mineral oil antifoam on the foam behavior, at two temperatures, of these solutions of sodium alkyl benzene sulfonate homologues where equilibrium values of surface tensions have been used to calculate B.

TABLE 18.3

Relationship between Bridging Coefficients and the Effectiveness of Mineral Oil-Hydrophobed Silica Antifoam after Cessation of Foam Generation in Solutions of Homologous[a] Sodium Alkylbenzene Sulfonate Blends

Chain Length	$B/(mN\ m^{-1})^2$	$\theta^*/deg.$	Antifoam Effectiveness/R_F[b]
		25°C	
C_9	+170	113	0.50
C_{10}	+241	121	0.33
C_{11}	+89	102	0.62
C_{12}	−25	85	0.94
C_{13}	−115	60	1.00
C_{14}	−183	0	0.90
		50°C	
C_{10}	+346	127	0.00
C_{11}	+169	109	0.28
C_{12}	+34	96	0.79
C_{13}	−61	79	0.92
C_{14}	−119	56	0.95

Source: Garrett, P.R., Moore, P.R. *J Colloid Interface Sci* 1993; 159: 214–225 [79].

[a] Foam generation by Ross-Miles. Mineral oil surface tension = 31 mN m⁻¹ at 25° and 29.5 mN m⁻¹. Each homologue is a blend of all possible isomers except the 1-phenyl isomer.

[b] R_F = F(t = 960 s)/F(t = 0 s) where F = volume of air in foam with antifoam/volume of air in foam without antifoam.

We see reasonable agreement with the expectation within the likely experimental error. It could be argued here, however, that slow antifoam action over a period of 960 s, where $R_F < 1$ for the lower homologues, is unlikely to concern antifoam effects in foam films since any antifoam drops will have long been removed by drainage from films over that time. If so, then we have evidence that antifoam action in Plateau borders may also concern a bridging-stretching process as we have proposed here.

18.7 ANTIFOAM ACTION IN NONAQUEOUS SYSTEMS

There would appear to have been few studies of the mode of action of antifoams in nonaqueous media. For the most part, such studies have been concerned with antifoams for the oil industry with problematic foams formed in crude oils [80] and lube oils [44,45,81]. As we have suggested here, antifoam action in nonaqueous foams by suitable antifoam oils can be realized by a bridging mechanism similar to that proposed for aqueous foams. However, significant differences in the specification of antifoam materials for use in a nonaqueous context exist. For example, the dielectric constant of typical hydrocarbons is almost two orders of magnitude less than that of water [82]. As a consequence, electrostatic effects arising from the separation of charges are likely to be weak or absent in hydrocarbons. This means that the repulsion contributions of overlapping electrical double layers to the positive disjoining pressures found in aqueous foam and pseudoemulsion films are also likely to be weak or absent. This in turn implies the absence of the need for any agency required for rupture of the relevant pseudoemulsion films. Antifoams for nonaqueous liquids are, therefore, simple liquids without the particles found in antifoams for aqueous liquids.

Such antifoams tend to be either PDMS oils or various derivatives of PDMS. They are applied at concentrations in the range of $(1–50) \times 10^{-3}$g dm^{-3} (see e.g., [34]) which is significantly lower than, for example, the typical concentrations of 0.1–0.2 g dm^{-3} of PDMS/hydrophobed silica [8] used in aqueous media. The solubilities of PDMS oils in, for example, hydrocarbons are also considerably higher than in water (where the latter is $<10^{-6}$ g dm^{-3} in the case of PDMS oils of molecular weights typical of those used in that context [83]). PDMS oils of the viscosity range 12.5–60 Pa are, for example, completely soluble in hexane [84]. Dissolution of PDMS often leads to profoaming behaviour rather than antifoam behavior. Therefore, alkoxy or perfluoro groups are often grafted onto PDMS chains to produce effective antifoams with low solubility in nonaqueous media.

The wetting behavior of antifoam liquids on the surfaces of nonaqueous foaming liquids has rarely been directly studied. Here we extend the definition of the possible spreading behavior, illustrated in Figure 18.7 for aqueous systems, to nonaqueous systems. Changes in the surface rheology and surface tension of crude oils upon addition of PDMS [85], alkoxy PDMS [34] and perfluoro PDMS oils [86] suggest either complete wetting or pseudo-partial wetting by these antifoam oils. Such studies often reveal profoaming behavior if the antifoam oils are present at concentrations below the solubility limit much as first described by Shearer and Akers [44,45] for PDMS in lube oils nearly 60 years ago. The effect of alkoxy PDMS antifoam oils on the foam behavior of crude oil by Mansur and coworkers [34] exemplifies aspects of the behavior of such antifoams. Results are shown in Figure 18.20. An alkoxy PDMS, which is too soluble, is ineffective as an antifoam and acts as a profoamer at all concentrations. Another alkoxy PDMS only functions as an antifoam at concentrations in excess of the solubility limit but shows no apparent profoaming behavior at concentrations below the solubility limit. There is of course the possibility that dissolved, surface active and profoaming antifoam may adversely affect antifoam effectiveness at concentrations above the solubility limit.

Mansur and coworkers [34] have measured entry, spreading, and bridging coefficients in these alkoxy PDMS + crude oil systems. As for the corresponding aqueous system, it is possible to define those coefficients where appropriate values of the relevant surface tensions are used. The bridging coefficient, for example, then becomes defined as

$$B = \sigma_{GL}^2 + \sigma_{AL}^2 - \sigma_{AG}^2 \tag{18.22}$$

where σ_{GL} is the gas-foaming liquid surface tension, σ_{AL} is the antifoam-foaming liquid surface tension and σ_{AG} is the antifoam-gas surface tension. For the limiting values of the range of admissible bridging coefficients for antifoam action we can, therefore, write

$$0 < B \leq 2\sigma_{AL}\sigma_{GL} \tag{18.23}$$

Mansur and coworkers [34] report positive entry coefficients, negative initial spreading coefficients, and large positive initial bridging coefficients for these alkoxy PDMS + crude oil systems. However, these alkoxy PDMS, with significant solubilities, reduced the surface tension of crude oil and sometimes produce profoaming effects at concentrations below the solubility limit as shown in Figure 18.20. Reduction of the air-crude oil surface tension simply makes the equilibrium spreading coefficient more negative than the initial spreading coefficient and the equilibrium bridging coefficient slightly less positive. All of this indicates that, at concentrations greater than the solubility limit, the less soluble alkoxy PDMS antifoam must form lenses at the crude oil-air surface and exhibit no tendency to spread (although adsorption from solution means that a monolayer of adsorbed antifoam is probably present together with lenses at equilibrium which implies pseudo-partial wetting). It seems that antifoam action in this case must derive from formation of unstable antifoam bridges in foam films where high values of B imply a bridging-stretching process as suggested by Mansur and coworkers [34].

(a) Volume air in foam/volume liquid

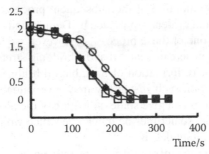

(b) Volume air in foam/volume liquid

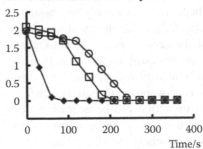

FIGURE 18.20 Plots of ratio of volume of air in foam/volume of liquid in foam as function of time, comparing two linear alkoxylated PDMS antifoams in crude oil; □ crude oil; ○ after addition of more soluble alkoxylated PDMS (with EO_{15}); ◆ after addition of less soluble alkoxylated PDMS (with EO_{19}). (a). concentration alkoxylated PDMS 2×10^{-2} g dm^{-3}. (b). concentration alkoxylated PDMS 5×10^{-2} g dm^{-3}. After (Reproduced from Fraga, A.K., Santos, R.F., Mansur, C.R.E. *J Appl Polymer Sci* 2012; 124(5): 4149–4156 [34].)

18.8 SUMMARIZING REMARKS

The main aspects of the present understanding of the mode of action of antifoams have been described here. Other important aspects such as the role of antifoam concentration on performance and the mechanism of deactivation in the case of some oil-particle mixed antifoams have been omitted because they are concerned mainly with consequences of antifoam action rather than causes.

As we have highlighted here, there are many aspects of understanding of antifoam mechanism which are incomplete. Even in the relatively simple case of the antifoam behavior of hydrophobic particles in aqueous liquids there are issues concerning the details of their mode of action as influenced by particle geometry. A recent paper by Morris et al. [30] even reports a simulation, for example, which suggests that foam film collapse by a bridging crystalline particle may be induced in the foam film rather than, as hitherto accepted, at the surface of the particle.

The presence of positive (usually electrostatic) contributions to the disjoining pressure isotherms of aqueous pseudoemulsion films is key to understanding not only the emergence of antifoam oil drops into air-water surfaces, but is the cause of the main distinguishing feature of the design of antifoams for aqueous and nonaqueous systems. The contribution of disjoining pressures to the emergence of oil drops into, for example, air-water surfaces, is often accounted for by a generalized entry coefficient. However, the usual definition of a generalized entry coefficient leads always to a negative value and no entry in the case of disjoining pressure isotherms dominated by a positive contribution at high film thicknesses. It takes no account of the attainment of a balance between the positive disjoining pressure and a negative external capillary pressure where entry gives rise to metastable pseudoemulsion films of finite thickness.

As mentioned here we have a thorough understanding of the stability of bridging oil drops, but little clear theoretical understanding of the subsequent processes resulting from instability. It is obvious from the experimental evidence, provided by Denkov and coworkers [8,52], that the bridging-stretching process must be one of those processes. However, we lack a theoretical understanding – even the simple analysis given here indicates that the bridging-stretching process may not necessarily represent a unique response to formation of unstable oil bridges in aqueous films. Simulation of the processes consequent upon such instability may give valuable insight concerning the role of capillarity and hydrodynamics and give some indication of the ultimate fate of the drop. Breakup of an oil drop may, for example, sometimes be inevitable which would have implication for continued antifoam activity by the resultant debris.

The role of the particles in the mixed oil-particle antifoam is now clear. However, the relationship between particle size, shape, contact angle, and effectiveness is only now becoming addressed (see e.g., [7]). The role of crystal habit in determination of crystal orientation at fluid-fluid surfaces can apparently, however, only be explored using the surface energy minimization approach of Morris et al. [32,33] for crystals with a maximum of six faces. Generalization to more complex crystals represents a future challenge. Disjoining forces are obviously important in the case of the action of particles in rupture of pseudoemulsion films. Unfortunately, the surface energy minimization approach uses the Surface Evolver of Brakke [31], which cannot apparently handle such forces. Clearly no lack of challenges here!

It is now clear that antifoam action can be dependent upon the extent to which the air-water surfaces during foam generation deviate from equilibrium. This derives from a failure of surfactant supply to maintain adsorption levels to rapidly expanding air-water surfaces. These effects are dependent upon the nature of the foam generation methodology. However, no suitable methodology exists for measurement of the extent to which deviations from equilibrium exist at, say, the top of foam columns during air entrainment by various foam generation techniques.

Far less attention has been paid to antifoam action in nonaqueous systems than in aqueous systems. Presumably this reflects foam problems of a less severe nature. Absence of significant electrostatic effects suggests, however, that such systems may present suitable models for study of liquid bridging mechanisms of antifoam action where the complications of disjoining forces and presence of particles are minimal or absent altogether.

REFERENCES

1. Garrett, P.R. *The Science of Defoaming; Theory, Experiment and Applications.* Vol. 155, Taylor & Francis, Boca Raton, Surfactant Sci. Series, 2013, Chpts. 8–11.
2. Garrett, P.R. (Ed.). *Defoaming: Theory and Industrial Applications.* Vol. 45, Marcel Dekker, N.Y., Surfactant Sci. Series, 1993, Chpts. 2–8.
3. Kerner, H.T., *Foam Control Agents.* Noyes Data Corp.: N.Y., 1976.
4. Zhang, H., Miller, C.A., Garrett, P.R., Raney, K.H. Mechanism for defoaming by oils and calcium soap in aqueous systems. *J Colloid Interface Sci* 2003; 263: 633–644.
5. Garrett, P.R., Ran, L. The effect of calcium on the foam behaviour of aqueous sodium alkyl benzene sulphonate solutions. 2. In the presence of triglyceride-based antifoam mixtures. *Colloids Surfaces A Physicochem Eng Aspects* 2017; 513: 402–414.
6. Garrett, P.R., Ran, L. The effect of calcium on the foam behaviour of aqueous sodium alkyl benzene sulphonate solutions. 3. The role of the oil in triglyceride-based antifoams. *Colloids Surfaces A Physicochem Eng Aspects* 2017; 513: 415–421.
7. Garrett, P.R., Ran, L. Morris, G.D.M. The effect of calcium on the foam behaviour of aqueous sodium alkyl benzene sulphonate solutions. 4. The role of particles in triglyceride-based antifoam mixtures. *Colloids Surfaces A Physicochem Eng Aspects* 2017; 513: 28–40.
8. Denkov, N.D. Mechanisms of foam destruction by oil-based antifoams. *Langmuir* 2004; 20: 9463–9505.
9. Denkov, N.D., Marinova, K., Denkov, N.D., Tcholakova, S.S. Mechanistic understanding of the modes of action of foam control agents. *Adv Colloid Interface Sci* 2014; 206: 57–67.
10. Garrett, P.R. The mode of action of antifoams. In: *The Science of Defoaming; Theory, Experiment and Applications*, Vol. 155. Taylor & Francis: Boca Raton, Surfactant Sci. Series, 2013, Chpt. 4, pp. 115–308.

11. Wasan, D.T., Christiano, S.P. Foams and Antifoams: A thin film approach. In: Birdi, K.S. (Ed.), *Handbook of Surface and Colloid Chemistry*, CRC Press: N.Y., 1997, Chp. 7.

12. Garrett, P.R., Ran, L. The effect of calcium on the foam behaviour of aqueous sodium alkyl benzene sulphonate solutions. 1. In the absence of antifoam. *Colloids Surfaces A Physicochem Eng Aspects* 2017; 513: 325–334.

13. Garrett, P.R., Gratton, P.L. Dynamic surface tensions, foam and the transition from micellar solution to lamellar phase dispersion. *Colloids Surfaces A Physicochem Eng Aspects* 1995; 103: 127–145.

14. Jha, B.K., Patist, A., Shah, D.O. Effect of antifoaming agents on the micellar stability and foamability of sodium dodecyl sulphate solutions. *Langmuir* 1999; 15: 3042–3044.

15. Nemeth, Zs., Racz, Gy., Koczo, K. Antifoaming action of polyoxyethylene-polyoxypropylene-polyoxyethylene-type triblock copolymers on BSA foams. *Colloids Surfaces A Physicochem Eng Aspects* 1997; 127: 151–162.

16. Sedev, R., Nemeth, Zs., Ivanova, R., Exerowa, D. Surface force measurement in foam films from mixtures of protein and polymeric surfactants. *Colloids Surf A Physicochem Eng Aspects* 1999; 149: 141–144.

17. Clark, D., Wilde, D., Wilson, D. Destabilisation of alpha lactalbumin foams by competitive adsorption of the surfactant Tween 20. *Colloids Surf A Physicochem Eng Aspects* 1991; 59: 209–223.

18. Marinova, K., Dimitrova, L.M., Marinov, R.Y., Denkov, N.D., Kingma, A. Impact of the surfactant structure on the foaming/defoaming performance of nonionic block copolymers in Na Caseinate solutions. *Bug J Phys* 2012; 39: 53–64.

19. Wasan, D., Nikolov, A.D., Huang, D.D., Edwards, D.A. Foam stability: Effects of oil and film stratification. In: Smith, D.H. (Ed.), *Surfactant Based Mobility Control*, Vol. 373. A.C.S. Symposium Series, American Chemical Society: Washington DC, 1988, p. 136.

20. Garrett, P.R. *The Science of Defoaming; Theory, Experiment and Applications*. Vol. 155. Taylor & Francis, Boca Raton, Surfactant Sci. Series, 2013, Chpts. 1–11, pp 572.

21. Dippenaar, A. The destabilization of froth by solids. I. The mechanism of film rupture. *Int J Mineral Processsing* 1982; 9: 1–14.

22. Frye, G.C.C., Berg, J.C. The antifoam action of solid particles. *J Colloid Interface Sci* 1989; 127: 222–238.

23. Aveyard, R., Binks, B.P., Fletcher, P.D.I., Peck, T.G., Rutherford, C.E. Aspects of aqueous foam stability in the presence of hydrocarbon oils and solid particles. *Ad Colloid Interface Sci* 1994; 48: 93–120.

24. Garrett, P.R. The effect of polytetrafluoroethylene particles on the foamability of aqueous surfactant solutions. *J Colloid Interface Sci* 1979; 69(1): 107–121.

25. Bartsch, O. On foam systems. *Kolloidchem Beihefte* 1924; 20: 1–49.

26. Dippenaar, A. Destabilization of froth by solids. II. The rate determining step. *Int J Mineral Processing* 1982; 9: 15–27.

27. Oliver, J.F., Huh, C., Mason, S.G. Resistance to spreading of liquids by sharp edges. *J Colloid Interface Sci* 1977; 59: 568–581.

28. Bayramu, E., Mason, S.G. Liquid spreading: Edge effect for zero contact angle. *J Colloid Interface Sci* 1978; 66: 200–202.

29. Garrett, P.R. Defoaming: Antifoams and mechanical methods. *Curr Opin Colloid Interface Sci* 2015; 20: 81–91.

30. Morris, G.D.M., Cilliers, J.J. Behaviour of a galena particle in a thin film, revisiting Dippenaar. *Int J Mineral Processing* 2014; 131: 1–6.

31. Brakke, K. The surface evolver. *Exp Math* 1992; 1: 141–165.

32. Morris, G.D.M., Neethling, S.J., Cilliers, J.J. A model for the behaviour of non-spherical particles at interfaces. *J Colloid Interface Sci* 2011; 354(1): 380–385.

33. Morris, G.D.M., Neethling, S.J., Cilliers, J.J. Modelling the self orientation of particles in a film. *Minerals Eng* 2012; 33: 87–92.

34. Fraga, A.K., Santos, R.F., Mansur, C.R.E. Evaluation of the efficiency of silicone polyether additives as antifoams in crude oil. *J Appl Polymer Sci* 2012; 124(5): 4149–4156.

35. Ross, S., Nishioka, G. Foaming behaviour of partially miscible liquids as related to their phase diagrams. In: *Foams; Proceedings of a Symposium Organized by the Soc Chem Ind Colloid and Surface Chem Group Brunel Univ in 1975* Akers, R.J. (Ed.), Academic Press: London, 1976, pp. 17–32.

36. Bonfillon-Colin, A., Langevin, D. Why do ethoxylated nonionic surfactants not foam at high temperature? *Langmuir* 1997; 13: 599–601.

37. Antonow, G.N. Surface tension at the limit of two layers. *J Chim Phys* 1907; 5: 372–385.

38. Bergeron, V., Fagan, M.E., Radke, C.J. Generalized entering coefficients: A criterion for foam stability against oil in porous media. *Langmuir* 1993; 9: 1704–1713.

39. de Feijter, J.A., Rijnbout, J.B., Vrij, A. Contact angles in thin liquid films. *J Colloid Interface Sci* 1978; 64: 158–268.

40. Bergeron, V., Radke, C.J. Disjoining pressure and stratification in asymmetric thin liquid films. *Colloid Polym Sci* 1995; 273: 165.

41. Bergeron, V., Radke, C.J. Equilibrium measurements of oscillatory disjoining pressures in aqueous foam films. *Langmuir* 1992; 8: 3020.

42. Lobo, L., Wasan, D.T. Mechanisms of Aqueous foam stability in the presence of emulsified non-aqueous-phase liquids: Structure and stability of the pseudoemulsion film. *Langmuir* 1993; 9: 1668–1677.

43. Garrett, P.R., Wicks, S.P., Fowles, E. The Effect of high volume fractions of latex particles on foaming and antifoam action in surfactant solutions. *Colloids Surfaces A Physicochem Eng Aspects* 2006; 282–283: 307–328.

44. Shearer, L.T., Akers, W.W. Foaming in lube oils. *J Phys Chem* 1968; 62: 1269–1270.

45. Shearer, L.T., Akers, W.W. Foam stability. *J Phys Chem* 1968; 62: 1264–1268.

46. Hadjiiski, A., Denkov, N.D., Tcholakova, S.S., Ivanov, I.B. In: Mittal, K., Shah, D. (Eds.), *Role of Entry Barriers in Foam Destruction by Oil Drops in Adsorption and Aggregation of Surfactants in Solution*, Vol. 109. Marcel Dekker: NY, Surfactant Sci. Series., 2002, Chpt. 23, pp. 465–500.

47. Ewers, W., Sutherland, K. The role of surface transport in the stability and breakdown of foams. *Aust J Sci Res* 1952; 5: 697–790.

48. Prins, A. Theory and practice of formation and stability of food foams. In: Dickinson, E. (Ed.), *Food Emulsions and Foams*, Vol. 58. Royal Soc. Chem. Special Publications: London, 1986, 30–39.

49. Garrett, P.R., Davis, J., Rendall, H.M. An experimental study of the antifoam behaviour of mixtures of a hydrocarbon oil and hydrophobic particles. *Colloids Surfaces A Physicochem Eng Aspects* 1994; 85: 159–197.

50. Brochard-Wyart, F., di Meglio, J., Quere, D., de Gennes, P. Spreading of nonvolatile liquids in a continuum picture. *Langmuir* 1991; 7: 335–338.

51. Bergeron, V. Cooper, P., Fischer, C., Giermanska-Kahn, J., Langevin, D., Pouchelon, A. Polydimethylsiloxane (pdms)-based antifoams. *Colloids Surfaces A Physicochem Eng Aspects* 1997; 122: 103–120.

52. Denkov, N.D., Cooper, P., Martin, J. Mechanisms of action of mixed solid-liquid antifoams. 1. Dynamics of foam film rupture. *Langmuir* 15, 8514–8529, 1999.

53. Garrett, P.R. Oils at Interfaces; Entry Coefficients, Spreading Coefficients, and Thin film forces. In: *The Science of Defoaming; Theory, Experiment and Applications* Vol. 155. Taylor & Francis: Boca Raton, Surfactant Sci. Series, 2013, Chpt. 3, pp. 100–101.

54. Ross, S., Nishioka, G. Experimental researches on silicone antifoams. In: *Emulsions, Latices, and Dispersions*, Becher, P., Yudenfreund, M. (Eds.), Marcel Dekker, NY, 1978, p. 237.

55. Trapeznikoff, A.A., Chasovnikova, L.V. Stabilisation of bilateral films by monolayers and thin films of poly(dimethylsiloxanes). *Colloid J USSR* 35: 926, 1973.

56. Ross, S. Inhibition of foaming II. A mechanism for the rupture of liquid films by antifoaming agents. *J Phys Colloid Chem* 1950; 54: 429–436.

57. Garrett, P.R. Preliminary considerations concerning the stability of a liquid heterogeneity in a plane-parallel liquid film. *J Colloid Interface Sci* 1980; 76(2): 587–590.

58. Denkov, N.D. Mechanisms of action of mixed solid-liquid antifoams. 2. Stability of oil bridges in foam films. *Langmuir* 1999; 15: 8530–8542.

59. Culick, F.E.C. Comments on a ruptured soap film. *J App Phys* 1960; 31: 1128–1129.

60. McEntee, W.R., Mysels, K.J. The bursting of soap films. I. An experimental study. *J Phys Chem* 1969; 73(9), 3018–3028.

61. Frye, G.C., Berg, J.C. Mechanisms for the synergistic antifoam action by hydrophobic solid particles in insoluble liquids. *J Colloid Interface Sci* 1989; 130(1): 54–59.

62. Arnaudov, L., Denkov, N.D., Surcheva, I., Durbut, P., Broze, G., Mehreteab, A. Effect of oily additives on foamability and foam stability. 1. Role of interfacial properties. *Langmuir* 2001; 17: 6999–7010.

63. Basheva, E.S., Stoyanov, S., Denkov, N.D., Kasuga, K., Satoh, N., Tsujii, K. Foam boosting by amphiphilic molecules in the presence of silicone oil. *Langmuir* 2001; 17: 969.

64. Simjoo, M., Rezaei, T., Andrianov, A., Zitha, P.L.J. Foam stability in the presence of oil: Effect of surfactant concentration. *Colloids Surfaces A Physicochem Eng Aspects* 2013; 438: 148–158.

65. Gao, F., Yan, H., Wang, Q., Yuan, S. Mechanism of foam destruction by antifoams: A molecular dynamics study. *Phys Chem Chem Phys* 2014; 16: 17231–17237.

66. Binks, B.P., Horozov, T.S. Colloidal particles at liquid interfaces: An introduction. In: Binks, B.P., Horozov, T.S. (Eds.), *Colloidal Particles at Liquid Interfaces*, Cambridge Univ. Press: Cambridge, 2006, Chpt. 1, p. 1–74.

67. Scheludko, A. Thin liquid films. *Adv Colloid Interface Sci* 1967; 1: 391–464.

68. Denkov, N.D., Tcholakova, S., Marinova, K.G., Hadjiiski, A. Role of spreading for the efficiency of mixed oil-solid antifoams. *Langmuir* 2002; 18: 5810–5817.

69. Koczo, K., Koczone, J., Wasan, D. Mechanisms for antifoaming action in aqueous systems by hydrophobic particles and insoluble liquids. *J Colloid Interface Sci* 1994; 166: 225–238.

70. Garrett, P.R. (Ed.). The mode of action of antifoams. In: *Defoaming: Theory and Industrial Applications*, Vol. 45. Marcel Dekker: N.Y., Surfactant Sci. Series, 1993, Chpt. 1, 1–117.

71. Marinova, K., Denkov, N.D., Branlard, P., Giraud, Y., Deruelle, M. Optimal hydrophobicity of silica in mixed oil-silica antifoams. *Langmuir* 2002; 18(9): 3399–3403.

72. Garrett, P.R. *The Science of Defoaming; Theory, Experiment and Applications*. Vol. 155. Taylor & Francis, Boca Raton, Surfactant Sci. Series, 2013, Chpt. 1, 1–31.

73. Nikolov, A.D., Wasan, D., Huang, D.D.W., Edwards, D.A. *Paper SPE 15443. Presented at 61st Annual Technical Conference and Exhibition of Soc Pet Engrs*, New Orleans, 1986.

74. Princen, H. Pressure/volume/surface area relationships in foams and highly concentrated emulsions: Role of volume fraction. *Langmuir* 1988; 4: 164–169.

75. Basheva, E.S., Ganchev, D., Denkov, N.D., Kasuga, K., Satoh, N., Tsujii, K. Role of betaine as foam booster in the presence of silicone oil drops. *Langmuir* 2000; 16: 1000–1013.

76. Farajzadeh, R., Andrianov, A., Krastev, R., Hirasaki, G., Rossen, W.R. Foam-oil interaction in porous media: Implications for foam assisted enhanced oil recovery. *Adv Colloid Interface Sci* 2012; 183–184: 1–13.

77. Wang, C., Fang, H., Gong, Q., Xu, Z., Liu, Z., Zhang, L., Zhang, Lu., Zhao, S. Roles of catanionic surfactant mixtures on the stability of foams in the presence of oil. *Energy Fuels* 2016; 30: 6355–6364.

78. Patist, A., Axelberd, T., Shah, D. Effect of long chain alcohols on micellar relaxation time and foaming properties of sodium dodecyl sulphate solutions. *J Colloid Interface Sci* 1998; 208: 259–265.

79. Garrett, P.R., Moore, P.R. Foam and dynamic surface properties of micellar alkyl benzene sulphonates. *J Colloid Interface Sci* 1993; 159: 214–225.

80. Garrett, P.R. *The Science of Defoaming; Theory, Experiment and Applications*. Vol. 155. Taylor & Francis, Boca Raton, Surfactant Sci. Series, 2013, Chpt.10.

81. Mcbain, J., Ross, S. Brady, A., Robinson, J., Abrams, I., Thorburn, R., Lindquist, C. *Foaming of aircraft engine oils as a problem in colloid chemistry*. National Advisory Committee for Aeronautics, Advanced restricted report NACA ARR No 4I05, 1944.

82. Lide, D.R. (Ed.), Handbook of chemistry and physics. CRC Press. Boca Raton, 85 Edn. 2004–2005.

83. Varaprath, S., Frye, C.L., Hamelink, J. Aqueous solubility of permethylsiloxanes (silicones). *Environ Toxicology Chem* 1996; 15(8): 1263–1265.

84. Callaghan, I.C., Hickman, S.A., Lawrence, F.T., Melton, P.M. Antifoams in gas-oil Separation. In: Karsa, D. (Ed.), *Industrial Applications of Surfactants*, Royal Society of Che. Special Publications 59, 48, 1987.

85. Callaghan, I., Gould, C., Hamilton, R., Neustadter, E. The relationship between the dilational rheology and crude oil foam stability I. preliminary studies. *Colloids Surfaces* 1983; 8: 17–28.

86. McKendrick, C., Smith, S., Stevenson, P. Thin liquid films in a non-aqueous medium. *Colloids Surfaces* 1991; 52: 47–70.

Section V

Applications

Applications

19 Foam Fractionation

Vamseekrishna Ulaganathan and Georgi Gochev

CONTENTS

19.1 INTRODUCTION

Foam fractionation is a technique which is used to separate the dissolved species of surface active molecules in aqueous phase. Floatation technique is the basis for foam fractionation where the former deals with macro species that needs to be separated and the latter with surface active molecules. It can be achieved by sparging air into the liquid containing surface active materials. An example of such a process can be seen in the cleaning of aquariums (see the pictures below) where protein-like waste substances are removed by sparging and discarding the generated foam [1] (Figure 19.1).

The history of foams shows that Schütz was probably the first who elaborated on the idea of foam fractionation in the late 1930s [2,3]. Later, Lemlich described and developed the process of foam fractionation with different methods such as batch, semibatch, and continuous foam fractionation [1–4].

In all the methods of foam fractionation employed, the underlying principle is to produce air bubbles in a solution to be foamed. It is required that the foam produced has to be relatively stable in order to be able to harvest it. The foam has to be allowed to drain to remove the interstitial fluid in between the bubbles. This is crucial because the dryer the foam, the better is the enrichment in the foam fractionation process [1,5].

Most of the studies on foam fractionation has been done on proteins and biosurfactants as they are generally abundant in food and bio technological processes [6–15]. Proteins are macromolecules which are amphiphilic in nature and their molecular conformation changes with solution conditions. This in turn would influence its surface activity which affects the foaming properties [8,16,17]. Therefore, for the separation process, it is best to choose the optimum foaming conditions. More details on the parameters influencing the foam fractionation process are given below.

Foam fractionation offers a cheap and eco-friendly method of separating the surface active material from a solution via an adsorption process. However, when separating proteins, there is a loss of its functionality as proteins undergo denaturation and structural changes upon adsorption [18–20].

19.2 MODES OF OPERATION

There are several modes of operation employed for foam fractionation which are similar to that of floatation. The broadly classified modes are (1) Batch mode, (2) Semi-batch mode, and (3) Continuous mode [1,4]. In all cases, air is sparged through a solution, B, with initial bulk concentration C_B. The produced foam is collected and then broken down to liquid and this is known as foamate F with concentration C_F. The enrichment ratio, E is given by

$$E = C_F/C_B \qquad (19.1)$$

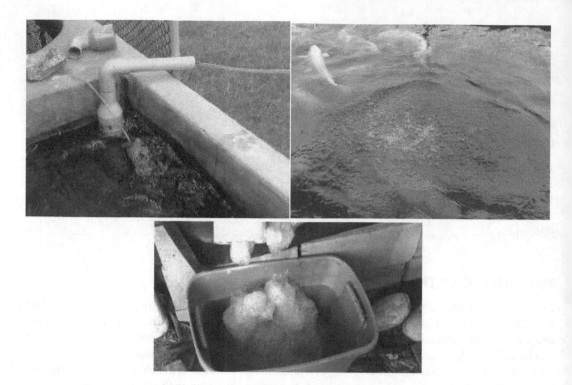

FIGURE 19.1 Aeration of the fish tank and collection of the foam containing surface active waste. (Taken from YouTube: https://www.youtube.com/watch?v=FE_iFrRR_dI).

In the *batch mode*, the foam in produced in a column by sparging into a fixed volume of solution. The foam is allowed to stand and drain after which the foamate is harvested. In the *semi-batch mode*, the foam is generated continuously for a single batch of feed solution and collected as it overflows. It is especially employed where the foaming is not desired in a process and the surface active material is removed by sparging until no significant foam could be further generated.

In the *continuous mode*, the solution is supplied into the tank and gas bubbles are generated at the same time at certain fixed flowrates. Thus, there is continuous influx of feed and the gas bubbles which generates mixing. The foam thus generated is collected at a certain height after certain drainage is allowed.

19.3   PARAMETERS AFFECTING THE ENRICHMENT

The bubble size has a major influence on the enrichment in the foamate [21]. Generally, the smaller the bubble size is, the slower is its rising velocity in the liquid [22]. Therefore, the more time that is available for the bubble to gather the surface active material, the more it is stable against coalescence [23]. The drainage of the liquid flowing through the bubble matrix is also slower for smaller bubbles [24]. This is because in a fixed cross sectional area of the fractionation column, the smaller the bubbles are, the greater is the total surface area offering resistance to the flow.

Another factor which retards the liquid drainage is the plateau borders. These plateau borders create capillary suction which acts against the liquid drainage due to gravity [2]. Therefore, the smaller the bubble size, the greater will be the liquid fraction in the foam. This might create a stable foam, but this is undesirable for enrichment because of the presence of interstitial liquid [5,25]. For more details on foam drainage, the reader is kindly asked to read the Sections 4.4 and 4.5.

Therefore, it is desirable to have bubble sizes which are large enough to allow quicker drainage in order to gain better enrichment. Note, however, the surface area to solution bulk volume is higher for smaller bubbles. Hence, the optimum bubble size has to be determined experimentally for each individual system.

Another factor which influences the enrichment is the gas flow rate. The gas is usually pumped through porous frit material which contains numerous capillary orifices [26]. The gas flow rate is directly proportional to the liquid flux carried into the foam. Though it is desirable to generate bubbles faster, the increase in liquid fraction in foam is counterproductive. Therefore, it is important to choose an optimum flowrate to reduce liquid flux as much as possible [1,25].

There are other parameters which are specific to each system such as pH and salt concentration. If the system contains proteins, they are very sensitive to the pH and ionic strength of their environment. Therefore, the surface activity of the protein is dependent of the solution pH which in turn would affect the foaming properties [16,17]. For a mixture of proteins in a solution with different isoelectric points this could be a useful parameter to separate the proteins from each other [27].

19.4   TECHNIQUES TO IMPROVE ENRICHMENT

There have been several attempts made to improve the enrichment over the course of improving the foam fractionation technique. One such techniques is connecting foam fractionation columns in series, therefore, the foamates collected are sequentially fractionated until the desired concentration is achieved. However, this technique has a disadvantage of requiring many fractionating columns [28,29]. By using a single column, the same efficiency is achieved by having a reflux of foamate. The collected foamate is reduced to liquid and fed back to the top of foam in the fractionation column. This would enhance the surface concentration in the foam which in turn would positively enhance the enrichment [30,31] (Figure 19.2).

Stevenson et al. [30] gave a model for predicting the enrichment of foamate with a reflux for a continuous foam fractionation column. Assuming a gas feed flow rate, v_g of 0.5 cm/s in the column as shown above, it generates an average bubble radius, $r_b = 0.5$ mm. For this illustration, the foam drainage parameters m = 0.016 and n = 2 were chosen. The simulation is presented below which shows the recovery fraction which is given by $C_F/(C_F + C_B)$ product rate which is the rate at which the foamate is produced. The effect of reflux ratio which is a ratio between volume of foamate fed back as reflux and total volume of foamate is shown below in Figure 19.3.

It can be noted that it is obvious that product rate is zero for reflux ratio which is 1. Whereas the recovery is maximum for the highest reflux ratio. This helps to choose an optimum reflux ratio with high recovery fraction and this model agrees with experimental observations. For more details, please read this article [30].

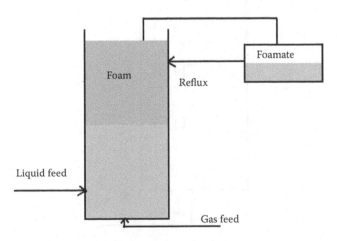

FIGURE 19.2   Schematic of a foam fractionation column with reflux. (Redrawn from Stevenson, P., Jameson, G.J. *Chem Eng Process: Process Intensification* 2007; 46: 1286–1291 [30].)

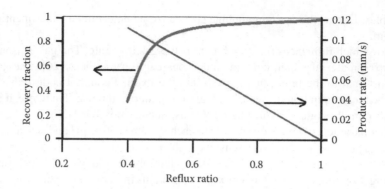

FIGURE 19.3 Simulation of a foam fractionation column with reflux. (Redrawn from Stevenson, P., Jameson, G.J. *Chem Eng Process: Process Intensification* 2007; 46: 1286–1291 [30].)

However, such reflux methods will only help with systems containing a low concentration of surface active material. In the case of high surfactant concentration, one would need to consider other methods that enhance liquid drainage [1].

Another method of improving the enrichment in a continuous mode was given by Boonyasuwat et al. [32]. The authors discuss a multistage fractionation process performed within a foam column with various parameters to select, such as number of trays to insert into the column, air flow rate, residence time, foam height, and surfactant concentration (see Figure 19.4). They performed foam fractionation of aqueous solutions of the anionic surfactant sodium dodecyl sulfate SDS and the cationic surfactant hexadecyl pyridinium chloride CPC. In Figure 19.5, some of the results are summarized.

For the two mentioned surfactants, an analysis of the influence of these system parameters on the enrichment ratio E of the surfactant in the foamate from the original solution is presented. As one can see, with increasing flow rate and initial bulk concentration the E decreases, while an increase is observed with increasing foam height and number of trays. In summary, we can state that the E

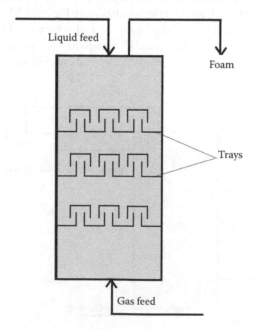

FIGURE 19.4 Schematic of a multistaged foam fractionation column with trays. (Redrawn from Boonyasuwat, S. et al. *Chem Eng J* 2003; 93: 241–252 [32].)

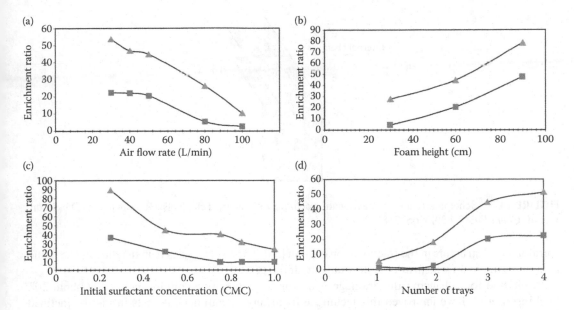

FIGURE 19.5 Enrichment ratio for aqueous (■) SDS and (▲) CPC solutions established in a multistage foam fractionation process as a function of different system parameters; (a) effect of air flow rate, (b) effect of foam height, (c) effect of initial surfactant concentration, (d) effect of the number of trays; base conditions were: air flow rate, 50 L/min; feed flow rate, 25 mL/min; foam height, 60 cm; surfactant feed concentration, 0.5CMC; temperature, 25°C; number of trays, 3. (Redrawn from Boonyasuwat, S. et al. *Chem Eng J* 2003; 93: 241–252 [32].)

is much better for CPC as compared to SDS. This, however, is no surprise because the CMC values of the two surfactants differ by a factor of 67, that is, the absolute bulk concentrations for the CPC are much lower than for the SDS solutions and hence depletion of the solution concentration due to material transferred into the foam must be much more efficient for CPC.

Some authors suggest using certain structures in the foam column in order to restrict the liquid flux while allowing the bubbles to rise. Bando et al. suggest the use of draft tubes inside the column through which bubbles would rise [33]. Tzubomizu et al. have used perforated plates for this purpose [34]. Wang et al. have successfully demonstrated enhanced enrichment using internal baffles mounted in the column [35] (see Figure 19.6). Yang et al. have shown improvements in enrichments with a

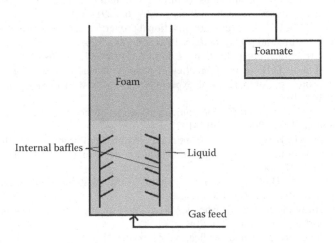

FIGURE 19.6 Schematic of a foam fractionation column with internal baffles. (Redrawn from Wang, L. et al. *J Food Eng* 2013; 119: 377–384 [35].)

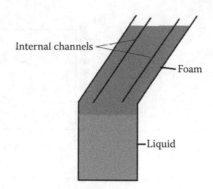

FIGURE 19.7 Schematic of a foam fractionation column with inclined channels. (Redrawn from Dickinson, J. et al. *Chem Eng Sci* 2010; 65: 2481–2490.)

spiral internal structure in the fractionation column [36]. These techniques help in reducing the liquid content in the foam which would allow quicker drainage.

A method for enhancing the drainage is having a fractionation column inclined at about 20°. Dickinson et al. have improved this technique by arranging parallel channels inside the inclined column. In this way, the interstitial liquid only needs to drain to the top of these channels and then flow back to the bottom liquid in the column (see Figure 19.7). It has been shown that in this way the enrichment can be increased four times more than with a vertical column [37].

REFERENCES

1. Stevenson, P., Li, X. *Foam Fractionation: Principles and Process Design*. CRC Press: Boca Raton, 2014.
2. Gochev, G., Ulaganathan, V., Miller, R. Foams. In: *Ullmann's Encyclopedia of Industrial Chemistry*, 2016, pp. 1–31. Wiley-VCH: Weinheim, Germany. doi: 10.1002/14356007.a11_465.pub2.
3. Schütz, F. Adsorption on foams. *Nature* 1937; 10: 629–630.
4. Lemlich, R. Adsorptive bubble separation methods—Foam fractionation and allied techniques. *Ind Eng Chem* 1968; 60: 16–29.
5. Leonard, R.A., Lemlich, R. A study of interstitial liquid flow in foam. Part I. Theoretical model and application to foam fractionation. *AIChE J* 1965; 11: 18–25.
6. Brown, A., Kaul, A., Varley, J. Continuous foaming for protein recovery: Part I. Recovery of β-casein. *Biotechnol Bioeng* 1999; 62: 278–290.
7. Brown, A., Kaul, A., Varley, J. Continuous foaming for protein recovery: Part II. Selective recovery of proteins from binary mixtures. *Biotechnol Bioeng* 1999; 62: 291–300.
8. Lee, S.Y., Morr, C.V., Ha, E.Y. Structural and functional properties of caseinate and whey protein isolate as affected by temperature and pH. *J Food Sci* 1992; 57: 1210–1229.
9. Burghoff, B. Foam fractionation applications. *J Biotechnol* 2012; 161: 126–137.
10. Uraizee, F., Narsimhan, G. Effects of kinetics of adsorption and coalescence on continuous foam concentration of proteins: Comparison of experimental results with model predictions. *Biotechnol Bioeng* 1996; 51: 384–398.
11. Chen, C.Y., Baker, S.C., Darton, R.C. Continuous production of biosurfactant with foam fractionation. *J Chem Technol Biotechnol* 2006; 81: 1915–1922.
12. Heyd, M., Franzreb, M., Berensmeier, S. Continuous rhamnolipid production with integrated product removal by foam fractionation and magnetic separation of immobilized Pseudomonas aeruginosa. *Biotechnol Prog* 2011; 27: 706–716.
13. London, M., Cohen, M., Hudson, P.B. Some general characteristics of enzyme foam fractionation. *Biochim Biophys Acta* 1954; 13: 111–120.
14. Li, R. et al. Pilot study of recovery of whey soy proteins from soy whey wastewater using batch foam fractionation. *J Food Eng* 2014; 142: 201–209.
15. Charm, S.E. et al. The separation and purification of enzymes through foaming. *Anal Biochem* 1966; 15: 498–508.

16. Phillips, L., Schulman, W., Kinsella, J. pH and heat treatment effects on foaming of whey protein isolate. *J Food Sci* 1990; 55: 1116–1119.

17. Zhang, Z., Dalgleish, D., Goff, H. Effect of pH and ionic strength on competitive protein adsorption to air/water interfaces in aqueous foams made with mixed milk proteins. *Colloids Surf B: Biointerfaces* 2004; 34: 113–121.

18. Barackov, I. et al. Investigation of structural changes of β-casein and lysozyme at the gas–liquid interface during foam fractionation. *J Biotechnol* 2012; 161: 138–146.

19. Clarkson, J., Z. Cui, Darton, R. Protein denaturation in foam: I. Mechanism study. *J Colloid Interface Sci* 1999; 215: 323–332.

20. Clarkson, J., Cui, Z., Darton, R. Protein denaturation in foam: II. Surface activity and conformational change. *J Colloid Interface Sci* 1999; 215: 333–338.

21. Du, L., Prokop, A., Tanner, R.D. Effect of bubble size on foam fractionation of ovalbumin. In: Adney, W.S., McMillan, J.D., Mielenz, J.R., Klasson, K.Th. *Biotechnology for Fuels and Chemicals*. Springer, 2002, pp. 1075–1091.

22. Tan, Y. et al. Bubble size, gas holdup and bubble velocity profile of some alcohols and commercial frothers. *Int J Miner Process* 2013; 119: 1–5.

23. Kirkpatrick, R., Lockett, M. The influence of approach velocity on bubble coalescence. *Chem Eng Sci* 1974; 29: 2363–2373.

24. Manev, E., Sazdanova, S., Wasan, D. Emulsion and foam stability—The effect of film size on film drainage. *J Colloid Interface Sci* 1984; 97: 591–594.

25. Stevenson, P. Hydrodynamic theory of rising foam. *Miner Eng* 2007; 20: 282–289.

26. Drenckhan, W., Saint-Jalmes, A. The science of foaming. *Adv Colloid Interface Sci* 2015; 222: 228–259.

27. Lockwood, C.E., Jay, M., Bummer, P.M. Foam fractionation of binary mixtures of lysozyme and albumin. *J Pharm Sci* 2000; 89: 693–704.

28. Darton, R., Supino, S., Sweeting, K. Development of a multistaged foam fractionation column. *Chem Eng Process: Process Intensification* 2004; 43: 477–482.

29. Morgan, G., Wiesmann, U. Single and multistage foam fractionation of rinse water with alkyl ethoxylate surfactants. *Sep Sci Technol* 2001; 36: 2247–2263.

30. Stevenson, P., Jameson, G.J. Modelling continuous foam fractionation with reflux. *Chem Eng Process: Process Intensification* 2007; 46: 1286–1291.

31. Martin, P. et al. Foam fractionation with reflux. *Chem Eng Sci* 2010; 65: 3825–3835.

32. Boonyasuwat, S. et al. Anionic and cationic surfactant recovery from water using a multistage foam fractionator. *Chem Eng J* 2003; 93: 241–252.

33. Bando, Y. et al. Development of bubble column for foam separation. *Korean J Chem Eng* 2000; 17: 597–599.

34. Tsubomizu, H. et al. Effect of perforated plate on concentration of poly (vinyl alcohol) by foam fractionation with external reflux. *J Chem Eng Japan* 2003; 36: 1107–1110.

35. Wang, L. et al. Enhancing the adsorption of the proteins in the soy whey wastewater using foam separation column fitted with internal baffles. *J Food Eng* 2013; 119: 377–384.

36. Yang, Q.-W. et al. Enhancing foam drainage using foam fractionation column with spiral internal for separation of sodium dodecyl sulfate. *J Hazard Mater* 2011; 192: 1900–1904.

37. Dickinson, J. et al. Enhanced foam drainage using parallel inclined channels in a single-stage foam fractionation column. *Chem Eng Sci* 2010; 65: 2481–2490.

20 Foams in Food

Deniz Z. Gunes, Jan Engmann, Cécile Gehin-Delval,
Christophe Schmitt, and Martin E. Leser

CONTENTS

20.1 GENERAL INTRODUCTION

20.1.1 BRIEF DESCRIPTION OF CONTEXT

Foam research related to the food industry is one of the most active fields in soft matter science, addressing questions and using methods which have important generic aspects and hence are relevant to many other industries [1–8]. The goal in most food foams applications is to generate structures aimed at providing the consumer maximal pleasure upon consumption, with consistent quality. To do so, those structures should be stable until the moment of consumption. Structure results from formulation and process combined. It will determine not only the product texture and stability, but also other properties such as flavor perception or its visual appearance (think of what makes a coffee or beverage cup your favorite!—see examples in Figure 20.1). There is also a strong demand from consumers to reduce the number of calories [36,141,209] and saturated fatty acid content [9] in food products and to replace these with nutritious ingredients [10–12] which offer them better control of satiety [13,14]. In this context, it can become challenging to control structure and stability [3,9,10,15], while always improving consumer pleasure.

In many processes, the aeration step is positioned after mixing of ingredients and after heating or cooling. This is true for example in most processes for ice cream [16,206], whipped cream [17–19], beverages, coffee (Figure 20.1), aerated chocolate (Figure 20.2) [20], dough [21], and so on. Alternatively, the aeration step may be performed at a very early stage and the foam subsequently incorporated into the final liquid matrix [22], possibly with further processing such as a temperature treatment step. In many products, like with cappuccino foams (Figure 20.1), the first and foremost category of attributes governing the sensory perception the consumer is aware of is certainly the textural one. It concerns an important part controlled by the rheology of the foam and the way it

FIGURE 20.1   Illustration of liquid foams in coffee and beverages. Left: picture of a Ristretto Ardenza Dolce Gusto coffee cup: a black coffee with a fine coffee cream on top. Right: picture of a Dolce Gusto Cappuccino with its generous milky foam.

FIGURE 20.2   Illustration of semi-solid to solid foams in food. Left: picture of tablet pieces of aerated chocolate (*Aero* from Nestlé). Right: picture of a spoon of chocolate mousse.

collapses in the mouth. The actual perception can, however, be strongly coupled to other attributes such as flavor and taste perception. Foam mouthfeel is not highly affected by the presence of small bubbles for low air volume fractions, however, larger bubbles can be perceived and control of size distribution is thereby important [23–26]. In contrast, highly aerated foams can possess high moduli without being dispersed in a highly viscous or elastic matrix (think of a milky froth—Figure 20.1). In those, the modulus is larger for foams with smaller bubbles [27–29]. In this chapter, we discuss the influence of foam characteristics on stability, rather than the stabilizing functionality of specific emulsifiers or specific food structures. We will address structural and stability characterization of food foams with macroscopic and microscopic tools. This task implies the ability to control and measure structural parameters in food foams at all scales, whereby the limits of current capabilities will be shown.

20.1.2 Formulation of Main Scientific Questions

One important reason why food foams are much more difficult to stabilize than emulsions, is that the surface tension is difficult to lower by a factor of more than 2 for foams when classical emulsifier types are used. When bare interfacial tensions are compared, the ratio between the interfacial tension of a drop to the surface tension of a bubble is 2 or 3. However, coverage of the water/oil interface with emulsifiers can lower the interfacial tension by a factor 10 or 100. In contrast, reaching a similar decrease factor for an air/water interface would mean surface pressure level near 70 mN/m. As a consequence, rare are the emulsifiers or particles that can induce levels of surface pressure so high that nonspherical shapes can be retained at long time scales for foams, with the exception of particles [30]. In other words, the driving force for Ostwald ripening (also called disproportionation) is generally more difficult to counteract for foams [5,31–33] in comparison to emulsions [34,35]. This will be discussed in more detail in this chapter. Another important reason of the difference is due to the much lower density of the dispersed phase in the case of a bubble as compared to a drop. It implies that more material needs to be exchanged in the case of an emulsion, to modify sizes from an initial distribution S_i to a final one S_f.

Scientists and engineers are constantly developing new food products by making use of material science and engineering rules and empirical knowledge. Still, there is a need to pose well defined questions that would allow effective correlation between either bulk or interfacial rheology properties and foam stability. In practice, a lot of the knowledge necessary to control foam structure and stability can only be empirically generated. Still, every scientific aspect investigated will make the developments more robust, more rapid, and more cost effective. This chapter summarizes material and guidelines of how to address structure related questions in food from a soft matter perspective, for developing more stable and functional structures [1,15].

The chapter is organized as follows. Processes to produce foamed structures are overviewed in 20.2. Structures generated in those processes are generally out of equilibrium, and usually involve molecular, colloidal, and macroscopic scales. The section "Destabilization phenomena and their characterization" highlights modern tools that have proven really useful in the characterization of foam structure and dynamics. The section "Interfacial and/or bulk stabilization of aerated structure: Stabilizing mechanisms," tackles the question of the relative importance of bulk or interfacial stabilization in traditional food systems and the tools to explore it. Generally speaking, the science of foams to date does not provide predictive theories of destabilization rates in most real systems, for example, the effect of interfacial and bulk viscoelasticity still needs to be established. There is certainly considerable complexity that comes from the coupling of the different destabilization mechanisms (Ostwald ripening, coalescence, drainage). To those add the difficulty to create small scale experiments that address the structures at the right time and length scales and to characterize foam properties like its rheology. We hope that the present chapter helps scientists working on food in identifying the important questions and the relevant methodologies for their specific targets.

20.2 FOAM GENERATION

The creation of aerated food structures is closely linked to stabilization of these structures since any newly created foam usually changes average bubble size, bubble size distribution, and gas content rapidly (unless the different destabilization mechanisms discussed above are sufficiently slowed down). In this section, we review different processes for gas incorporation and bubble expansion and discuss their interplay with structural stabilization.

Campbell and Mougeot [36] have reviewed both artisanal and industrial food aeration processes and their basic principles. Here, we update and supplement their overview, considering also some aeration technologies outside the food sector and discussing more specifically the role of food matrix rheology in structure formation.

A basic distinction for creating foam structures can be made between

- *Incorporation* of gas bubbles (a topological structure change which affects the number of bubbles as well as gas content and bubble sizes) by subdivision of a gas-liquid interface or by gas bubble nucleation and
- *Growth* of gas bubbles (which leaves the foam topology unchanged but affects both gas content and bubble size).

Incorporation of gas bubbles into food structures often takes place by the mere presence of gas cells/pores as part of the ingredients (e.g., porous powder particles, which may be soluble or insoluble) and the interstitial air between particles incorporated when they are mixed (e.g., batters or bread dough [36,37]). The number of pores is limited by geometric factors (e.g., the number of distinct cavities between ingredient particles is linked to the number of particles).

Additional incorporation of gas volume is often achieved through rapid agitation of the food matrix in contact with the atmosphere or, less frequently, by a gas reservoir under positive pressure. The macroscopic surface area of the food matrix and its rate of turnover during agitation determine the *amount* of gas incorporated, whereas the shape of this surface, specific patterns of turnover, and pressure can influence the size of the incorporated gas bubbles [38,39]. Gas incorporation is often combined in a single unit operation with mixing of ingredients to develop an aerated food structure, the classic example being doughs and batters. "Whipping" of low viscosity food matrices achieves rapid and often locally turbulent flow conditions near the gas/liquid or semisolid interface and hence increases the incorporation of small gas bubbles. High speed agitation can also lead to further subdivision of gas bubbles when the local shear stress exceeds a critical interfacial stress required to break up a bubble [40–45]. Most traditional foaming processes employed in home or restaurant kitchens incorporate gas bubbles by some sort of whipping or shaking. The rheological properties of the matrix have a strong effect on the shape of the surface and local flow patterns and are, therefore, important in determining the amount of gas incorporated and the bubble size. Typical rheology modifiers also often affect bubble rise/drainage and coalescence, making it a challenge to assess their effects systematically and selectively [46]. Gas incorporation may also be performed in combination with strong temperature-induced changes of rheological properties, as, for example, during the freezing of ice cream mix in scraped surface heat exchangers [47].

A more controlled gas incorporation (gas content and bubble size) than by open surface aeration can be achieved by *injection of a gas stream* into a liquid phase through single or multiple nozzles or pores [46,48–53]. The degree of control depends on the rate of gas injection and on the variability of the flow conditions near the injection point since the break up of a gas jet into individual bubbles depends on the balance between local stresses in the fluid and interfacial stresses [54,55]. Gas injection into a stream of material flowing at steady state is likely to give more consistent results than injection into a stirred tank with strong transients. The rheological properties of the matrix have, of course, a strong effect on the stresses near the gas/liquid interface and, therefore, affect the break up conditions [55]. Gas stream injection is used primarily in industrial food aeration processes, but also in milk frothing devices (steam injection), for example, for coffee beverage preparation.

Finally, incorporation of very large gas cells can be achieved through folding processes of solid and semisolid food matrices, as, for example, in puff pastry where the creation and closure of open surfaces is specifically controlled with processing equipment.

Homogeneous nucleation of gas bubbles from dissolved gas inside a liquid phase requires a high degree of supersaturation and, therefore, a strong change in solute concentration (e.g., by CO_2 production from yeast or chemical leavener), pressure or temperature as well as the absence of heterogeneous nucleation sites [26,56]. Nucleation of gas bubbles is particularly relevant in beverages (e.g., carbonated soft drinks, beer, champagne). Recently, such a process was shown to be as well significant for espresso coffee foams [57]. The driving forces for nucleation are strongly linked to those for bubble *growth*, which we will now discuss. A specific case worth mentioning is that of bubble size control (before destabilization) in beer and champagne. Bubble size is tuned by controlling detachment at microstructured glass wall, the gas oversaturation level, and liquid height which will determine its expansion rate and final size [58].

The degree of foaming/aeration (average gas content in the whole matrix) achievable via gas incorporation is usually quite moderate except for some highly stable foams (e.g., egg white foams) since conditions favoring strong incorporation of gas bubbles (high local shear rate) usually favor destabilization processes as well. High gas contents ($\phi > 50\%$) with reasonable lifetimes, therefore, often require some form of *bubble growth process* with simultaneous stabilization. The theoretical potential of different bubble "growth drivers" can be easily estimated by considering the ideal gas law ($P\,V = n\,R\,T$) for a single bubble. Expressing this as $V = n\,R\,T/P$, it is seen that the total gas "overrun" (V_{gas}/V_{matrix}) is directly proportional to the absolute temperature T and inversely proportional to the (absolute) pressure P in the product. The molar amount of gas n is different for each bubble (except for monodispersive foams). For the case of very small gas bubbles and high interfacial tension, the excess (Laplace) pressure inside the bubbles needs to be accounted for when making such estimates.

The absolute temperature T can usually not be changed by more than 30% without evaporation or solidification of the food matrix (unless gas is incorporated directly at extremely low temperatures, as e.g., celebrated by "molecular gastronomists," e.g., Barham [59]). Changes in pressure and gas content inside the gas cells/bubbles are, therefore, the most important drivers for bubble growth. Temperature, however, also influences the chemical potential of solutes and gas molecules and, therefore, has an additional effect on bubble growth besides the thermal expansion of the gas (as well through its effect on the matrix rheology). This is discussed in the next paragraph.

The content of each type of gas molecule* in a bubble is linked to its local concentration in the surrounding liquid phase by a chemical equilibrium unless a low permeability gas barrier (e.g., through local gelation of macromolecules such as proteins) has been formed at the interface. Any change in chemical potential difference for a given molecular species (caused by a concentration increase in the liquid phase, a temperature or pressure change) can shift this equilibrium and drive gas molecules toward and into the bubbles. Carbon dioxide (CO_2) produced by a chemical reaction (e.g., baking soda) or biochemical reaction (e.g., baker's yeast) leads to supersaturation and diffusion towards the bubbles. Since this process is diffusive, the *kinetics of bubble expansion* (e.g., the rate at which the total overrun increases) can be governed by the mobility of the gas molecules in the phase surrounding the bubbles. The fastest expansion mechanism achieved by a chemical potential difference is *water evaporation* where the liquid/gas equilibrium of water is shifted via a rapid temperature or, less conventional, pressure change. Since water is relatively abundant in most food matrices, diffusion is less of a controlling factor. Water evaporation, thermal expansion, and desorption of CO_2 all contribute to bubble growth in baking processes [60–64].

Pressure can be used to shift chemical equilibria, but also to directly grow gas bubbles. In contrast to absolute temperature changes, relative pressure shifts by one or two orders of magnitude are relatively easy to realize without affecting food matrix properties. If gas is incorporated under positive pressure (e.g., in high pressure mixing of dough), the subsequent decrease of pressure to atmospheric conditions leads to rapid bubble growth [65]. Inversely, a partial vacuum can be used [20], in which case the structure needs to be well set (solidified) to avoid collapse when atmospheric pressure is attained. In a continuous process (where pressure release coincides with flow), high pressures also imply high *pressure gradients* along the process line, which inevitably lead to high shear stresses (consider, e.g., the stress balance for a simple cylindrical pipe, where the stress at the wall is proportional to the pressure gradient). High shear stress leads to high shear and flow rates. This can be desirable (e.g., to achieve additional bubble break up during flow), but in the case of already high gas content, it may also promote bubble coalescence. It can also be noted that very high pressure levels (of order 100 MPa) can also induce molecular level changes in the matrix [66].

If solute diffusion toward the bubbles is not limiting the bubble growth rate, high interfacial dilatational viscosity can do so. Both interfacial and bulk rheology may slow down the bubble growth process. Combinations of pressure, temperature, and mass transfer effects to produce bubble growth are common and, for example, used in the aeration of cereal matrices, which can undergo a rapid change in rheological properties as a function of water content and temperature and, therefore, allow rapid setting of a structure [67–75]. The type of process and complex rheological nature of the matrix has analogies in synthetic polymer foaming, where many basic analyses of bubble growth processes have been performed [76,77].

---

* Mostly N_2, O_2, CO_2, H_2O, N_2O (beer widget); there are not many other gases that are nontoxic, safe, and economically feasible for use in food.

As stated earlier, stabilization of a foam structure by formation of interfacial and/or bulk structures needs to occur during foam formation if coarsening and loss of gas content (by coalescence) are to be avoided. Bubble growth processes usually compete with stabilization since whatever slows down or prevents coarsening by reducing molecular mobility at or near the liquid/gas interfaces has the potential to also reduce or slow down bubble growth. In baking of bread, for example, the expansion of a loaf by water evaporation is eventually stopped by the formation of an outer crust, the solidification of the crumb, the creation of an open, interconnected pore structure or a combination of those effects [67]. Without these effects, a bread loaf would simply collapse as soon as the temperature is lowered again due to contraction and the open pore structure. Classical transport models [78–80] have been used to analyze this interplay and how it depends on both material properties and geometric factors. Aeration with simultaneous freezing of the structure is also performed in classical ice cream processing (scraped surface heat exchangers) [43,47]. The "competition" between bubble growth and foam structure setting is a generic feature of foams and not limited to food systems: see for example, the extensive literature on synthetic polymer foaming [81–83] or the emerging field of biomaterials [73,84].

Given this complex interplay between foam structure formation, structure evolution, and (for "solid" foams) structure setting, we may conclude that a precise prediction of foam structure development is rarely possible, even when the basic material properties (bulk rheological properties, diffusivity, interfacial rheology) have been very well characterized. A good understanding of the governing mechanisms and material properties allows, however, to identify effective design rules for formulation and process conditions and to obtain good control in producing attractive food foams.

20.3 DESTABILIZATION PHENOMENA AND THEIR CHARACTERIZATION

20.3.1 FOAM STRUCTURE CHANGES (COLLAPSE, COARSENING, DRAINAGE)

In this section, we will try to highlight techniques that are either well established or promise to be particularly insightful for the understanding of physical stability of food foams. Obviously, all techniques that have been proven useful for nonfood foams should be considered at least extensible to food foams. For most food foams, visual observation combined with image analysis is certainly the first characterization step (when possible). We usually access only the external layers (top, bottom or side), which can be representative of the whole foam [85], at least from the qualitative behavior point of view. The possible invasive role of container walls or the interface at the boundary with the food system considered needs to be addressed. For food foams, the complexity can even be larger than for nonfood, due to the fact that dispersions involve structures with many refractive indices as well as light absorption spectra. It implies that the commonly used techniques always need to be questioned.

For example, the dynamic evolution profile of *the liquid fraction versus altitude* in a collapsing foam is still a challenge to correctly measure for any foam. The time evolution of the liquid fraction can be studied using electric conductimetry profiling, usually done at a scale larger than 1 cm. Recalibration of the conductivity of the drained liquid versus the conductivity of the liquid is usually required for food foams. When the number of electrodes is high enough, one can speak of conductimetric tomography of foams [86], whereby scales down to 1 mm can be explored. In any case, conductimetry is useful to monitor the loss of liquid from the foam by the combination of drainage, coarsening, and coalescence [87,88]. Turbidimetry will be in such case a complementary technique for comparing the stability in time of different foams [89], as it is sensitive to bubble size and liquid fraction [90], even when the continuous phase is turbid [91]. The resolution, when detecting light in transmission, is limited by the sample thickness; however, it can be quite smaller in backscattering configuration. Magnetic Resonance Imaging (MRI) gives access to time-lapse evolution in case of foam collapse not faster than at the scale of a few minutes [92], hence can address only slow phenomena. Nuclear Magnetic Resonance (NMR) was proven relevant for the characterization of size in the case of emulsions and foams (up to 50% air fraction or so), implying important assumptions on the size distribution [93,94].

Two-dimensional (2D) foams find a high number of supporters amongst researchers dealing with more fundamental aspects [95,96], and the approach certainly has advantages for applied research as well [97]. Let us cite here a few reasons why, starting from the possibility of full observation of the foam structure evolution. Fundamental studies are possible in 2D, even if the phenomena studied may follow laws somewhat different than in 3D, as established for Ostwald ripening [98,99,210] and drainage [100,101]. Investigations in 2D allow to decouple destabilization driving forces, for example, by suppressing drainage, even when they are coupled in real foams. A particularly important advantage when working with 2D foams is the possibility to tune liquid fraction. One way is to use gravity to establish a liquid fraction profile by inclining the 2D foam at a known angle [96]. Another is to use wall wetting properties to obtain a uniform liquid fraction of tunable value [102]. 2D configuration was used to show that foams with multilamellar emulsifier structures can become more stable as they reinforce with time [97], with a correlation with an increase of interfacial rheology properties [103]. We will mention more particularly the use of a Thin Film Balance apparatus (TFB) for thin film characterization in the next section. Conductimetry is another technique demonstrated to measure single film drainage kinetics [104].

Regarding *Ostwald ripening*, a combination of methods to quantify the effect of the single interface, of the thin film, and of the contribution of all effects including bulk would be ideal. The first one can be done by measuring the dissolution rate of a single bubble into a solution of known chemical potential of dissolving gas [105,207]. The second may be measured by monitoring the shrinkage rate of a bubble below a free surface [106]. The third may be characterized by following the mean bubble size in the foam or by observing 2D foam evolution, whereby changes due to Ostwald ripening or coalescence can be distinguished [97,107]. The dynamics at colloidal scale may be different from the dynamics at molecular diffusion scale, which makes the problem of Ostwald ripening a multiscale, complex one whose dominating parameters are not intuitive. Measuring the *film thickness as a function of disjoining pressure* (using the Thin Film Balance (TFB) [108,109] apparatus) can be related to actual thickness values within a foam, which are related to foam stability [110]. Though TFB rarely allows direct prediction of foam stability [111,112], it can even give access to important film structural characteristics such as the specific molecular and colloidal organization therein [113–117]. Its importance must be emphasized in particular in the presence of *confinement induced effects* [29,103,115–117], for example, at the scale of the thin film separating two bubbles. Those can affect the film thickness profile versus pressure applied, its effective permeability to gas diffusion, and its rheology. This is why the foam science community has defined and used empirical approaches to determine certain parameters, like thin film permeability with respect to gas diffusion. That is justified only due to the presence of films which are complex at structural and compositional levels, preventing us from modeling their effects on permeability coefficients [106,118]. The same holds for the rheology of the thin film, which is often equally complex to understand and characterize. The "bubble diminishing method" nicely allows measuring effective surface and film properties affecting foam stability with respect to coarsening [87,118]. In using that instrument, one needs to be cautious in making sure that the creation of the thin film between the bubble and the free interface and the foam creation are compared within the same time scale. In principle, it should enable one to determine steric contributions to stability versus rheological ones resulting from the thin film rheology. An improvement of the method to control the age of the interfaces independently from the film creation time is desired [97].

Probing the *dynamics within a foam* is possible by using Diffusing Wave Spectroscopy (DWS). It is the extension of the classical (single) dynamic light scattering principle to highly multiple scattering media. Because of its noninvasive nature, it is a method of choice for measuring rates of bubble rearrangements in foams. In a few words, the technique is based on the measurement of the characteristic time of decorrelation of the autocorrelation function of the scattered light intensity of a laser beam. It can be coupled to a flow apparatus [119]. One inherent difficulty that needs to be addressed when applying DWS to food foams, is that the nongaseous medium is often turbid, hence can contribute to the DWS signal nontrivially. That contribution would be problematic if its associated time scale was shorter than that of the foam dynamics of interest. The same reason makes it challenging to use tracer scatterers-based DWS [120] (as in microrheology). New developments of turbid media dynamic light scattering signal analysis [121] are aimed at improving the issues just cited.

Apparatuses developed to *work with a few bubbles* allow focusing on a particular question at the smallest scale, for example, film drainage [122] or topological change (T1 type of rearrangement) upon stress application [123]. Such experiments, when combined with more classic ones to characterize foam stability, allowed showing that coalescence is driven by rearrangement events for many foam types. Topological rearrangements, which can be driven by, for example, Ostwald ripening [107] or drainage, will induce film rupture if the local liquid fraction is below a certain critical value [124]. Coalescence driven by pressure changes within the process [125] may also be a related phenomenon.

20.3.2 OBSERVING STRUCTURES WITHIN THE FOAM AND AT INTERFACES

Observing the real food foam, that is, ensuring on one side that it is *not producing too strong artifacts*, and on the other side, that it is obtaining information below the scale of the bubble, is a difficult task with food materials. Fundamentals of *confocal microscopy* were consolidated to render possible the analysis of complex food systems [126]. An estimate of bubble or drop size distribution can be obtained even in real food formulations [127]. For many decades, considerable knowledge was gained based on rather extensive use of optical and *electron microscopy* [128]. A good example is ice cream stability, whereby correlation was established between fat agglomeration status and product stability [129]. There is real difficulty in characterizing structures at minus temperatures or in using microscopy while not inducing significant artifacts due to sample preparation (freezing, cutting, recrystallizing, shining too strong an electron beam, etc.). Even when the presence of certain structures is shown to correlate well with higher stability, underlying physical explanations are in most cases to be further established. An example is the role of the partially coalesced network of fat drops in whipped creams [17–19]. The building of a robust approach would combine the use of model systems, whereby structures would be controlled at all multiple scales and compared with the study of structural evolution of real foams.

The variety and potential of techniques that can be applied to the structure and dynamics characterization of model foams is much higher than those that can apply to real foams. *X-ray tomography* has been shown to be powerful for the investigation of aerated food [130–134], with a resolution better than 10 μm for a pixel. With the development of phase contrast imaging, the technique seems promising to not only distinguish between air and other phases, but also between different condensed phases in a complex structure. The description and discussion of ice cream structures subjected to heat shocks can be found in [5]. There is still significant room for progress in data analysis before achieving quantitative structure characterization with X-ray tomography [5,131].

X-ray scattering can be applied to foams to reveal complex structures [133], even permitting the investigation of molecular structures of emulsifier-laden interfaces [135]. Food formulations very often imply structures that possess many organization levels of matter below the micron scale. Protein stabilized films have been shown to be sometimes very heterogeneous, despite the size of proteins being of the order of a few nanometers [115,116]. Typical evidence for that character is found in the structure of thin films. They need to be investigated using the many different well-established techniques [114] such as the thin film balance, X-ray and neutron scattering from interfaces, and so on. *Neutron scattering* has been shown to be very useful for investigating complex structures at interfaces [7,132,136]. *Atomic Force Microscopy* [137–139], *dynamic surface tension* [140,141] or *rising bubble velocity* time dependency [142] were shown to bring insight into competitive adsorption and structural dynamics at interfaces. Protein conformation and interactions at interfaces can be probed by using *circular dichroism* and *dual polarization interferometry* [143].

20.4 INTERFACIAL AND/OR BULK STABILIZATION OF AERATED STRUCTURES: STABILIZING MECHANISMS

20.4.1 INTRODUCTORY CONSIDERATIONS

Most food foams are stabilized by a combination of bulk and interfacial effects, which will exclude examples cited above as baked products or aerated chocolate (where bulk stabilization suffices—see Figure 20.2). We are dealing here with systems where the different destabilization driving forces, as well as stabilizing factors, are known. They usually contain important kinetic (e.g., viscous) as well as thermodynamic (e.g., elastic or surface tension) effects. Examples of pure bulk stabilization less obvious than cited above would be aerated gel structures, for example, made of hydrocolloids such as food proteins [144] (like in chocolate mousse; Figure 20.2) or gelatin [145]. It may appear obvious but let us recall that when dealing with stability questions, we need to involve time scale considerations as soon as the system considered is not fully stable. If we address structural stability during the storage of soft food systems (i.e., having their linear elastic modulus at the low frequency limit in the order of 100 Pa or lower), as explained before it cannot be solved by use of bulk elasticity alone [146]. Destabilization mechanisms can either take place according to mechanisms present throughout the product lifetime like in ice cream [5] or their effects can develop dramatically only from a certain point on [124]. Note that even ice cream cannot be considered as having fully efficient bulk stabilization over the shelf lifetime [5], part of the issue rising from the difficulty to maintain constant cold temperatures at acceptable cost. In the case of ice cream, the whole problem becomes a level more complicated because of the coupling of the aerated structure destabilization to the coarsening of the ice crystals structure (by Ostwald ripening or due to temperature gradients).

In this section, rather than discussing destabilization mechanisms one by one, we will address a few effects (or parameters) and specify their role in the different mechanisms (coalescence, Ostwald ripening, drainage). In the food industry, it is often possible to generate efficient stabilizing structures in process [23,133,134]. The use of a special stabilizer in the list of possible ingredients is a suitable option in case of obvious drastic stability improvement (thereby bringing a significant value to the consumer benefit or commercially); or when its use would actually be based on a generic functionality, that is, for different product categories. We are here not aiming at being comprehensive. Now, dealing with structures in the presence of interfaces is a level more complex than dealing with bulk structures [147,208]. While much of applied research even to describe only foamability is done empirically [144,148,149], we want to stress here that existing tools can significantly accelerate the design of more stable structures and more efficient processes. We will see that interactions can be tuned to generate metastable composite structures at interfaces or in thin films, which can efficiently stabilize food foams as will be described later.

20.4.2 ROLE OF INTERFACES

Within *interfacial stabilizing effects*, we may distinguish between repulsive interactions between surfaces due to *electrostatic or steric* origin [110,115,150,151], and *interfacial rheology* effects (in shear, dilatation, and bending modes). The first type of interactions is thought to be very important for *foamability* and short term stability of liquid protein solutions for making froths of milky beverages [110,115]. Repulsive effects mentioned before impart sufficient stability to bubbles against coalescence in the young foam. Surface charge effects are, however, not sufficient to stabilize against coarsening even against short term coarsening in common protein foams [152], *a fortiori* not efficient long term either. Whether bulk or interfacial effects dominate the behavior depends on where the dominating colloidal interactions reside [29,146,153]. Food foams that exhibit long term stability, that is, beyond a day up to months, in practice need to be stabilized with strong rheological contributions from the interface, the bulk or the film between bubbles [146].

Considering *foamability*, parameters that determine the adsorption kinetics of proteins are relatively high in number, hence systematic investigations are needed to probe their contributions [154]. It was shown that the zeta potential of protein-laden bubbles [155] can be non-zero even before protein adsorption because of ion (hydroxyl, multivalent) adsorption to the interface [156–158]. It may in practice result in adsorption kinetics with two regimes, with the development of a slow adsorption mode as the interface effective charge increases with surface coverage, as, for example, for oil-water interfaces [159]. Theories have been developed which attempt to take into account ion adsorption and image charges near interface [158,160]. However, for food systems based on proteins or other type of molecules involved in complex assemblies, quantitative model development is not yet within reach. For example, desorption probability [154] is usually not predicted from molecular theories or simulation. Furthermore, ion-molecule interactions often add to the complexity level.

The stability of protein-based liquid foams will critically depend on foam height, which sets the gravity pressure counteracted by the *disjoining pressure of thin films*. Because of the time needed for the foam to drain, in practice these counteracting forces generally do not balance before destabilization by coalescence and disproportionation. The film thickness profile versus altitude at a given age of the foam will thus be the result of drainage [161] and the coupled mechanisms mentioned earlier. For compressed emulsions, catastrophic destabilization by coalescence occurs when the osmotic pressure applied on the film overcomes the disjoining pressure between drops [162]. Such behavior can be fully rationalized by quantifying the contribution of electrostatic interactions between drops or bubbles, for example, in the case of protein stabilized emulsions [151]. Similarly, the critical disjoining pressure for inducing film rupture correlates well with foam stability [163].

Most food products not only contain proteins as surface active agents, but also *smaller molecular weight surface active species*, such as protein hydrolysates or lipids. Laws of thermodynamics impose that they will be engaged in a competition with proteins for adsorption at the interface. The smaller molecules can displace proteins from interfaces if sufficiently concentrated as nicely shown by several groups [51,164] or on the contrary complex with them to increase stability [165]. Protein hydrolysates alone [166] as well as peptide solutions [167] were shown to possess important foaming properties. As already noted, in their competition for occupation of the interface, energy barriers often take a role and influence the effective interaction between a protein and the—possibly covered—interface. The interaction potential between a protein and an air-water interface comprises a component which depends on the effective dielectric constant of the protein, making it protein dependent (size and type) [168,169].

Let us now consider *interfacial stabilization against Ostwald ripening*. If bubble interfaces are not far apart within the foam, Ostwald ripening will slowly coarsen the structure unless the force driving that mechanism is counteracted. The opposing contribution to the free energy of the system needs to be of similar amplitude as the interfacial energy reduction prevented. It is now common knowledge that bubble dissolution can be arrested by a layer of hard particles adsorbed at its interface (by the so called Pickering effect) [170]. Very recent experimental and theoretical work evidenced that the stability of bubbles covered with rigid spheres can be accounted for only by the cost of interfacial energy to expel the particles from the interface [171]. It is not yet clear whether that simple explanation of bubble stability holds for other systems involving, for example, softer particles or mixed interfacial stabilizers. Another possibility for suppressing disproportionation would be an annihilation of the driving force, that is, suppressing the Laplace pressure differences between bubbles. It would mean that every bubble would possess the same mean curvature, which could be a zero mean curvature [30] (thereby also suppressing the chemical potential difference between bubbles and the head space). It would mean that the average curvature would be the result of position at surface and maybe the shape of the particles as well. In practice, there has not been yet experimental demonstration that foams stable to disproportionation could be produced without the need to first have a jammed structure at the interface, for example, with particles. Foams with zero Laplace pressure were also not produced as we far as literature lets us know. A well known example of food foams where suppression of Ostwald ripening and coalescence was achieved were made using *highly elastic interfaces* such as hydrophobin class II monolayer covered bubbles [172,173]. Those interfaces were shown to resist bubble dissolution to high compressional modulus once a large surface coverage was achieved, combined with a large elastic resistance to bending. The latter is a condition to prevent the bubble surface from buckling, hence the bubble from losing volume without losing surface area by making wrinkles and folds. It is natural to assume that there must be important lateral interactions between the polar parts of the adsorbed hydrophobins which are responsible for such high levels of bending modulus [174]. This property has been shown to be a remarkable feature of hydrophobin class II covered interfaces. It can be noted that the field of bubbles as ultrasound contrast agents is using high elastic modulus interfaces that can stabilize bubbles against dissolution for remarkably long times [175]. Shelf stability time scales are, however, one or more orders of magnitude larger, which makes it compulsory to have interfacial structures with characteristic relaxation times at least comparable to the shelf life. Another class of particles demonstrated to bring remarkable stability to foams is ethyl cellulose particles, which are much larger in average diameter. Their surface chemistry is difficult to control, but when adequately done, they form packed structures at interfaces, allowing bubbles to retain nonspherical shapes in water for long times [176]. The presence of dairy proteins added to a highly stabilizing agent such as a hydrophobin class II or cellulose particles was seen to have different possible effects. In the case hydrophobin class II mixed with dairy proteins, stability versus Ostwald ripening was decreased by surface fluidization in comparison to the pure hydrophobin II stabilized bubbles [31]. In contrast, stability of the mixed cellulose-ethyl cellulose systems with dairy proteins was increased [177], possibly due to an increased interfacial packing efficiency. The difference in behavior modification between the two cases can be interpreted as a difference in the degree of specificity of interactions within the interfacial layer.

20.4.3 BULK EFFECTS

Within bulk stabilizing effects, let us consider first the role of the *rheology of the continuous phase*. Viscous effects at the smallest gas diffusion scale will have nearly linear effects regarding the rate of gas volume exchange by Ostwald ripening. We do not discuss here as much as they would deserve to be effects using gas solubility or mixtures of gases of different solubilities. Let us, however, recall that academic research has helped the beer industry with making much more stable beer froths by playing with mixtures of gases of different solubilities [33], adapting smartly an idea previously demonstrated to work against the Ostwald ripening in emulsions [34,35]. The bulk rheology obviously also affects the drainage kinetics, as well as the coalescence probability by their role in the shape fluctuations of a thin film. For relatively wet foams (most bubbles are apart, i.e., do not share interface via a thin film separating them), the rheology of the bulk phase as can be measured with large gap rheometers will have a very important role. Indeed, while surface activity is very often important for food foam generation, bulk rheology often dominates the stabilization [17,18,127,178] with many products like marshmallows or aerated chocolate being commonly fully bulk stabilized [23,133,134]. The bulk modulus is thereby high enough to prevent the foam from increasing the air fraction by gravity effects. The yield stress value needed for the foam to be stable against gravity is proportional to the foam height, and of order 1000 Pa for a 10 cm high foam. It was, however, proven that the question of bulk rheology effects is much more complex than currently understood [179,200]. Indeed, the rise of a sphere in a yield stress fluid can strongly depend on its microstructure, that is, the causal analysis of structural evolution of a dispersion of bubbles in such a matrix needs to incorporate time dependent evolution of microstructure [179]. Even if a foam is stabilized against drainage by bulk effects, that may not guarantee its shelf stability against Ostwald ripening, which may slowly coarsen the structure. This is because from the theory side, it is not well understood which level of bulk modulus is needed to arrest Ostwald ripening. Predictions by Meinders et al. regarding the needed bulk moduli are below the levels practically observed that imply foam stabilization [146].

Let us cite some recent examples of bulk stabilizing structure design, which are or could be relevant to food products. Following van Aken et al. [211], Patel et al. [180] showed foam stabilization by emulsion droplets and hydrocolloid complexes as surface active agents without the use of classical molecular surfactants. The structure obtained can be considered similar to whipped cream, except for the partially coalesced structures, which are not necessarily present. Salonen et al. [181] showed that the foam stabilization is higher when the oil fraction is higher due to droplets being located in Plateau borders and films, thereby increasing their effective diameter and thickness.

20.4.4 CONFINED STRUCTURES: INTERFACE/BULK EFFECTS

The *structure and the rheology of the thin films* separating bubbles involve interfacial and continuous phase rheology effects, which are relevant to a number of food foams. Even for systems with established important interfacial stabilization like for hydrophobin (type II) foams [172], there are strong hints of other significant contributions. For example, aggregates are usually destroyed during preparation of the mother hydrophobin solution, but they form consequently from the high shear forces applied in the foaming process [31]. Their presence, including the possibility of multilayers, prevents neighboring bubbles from getting close, thereby contributing to stabilization by a partial *bulk (or steric) aspect* [31]. The same type of questions can be asked about foams stabilized solely by ethyl cellulose particles [176]. Those were shown to be superstable in regimes where the liquid draining from the foam was nearly devoid of particles. It was even when they were in large excess versus the created air-water surface area, that is, Plateau borders and foam lamellae were thickened by particles. Advanced functional shapes such as "lumpy microrods" were developed in order to bring steric stabilization to foams [182]. Those features can be expected to be particularly important to avoid foam collapse during processing by helping to maintain larger separation levels between bubbles.

A large proportion of food foams are stabilized by relatively *thick structures adsorbed or formed at interfaces.* We are talking here about structures which can consist of tens to thousands of molecular diameters. Protein-polysaccharide complexes are the typical result of food engineering based on colloid science, which can form thick structures around bubbles helping their stabilization against coarsening [88,183,184]. It should be noted that complex films do not necessarily stabilize against coalescence, since certain amphiphilic structure types were demonstrated to easily rupture [185]. Even when bringing high stability against coalescence at rest, such films also display high tendency to rupture under applied macroscopic deformation. They are unable to respond to rapid deformation without breaking, even if they are the product of dynamic self-assembly of small molecules like polyglycerol ester [97]. Similar behavior was observed in aerated casein gels and emulsion gels, whereby aerated casein emulsions gels display higher structural stability upon deformation [186]. The same could be said about partially coalesced fat networks, which are held responsible for the high stability of whipped creams [17,18].

The challenge to understand what parameters govern structure formation in thin films form is still very much present, particularly in the light of new insight into the importance of *confinement induced effects.* Given the out of equilibrium character of most food structures [4,18,97,140,187], they will depend on the path of formation of colloidal structure formation in the bulk and the thin film. It was established using TFB that the confinement effects can even affect structures in protein stabilized foams at relatively large scales, that is, comparable to the film thickness [103,115,116]. This will impact the rheological properties of the film and its dynamics and stability. Food foams such as based on proteins are in many aspects similar to other, nonfood foams such as those stabilized by polyelectrolyte complexes [188–190]. In the case of protein foams, recent results from the same group (as for [188–190]) brought evidence of aggregate structure formation within the film under conditions which stabilize the foams. Those aggregates impact the film thinning under gravitational pressure, as well as its ability to resist against rupture. Since not long ago, modeling work aims at understanding the influence of (nonsurface active) suspended particles on foam drainage [191].

In this context, small gap rheology is of very high relevance to the understanding of foam stability. It is because the thin film rheology will determine the drainage characteristics as well as its ability to resist against rupture under fast events such as T1 type of bubble rearrangements [122,124]. Low gap rheometers such as developed by Clasen and McKinley [192,193] are instruments of choice to tackle those types of issues. Proteins are macromolecules carrying electrostatic charges. It can be noted that complex collective effects even take place in single foam layers, whereby the free surface boundary also plays a role like for champagne [194].

20.4.5 Role of a Complex Chemistry and Physical Chemistry

From a general point of view, food foams are commonly very sensitive to physical chemistry, that is, to pH, ionic strength, concentration of stabilizer, as is, for example, the case for protein-based foams [29,142]. For example, the importance of *multivalent ion concentration* is high, such as of calcium ions or copper ions [29,195], which holds as well for trivalent ions [196]. Their influence is expected to strongly impact on interactions and structure development within thin films. Such interactions were proven to impart higher resistance of foam films versus destabilization due to competitive adsorption with small emulsifiers [196]. However, because such foam films can reach high rheological properties [29], they can also suffer from a poor ability to undergo bubble rearrangements without rupturing irreversibly [124]. Furthermore, protein-polysaccharide interactions under well controlled pH conditions favoring electrostatic complex formation were shown to induce strong viscoelastic film formation [88,184,190]. Proteins are also famous for interacting with lipidic molecules [165]. Furthermore, different phases can be engaged in competition for occupying interfaces. For example, competition between fat crystals, oil drops, and proteins at the water/air interface takes place in whipped creams and ice cream, thereby influencing the partial coalescence behavior [17–19].

The very *special case of coffee foams* (illustration in Figure 20.1) needs to be considered separately. The principle reason is the *chemical reactivity* of part of its components, which induces covalent bond creation through, for example, Maillard type of reactions [197]. The importance of exposure to the air above the cup comes on top, resulting in unique physical complexity. The stabilization and destabilization mechanisms involved often include irreversible structure formation in between bubbles as a result of phenomena that are to date not very well understood. An illustrative example is that of espresso coffee crema, which contain lipidic components that can have detrimental effects on crema stability [57]. To that complexity adds the variability of the water hardness [198] and other parameters of coffee preparation linked to the beverage fabrication parameters (e.g., heating, pressure, cooling) [199]. It seems that the stability of the crema is rather strongly related to interfacial rheology properties of the bubbles. The role of proteins and the melanoidin components, which are products of Maillard reactions, is believed to be critical in the stability [197]. Contribution of solid particles, which are known to adsorb to interfaces in the crema, particularly for *Arabica* rich coffee, was shown [57]. They are presumed to have at least a role in slowing down the drainage in the crema by occupying significant volume in the Plateau borders despite being not fully covering bubble interfaces. High molecular weight components of espresso coffee have been shown to contribute particularly to foam stability [200]. Coffee foam stability, therefore, should be looked at under various angles, and better understanding may be achieved only if further work aims to decouple the diverse contributions.

20.4.6 Additional Remarks: Sources of Uncontrolled Variability

To take a few examples of uncontrolled causes of foam destabilization, let us consider beverage foams. They undergo destabilization with the contribution of certain seldom explored mechanisms, such as local rapid drying of bubble surface by convection currents of air, leading to frequent bubble bursts. There is some evidence in the literature of the importance of this mechanism in actual foam destabilization, for example, in coffee espresso crema [57] or in the top bubble layer in a glass of champagne [201]. One can foresee that in many food foams, the level of importance should strongly depend on the type of structures in the thin film between the bubbles and the air above the cup. The frequency of these bursts can be measured by acoustic methods [202]. We stress the idea that model systems can be very useful to investigate questions in food science. Furthermore, there is special care to take with respect to proper validation preparation protocol and ensuring the constant character of the sources. This is of particular importance when dealing with systems comprising proteins and their hydrolysates or structures where reactions can occur. It is also true of any system where minor components are suspected to have non-negligible effects.

20.5   CONCLUSIONS

The creation of stable foamed structures with optimized process, while targeting a specific overrun range, often remains a challenge for the food industry. The challenge will remain, if not increased in importance, as new formulations and new kinds of textures will be developed. A large number of them should be overcome by ensuring good systematics in food engineering. The deployment of structure design principles across several product categories requires generic understanding and engineering of aerated foods.

Now from a wider angle, the science and technology of foams is facing challenges at several levels, from the very fundamental to the very applied. Even for model foams, the task to break down the scientific questions included in foam physics to a number of questions that can be addressed experimentally and modeled theoretically is not a simple one. Understanding the fundamentals of foam destabilization mechanisms is still a theoretical challenge, and an even more difficult one when one considers the coupling of the phenomena involved. The structure complexity in food foams is not only in the aerated structure, but also in the structure of the continuous phase. The latter is often as complex and as rich in organizational levels and dynamics as the aerated structure itself (think of chocolate mousse, ice cream, coffee crema). The same could be said about interfaces and their often slow "out of equilibrium" dynamics.

Another step in developing an applied foam science will come from more effective down-scaling of factory questions. By down-scaling, we mean the translation of large scale phenomena to small scale, while integrating correctly all implications of dimensional changes (by dimensional analysis). Structure development phenomena are most often so complex and numerous in a single process that such an approach cannot involve full exploration of all of them. Still, applying a common chemical engineering approach is only possible by identifying the most important phenomena involving small scale structures. Down-scaling of physical stability questions should reach to the level of a few bubbles. Well controlled small scale experiments as, for example, microfluidic methods [97,102,203,204] are expected to help in that respect, toward better understanding the relationship between process and food structure. It would, for example, consist of the investigation of the stability of model systems with tailored interfacial or bulk rheology properties, as done in rheology studies of model foams [29]. Using the same tools and science as for the down-scaling, the solutions to the problems eventually need to be scaled up. Those tools can only complement the classical soft matter techniques proven to study colloidal and molecular structures in bulk and at interfaces. Generalizing such approaches still needs a higher level of engineering and science interplay, as well as a higher level of cross-disciplinary collaboration between players from all backgrounds.

REFERENCES

1. Basheva, E.S., Kralchevsky, P.A., Christov, N.C., Danov, K.D., Stoyanov, S.D., Blijdenstein, T.B.J., Kim, H.-J., Pelan, E.G., Lips, A. Unique properties of bubbles and foam films stabilized by HBFII hydrophobin. *Langmuir* 2011; 27(6): 2382–2392.
2. Ferri, J.K., Gorevski, N., Kotsmar, C.S., Leser, M.E., Miller, R. Desorption kinetics of surfactants at fluid interfaces by novel coaxial capillary pendant drop experiments. *Colloids Surf A* 2008; 319(1–3): 13–20.

3. Marze, S. Bioaccessibility of nutrients and micronutrients from dispersed food systems: Impact of the multiscale bulk and interfacial structures. *Crit Rev Food Sci Nutr* 2013; 53(1): 76–108.
4. Van Aken, G.A., Blijdenstein, T.B.J., Hotrum, N.E. Colloidal destabilisation mechanisms in protein-stabilized emulsions. *Curr Opin Coll Interface Sci* 2003; 8(4–5): 371–379.
5. Guzman-Sepulveda, J.R., Douglass, K.M., Amin, S., Lewis, N.E., Dogariu, A. Passive optical mapping of structural evolution in complex fluids. *RSC Adv* 2015; 5(7): 5357–5362.
6. Webster, A.J., Cates, M.E. Stabilization of emulsions by trapped species. *Langmuir* 1998; 14(8): 2068–2077.
7. Lopez-Rubio, A., Gilbert, E.P. Neutron scattering: A natural tool for food science and technology research. *Trends Food Sci Technol* 2009; 20(11–12): 576–586.
8. Ubbink, J., Burbidge, A., Mezzenga, R. Food structure and functionality: A soft matter perspective. *Soft Mat* 2008; 4(8): 1569–1581.
9. Wassell, P., Bonwick, G., Smith, C.J., Almiron-Roig, E., Young, N.W.G. Towards a multidisciplinary approach to structuring in reduced saturated fat-based systems—A review. *J Food Sci Technol* 2010; 45: 642–655.
10. Sensoy, I. A review on the relationship between food structure, processing, and bioavailability. *Crit Rev Food Sci Nutr* 2014; 54(7): 902–909.
11. Norton, J.E., Wallis, G.A., Spyropoulos, F., Lillford, P.J., Norton, I.T. Designing food structures for nutrition and health benefits. *Annu Rev Food Sci Technol* 2014; 5(1): 177–195.
12. Lesmes, U., McClements, D.J. Structure-function relationships to guide rational design and fabrication of particulate food delivery systems. *Trends Food Sci Technol* 2009; 20(10): 448–457.
13. Mackie, A.R., Rafiee, H., Malcolm, P., Salt, L., van Aken, G. Specific food structures supress appetite through reduced gastric emptying rate. *Am J Physiol—Gastrointest Liver Physiol* 2013; 4(11): G1038–G1043.
14. Juvonen, K.R. et al. Structure modification of a milk protein-based model food affects postprandial intestinal peptide release and fullness in healthy young men. *J Nutr* 2011; 106(12): 1890–1898.
15. Campbell, A.L., Stoyanova, S.D., Paunov, V.N. Fabrication of functional anisotropic food-grade micro-rods with micro-particle inclusions with potential application for enhanced stability of food foams. *Soft Mat* 2009; 5: 1019–1023.
16. Goff, H.D. Partial coalescence and structure formation in dairy emulsions. *Adv Exp Med Biol* 1997; 415: 137–148.
17. Haedelt, J., Pyle, D.L., Beckett, S.T., Niranjan, K. Vacuum-induced bubble formation in liquid-tempered chocolate. *J Food Sci* 2005; 70(2): E159–E164.
18. Fameau, A.-L., Cousin, F., Navailles, L., Nallet, F., Boué, F., Douliez, J.-P. Multiscale structural characterizations of fatty acid multilayered tubes with a temperature-tunable diameter. *J Phys Chem B* 2011; 115(29): 9033–9039.
19. Hotrum, N.E., Cohen Stuart, M.A., Van Vliet, T., Avino, S.F., Van Aken, G.A. Elucidating the relationship between the spreading coefficient, surface-mediated partial coalescence and the whipping time of artificial cream. *Colloids Surf A* 2005; 260(1–3): 71–78.
20. Mills, E.N.C., Wilde, P.J., Salt, L.J., Skeggs, P. Bubble formation and stabilization in bread dough. *Food and Bioproducts Processing: Transactions of the Institution of Chemical Engineers, Part C* 2003; 81(3): 189–193.
21. German, J.B., McCarthy, M.J. Stability of aqueous foams—Analysis using magnetic-resonance imaging. *J Agric Food Chem* 1989; 37(5): 1321–1324.
22. Saggin, R., Leser, M., Bezelgues, J.-B., Livings, S., Sher, A.A. Generating foam with desired texture, stability and bubble size distribution. US7537138 B2.
23. Ferrari, M., Navarini, L., Liggieri, L., Ravera, F., Liverani, F.S. Interfacial properties of coffee-based beverages. *Food Hydrocoll* 2007; 21: 1374–1378.
24. Piazza, L., Gigli, J., Bulbarello, A. Interfacial rheology study of espresso coffee foam structure and properties. *J Food Eng* 2008; 84: 420–429.
25. Chavez-Montes, B.E., Choplin, L., Schaer, E. Rheological characterization of wet food foams. *J Texture Stud* 2007; 38: 236–252.
26. Tcholakova, S., Denkov, N.D., Ivanov, I.B., Campbell, B. Coalescence in β-lactoglobulin-stabilized emulsions: Effects of protein adsorption and drop size. *Langmuir* 2002; 18(23): 8960–8971.
27. Kogan, M., Ducloué, L., Goyon, J., Chateau, X., Pitois, O., Ovarlez, G. Mixtures of foam and paste: Suspensions of bubbles in yield stress fluids. *Rheol Acta* 2013; 52(3): 237–253.
28. Salonen, A., Lhermerout, R., Rio, E., Langevin, D., Saint-Jalmes, A. Dual gas and oil dispersions in water: Production and stability of foamulsion. *Soft Mat* 2012; 8: 699.

29. Lexis, M., Willenbacher, N. Relating foam and interfacial rheological properties of β-lactoglobulin solutions. *Soft Mat* 2014; 10(48): 9626–9636.

30. Abkarian, M., Subramaniam, A.B., Kim, S.-H., Larsen, R.J., Yang, S.-M., Stone, H.A. Dissolution arrest and stability of particle-covered bubbles. *Phys Rev Lett* 2007; 99(18): 188301.

31. Du, Z., Bilbao-Montoya, M.P., Binks, B.P., Dickinson, E., Ettelaie, R., Murray, B.S. Outstanding stability of particle-stabilized bubbles. *Langmuir* 2003; 19: 3106–3108.

32. Dutta, A., Chengara, A., Nikolov, A.D., Wasan, D.T., Chen, K., Campbell, B. Destabilization of aerated food products: Effects of Ostwald ripening and gas diffusion. *J Food Eng* 2004; 62(2): 177–184.

33. Kabalnov, A.S., Pertzov, A.V., Shchukin, E.D. Ostwald ripening in two-component disperse phase systems: Application to emulsion stability. *Coll Surf* 1987; 24: 19–32.

34. Liger-Belair, G., Seon, T., Antkowiak, A. Collection of collapsing bubble driven phenomena found in champagne glasses. *Bubble Sci Eng Technol* 2012; 4(1): 21–34.

35. Fainerman, V.B., Miller, R., Ferri, J.K., Watzke, H., Leser, M.E., Michel, M. Reversibility and irreversibility of adsorption of surfactants and proteins at liquid interfaces. *Adv Colloid Interface Sci* 2006; 123–126(16): 163–171.

36. Allais, I., Roch-Boris, E.G., Gros, J.B., Trystram, G. Interfacial and foaming properties of some food grade low molecular weight surfactants. *Colloids Surf A* 2008; 331: 56–62.

37. Baker, J.C., Mize, M.D. The origin of the gas cells in bread dough. *Cereal Chem* 1941; 19: 84–94.

38. Allais, I. et al. Influence of egg type, pressure and mode of incorporation on density and bubble distribution of a lady finger batter. *J Food Eng* 2006; 74(2): 198–210.

39. Salerno, A., Oliviero, M., Di Maio, E., Iannace, S., Netti, P. Design of porous polymeric scaffolds by gas foaming of heterogeneous blends. *J Mat Sci* 2009; 20(10): 2043–2051.

40. Mack, S., Hussein, M.A., Becker, T. On the theoretical time-scale estimation of physical and chemical kinetics whilst wheat dough processing. *Food Biophys* 2013; 8(1): 69–79.

41. Trinh, L., Lowe, T., Campbell, G.M., Withers, P.J., Martin, P.J. Bread dough aeration dynamics during pressure step-change mixing. *Chem Eng Sci* 2013; 101: 470–477.

42. Horvat, M., Schuchmann, H.P. Investigation of growth and shrinkage mechanisms in vapor-induced expansion of extrusion-cooked corn grits. *Food Bioprod Techn* 2013; 6(12): 3392–3399.

43. Padiernos, C.A., Lim, S.-Y., Swanson, B.G., Ross, C.F., Clark, S. High hydrostatic pressure modification of whey protein concentrate for use in low-fat whipping cream improves foaming properties. *J Dairy Sci* 2009; 92(7): 3049–3056.

44. Delaplace, G., Copenolle, P., Cheio, J., Ducept, F. Influence of whip speed ratios on the inclusion of air into a bakery foam produced with a planetary mixer device. *J Food Eng* 2012; 108(4): 532–540.

45. Chesterton, A.K.S., De Abreu, D.A.P., Moggridge, G.D., Sadd, P.A., Wilson, D.I. Evolution of cake batter bubble structure and rheology during planetary mixing. *Food Bioprod Proc* 2013; 91(3): 192–206.

46. Narchi, I., Vial, C., Djelveh, G. Influence of bulk and interfacial properties and operating conditions on continuous foaming operation applied to model media. *Food Res Int* 2007; 40(8): 1069–1079.

47. Netti, P. Biomedical foams for tissue engineering applications. *Biomedical Foams for Tissue Engineering Applications*. Elsevier, 2009.

48. Bindler, U., Müller, M., Rieck, H. Process and installation for preparation of choclate-mass containing gas bubbles. *Eur. Pat. 724836*, 1996.

49. Venerus, D. Modeling diffusion-induced bubble growth in polymer liquids. *Cell Polym* 2003; 22(2): 89–101.

50. Müller-Fischer, N., Windhab, E.J. Influence of process parameters on microstructure of food foam whipped in a rotor-stator device within a wide static pressure range. *Coll Surf A* 2005; 263(1–3): 353–362.

51. Gunning, A.P., Kirby, A.R., Mackie, A.R., Kroon, P., Williamson, G., Morris, V.J. Watching molecular processes with the atomic force microscope: Dynamics of polymer adsorption and desorption at the single molecule level. *J Microsc* 2004; 216(1): 52–56.

52. Müller-Fischer, N. et al. Dynamically enhanced membrane foaming. *Chem Eng Sci* 2007; 62(16): 4409–4419.

53. Massey, A.H. et al. Air inclusion into a model cake batter using a pressure whisk: Development of gas hold-up and bubble size distribution. *J Food Sci* 2001; 66(8): 1152–1157.

54. Fan, J. et al. Model for the oven rise of dough during baking. *J Food Eng* 1999; 41(2): 69–77.

55. Müller-Fischer, N., Suppiger, D., Windhab, E.J. Impact of static pressure and volumetric energy input on the microstructure of food foam whipped in a rotor-stator device. *J Food Eng* 2007; 80(1): 306–316.

56. Blander, M., Katz, J.L. Bubble nucleation in liquids. *AIChE J* 1975; 21(5): 833–848.

57. Ding, J., Tsaur, F.W., Lips, A., Akay, A. Acoustical observation of bubble oscillations induced by bubble popping. *Phys Rev E* 2007; 75(4): art. no. 041601.
58. Liger-Belair, G., Jeandet, P. Capillary-driven flower-shaped structures around bubbles collapsing in a bubble raft at the surface of a liquid of low viscosity. *Langmuir* 19, 5771–5779, 2003.
59. Barham, P. *The Science of Cooking*. Springer, 2001.
60. Bloksma. Rheology of the breadmaking process. *Cereal Foods World* 1990; 35(2): 228–244.
61. Babin, P., Della Valle, G., Chiron, H., Cloetens, P., Hoszowska, J., Pernot, P., Réguerre, A.L., Salvo, L., Dendievel, R. Fast X-ray tomography analysis of bubble growth and foam setting during breadmaking. *J Cereal Sci* 2006; 43(3): 393–397.
62. Leung, S.N., Park, C.B., Xu, D., Li, H., Fenton, R.G. Computer simulation of bubble-growth phenomena in foaming. *Ind Eng Chem Res* 2006; 45(23): 7823–7831.
63. Sumnu, G. A review on microwave baking of foods. *Int J Food Sci Tech* 2001; 36(2): 117–127.
64. Della Valle, G., Chiron, H., Cicerelli, L., Kansou, K., Katina, K., Ndiaye, A., Whitworth, M., Poutanen, K. Basic knowledge models for the design of bread texture. *Trends Food Sci Techn* 2014; 36(1): 5–14.
65. Chin, N.L., Martin, P.J., Campbell, G.M. Aeration during bread dough mixing: I. Effect of direction and size of a pressure step-change during mixing on the turnover of gas. *Food Bioproducts Process* 2004; 82(4 C): 261–267.
66. Barigou, M., van Ginkel, M. *Colloids Surf A* 2007; 311: 112–123.
67. Dobraszczyk, B.J. The physics of baking: Rheological and polymer molecular structure-function relationships in breadmaking. *J Non-Newt Fluid Mech* 2004; 124(1–3): 61–69.
68. Della Valle, G., Vergnes, B., Colonna, P., Patria, A. Relations between rheological properties of molten starches and their expansion behaviour in extrusion. *J Food Eng* 1997; 31(3): 277–296.
69. Rohenkohl, H., Kohlus, R. Foaming of ice cream and the time stability of its bubble size distribution. In: Campbell, G.M. (Ed.), *Bubbles in Food 2*, Eagan Press, St. Paul, 2008.
70. Alavi, S.H., Rizvi, S.S.H., Harriott, P. Process dynamics of starch-based microcellular foams produced by supercritical fluid extrusion. *Food Res Int* 2003; 36(4): 309–330.
71. Arhaliass, A., Bouvier, J.M., Legrand, J. Melt growth and shrinkage at the exit of the die in the extrusion-cooking process. *J Food Eng* 2003; 60(2): 185–192.
72. Boischot, C., Moraru, C.I., Kokini, J.L. Factors that influence the microwave expansion of glassy amylopectin extrudates. *Cereal Chem* 2003; 80(1): 56–61.
73. Narchi, I., Vial, C., Djelveh, G. Effect of protein-polysaccharide mixtures on the continuous manufacturing of foamed food products. *Food Hydrocoll* 2009; 23(1): 188–201.
74. Garstecki, P., Fuerstman, M.J., Stone, H.A., Whitesides, G.M. Formation of droplets and bubbles in a microfluidic T-junction—Scaling and mechanism of break-up. *Lab Chip* 2006; 6(3): 437–446.
75. Turbin-Orger, A., Boller, E., Chaunier, L., Chiron, H., Della Valle, G., Réguerre, A.-L. Kinetics of bubble growth in wheat flour dough during proofing studied by computed X-ray micro-tomography. *J Cereal Sci* 2012; 56(3): 676–683.
76. Mills, N.J. *Polymer Foams Handbook*. Butterworth-Heinemann, 2007.
77. Lee, S.-T., Park, C.B., Ramesh, N.S. *Polymeric Foams: Science and Technology*. CRC Press, 2006.
78. Fan, J., Mitchell, J.R., Blanshard, J.M.V. Computer simulation of the dynamics of bubble growth and shrinkage during extrudate expansion. *J Food Eng* 1994; 23(3): 337–356.
79. Zhang, J., Datta, A.K. Mathematical modeling of bread baking process. *J Food Eng* 2006; 75(1): 78–89.
80. Chiotellis, E., Campbell, G.M. Proving of bread dough. *Food Bioprod Proc* 2003; 81(3): 194–216.
81. Skurtys, O., Bouchon, P., Aguilera, J.M. Formation of bubbles and foams in gelatine solutions within a vertical glass tube. *Food Hydrocoll* 2008; 22(4): 706–714.
82. Klempner, D., Sedijarevic, V. *Polymeric Foams and Foam Technology*. Hanser, München: Germany, 2004.
83. Mezdour, S., Balerin, C., Aymard, P., Cuvelier, G., Ducept, F. Effect of rheology of the continuous phase on foaming processes: Viscosity-temperature impact. In: Campbell, G.M. (Ed.), *Bubbles in Food 2*, Eagan Press, St. Paul, 2008.
84. Robin, F., Dubois, C., Pineau, N., Labat, E., Théoduloz, C., Curti, D. Process, structure and texture of extruded whole wheat. *J Cereal Sci* 2012; 56(2): 358–366.
85. Germain, J.C., Aguilera, J.M. Identifying industrial food foam structures by 2D surface image analysis and pattern recognition. *J Food Eng* 2012; 111(2): 440–448.
86. Kostoglou, M., Georgiou, E., Karapantsios, T.D. A new device for assessing film stability in foams: Experiment and theory. *Colloids and Surfaces A* 2011; 382: 64–73.
87. Minor, M., Vingerhoeds, M.H., Zoet, F.D., de Wijk, R., van Aken, G.A. Preparation and sensory perception of fat-free foams—effect of matrix properties and level of aeration. *Int J Food Sci Technol* 2009; 44: 735–747.

88. Schmitt, C., Palma Da Silva, T., Rami-Shojaei, C.B.S., Frossard, P., Kolodziejczyk, E., Leser, M.E. Effect of time on the interfacial and foaming properties of β-lactoglobulin/acacia gum electrostatic complexes and coacervates at pH 4.2. *Langmuir* 2005; 21(17): 7786–7795.

89. Meagher, A.J., Mukherjee, M., Weaire, D., Hutzler, S., Banhart, J., Garcia-Moreno, F. Analysis of the internal structure of a monodisperse liquid foam by X-ray tomography. *Soft Mat* 2011; 7(21): 9881–9885.

90. Nikolov, A.D., Wasan, D.T. Ordered micelle structuring in thin films formed from anionic surfactant solutions. I. Experimental. *J Colloid Interface Sci* 1989; 133(1): 1–12.

91. Gardiner, B.S., Dlugogorski, B.Z., Jameson, G.J. Coarsening of two- and three-dimensional wet polydisperse foams. *Philos Mag A* 2000; 80(4): 981–1000.

92. Romoscanu, A.I., Fenollosa, A., Acquistapace, S., Gunes, D., Martins-Deuchande, T., Clausen, P., Mezzenga, R., Nydén, M., Zick, K., Hughes, E. Structure, diffusion, and permeability of protein-stabilized monodispersed oil in water emulsions and their gels: A self-diffusion NMR Study. *Langmuir* 2010; 26(9): 6184–6192.

93. Holland, D.J., Blake, A., Tayler, A.B., Sederman, A.J., Gladden, L.F. A Bayesian approach to characterising multi-phase flows using magnetic resonance: Application to bubble flows. *J Magn Reson* 2011; 209(1): 83–87.

94. Enríquez, O.R., Hummelink, C., Bruggert, G.-W., Lohse, D., Prosperetti, A., van der Meer, D., Sun, C. Growing bubbles in a slightly supersaturated liquid solution. *Rev Sci Instrum* 2013; 84: 065111.

95. Amin, S., S. Blake, Kenyon, S.M., Kennel, R.C., Lewis, E.N. A novel combination of DLS-optical microrheology and low frequency Raman spectroscopy to reveal underlying biopolymer self-assembly and gelation mechanisms. *J Chem Phys* 2014; 141: 234201.

96. Duri, A., Sessoms, D.A., Trappe, V., Cipelletti, L. Resolving long-range spatial correlations in jammed colloidal systems using photon correlation imaging. *Phys Rev Lett* 2009; 102(8), art. no. 085702.

97. Farajzadeh, R., Krastev, R., Zitha, P.L.J. Foam film permeability: Theory and experiment. *Adv Colloid Interface Sci* 2008; 137: 27–44.

98. MacPherson, R.D., Srolovitz, D.J. The von Neumann relation generalized to coarsening of three-dimensional microstructures. *Nature* 2007; 446(7139): 1053–1055.

99. Hutzler, S., Cox, S.J., Wang, G. Foam drainage in two dimensions. *Colloids Surf A* 2005; 263: 178–183.

100. Tong, M., Cole, K., Neethling, S.J. Drainage and stability of 2D foams: Foam behaviour in vertical Hele-Shaw cells. *Colloids Surf A* 2011; 382: 42–49.

101. Hartel, R.W. Ice crystallization during the manufacture of ice cream. *Trends Food Sci Technol* 1996; 7(10): 315–321.

102. Karapantsios, T.D., Papara, M. On the design of electrical conductance probes for foam drainage applications: Assessment of ring electrodes performance and bubble size effects on measurements. *Colloids Surf A* 2008; 323: 139–148.

103. Saint-Jalmes, A., Peugeot, M.-L., Ferraz, H., Langevin, D. Differences between protein and surfactant foams: Microscopic properties, stability and coarsening. *Colloids Surf A* 2005; 263(1–3): 219–225.

104. Duplat, J., Bossa, B., Vermillaux, E. On two-dimensional foam ageing. *J Fluid Mech* 2011; 673: 147.

105. Liger-Belair, G., Seon, T., Antkowiak, A. Collection of collapsing bubble driven phenomena found in champagne glasses. *Bubble Sci Eng Technol* 2012; 4(1): 21–34.

106. Durian, D.J. Bubble-scale model of foam mechanics: Melting, nonlinear behavior, and avalanches. *Phys Rev E* 1997; 55: 1739–1751.

107. Bohn, S. Bubbles under stress. *Eur Phys J E* 2003; 11(2): 177–189.

108. Goff, H.D. Instability and Partial Coalescence in Whippable Dairy Emulsions. *J Dairy Sci* 1997; 80(10): 2620–2630.

109. Scheludko, A., Exerowa, D. Über den elektrostaschen Druck in Schaumfilmen aus wässerigen Elektrolytlösungen. *Kolloid Zeitschrift* 1959; 165: 148.

110. Wasan, D.T., Nikolov, A.D. Foams and emulsions: The importance of structural forces. *Aust J Chem* 2007; 60(9): 633–637.

111. Bergeron, V., Langevin, D., Asnacios, A. Thin-film forces in foam films containing anionic polyelectrolyte and charged surfactants. *Langmuir* 1996; 12(6): 1550–1556.

112. Fortuna, I., Thomas, G.L., De Almeida, R.M.C., Graner, F. Growth laws and self-similar growth regimes of coarsening two-dimensional foams: Transition from dry to wet limits. *Phys Rev Lett* 2012; 108(24): art. no. 248301.

113. Blomqvist, B.R., Ridout, M.J., Mackie, A.R., Wärnheim, T., Claesson, P.M., Wilde, P. Disruption of viscoelastic β-lactoglobulin surface layers at the air-water interface by nonionic polymeric surfactants. *Langmuir* 2004; 20(23): 10150–10158.

114. Ivanov, I.B., Radoev, B., Manev, E., Scheludko, A. Theory of the critical thickness of rupture of thin liquid films. *Trans Faraday Soc* 1970; 66: 1262–1273.
115. Emile, J., Pezennec, S., Renault, A., Robert, E., Artzner, F., Meriadec, C., Faisant, A., Meneau, F. Protein molecule stratification inside a single curved film: Evidence from X-ray scattering. *Soft Mat* 2011, 7(19): 9283–9290.
116. Denkov, N.D., Yoshimur, H., Nagayama, K., Kouyama, T. Nanoparticle arrays in freely suspended vitrified films. *Phys Rev Lett* 1996; 76(13): 2354–2357.
117. Jones, S.F., Evans, G.M., Galvin, K.P.U. Bubble nucleation from gas cavities: A review. *Adv Colloid Interface Sci* 1999; 80: 27–50.
118. Platikanov, D., Nedyalkov, M., Nasteva, V. Line tension of Newton black films. II. Determination by the diminishing bubble method. *J Colloid Interface Sci* 1980; 75(2): 620–628.
119. Pine, D.J., Weitz, D.A., Chaikin, P.M., Herbolzheimer, E. Diffusing wave spectroscopy. *Phys Rev Lett* 1988; 60: 1134.
120. Biance, A.-L., Cohen-Addad, S., Höhler, R. Topological transition dynamics in a strained bubble cluster. *Soft Mat* 2009; 5: 4672–4679.
121. Trater, A.M., Alavi, S. Use of non-invasive X-ray microtomography for characterizing microstructure of extruded biopolymer foams. *Food Res Int* 2005; 38: 709–719.
122. Biance, A.L., Delbios, A., Pitois, O. How topological rearrangements and liquid fraction control liquid foam stability. *Phys Rev Lett* 2011; 106: 068301.
123. Bolliger, S., Goff, H.D., Tharp, B.W. Correlation between colloidal properties of ice cream mix and ice cream. *Int Dairy J* 2000; 10(4): 303–309.
124. Geoffrey, C.-P. (Ed.). *Food Science and Technology.* Wiley-Blackwell, 2009.
125. Raikos, V., Campbell, L., Euston, S.R. Effects of sucrose and sodium chloride on foaming properties of egg white proteins. *Food Res Int* 2007; 40: 347–355.
126. Zhai, J.L., Day, L., Aguilar, M.-I., Wooster, T.J. Protein folding at emulsion oil/water interfaces. *Curr Opin Colloid Interface Sci* 2013; 18(4): 257–271.
127. Entwistle, A. The effects of total internal reflection on the spread function along the axis in confocal microscopy. *J Microsc* 1995; 180(2): 148–157.
128. Durian, D.J., Weitz, D.A., Pine, D.J. Dynamics and coarsening in three-dimensional foams. *J Phy: Condens Matter 2* 1990; 069: SA433–SA436.
129. Ivanov, I.B. (Ed.) *Thin Liquid Films: Fundamentals and Applications.* Marcel Dekker, Inc.: New York, 1988.
130. Maestro, A., Drenckhan, W., Rio, E., Höhler, R. Liquid dispersions under gravity: Volume fraction profile and osmotic pressure. *Soft Mat* 2013; 9: 2531.
131. Curschellas, C., Kohlbrecher, J., Geue, T., Fischer, P., Schmitt, B., Rouvet, M., Windhab, E.J., Limbach, H.J. Foams stabilized by multilamellar polyglycerol ester self-assemblies. *Langmuir* 2013; 29(1): 38–49.
132. Penttilä, P.A., Suuronen, J.-P., Kirjoranta, S., Peura, M., Jouppila, K., Tenkanen, M., Serimaa, R. X-ray characterization of starch-based solid foams. *J Mater Sci* 2011; 46(10): 3470–3479.
133. Terriac, E., Emile, J., Axelos, M.A.V., Grillo, I., Meneau, F., Boué, F. Characterization of bamboo foam films by neutron and X-ray experiments. *Colloids Surf A* 2007; 309(1–3): 112–116.
134. Miller, R., Ferri, J.K., Javadi, A., Krägel, J., Mucic, N., Wüstneck, R. Rheology of interfacial layers. *Colloid Polym Sci* 2010; 288(9): 937–950.
135. Lau, K., Dickinson, E. Structural and rheological properties of aerated high sugar systems containing egg albumen. *J Food Sci* 2004; 69(5): E232–E239.
136. Zúñiga, R.N., Aguilera, J.M. Structure-fracture relationships in gas-filled gelatin gels. *Food Hydrocoll* 2009; 23(5): 1351–1357.
137. Gunning, P.A., MacKie, A.R., Gunning, A.P., Wilde, P.J., Woodward, N.C., Morris, V.J. The effect of surfactant type on protein displacement from the air–water interface. *Food Hydrocoll* 2004; 18(3): 509–515.
138. Campbell, G.M., Mougeot, E. Creation and characterisation of aerated food products. *Trends Food Sci Technol* 1999; 10: 283–296.
139. Carrera Sanchez, C., Rodrıguez Patino, J.M. Interfacial, foaming and emulsifying characteristics of sodium caseinate as influenced by protein concentration in solution. *Food Hydrocoll* 2005; 19: 407–416.
140. Van Der Sman, R.G.M., Van Der Goot, A.J. The science of food structuring. *Soft Mat* 2009; 5(3): 501–510.
141. Aguilera, J.M. Why food micro structure? *J Food Eng* 2005; 67(1–2): 3–11.

142. Ulaganathan, V., Krzan, M., Lotfi, M., Dukhin, S.S., Kovalchuk, V.I., Javadi, A., Gunes, D.Z., Gehin-Delval, C., Malysa, K., Miller, R. Influence of β-lactoglobulin and its surfactant mixtures on velocity of the rising bubbles. *Colloids Surf A* 2014; 460: 361–368.

143. Schmitt, C., Bovay, C., Frossard, P. Kinetics of formation and functional properties of conjugates prepared by dry-state incubation of β-lactoglobulin/acacia gum electrostatic complexes. *J Agric Food Chem* 2005; 53(23): 9089–9099.

144. Shimoyama, A., Kido, S., Kinekawa, Y.-I., Doi, Y. Guar foaming albumin: A low molecular mass protein with high foaming activity and foam stability isolated from guar meal. *J Agric Food Chem* 2008; 56: 9200–9205.

145. Kloek, W., Van Vliet, T., Meinders, M. Effect of bulk and interfacial rheological properties on bubble dissolution. *J Colloid Interface Sci* 2001; 237(2): 158–166.

146. Weaire, D., Pageron, V. Frustrated froth: Evolution of foam inhibited by an insoluble gaseous component. *Philos Mag Lett* 1990; 62(6): 417–421.

147. Maldonado-Valderrama, J., Langevin, D. On the difference between foams stabilized by surfactants and whole casein or β-casein. Comparison of foams, foam films, and liquid surfaces studies. *J Phys Chem B* 2008; 112: 3989–3996.

148. Eisner, M.D., Wildmoser, H., Windhab, E.J. Air cell microstructuring in a high viscous ice cream matrix. *Colloids Surf A* 2005; 263: 390–399.

149. Phianmongkhol, A., Varley, J. ζ potential measurement for air bubbles in protein solutions. *J Colloid Interface Sci* 2003; 260: 332–338.

150. Rullier, B., Axelos, M.A.V., Langevin, D., Novales, B. β-Lactoglobulin aggregates in foam films: Correlation between foam films and foaming properties. *J Colloid Interface Sci* 2009; 336(2): 750–755.

151. Tcholakova, S., Denkov, N.D., Sidzhakova, D., Ivanov, I.B., Campbell, B. Effects of electrolyte concentration and pH on the coalescence stability of β-lactoglobulin emulsions: Experiment and interpretation. *Langmuir* 2005; 21(11): 4842–4855.

152. Wierenga, P.A., van Norél, L., Basheva, E.S. Reconsidering the importance of interfacial properties in foam stability. *Colloids Surf A* 2009; 344(1–3): 72–78.

153. Marinova, K.G., Basheva, E.S., Nenova, B., Temelska, M., Mirarefi, A.Y., Campbell, B., Ivanov, I.B. Physico-chemical factors controlling the foamability and foam stability of milk proteins: Sodium caseinate and whey protein concentrates. *Food Hydrocoll* 2009; 23(7): 1864–1876.

154. De Jongh, H.H.J., Kosters, H.A., Kudryashova, E., Meinders, M.B.J., Trofimova, D., Wierenga, P.A. Protein adsorption at air-water interfaces: A combination of details. *Biopolymers* 2008; 74(1–2): 131–135.

155. Sceni, P., Wagner, J.R. Study on sodium caseinate foam stability by multiple light scattering. *Food Sci Tech Int* 2007; 13(6): 461–468.

156. Li, C., Somasundaran, P. Reversal of bubble charge in multivalent inorganic salt solutions-Effect of aluminum. *J Colloid Interface Sci* 1992; 148(2): 587–591.

157. Marinova, K.G., Alargova, R.G., Denkov, N.D., Velev, O.D., Petsev, D.N., Ivanov, I.B., Borwankar, R.P. Charging of oil-water interfaces due to spontaneous adsorption of hydroxyl ions. *Langmuir* 1996; 12(8): 2045–2051.

158. Luo, G., Malkova, S., Yoon, J., Schultz, D.G., Lin, B., Meron, M., Benjamin, I., Vanýsek, P., Schlossman, M.L. Ion distributions near a liquid liquid interface. *Science* 2006; 311(5758): 216–218.

159. Donsmark, J., Rischel, C. Fluorescence correlation spectroscopy at the oil—Water interface: Hard disk diffusion behavior in dilute β-lactoglobulin layers precedes monolayer formation. *Langmuir* 2007; 23(12): 6614–6623.

160. Karraker, K.A., Radke, C.J. Disjoining pressures, zeta potentials and surface tensions of aqueous non-ionic surfactant/electrolyte solutions: Theory and comparison to experiment. *Adv Colloid Interface Sci* 2002; 96(1–3): 231–64.

161. Mader, K., Mokso, R., Raufaste, C., Dollet, B., Santucci, S., Lambert, J., Stampanoni, M. Quantitative 3D characterization of cellular materials: Segmentation and morphology of foam. *Colloids Surf A* 2012; 415: 230–238.

162. Pinzer, B.R., Medebach, A., Limbach, H.J., Dubois, C., Stampanonib, M., Schneebeli, M. 3D-characterization of three-phase systems using X-ray tomography: Tracking the microstructural evolution in ice cream. *Soft Mat* 2012; 8: 4584.

163. Exerowa, D., Kolarov, T., Pigov, I., Levecke, B., Tadros, T. Interaction forces in thin liquid films stabilized by hydrophobically modified inulin polymeric surfactant. 1. foam films. *Langmuir* 2006; 22(11): 5013–5017.

164. Cooper, D.J., Husband, F.A., Mills, E.N.C., Wilde, P.J. Role of beer lipid-binding proteins in preventing lipid destabilization of foam. *J Agric Food Chem* 2002; 50: 7645–7650.

165. Morris, V.J., Gunning, A.P. Microscopy, microstructure and displacement of proteins from interfaces: Implications for food quality and digestion. *Soft Mat* 2008; 4: 943–951.

166. Dimitrijev-Dwyer, M., He, L., James, M., Nelson, A., Wang, L., Middelberg, A.P.J. The effects of acid hydrolysis on protein biosurfactant molecular, interfacial, and foam properties: PH responsive protein hydrolysates. *Soft Mat* 2012; 8(19): 5131–5139.

167. Dexter, A.F., Malcolm, A.S., Middelberg, A.P.J. Reversible active switching of the mechanical properties of a peptide film at a fluid-fluid interface. *Nat Mater* 2006; 5(6): 502–506.

168. Roth, C.M., Lenhoff, A.M. Electrostatic and van der Wals contributions to protein adsorption: Computation of equilibrium constants. *Langmuir* 1993; 9(4): 962–972.

169. Wierenga, P.A., Meinders, M.B.J., Egmond, M.R., Voragen, A.G.J., De Jongh, H.H.J. Quantitative description of the relation between protein net charge and protein adsorption to air-water interfaces. *J Phys Chem B* 2005; 109(35): 16946–16952.

170. Luck, P.J., Foegeding, E.A. The role of copper in protein foams. *Food Biophys* 2008; 3: 255–260.

171. Murray, B.S., Durga, K., Yusoff, A., Stoyanov, S.D. Stabilization of foams and emulsions by mixtures of surface active food-grade particles and proteins. *Food Hydrocoll* 2011; 25: 627–638.

172. Gonzenbach, U.T., Studart, A.R., Tervoort, E., Gauckler, L.J. Stabilization of foams with inorganic colloidal particles. *Langmuir* 2006; 22: 10983–10988.

173. Wang, Y., Bouillon, C., Cox, A., Dickinson, E., Durga, K., Murray, B.S., Xu, R. Interfacial study of class II hydrophobin and its mixtures with milk proteins: Relationship to bubble stability. *J Agric Food Chem* 2013; 61: 1554–1562.

174. Kasapis, S., Norton, I., Ubbink, J. *Modern Biopolymer Science—Bridging the Divide between Fundamental Treatise and Industrial Application*. Elsevier, 2009.

175. Pu, G., Borden, M.A., Longo, M.L. Collapse and shedding transitions in binary lipid monolayers coating microbubbles. *Langmuir* 2006; 22(7): 2993–2999.

176. Taccoen, N., Lequeux, F., Gunes, D.Z., Baroud, C.N. Probing the mechanical strength of an armored bubble and its implication to particle-stabilized foams. *Phys Rev X* 2016; 6: 011010.

177. Radulova, G.M., Golemanov, K., Danov, K.D., Kralchevsky, P.A., Stoyanov, S.D., Arnaudov, L.N., Blijdenstein, T.B.J., Pelan, E.G., Lips, A. *Langmuir* 2012; 28: 4168–4177.

178. Rio, E., Drenckhan, W., Salonen, A., Langevin, D. Unusually stable liquid foams. *Adv Colloid Interface Sci* 2014; 205: 74–86.

179. Emady, H., Caggioni, M., Spicer, P. Colloidal microstructure effects on particle sedimentation in yield stress fluids. *J Rheol* 2013; 57(6): 1761–1772.

180. Patel, A.R., Drost, E., Blijdenstein, T.B.J., Velikov, K.P. Stable and temperature-responsive surfactant-free foamulsions with high oil-volume fraction. *Chemphyschem* 2012; 13: 3777–3781.

181. Drenckhan, W., Langevin, D. Monodisperse foams in one to three dimensions. *Curr Opin Coll Interface Sci* 2010; 15: 341.

182. Cox, A.R., Aldred, D.L., Russell, A.B. Exceptional stability of food foams using class II hydrophobin HFBII. *Food Hydrocoll* 2009; 23: 366–376.

183. Drenckhan, W., Cox, S.J., Delaneya, G., Holste, H., Weaire, D., Kern, N. Rheology of ordered foams—on the way to discrete microfluidics. *Colloids Surf A* 2005; 263: 52–64.

184. Perez, A.A., Carrara, C.R., Sánchez, C.C., Santiago, L.G., Rodríguez Patino, J.M. Interfacial and foaming characteristics of milk whey protein and polysaccharide mixed systems. *AIChE J* 2010; 56(4): 1107–1117.

185. Mileva, E., Exerowa, D. Amphiphilic nanostructures in foam films. *Curr Opin Colloid Interface Sci* 2008; 13: 120–127.

186. Dickinson, E. Structure formation in casein-based gels, foams, and emulsions. *Colloids Surf A* 2006; 288: 3–11.

187. Jung, J.-M., Gunes, D.Z., Mezzenga, R. Interfacial activity and interfacial shear rheology of native β-lactoglobulin monomers and their heat-induced fibers. *Langmuir* 2010; 26(19): 15366–15375.

188. Asnacios, A., Bergeron, V., Langevin, D., Argillier, J.-F. Anionic polyelectrolyte-cationic surfactant interactions in aqueous solutions and foam films stability. *Revue de l'Institut Francais du Petrole* 1997; 52(2): 139–144.

189. Stubenrauch, C., Albouy, P.-A., Klitzing, R.V., Langevin, D. Polymer/surfactant complexes at the water/air interface: A surface tension and X-ray reflectivity study. *Langmuir* 2000; 16(7): 3206–3213.

190. Guillot, S., McLoughlin, D., Jain, N., Delsanti, M., Langevin, D. Polyelectrolyte-surfactant complexes at interfaces and in bulk Protein-polysaccharide interactions at fluid interfaces. *J Phys Condens Matter* 2003; 15(1): S219–S224.

191. Haffner, B., Khidas, Y., Pitois, O. Flow and jamming of granular suspensions in foams. *Soft Mat* 2014; 10(18): 3277–3283.

192. Clasen, C., McKinley, G.H. Gap-dependent microrheometry of complex liquids. *J Non-Newton Fluid Mech* 2004; 124(1–3 SPEC. ISS.): 1–10.

193. Clasen, C., Gearing, B.P., McKinley, G.H. The flexure-based microgap rheometer (FMR). *J Rheol* 2006; 50(6): art. no. 008606JOR, 883–905.

194. Curschellas, C., Gunes, D.Z., Deyber, H., Watzke, B., Windhab, E., Limbach, H.-J. Interfacial aspects of the stability of polyglycerol ester covered bubbles against coalescence. *Soft Mat* 2012; 8: 11620–11631.

195. Sarker, D.K., Wilde, P.J., Clark, D.C. Enhancement of the stability of protein-based food foams using trivalent cations. *Colloids Surf A* 1996; 114: 227–236.

196. Mackie, A.R., Gunning, A.P., Wilde, P.J., Morris, V.J. Orogenic displacement of protein from the air/water interface by competitive adsorption. *J Colloid Interface Sci* 1999; 210: 157.

197. Navarini, L., Rivetti, D. Water quality for espresso coffee. *Food Chem* 2010; 122: 424–428.

198. Haedelt, J., Beckett, S.T., Niranjan, K. Bubble-included chocolate: Relating structure with sensory response. *J Food Sci* 2007; 72(3): E138–42.

199. d'Agostina, A., Boschin, G., Bacchini, F., Arnoldi, A. Investigations on the high molecular weight foaming fraction of espresso coffee. *J Agric Food Chem* 2004; 52: 7118–7125.

200. Mackie, A.R., Gunning, A.P., Ridout, M.J., Morris, V.J. Gelation of gelatin: Observation in the bulk and at the air-water interface. *Biopolymers* 1998; 46(4): 245–252.

201. Illy, E., Navarini, L. Neglected food bubbles: The espresso coffee foam. *Food Biophys* 2011; 6: 335–348.

202. Rühs, P.A., Affolter, C., Windhab, E.J., Fischer, P. Shear and dilatational linear and nonlinear subphase controlled interfacial rheology of β-lactoglobulin fibrils and their derivatives. *J Rheol* 2013; 57(3): 1003–1022.

203. Roth, A.E., Jones, C.D., Durian, D.J. Bubble statistics and coarsening dynamics for quasi-two-dimensional foams with increasing liquid content. *Phys Rev E* 2013; 87: 042304.

204. Jang, W., Nikolov, A., Wasan, D.T., Chen, K., Campbell, B. Prediction of the bubble size distribution during aeration of food products. *Ind Eng Chem Res* 2005; 44(5): 1296–1308.

205. Jin, H., Zhou, W., Cao, J., Stoyanov, S.D., Blijdenstein, T.B.J., de Groot, P.W.N., Arnaudov, L.N., Pelan, E.G. Super stable foams stabilized by colloidal ethyl cellulose particles. *Soft Mat* 2012; 8: 2194.

206. X.E., Pei, Z.J., Schmidt, K.A. Ice cream: Foam formation and stabilization-a review. *Food Rev Int* 2010; 26(2): 122–137.

207. Epstein, P.S., Plesset, M.S.J. On the stability of gas bubbles in liquid-gas solutions. *Chem Phys* 1950; 18: 1505.

208. Ducloué, L., Pitois, O., Goyon, J., Chateau, X., Ovarlez, G. Rheological behaviour of suspensions of bubbles in yield stress fluids. *J Non-Newton Fluid Mech* 2015; 215: 31–39.

209. Rodríguez-García, J., Salvador, A., Hernando, I. Replacing fat and sugar with inulin in cakes: Bubble size distribution, physical and sensory properties. *Food Bioprocess Technol* 2013; 7(4): 964–974.

210. Marchalot, J., Lambert, J., Cantat, I., Tabeling, P., Jullien, M.-C. 2D foam coarsening in a microfluidic system. *Europhys Lett* 2008; 83(6): 64006.

211. van Aken, G.A. Aeration of emulsions by whipping. *Coll Surf A* 2001; 190(3): 333–354.

21 The Froth in Froth Flotation

Rossen Sedev and Jason N. Connor

CONTENTS

Froth flotation is a highly selective method of separating value ore particles from gangue solids which has made it the most widely used process in minerals beneficiation. The process involves bubble-particle attachments in a vigorously stirred concentrated pulp followed by further transport of the stable aggregates to the froth layer. Extensive research has been directed at the processes in the pulp layer as particles and bubbles meet, collide, and eventually attach in this zone of the flotation cell. The physicochemical processes related to the froth layer are less well understood and, therefore, are emphasized in this chapter.

21.1 INTRODUCTION

Froth flotation is a flexible and efficient method for ore beneficiation that has been around for more than a century and is still the workhorse of the minerals industry [1]. The economic significance of froth flotation is enormous given the high volumes of ores processed. It is also having a major environmental impact due the significant use of water and the disposal of large volumes of tailings. Originally, froth flotation was designed for dealing with metal sulfide ores, but today it is also used with metal oxide ores, in coal production [2], and for bitumen recovery from oil sands [3]. Variants of froth flotation are also employed in the treatment of wastewater [4] and deinking [5].

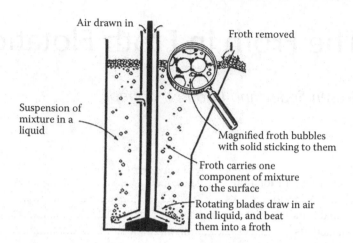

FIGURE 21.1 A historic schematic of the flotation process (Potter-Delprat process [6]). The pulp is contained in a mechanically agitated tank. Air is entrained and dispersed at the bottom of the flotation cell. Bubbles collect hydrophobic particles and carry them to the froth layer. Hydrophilic particles remain in the pulp.

Froth flotation is a highly selective process for separating valuable mineral grains from a mixture of particles obtained by grinding a suitable ore. The particles are suspended in an aqueous environment to create a pulp; a series of flotation reagents are added to achieve the desired conditioning and then air is pumped through the flotation cell which is stirred continuously. A historical yet valid depiction of the process is shown in Figure 21.1 [6]. The small bubbles introduced at the bottom of the flotation cell rise through the pulp and collect hydrophobic particles only by virtue of adhesion. The particle-laden bubbles arrive at the top of the vessel and form a froth layer, that is, a foam layer loaded with solid particles. The froth is then removed (or overflows naturally) and this concentrate containing a higher grade of the valuable mineral is further processed by other metallurgical methods.

Pulp conditions are of utmost importance in froth flotation as this is where bubbles and particles mix, collide, and eventually attach. Flotation is most efficient for particle sizes in the optimal range of approximately 10–200 μm. Ore particles are heavy and to keep them suspended the cell is stirred vigorously. The bubbles introduced at the bottom of the column are typically 1–2 mm in size and they float naturally through buoyancy. The mixing conditions in the flotation cell are critical for maximizing the number of bubble-particle collisions. The chemical and electrochemical conditions in the pulp are equally important as they determine the physicochemical state of the solid and liquid surfaces (adsorption, coverage, charge, etc.). Pulp parameters must be adjusted precisely so that value particles are hydrophobic enough to float efficiently while gangue particle are hydrophilic and remain in the pulp (occasionally the gangue particles are floated in a process known as reverse flotation). The selectivity and efficiency of froth flotation are largely determined by creating and maintaining optimal chemical and mechanical conditions in the pulp.

The bubbles loaded with particles are gradually incorporated in the froth layer situated atop the pulp layer (Figure 21.1). The froth (i.e., a foam containing solid particles) must be stable enough to avoid releasing the value particles back into the pulp. On the other hand, it should not be particularly stable as this may adversely affect the subsequent transport and processing of the concentrate. Also, small and hydrophilic particles are always entrained and reach the froth layer thus decreasing the grade of the concentrate. It is desirable to maintain froth conditions allowing these unwanted particles to drop back into the pulp. The amount and stability of the froth are manipulated through the use of frothers, but frothing is influenced by the presence of particles as well as other reagents. The pulp and froth processes are consecutive, therefore, froth behavior is crucial to the overall success of ore separation through flotation.

In spite of the extensive research conducted on most aspects of froth flotation and detailed investigations of specific processes, the operation of flotation plants is still based mostly on empirical knowledge. The detailed scientific understanding of flotation remains incomplete. At the same time, flotation plants process very large amounts of ore which makes them both inert and expensive systems to control. The quality of the ore varies significantly which means that comprehensive laboratory results may not be available to guide daily decisions. Exploratory measurements are usually done with small volume cells, but flotation outcomes are not easy to scale up. Additionally, the minerals industry faces new and enduring problems: the grade of the ores processed is continuously declining, the use of water is becoming more restrictive, and environmental standards are getting stricter. These industrial challenges can only be resolved with multidisciplinary and innovative research and development activities.

In this chapter, we outline the processes involved in froth flotation based on the physicochemical principles involved. The emphasis is on various aspects frothing and froth behavior in the wider context of froth science. This chapter should be seen as an up-to-speed introduction to the extensive literature on various aspects of froth flotation: concise reviews of froth flotation [7] and bioflotation [8], extensive reviews of froth flotation [1,4,9–13], physicochemical aspects of flotation [14,15], flotation of coal [2], and flotation reagents [16].

21.2 PHYSICAL CHEMISTRY OF FLOTATION

The basic idea of froth flotation is to achieve selective attachment of the value particles suspended in the pulp to the rising bubbles (Figure 21.2). The stable bubble-particle aggregates travel upward through the pulp, which is stirred vigorously and unevenly, to reach the froth layer at the top of the cell (Figure 21.1). This concentrate is taken for further processing. The grade of the concentrate is improved because the gangue particles are hydrophilic and mostly remain in the pulp (Figure 21.2). The rejected gangue particles (called tailings) leave the cell from the lower end (not shown in Figure 21.1) for further processing or disposal.

Relatively few minerals (such as talc) are hydrophobic enough to be floated in their natural state. Most minerals enriched through flotation are made sufficiently hydrophobic through the use of collectors (surfactants that adsorb on the solid surface and render it more hydrophobic). The froth

FIGURE 21.2 Valuable (hydrophobic) particles are captured and carried up by bubbles while gangue (hydrophilic) particles remain in the pulp. The pulp is energetically mixed at all times to maintain the suspension of the solids and increase the number of bubble-particle collisions.

layer is created and maintained by adding frothers (surfactants that stabilize foams) to the pulp. The unwanted particles (gangue) are naturally hydrophilic or their hydrophobicity is reduced by the addition of suitable depressants and adjusting the pH and Eh (pulp potential) of the pulp. Additional reagents are widely used to tune up the performance of the flotation process.

The above description implies that adhesion between solids and gas plays a major role and indeed this is the case for froth flotation. The condition of the interfaces involved (solid-air, water-air, and solid-water) largely determines the adhesion and is manipulated through the use of surfactants. Thus, the behavior of surfactants and particles at interfaces is of key relevance. Both pulp and froth are dispersive multiphase systems (the terms "pulp phase" and "froth phase" are inexact but widely used). The stability of a dispersive system is often related to the stability of the liquid films intervening between different phases (e.g., between a bubble and a particle in the pulp or between two bubbles in the froth). This brings into focus the topic of surface forces which determine the stability of thin liquid films.

In the following three subsections, these key concepts are briefly presented. As usual, equilibrium properties are invoked because they are easier to understand and measure in comparison with dynamic ones. However, froth flotation is a dynamic process involving complex heterogeneous systems. The pulp is constantly agitated, turbulence is very significant (especially near the baffles) but varies widely with location, and air is bubbled through continuously. These highly dynamic mechanical conditions are coupled with the physicochemical processes at the interfaces and make the interpretation and modeling of the overall flotation process quite challenging.

21.3 WETTABILITY AND ADHESION

The selectivity of froth flotation is based on the different wettabilities of the solid particles in the pulp. The wettability of a smooth solid surface is determined by the molecular interactions between the solid and the liquid in a given fluid environment. The most common measure of hydrophobicity is the contact angle, that is, the angle at which the water-air surface meets the solid surface (Figure 21.3). In the simplest thermodynamic model, an ideal flat smooth surface (S), which is also rigid and inert, comes into contact with two fluids—air (A) and water (W)—Figure 21.3a.

In this ideal case, the equilibrium contact angle, θ_E, assumes a unique value prescribed by the Young Equation [17]:

$$\cos \theta_E = \frac{\gamma_{SA} - \gamma_{SW}}{\gamma_{WA}}$$

(21.1)

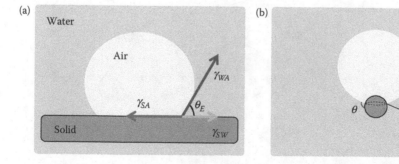

FIGURE 21.3 Contact angle: (a) The equilibrium contact angle, θ_E, is entirely prescribed by the three interfacial tensions: solid-air (γ_{SA}), solid-water (γ_{SW}), and water-air (γ_{WA}); (b) In most real systems, the static contact angle, θ, can adopt any value between the advancing contact angle, θ_A, and the receding contact angle, θ_R.

This equilibrium contact angle is defined by the three interfacial tensions (γ_{SA}, γ_{SW}, and γ_{WA}) which represent the interactions between the three pairs of bulk phases (S, W and A) and it is, therefore, a specific property of the three phase system. The crucial role of the contact angle in capillarity stems from the fact that it provides the boundary condition for the fluid interface and thus affects critically the capillary force acting in the system. Hence it determines how strongly particles attach to bubbles (Figure 21.3b).

The work of adhesion, W_A, defined as the difference between the interfacial free energy per unit area before (γ_{SA} and γ_{WA}) and after contact (γ_{SW}) between the solid and water, is directly related to the contact angle:

$$W_A \equiv \gamma_{SA} + \gamma_{WA} - \gamma_{SW} = \gamma_{WA}(1 + \cos\theta_E) \tag{21.2}$$

This energy of adhesion between solid and water depends on the surface tension of water, γ_{SA}, and the hydrophobicity of the solid surface expressed through the contact angle, θ_E. The work of adhesion between solid and air, W_A^{air}, can be expressed in exactly the same manner (the relevant angle in this case is $180° - \theta_E$, see Figure 21.3a):

$$W_A^{air} = \gamma_{WA}(1 - \cos\theta_E) \tag{21.3}$$

The works of adhesion of the solid surface to water and to air are always in competition because their sum is constant and equal to $2\gamma_{WA}$ (a quantity known as work of cohesion):

$$W_A + W_A^{air} = 2\gamma_{WA} \tag{21.4}$$

Thus, hydrophobic surfaces (with larger θ_E) adhere weakly to water and, therefore, strongly to air. In the pulp, hydrophobic particles are much more likely to attach to a bubble as sketched in Figure 21.3b. On the contrary, particles with hydrophilic surfaces (i.e., smaller θ_E) adhere stronger to water and much less to bubbles. In the limiting case of a completely wettable particle ($\theta_E = 0$), a thin water film covering its surface will be stable in air and a direct bubble-particle contact is avoided. Hence the selectivity of the flotation process comes down primarily to the value of the contact angle which quantifies the hydrophobicity of the surface.

In real systems, the static contact angle rarely adopts an equilibrium value. Instead, a whole range of static contact angles is observed. This contact angle hysteresis can be due to surface roughness or chemical heterogeneity of the solid surface [17], but also ongoing surface modification (very few solid surfaces are truly inert when immersed in a liquid). The upper and lower limits of the hysteresis interval are designated as advancing and receding contact angles—θ_A and θ_R, respectively. The static contact angle observed in an overwhelming majority of experiments is likely to adopt an intermediate value θ such that $\theta_R \leq \theta \leq \theta_A$. The intermediate values in this interval are set by external conditions and, therefore, are not representative of the wettability in this system. The equilibrium contact angle, θ_E, is a thermodynamic quantity, but is rarely realized even in static experiments. The advancing static contact angle, θ_A, is often treated as a quantity representative of the adhesion in the system [18]. The receding contact angle, θ_R, is far too often ignored with no solid validation. For mineral particles (usually heterogeneous), contact angle hysteresis can be very significant, therefore, relying on a single value for the contact angle can be misleading.

Particles typically encountered in froth flotation (size between 20 and 200 µm) are very large in comparison to the thickness of the interface which is on the order of the molecular size. Therefore, the curvature of their surfaces is too small to affect the contact angle through second-order effects (e.g., line tension). Documented correlations between static contact angle and particle size should be considered in relation to the degree of liberation, that is, what fractions of the bulk chemical components are exposed on the surface of a heterogeneous particle. Surface heterogeneity affects both the equilibrium contact angle and the hysteresis range [17].

If external forces drive the static contact angle below θ_R or above θ_A, the contact line (or three phase contact line) recedes or advances over the solid. The dynamic contact angle is generally a function of contact line speed and mode of dissipation [17].

Unfortunately, the particles typically encountered in froth flotation vary in size from small (100 µm) to very small (below 10 µm), and are typically irregular in shape, so most of the standard methods for contact angle measurement are not applicable. Instead, liquid imbibition into packed beds of particles is examined (under static or dynamic conditions) and the contact angle is obtained indirectly [19].

Particle-to-bubble attachments in the pulp occur under highly turbulent conditions, therefore, the static contact angle is likely to vary within the hysteretic range. This is almost universally disregarded though it can be a misleading approximation. For instance, immediately after the liquid film between a bubble and a particle ruptures, the newly formed contact line recedes over the solid surface and the advancing contact angle could be quite irrelevant. Furthermore, given the high speed of contact line expansion, most certainly a dynamic contact angle should be considered. Once the contact line (contact perimeter) has expanded to provide adequate adhesion, the bubble-particle aggregate is formed. It must then travel upward in a turbulent environment to reach the froth layer. This is possible only if it is stable enough to resist the external stresses. One can image that the hysteresis interval of the contact angle, which is almost never assessed, actually plays a major role in providing stability to the bubble-particle aggregate.

21.4 SURFACTANTS AND SMALL PARTICLES AT INTERFACES

21.4.1 SURFACTANTS

Pure water does not foam. The reason is that energetic agitation may produce many bubbles, but they are very unstable and coalesce rapidly. From a thermodynamic point of view, the dispersive system has a much larger interfacial area than the two separate immiscible phases, and therefore, larger interfacial free energy. Stable foams can be obtained only in the presence of surfactants.

Surfactants have amphiphilic molecules, that is, one part of the molecule is polar and water soluble (hydrophilic), while another one is nonpolar and soluble in oil only (hydrophobic). Owing to this duality, surfactants adsorb strongly at interfaces and decrease the surface (interfacial) tension even at relatively low bulk concentrations. The equilibrium properties of surfactant-laden surfaces are summarized in their adsorption isotherms, that is, surface (interfacial) tension versus bulk surfactant concentration dependences. Surfactant molecules can be ionic or nonionic but always contain both polar groups ($-COOH$, $-OH$, $-NH_2$, $-(CH_2CH_2O)_n-$, etc.) and nonpolar groups ($-CH_2$, $-CF_2-$, $-(CH_2CH(CH_3)O)_n-$, etc.).

Depending on their molecular structure, surfactants can be more water soluble or oil soluble and that will affect their action. The most popular classification criterion is the hydrophile-lipophile balance (HLB). According to Davis [20], HLB can be calculated by summing tabulated group contributions, B_i:

$$HLB = 7 + \sum_i B_i \tag{21.5}$$

Surfactants with high HLB are more prominent on the water side of the water-oil (or water-air) interface when adsorbed. Because of lateral steric interactions, they favor a convex interface (from the water side) and thus promote the formation of oil-in-water (O/W) emulsions. Surfactants with low HLB do exactly the opposite and thus foster the formation of W/O emulsions.

Surfactant layers exhibit rich dynamic behavior. At equilibrium, a fixed amount of surfactant is adsorbed over a given surface area (in accordance with the adsorption isotherm). If the area in question is increased, the surface concentration of surfactant molecules drops, the surface tension increases, and thus the tendency of the area to shrink increases. If the area is decreased, the opposite happens, but now the lower surface tension favors area expansion. In other words, the surfactant presence imparts some elasticity to the liquid surface. In a useful analogy, the surfactant layer is approximated to an elastic body with a Young modulus, E ($=x\partial\sigma/\partial x$, σ is stress and dx/x is Cauchy strain). The Gibbs elasticity of the surfactant layer is defined as [20] (γ is surface tension and A is area of the layer):

$$E = A\frac{\partial\gamma}{\partial A} \tag{21.6}$$

This elasticity is entirely due to the presence of surfactant molecules and is affected by their bulk concentration. Elasticity is considered a major factor in the stability of interfaces and disperse systems.

When subjected to an external disturbance, a surfactant layer reacts by redistributing itself along the interface to maintain uniformity. For instance, if an equilibrated surfactant layer is compressed in the direction of a liquid flow, regions of higher and lower surface tension will be created due to the different surface coverage. The surface pressure (the difference in surface tension) will immediately drive surfactant molecules toward the region of lower surface concentration, that is, oppose the external disturbance. The surface redistribution of surfactant can be very significant and entrain the surface layers of the underlying liquid. The effect is known as the Marangoni effect [17,20] and it often plays a central role in dynamic situations. An important example of the Marangoni effect is the effective immobilization of a liquid surface (zero fluid velocity) by minute amounts of surfactant.

Some authors [21] lump together the effect of surface elasticity and the Marangoni effect, but a stricter analysis [22] shows that Gibbs elasticity relates to slow interfacial expansion circumstances, while the Marangoni effect is considered in situations of rapid interfacial expansion. More generally, surfactant layers are viscoelastic and the viscoelastic modulus, E_{VE}, reflects that [23]:

$$E_{VE} = E + i\omega\eta \tag{21.7}$$

The real part is the dilational elasticity as given by Equation 21.6 and the imaginary part is related to the surface viscosity, η (i is imaginary unit and ω is angular frequency of the external disturbance).

The presence of surfactants is of crucial importance to the stability of dispersive systems and foams, in particular. Surfactant science is a vast subject of its own [24,25].

21.4.2 SMALL PARTICLES

Small particles of intermediate wettability attach strongly to fluid interfaces. If a spherical particle traverses an undisturbed liquid surface its interfacial energy, F_{int}, can easily be calculated from spherical geometry and the interfacial tensions γ_{SW}, γ_{SA}, and γ_{WA}. As an example, the F_{int} for a model hydrophobic particle crossing the water surface is plotted in Figure 21.4a. The running contact angle, α, is merely a measure of the particle position with respect to the interface. The total F_{int} is the sum of three components (one for each of the three interfaces) and has a minimum at the equilibrium contact angle (120° in this case).

This thermodynamic approach also estimates the energy of detachment, ΔF, for a particle of radius r, trapped at the interface [26]:

$$\Delta F = \pi r^2 \gamma_{WA}(1 \pm \cos\theta)^2 \tag{21.8}$$

The positive sign is for a particle going into the aqueous phase and the negative sign is for a particle moving into the gas phase. The free energy of detachment is plotted in Figure 21.4b as a function of the contact angle, θ. It is clear that completely hydrophilic particles ($\theta = 0$) will remain in water, while completely hydrophobic particles ($\theta = 180°$) will stay in air (such particles would be labeled

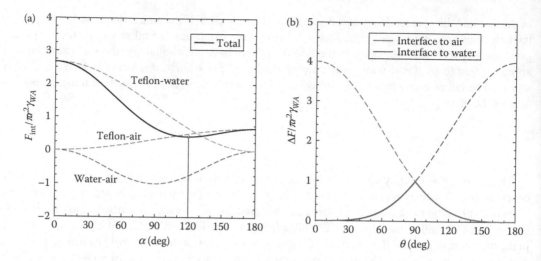

FIGURE 21.4 A single spherical particle (size 100 μm) at an interface: (a) Interfacial free energy of a Teflon particle positioned at different locations across a flat water-air interface (cos α = –h/r, h is elevation of the center of gravity above the interface and r is particle radius); (b) Energy required to detach a Teflon particle, trapped at the water-air interface, and relocate it into a neighboring phase (air or water).

superhydrophobic as the largest possible water contact angle on a smooth solid surface is about 125°). Particles with equal affinity to both phases (i.e., $\theta = 90°$) are attached the strongest—Figure 21.4b.

It is worth noting that this maximum detachment energy is very large on the molecular scale and, therefore, small particles are de facto anchored at the interface [26]. This is one key difference from surfactants—small particles attach at interfaces irreversibly. Another difference is that most particles do not have clearly segregated hydrophilic and hydrophobic sides (with the notable exception of Janus particles). Finally, small particles are larger than typical surfactant molecules and their diffusion is much slower.

The interest in particles at interfaces is growing rapidly. The presence of small particles in froths can vary from insignificant to dominant. In real systems, both particles and surfactants are often present and complex mutual influences can develop.

21.5 THIN LIQUID FILMS AND SURFACE FORCES

Bubbles can be formed by shaking pure water, but they are very unstable and coalesce within seconds after stopping the mechanical action. On a microscopic level, the water films intervening between the bubbles are unstable—they drain fast and burst. Thus, thin liquid films play a key role in the stability of disperse system as their stability determines the outcome of an encounter between two bubbles, two particles or a bubble and a particle. A thin liquid film is a film which is thin enough so that the two interfaces are not independent and exert a force on each other [27–29]. This surface force (or disjoining pressure, Π) is the pressure difference between the liquid in the thin liquid film, $P(h)$, where h is the film thickness, and the liquid in the bulk, $P(\infty)$:

$$\Pi = P(h) - P(\infty) \tag{21.9}$$

A thermodynamic consideration links the disjoining pressure, Π, to the free energy of interactions in the film per unit area, F, as:

$$\Pi = -\left(\frac{\partial F}{\partial h}\right)_{T,V,\mu_i} \tag{21.10}$$

The disjoining pressure can be measured experimentally and is a key concept in understanding the stability of disperse systems through the different types of interfacial interactions across thin liquid films.

The disjoining pressure, Π, is a thermodynamic quantity, but in practice, it is modeled as a sum of components arising from different types of interactions [27–29]. In relation to froth flotation, three types of surface forces are usually discussed (Figure 21.5) and the disjoining pressure is

$$\Pi = \Pi_{VW} + \Pi_{EL} + \Pi_H \tag{21.11}$$

The van der Waals component, Π_{VW}, is due to van der Waals interactions (Figure 21.5a) and is relatively short-ranged. The sign and magnitude at a given film thickness are determined by the Hamaker constant which is tabulated for many materials. The electrostatic component, Π_{EL}, is due to the overlap of the electrical double layers carried by both interfaces (Figure 21.5b) and its range of action can extend up to about 100 nm. Its magnitude is a function of the surface charge density at the interfaces and also the ionic strength of the aqueous solution. The van der Waals and electrostatic interactions, known together as DLVO interactions, are relatively well understood, especially in model systems [27–30].

In a symmetric system, for example, a foam film, van der Waals forces are always attractive and electrostatic forces are always repulsive. In an asymmetric system, for example, particle-water-air, the signs of both Π_{VW} and Π_{EL} can be either positive or negative. The calculation of the electrostatic component, in both symmetric and asymmetric cases, depends on the boundary conditions at the interfaces. The constant potential and constant charge approximations provide a useful range of behavior unless a more detailed charge regulation model is available.

The hydrophobic component, Π_H, is often invoked when resolving bubble-particle interactions because the particles of interest in flotation are hydrophobic. The hydrophobic interaction is always attractive, that is, destabilizes the thin water film, and its range of action can be very large.

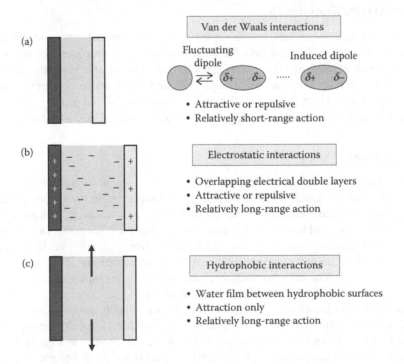

FIGURE 21.5 Surface forces between a solid and a bubble separated by an aqueous film. Surface forces most relevant to froth flotation: (a) van der Waals interactions; (b) Electrostatic interactions; and (c) Hydrophobic interactions.

A simplistic explanation of Π_H can be outlined as follows. An aqueous film can be force spread on a single hydrophobic surface (e.g., by spin-coating) but it is metastable and will rupture and retreat at the first opportunity (e.g., film puncture near a defect or created by an external influence). In a similar manner, an aqueous film is metastable when trapped between two hydrophobic surfaces as in Figure 21.5c (note that air is hydrophobic; if a water droplet is sitting on top of a film of air—superhydrophobic surfaces provide a good approximation to this situation—the contact angle is 180°). This aqueous film displays a tendency to leave the gap between the two surfaces and that translates into the specific force of hydrophobic attraction. Empirical expressions for Π_H are available, but the details of hydrophobic attraction are still being debated [31–34].

Surface forces are studied intensively as they provide deep insights into the interactions in various disperse systems. However, a cohesive understanding of the role of surface forces in the subprocesses of flotation is still lacking.

The features of surface forces mentioned above related to equilibrium conditions. Under dynamics conditions, the influence of surface forces overlays with the drainage of the liquid film. The drainage of liquid films between solid surfaces is relatively well understood. The liquid film drains in accordance with the laws of hydrodynamics and the conditions at the surfaces. At sufficiently small thicknesses, the film may be affected by surface forces, including becoming unstable. If stable, the liquid film thins down to thicknesses where the molecular structure of the liquid comes into play (oscillatory surface forces have been detected and described in this region). When the liquid film is flanked by one or two fluid phases, drainage becomes more complex as various deformations are possible (e.g., dimple—a growth in thickness in the inner part of a thinning liquid film and wimple—a multiple curvature inversion). A hydrodynamic model incorporating the role of surface forces and flow conditions at the interfaces has been developed by Chan and coauthors [35].

21.6 FLOTATION REAGENTS

The natural wettability contrast between valuable mineral particles and unwanted gangue particles is rarely sufficient, except for very hydrophobic minerals like talc or graphite. The action of additives is not always specific enough and that necessitates the use of additional reagents. The list is further complemented by chemicals employed to adjust pH and Eh or enhance/suppress the effects of reagents already in use. Thus, a number of reagents are always added to the pulp. Based on their roles, flotation reagents are broadly classified as collectors, frothers, and modifiers [36].

21.6.1 COLLECTORS

The hydrophobicity of the valuable particles is the major factor in froth flotation. A certain level of hydrophobicity (large enough water contact angle) is necessary for the successful capture of the particles in the pulp and their transport to the froth layer. In most cases, the natural hydrophobicity of the minerals is too low and has to be modified by using suitable collectors. An example of floating the sulfide minerals chalcopyrite and pyrite in laboratory conditions [37] is shown in Figure 21.6. Their flotation recovery increases in time in an exponential manner and it happens faster at higher collector concentrations. The collector used produces better results with chalcopyrite.

It was shown, using UV spectroscopy, that in both cases, the amount of collector adsorbed on the mineral particles improved at higher bulk collector concentrations. This in turn made the chalcopyrite and pyrite particles more hydrophobic and, therefore, more floatable (recoverable).

The outcomes of a laboratory microflotation of quartz are plotted in Figure 21.7 as flotation recovery versus collector concentration. Recovery increases in a typical S-shape manner—slowly at low concentrations of the collector (n-dodecylamine) and then rising sharply at the so called critical flotation concentration, CFC—Figure 21.7a. The CFC can be considered as the minimum collector concentration at which recovery is significant and is, therefore, an indicator of collector efficiency

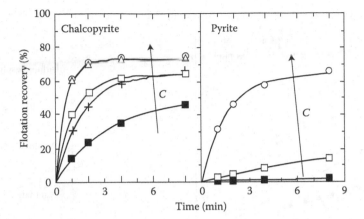

FIGURE 21.6 Flotation of chalcopyrite and pyrite with O-isobutyl-N-ethoxycarbonyl thionocarbamate in a microflotation Hallimond tube (solids 2 g/L, pH 7). The lines are first-order kinetics fits [37]. The arrows point at the increasing collector concentration.

[38]. The flotation recovery of quartz with alkyl ammonium acetates of different chain length is shown in Figure 21.7b. The curves obtained with different collectors are similar but shifted with respect to CFC. The logarithm of CFC decreases linearly with the length of the alkyl chain. Thus, collectors with longer hydrophobic chains are more effective and can be used at lower concentrations. The slope of the dependence gives the gain in free energy when removing a single CH_2 segment from the solution (about 1 kT) [39].

Collectors are surfactants that impart hydrophobicity to the valuable mineral particles by adsorbing on their surfaces. Adsorption is determined by the surface chemistry of the solid and because of that, the sulfides and nonsulfides groups of minerals are considered separately [36]. Sulfide minerals are bonded by strong bonds (covalent or metallic) and have low solubility in water, higher natural hydrophobicity (weakly hydrated surfaces, low hydrogen bonding capacity), and a pronounced affinity for sulfur containing ligands. Their pulp chemistry is strongly affected by the electrochemical affinities of the species present on the solid surface and in the aqueous solution. The nonsulfide

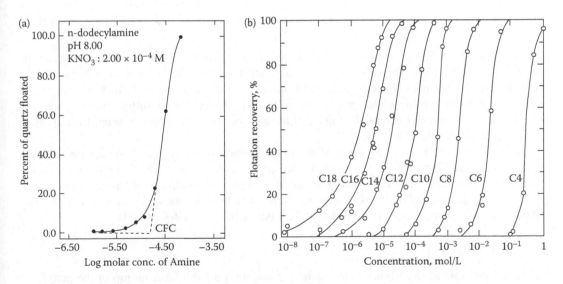

FIGURE 21.7 Microflotation of quartz in a Hallimond tube (a) With n-dodecylamine as a collector (solids 1 g, 0.2 mM KNO_3, pH 8) [38]; (b) With alkyl ammonium acetates of different chain length.

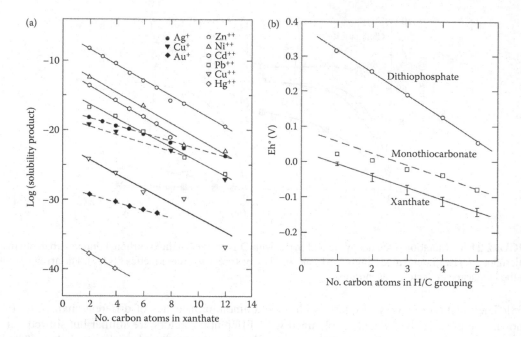

FIGURE 21.8 (a) Solubility products of some metal xanthates as function of the number of carbon atoms; (b) Standard redox potentials, E_h^0, for homologous series of sulfydryl collectors as a function of number of carbon atoms in the substituent group [36].

minerals typically include ionic bonding and a higher solubility in water, lower natural hydrophobicity (hydrated surfaces, stronger hydrogen bonding), and an affinity for oxygen containing ligands. Their pulp chemistry is determined mostly by ion exchange and electrostatic interactions.

The variety of collectors in use is wide [36]. The collectors used in the flotation of sulfide minerals almost always contain a sulfur atom in their molecules and are anionic or nonionic. Typical dosages vary from 10 to 100 g per metric ton of dry ore. It has been shown that the basic bonding group is a C=S, P=S or C–S–. The metal cation (M^{z+}) in the mineral and the collector (X^-) interact as follows:

$$M^{z+} + zX^- \leftrightarrow MX_z \tag{21.12}$$

The driving force for adsorption is due to the reactivity of the collector and the solubility of the mineral-reagent complex. The solubility of metal xanthates (an example is shown in Figure 21.8a) and dithiophosphates quickly decreases with the number of carbon atoms in the hydrophobic group. Longer chains favor a stronger adsorption of the collector molecules. Many sulfide mineral collectors are reactive and their redox potentials (Figure 21.8b) must be taken into account, hence the important role of the pulp potential, Eh.

The collectors used in the flotation of nonsulfide minerals are typically ionic, often contain oxygen atoms, and bonding occurs through the structures C–O, N–O, S–O, or P–O. Examples include fatty acids, alkyl/aryl sulfonates, and various amines with relatively long alkyl chains (12–22 carbon atoms). Because of the weaker, nonspecific bonding between these collectors and nonsulfide minerals, typical dosages vary from 100 to 1000 g per metric ton of dry ore [7].

21.6.2 FROTHERS

The functional role of the frother is to create and maintain a froth layer on top of the pulp layer until the former is removed to launders. When introduced in the flotation system, frothers lower the surface tension of the aqueous phase, stabilize small bubbles, and generate a froth layer.

The most commonly used frothers are short aliphatic alcohols, low molecular weight poly(ethylene) glycol, poly(propylene) glycols and their monoethers, and alkoxy compounds [36]. Methyl Isobutyl Carbinol (MIBC) is very popular and is often taken as a standard frother. Alcohol and glycol frothers are quite different due to differences in water solubility, surface activity, and diffusion coefficients. Hydrocarbons, such as fuel oil and kerosene, are often added to alcohols in order to manipulate the structure and stability of the froth. Typical frother dosages are low 10–50 g per metric ton, but also vary significantly (0–300 g per metric ton [7]).

The consistent measurement of the frothing ability of frothers and their classification are still being examined. Bikerman [40] proposed the unit of foaminess, Σ, which is the ratio of the average steady-state froth volume, V_F, and volumetric gas flow rate, Q:

$$\Sigma = \frac{V_F}{Q} \cong \tau_F \tag{21.13}$$

Obviously, this ratio can be interpreted as the average time that gas spends in the foam. More recently, the above ratio, renamed as the dynamic stability factor [41], was examined in the case of ore flotation under laboratory conditions. The dynamic froth height, h, followed an exponential growth in time:

$$h = h_{max}\left[1 - \exp\left(\frac{-t}{\tau}\right)\right] \tag{21.14}$$

The volume fraction of air overflowing the flotation cell was measured (using image processing) and predicted correctly using h_{max} [41].

Laskowski [42] mapped a variety of frothers based on their molecular mass and HLB—Figure 21.9. In the shaded space, more selective frothers are positioned to the left while stronger frothers cluster to the right side. Arguably, selective frothers should be used in the flotation of very fine particles, whereas stronger ones would provide higher recoveries in the floating of coarser particles [42].

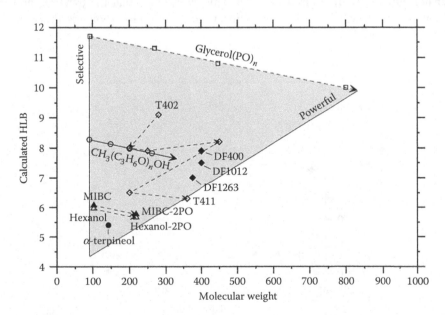

FIGURE 21.9 HLB versus molecular mass diagram for flotation frothers [42]. Frothers are classified as more selective or more powerful.

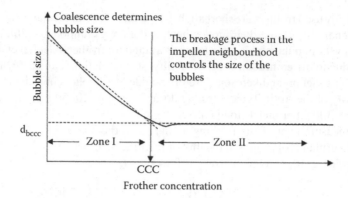

FIGURE 21.10 Definition of the critical coalescence concentration (CCC) [43].

The mechanical dispersion of air into small bubbles is assisted by the presence of frothers. The generic relation between bubble size and frother concentration is shown in Figure 21.10 [43]. In zone 1, bubble size decreases with frother concentration because increasing amounts of surfactant prevent bubble coalescence. In zone 2, coalescence is completely suppressed and the dispersion of the bubbles is determined by the hydrodynamic conditions in the cell. The boundary between these two zones determines the critical coalescence concentration, CCC.

The CCC values appear to be independent of the type of machine used for dispersing the air and the operating conditions (air flow rate and impeller speed) in the cell [43]. If frother concentration is normalized with CCC, then bubble size versus relative frother concentration curves for different frothers collapse on a master curve [42]. Therefore, the CCC can be used to classify frothers, for example, by using DFI (defined later by Equation 21.16) versus CCC plot—Figure 21.11.

In this space, frothers positioned in the upper left corner (high DFI and low CCC) are very powerful ones; the ones situated in the bottom right corner (low DFI and high CCC) are weaker and more selective [42]. Revealingly, no correlation between the above diagram and the equilibrium surface tension of aqueous frothers was found. This emphasizes that the dynamic properties of surfactant layers are more relevant than equilibrium ones in highly dynamic processes such as frothing and froth flotation.

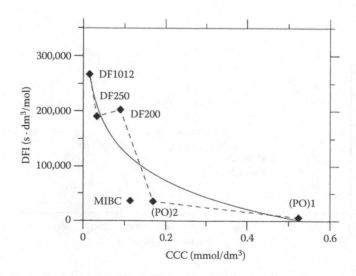

FIGURE 21.11 DFI versus CCC diagram for flotation frothers [44].

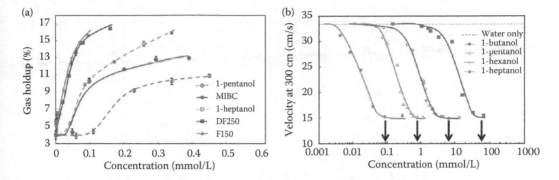

FIGURE 21.12 Effect of frother concentration on: (a) Gas holdup (volume fraction of air); (b) Single bubble velocity (dB = 1.45 mm, h = 300 cm) [45].

Frothers diminish the bubble size and increase the gas holdup (volume fraction of air) in flotation. Finch and coworkers [45] were able to correlate gas holdup and the rise velocity of single bubbles—Figure 21.12. While all frothers tested decreased the bubble size to about 0.8 mm above their respective CCC, gas holdup was different—Figure 21.12a. The rise of individual bubbles was followed in a 4 m tall cell and their velocity is shown in Figure 21.12b as a function of frother concentration.

The concentration to reach minimum velocity, concentration of minimum velocity (CMV), at a given height in the cell (arrows in Figure 21.12b) ranks the frothers in exactly the same way as gas holdup. It also clearly illustrates the role of the alkyl chain length in a homologous series of alcohols—the CMV decreases by approximately an order of magnitude for each additional methylene group.

Malysa and coworkers [46,47] argued that Bikerman's definition (Equation 21.13) should be refined and introduce the retention time, τ, defined as (V_G is total gas volume in both foam and solution, and Q is volumetric gas flow rate):

$$\tau = \frac{V_G}{Q} \tag{21.15}$$

This retention time is a foamability parameter independent of the gas flow rate and the dimensions of the measuring device. Its physical meaning is the average time needed for a unit gas volume to pass through the system. The retention time τ was used to define the dynamic foamability index, DFI, as [47]

$$\text{DFI} = \left(\frac{\partial \tau}{\partial c}\right)_{c=0} \tag{21.16}$$

While τ is an exponentially saturating function of the bulk frother concentration, the DFI is a parameter independent of surfactant concentration and characteristic of the given frother. At low frother concentrations (typical for froth flotation), both τ and DFI give the same results.

Malysa and coworkers measured the surface elasticity for aqueous solutions of two homologous series of surfactants: fatty acids and linear alcohols [46]. The modulus of elasticity was termed effective as it was determined at surface coverages lower than the equilibrium ones—a situation representative of the conditions in dynamic foams. The combined results in Figure 21.13 demonstrate that retention time, τ, correlates very well with the effective modulus of elasticity, E_{eff}.

This comparison shows that the stability of wet foams correlates with the effective elasticity. The close agreement between effective surface elasticity and retention time (i.e., foamability) as functions of both concentration and alkyl chain length (number of carbon atoms) shows that effective elasticity determines the formation and stability of wet foams [46]. It confirms again that dynamic foamability does not necessarily follow equilibrium surface activity. Finally, in wet foams liquid films, rupture

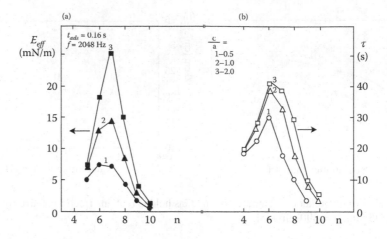

FIGURE 21.13 Properties of aqueous n-alcohols at different relative surfactant concentrations: (a) Effective modulus of elasticity; (b) Retention time [46].

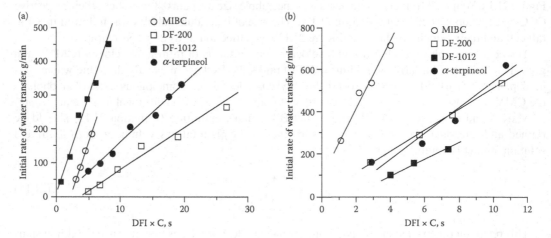

FIGURE 21.14 Initial rates of water transfer from the aqueous solution to the froth layer as a function of frother concentration, C, for four different frothers: (a) in a foam (in the absence of particles); (b) in the presence of hydrophobic coal particles [48].

at relatively large thicknesses and the role of surface forces is insignificant [46]. It should be noted that the above considerations apply to surfactant solutions in the absence of any particles.

In froth flotation, it is mainly the presence of a frother that gives rise to a foam column on top of the pulp layer. This foam contains water and it is well established that the amount of water transferred to the froth layer is proportional to the concentration of the frother in the pulp. The initial rate of water transfer is a linear function of the frother concentration—Figure 21.14, and the slope is characteristic of the frother used [48].

However, the performance of the frothers in a foam (Figure 21.14a), that is, in the absence of particles, is different from their performance when hydrophobic particles are added, that is, in a froth (Figure 21.14b). Furthermore, the changes were different depending on particle hydrophobicity [48].

21.6.3 Modifiers (Regulators)

A number of additional modifiers (or regulators) are also used in froth flotation. Their importance is proportional to the mineralogical complexity of the system [36]. In many cases, they improve the selectivity of collector adsorption and are relatively more important in nonsulfide flotation. From a molecular point of view, modifiers can be grouped into (i) inorganics, (ii) small organics and oligomers, and (iii) polymeric molecules. From a practical point of view, modifiers can affect almost any parameter in the system and sometimes can have multiple effects. Activators are added to the flotation system because they improve collector attachment on the surface of the value particles. For instance, sphalerite flotation with a xanthate as a collector is not possible. However, after the addition of copper sulfate (activator), copper ions exchange with zinc ions on the surface of the mineral and flotation with xanthate becomes possible [7]. The sulfidization of metal-containing ores in an oxygen-rich environment promotes the efficiency of sulfide collectors. In some cases, the deposition of a metal hydroxide on the surface of mineral particles paves the way to efficiently adsorb collector.

Depressants are employed to decrease the hydrophobicity of unwanted particles which are hydrophobic enough to attach to the bubbles under given conditions [7]. Soluble polymers are used to adsorb on the surface of certain particles and thus block the possible adsorption of a collector. Oxidizing agents are used to prevent collector adsorption by oxidizing the mineral surface.

Dispersants and flocculants are also used at different stages of the ore treatment [7].

21.7   ROLE OF pH AND Eh

Various modifiers are used to control pH and Eh in the flotation cell. The pH of the pulp affects the ionization of groups on the surface of the solid particles and also the ionic equilibria involving flotation reagents. In the example shown in Figure 21.15a, the recovery of mica with dodecylamine as a collector peaks at pH 8. It was shown that the maximum is not due to a minimum in electrostatic interactions as both water-air and mica-solution interfaces were positively charged and the Hamaker constant was negative. The reason was that the hydrophobic monolayer of dodecylamine was most compact at pH 8. At lower pH, the layer is less tightly packed; at higher pH, multilayer adsorption occurs. In both cases, the hydrophobicity of the particle surface decreases and so does flotation recovery [49]. Common pH regulators include lime, soda ash, caustic, sulfuric acid, and hydrochloric acid. The manipulation and control of pH in the pulp is not only necessary to optimize the flotation of the mineral particles, but it is also relevant in the recycling and disposal of used water.

The pulp potential (Eh) is an important electrochemical parameter that has been correlated with flotation outcomes [51]. Eh is a redox potential, but its meaning when several redox couples are

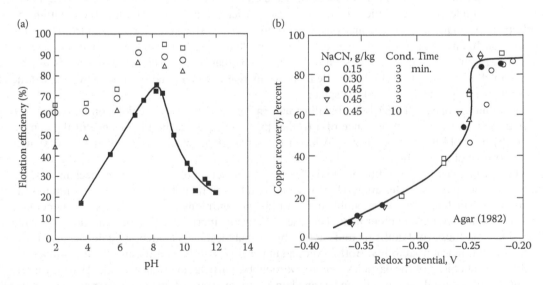

FIGURE 21.15   Flotation recovery: (a) Mica with a 1 mM dodecylamine collector as a function of pH [49]; (b) Copper from a chalcopyrite-pentalandite ore as a function of the redox potential Eh [50].

present (known as mixed potential) is less clear. In practice, it is measured as the potential of a platinum wire inserted in the pulp with respect to a saturated calomel electrode or silver/silver chloride electrode (the results are usually reported on the standard hydrogen electrode scale). The role of Eh is crucial whenever a redox reaction forms the basis of the interaction between reagent and ore particles. In the example shown in Figure 21.15b, the recovery drops at lower potentials because the collector is unable to oxidize under reducing conditions [50]. Adjusting and monitoring the pulp potential is a requirement in such cases.

21.7.1 SELECTION OF REAGENTS

It has been recognized that the current understanding of the processes in flotation is insufficient to formulate a general approach to the selection of reagents [36]. The number of reagents and variable flotation conditions further make the selection difficult. The reagent choice is largely guided by the chemistry of the ore to be processed. It is also influenced by the effects on the whole circuit rather than just the flotation cells. This becomes even more pressing in view of stricter environmental requirements as well as tighter access to fresh water. Reagents' price and availability can have an otherwise unexpected influence on the selection. Finally, the choice is often guided by local knowledge, that is, previous experience or seasonal changes.

21.8 FLOTATION KINETICS

Froth flotation is a dynamic process and given its complexity and the large tonnage of ore processed in flotation plants, the kinetic aspects are of high interest. Particle size plays a key role as it affects the dynamics of particle movement which is highly uneven in the flotation cell. Froth flotation is viewed as a first-order kinetic process and various models have been proposed. The state and stability of the froth are critical for understanding particle movement and particle entrainment, in particular. Froth management is necessary for maximizing the grade of the concentrate.

21.8.1 PARTICLE SIZE

The role of particle size is critical for successful and efficient flotation. It is not possible to develop a meaningful understanding of the mineralogy of the particles and their behavior in the flotation cell without a systematic size-by-size analysis. The optimal particle size range for efficient froth flotation is between 10 and 150 μm [14]. Two examples of particle recovery as a function of particle size are shown in Figure 21.16. It is clearly seen in Figure 21.16a that the recovery of chalcocite is highest at intermediate particle sizes [52]. Particle recovery increases when flotation time increases, but the optimal size range remains about the same.

The influence of particle size is further illustrated in Figure 21.16b with the flotation of chalcopyrite without collector but in the presence of a frother [53]. In this case, the importance of the contact angle (hydrophobicity) is emphasized. A larger contact angle improves recovery and also expands the size range of floatability.

The optimal size range of floatability is limited by the action of two different mechanisms. On one hand, very large particles are too heavy to be lifted by small bubbles (typically 1–2 mm in size) or prone to detachment from the bubble under turbulent conditions. These effects set an upper size limit to floatability. On the other hand, because of their low inertia, very small particles follow the flow lines and thus avoid collision and capture by the bubbles. This tendency establishes the lower size limit to efficient floatability. Both lower and upper limits of floatability are strongly influenced by the hydrophobicity of the particles—more hydrophobic particles are easier to float. Hydrodynamic conditions inside the flotation cell can be tuned up for the recovery of fine or coarse particles but not both, as the requirements are conflicting.

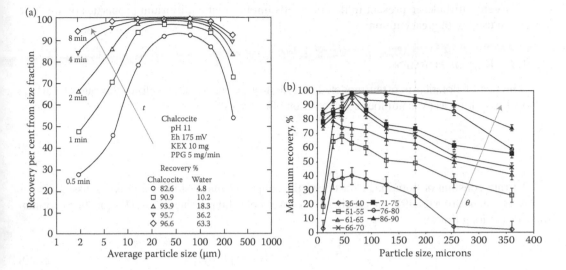

FIGURE 21.16 Recovery versus particle size: (a) Influence of time of flotation, t, (values are listed within the plot) for chalcocite floated with potassium ethyl xanthate [52]; (b) Influence of hydrophobicity (i.e., water contact angle, θ, (values are listed within the plot) for chalcopyrite floated without collector [53].

It is tempting to float larger particles as the communition of ores uses very large amounts of energy. However, the benefits of doing this may be lost as the grade of the ores processed is steadily declining over the years. Often the valuable mineral is dispersed inside the gangue matrix, therefore, finer grinding is necessary to liberate the target material.

21.8.2 TURBULENT MIXING

Turbulent mixing inside the flotation cell is necessary to maintain the particles suspended, but also to maximize the number of bubble-particle contacts. Turbulence is highest near the impeller and quickly decreases away from it [54,55]. A quiescent zone is also needed as intense mixing is likely to destroy bubble-particle aggregates as well as affect the froth layer. In both cases, the effect on flotation recovery is detrimental. The optimal distribution of turbulence within a flotation cell is a central problem in the scale up of froth flotation.

Viscous dissipation is the process of converting mechanical energy into heat. The rate of viscous dissipation, D, is related to the viscous force, f_μ (given by Newton's law), and the fluid velocity, u:

$$D = f_\mu u \cong \mu \left(\frac{\partial u}{\partial z}\right)^2 V \tag{21.17}$$

It is proportional to the liquid's dynamic viscosity, μ, the squared velocity gradient normal to the direction of flow, and the volume of the system, V. In engineering, the more relevant quantity is dissipation rate per unit mass, ε:

$$\varepsilon = \frac{D}{m} = \frac{\mu}{\rho} \left(\frac{\partial u}{\partial z}\right)^2 \tag{21.18}$$

The energy dissipation in flotation has major implications in both modeling the kinetics of flotation as well as designing flotation cells. Recent experimentation with the hydrodynamics in a pilot-scale flotation cell showed that the size of the turbulent zone is of key relevance to ore recovery [56].

The intense turbulence present in flotation cells implies that equilibrium properties of any type should be used with great caution.

21.8.3 RATE OF FLOTATION

The kinetics of flotation is often modeled as a first-order reaction, therefore, the rate of particle removal is written as (N is number concentration of particles):

$$-\frac{dN}{dt} = kN \tag{21.19}$$

The rate constant of flotation, k, has been the subject of many studies and is discussed further below. By integrating (21.19), one obtains the recovery of solid particles, R, as (R_{max} is the maximum recovery at long times):

$$R = \frac{N_0 - N}{N_0} = R_{max}[1 - \exp(-kt)] \tag{21.20}$$

Equation 21.20 is often obeyed and almost universally used in practice. The modeling of the flotation rate constant, k, is based on the understanding of flotation as a sequence of elementary processes.

21.9　ELEMENTARY STEPS OF FLOTATION

The overall froth flotation process is seen as proceeding through several consecutive steps:

1. A particle collides with a bubble;
2. Under favorable conditions, the particle adheres to the bubble;
3. The bubble-particle aggregate is carried upward in the flotation cell and, if stable enough, survives the turbulent environment;
4. The particle-bubble aggregate joins the froth layer and the value particles remain there until removed mechanically.

The mechanisms of flotation have been discussed for a long time and the main elementary steps are widely recognized—Figure 21.17. A collision between the particle and a bubble is a necessary condition. Not every approach leads to collision and collision efficiency, E_1, is defined as the probability of a collision occurring. When the particle comes close to the bubble, two extreme scenarios are possible. If the particle is moving toward the apex region of the bubble (small collision angle), it may bounce back from the liquid-air interface and the time of contact (impact time) is very short—only a few milliseconds. At larger collision angles, the particle slides around the bubble and the contact time (sliding time) is longer—about 20–30 ms. Intermediate behavior is also possible.

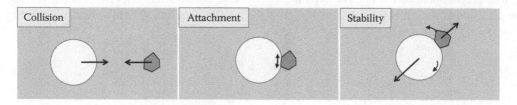

FIGURE 21.17 Elementary steps of flotation in the pulp zone: (i) Collison between a particle and a bubble; (ii) Attachment of the particle to the bubble; (iii) Stability of the bubble-particle aggregate during upward travel to the froth zone.

Once in close proximity, the particle and the bubble must attach, that is, form a stable aggregate. Physically the thin aqueous film between them must drain and rupture so that a three phase contact is established. The contact perimeter (closed contact line) then must expand quickly enough to provide the capillary force needed for attachment. The time required for these processes to occur is known as induction time. Normally film thinning is the slowest process and it defines the induction time. Not every bubble-particle encounter leads to capture and capture efficiency, E_2, is defined as the probability of a successful attachment. It has been documented that multiple particles can attach to a single bubble, but this complication is usually disregarded.

After the bubble-particle aggregate is formed it floats through the pulp layer in order to reach the forth layer. The trajectory is nonlinear in the continuously stirred flotation cell and turbulent conditions challenge the stability of the aggregate. In fact, stability efficiency, E_3, must be considered to account for the probability of detachment $1 - E_3$.

The three elementary steps of flotation sketched in Figure 21.17 are seen as consecutive and independent, and thus the overall collection (or capture) efficiency, E_C, of bubble-particle encounters in the pulp zone is written as:

$$E_C = E_1 E_2 E_3 \tag{21.21}$$

It has been recognized that the hydrodynamic conditions within the flotation cell are crucially important and must be included in a realistic description of flotation. In the Wark flotation model (named after the Ian Wark Research Institute, University of South Australia, Adelaide), this has been incorporated by expressing the rate constant of flotation, k, as follows [57,58]:

$$k = MTE_C \tag{21.22}$$

The mechanical term, M, is related to gas flow rate, Q, bubble diameter, d_B, and volume of the flotation cell, V_{cell}:

$$M = 2.39 \frac{Q}{d_B V_{cell}} \tag{21.23}$$

The turbulence term, T, is related to viscous dissipation per unit mass, ε, bubble diameter, d_B, bubble velocity, u_B, kinematic viscosity, ν, air density, ρ_A, and liquid density, ρ_W:

$$T = \frac{0.33 \varepsilon^{4/9} d_B^{7/9}}{u_B \nu^{1/3}} \left(\frac{\rho_W - \rho_A}{\rho_W} \right)^{2/3} \tag{21.24}$$

The collision efficiency, E_1, is largely determined by bulk hydrodynamics and is calculated assuming potential flow around the bubble and a mobile liquid-gas interface (the upward movement of the bubble means that fluid flow pushes any adsorbed surfactant molecules to the back of the bubble; if the apex region is severely depleted it becomes mobile). The attachment efficiency, E_2, is dominated by interfacial behavior and is calculated using the contact angle of the particle and equating the sliding time to the induction time. The stability efficiency is affected by both hydrodynamics and interfacial behavior and is calculated using the Schulze model [59], which compares the attachment force with the mechanical stress experienced by the bubble-particle aggregate.

Two examples of applying this model are illustrated in Figure 21.18. The Wark model (shown with a solid line) works well for laboratory flotation experiments, as demonstrated with the flotation of hydrophobized quartz particles in a Rushton cell [57]—Figure 21.18a. The model has also been tested with data about the performance of a large operating flotation plant [60]. The correspondence was very good for all 12 cells in the rougher section of the plant—Figure 21.18b. Thus, a model

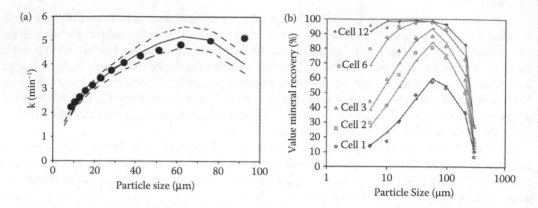

FIGURE 21.18 Application of the Wark flotation model—experimental and calculated flotation rate constants: (a) Flotation of hydrophobic quartz particles in a Rushton flotation cell (2.25 L, 0.1 M KNO3, pH 5.6, flotation time 8 min) [57]; (b) Recovery in the rougher flotation section of an operating plant [60].

based on the physical properties of the system was able to predict the performance of a real plant and explicitly assess the importance of particle size, particle contact angle, and cell hydrodynamics.

The JKMRC model (named after the Julius Kruttschnitt Mineral Research Centre, University of Queensland, Australia) is also based on Equation 21.19 but takes a more down to earth approach. Based on extensive empirical evidence, the flotation rate constant, k, is expressed as [61]

$$k = PS_B R_F \tag{21.25}$$

In this equation, P is particle floatability (dimensionless), S_B is bubble surface area flux, and R_F is froth recovery. Equation 21.25 is considered to be independent of cell size and operating parameters—an example with two largely different flotation cells is presented in Figure 21.19.

The JKMRC model has been incorporated into a software package named JKSimFloat [63]. In a recent version (6.1 Plus), the model works as follows. A value of S_B is obtained through empirical

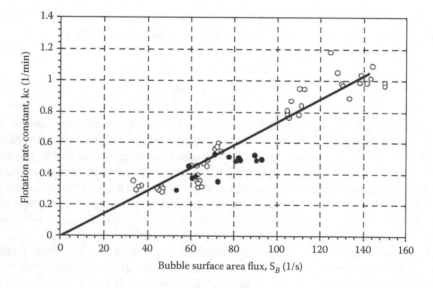

FIGURE 21.19 Flotation rate constant, k, as a function of the bubble surface area flux, S_B, for two different flotation columns (○ 50 L laboratory cell; ●: 1.1 m³ pilot cell) [62].

measurements. For each class of particles, froth recovery, F_R, is calculated as (ϕ is fraction of nondraining particles)

$$R_F = 1 - (1 - \phi)P_D \tag{21.26}$$

In this expression, the probability of detachment, P_D, is linked to the froth residence time, τ_F, via an empirical parameter, β:

$$P_D = 1 - \exp(-\beta \tau_F) \tag{21.27}$$

Entrainment, *Ent*, defined as the ratio of entrained particles and entrained water ($Ent = R_P/R_W$) is also determined empirically. The iteration cycle starts with a setting of 10% feed reaching the concentrate. The flow rate of water is determined from the entrainment model and the cell residence time, τ, is calculated as (ε_G is volume fraction of the gas, V is pulp volume, Q_T is flow rate of the tailings)

$$\tau = \frac{(1 - \varepsilon_G)V}{Q_T} \tag{21.28}$$

Then the rate constant for each particle class is calculated via Equation 21.25 and used to obtain the recovery for that class as [61]

$$R = \frac{k\tau}{1 + k\tau} \tag{21.29}$$

The process is iterated until the total mass of particles converges. The software has been successfully applied in a number of flotation operations and offers significant flexibility in modeling flotation circuits.

21.10 STABILITY OF FOAMS AND FROTHS

The properties and behavior of the froth (foam with solid particles) are relevant in forth flotation, but it is useful to consider a foam and then the effects due to the presence of small particles. A foam is composed of thin foam films (liquid lamellae) delineated by liquid channels (Plateau borders) which in turn meet in vertices (nodes). In a wet foam (high volume fraction of water), the foam films are relatively thick and the bubbles are approximately spherical. In a dry foam (low volume fraction of water), the liquid films are much thinner and the bubbles are polyhedral.

Foams obtained through mechanical agitation of a pure liquid are unstable. The reason is that the additional surface area created in the mixing process carries an excess surface free energy and the system, when left on its own, quickly separates into two homogeneous phases. Foams are stabilized by the addition of surfactants. Surfactants adsorb at the interface and decrease surface tension, but even more importantly, provide mechanical resilience under dynamic conditions (Section 12.4.1). Small particles of intermediate hydrophobicity are also effective stabilizers because in many respects they act like surfactants (Section 21.4.2). However, the strong attachment of particles at liquid interfaces (Equation 21.8) does not immediately explain the stability of particle-stabilized foams and emulsions [64]. In the special case of very high surface coverages (beyond a full monolayer), the stratified surfaces mechanically compress against each other and thus prevent the coalescence of bubbles (or emulsion droplets), but the fluid nature of the interfaces has been compromised. At lower surface coverages, a capillary pressure arising from the presence of particles at the surfaces of the liquid film interfaces must be considered [65]. This capillary pressure, due to the menisci formed between the particles, stabilizes the film for contact angles below 90°. The effect is quantified by the maximum capillary pressure, which is larger for smaller particles and larger contact angle hysteresis [64,65].

The lifetime of a foam is very different depending on the amount of liquid (volume fraction ϕ). In wet foams (ϕ is about 0.3), the thickness of the foam films is large and the main mechanism of destruction is liquid drainage. In dry foams (ϕ below 0.07), most of the liquid has already drained and the stability of the system is provided mainly by the stability of the thin liquid films.

The drainage of liquid from a wet foam is modeled as proceeding through the Plateau borders only—Figure 21.6a. The area of the cross section of the plateau border, A, is approximated as the intersection of three cylindrical surfaces (Figure 21.20b). The radius of curvature of the liquid-air interface, R, is constant along the plateau border (cylindrical symmetry).

The flow is modeled as steady, unidirectional, and sufficiently slow, using a Darcy-type equation. The hydrostatic pressure, P_H ($= -\rho g x$, ρ is liquid density, g is gravitational acceleration, and x is vertical length of the Plateau border), and capillary pressure, P_C ($= \gamma/R = C\gamma/\sqrt{A}$; γ is surface tension and A is area of the cross section of the Plateau border) are taken into account. Finally, using the material balance of the liquid, the foam drainage equation is obtained as (μ is viscosity of the liquid)

$$\frac{\partial A}{\partial t} + \frac{1}{50\mu}\frac{\partial}{\partial x}\left(\rho g A^2 - \frac{C\gamma}{2}A^{1/2}\frac{\partial A}{\partial x}\right) = 0 \tag{21.30}$$

There is a broad resemblance between viscous flow in foams and in porous bodies. However, in foams (and froths), the dimensions of the liquid channels vary with liquid content and this makes Equation 21.30 more complicated [66]. The drainage equation has been examined for a variety of dynamic and physicochemical conditions [67].

In dry foams, the drainage is practically over and the stability of the foam is largely governed by the stability of the foam films. These films are typically thin and surface forces play a decisive role. An extensive treatment of foam films and foams is given in Exerowa and Kruglyakov's monograph [28]. The froths encountered in flotation are relatively wet and the stability mechanisms for dry foams are of secondary relevance.

The froth layer in a flotation cell is a foam layer that contains many particles. Ideally, only valuable particles should reach the froth via bubble-particle aggregates, but the situation is more complicated (Section 4.7). Based on static considerations, hydrophilic particles in the froth are likely to avoid the interfaces and follow the liquid in the Plateau borders and nodes. On the contrary, hydrophobic particles should be attached to the interfaces, mostly the foam films because of their large surface

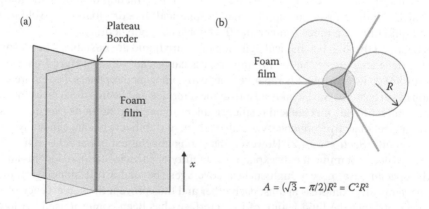

FIGURE 21.20 Foam drainage: (a) Plateau borders are the main channels draining the liquid; (b) Cross section of a Plateau border and its area A (R is radius of curvature of the liquid-air interface).

area. However, the forth is not very stable and intense bubble bursting and coalescence happen within the froth layer. Thus, some hydrophobic particles will be present in the Plateau borders and nodes of the froth. All unattached particles, regardless of their hydrophobicity, move with the liquid but the overall motion is complicated. The liquid-gas interfaces in foams are typically immobile because of the presence of surfactants and the Marangoni effect. A velocity profile develops in the Plateau borders and particles will spread accordingly (Plateau border dispersion). When reaching a node, particles may take different directions (geometric dispersion). Finally, particles are heavy and have a tendency to settle under gravity.

In an elegant model study, Horn and coworkers [68] studied the drainage of single vertical foam films stabilized with polypropylene glycol and loaded with submicron size silica particles. They varied the hydrophobicity of the particles and tracked the evolution of the film profile during drainage using interferometry. Their results are summarized in Figure 21.21a. In the absence of particles, the films were metastable and drained with seconds, that is, quickly. The presence of hydrophilic particles (small water contact angle) did not make much of a difference. This is natural as hydrophilic particles would not attach to the water-air interface and will drain as a suspension with the core of the film. Hydrophilic particles could slow down film drainage through the increase of bulk viscosity at high particle concentrations. This physical mechanism is exploited in flotation practice when wash water is added to the froth layer. The water is carefully added on top of or inside the froth so that liquid drainage is maintained so that hydrophilic (gangue) particles are removed as completely as possible.

Moderately hydrophobic particles present in the liquid film provide the strongest effect—they slow down film thinning significantly and at the same time do not destabilize the foam films—Figure 21.21a. This regime corresponds to the optimal flotation scenario where value particles are strongly adhering to bubbles and after reaching the froth layer remain attached to the liquid-air interfaces until removed as a concentrate.

Finally, highly hydrophobic particles also reduce the rate of film drainage, but simultaneously reduce film stability—Figure 21.21a. The dissimilar action of moderately and strongly hydrophobic particles was explained as due to the different particle rearrangement at the surface [68]. A monolayer of particles is able to sustain a continuous liquid film only at intermediate hydrophobicity—Figure 21.21b. Highly hydrophobic particles break the film by bridging the two film surfaces as sketched in Figure 21.21b. This explanation is in line with the concept of a maximum capillary pressure stabilizing a foam film with particles at the interfaces [64]. It also fits with the general observation that small hydrophobic particles exhibit anti-foaming properties.

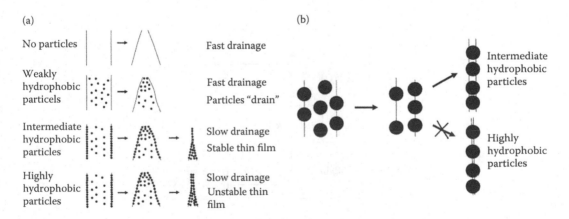

FIGURE 21.21 Drainage mechanisms in foams and froths [68]: (a) Influence of particle hydrophobicity; (b) Difference between the behavior of moderately hydrophobic and highly hydrophobic particles.

21.11   FROTH RECOVERY AND AIR RECOVERY

For a while, flotation models have focused mainly on the processes in the pulp and the role of the froth has been somewhat neglected. This has been recognized as a weakness in modeling flotation and efforts are being made to redress the situation. Froth recovery is the fraction of particles in the concentrate with respect to the particles entering the froth layer. The complete flowchart of froth recovery is given in Figure 21.22 and includes froth recovery, R_F, which is missing from the traditional scheme in Figure 21.17. The overall flotation recovery, R, is determined by recovery in the collection zone, R_C, and recovery in the froth zone, R_F. Because these two steps are consecutive and independent, $R = R_C R_F$.

Froth recovery is affected by the type and concentration of frother used, gas flow rate, speed and design of the impeller, and the depth of the froth layer.

The importance of air recovery stems from the energetic coalescence in the froth—under typical conditions each bubble passing through the froth coalesces about 15 times, yet particle-laden films are quite stable [66]. Air recovery is the ratio of air in the surviving bubbles and air entering the pulp. It rarely exceeds 0.50 in cleaner flotation cells and is lower (0.05–0.20) in rougher and scavenger cells. Air recovery reflects the balance between particles recovered from froth films and Plateau borders and affects the boundary conditions for modeling froth motion [66].

The role of the froth in froth flotation is to keep the valuable particles there until they are removed. The bubble-particle aggregates arriving at the bottom of the froth carry mineral particles but also some gangue particles. These unwanted particles are hydrophilic and the main reason they reached the froth layer is entrainment, that is, small particles are carried out by the upward flow of water. These particles enter the liquid phase of the froth but, given sufficient time, will drain down to the pulp. In practice, this process is encouraged by spraying or injecting additional water in the froth as discussed in Section 4.5.

In total, the mineral recovery includes particle floated by genuine flotation (flotation due to selective attachment to bubbles), locked particles, and entrainment.

21.11.1   PARTICLE ENTRAINMENT

Particle entrainment into the froth layer is unavoidable and must be controlled in order to keep the grade of the concentrate high. The recovery of gangue particles is directly related to the recovery of water as illustrated in Figure 21.23a. The recovery of silica increases linearly with water recovery, except at very low values of water recovery. The slope of the dependence decreases with increasing particle size. The extrapolated intercept is zero for fine particles but increasingly negative for larger ones [69].

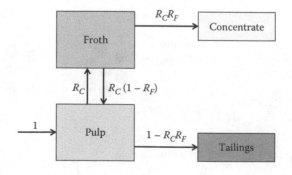

FIGURE 21.22   The complete description of froth flotation includes the process in both the pulp and froth layers. The total recovery is $R_C R_F$ (R_F is recovery in the collection zone and R_F is the recovery in the froth zone).

Neethling and Cilliers [71] developed a model which is based on the foam drainage equation and furthermore accounts for the presence and motion of particles, including the effects of dispersion and settling. The increasing water recovery was simulated by increasing the gas flow rate and a simulation result is given in Figure 21.23b. The model reproduces all key features found in the experiments. In a further development, Neethling and Cilliers [70] considered the role of particle size. They emphasized that in the froth the bubble size increases sharply from about 1 mm at the pulp–froth boundary to about 1 cm at the top of the froth column. This is coupled with a very substantial difference in water speed. The sedimentation of particles is strongly dependent on their size. Their simulation reproduced well the experimentally established dependence—Figure 21.24.

Various mechanisms for particle entrainment have been suggested. In a graphic summary, Konopacka and Drzymala [74] depict six different scenarios for mechanical flotation (a descriptive if less common term for particle entrainment)—Figure 21.25: (a) hydrophilic particles entrained by bubble-particle aggregates without attaching to any interface; (b) hydrophilic particles entrapped by flocculating bubble-particle aggregates; (c) hydrophilic particles lifted by such flocks; (d) slimes (very fine hydrophilic particles) attached to the surface of the value particles; (e) both hydrophilic and hydrophobic particles carried up in the wakes of bubbles; (f) particles strongly attracted to the

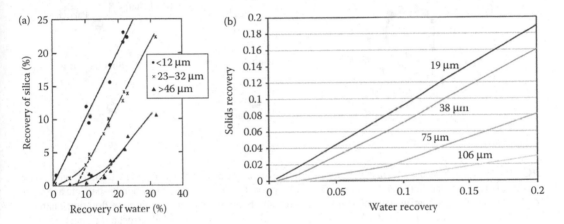

FIGURE 21.23 Particle recovery as a function of water recovery: (a) Experimental results for silica [69]; (b) Simulated results [70].

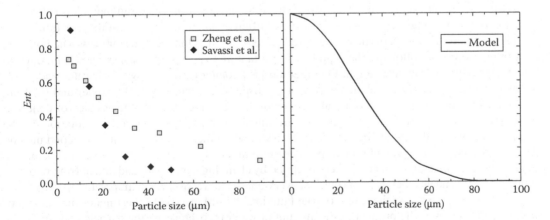

FIGURE 21.24 Entrainment factor, *Ent*, as a function of particle size: (left) Experimental results from Zheng et al. [72] and Savassi et al. [73]; (right) Model calculations by Neethling and Cilliers. (Adapted from Neethling, S.J., Cilliers, J.J. *Int J Miner Proc* 2009; 93(2): 141–148 [70].)

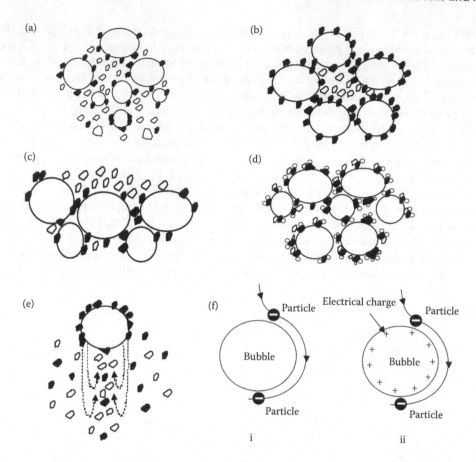

FIGURE 21.25 Possible mechanisms of gangue particle entrainment: (a) Upward transfer with Plateau borders, (b) Entrapment between bubble-particle aggregates; (c) Upward transfer by bubble-particle aggregates, (d) Slime coating on floating particles; (e) Following the bubble wake; (f) Contactless flotation enabled by attractive surface forces ((i) van der Waals forces only; (ii) van der Waals and electrostatic forces).

bubble without breaking the thin liquid film (such contactless flotation is possible if strong attractive surface forces are present in the system).

While this comprehensive list includes all suggested mechanisms of entrainment, it is a daunting task to obtain experimental evidence allowing to differentiate between some of these options.

The distinction between particles recovered by selective attachment to the bubbles (true flotation) and by entrainment (following the liquid) is important and various approaches have been devised for its experimental quantification. One method [75] compares the recovery of water and solids in two separate flotation experiments: (i) from a suspension containing only frother and (ii) from a suspension containing both frother and collector. It is assumed that, in the absence of collector, all particles in the froth have arrived via entrainment. The recovery due to true flotation can be estimated from the difference in particle recovery between the two experiments. A second method [76] was based on the assumption that particle entrainment does not take place in a dry foam. In a series of experiments, the water recovery is varied by changing froth depth and rate of froth removal. The linear dependence between particle recovery and water recovery is extrapolated to zero water recovery in order to assess the extent of true flotation. A third method [77] estimates the amount of entrained particles as the product of the amount of water recovered and the particle concentration, which is assumed to be identical to that in the pulp. George et al. [78] have suggested a fourth method, which compares the flotation of silica and alumina particles in the presence of a cationic surfactant (collector) and a frother. The silica particles, being negatively charged, adsorb the cationic collector and float by true flotation. The alumina particles are charged positively and reach the forth by entrainment only. The results obtained using the above four methods were highly consistent.

21.12   PRACTICAL ASPECTS OF FROTH IN INDUSTRIAL FLOTATION CELLS

The apparent simplicity of froth flotation (Figure 21.1) is deceptive. The process is complex both mechanically and chemically, as it includes zones of very different turbulence and a number of reagents as well as modifiers. The number of variables is thus large and the access to industrial flotation for chemical and physical probing rather limited. In froth flotation, both pulp and froth zones play important roles and they should not be decoupled from one another in the holistic modeling of the process (Figure 21.22). In some cases, tracing measurements have shown that most upgrading occurs in the froth. Frothing should be taken into account in the design of column flotation cells [79]. The processes occurring during froth flotation are notoriously difficult to scale up which explains the efforts to use model columns of various sizes and the general skepticism in considering laboratory based results. At the same time, experimental and theoretical research has revealed most of the key processes on length scales spanning from molecular size to large industrial flotation cells and columns. However, the declining grade of ores and the shift to ore bodies such as oxides will bring new challenges. A deeper understanding of the froth layer is likely to play an inherent role in the overall recovery strategy. Elucidating further the details of froth flotation as well as modeling the whole process will require truly multidisciplinary collaborative efforts.

ACKNOWLEDGMENTS

The authors warmly acknowledge the knowledge and support received from colleagues and friends, with special thanks to Grant Small for reading the draft manuscript.

REFERENCES

1. Fuerstenau, M.C., Jameson, G., Yoon, R.-H. *Froth Flotation: A Century of Innovation.* SMME: Littleton, Colorado, 2007.
2. Laskowski, J. *Coal Flotation and Fine Coal Utilization*, 1st ed. Elsevier: Amsterdam, 2001. p xiv, 368 p.
3. Rao, F., Liu, Q. Froth treatment in Athabasca oil sands bitumen recovery process: A review. *Energy & Fuels* 2013; 27(12): 7199–7207.
4. Wang, L.K., Shammas, N.K., Selke, W.A., Aulenbach, D.B. *Flotation Technology Volume 12*, 1st ed. Humana Press: Totowa, NJ, 2010.
5. Bajpai, P. *Recycling and Deinking of Recovered Paper*, 1st ed. Elsevier: Amsterdam, 2014. p vii, 304 p.
6. Froth Flotation Process. www.powerhousemuseum.com.
7. Klimpel, R.R. Froth flotation. In: Meyers, R.A. (Ed.), *Encyclopedia of Physical Science and Technology*, 3rd ed. Academic Press: Cambridge, MA, 2003, pp. 219–234.
8. Hanumantha Rao, K., Subramanian, S. Bioflotation and bioflocculation of relevance to minerals bioprocessing. In: Donati, E.R., Sand, W., (Eds.), *Microbial Processing of Metal Sulfides*. Springer: Berlin, 2007, pp. 267–286.
9. Fuerstenau, M.C., Han, K.N. *Principles of Mineral Processing.* SME: Littleton, Colorado, 2003.
10. Ramachandra Rao, S. *Surface Chemistry of Froth Flotation*, 2nd ed. Springer: New York, 2004, Vol. 1.
11. Alexander, D., Manlapig, E., Bradshaw, D., Harbort, G. Froth flotation. In: Wills, B.A., Napier-Munn, T.J., (Eds.), *Wills' Mineral Processing Technology*, 7th ed. Butterworth-Heinemann: Oxford, UK, 2006, pp. 267–352.
12. Marvros, P., Matis, K.A. *Innovations in Flotation Technology.* Springer: Berlin, 1992.
13. Miller, J.D., Parekh, B.K. *Advances in Flotation Technology.* Society for Mining, Metallurgy, and Exploration: Littleton, CO, 1999. p xxxiii, 463 p.
14. Ives, K.J. *The Scientific Basis of Flotation.* Martinus Nijhoff: The Hague, 1984. p vi, 429 p.
15. Nguyen, A.V., Schulze, H.J. *Colloidal Science of Flotation.* Marcel Dekker: New York, 2004. p xix, 850 p.
16. Bulatovic, S.M. *Handbook of Flotation Reagents.* Elsevier: Amsterdam, 2007, p. 446.

17. de Gennes, P.-G., Brochard-Wyart, F., Quéré, D. *Capillarity and Wetting Phenomena: Drops, Bubbles, Pearls, Waves*. Springer: New York, 2004.

18. Neumann, A.W., David, R., Zuo, Y. *Applied Surface Thermodynamics*, 2nd ed. Taylor & Francis: Boca Raton, 2011. p xxii, 743 p.

19. Neumann, A.W., Good, R.J. Techniques of measuring contact angles. In: Good, R.J., Stromberg, R.R. (Eds.), *Surface and Colloid Science: Volume 11: Experimental Methods*. Springer: Boston, MA, 1979, pp. 31–91.

20. Adamson, A.W., Gast, A.P. *Physical Chemistry of Surfaces*, 6th ed. Wiley: New York, 1997. p xxi, 784 p.

21. Pugh, R.J. Foaming, foam films, antifoaming and defoaming. *Adv Colloid Interface Sci* 1996; 64: 67–142.

22. Edwards, D.A., Brenner, H., Wasan, D.T. *Interfacial Transport Processes and Rheology*. Butterworth-Heinemann: Boston, 1991. p xvii, 558 p.

23. Miller, R., Liggieri, L. *Interfacial Rheology*. Brill: Leiden, 2009.

24. Holmberg, K., Jönsson, B., Kronberg, B., Lindman, B. *Surfactants and Polymers in Aqueous Solution*. 2nd ed. John Wiley & Sons: Chichester, 2002.

25. Dukhin, S.S., Kretzschmar, G., Miller, R. *Dynamics of Adsorption at Liquid Interfaces: Theory, Experiment, Application*. Elsevier: Amsterdam, 1995. p xviii, 581 p.

26. Binks, B.P., Horozov, T. *Colloidal Particles at Liquid Interfaces*. Cambridge University Press: Cambridge, 2006.

27. Israelachvili, J.N. *Intermolecular and Surface Forces [Online]*, 3rd ed. Academic Press: Burlington, MA, 2011, p. 674.

28. Exerowa, D.R., Kruglyakov, P.M. *Foam and Foam Films: Theory, Experiment, Application*. Elsevier: Amsterdam, 1998. p xxii, 773 p.

29. Butt, H.-J.R., Kappl, M. *Surface and Interfacial Forces*. Wiley-VCH: Weinheim, 2010. p xv, 421 p.

30. Elimelech, M., Jia, X., Gregory, J., Williams, R. *Particle Deposition & Aggregation Measurement, Modelling and Simulation*. Elsevier: Burlington, 1998, p. 459.

31. Christenson, H.K., Claesson, P.M. Direct measurements of the force between hydrophobic surfaces in water. *Adv Colloid Interface Sci* 2001; 91(3): 391–436.

32. Attard, P. Nanobubbles and the hydrophobic attraction. *Adv Colloid Interface Sci* 2003; 104(1–3): 75–91.

33. Mishchuk, N.A. The model of hydrophobic attraction in the framework of classical DLVO forces. *Adv Colloid Interface Sci* 2011; 168(1–2): 149–166.

34. Tabor, R.F., Grieser, F., Dagastine, R.R., Chan, D.Y.C. The hydrophobic force: Measurements and methods. *Phys Chem Chem Phys* 2014; 16(34): 18065–18075.

35. Chan, D.Y.C., Klaseboer, E., Manica, R. Film drainage and coalescence between deformable drops and bubbles. *Soft Mat* 2011; 7(6): 2235–2264.

36. Chander, S., Nagaraj, D.R. Flotation reagents. In: *Techniques and Theory of Separation by Flotation*. Elsevier, 2007, pp. 1–14.

37. Fairthorne, G., Fornasiero, D., Ralston, J. Interaction of thionocarbamate and thiourea collectors with sulphide minerals: A flotation and adsorption study. *Int J Miner Proc* 1997; 50(4): 227–242.

38. Bleier, A., Goddard, E.D., Kulkarni, R.D. Adsorption and critical flotation conditions. *J Colloid Interface Sci* 1977; 59(3): 490–504.

39. Fuerstenau, D.W., Healy, T.W., Somasundaran, P. The role of the hydrocarbon chain of alkyl collectors in flotation. *Trans Soc Min Eng AIME* 1964; 229: 321–24.

40. Bikerman, J. *Foams*, vol. 10. Springer-Verlag: Berlin, 1973.

41. Barbian, N., Ventura-Medina, E., Cilliers, J.J. Dynamic froth stability in froth flotation. *Miner Eng* 2003; 16(11): 1111–1116.

42. Laskowski, J.S. Testing flotation frothers. *Physicochem Probl Miner Proc* 2004; 38: 13–22.

43. Grau, R.A., Laskowski, J.S., Heiskanen, K. Effect of frothers on bubble size. *Int J Miner Proc* 2005; 76(4): 225–233.

44. Laskowski, J.S., Tlhone, T., Williams, P., Ding, K. Fundamental properties of the polyoxypropylene alkyl ether flotation frothers. *Int J Miner Proc* 2003; 72(1–4): 289–299.

45. Tan, Y.H., Rafiei, A.A., Elmahdy, A., Finch, J.A. Bubble size, gas holdup and bubble velocity profile of some alcohols and commercial frothers. *Int J Miner Proc* 2013; 119: 1–5.

46. Małysa, K. Wet foams: Formation, properties and mechanism of stability. *Adv Colloid Interface Sci* 1992; 40: 37–83.

47. Czarnecki, J., Małysa, K., Pomianowski, A. Dynamic frothability index. *J Colloid Interface Sci* 1982; 86(2): 570–572.

48. Melo, F., Laskowski, J.S. Effect of frothers and solid particles on the rate of water transfer to froth. *Int J Miner Proc* 2007; 84(1–4): 33–40.

49. Pugh, R.J., Rutland, M.W., Manev, E., Claesson, P.M. Dodecylamine collector—pH effect on mica flotation and correlation with thin aqueous foam film and surface force measurements. *Int J Miner Proc* 1996; 46(3): 245–262.

50. Chander, S. A brief review of pulp potentials in sulfide flotation. *Int J Miner Proc* 2003; 72(1–4): 141–150.

51. Ross, V.E., Van Deventer, J.S.J. The interactive effects of the sulphite ion, pH, and dissolved oxygen on the flotation of chalcopyrite and galena from Black Mountain ore. *J South African Inst Min Metall* 1985; 85(1): 13–21.

52. Trahar, W.J. A rational interpretation of the role of particle size in flotation. *Int J Miner Proc* 1981; 8(4): 289–327.

53. Muganda, S., Zanin, M., Grano, S.R. Influence of particle size and contact angle on the flotation of chalcopyrite in a laboratory batch flotation cell. *Int J Miner Proc* 2011; 98(3–4): 150–162.

54. Newell, R., Grano, S. Hydrodynamics and scale up in Rushton turbine flotation cells: Part 1—Cell hydrodynamics. *Int J Miner Proc* 2007; 81(4): 224–236.

55. Amini, E., Bradshaw, D.J., Xie, W. Influence of flotation cell hydrodynamics on the flotation kinetics and scale up, Part 1: Hydrodynamic parameter measurements and ore property determination. *Miner Eng* 2016; 99: 40–51.

56. Tabosa, E., Runge, K., Holtham, P. The effect of cell hydrodynamics on flotation performance. *Int J Miner Proc* 2016; 156: 99–107.

57. Pyke, B., Fornasiero, D., Ralston, J. Bubble particle heterocoagulation under turbulent conditions. *J Colloid Interface Sci* 2003; 265(1): 141–151.

58. Ralston, J., Miller, J.D., Rubio, J. *Flotation and Flocculation*. Snap Printing: Adelaide, 2003.

59. Schulze, H.-J., Stöckelhuber, W. Flotation as a heterocoagulation process. In: Stechemesser, H., Dobiás, B. (Eds.), *Coagulation and Flocculation*, 2nd ed. Taylor & Francis: Boca Raton, FL, 2005, pp. 455-.

60. Ralston, J., Fornasiero, D., Grano, S., Duan, J., Akroyd, T. Reducing uncertainty in mineral flotation—Flotation rate constant prediction for particles in an operating plant ore. *Int J Miner Proc* 2007; 84(1–4): 89–98.

61. Harris, M.C., Runge, K.C., Whiten, W.J., Morrison, R.D. Jksimfloat as a practical tool for flotation process design and optimization. In: *Mineral Processing Plant Design, Practice and Control*, 2002, SME: Vancouver, pp. 461–478.

62. Hernández, H., Gómez, C.O., Finch, J.A. Gas dispersion and de-inking in a flotation column. *Minerals Engineering* 2003; 16(8): 739–744.

63. Schwarz, S., Richardson, J.M. Modeling and simulation of mineral processing circuits using Jksimmet and Jksimfloat. In: *SME Annual Meeting*, 2013, SME: Denver, CO, pp. 13–120.

64. Kaptay, G. On the equation of the maximum capillary pressure induced by solid particles to stabilize emulsions and foams and on the emulsion stability diagrams. *Colloids Surfaces A* 2006; 282–283: 387–401.

65. Denkov, N.D., Ivanov, I.B., Kralchevsky, P.A., Wasan, D.T. A possible mechanism of stabilization of emulsions by solid particles. *J Colloid Interface Sci* 1992; 150(2): 589–593.

66. Cilliers, J. The froth in column flotation. In: Fuerstenau, M.C., Jameson, G., Yoon, R.-H. (Eds.), *Froth Flotation: A Century of Innovation*. SME: Littleton, CO, 2007, pp. 708–729.

67. Kruglyakov, P.M., Karakashev, S.I., Nguyen, A.V., Vilkova, N.G. Foam drainage. *Curr Opin Colloid Interface Sci* 2008; 13(3): 163–170.

68. Tan, S.N., Yang, Y., Horn, R.G. Thinning of a vertical free-draining aqueous film incorporating colloidal particles. *Langmuir* 2010; 26(1): 63–73.

69. Engelbrecht, J.A., Woodburn, E.T. The effects of froth height aeration rate and gas precipitation on flotation. *J South Afr Inst Min Metall* 1975; 76(3): 125–132.

70. Neethling, S.J., Cilliers, J.J. The entrainment factor in froth flotation: Model for particle size and other operating parameter effects. *Int J Miner Proc* 2009; 93(2): 141–148.

71. Neethling, S.J., Cilliers, J.J. The entrainment of gangue into a flotation froth. *Int J Miner Proc* 2002; 64(2–3): 123–134.

72. Zheng, X., Johnson, N.W., Franzidis, J.P. Modelling of entrainment in industrial flotation cells: Water recovery and degree of entrainment. *Miner Eng* 2006; 19(11): 1191–1203.

73. Savassi, O.N., Alexander, D.J., Franzidis, J.P., Manlapig, E.V. An empirical model for entrainment in industrial flotation plants. *Miner Eng* 1998; 11(3): 243–256.

74. Konopacka, Z., Drzymala, J. Types of particles recovery—Water recovery entrainment plots useful in flotation research. *Adsorption* 2010; 16(4): 313–320.

75. Trahar, W.J., Warren, L.J. The flotability of very fine particles—A review. *Int J Miner Proc* 1976; 3(2): 103–131.
76. Warren, L.J. Determination of the contributions of true flotation and entrainment in batch flotation tests. *Int J Miner Proc* 1985; 14(1): 33–44.
77. Ross, V.E. Flotation and entrainment of particles during batch flotation tests. *Miner Eng* 1990; 3(3): 245–256.
78. George, P., Nguyen, A.V., Jameson, G.J. Assessment of true flotation and entrainment in the flotation of submicron particles by fine bubbles. *Miner Eng* 2004; 17(7–8): 847–853.
79. Flint, I.M., Burstein, M.A. Froth processes and the design of column flotation cells. In *Techniques and Theory of Separation by Flotation*. Academic Press: Cambridge, MA, 2000, pp. 1521–1527.

22 Fire Fighting Foams

Dirk Blunk

CONTENTS

22.1 INTRODUCTION

Fire fighting foam is an indispensable operating resource for firefighters. It possesses many advantages during the emergency action. The primary effect of the foam blanket is that it separates the burning material from the surrounding atmosphere and in this way interrupts the burning process by cutting off the supply of the reaction area (i.e., the flames) from inflammable gases or vapor. Furthermore, due to its insulating and scattering properties, it protects the material underneath the foam blanket against heat in the form of thermal radiation or convection. Closely related is a cooling effect which arises from the large heat capacity and heat of evaporation of the water content of the foam [1–5].

Apart from these direct extinguishing properties, firefighting foams may also be used to cover combustible liquids in order to prevent the formation of explosive gas compositions in the atmosphere, to squeeze dangerous gases out of rooms or compartments or to protect objects from heat irradiation.

Last but not least, firefighting foam can easily be prepared at the site of deployment simply from water, air, and a foaming agent (Figure 22.1).

22.2 TECHNICAL TERMS IN RELATION TO FIRE FIGHTING FOAMS

22.2.1 ALCOHOL RESISTANCE

Common firefighting foams and also aqueous film forming foams (AFFF) are almost immediately destroyed if they come into contact with alcohols or other polar (water miscible) liquids. Amongst

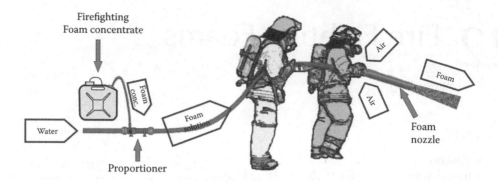

FIGURE 22.1 Typical setup at an incident area with foam application.

others, the family of burnable polar liquids comprises compounds such as conventional alcohol (ethanol), methanol, propanol, isopropanol, and acetone. The reason for the destructive effect of such polar liquids is mainly their miscibility with water which leads to an immediate loss of water from the foam lamella in contact with the polar liquid and conversely the diffusion of the polar liquid into the foam solution in the contact region. This destroys the sensitive physical conditions for the stability of the foam. It is self-evident that also a water film, the characteristic feature of AFFF, cannot form on top of a water miscible liquid. Hence, these and other liquids cannot be extinguished with conventional types of firefighting foam.

In order to achieve an effective extinguishing result, the exchange between the two media (foam and burnable polar liquid) should be minimized. This is achieved by certain polymeric components in specialized so called alcohol resistant firefighting foams which are consequently marked with the extension "AR" (e.g., AFFF-AR). The polymeric components most often are xanthan or rhamsan gum. These components form a highly viscous gel in the contact region between the foam layer and the polar burnable liquid. This gel layer separates and in this way protects the foam from the polar liquid.

22.2.2 APPLICATION RATE

The application rate is defined as the volume of finished foam which is applied to a unit area in a certain time. The resulting unit is, for example, $L/(m^2*min)$. This is a very significant number since almost any foam which is applied with an overwhelming huge application rate can extinguish any fire. Such unreal application rates may occur, for example, in small sized fire extinguishing tests and result in an incorrect picture of the performance of a fire fighting foam. In reality, the application rate is determined and limited by the size and accessibility of the incident area, by the available firefighting technology, the supply chain, and so on.

Below a critical application rate, a fire cannot be extinguished since the foam brakes down faster than it can be applied due to the heat of the fire and the chemical or physical interaction of the foam with the burning material [6–10]. Critical (minimum) application rates have been determined by extinguishing experiments for emergency situations, types of foam, and types of the burning liquid.

22.2.3 DESTRUCTION RATE

The destruction rate is defined as the amount of firefighting foam which is destroyed immediately when it hits the fire. Apart from the heat of the fire, this can also be caused by chemical reactions or by physical processes (e.g., mixing) of the foam solution with the burning material or other liquids/gases from the surroundings. This is especially important with burnable polar liquids, for example, alcohols (cf. alcohol resistance).

Generally, the calculation of the required foam quantities is based on a destruction rate of 50%, but in some cases, for example, in very hot mineral oil fires, the destruction rate can be as high as 70%.

22.2.4 DRAINAGE RATE

The drainage rate defines the amount of aqueous solution which drains from the foam in a given time. It is often given as the half-life or 50% drain time, or quarter life or 25% drain time, respectively, which means that 50% (or 25%, respectively) of the foam solution drains out at the specified time. The drainage rate is most easily measured with a large measuring jar standing on a balance. The complete amount of foam solution is given by its weight, while the volume of the solution draining out of the foam can easily be measured with the scaling of the measuring jar over time.

Generally, the water content of the foam is very important with respect to its cooling ability and its durability, as well as its resistance and protective abilities against heat [11,12]. Since a lower drainage rate means a longer lasting, higher amount of water within the foam, this is equivalent to a higher durability, heat resistance, and so on.

Especially for AFFF, the drainage rate also has an important influence on the fluidity and mobility of the foam spreading across the fuel surface, as the draining solution supports the water film between the foam and the fuel and creates the flow at the rim of the foam blanket.

22.2.5 EXPANSION RATE

The expansion rate is defined as the ratio of the volumes of the final foam to the foam solution. The expansion rate mainly depends on the type and concentration of the surfactants in the foam solution, that is, the chemical composition of the foam concentrate and the mode of foam generation (e.g., type of foam nozzle, foam generator etc.). Further factors influence the expansion rate as well, for example, the flow rate of the foam solution at the Venturi nozzle or the amount and type of salts or minerals in the water. Calcium or magnesium ions in hard water or the salt load in sea water can especially influence the expansion rate. Impurities in the ingested air and its temperature may also influence the expansion rate. Examples for this are sooty particles or gaseous components in the aspired air, for example, acidic flue gas components such as CO_2, HCl, HCN, NO_x or SO_x.

22.2.6 FOAM CONCENTRATE

For the formation of firefighting foam, water is mixed in the correct proportion with a foam concentrate at the place of incidence. Such foam concentrates are liquids which are supplied from a manufacturer and contain a blend of surfactants and chemical auxiliaries.

22.2.7 FOAM SOLUTION

The ready-to-use solution of a foam concentrate in water is called the foam solution.

22.2.8 MIXING RATE, DILUTION RATE OR PROPORTIONING RATE

The mixing rate describes the proportion of the foam concentrate in the foam solution. The most common mixing rates are 6%, 3% or 1%. For example, a concentrate with a required mixing rate of 3% needs 3 liters of foam concentrate added to 97 liters of water (assuming ideal miscibility). It has to be considered that under operating conditions and with respect to the available technique the accuracy of the proportioning might not be ideal.

At present, concentrates with 6% admixture rate are becoming rarer and 3% concentrates are the most often used type. Currently, concentrates with 1% admixture or even less are becoming more and more important. Their main advantage is that the emergency forces can produce a larger amount of finished foam out of a given amount of foam concentrate. This is not only an aspect of safety but has also technical consequences. For example, emergency vehicles can be designed with a lower load capacity or space needed for the foam concentrate. In marine use cases volume and weight also play important roles so that concentrates with a 1% mixing rate have indispensable advantages. However, technically, their application is often limited by the type of the foam proportioner which is available, since the exact setting of the low proportioning rate may cause problems especially with passive foam inductors.

22.3   CHEMICAL COMPONENTS IN FIRE FIGHTING FOAMS

Generally, firefighting foam concentrates consist of a blend of different surfactants, mainly designed for an optimal foam formation and durability. A selection of commonly used surfactants is shown in Figure 22.2.

To achieve special properties, further additives are used. For example, in AFFF polyfluorinated surfactants (fluorotelomers) are added as film forming components in addition to the foaming surfactants. Alcohol resistance is introduced by the addition of xanthan or rhamsan and the water drainage properties of the foam are tuned by the additives 1,2-ethandiol (glycol), triethylene glycol or 2-(2-butoxyethoxy)ethanol (n-butyldiglycol). Furthermore, anti-corrosive agents, preservatives, and other chemical auxiliaries may be included.

22.4   CLASSES OF FIRE

Fire can be divided into several classes with respect to the burning medium. Although this classification is almost internationally homogeneous, some national deviations exist.

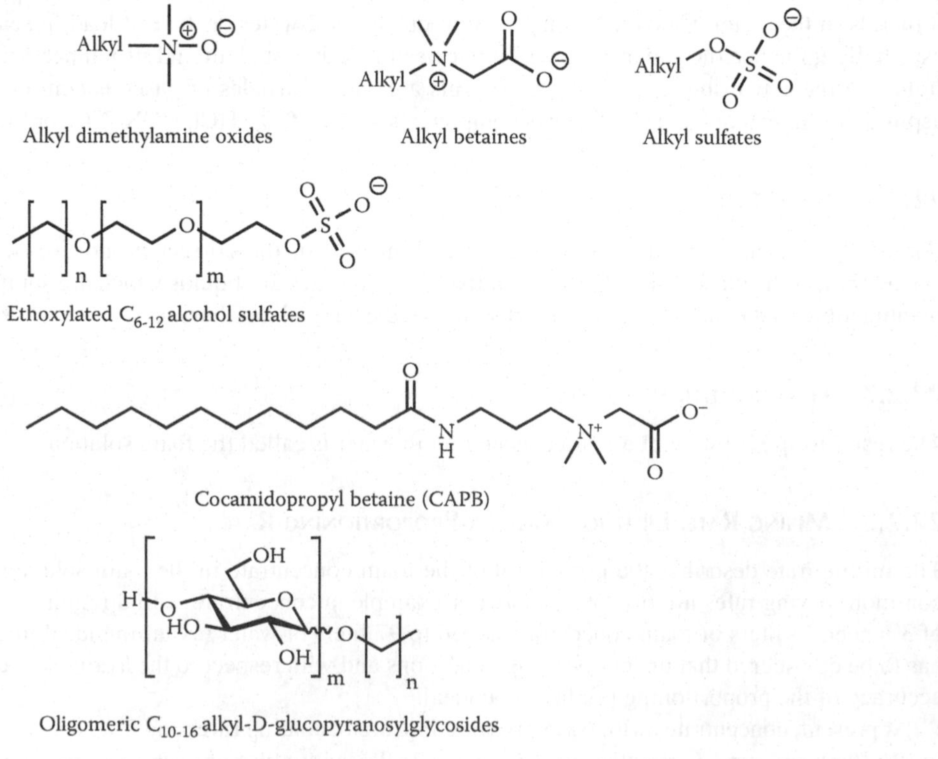

Alkyl dimethylamine oxides

Alkyl betaines

Alkyl sulfates

Ethoxylated C_{6-12} alcohol sulfates

Cocamidopropyl betaine (CAPB)

Oligomeric C_{10-16} alkyl-D-glucopyranosylglycosides

FIGURE 22.2   Typical surfactants used as foaming components in fire fighting foam concentrates.

Class A fires are burning solids, for example paper, wood, coal, cloth, fabric or plastic. Often such materials are also glowing or smoldering which makes the process more complicated since the extinguishing medium cannot easily infiltrate the inside of the glowing material. However, in these cases a low concentrated aqueous solution of a fire fighting foam concentrate may also help; this is typically called "wet water."

Class B fires are burning liquids, for example, gasoline, petrol oil, paint or alcohols. Sometimes plastic which liquifies in the heat is also counted as this type of fire, although such fires are typically better described as class A.

In general, it is not the liquid itself which burns, but the vapor or gaseous phase above the liquid.

In Europe and Australia, class C fires are defined to be burning gases, such as methane, propane, butane, illuminating gas or likewise. In the U.S., class C fires are so called electrical fires, which are in principle class A fires where conducting extinguishing agents like water should not be used.

Class D fires are burning metals. Most often this relates to magnesium, titanium or aluminum, but also lithium, sodium or potassium.

Class F fires are burning cooking oils or fats or grease. In the U.S., this type of fire is referred to as class K.

In general, fire extinguishing foams are not suitable for fires of the classes C or D, that is, burning gases or metals. In the other cases, fire extinguishing foams can be used, though specialized foam types may be appropriate.

22.5 FOAM TYPES

The different types of fire extinguishing foam are mainly characterized by their volume to liquid ratio. The expansion ratio is defined as the volume of the foam divided by the volume of the application solution from which it is formed (cf. section 2).

22.5.1 HIGH EXPANSION FOAM

High expansion foam has a high volume with a low liquid content. It is, therefore, a very dry foam. Its expansion ratio is $>=200$ [13], typically between 750 and 1000. Due to its low water content, high expansion foam has only a very limited cooling ability and its resistance to heat and flowability is also rather weak.

Since this type of foam is very light weight, it cannot be thrown as it is blown away by light wind. Thus, its applicability outdoors is very dependent on the weather conditions, in particular air movements.

High expansion foam is mainly used in closed rooms, for example, aircraft hangars, stockrooms, mines, ships or in large compartments as filling foam and its effect is mainly achieved by the substitution of the gas phase. Its characteristic impact is the immediate elimination of flames, but due to its low water content, embers remain and have to be treated afterward with spray jet or wetting agent. The latter is water which contains surfactants at a lower dosage than for foam formation to reduce the surface tension and enhance its wetting abilities.

For the formation of high expansion foam, a special generator is needed which mainly consists of a large fan pushing an air stream into a flow channel with a foaming sieve onto which the application solution is sprayed. Thus, in contrast to the other foam types, high expansion foam is not thrown with a nozzle but is blown onto or into the object.

22.5.2 MEDIUM EXPANSION FOAM

Medium expansion foam has an expansion ratio of $>=20$ and <200 [14]. It is mainly used for the suppression of liquid fires and for the flooding of rooms. Furthermore, it is applied on spilled burnable liquids to prevent ignition of the puddle or to prevent the formation of explosive vapor compositions.

Due to its low content of aqueous solution, this type of foam does not develop a significant cooling effect. The effectivity of medium expansion foam is higher than with low expansion foam (see below) because larger areas can be covered with respect to the same amount of foam solution due to the higher expansion rate. The foam blanket of medium expansion foam is usually thicker than that of low expansion foam and the bubbles are not so mobile with respect to each other due to the lower water content and the thinner lamella between the bubbles. As a consequence, medium expansion foam blankets have a lower permeability than conventional low expansion foam blankets. The latter does not hold for so called aqueous film forming variants of low expansion foams (AFFF) which form an additional water film as a highly effective barrier between the foam blanket and the burnable liquid.

The typical throw distance of medium expansion foam is 6–10 m with conventional foam nozzles or about 25 m with special foam nozzles. The latter, however, have a lower expansion rate, that is, they produce a slightly heavier, that is, "wetter" medium expansion foam.

22.5.3 LOW EXPANSION FOAM

Low expansion foam has an expansion ratio of <20 [15].

Due to its high water content, low expansion foam is heavy and can be thrown over large distances. Throwing distances of more than 25–35 m are possible with hand held foam nozzles and large airport fire engines are able to throw low expansion foam to distances of about 50–90 m. The high throwing distances are one of the main advantages of low expansion foam. With this, it is possible to remain a safe distance from the place of incidence, for example, in the case of tank fires, leaking technical or infrastructure installations or burning airplanes. Furthermore, the high water content of low expansion foam is responsible for its comparatively good cooling abilities. As another result of its high water content, it can successfully be applied to solid fires (fire class A). Furthermore, low expansion foam is used for covering endangered areas with a foam blanket which cools down the coated facilities and protects them from heat radiation.

The required amount of foam concentrate necessary for the formation of a certain volume of low expansion foam is comparably high in comparison with high or medium expansion foam. This means that the logistics and provision of enough foam concentrate is of particular importance in the case of an emergency before starting the foam attack.

22.6   AQUEOUS FILM FORMING FOAM (AFFF)

For various reasons, for example, accidents, lightning strikes, attacks or combat operations, large quantities of liquids, for example, fuels, solvents or chemicals, can be set on fire. Such liquid fires are especially dangerous due to their large areas, enhanced mobility, and unpredictability as well as for their enormous heat irradiation. Furthermore, such fires often occur in difficult locations, for example, industrial plants, refineries, tank storage facilities, airports, marine engine rooms or oil rigs. Such fires are difficult to control with conventional extinguishing agents, including "normal" extinguishing foams, since these extinguishing agents can only act where they are applied and often the application of the extinguishing agent is severely hampered by the enormous heat irradiation or complex local situation. Another argument may be the necessity to knock down the fire in a very short time, for example, in the case of a burning passenger airplane.

Conventional fire extinguishing foams are often not adequately capable of extinguishing burning organic liquids (class B fires) under such conditions and hence, high performance extinguishing foams are needed for the suppression of large class B fires. In such cases, AFFF are used [16]. An AFFF is designed in such a way that a thin water film is formed between the burning liquid and the foam blanket, cf. Figure 22.3.

This water film formed by the aqueous solution of AFFF gives this special type of firefighting foam decisive advantages in the case of class B fires over conventional extinguishing foams. The thin water film cools the burning goods and constitutes an additional and effective vapor barrier so that

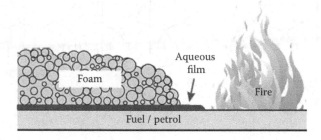

FIGURE 22.3 Formation and situation of the name giving water film between the foam layer and the combustible liquid (e.g., fuel). The water film is typically 0.01–0.03 mm thick and continuously supported by drainage water from the foam layer.

the combustible liquid can no longer pass into the gas phase or cross the foam layer. Consequently, the fire is knocked down much faster [17]. Furthermore, the water film makes it much easier for the extinguishing foam to disseminate on the burning goods. This makes the foam blanket more agile and the extinguishing effect much faster. Additionally, the good spreading ability of the foam blanket allows the extinguishing of liquid fires, where the foam cannot be applied onto the whole burning surface. This plays a crucial role, for example, if the area is too large, comparted with barriers or in angular and labyrinthine compartments such as engine rooms or industrial facilities.

The water film also promotes the self-healing properties of the foam blanket. This is a remarkable safety feature since the foam blanket closes holes autonomously when it gets damaged by falling debris or by firefighters or victims who may have to walk through it. Furthermore, the water film remarkably prolongs the burn back time. The burn back time is a measure with respect to the reignition of the combustible liquid at hot surfaces.

Up until now, in the year 2017, without exception, all commercially available AFFF agents contain polyfluorinated surfactants. Initially, perfluorooctanesulfonic acid (PFOS, cf. Figure 22.4) was used in AFFF, but this chemical is now already banned in numerous countries, mainly as a result of its listing as a persistent organic pollutant in the Stockholm convention [18]. Furthermore, perfluorooctanoic acid (PFOA, cf. Figure 22.4) and its salts were used, but are also out of use in firefighting foam concentrates and are about to be banned completely as they are identified as substances of very high concern (SVHCs) due to their reprotoxicant, persistent, bioaccumulating, and toxic properties [19]. In general, such per- or polyfluorinated chemicals (PFCs) are harmful and environmentally hazardous [20].

Due to the legal restrictions, PFOS, PFOA, and its salts have been replaced in AFFF by fluorotelomer surfactants as shown in Figure 22.5. The problem with these substitutes is that they do not solve the environmentally toxicological core problem, as in nature, these compounds are again degraded to persistent products as perfluoroalkyl carboxylates (PFCAs).

One possible environmentally advantageous alternative to the hazardous fluorinated surfactants are siloxane surfactants. Newly developed surfactants of this type are capable of exhibiting similar positive properties with respect to fire extinguishing properties as the established polyfluorinated surfactants, but are not persistent or bioaccumulative [3–5].

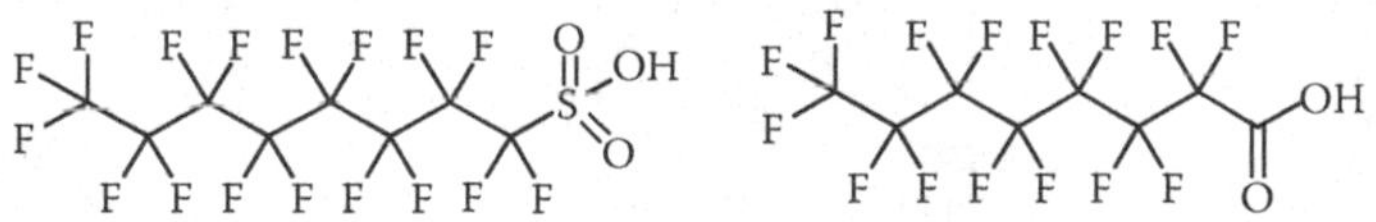

FIGURE 22.4 Chemical structures of perfluorooctanesulfonic acid (PFOS) and perfluorooctanoic acid (PFOA).

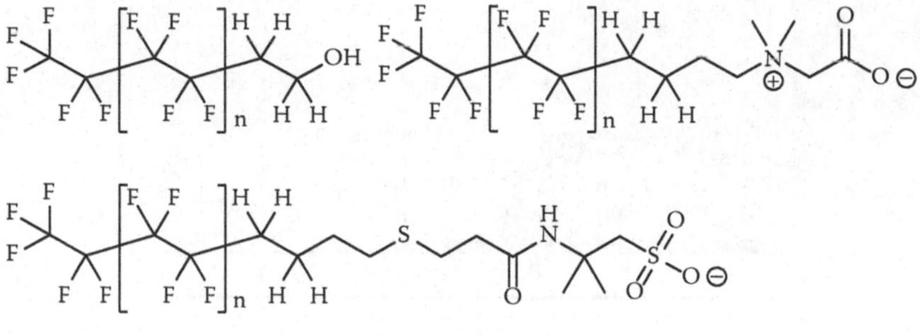

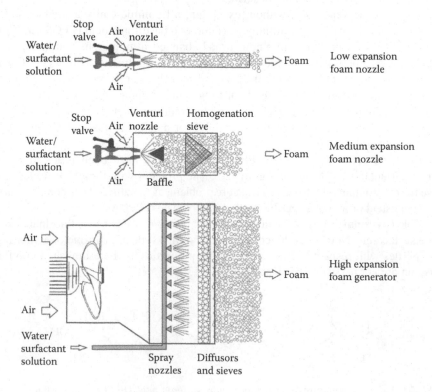

FIGURE 22.5 Fluorotelomer alcohol and two examples of fluorotelomer surfactants used in present day AFFF.

22.7 FOAMING TECHNIQUES

Common construction principles for low, medium, and high expansion foam devices are shown in Figure 22.6.

Originally and still most often today, firefighting foam is produced at the place of incidence in a foaming device which works with the principle of a Venturi nozzle. The foam solution is delivered with sufficient pressure into the Venturi nozzle, sucking in surrounding air and intimately mixing both phases to form the foam. Afterward, the foam can be further processed, for example, by pressing it through a homogenation sieve.

High expansion foam is produced slightly differently. Here, a powerful fan creates a stream of air in which the foam solution is sprayed through several spray nozzles onto diffusors and sieves,

FIGURE 22.6 Sketches of the principal construction of foam generation devices for different types of foam.

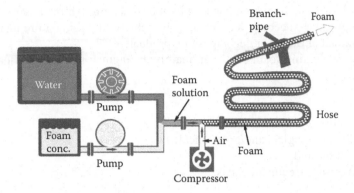

FIGURE 22.7   General setup for the generation and application of compressed air foam (CAF).

creating a very light, high expansion foam. Usually this device is placed outside the incident area and the foam is directed to the fire place through an air pipe.

Although compressed air foam (CAF) was originally developed in the 1930s–1940s, it again constitutes a currently emerging trend. In CAF, the foam solution is not foamed by ambient air at the nozzle but with pressurized air, for example, in the fire engine. The expansion rate can be tuned from about 7 (wet, heavy, low expansion foam) to 30 (dry, light foam). The general setup of a CAF system is shown in Figure 22.7. The advantage of CAF is mainly its large possible throwing distance and height. Furthermore, since the air needed for the foaming is not aspired at the place of the fire, CAF can be used to tackle interior fires with an inside attack. However, the systematic investigation and experience with modern CAF are only at their infancy. Possible dangers are the application of foam which is too dry, looking stable and dense and thus providing a false sense of safety. Underneath such dry foams the fire may propagate unperceived. Another problem is the technical complexity of the CAF generation and the extensive incorporation of highly pressurized gas/water mixtures, that is, with high energy content, in large pipe systems. This creates new dangers for the action forces.

22.8   ENVIRONMENTAL ASPECTS

Firefighting foam concentrates are a blend of different chemicals, mainly surfactants, designed to decrease the surface tension of water and to build foams. As such it is clear that these products should not be released into the environment [21–24]. Yet they are intended to be used in emergency cases and the place of incidence is not by choice. The situation at such emergency places can be manifold: from incidences in free nature to industrial plants, from an intact infrastructure to devastated places, everything is possible and occurs in real life. Very often firefighters have to make difficult decisions based on an assessment of the associated risks but with incomplete data and form an overview within a few seconds.

Aside from the immediate threat from the fire, environmental and toxicological aspects of the fire extinguishing agents are within the focus of public interest.

One problem is that the foam solution can be pressed into the fresh water supply infrastructure by the powerful pumps of the fire engines. This is, for example, possible if the pump system and water distribution is connected in a false way at a confusing scene of emergency. Technical appliances such as back-pressure valves should avoid such incidents, but they nevertheless may occur.

The other and bigger problem is the forge water. The forge water is loaded with the surfactants from the foam concentrate and the decomposition products of these chemicals and, of course, with the combustion products of the burning materials. The latter can contain highly toxic compounds as, for example, polychlorinated or polybrominated dibenzodioxines or –furanes (PCDD/F), polycyclic aromatic hydrocarbons (PAH) or dioxine-like polychlorinates biphenyls. However, this partition of the noxious contaminants cannot be influenced by the choice firefighting foam, except for the fact that a faster extinguishing success – and this may be achieved by using a fire fighting foam – also minimizes the amount of the other decomposition products.

Regarding the firefighting foams, one can mainly distinguish five different problematic parts. These are the

- Acute toxicity,
- Chronic toxicity,
- Decomposition (biological and/or chemical),
- Bioaccumulation, and
- Behavior in sewage treatment plants,

of the various chemical components of the foam concentrate.

A special problem with firefighting foams is that the place and the circumstances of their applications are not always under complete control. In many countries, fire workers are required or obliged to collect the forge water if they apply firefighting foam and to release it only after an analysis under controlled conditions or in further dilution to the sewer system. In other cases, the forge water must be treated, for example, with absorption filters or is brought to special waste incineration plants.

In cases where it is not possible to collect the forge water, it may drain into the soil reaching the ground water or other natural water like rivers, streams, and lakes. Incidents on ships and oil rigs or in harbors may lead to a contamination of the sea.

All of this makes clear that firefighting foam concentrates should at best contain only nontoxic, degradable surfactants as far as possible. Much literature has been published already concerning the toxicity of surfactants in general [21–37].

Furthermore, due to the intended special use of firefighting foam in emergency cases, one has to assess the acute toxic impacts differently from the chronic toxicity, persistency, and bioaccumulation. Under bad circumstances, especially if the forge water cannot be collected properly for any reason, the acute toxicity of firefighting foams may lead to a localized die-off of aquatic populations. This harm must be balanced against the risks for fire victims, action forces, and property damage. A timely limited environmental damage due to acute toxicity might be accepted under certain circumstances, while over decades, lasting burdens due to chronic toxicity, persistency, and bioaccumulation are probably unacceptable. However, it is also important to remember that the products resulting from the burning material can also be vast, dangerous, acutely and/or chronically toxic, persistent, and bioaccumulative. In such cases, a fast and save extinguishing success due to the employment of firefighting foam might be less harmful in the consequence and minimize the total damage. On the other hand, the surfactants in the firefighting foam solution may also solubilize harmfully but normally insoluble fire by-products and make them effective in this way.

Unfortunately, incidents as described above are sometimes unavoidable and the decision as to whether firefighting foam is to be applied must be made by the command staff often in a very short time, under pressure, and probably with incomplete knowledge about the damage situation. Hence, it is obvious that firefighting foam can reach into the environment and should be as harmless as possible.

It is clear that firefighting foams are not necessarily the first choice in every fire incident, especially not for relatively small fires which can be controlled by other extinguishing techniques. Therefore, the highest possible level of training for the fire fighters and proper, justified standards are indispensable.

But summarizing the rules of engagement in many developed countries and balancing the risks of the application of firefighting foam against the dangers of the fire and other extinguishing techniques, one may conclude that the application of firefighting foam is reasonable and probably often the best choice in certain situations. The only exceptions are the fluorinated foams (fluorinated AFFF) due to their inherent negative environmental properties (persistency, bioaccumulative properties, toxicity). Here, environmentally sound alternatives are urgently needed and are under promising development.

In general, firefighting foams are an indispensable tool for fire fighters to tackle the most difficult fire incidents. They allow an optimal extinguishing success, save lives, protect the victims, action forces, and goods, and thus are of the utmost value for the safety of our society.

ACKNOWLEDGMENT

The author gratefully acknowledges the financial funding and partial experimental support of our research on nontoxic and environmentally sound fire fighting foams of the following institutions: Bundeswehr Research Institute for Protective Technologies and NBC-Protection, Munster, Germany, European Union European Regional Development Fund, European Commission, Brussels, Belgium, State Government of North Rhine-Westphalia, Düsseldorf, Germany and Fraport Environmental Fund, Frankfurt am Main, Germany.

REFERENCES

1. Tabar, D.C. *Plant/Oper Prog* 1989; 8: 218.
2. ICAO. *Airport Services Manual, Part 1, Rescue and Fire Fighting*, International Civil Aviation Organization, 1990.
3. Blunk, D., Hetzer, R., Wirz, K. *Crisis Prevention* 2014, 38.
4. Hetzer, R.H., Kümmerlen, F., Blunk, D. Fire testing of experimental siloxane-based AFFF: Results from new experiments. In: *Suppression, Detection and Signaling Research and Applications Symposium (SUPDET 2015)*, Orlando, Florida, U. S. A., 2015, http://www.nfpa.org/~/media/files/news-and-research/resources/research-foundation/symposia/2015-supdet/2015-papers/supdet2015hetzer.pdf?la=en.
5. Hetzer, R.H., Kümmerlen, F., Wirz, K., Blunk, D. Fire testing a new fluorine-free AFFF based on a novel class of environmentally sound high performance siloxane surfactants. In: *International Association for Fire Safety Science (IAFSS) Symposium*, Canterbury, New Zealand, 2014, www.iafss.org/publications/fss/11/1261/view/fss_11-1261.pdf.
6. French, R.J., Hinkley, P.L., Fry, J.F. *Chem Ind (London, U. K.)* 1956; 75: 260.
7. French, R.J., Hinkley, P.L. *J Appl Chem* 1954; 4: 513.
8. Clark, N.O., Thornton, E., Lewis, J.A. *J Inst Pet* 1947; 33: 192.
9. HM Fire Service Inspectorate Publications Section. *Fire Service Manual - Volume 2 - Operational - Firefighting Foam.* The Stationery Office: London, 1998.
10. Sharma, T., Chimote, R., Gupta, S., Singh, J. *Fire Saf Sci* 1994; 4: 865.
11. French, R.J. *J Appl Chem* 1952; 2: 60.
12. Thomas, P.H. *J Appl Chem* 2007; 9: 265.
13. DIN 1568-2: Feuerlöschmittel – Schaummittel – Teil 2: Anforderungen an Schaummittel zur Erzeugung von Leichtschaum zum Aufgeben auf nicht-polare (mit Wasser nicht mischbare) Flüssigkeiten; Deutsche Fassung EN 1568-2:2008, DIN Deutsches Institut für Normung e.V., 2008.
14. DIN 1568-1: Feuerlöschmittel – Schaummittel – Teil 1: Anforderungen an Schaummittel zur Erzeugung von Mittelschaum zum Aufgeben auf nicht-polare (mit Wasser nicht mischbare) Flüssigkeiten; Deutsche Fassung EN 1568-1:2008, DIN Deutsches Institut für Normung e.V., 2008.
15. DIN 1568-3: Feuerlöschmittel – Schaummittel – Teil 3: Anforderungen an Schaummittel zur Erzeugung von Schwerschaum zum Aufgeben auf nicht-polare (mit Wasser nicht mischbare) Flüssigkeiten; Deutsche Fassung EN 1568-3:2008, DIN Deutsches Institut für Normung e.V., 2008.
16. Hetzer, R.H., Schönherr, E., Mickeleit, M. *vfdb - Zeitschrift für Forschung, Technik und Management im Brandschutz*, Ebner Verlag GmbH & Co KG, Bremen, Germany, 2014; 4: 171.
17. Magrabi, S.A., Dlugogorski, B.Z., Jameson, G.J. *Fire Saf J* 2002; 37: 21.
18. Stockholm Convention on Persistent Organic Pollutants, approved on 22.05.2001, in force since 17.05.2004, with amendments 2009, 2011 und 2013, Stockholm Convention, 2009.
19. Annex XV Restriction Report - Proposal for a Restriction, ECHA European Chemicals Agency, 2014.
20. Arenholz, U. Sabine Bergmann, K. Bosshammer, D. Busch, K. Dreher, W. Eichler, K.-J. Geueke, G. et al. In: *Verbreitung von PFT in der Umwelt - Ursachen – Untersuchungsstrategie – Ergebnisse – Maßnahmen.* LANUV-Fachbericht: Recklinghausen, 2011.
21. Jardak, K., Drogui, P., Daghrir, R. *Environ Sci Pollut Res* 2016; 23: 3195.
22. Rebello, S., Asok, A.K., Mundayoor, S., Jisha, M.S. *Environ Chem Lett* 2014; 12: 275.
23. Blasco, J., Hampel, M., Moreno-Garrido, I. *Compr Anal Chem* 2003; 40: 827.

24. Scott, M.J., Jones, M.N. *Biochim Biophys Acta* 2000; 1508: 235.
25. Gloxhuber, C. *Arch Toxicol* 1974; 32: 245.
26. Boethling, R.S. *Water Res* 1984; 18: 1061.
27. Yoshimura, K. *J Am Oil Chem Soc* 1986; 63: 1590.
28. Lewis, M.A. *Ecotoxicol Environ Saf* 1990; 20: 123.
29. Lewis, M.A. *Water Res* 1991, 25, 101.
30. Lewis, M.A. *Water Res* 1992, 26, 1013.
31. Boethling, R.S. In: Cross, J. Singer, E.J. (Eds.), *Surfactant Sci Ser*, Marcel Dekker, Inc., New York, Basel, Hong Kong, 1994, Vol. 53, pp. 95.
32. Drobeck, H.P. In: Cross, J. Singer, E.J. (Eds.), *Surfactant Sci Ser*, Marcel Dekker, Inc., New York, Basel, Hong Kong, 1994, Vol. 53, pp. 61.
33. Yeh, D.H., Pennell, K.D., Pavlostathis, S.G. *Water Sci Technol* 1998; 38: 55.
34. Cserháti, T. Forgács, E. Oros, G. *Environ Int* 2002; 28: 337.
35. Ying, G.-G. *Environ Int* 2006; 32: 417.
36. Heinze, J. *Environ Int* 2007; 33: 272.
37. Xu, L., Shi, Y., Cai, Y. *Water Res* 2013; 47: 715.

23 Metallic Foams

F. García-Moreno

CONTENTS

23.1 INTRODUCTION

Solid foams are light weight cellular materials inspired by nature. Wood, bones or sea sponges are some well known exemplars. In fact, solid metallic foam structures are the conserved image of the corresponding liquid metallic foam. Scientific attempts to improve foam quality have to concentrate on the foam physics, that is, bubble formation, foam nucleation, growth, stability, development, and gas diffusion in the liquid state where the foam structure evolves [1–13]. The engineering and industrial approach is to scale up the processes and provide reliable conditions for serial productions at acceptable quality and cost levels [14–17]. Depending on the production method, the foam structure is more or less homogeneous and comprises different characteristic features that determine its properties, and therefore, the fields of application [18].

Some applications based on foamed materials such as polymeric foams, porous concrete or food foams are quite popular in society. Several applications of ceramic foams are also well known in industry [19], but metal foams and their applications are still quite unfamiliar. The reason is they are still not wide spread, although they have a very high potential and a large number of applications already exist on the market.

Several reviews about the applications of metallic foams are available in the literature [18,20,21], but in the past years, new applications and application fields have emerged. These applications depend strongly on the foam type and properties, that is, on the manufacturing process. There is a large number of manufacturing processes divided into two main families: the melt and the powder metallurgical route. All these production methods are extensively described in literature [18,22]. There are further manufacturing methods which are not directly connected with foaming in the strict sense of the word, but lead to a foam similar structure, like foam replication or casting, coating of polymeric foam, place holder methods, and so on. This is especially the case for open cell metal foams, which should be called metal sponges more precisely speaking, which can obviously not be foamed directly as the gas will leak [23]. On the other hand, these products possess a foam-like structure, therefore, we will consider them for applications. In contrast, other kinds of cellular or porous materials like lattices, fibers, honey combs, and so on, will not be taken into account.

A large number of companies made a bet for the future and started producing and commercializing metallic foams in the past two decades, although several closed, stopped the production field or

445

concentrated on their core business, especially during the worldwide economic crisis starting in 2007. Their major problem was caused not by the quality or properties of their products, but mainly by two factors: a lack of effective marketing for the small metallic foam market and the price, that is, their benefit margin. Some of the most relevant foam producing companies at present are: Havel Metalfoam, Cymat, Aluinvent, m-pore (Mayser), pohltec metalfoam, Recemat, Alantum, Foamtech, ERG, and so on. A more detailed list can be found on the internet at www.metalfoam.net [24]. Beside the company names, some trade names have been established as being characteristic for a production method such as Fominal (Frauenhofer), Alporas (Shinko Wire, now Foamtech), Alusion (Cymat), Aluhab (Aluinvent), and Duocel® (ERG).

The main applications of metal foams are divided into structural and functional and are based on several excellent properties of the material [18]. We will review applications of closed, partially open, and open cell metal foams. Open structures allow for media to penetrate through and provide interesting properties for functional applications. For serial applications, the production processes should be reliable, reproducible, and the final product should have a reasonable price. In the next paragraphs, we will first consider the most relevant properties of metal foams and discuss some general properties and cost considerations. Then we will review the most important fields of applications and refer to some prominent examples of prototypes, products, and serial applications. In fact, a large number of applications are hard to classify in a certain category, as several properties are apparent for it. In this case, the most relevant and characteristic property will be considered.

23.2 PROPERTIES

Applications of metal foams are strongly linked to the properties which such kind of materials can offer and especially to those which are excellent or even unique. Some of the properties are obviously related to those of the matrix metal itself, for example, temperature or corrosion resistance, while others appear only in combination with the cellular structure, for example, large surface area or damping. Therefore, the different types of mechanical, functional, and other properties are reviewed here.

23.2.1 MECHANICAL PROPERTIES

The mechanical properties of metal foams are, of course, correlated to the ones of the bulk phase metal, but in a reduced manner. The dominating factors here are the density and the structure itself. The foam structure is the characteristic feature of foams. Mechanical properties differ and are influenced by quality in the sense of cell connectivity, cell roundness and diameter distribution, fraction of the solid contained in the cell nodes, edges or the cell faces, and so on. Gibson and Ashby proposed a simple beam model of a cubic cell for describing the mechanical response of foams [25]. Although this model is based on regular cellular structures, so called lattices, it provides quite realistic results. Note that foam cells do not have eight neighboring cells, but usually have around 14 [26]. This model and experimental results show the strong dependence of the mechanical properties on the foam's relative density [25,27–31]. The differences between open and closed cell foams can be comprehended in the sense of material distribution.

A certain mechanical property $P*$ of the foam should be evaluated in accordance with the weight, that is, density. Therefore, we should talk about high specific stiffness or moduli or about mass-related properties, as $P*$ will usually never be as good as the one for the same bulk samples of the same volume P_s. However, there are exceptions like damping, energy absorption, and so on. Accordingly, the relative property is proposed in dependence of the relative density ρ/ρ_s, with the coefficient a and the structural constant k, as

$$\frac{P*}{P_s} \approx k\left(\frac{\rho*}{\rho_s}\right)^a \tag{23.1}$$

where a and the structural constant k depend on the property, on the foam's structure, and on the type of deformation.

One of the most important mechanical properties for structural applications is the stiffness or Young's modulus. In our case, as already mentioned, we have to look for the relative modulus according to Equation 23.1. The Young's modulus of a foam E^* corresponds to the slope of the compression strength in dependence of the strain ε in the elastic regime, that is,

$$E^* = \frac{\sigma}{\varepsilon}. \tag{23.2}$$

Evans et al. and Banhart showed that the stiffness of a flat panel of a given weight is inversely proportional to its density [18,32]. Ashby et al. [22] proposed Equation 23.1 for the relative Young's modulus

$$\frac{E^*}{E_s} \approx k \left(\frac{\rho^*}{\rho_s} \right)^2 \tag{23.3}$$

where E_s is the modulus of the solid bulk material and $k \approx (0.1\text{--}4)$.

Based on their model, they further proposed a specific compressive strength of the foam in the elastic regime and came to the following approximation

$$\frac{\sigma_{el}^*}{E_s} \approx 0.05 \left(\frac{\rho^*}{\rho_s} \right)^2 \tag{23.4}$$

and similarly in the plastic regime

$$\frac{\sigma_{pl}^*}{\sigma_{ys}} \approx 0.3 \left(\frac{\rho^*}{\rho_s} \right)^{3/2} \tag{23.5}$$

with σ_{el}^* being the compressive strength in the elastic regime and σ_{pl}^* in the plastic regime, respectively [22].

Considering the models and experimental results obtained by Gibson [25], the compressive strength of an open cell foam can be summarized as

$$\frac{\sigma_{pl}^*}{\sigma_{ys}} \approx k \left(\frac{\rho^*}{\rho_s} \right)^{3/2} \tag{23.6}$$

and of the closed cell foam to

$$\frac{\sigma_{pl}^*}{\sigma_{ys}} \approx k \left[0.5 \left(\frac{\rho^*}{\rho_s} \right)^{2/3} + 0.3 \left(\frac{\rho^*}{\rho_s} \right) \right] \tag{23.7}$$

with $k = (0.1\text{--}1.0)$.

The tensile strength σ_{ts}^* is a weak property of foams in general, as the resistance against a propagating crack mainly depends on the weakest link of the material, which is given here as a thin cell wall or Plateau border, which can break easily. σ_{ts}^* is density-independent, and follows the relation

$$\frac{\sigma_{ts}^*}{\sigma_{ys}} \approx k' \tag{23.8}$$

with $k' = (1.1-1.4)$ [22].

A metal foam can suffer from fatigue under compression or tension load, but mostly it is a combination of both due to bending and sometimes even torsion. Therefore, fatigue is mostly limited by the tensile strength of metallic foams [33–37]. Many applications, especially in the automotive industry, suffer a large number of cyclic loads, therefore, it is very important to guarantee no failure during the operative lifetime. Fatigue can be represented by the number of cycles which a foam can withstand at a certain load without structural degradation. Since cracks growth in a catastrophic manner, a structural failure will be the result.

Most probably the best and more characteristic mechanical property of metal foams is their energy absorption capability [38,39]. Foams can absorb a maximum of mechanical energy without exceeding a certain stress limit σ_D due to plastic, irreversible deformation over a large strain range. This property and isotropy make them almost ideal crash absorbers. The maximum energy W_{max} absorbed by the plastic deformation of a metal foam is the integral of the stress-strain curve up to the stress limit and is given by Gibson and Ashby as [25]

$$\frac{W_{max}}{E_s} = \frac{\sigma_D}{E_s}\left\{1 - 3.1\left(\frac{\sigma_D}{E_s}\right)^{2/3}\right\}. \tag{23.9}$$

The mechanical properties of metal foams can be improved, similar to those for bulk metals, by different hardening mechanisms such as alloy composition, grain refinements, heat treatments, and so on. [35]. Furthermore, structural designs and property optimizations are possible depending on the application and practical aspects [38].

Integral foam structures show a density and property gradient, usually with a denser skin and a lighter core similar to a bone structure, but are difficult to manufacture [40,41]. A further interesting optimization of a panel is the sandwich structure composed of two bulk face sheets and a lighter metal foam core as will be presented later [15,42,43].

23.2.2 FUNCTIONAL PROPERTIES

There is a wide range of functional properties of metal foams originating from their cellular character. This implies a large surface area and number of cells. Very often the combination of functional properties such as acoustic [44–49], thermal [50–55], electrical [21] or chemical resistance with mechanical properties like strength or stiffness allows interesting new applications [32].

Foams, in general, are known for their thermal insulation properties, especially for ceramic, glass, and polymeric foams. They are the basis for a large number of applications related to thermal control. Thermal conduction has to percolate through the matrix network, radiation between cell walls is hindered due to the large number of cells, and gas convection is disabled due to the small gas volume in each separated cell. Metal foams' thermal conductance is less than that of the corresponding bulk metal, but in absolute values, they cannot compete with polymeric foams in terms of thermal insulation.

On the other hand, open cell metal foams have a large surface area that can be used in combination with a good conductivity of the matrix metal (e.g., Cu or Al) for a complete opposite purpose, namely heat dissipation or passive cooling [53]. In fact, Al and Cu foams can conduct temperature very well through the matrix, therefore, such foams are used for heat exchangers. The relative thermal conductivity in metal foams λ^*/λ_S can be expressed in a simple model as a function of the relative density

$$\frac{\lambda^*}{\lambda_s} = \left(\frac{\rho^*}{\rho_s}\right)^a \tag{23.10}$$

with $a = (1.65-1.8)$ [22]. Further advanced models are summarized in the literature [55].

Metals are also known to be good electrical conductors. Again, the combination with a large foam surface allows their utilization as electrodes in batteries or for other electrochemical purposes. The relative electrical resistivity R^*/R_s is given by

$$\frac{R^*}{R_s} = \left(\frac{\rho^*}{\rho_s}\right)^{-a} \tag{23.11}$$

with $a = (1.6-1.8)$ [22].

Polymeric foams are good for acoustic and vibration damping over a large frequency range due to the cellular structure and the same property is present in metallic foams [45,48,56]. Additionally, electromagnetic shielding can be provided due to their metallic nature [15].

23.2.3 OTHER PROPERTIES

The foam structure, and accordingly the foam properties, are usually described as isotropic. But obviously, the number of cells has to be large enough to be able to consider the foam as a continuum. This is why, for example, for compression tests, the German Institute for Standardization (DIN) standard 50134 "*Compression test of metallic cellular materials*" requests that the number of cells in each direction of the sample tested has to be equal to or larger than 10. Most other cellular materials, for example, lattices or honey combs, are obviously anisotropic in structure and in their properties. Anisotropy is an important advantage of metal foams for certain purposes, although some anisotropies in the structure can also be favorable for specific applications and occur depending on the production method [21,57].

Recycling is becoming more and more important in our daily live and is becoming a required property for advanced and new materials. Obviously, the main component in a metallic foam is air, corresponding to 50–90 vol.% of the material, depending on the porosity. The rest is the metal matrix, which can be recycled as well as the metal material itself, among which aluminium alloy foams are the most prominent candidates. Additives like the blowing agents (TiH_2, ZrH_2, Ca_2CO_3, etc.) are usually hydrides or oxides which lose the gaseous component during the foaming process when the metal reacts with the matrix material and can be considered part of the alloy due to their metallic nature and low amount (0.2–1 wt.%) [58,59]. There are exceptions, however, such as gas injected Al foams stabilized by large amounts (5–20 vol.%) of ceramic particles (usually SiC) commercially known as Cymat or Alusion foams [60].

The possibility to adjust the metallic foam density in a large range from 0.05 to 10 g/cm^3 depending on the porosity and matrix metal allows for properties like buoyancy, which can be tailored, varying the material density, and they can be used, for example, for ships, floaters or sea markers. It can be also combined with other properties like high temperatures or corrosion resistance, where standard polymeric foams cannot be used.

Metals such as Ti or Mg are biocompatible or biodegradable, respectively. These properties also apply to their foamed structures, being additionally able to adjust the Young's modulus by adjusting their density and offering a large surface, especially in the case of metal sponges, for osseointegration.

23.3 COSTS AND FEASIBILITY CONSIDERATIONS

A large number of successful prototypes have already demonstrated that there is a very wide range of possible applications. Depending on the desired performance and manufacturing feasibility, different production methods are considered. In most applications, metal foams are as good as or better than their competitors [18]. But, of course, one key point for industrialization breakthrough is still the price of the product [16].

Additionally, when it comes to costs and system integration, bifunctional or even multifunctional applications are in favor [18,32]. Therefore, for real applications, cost effective production methods and multifunctionality need to be combined as priority goals [61]. A multipurpose approach could be, for example, an application as a light weight, stiff, and recyclable material with vibration damping properties and very high volume-effective crash absorption protection for electric cars. In such cases, the balance between costs and benefits are not in favor of metal foams. But sometimes other aspects such as innovation, image, marketing, prestige or strategical aspects can play an important role. Depending on the application field, the price may then play a less important role, which is the case in the fields of arts, design, medicine or sports.

More important even is to find applications where the properties provided by the metal foam are unique when compared to other materials and the price is only secondary, for example, the best crash absorption behavior in a minimal space and weight to improve safety.

For mobility, that is, applications for automotive, railway or aerospace industries, light weight plays an enormously important role, as in these cases saving weight leads to a large saving of energy due to the continuous acceleration and breaking cycles of the mass. This point seems to be particularly important for actual electro cars and buses, where the batteries still do not provide a sufficient operation range.

Applications in aerospace could be the field where multifunctionality offers the most financial advantages. Thinking, for example, about a plane fuselage which is made traditionally from Al sheets with welded or riveted struts, it could be replaced by a foamed, curved sandwich structure. A metallic foam sandwich will provide additional noise and vibration damping of the turbines and a certain additional temperature insulation compared with the traditional metal sheets. It can be also recycled and it is nonflammable. Saving 1 kg in a plane fuselage corresponds roughly to 1 Million € savings over the whole life of the plane (25 years). And considering that the demand and price for fossil fuels will increase strongly in the future, the advantages of light weight materials will increase also. But the aeronautic industry seems to still have some reservations about the application of metallic foams, most probably due to the long and expensive validation procedure of new materials.

For small or medium sized applications, it is difficult to fix an actual price for metal foams, as it depends mostly on the foam density, shape, degree of complexity, number of units of the desired product, and so on. In case of foams made by following the powder metallurgical route (not Cymat), the price of the raw material is $\sim$3 €/kg for the Al powder and <1 €/kg for the blowing agent (TiH$_2$) compared to the weight of the foam. But foam prices should not be given in a price-per-weight format, as this will suggest an expensive product compared to bulk material, but better in price-per-volume or -area. As a rough estimation, we can find actually Al foam panels on the market for prices in the range of 60–200 €/qm, depending on the type of foam, panel thickness, and so on. A diagram of costs related to material density and modulus for different bulk and foamed metals can be found in the literature, although the given prices are not the actual prices anymore [22].

23.4 STRUCTURAL APPLICATIONS

Light weight materials with high stiffness are often desired for applications. Standard products in the market such as honeycomb panels use a cellular structure as the core and brazed or glued face sheets to provide the desired properties. They are usually cheap, but have some disadvantages as they cannot be curved, resist high temperatures due to the glue or cannot be recycled. Aluminum foam sandwiches (AFS) and steel aluminum foam sandwiches (SAS) are promising products for structural applications and are already on the market [15,62–64].

AFS panels are used as support frames, for example, for solar panels, mirrors, so on, and everywhere, where light and rigid metallic panels are needed. Most of the customers prefer to provide the material as panels and manufacture their own products. They often even keep their innovative field of application a secret to assure a competitive advantage for their products on the market.

A good example are industrial machines, where foam-filled beams and columns are stiff but light. With reduced inertia, they can be moved quickly and positioned precisely. Examples are drilling, milling, textile, cutting, printing, pressing or blanking industrial machines. Additionally, damping of the system and of, for example, an additional vibration tool can improve the performance in precision of positioning and wear, reduce fatigue problems, and increase the operational lifetime. An example of an application for a high-speed milling machine of Niles-Simmons in cooperation with the Fraunhofer-Institut für Werkzeugmaschinen und Umformtechnik (IWU) in Chemnitz, Germany is given in Figure 23.1a. The sliding bed is made of 11 welded AFS parts and the construction is 28% lighter than the cast part with the same stiffness but improved vibration damping. Around 15 parts per year are manufactured. Figure 23.1b shows a beam of a textile machine filled with Alporas foam produced by the Au Metallgießerei in Sprockhövel, Germany. This part is 1590 mm × 280 mm × 160 mm in size and provides 60% of reduction in amplitude at the resonance frequency. The production is ~1000 pieces per year. A similar hybrid material concept was applied to a tool column prototype of the Technical University Prague for a cutting machine (model Prisma S) from TOS Varnsdorf s.a., Czech Republic, in which an Alporas foam core is integrated.

The automotive industry wants to benefit from the developments of new light weight materials. For large serial productions, not only material properties and costs are important, but also other engineering and strategical aspects play an important role, for example, the necessity of redesigning other components, the system integration capability or the number of available suppliers [65].

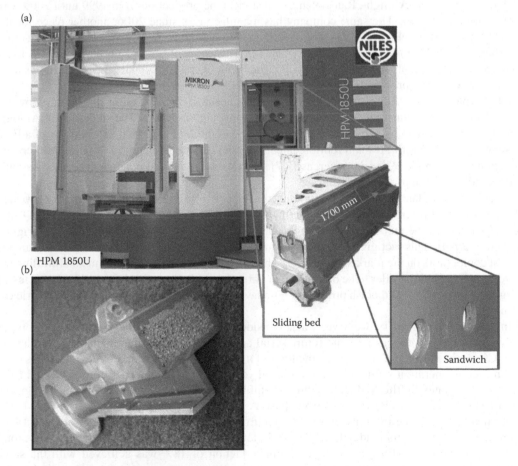

FIGURE 23.1   (a) Sliding bed of a milling machine made of a welded aluminum foam sandwich (Courtesy of Th. Hipke, IWU Chemnitz, Germany.) and (b) beam of a textile machine filled with Alporas foam. (Courtesy of the Au Metallgießerei, Germany.)

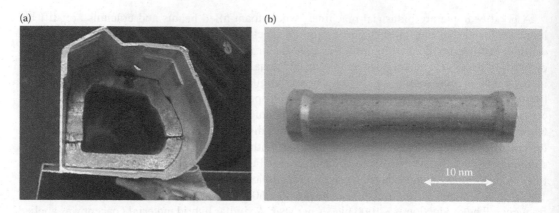

FIGURE 23.2   (a) Al foam part for the Ferrari 360 and 430 Spider and (b) small crash absorbing element for the Audi Q7. (Courtesy of Alulight, Ranshofen, Austria.)

In addition to a large number of prototypes several large series went into production. As an example (Figure 23.2a), the door sill was filled with Al foam to reinforce the frame and increase stiffness as well as the behavior in case of a side crash in high premium cars such as the Ferrari 360 and 430 Spider. The company Alulight, Ranshofen, Austria, held the production from 1999 until 2009 with ~5000 pieces per year. The same company has manufactured, since 2006, another piece for the Audi Q7 sport utility vehicle (SUV) vehicle in a complete automated series, namely a small 7 g crash absorber (Figure 23.2b) placed at the safety net [14]. This element is able to protect the passengers from the load in case of accident or strong braking. With over 100,000 parts per year, this is until now the largest serial production of metallic foams in the automotive industry.

Also exploiting the energy absorption capability of metallic foams, Foamtech, Korea, provided crash elements for a guardrail at the Massan-Chanwon Bridge in Korea. The suitability of Al foam as a protection element in A-pillars was also analyzed in the literature [66]. An A-pillar of a Ford passenger car filled with Al foam reinforcement provides an improvement of 30% in crash energy absorption with only a 3% weight increase [67]. A redesign of the components instead of a simple filling could even improve theses values.

Since 2003, Pohltec metalfoam has produced a support of a working platform made of welded AFS parts for mobile cranes for Teupen, Germany (Figure 23.3). This is a serial production of ~100 pieces/year where the metallic foam component is saving around 95 kg compared to the original steel counterpart. This fact gives the company and the end user an important strategical advantage against competitors on the market, namely, it is the only crane with a range of 25 m and fulfilling the 3.5 tons weight limit in order to be driven with a normal driving license. AFS panels are also tested as light weight support and crash protection for heavy battery modules at the bottom of electric car or bus bodies.

In the frame of mobility, we have to also consider the railway industry. Promising prototypes evolved in the past years as possible future serial application. AFS foam panels delivered by the IWU Chemnitz, Germany, are used in the floor of a wagon of the metro in Peking which has been in continuous operation without issue for several years. A train front structure was welded from curved AFS plates by the Wilhelm Schmidt GmbH in cooperation with the Brandenburgische Technische Universität (BTU) in Cottbus [68]. A more prominent and recent prototype is the power head cover of the intercity-express (ICE) train fabricated by Voith Engineering and IWU, Chemnitz, Germany. It is made of welded AFS plates and carbon fibers in the front, with a total length of around 6 m (Figure 23.4). A weight reduction of 18% was achieved with the same stiffness, improving vibration damping and reducing the manufacturing steps compared to the traditional construction procedure. These examples clearly show the advantages of metallic foams or AFS panels against honeycomb panels, namely when curved sections are required.

(a) (b)

FIGURE 23.3 (a) Support of a working platform made of welded AFS made by Pohltec metalfoam, Germany, and (b) mobile crane vehicle from Teupen, Germany, where the support is applied. (Courtesy of Pohltec metalfoam.)

Further railway applications can also be mentioned, for example, for the Combino tram in Budapest, where Alulight, Austria, delivers a foam block placed behind the bumper to absorb energy in the case of collision with cars, trying to avoid expensive write-offs (Figure 23.5a). A similar crash absorption box can be found in the Sprinter Light Train (SLT) in Holland, where ~500 pieces per year have been delivered since 2008 by Gleich, Germany, and the Slovak Academy of Sciences (SAS) in Bratislava (Figure 23.5b).

(a) (b)

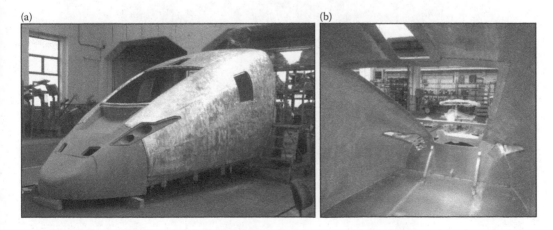

FIGURE 23.4 Prototype of German high velocity train ICE made of welded aluminium foam sandwich. View of the locomotive from (a) outside and (b) from inside. (Courtesy of Th. Hipke, IWU Chemnitz and Voith Engineering, Chemnitz, Germany.)

(a)   (b)

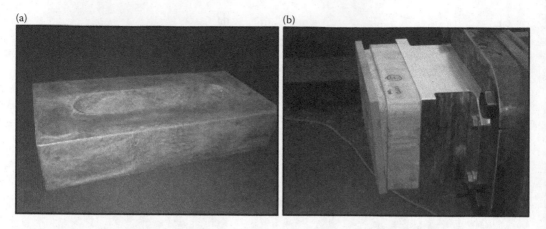

FIGURE 23.5 Crash absorbers (a) of the Combino tram in Budapest produced by Alulight, Austria, and (b) of the Sprinter Light Train in Holland produced by Gleich and SAS Bratislava. (Courtesy of Alulight, Gleich and SAS Bratislava.)

Crash absorption capability is of special interest for military applications related to the armored protection of vehicles or blast mitigation. These types of applications are obviously mostly confidential and it is difficult to gain detailed information about them. Tests of metallic foams for protection and crash absorption at the bottom of helicopters and of blast mitigation systems are known. Figure 23.6 shows a foam block from Aluinvent with explosives to test its blast mitigation capability. Duocel aluminum foam provides critical blast energy absorption for crew protection in the cabin retrofitted in the family of military Medium Tactical Vehicles and in cabin seat mounts [69].

The ship building industry is considering metal foam applications concerning weight reduction and distribution, that is, upper structures should be as light as possible to increase load capability. In 2009, a project started where the development of a new type of inland area cargo vessel for weight reduction in ships was the aim [70]. Prototypes of sea markers made of metallic foam showed that in the case of damage due to ship contact or ice pressure, they still continue floating, which is an important advantage for security reasons.

The aeronautic industry considered metallic foams for protection against bird strikes in planes [71]. Protection against micrometeoroids in satellites or in the space station are also of great interest,

FIGURE 23.6 Block of Aluinvent continuously casted aluminium foam with 85 mm thickness and 15 mm sized bubbles for blast mitigation test with explosives shortly before detonation. (Courtesy of N. Babcsán, Aluinvent.)

especially due to their isotropic properties [72,73]. Pohltec metalfoam manufactured a prototype of a conical adapter (>4 m diameter) for the Ariane rocket with welded, curved AFS panels, which performed better than standard materials due to additional vibration damping and demonstrating again the feasibility of curved and 3D structures.

For some sports equipment, the unique properties of metallic foams become more relevant, as price usually plays a secondary role here. For example, some golf putters use the damping properties of a metal foam structure to improve the strike control [21]. Some areas, like shinbone protectors for football players or helmets, could also be exploited. For example, Alcarbon, Germany, combines the properties of aluminum foams with carbon fibers for sports articles [74].

23.5 FUNCTIONAL APPLICATIONS

A wide palette of functional applications based on metallic foams can be found on the market, with most of these related to the corresponding functional property. Again, a multipurpose approach has the best chance to produce a competitive or unique product.

Ceilings in auditoriums or large rooms are very often planked with perforated metal sheets for sound control. As an alternative to this traditional construction, applications of metallic foam panels for sound absorption are already available on the market, offered by different companies [60,63,75]. At the foam surface, the sound waves are guided and redirected to the foam interior, where they are caught and damped after several reflections. The pore size distribution and the different orientations allow for a very effective damping over a broad frequency spectrum. These applications could be also considered as architectural, but we include them here as their main function is sound absorption. They combine the advantage of light weight, self-supporting capability of large metallic foam panels made of open cells or just sliced closed cell foams with a design component. Figure 23.7 shows the ceilings of an audience hall and a restaurant covered by Alusion foam provided by Cymat, Canada.

Further sound absorption applications made by Alporas foams provided by Shinko Wire, Japan, can be found in train rails, metro tunnels, elevators, on the underside of an elevated highway, and so on. [76]. Newer products also based in Alporas foams were developed by Foamtech, Korea, and applied on concert halls, conference rooms, auditoriums, sports centers, and the walls and ceilings of the machinery and operating rooms of industrial plants as a nonflammable, acoustic absorbent [75]. Further applications can be found in the ship industry, where, for example, prevention of noise from the engine room, combustive exhaust pipe, and air cleaning system is provided by foamed inside walls between cabins [75]. Metallic foams are also used by Foamtech for sound absorption in metro tunnels, in the railway, and on the tunnel and station walls, where they have to support high air pressure changes and vibration while still being nonflammable [75].

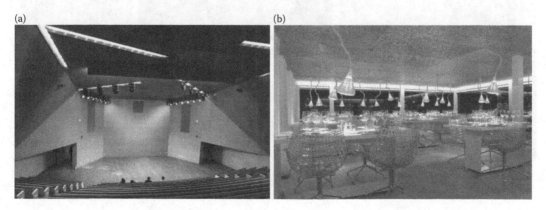

FIGURE 23.7 Ceiling of (a) audience hall and (b) restaurant covered by Alusion foam for sound control. (Courtesy of Alusion.)

We can find aerodynamic noise reduction prototypes in the railway industry, for example, in pantographs for the Shinkansen train in Japan [77,78]. There, open cell Al foam is used for shape-smoothing of the panhead and its support and covering. Applications such as aerodynamic noise reduction of jet turbines by airplanes are also under discussion [79].

Due to the high temperature and chemical resistance of metal foams, and combined with sound absorbing properties, open cell foams from Alantum have being applied as silencers, mufflers, diesel particulate filters, selective catalysis reduction, and so on. [23,80]. Exxentis, Switzerland, and Mott Corporation, USA, offer filters for the filtration of gases and liquids, flow control, diffusion, sparging, fluidizing, venting, and wicking [81,82]. Duocel, Recemat, Inco, and Alantum have open cell metal foams as a catalyst carrier, for filtration fluid control or antisloshing purposes in their portfolios [23,69,83,84]. Even nanoporous gold foams find applications as catalysts [85].

A strong market for metallic sponges, with a large number of different products, is represented by heat exchangers. The very good thermal conductivity of metals and the large area of metal sponges provide very effective passive cooling [32,53,69,83]. As an example, Figure 23.8 shows a large number of different heat exchangers from m-pore, Germany [86].

ERG also produces heat exchanger from Duocel (Al- and Cu-based open cell foams) [69]. They have built different devices, for example, a small heat exchanger to thermally stabilize the lens of electron scanning microscopes or thermal energy absorbers for medical laser applications. Furthermore, the breather plug used in the Lockheed Martin F-22 fighter aircraft for pressure release during rapid altitude changes, electromagnetic shielding protection, and moisture wicking is made of Duocel [69]. An ERG heat exchanger made of Al foam was used as heat exchange media and a support matrix of granulated chemicals consisting of multiple layers of amine-based filter beds to remove carbon dioxide and moisture in the Space Shuttle, and it is also used in the International Space Station. After reaching its full absorption capacity, the assembly rotates to emit the CO_2 and moisture into space, after which it can rotate back toward the interior and continue filtration, facilitating uninterrupted CO_2 removal [87].

A special application of heat exchangers is passive thermal cooling, a field where the demand of very effective heat sinks is increasing from day to day due to the rapidly growing performance of computers and mobile electronic devices. Today, standard bulbs are being replaced by modern, powerful LED lamps

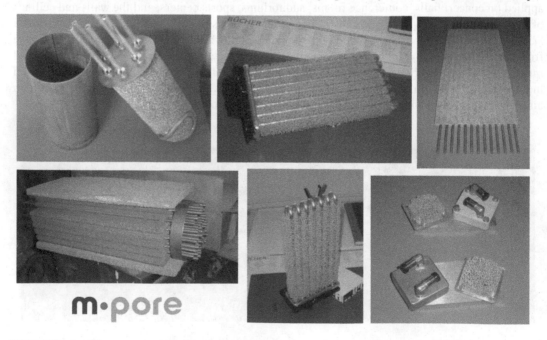

FIGURE 23.8 Diverse types of heat exchangers based on Al and Cu foams. (Courtesy of m-pore and Mayser GmbH.)

(a)   (b)

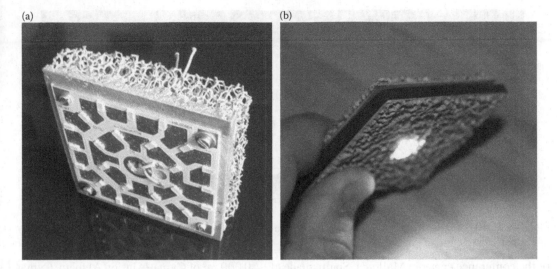

FIGURE 23.9 Passive thermal cooling of LED lamps, (a) Al open cell foam from m-pore (Courtesy of m-pore) and (b) cut AFS from the Technical University Berlin.

(see Figure 23.9). This power has to be cooled down to assure light-emitting diode (LED) efficiency and protect the electronics. Here, where a "noisy" fan cannot be installed, foamed passive cooling devices have an opportunity. These applications can combine their functionality with an innovative design.

Due to the high melting temperature and thermal conductivity of metals, they can be used for heat and fire resistance [36]. Alulight provided Al alloy foam panels for the protection of concrete against flames in fire houses, thus dissipating the heat over large surfaces. Similarly, heat dissipation was used by Pohltec metalfoam on AFS panels cut and closed by hot forming and covered by a ceramic plasma coating as barbecue and cooking plates, providing convenient, homogeneous temperature distribution [15]. Duocel aluminum foam was used as the primary component in satellite cryogenic tanks to provide uniform heating and cooling. Space based infrared optics using solid cryogenic coolers also utilize Duocel aluminum foam as an isothermalizer and baffle structure. Keeping a solid cryogen at a uniform temperature gives the infrared optics a longer useful lifetime [69].

Al-, Zn, and Ni sponges are used for electrochemical applications, where the large surface offers good advantages during operation, for example, in water purification where ions react with the matrix material or as electrodes for Zn- and Ni-based batteries [23,83,84,88–92]. The application of Inco Ni foams as an anode in batteries led to a production volume of about 3,000,000 m^2 foam per year.

Biomedical applications are another field of interest. Here high quality products based on Ti foams provide excellent biocompatibility properties. The trend goes in the direction of porous structures to improve osseointegration as is shown in Figure 23.10, where a tomography of a Ti-based dental implant shows its porous structure. Although Ti is quite difficult to foam, foam-like structures can be created through different production methods. This field has to be exploited more in the future.

23.6 ARCHITECTURAL APPLICATIONS

The organic cellular structure of metallic foams makes their surfaces unique and very attractive. Closed cell foams with a bubbly surface, cut closes cell foams with a first open cell layer or open cell foams, which are more or less translucent depending on the pore size and thickness of the material, have become in the last decade very interesting for architectural applications. These types of applications, especially as facades, are very appealing for metallic foam producers, as usually large areas are required promising good benefits.

In 2003, the architect Slawomir Kochanowicz covered his office building in Bochum, Germany, with closed cell Alporas Al-foam. Further facades were produced in the next years, such as the one

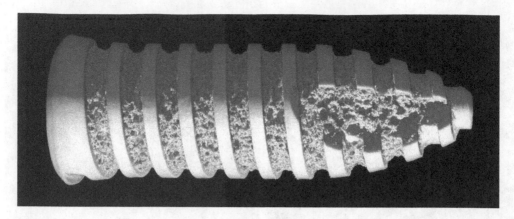

FIGURE 23.10 Tomography of a Ti-based porous dental implant for osseointegration improvement. (Courtesy of L.-Ph. Lefevre.)

of the conference center in Mallorca, Spain, made of ~20,000 m² of Cymat/Alusion Al foam. Cymat also provided cladding of the exterior facade of a 12,000 m² building in downtown Tbilisi, Georgia.

Not only facades, but other architectural applications were realized with Alusion foam, for example, the entrance of a souterrain or the support structure of a clarion as part of a monument as shown in Figure 23.11a and b, respectively. The memorial of Service Employees International Union, dedicated to its members lost in the World Trade Center attack on September 11, 2001, designed by the architects Furnstahl and Simon, is made of Alusion foam.

In 2003, Kauffmann, Theilig, and Partner and Markgraph created for Daimler Chrysler a stand at the international auto show in Geneve, Switzerland, from Alporas foam blocks to represent their advanced technologies. Ferrari, Audi, and others also presented stands with metal foam components (Figure 23.12). The company Aluinvent has manufactured a reception desk for Market Zrt. Budapest, Hungary (Figure 23.13). Pohltec metalfoam has design tiles made of a molten AFS surface in their portfolio for interior architectural purposes (Figure 23.14).

(a)   (b)

FIGURE 23.11 Translucent entrance of a souterrain and monument made of Alusion foam. (Courtesy of Alusion.)

FIGURE 23.12 Stand for convention center with Alusion foam structure. (Courtesy of Alusion.)

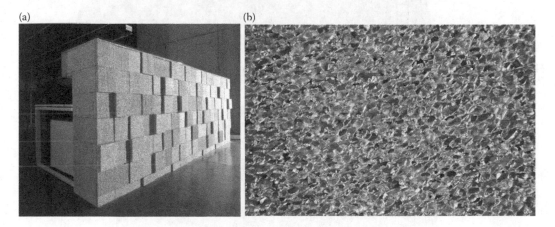

FIGURE 23.13 (a) Reception desk from Aluinvent made of Aluhab foam with bubble size of 3 mm built for Market Zrt. Budapest and (b) detail of the foam surface structure. (Courtesy of N. Babcsán, Aluinvent.)

23.7 DESIGN, ART, AND DECORATION

The transition from architectural applications to design, art, and decoration is very smooth and seamless, and it is often very difficult to distinguish between them, as in most cases a combination of these aspects is present. We can speak again about multifunctional applications. The immense variety of applications and projects already existing makes it impossible to review all types, but a few examples can be mentioned. The peculiar, sophisticated surface of the cellular metal foam structure is always in the foreground, while the structural and other properties stay in the background.

Several artists and designers use metallic foams for decorations and artistic compositions. The designer Gerd Kaden produced several compositions of wood and metallic foam to show the contrast between natural and man-made cellular materials (Figure 23.15). Figure 23.16a shows the support of a statue made of Alusion foam and Figure 23.16b shows the head of Pythagoras foamed in a mold with Alulight aluminum foam by F. Simančík from the Slovak Academy of Science in Bratislava. A design chair from welded Aluhub foam was produced at the Imperial College London by Andor Ivan (Figure 23.17a). An artistic light scattering effect of a designer lamp manufactured with m-pore open cell foam can be observed in Figure 23.17b. Gold and silver foams are even used for jewelry, demonstrating the diversified fields of applications based on metallic foams existing on the market.

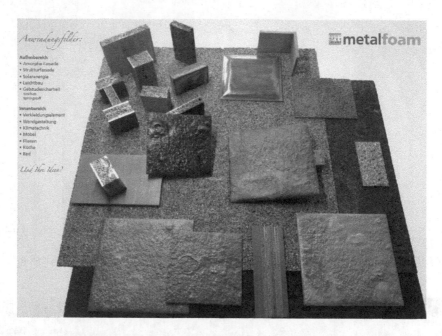

FIGURE 23.14 Different AFS, cut AFS, and tiles made of AFS for internal architectural applications. (Courtesy of Pohltec metalfoam.)

FIGURE 23.15 Artistic compositions of the designer Gerd Kaden created from wood and metallic foams.

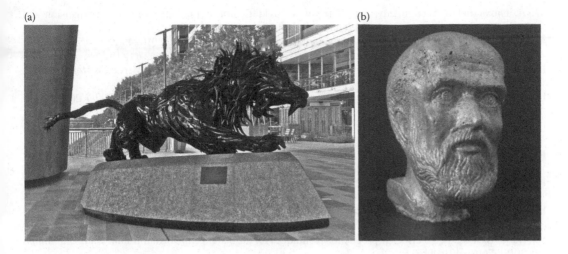

FIGURE 23.16 Statue support made of Alusion foam. (Courtesy of Alusion.) and Pythagoras head made of Alulight metal foam for the Slovak Academy of Sciences. (Courtesy of F. Simančík, SAS.)

FIGURE 23.17 (a) Welded design chair made of Aluhab. (Courtesy of N. Babcsán, Aluinvent.) and (b) art composition with metal sponge and light scattering effect from m-pore. (Courtesy of m-pore.)

REFERENCES

1. Byakova, A.V. et al. Influence of wetting conditions on bubble formation at orifice in an inviscid liquid: Mechanism of bubble evolution. *Colloids Surf A, Physicochem Eng Aspects* 2003; 229(1–3): 19–32.
2. Mao, D., Edwards, J.R., Harvey, A. Prediction of foam growth and its nucleation in free and limited expansion. *Chem Eng Sci* 2006; 61(6): 1836–1845.
3. Rack, A. et al. Early pore formation in aluminium foams studied by synchrotron-based microtomography and 3-D image analysis. *Acta Mater* 2009; 57(16): 4809–4821.
4. Asavavisithchai, S., Kennedy, A.R. In-situ oxide stabilization development of aluminum foams in powder metallurgical route. *High Temp Mater Process* 2011; 30(1–2): 113–120.
5. Garcia-Moreno, F. et al. Metal foaming investigated by x-ray radioscopy. *Metals* 2011; 2(1): 10–21.
6. Wang, Y.W. et al. Bubble nucleation of PM Al-9Si foam. *Adv Mater Res* 2011; 183–185: 1682–1686.

7. Asavavisithchai, S., Kennedy, A.R. The role of oxidation during compaction on the expansion and stability of a foams made via a PM route. *Adv Eng Mater* 2006; 8(6): 568–572.
8. Liu, X. et al. Foam stability in gas injection foaming process. *J Mater Sci* 2010; 45(23): 6481–6493.
9. Vandewalle, N. et al. Foam stability in microgravity. *J Phys Conf Ser* 2011; 327(1): 012024–0.
10. Körner, C., Arnold, M., Singer, R.F. Metal foam stabilization by oxide network particles. *Mater Sci Eng A* 2005; 396(1–2): 28–40.
11. Mukherjee, M. et al. Al and Zn foams blown by an intrinsic gas source. *Adv Eng Mater* 2010; 12(6): 472–477.
12. Mukherjee, M., Garcia-Moreno, F., Banhart, J. Solidification of metal foams. *Acta Mater* 2010; 58(19): 6358–6370.
13. Garcia-Moreno, F., Solorzano, E., Banhart, J. Kinetics of coalescence in liquid aluminium foams. *Soft Matter* 2011; 7(19): 9216–9223.
14. Schäffler, P. et al. Alulight metal foam products. In: Lefebvre, L.P., Banhart, J., Dunand, D.C. (Eds.), *Porous Metals and Metallic Foams: Metfoam*. The Japan Institute of Metals: Kyoto, Japan, 2008.
15. Banhart, J., Seeliger, H.-W. Recent trends in aluminum foam sandwich technology. *Adv Eng Mater* 2012; 14(12): 1082–1087.
16. Banhart, J. et al. Cost-effective production techniques for the manufacture of aluminium foam. *Aluminium* 2000; 76(6): 491–496.
17. Kevorkijan, M. Cost effective foaming with CaCO3. *Aluminium* 2010; 12: 59–65.
18. Banhart, J. Manufacture, characterisation and application of cellular metals and metal foams. *Prog Mater Sci* 2001; 46(6): 559–632.
19. Scheffler, M., Colombo, P. *Cellular Ceramics: Structure, Manufacturing, Properties and Applications*. WILEY-VCH: Weinheim, 2005, p. 645.
20. Davies, G.J., Zhen, S. Metallic foams: Their production, properties and applications. *J Mater Sci* 1983; 18(7): 1899–1911.
21. Nakajima, H. Fabrication, properties and application of porous metals with directional pores. *Prog Mater Sci* 2007; 52(7): 1091–1173.
22. Ashby, M.F. et al. *Metal Foams: A Design Guide*. Butterworth-Heinemann: Boston, 2000.
23. Kim, S., Lee, C.-W. A review on manufacturing and application of open-cell metal foam. *Procedia Mater Sci* 2014; 4: 305–309.
24. www.metalfoam.net. www.metalfoam.net. 2015 [cited 2015 27.07.2015]; Available from: www.metalfoam.net
25. Gibson, L., Ashby, M. *Cellular Solids: Structure and Properties*. Cambridge University Press: Cambridge, UK, 1997.
26. Weaire, D., Hutzler, S. *The Physics of Foams*. Oxford University Press: Oxford, USA, 1999.
27. Hall, I.W., Guden, M., Yu, C.-J. Crushing of aluminum closed cell foams: Density and strain rate effects. *Scr Mater* 2000; 43(6): 515–521.
28. Olurin, O.B., Fleck, N.A., Ashby, M.F. Deformation and fracture of aluminium foams. *Mater Sci Eng A* 2000; 291(1–2): 136–146.
29. Kennedy, A.R. Aspects of the reproducibility of mechanical properties in Al based foams. *J Mater Sci* 2004; 39(9): 3085–3088.
30. Ramamurty, U., Paul, A. Variability in mechanical properties of a metal foam. *Acta Mater* 2004; 52(4): 869–876.
31. Beals, J.T., Thomson, M.S. Density gradient effects on aluminium foam compression behaviour. *J Mater Sci* 1997; 32(13): 3595–3600.
32. Evans, A.G., Hutchinson, J.W., Ashby, M.F. Multifunctionality of cellular metal systems. *Prog Mater Sci* 1998; 43(3): 171–221.
33. Sugimura, Y. et al. Compression fatigue of a cellular Al alloy. *Mater Sci Eng A* 1999; 269(1–2): 38–48.
34. Harte, A.-M., Fleck, N.A., Ashby, M.F. Fatigue failure of an open cell and a closed cell aluminium alloy foam. *Acta Mater* 1999; 47(8): 2511–2524.
35. Lehmhus, D. et al. Influence of heat treatment on compression fatigue of aluminium foams. *J Mater Sci* 2002; 37(16): 3447–3451.
36. Kolluri, M. et al. Fatigue of a laterally constrained closed cell aluminum foam. *Acta Mater* 2008; 56(5): 1114–1125.
37. Harte, A.-M., Fleck, N.A., Ashby, M.F. The fatigue strength of sandwich beams with an aluminium alloy foam core. *Int J Fatigue* 2001; 23(6): 499–507.
38. Hanssen, A.G., Langseth, M., Hopperstad, O.S. Optimum design for energy absorption of square aluminium columns with aluminium foam filler. *Int J Mech Sci* 2001; 43(1): 153–176.

39. Ramachandra, S., Sudheer Kumar, P., Ramamurty, U. Impact energy absorption in an Al foam at low velocities. *Scr Mater* 2003; 49(8): 741–745.

40. Körner, C. *Integral Foam Molding of Light Metals: Technology, Foam Physics and Foam Simulation.* Springer-Verlag: Berlin Heidelberg, 2008.

41. Hartmann, J., Trepper, A., Körner, C. Aluminum integral foams with near-microcellular structure. *Adv Eng Mater* 2011; 13(11): 1050–1055.

42. Lehmhus, D. et al. Influence of core and face sheet materials on Quasi-static mechanical properties and failure in aluminium foam sandwich. *Adv Eng Mater* 2008; 10(9): 863–867.

43. Lies, C., Hohlfeld, J., Hipke, T. Adhesion in sandwiches with aluminum foam core. In: Lefebvre, L.P., Banhart, J., Dunand, D.C. (Eds.), *Porous Metals and Metallic Foams: Metfoam 2007*, 2008, pp. 31–34.

44. Han, F. et al. Acoustic absorption behaviour of an open-celled aluminium foam. *J Phys D Appl Phys* 2003; 36(3): 294.

45. Banhart, J., Baumeister, J., Weber, M. Damping properties of aluminium foams. *Mater Sci Eng A* 1996; 205(1–2): 221–228.

46. Avilova, G.M., Grushin, A.E., Lebedeva, I.V. Sound insulation of foam shells. In: *XIII Session of the Russian Acoustical Society*, Moscow, 2003.

47. Byakova, A. et al. Closed-cell aluminum foam of improved sound absorption ability: Manufacture and properties. *Metals* 2014; 4(3): 445–454.

48. Jiejun, W. et al. Damping and sound absorption properties of particle reinforced Al matrix composite foams. *Compos Sci Technol* 2003; 63: 569–574.

49. Kádár, C. et al. Acoustic emission measurements on metal foams. *J Alloys Compd* 2004; 378(1–2): 145–150.

50. Fiedler, T. et al. Non-linear calculations of transient thermal conduction in composite materials. *Comput Mater Sci* 2009; 45(2): 434–438.

51. Fiedler, T. et al. Lattice Monte Carlo and experimental analyses of the thermal conductivity of random-shaped cellular aluminum. *Adv Eng Mater* 2009; 11(10): 843–847.

52. Fiedler, T. et al. Computed tomography based finite element analysis of the thermal properties of cellular aluminium. *Mater.wiss Werkst.tech* 2009; 40(3): 139–143.

53. Lu, T.J., Stone, H.A., Ashby, M.F. Heat transfer in open-cell metal foams. *Acta Mater* 1998; 46(10): 3619–3635.

54. Lu, T.J., Chen, C. Thermal transport and fire retardance properties of cellular aluminium alloys. *Acta Mater* 1999; 47(5): 1469–1485.

55. Solórzano, E. et al. An experimental study on the thermal conductivity of aluminium foams by using the transient plane source method. *Int J Heat Mass Transf* 2008; 51(25–26): 6259–6267.

56. Golovin, I.S., Sinning, H.R. Damping in some cellular metallic materials. *J Alloys Compd* 2003; 355(1–2): 2–9.

57. Banhart, J., Baumeister, J. Deformation characteristics of metal foams. *J Mater Sci* 1998; 33(6): 1431–1440.

58. Rasooli, A. et al. Kinetics and mechanism of titanium hydride powder and aluminum melt reaction. *Int J Miner Metal Mater* 2012; 19(2): 165–172.

59. Jiménez, C. et al. Partial decomposition of TiH_2 studied in situ by energy-dispersive diffraction and ex situ by diffraction microtomography of hard X-ray synchrotron radiation. *Scr Mater* 2012; 66(10): 757–760.

60. Cymat. http://www.cymat.com/. 2015 [cited 2015 20.07.2015]; Available from: http://www.cymat.com/

61. Stöbener, K. et al. Forming metal foams by simpler methods for cheaper solutions. *Metal Powder Report* 2005; 60(1): 12–16.

62. Neugebauer, R., Hipke, T. Machine tools with metal foams. *Adv Eng Mater* 2006; 8(9): 858–863.

63. Metalfoam, P. http://metalfoam.de/. 2015 [cited 2015 27.07.2015]; Available from: http://metalfoam.de/

64. IWU. http://www.iwu.fraunhofer.de/. 2015 [cited 2015 27.07.2015]; Available from: http://www.iwu.fraunhofer.de/

65. Baumeister, J., Banhart, J., Weber, M. Aluminium foams for transport industry. *Mater Design* 1997; 18(4–6): 217–220.

66. Kretz, R., Götzinger, B. Energy absorbing behaviour of aluminium foams: Head impact tests on an A-pillar of a passenger car. In: Banhart, J., Ashby, M.F. and Fleck, N.A. (Eds.), *Cellular Metals and Metal Foaming Technology: Metfoam.* MIT Publishing: Bremen, 2001.

67. Hanssen, A.G. et al. Optimisation of energy absorption of an A-pillar by metal foam insert. *Int J Crashworthiness* 2006; 11(3): 231–241.

68. Viehweger, B., Sviridov, A. Frontmodule für Schienenfahrzeuge aus Aluminiumschaumsandwich – Fertigungstechnologien zur Bauteilherstellung. *Forum der Forschung* 2007; 20: 69–72.

69. Duocel. http://www.ergaerospace.com/index.html. 2015 [cited 2015 22.07.2015]; Available from: http://www.ergaerospace.com/index.html

70. https://www.martec-era.net/. https://www.martec-era.net/. 2015 [cited 2015 27.07.2015]; Available from: https://www.martec-era.net/

71. Hanssen, A.G. et al. A numerical model for bird strike of aluminium foam-based sandwich panels. *Int J Impact Eng* 2006; 32(7): 1127–1144.

72. Thoma, K., Wicklein, M., Schneider, E. New protection concepts for meteoroid/debris shields. In: *European Conference on Space Debris*. ESA: Darmstadt, Germany, 2005.

73. Ryan, S., Christiansen, E., Lear, D. *Shielding against micrometeoroid and orbital debris impact with metallic foams*. NASA Bienial reports. 2011. p. 267–269.

74. http://www.alcarbon.de/. http://www.alcarbon.de/. 2015 [cited 2015 27.07.2015]; Available from: http://www.alcarbon.de/

75. Foamtech. http://www.foamtech.co/kreng02/. 2013 [cited 2013 20.02; accessed:]. Available from: http://www.foamtech.co/kreng02/

76. Miyoshi, T. et al. ALPORAS aluminum foam: Production process, properties, and applications. *Adv Eng Mater* 2000; 2(4): 179–183.

77. IKEDA, M. et al. Aerodynamic noise reduction in pantographs by shape-smoothing of the panhead and its support and by use of porous material in surface coverings. *Q Rep RTRI* 2010; 51(4P): 220–226.

78. Mitsumoji, T. et al. Aerodynamic noise reduction of a pantograph panhead by applying a flow control method. In: Nielsen, J.C.O. et al. (Eds.), *Noise and Vibration Mitigation for Rail Transportation Systems*. Springer: Berlin Heidelberg, 2015. pp. 515–522.

79. Paun, F., Gasser, S., Leylekian, L. Design of materials for noise reduction in aircraft engines. *Aerosp Sci Technol* 2003; 7(1): 63–72.

80. Alantum. http://www.alantum.com/. 2015; Available from: http://www.alantum.com/en/home.html

81. Exxentis. http://www.exxentis.co.uk/index.html. 2018 [cited 2018 20.04.2018]; Available from: http://www.exxentis.co.uk/.

82. Corporation, M. http://www.mottcorp.com/products/. 2015 [cited 2015 24.07.2015]; Available from: http://www.mottcorp.com/products/

83. Recemat. http://www.recemat.nl/eng/. 2015 [cited 2015 24.07.2015]; Available from: http://www.recemat.nl/eng/

84. Paserin, V. et al. The chemical vapor deposition technique for Inco nickel foam production–manufacturing benefits and potential applications. In: Banhart, J. and Fleck, N.A. (Eds.), *Cellular Metals and Metal Foaming Technology*. MIT-Verlag: Berlin, Germany, 2003.

85. Wittstock, A. et al. Nanoporous gold catalysts for selective gas-phase oxidative coupling of methanol at low temperature. *Science*, 2010; 327(5963): 319–322.

86. m-pore. http://www.m-pore.de/. 2015 [cited 2015 24.07.2015]; Available from: http://www.m-pore.de/

87. Raval, S. spacesafetymagazine 2013 [cited 2015 23.07.2015]; Available from: http://www.spacesafetymagazine.com/aerospace-engineering/spacecraft-design/camras-nasas-co2-moisture-removal-system-ready-final-tests/

88. Montillet, A., Comiti, J., Legrand, J. Application of metallic foams in electrochemical reactors of filter-press type Part I: Flow characterization. *J Appl Electrochem* 1993; 23(10): 1045–1050.

89. Cognet, P. et al. Application of metallic foams in an electrochemical pulsed flow reactor Part II: Oxidation of benzyl alcohol. *J Appl Electrochem* 1996; 26(6): 631–637.

90. Treviño, P., Ibanez, J.G., Vasquez-Medrano, R. Chromium (VI) reduction kinetics by zero-valent aluminum. *Int J Electrochem Sci* 2014; 9: 2556–2564.

91. Langlois, S., Coeuret, F. Flow-through and flow-by porous electrodes of nickel foam. II. Diffusion-convective mass transfer between the electrolyte and the foam. *J Appl Electrochem* 1989; 19(1): 51–60.

92. Paserin, V. et al. CVD technique for Inco nickel foam production. *Adv Eng Mater* 2004; 6(6): 454–459.

24 Ceramic Foams

F. A. Costa Oliveira

CONTENTS

24.1 INTRODUCTION

Many materials have a cellular structure: an assembly of prismatic or polyhedral cells with solid edges and faces packed together to fill space. Nature and man both exploit the remarkable properties of cellular solids. To those who study nature, they are the structural materials of their subject: cork, balsa wood, coral, and trabecular bone. To the engineers, they are of upmost importance in building light weight structures, for energy management, for thermal insulation, for filtration, and so on.

Ceramic foams constitute a specific class of materials containing a high level of porosity (typically greater than 70 vol%) which are characterized by the presence of a recognizable "cell," that is an enclosed empty space possessing faces and solid edges (known as struts). The faces can be either fully solid or void, given a closed cell or an open cell material, respectively.

Examples of cellular ceramic structures include lattice structures, honeycombs, assembled hollow spheres, foams, and sponges. Open cell structures allow fluid transport, whereas closed cell ones are used for thermal insulation purposes.

Recent advances in techniques for foaming ceramics have led to their intense study, enabling the near-net shape fabrication of light weight, inorganic foam structures with tailored thermal, elastic, mechanical, electrical, and magnetic characteristics for high temperature applications. Modern imaging and analysis techniques allow their properties to be understood in greater detail. Their cellular structure gives them unique properties, such as low density, high specific strength, high permeability, high thermal shock resistance, low specific heat, and high thermal insulation, that are exploited in a variety of applications, such as filters for molten metals, hot gases and ion exchange, catalyst supports, heat exchangers, and scaffolds for cell growth. Cellular properties are very much dependent upon the material from which the foam is made, the cell topology and shape, and the relative density (ratio of the density of the foam and that of the solid from which it is made). Typically, one can properly speak of foams only when the relative density of the material is ≤ 0.3.

Typical examples of engineered cellular materials include cordierite and aluminum honeycombs, open cell nickel foam, closed cell glass foam, open cell zirconia foam, hydroxyapatite and collagen-based foams, silica foam, silicon carbide foam, alumina foam, mullite foam, and cordierite foam among others. The different properties required of cellular ceramics imply that a range of processing routes is needed to fabricate them; there is no single route to yield all the necessary structures. First, a brief overview of the main techniques available to fabricate ceramic foams will be presented, with particular attention being paid to the replication techniques (e.g., using a polymer as a template) and to the direct foaming methods. A comparison between the properties of the foams produced by different routes and the models available to predict their bulk properties will be discussed. Finally, the main applications for ceramic foams including metal casting and hot gas filtration, electrodes for electrochemical processes, space appliances, heat sinks, kiln furniture, catalyst processes, solar radiation conversion, interpenetrating composites, scaffolds made of calcium phosphate-based foams for bone tissue regeneration and porous-medium burners will be highlighted.

24.2 MANUFACTURING

In this section, the main manufacturing strategies, namely replication, sacrificial template, and direct foaming processes will be briefly described.

24.2.1 REPLICATION TECHNIQUES

The replication of polymer foams was one of the first fabricating techniques developed for producing ceramic foams with controlled macroporosity and the method was patented by Schwartzwalder and Somers [1]. It is still common in the industry because of its simplicity and affordability. The process involves coating a flexible, open cell polymer foam with a ceramic slurry. After removal of the excess of slurry by squeezing and subsequent drying, the polymer is burned out (typically at 500°C) and the ceramic is sintered at an adequate temperature. The solid content in the slurry lies in the range 50–70 wt% and the suspension must be thixotropic, that is, the viscosity decreases with time at a fixed shear rate in order to avoid excessive drainage.

Replication of polymeric foams offers a simple, inexpensive, and versatile way of producing ceramic foams with pore size in the millimeter range (Figure 24.1a). However, the quantity and toxicity of the gases released during polymer burnout, for example, hydrogen cyanide in the case of polyurethane, makes scrubbing of the waste gases necessary.

24.2.2 SACRIFICIAL TEMPLATE TECHNIQUES

The sacrificial template method consists of coating a spherical polymer particle (e.g., polystyrene) by a slurry or a gel solution of ceramic material. After drying, a donut-type building block is obtained. The structure is then calcined at a suitable temperature. Hereafter, the hollow cores are packed together in a mold and joined by a second slurry casting. Finally, the components are dried, calcined, and sintered [2]. This method is one of the most widely used for making ceramic hollow spheres. The relative density (density of foam/density of the hollow sphere wall material) is typically in the range of 9%–24%.

FIGURE 24.1   Cordierite ($Mg_2Al_4Si_5O_{18}$) foams fabricated by the replication (a) and the direct foam (b) methods.

Hollow spheres have found applications as refractory thermal insulation, kiln furniture, lightweight composites, fiber optic sensors, drug delivery capsules, gas and chemical storage, and so on.

24.2.3   DIRECT FOAMING TECHNIQUES

The direct foaming method alloys the manufacturing of ceramic foams possessing dense struts. This method consists of preparing a stable, well-dispersed, high solid content, aqueous ceramic slurry that also incorporates an organic monomer together with an initiator and a catalyst to provide *in situ* polymerization. A foaming agent (surfactant) is added to the slurry (required to stabilize the presence of bubbles) and a high-shear mixer is used to provide mechanical agitation which results in the formation of a wet ceramic foam that is cast into an appropriate mold. After polymerization, the structure is strong enough to be demolded and transferred to an oven for drying. The polymeric binder is burnt out and the foam is sintered under controllable conditions to yield a strong highly porous ceramic since they do not possess hollow struts [3]. Gel casting is another alternative method of fabricating ceramic foams by direct foaming [4,5].

Another direct foaming route similar to the existing ones was developed which consists of premixing the ceramic precursor (30–48 wt%) to two component isocyanate-polyolsystem [6,7]. After obtaining homogeneous mixtures, a surfactant, a catalyst, and water (blowing agent) were added to the polyol mixture in adequate proportions. The two components were then blended by high shear mixing at 9000 rpm for about 1 min. The mixture was poured into a polyethylene mold, allowed to grow freely, and to cure for a period of 24 h at room temperature to ensure that the reaction was completed. After burnout of the polymer at 500°C and sintering of the ceramic at 1300°C, cordierite foams with densities in the range of 150–410 kg m^{-3} having porosities as high as 90% were obtained (Figure 24.1b).

24.3   PROPERTIES

24.3.1   STRUCTURE

In 1665, Robert Hooke examining a section of cork in his microscope used the term "cell" to describe its structure. Two centuries later, in 1887, Sir William Thomson (later Lord Kelvin) identified the space-filling unit cell which minimizes surface area per unit volume as the tetrakaidecahedron (a 14-side polyhedra containing 6 square faces and 8 hexagonal ones) with slightly curved faces (Figure 24.2a). In 1993, Weaire and Phelan, at Trinity College in Dublin, discovered that space filling was 0.3% more efficient with an array of 6 polyhedra with 14 faces and 2 polyhedra with 12 faces rather than with the Kelvin structure (Figure 24.2b). The Weaire-Phelanstructure is observed on a class of chemical compounds known as clathrates and in some liquid crystals. This structure inspired the

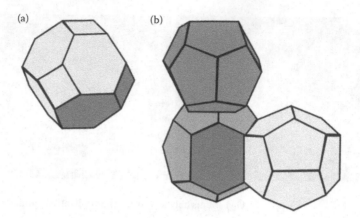

FIGURE 24.2   Kelvin (a) and Weaire-Phelan (b) structures.

construction of the "Water cube" building where the swimming competitions took place at the 2008 Beijing Olympics.

Today, most researchers use both computer tomography and magnetic resonance techniques to image 3D foams. Based on these images, numerical simulations using, for instance, finite element analysis are carried out to get information on a foam's behavior on localized volumes, the so called local effect approach.

Due to the complex nature of the three-dimensional structure of ceramic foams and their wide variety of internal structures, no single analytical technique available provides an accurate and complete description of their morphology. Combining information from various techniques seems to be the only way to obtain an overall picture of the real 3D architecture of these materials.

2D image analysis provides useful numerical information on pore characteristics such as the perimeter, the pore diameter, the porosity, and the shape factor of the pores [8,9].

For many applications, foam performance is directly influenced by cell size. For more than 25 years, the cellular structure and the cell size have been defined by the unit ppi (pores per inch). The number of pores is counted over a standard length of one inch.

The ceramic foam morphology is dependent on the fabrication technique used. Since the resulting ceramic foam is a direct replica of the original foam, the polymer foam structure and pore size are critical in determining its properties, for instance, its density and permeability. The main disadvantage of this process is that the ceramic struts are hollow and possess sharp edges (Figure 24.3a) because during sintering, the polymer is burnt out and totally decomposes. In addition, a large number of longitudinal cracks are formed upon firing due to the shrinkage and gas evolution resulting from both the combustion of the polymer and the thermal expansion mismatch between the

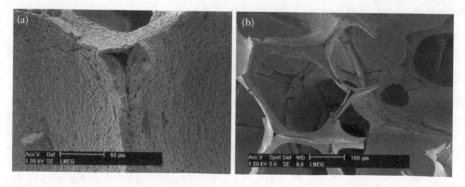

FIGURE 24.3   Typical features of cordierite foams obtained by the replication technique: triangular hollow in a strut (a) and cracking of the ceramic network (b) both due to volatilization of the polymer foam template.

ceramic and the polymer (Figure 24.3b). These defects significantly lower the strength of the final product. In addition, the replica technique can only produce open cell structures while the direct foaming techniques allow fabricating either open or closed cell foams. In open cell foams, the cells are interconnected such that there is a continuous pore phase and the solid is an interconnected array of struts. When the ceramic material is present only in the struts, these foams are referred to as reticulated foams. On the other hand, the cell windows can be either open or closed. When some cells present the windows closed with a thin membrane they are of the semiclosed cell type (as shown in Figure 24.1a). Closed cell foams have a thin membrane of solid in the cell faces sealing them from neighboring cells. In addition, ceramic foams also contain flaws formed during polymer removal due to the thermal expansion mismatch of the ceramic and polymer materials (Figure 24.3b).

24.3.2 MECHANICAL PROPERTIES

Modeling of mechanical properties of bulk foams is usually made using Gibson-Ashbyand Zhang approaches derived using a cubic unit cell and a Kelvin cell, respectively. Basically, each foam property is proportional to the relative density, that is the ratio of the bulk density and the density of the material from which they are made. Both the geometric constants and the power exponents are generally in good agreement with the experimental ones [10].

Sometimes ceramic foams are too weak to be used in any application. Hence, in order to improve their strength, they must be submitted to a dip-coating process, which consists of dipping the sintered foams into a suitable ceramic aqueous slurry [11]. In this case, different aqueous slurries with solid contents ranging from 30 to 60 wt% were prepared using the cordierite ceramic precursor. To provide well dispersed ceramic particles, the slurries were milled for 24 hours in a ball mill. A vacuum degassing was applied during the dip-coating of the sintered foams to remove air bubbles from the aqueous slurries. After drying, the coated ceramic foams were sintered under the same conditions described above.

Foams prepared by the direct foaming method (uncoated foams) presented strut thicknesses in the range of 60–70 μm and a cell mean diameter around 549 ± 99 μm. By increasing the solid content of the dip-coating slurry, an increase in the amount of ceramic material deposited in the ceramic foam walls was observed, resulting in thicker struts (70–90 μm for the 30% solid content slurries and 100–115 μm for the 60% ones) and smaller cells (483 ± 93 and 456 ± 84 μm, for the 30% and 60% solid content, respectively). The uncoated cordierite foams were very fragile compared to the coated ones, showing compressive strengths lower than 0.1 MPa. Such low strength is associated with the poor densification of the struts. On the other hand, the compressive strength was increased by a factor of 3 (0.3 MPa for the 30% solid content slurry) to 10 (1.0 MPa for the 60% solid content slurry) as the relative density of foams increased from 0.06 (uncoated foams) to 0.08 (30% solid content slurry) and 0.18 (60% sold content slurry). This effect of relative density on foam strength is in agreement with the predictions of the model developed by Gibson and Ashby [12]. In the case of cordierite foams fabricated by the replication method, the compressive strength of these foams increased (from 0.1 to 2 MPa) with increasing solids volume fraction [13]. These data are in agreement with the predictions of the model developed by Gibson and Ashby. However, the exponent predicted by the model was half of the measured one (≈3) over the range of relative densities investigated (80%–90%). Such discrepancy appears to be related to several factors, including the morphological differences in the structural unit of the developed foams with respect to a cubic open cell foam and to the presence of both open and closed cells. Indeed, the cordierite foams produced were of semiclosed cell type.

Very limited information is available on the properties of ceramic hollow spheres. In the case of alumina foams with 10% relative density, the compressive strength of individual hollow spheres was in the range of 25–31 MPa, whereas compressive strength of foams made of hollow spheres was in the range of 1–4 MPa [14].

Typically, the bending strength of foams fabricated using direct foaming techniques varies between 2 and 26 MPa for alumina foams of relative density ranging from 8% to 30%. The higher strength magnitudes obtained for this material compared to other conventional methods are not only due to the less flawed structure, but also to the dense struts and walls produced [3].

The elastic properties of open cell cordierite foams obtained by the replication technique were measured as a function of their relative densities by the impulse excitation method [15]. The values of Young's modulus (1–5 GPa) and the shear modulus (0.6–2.5 GPa) depended on the relative density in accordance with the available theoretical models. The values of Young's modulus and shear modulus were found to be dependent upon both the relative density and the microstructure of the cell structure with the elastic moduli increasing with decreasing of porosity. The density exponent was found to be in reasonably good agreement with theoretical models that considered cell strut bending as a major deformation mode.

24.3.3 THERMAL PROPERTIES

The thermal shock resistance of open cell cordierite foams was assessed by the change in flexural strength for specimens quenched into water [16]. The intent was to simulate very drastic temperature changes. The results of these experiments indicated a gradual loss in strength with increasing ΔT (quenching temperature difference), rather than an abrupt change observed for dense samples of the same material, as illustrated in Table 24.1. This implies a damage accumulation process. Indeed, the foam retained about half of its original flexural strength after water quenching with a temperature difference of 600 K, implying that their thermal shock behavior is better than predicted by the Gibson-Ashby model. On the other hand, a sudden drop in flexural strength was observed at $\Delta T = 325$ K for the same material in the dense form, as a result of thermal stresses leading to the propagation of macrocracks within the material at the critical condition. A similar pattern was observed for open cell alumina-mullite material [10].

24.4 NEW DEVELOPMENTS AND SUCCESSFUL APPLICATIONS

Taking advantage of their high permeability, low mass, and high thermal insulation, ceramics foams are mainly used in liquid metal filtration, hot gas (particulate) filtration, kiln furniture, catalyst processes, solar radiation conversion, 3D interpenetrating composites, biomedical and implant technology, and porous burners, among others [18].

24.4.1 FILTRATION

The use of ceramic foams in liquid metal filtration allows removal of inclusions, reduces trapped gas, and provides laminar flow. An example of a typical filter is shown in Figure 24.4. Since 1974, open cell alumina foams are being used by the aircraft industry for considerable reduction of aluminum casting scrap.

TABLE 24.1
Thermal Shock Resistance of Dense and Open Cell Cordierite Materials Quenched in Water

	Model	Estimated Value (K)	Experimental Value (K)
Hasselman [17]	$R = \dfrac{\sigma_r(1-\nu)}{E\alpha} = \Delta T_c$	128	325
Gibson and Ashby [12]	$R_b = R_s \cdot \left(\dfrac{C_2}{C_1}\right) \cdot \left(\dfrac{\rho}{\rho_s}\right)^{-0.5}$	232	>350

Note: $C_1 = 0.5$; $C_2 = 0.16$

FIGURE 24.4   Silicon carbide filters used in iron casting ($45 \times 45 \times 20$ mm^3).

Multifunctional filters based on catalytic ceramic foams are under development for application in waste incineration plants for abatement of fly ashes, NO_x, VOCs, and dioxins [19]. In this application, the filter is used to host a catalyst capable of promoting the abatement of gaseous pollutants inserted in Goretex filter bags. As these bags typically work at 200–210°C, innovative catalysts active in such temperature range for the selective catalytic reduction of NO_x with ammonia and simultaneous combustion of volatile organic compound (VOCs) must be developed and applied over the foam structure to satisfy the requirements of this application.

24.4.2   ELECTRODES

Reticulated vitreous carbon (RVC) foams are being employed as porous electrodes for electrochemical processes [20]. RVC is regarded as a useful electrode material, particularly where high current densities, low electrical/fluid flow resistance, and minimal cell volume loss to electrodes are required. Other RVC applications include: high temperature insulation, filters, storage batteries, scaffolds for biological cell growth, the manufacture of semiconductors, and acoustic appliances.

24.4.3   SPACE APPLIANCES

Ultramet open cell foam-based propellant injectors are being tested for burning hydrogen/air mixtures at the Edwards Air Force Research Laboratory in the USA [21].

DUOCEL SiC foams with 100 ppi and 10% density from ERG Aerospace (USA) are employed as hot gas filters for missile propulsion systems [22].

Silica aerogel is a unique silicon-based substance nicknamed "solid smoke" because it contains 99.8% air. It is one of the lightest solid materials known. It was used as insulation on the Sojourner Mars rover which landed on Mars on July 4th, 1997.

24.4.4   HEAT SINKS

Silicon carbide foams (both 100 and 300 ppi) from Ultramet are also applied in high performance liquid cooled heat sinks for removing heat from high power (up to 1000 W cm^{-1}) electronic components [23].

24.4.5   KILN FURNITURE

Ceramic foams used as kiln furniture has several advantages compared to their dense counterparts including longer life, better uniformity of the atmosphere surrounding the fired components, reduction of frictional forces generated during parts shrinkage, chemical inertness, and cost benefits [24]. The 60%–80% lower mass compared to dense kiln furniture results in much less material to heat in each thermal cycle and lower energy costs.

24.4.6   CATALYST PROCESSES

More than 80% of all base chemicals and products are produced or converted using a catalyst. Most of these processes are heterogeneously catalyzed (meaning that the physical condition of the solid catalyst is different of the liquid or gas starting material/product) at temperatures in the range 100–600°C. A well known example is the three-way catalyst used in automotive exhaust gas cleaning since 1974 relying on cordierite honeycombs coated with a so called wash coat consisting of γ-alumina stabilized with ceria containing a catalytically active material, typically Pt and Rh.

From the view point of catalytic performance, ceramic foams can be superior in some applications using honeycombs or packed-bed catalysts. Indeed, ceramic foams with adequate mechanical properties offer some advantages, in particular excellent mixing and mass and heat transfer in the radial directions owing to the flow tortuosity within the foam structure. However, catalytic applications of ceramic foams are still restricted to laboratory-scale experiments.

Cordierite foams developed at INETI—Instituto Nacional de Engenharia e Tecnologia Industrial (now LNEG) [25] were used as substrates for catalytic combustion of NO_x [26] and toluene, a common VOC used in the paper industry [27]. For this purpose, a new method of coating ceramic foams by a zeolite-based slurry has been successfully developed and patented [28,29]. The light off temperature of toluene was reduced by about 15°C when increasing the foam porosity from 70% to 90%. On the other hand, the Pt catalyst directly impregnated on the cordierite foams (PtCF) has a light off temperature $T_{50\%}$ higher (~50°C) than PtMFI coated ones, which attests to the improving effect of zeolite on the catalytic active phase [27]. PtMFI catalyst was prepared by ionic exchange in pentasil zeolite—also known as ZSM-5, framework type MFI—(Si/Al = 15, Zeolyst Corp.) in order to introduce 0.1 wt% of platinum.

24.4.7   SOLAR RADIATION CONVERSION

Many successful tests have shown the feasibility of using ceramic foam absorbers in high temperature applications. As ceramics offer high temperature resistance, they play a crucial role in the design of volumetric absorbers.

Advanced volumetric absorber silicon carbide foams (80/20 ppi) were employed to absorb concentrated solar radiation and to transfer the energy to a fluid passing through its open cells [30]. Typically, air is forced through the pores of the material and is heated to about 700°C. This hot air is then used to generate steam for a conventional steam turbine process. Due to its dark color, silicon carbide has been used for this purpose since it reaches weighted solar absorption values of 0.9.

Taking advantage of a visible light induced photocatalytic activity phenomenon, carbon-doped TiO_2 hollow spheres are being proposed for water disinfection purposes [31].

24.4.8   INTERPENETRATING COMPOSITES

The infiltration of ceramic foams offers the potential for producing tailored interconnectivity interpenetrating composites. In fact, ceramic reinforced aluminum matrix composites (AMCs) have received substantial attention from the automotive and aerospace industries due to their light weight, high elastic modulus, improved strength, and good wear resistance. The latter is one the most attractive properties of the AMCs in applications including automotive brake systems, callipers, pistons, cylinder liners, connecting rods, and turbine compressors. A pressureless infiltration technique was developed to produce metal-ceramic interpenetrating composites consisting of 3D-continuous matrices of discrete metal and ceramic phases by infiltrating molten aluminum alloys into a range of gel-cast ceramic foams produced from alumina, mullite, and spinel [32].

24.4.9 BIOMEDICAL APPLICATIONS

Bone grafts are necessary in orthopedic surgery for filling bone cavities, treatment of nonunion, and replacement of bone lost during trauma and tumor removal. Porous calcium phosphate-based ceramics are attractive for use as synthetic bone grafts allowing successful tissue ingrowth, which further enhances the implant-tissue attachment.

Hydroxyapatite (HA) bioceramic foams with interconnected porosity and controlled pore sizes required to simulate natural bone tissue morphology were fabricated by the replication technique through impregnation of reticulated polymeric foams with ceramic slips (Figure 24.5) [33]. Open cell HA foams with porosities of about 80% were obtained, that is, 30% higher than that determined for commercial ones (50%). Many of the commercial foam cells approach 500 μm in diameter whereas the developed foam cell size ranged from 300 up to 500 μm. Indeed, it is generally accepted that a minimum pore size of 100 μm is necessary for the porous implant materials to function properly and a pore size greater than 200 μm is an essential requirement for osteoconduction.

The ultimate compressive strength of the developed foams (1–2 MPa) was found to be higher than that recorded for the commercial ones (0.7 MPa), indicating that these foams can be more easily modeled by the surgeons than the ones in the market. Both the Young's modulus (2–3 GPa) and the compressive strength of the developed foams were found to increase with an increase of the relative density, in accordance with the predictions of available micromechanical models. Due to high cost of the HA powder, a fully automated synthesis process based on the chemical precipitation method was used to produce hydroxyapatite (HA) powders, with a molar ratio of Ca/P = 1.67 ($Ca_{10}(PO_4)_6(OH)_2$), for biomedical applications [34,35]. Results showed that the pH of the reaction is the main control parameter affecting the Ca/P molar ratio of the powders. The source of water also influences the phases formed upon calcination at 1000°C [36]. These powders were found to be suitable for producing HA foams by the replication technique.

FIGURE 24.5   HA foams fabricated by the replication technique ($125 \times 25 \times 12$ mm^3).

FIGURE 24.6 Porous gas burner manufactured by ECO Ceramics BV prior to (a) and during operation (b) in radiant mode (450 × 450 × 10 mm³).

24.4.10 POROUS BURNERS

Mullite ($3Al_2O_3.2SiO_2$) foams developed by ECN—Energy research Centre of the Netherlands (Petten, The Netherlands) and manufactured by ECO-Ceramics BV (Velsen-Noord, The Netherlands) are used in flat flame premixed surface burners due to their low density, high thermal stability, low permeability, and excellent thermal shock resistance. These burners are characterized by low emissions (NO_x < 20 mgkWh^{-1}; CO < 3 mgkWh^{-1}), low noise, large modulating range (100–2000 kW m^{-2}), high radiation efficiency in the range of 100 up to 600 kW m^{-2}, and being suitable for all type of gases such as natural gas, propane, butane, coke oven gas, landfill gas, hydrogen/ methane mixtures containing up to 75% H_2. Common applications include both domestic appliances such as hot water boilers and gas fired imitation fireplaces and industrial ones such as boosters for hot air generation for heating large buildings and for paper, textile, and glazes drying. At 30% excess air, the maximum pressure drop across the ceramic foam tile is below 200 Pa at 2000 kW m^{-2}. A significant decrease in NO_x emissions is achieved in radiant mode (200–600 kW m^{-2}). Such burners produce a flame very near to the burner support, which operates at a high temperature (up to about 1100°C for an air to fuel ratio n = 1.3) and radiates a part of the heat released by combustion to the appliance (Figure 24.6). Hence, the combustion temperature and thus the NO_x emissions decrease.

The damage imposed on open cell mullite ceramic foams was evaluated in premixed radiant gas burners under controllable operating conditions. After exposure to the prevailing combustion environment, foams suffered moderate strength degradation as a result of thermal stresses being imposed on the material during service. Damage in mullite foams was mainly localized at the top layer of the burners where higher temperatures and steeper thermal gradients were imposed on the material. There was also evidence of chemical attack during combustion although thermal shock measurements suggest that damage sustained by the foams resulted mainly from thermal shock rather than chemical degradation [37,38].

Another approach concerns the development of volumetric porous radiant burners, which completely trap the combustion flame inside the porous media and have the same benefits as radiant surface burners, being capable of operating at significantly higher heat loads (up to 8000 kW m^{-2} under atmospheric pressure for methane/air mixtures) [39].

24.4.11 CO₂ SEQUESTRATION

Solar thermochemical CO_2 splitting has been investigated in a solar cavity-receiver containing a reticulated porous ceramic (RPC) foam made of pure CeO_2 [40]. The results obtained so far at a laboratory scale are quite promising. The superior reactor performance was attributed to the relatively large density and macroporosity of the RPC, which enabled high mass loading and volumetric absorption of concentrated solar radiation. Indeed, solar-driven thermochemical cycles based on metal oxide redox reactions are known to be able to split H_2O and CO_2 to produce H_2 and CO (syngas), the precursors to the catalytic synthesis of conventional liquid fuels for the transportation sector [41].

24.5 SUMMARY

Because of their particular structure, ceramic foams display a wide variety of specific properties which make them suitable for various engineering applications. Not surprisingly, new applications for ceramic foams are currently under development.

Advanced ceramic foams must be properly engineered and integrated within each system to fulfill the requirements of a given application.

Component design includes selecting the adequate material, cell size, relative density, and fabrication technique to ensure optimal performance. The fabrication technique influences the foam properties since it affects the morphology, which is the flaw population of the ceramic material that constitutes the foam.

Novel fabrication techniques and developments in established manufacturing technologies should allow the production of components with improved reliability and properties, which can be optimized for a targeted appliance. Hence, there are increasing efforts to use ceramic foams in different fields as engineers become aware that these materials are especially suited for fulfilling some of the unique requirements of advanced technological developments.

ACKNOWLEDGMENTS

The author would like to acknowledge the contributions made by his coworkers for the research work that he has been carrying out in the field of ceramic foams in the past 12 years, namely José Maria Ferreira (UA), Susana Olhero (UA), Eduardo Pires (Ceramed), Elisa Costa (CATIM), Claudia Ranito (UNL), João Paulo Borges (UNL), Maria Fátima Vaz (IST), Jorge Fernandes (IST), Elisabete Silva (IST), Maria Filipa Ribeiro (IST), João Miguel Silva (ISEL), Fernando Ramôa Ribeiro (IST), João Bordado (IST), Susana Dias (LNEG), Carmen Rangel (LNEG), Carlos Henriques (IST), Rita Catalão (IST), Maria José Matos (LNEG), Nuno Correia (IST), Diamantino Dias (Rauschert Portuguesa Lda.). The financial support of FCT—Fundação para a Ciência e Tecnologia and EC—Brite Euram III programme is also acknowledged.

REFERENCES

1. Schwartzwalder, K., Somers, A.V. US Patent 3090094, 1963.
2. Luyten, J., Mullens, S., Cooymans, J., De Wilde, A.M., Thijs, I., Kemps, R. Different methods to synthesize ceramic foams. *J Eur Ceram Soc* 2009; 29: 829–832.
3. Sepulveda, P., Binner, J.G.P. Processing of cellular ceramics by foaming and *in situ* polymerisation of organic monomers. *J Eur Ceram Soc* 1999; 19: 2059–2066.
4. Janney, M.A., Marietta, M. Methods for moulding ceramic powders. US Patent 4894194, 1990.
5. Omatete, O.O., Janney, M.A., Strehlow, R.A. Gel-casting—A new ceramic forming process. *Ceram Bull* 1991; 70: 1641–1649.
6. Silva, E., Correia, N., Silva, J.M., Oliveira, F.A.C., Ribeiro, F.R., Bordado, J.C., Ribeiro, M.F. Produção de espumas cerâmicas de cordierite por polimerização *in situ*. PT Patent 103532, 2007 (in portuguese).
7. Silva, E., Correia, N., Silva, J.M., Oliveira, F.A.C., Ribeiro, F.R., Bordado, J.C., Ribeiro, M.F. Processing of cordierite foams by direct foaming. *Polymery* 2007; 52: 351–356.
8. Matos, M.J., Dias, S., Oliveira, F.A.C. Macrostructural changes of polymer-replicated open-cell cordierite-based foams upon sintering. *Adv Appl Ceram: Struct Funct Bioceram* 2007; 106: 209–215.
9. Matos, M.J., Vaz, M.F., Fernandes, J.C., Oliveira, F.A.C. Structure of cellular cordierite foams. *Key Eng Mater* 2004; 455–456: 163–167, Tech Publications Ltd.: Switzerland.
10. Brezny, R., Green, D.J. *Materials Science and Technology – A Comprehensive Treatment, Vol. 11, Structure and Properties of Ceramics*. VCH: Germany, 1994, pp. 463–516.
11. Silva, E., Silva, J.M., Oliveira, F.A.C., Ribeiro, F.R., Bordado, J.C., Ribeiro, M.F. Strength improvement of cordierite foams by a dip coating method. *Mater Sci Forum* 2008; 587–588: 123–127, Trans Tech Publications Ltd.: Switzerland.
12. Gibson, L.G., Ashby, M.F. *Cellular Solids, Structure & Properties*. Cambridge Univ. Press: Cambridge, U.K., 1997.
13. Oliveira, F.A.C., Dias, S., Vaz, M.F., Fernandes, J.C. Behaviour of open-cell cordierite foams under compression. *J Eur Ceram Soc* 2006; 26: 179–186.

14. Thijs, I., Luyten, J., Mullens, S. Producing ceramic foams with hollow spheres. *J Am Ceram Soc* 2003; 87: 170–172.

15. Oliveira, F.A.C. Elastic moduli of open-cell cordierite foams. *J Non-Cryst Solids* 2005; 351: 1623–1629.

16. Oliveira, F.A.C., Dias, S., Fernandes, J.C. Thermal shock behaviour of open-cell cordierite foams. *Mater Sci Forum* 2006; 514–516: 764–767, Trans Tech Publications Ltd.: Switzerland.

17. Hasselman, D.P.H. Strength behavior of polycrystalline alumina subjected to thermal shock. *J Am Ceram Soc* 1970; 53: 450–455.

18. Scheffler, M., Colombo, P. (Eds.) *Cellular Ceramics, Structure, Manufacturing Properties and Applications*. Wiley-VCH GmbH & Co.KgaA: Weinheim, 2005.

19. Fino, D., Russo, N., Saracco, G., Specchia, V. A multifunctional filter for the simultaneous removal of fly-ash and NO_x from incinerator flue gases. *Chem Eng Sci* 2004; 59: 5329–5336.

20. Friedrich, J.M., Ponce-de-León, C., Reade, G.W., Walsh, F.C. Reticulated vitreous carbon as an electrode material. *J Electroanalytical Chem* 2004; 561: 203–217.

21. http://ultramet.com/refractory-open-cell-foams-carbon-ceramic-and-metal/fuel-injectors/ (Accessed on) 2018/04/16).

22. http://ergaerospace.com/applications/duocel-foam-high-temperature-filters/ (Accessed on 2018/04/16).

23. https://ultramet.com/refractory-open-cell-foams-carbon-ceramic-and-metal/thermal-management/ (Accessed on 2018/04/16).

24. http://seleeac.com/kiln-furniture/ (Accessed on 2018/04/16).

25. Oliveira, F.A.C., Dias, S., Mascarenhas, J., Ferreira, J.M.F., Olhero, S., Dias, D. Fabrication of cellular cordierite foams. *Key Eng Mater* 2004; 455–456: 177–181, Trans Tech Publications Ltd.: Switzerland.

26. Dias, S., Oliveira, F.A.C., Henriques, C., Ribeiro, F.R., Rangel, C.M., Ribeiro, M.F. Selective catalytic reduction of NO_x over zeolite-coated cordierite-based ceramic foams: Water deactivation. *Mater Sci Forum* 2008; 587–588: 810–814, Trans Tech Publications Ltd.: Switzerland.

27. Ribeiro, F., Silva, J.M., Silva, E., Vaz, M.F., Oliveira, F.A.C. Catalytic combustion of toluene on Pt zeolite coated cordierite foams. *Catalysis Today* 2011; 176: 93–96.

28. Silva, E., Silva, J.M., Vaz, M.F., Oliveira, F.A.C., Ribeiro, M.F. Cationic polymer surface treatment for zeolite washcoating deposited over cordierite foam. *Mater Lett* 2009; 63: 572–574.

29. Silva, E., Silva, J.M., Oliveira, F.A.C., Ribeiro, F.R., Vaz, M.F., Ribeiro, M.F. Método e aplicação de revestimento à base de zeólitos sobre espumas cerâmicas. PT Patent 104022, 2009 (in Portuguese).

30. Fend, T., Pitz-Paal, R., Reutter, O., Bauer, J., Hoffschmidt, B. Two novel high-porosity materials as volumetric receivers for concentrated solar radiation. *J Sol Energy Mater Sol Cells* 2004; 84: 291–304.

31. Shi, J.-W., Zong, X., Wu, X., Cui, H.-J., Xu, B., Wang, L., Fu, M.-L. Carbon-doped titania hollow spheres with tunable hierarchical macroporous channels and enhanced visible light-induced photocatalytic activity. *Chem Cat Chem* 2012; 4: 488–491.

32. Liu, J., Binner, J., Higginson, R. Dry sliding wear behaviour of co-continuous ceramic foam/aluminium alloy interpenetrating composites produced by pressureless infiltration. *Wear* 2012; 276–277: 94–104.

33. Ranito, C.M., Oliveira, F.A.C., Borges, J.P. Hydroxyapatite foams for bone replacement. *Key Eng Mater* 2005; 284–286: 341–344, Trans Tech Publications Ltd.: Switzerland.

34. Ranito, C.M., Nogueira, C.A., Domingues, J., Pedrosa, F., Oliveira, F.A.C. Optimization of the synthesis of hydroxyapatite powders for biomedical applications using Taguchi's method. *Mater Sci Forum* 2006; 514–516: 1025–1028, Trans Tech Publications Ltd.: Switzerland.

35. Dias, S., Lourenço, V., Nogueira, C.A., Oliveira, F.A.C. Contrasting morphology of calcium phosphate powders prepared by the wet method using poly (ethylene glycol). *Mater Sci Forum* 2008; 587–588: 32–36, Trans Tech Publications Ltd.: Switzerland.

36. Lourenço, V., Dias, S., Nogueira, C.A., Oliveira, F.A.C. Formation of secondary phases on hydroxyapatite powders for bone regeneration: Effect of water source. *Mater Sci Forum* 2008; 587–588: 37–41, Trans Tech Publications Ltd.: Switzerland.

37. Oliveira, F.A.C. Damage assessment of mullite ceramic foams in radiant gas burners. *Adv Mater Sci Forum* 2008; 587–588: 99–103, Trans Tech Publications Ltd.: Switzerland.

38. Alliat, I., Olagnon, C., Chevalier, J., Oliveira, F.A.C. How to predict the lifetime of ceramic foam for radiant burners? In: *Proceedings of IGRC 2001-International Gas Research Congress*, 2001, Amsterdam, The Netherlands.

39. http://www.gogas.com/GoGaS_Ceramic_Burner.pdf (Accessed on 2018/04/16).

40. Furler, P., Scheffe, J., Gorbar, M., Moes, L., Vogt, U., Steinfeld, A. Solar thermochemical CO_2 splitting utilizing a reticulated porous ceria redox system. *Energy Fuels* 2012; 26: 7051–7059.

41. Smestad, G.P., Steinfeld, A. Review: Photochemical and thermochemical production of solar fuels from H_2O and CO_2 using metal oxide catalysts. *Ind Eng Chem Res* 2012; 51: 11828–11840.

25 Ceramic Foams via a Direct Foaming Technique

Philip N. Sturzenegger and Urs T. Gonzenbach

CONTENTS

25.1 INTRODUCTION

Although the synthesis of porous ceramics has been a growing field of scientific research since the late 1990s, it is just in recent years that foamed ceramic products have found their way to industrial applications. On one hand, this is astonishing because in energy intensive industries like fuel refinement or cement and metal production there is a clear need for effective high temperature insulation materials. Here, porous ceramics could be a suitable substitute for mineral fiber-based products that suffer from decreasing market acceptance due to real and potential health issues. On the other hand, everyone who was ever involved in the development of ceramic foams knows all too well that the road from unveiling the basic laws of foam stabilization to the manufacture of high quality foam products can be long and stony. There exist three basic strategies to manufacture ceramic foams [1]. Coating natural or synthetic foam structures with a ceramic suspension followed by removal of the template is called the replica method. Thermal and chemical transformation of a sponge into ceramic foam also belong to this category. Replica foaming typically results in open porous ceramic parts with intermediate to high porosity. Such foams are, for example, applied as filters for molten metals or catalyst carriers. Sacrificial templating summarizes methods where pores are formed by blending a ceramic suspension with any sort of a pore template followed by removal of the template after drying or while sintering the ceramic. Examples for pore templates are sawdust or polymeric beads. Low to intermediately high porosity foams are produced with this method. It allows for a good control over the porosity and shape of the pores. Under direct foaming we understand techniques where a ceramic suspension is directly foamed by incorporating bubbles using mechanical or chemical means. Generally, the bubbles are stabilized with the help of surfactants that adsorb to the air-water interface and reduce the free energy of the system. Directly foamed ceramic parts cover the whole range from low to very high porosity and show mainly closed pores. However, especially at high pore volume fractions, surfactant-stabilized foams usually exhibit pore openings and as a result connective channels throughout the foamed part. This is especially disadvantageous for thermal insulation applications where pore connectivity decreases insulation performance. Pore windows develop because surfactant-stabilized foams typically suffer from degradation mechanisms like drainage, bubble coalescence or Ostwald ripening. These processes cause not only open porosity at high air contents, but generally limit the control over the microstructure of directly foamed parts. A much more effective bubble stabilization is achieved through the adsorption of fine ceramic particles to the air–water interface [2]. If particles exhibit intermediate wettability with a contact angle between 70° and 86° at the air water interface, they cover the surface of bubbles during foaming and lead to wet foams of excellent stability [3]. In this contribution, we discuss the properties of high porosity, closed cell α-Al_2O_3 ceramic foam that is produced via the promising technique of direct foaming using *in situ* hydrophobized particles as foam stabilizers [4–7]. The term 'promising' is used here because the method of particle stabilization allows for processing extremely stable wet foams from a broad variety of ceramic materials. While the unprecedented wet foam stability assures good reproducibility and an excellent control over end product properties, the versatility of the method in respect to the use of raw materials allows broadening the scope of possible applications. In addition, the range of foam porosities that can be covered reaches from a low 30 up to 95 vol%. The foam porosity P_f is defined herein according to Equation 25.1 and is calculated from the foam density ρ_f and the theoretical density of the raw material ρ_d.

$$P_f = \left(1 - \frac{\rho_f}{\rho_d}\right) * 100\% \qquad (25.1)$$

In the following paragraphs, the technology of *in situ* particle hydrophobization is presented and the processing route to high porosity ceramics discussed. After the description of the foam properties, we report on a specific application before we complete the contribution by mentioning other foam products that are processed using the same technology.

25.2 *IN SITU* HYDROPHOBIZATION TECHNOLOGY

The surface of ceramic particles is typically hydrophilic and thus in suspension well wetted by water. Under such conditions, the particles are not likely to adsorb to the air-water interface and stabilize foams efficiently. As a consequence, no or just small volumes of air are stabilized in such a system. An efficient particle stabilization of air bubbles requires partial dewetting of the particles at the air-water interface. Therefore, the *in situ* hydrophobization technology is applied to partially hydrophobize the surface of the particles directly in suspension by adsorbing short-chain amphiphilic molecules [5,6]. It was shown that suitable amphiphiles for surface modification are comprised of 3 to 8 carbon atoms. The hydrophilic head group of the molecules is chosen such that it binds physically or chemically to the particle surface whereas the hydrocarbon tail decreases the hydrophilicity of the surface with increasing concentration of the amphiphiles [8]. The required amphiphile concentration in respect to powder weight varies typically between 0.05% and 0.5%. The higher the solids loading of the suspension and the molecular weight of the hydrocarbon chain, the lower is the required amphiphile concentration to later on produce stable foam. The surface-modified particles promote the formation of small bubbles during foaming and effectively stabilize foam by almost completely occupying the air–water interface [6]. Although the surface modification leads to some viscosity increase, the technique still allows for processing of suspensions at high solids concentration. This fact is decisive if the final goal is to produce porous ceramics as opposed to creating wet foam.

Figure 25.1 shows the process flow chart for the manufacture of high porosity closed cell foams via direct foaming using *in situ* hydrophobized particles as foam stabilizers.

In a first step, a stable suspension is prepared and ball milled if necessary. The solids loading of the suspension is ideally high to minimize shrinkage of the part during drying and sintering. The upper limit of solids concentration is defined by the suspension viscosity as follows: a suitable

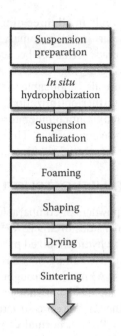

FIGURE 25.1 Process flow chart for the manufacture of high porosity closed cell foams via direct foaming using *in situ* hydrophobized particles as foam stabilizers.

suspension for foaming is very pourable with a viscosity in the range of 0.1–1 Pas at a shear rate of 100/s. In a subsequent step, the surface of the particles is partially hydrophobized adding short chain amphiphilic molecules under constant stirring. From this point on, the suspension effectively stabilizes air, therefore, unwanted air intake while stirring needs to be avoided. Thereafter, the formulation is finalized by adjusting the pH and diluting the suspension to the target solids loading. Now, the suspension is foamed by introduction of air either by physical or chemical means. This step transforms the pourable suspension into a stiff, viscoelastic wet foam. The most important wet foam properties are stability, porosity, bubble size, and foam stiffness. Stable foam maintains the original air content and bubble size over the whole course of ceramic manufacture. The wet foam stability, porosity, bubble size, and stiffness can be tailored to meet the requirements of specific applications. In practice, wet foam optimization is done by running well defined experimental series to test the impact of, for example, the solids loading and the concentration of surface modifier. At a defined solids loading, an increase in the surface modifier concentration first increases the air content to a high plateau value. Under further modifier addition, a decrease in the bubble size is generally observed, together with an increase in foam stiffness. Porosity, bubble size, and foam stiffness can be varied in a relatively wide range [4], but they are not completely independent. Stiff foams usually exhibit either air contents higher than approximately 75 vol% or small bubble sizes. In contrast, pourable foams are realized at intermediate air content of below 65 vol% or rather big bubble sizes. Once the required properties are achieved, the foam is shaped. Common formats are cylinders, sheets or bricks. But complex shapes like cups and jugs or special applications like the coating of parts by thin foam layers can also be realized. In a next step, the shaped parts need to be dried. It is advantageous to dry foams in the controlled atmosphere of a climate chamber where the temperature and the relative humidity can be defined. Drying is a crucial part in the fabrication process. Because wet and semiwet foams are mechanically delicate structures, inappropriate drying parameter settings easily lead to crack formation. Mastering the drying process gets increasingly difficult the bulkier and more voluminous a foam part is. In the majority of cases, the very origin of crack formation is the shrinkage of the foams during drying. Therefore, excessive friction or even sticking of the foam to the mold or support materials needs to be avoided. In addition, the drying parameters should be set such that the foam part can shrink evenly in all three dimensions. Early and complete drying of the foam surface while the core of the part is still wet must be avoided by all means. Binders are often found to diminish the tendency to crack formation because they increase the wet foam stiffness and reduce the moisture gradients during drying. Green foams can then be sintered. In general, sintering programs for dense ceramics are also applicable for firing of their porous counterparts. At least they do serve as a good basis for further optimization. Again, the lower mechanical strength of foams has to be borne in mind. The maximum temperature gradients tolerated by a porous body are always lower compared with the dense counterpart. That is why it is advisable to sinter foamed parts at lower heating rates of typically less than 3°C/min.

25.3 FOAM APPEARANCE AND MICROSTRUCTURE

On the basis of CT3000SG or alumina powders of comparable quality, foamed parts with porosities as high as 94 vol% and corresponding densities of 239 kg/m^3 can be manufactured by applying the technology of direct foaming with *in situ* hydrophobized particles as foam stabilizers. An example of an α-Al_2O_3 foam with 91 vol% porosity, 360 kg/m^3 density is presented in Figure 25.2a. The dimensions of the part are $222 \times 112 \times 63$ mm^3. Throughout the whole volume, the material is homogeneously structured and neither air inclusions nor gradients in the pore size can be made out. An electron micrograph showing the regular pore structure of the foam is displayed in Figure 25.2b. The pore size typically closely follows a normal distribution. The range of pore sizes for the product depicted in Figure 25.2 is 10–300 μm with a median pore size of 130 μm determined from the pore volume distribution. Such pore size distributions are typical for foams processed via direct foaming. If one aims for foams with monosized pores, other technologies like sacrificial templating appear more suitable. Nevertheless, porosities above 74 vol% can only be realized with at least two pore sizes, therefore, having a pore size distribution. A feature that is unique to the applied processing route is the mainly closed pore structure at air contents as high as 94 vol%. Such regular, closed pore microstructures can reproducibly be obtained only from a processing route that results in extremely stable wet foam. Because angular pore windows are not typical for open porous foams, it is very probable that at least some holes in the pore walls of Figure 25.2b are artifacts from sample preparation. The pore walls separating single cells are all of similar dimensions. Electron microscopy images recorded at higher magnification reveal cell wall thicknesses in the range of 1–3 μm. High resolution micrographs further show that the cell walls of the foams are built up by a single layer of α-Al_2O_3 grains. Such a homogeneous microstructure is expected to result in homogeneous material properties.

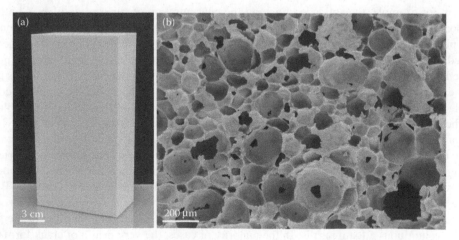

FIGURE 25.2 (a) High porosity α-Al_2O_3 with 91 vol% porosity and 360 kg/m^3 density. (b) Scanning electron microscopy image showing the pore structure of the foam.

25.4   MECHANICAL CHARACTERIZATION

The mechanical characterization of cellular materials as they are obtained from direct foaming processes is a relatively new field in research and development. Standard measurement procedures do not yet exist and examples from literature are scarce. For an in-depth discussion about advantages and disadvantages of three mechanical testing methods used for the characterization of high-porosity foams, the authors refer to the work of Seeber et al. [9]. The data discussed here with a focus on the material behavior under load originate from this publication.

The mechanical properties of high porosity α-Al$_2$O$_3$ foams are exemplified by testing samples of 90.9 vol% porosity, 360 kg/m^3 density and of 94.6 vol% porosity, 216 kg/m^3 density in compression and three-point bending mode. For compression experiments, cylindrical samples of 15 mm diameter and 20 mm height were cut. Rods of $10 \times 10 \times 80$ mm^3 were prepared for three-point bending. Typical stress strain curves recorded in compression and in bending mode and Weibull distributions from at least 10 runs are plotted in Figure 25.3. Under compression, the stress increases linearly with strain for both types of foam until an apparent yield stress is observed. Already in this linear regime, some single strut fracture can occur at the contact area between foam and sample fixation. Against any expectation, a further increase of the strain leads not to catastrophic failure of the specimen, but to a saw-tooth shaped progression of the stress-strain curve. Whereas for the 216 kg/m^3 sample the saw-tooth shape is not very pronounced, a significant stress increase is recorded between 1.3% and 3.3% strain for the sample with 360 kg/m^3 density. This increase is not steady but is interrupted by minor stress drops caused by joint breaking of numerous struts in planar zones oriented essentially in parallel and located mainly but not exclusively close to the sample fixations. From 3.3% to 4.2% strain, a net drop in compressive stress is observed. Such drops typically go together with the spalling of debris from the sample areas close

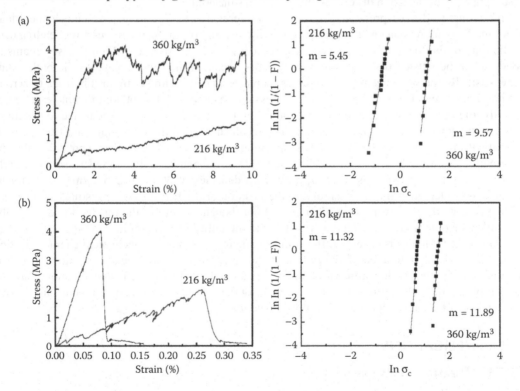

FIGURE 25.3   (a) Typical stress–strain curves from compression tests on α-Al$_2$O$_3$ foams with 360 and 216 kg/m^3 density (left) and Weibull distributions of at least 10 runs (right). (b) Typical stress-strain curves from three-point bending tests and corresponding Weibull distributions. (The graphs are adapted from Seeber, B.S.M., Gonzenbach, U.T., Gauckler, L.J. *J Mater Res* 2013; 28: 2281–2287 [9].)

TABLE 25.1

Results for Compressive and Three-Point Bending Measurements of α-Al$_2$O$_3$ Foams with 360 and 216 kg/m^3

Foam properties	Density (kg/m³)	360	216
	Porosity (vol%)	90.9	94.6
Compression	Apparent strength (MPa)	3.00 ± 0.38	0.46 ± 0.10
	Apparent strain (%)	1.25 ± 0.18	0.51 ± 0.20
	Weibull m	9.57	5.45
	Weibull σ_0	3.16	0.5
Three-Point bending	Stress at break (MPa)	4.55 ± 0.45	1.96 ± 0.20
	Strain at break (%)	0.093 ± 0.015	0.190 ± 0.033
	Weibull m	11.89	11.32
	Weibull σ_0	4.75	2.05

to the sample fixations. It is very interesting to note that this pattern is repeated for further increase in strain and that the specimen can repeatedly be loaded with the original maximum stress level. This result signifies that there is basically no macro crack propagation in high-porosity alumina foams when loaded under compression. Instead, compressive stresses just cause local damage in the area of impact. Apparent compressive strength values for both samples are listed in Table 25.1. With Weibull parameters m of 5.45 for 216 kg/m^3 and 9.57 for 360 kg/m^3 foams, the reliability of data from compression testing is acceptable or even high in the case of the less porous material.

In contrast to the compression test, where compression stresses dominate the mechanical load on a sample, compression and tensile stresses symmetrically occur during three-point bending tests in the top and bottom half of the specimen, respectively. The response of high-porosity foams to bending can be studied from the stress–strain curves plotted in Figure 25.3b. Again, both foams show basically the same behavior. The stress increases linearly with strain until it drops to zero at the maximum bearable strain. Whereas the stress-strain curve for the 360 kg/m^3 sample increases steadily, the one for the 216 kg/m^3 foams shows multiple stress drops that occur due to failure of struts at the contact line specimen/bearing. The result shows that the foams fail catastrophically if the critical strain is exceeded and indeed the specimens are observed to completely break in half. The failure mode signifies break due to tensile stresses. With 4.6 and 2.0 MPa, these stresses are clearly above expectations for porous α-Al$_2$O$_3$ parts with densities of 360 and 216 kg/m^3, respectively. Because no indentation marks are found at the position of the bearings after testing, the increase of contact area during testing is very small and not thought to impact the results. In addition, the Weibull distribution of the results proves that the reliability of three-point bending data of high porosity α-Al$_2$O$_3$ foams is excellent. In summary, the results from the mechanical characterization show a remarkable tolerance of the porous material against macro crack propagation which is surprising given the brittle nature of the raw material. Only under tensile load is unambiguous and catastrophic failure observed. The good reliability of the obtained stress and strain data are a direct consequence of the homogeneous foam microstructure.

25.5 APPLICATIONS

25.5.1 THERMAL INSULATION

With the combination of high air contents and mainly closed porosity, the presented alumina foams are well suited for thermal insulation applications at temperatures above 1350°C. The graph in Figure 25.4 displays thermal conductivity data for α-Al$_2$O$_3$ foam with 90.7 vol% porosity and

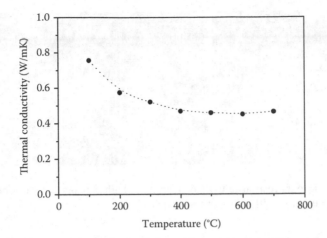

FIGURE 25.4   Thermal conductivity as function of temperature for α-Al$_2$O$_3$ alumina foam with 90.7 vol% porosity and 370 kg/m^3 density determined by the laser flash method.

370 g/m^3 density as obtained by the laser flash method. The thermal conductivity decreases with temperatures up to 400°C where it levels off at 0.46 W/mK. Typical values for dense alumina are in the order of 36 W/mK [10].

25.5.2 ELECTRICAL INSULATION

Processing α-Al$_2$O$_3$ into a high-porosity foam transforms this material not only into a good thermal insulator in the top temperature range, but also in a good electrical insulator. The static relative permittivity and the dielectric loss measured at 5.5 GHz for 89.5 vol% porous alumina are listed in Table 25.2, together with data for SiO$_2$, dense Al$_2$O$_3$, and poly(tetrafluoroethylene) PTFE, a material that is commonly used for high performance insulation applications.

25.5.3 MACHINABILITY OF SINTERED FOAMS

Owing to the homogeneous microstructure and excellent crack tolerance, the presented foams can readily be machined to high precision. In contrast to machining fiber-based insulation materials, the dust produced by cutting alumina foams is harmless and does not require the use of special dust control measures. These features facilitate a low cost manufacture of prototypes or a small series production. If nails and screws are driven into the material, no macro cracks are formed. The structure is just locally deformed. Prior to placing bigger screws, a hole is drilled into the material. Examples are presented in Figure 25.5a. The high porosity allows for cutting the material essentially without using cooling liquids. An unshaped sintered foam can, for example, be cut to a brick of

TABLE 25.2

Electric Properties of SiO$_2$, Al$_2$O$_3$, Poly(tetrafluoroethylene) PTFE, and High Porosity α-Al$_2$O$_3$

Material	Density (kg/m³)	Porosity (vol%)	Static Relative Permittivity	Dielectric Loss (10^{-4})
SiO$_2$ [11]	2240	None	3.9	
Al$_2$O$_3$ [11]	3980	None	11	
PTFE [12]	2350	None	2.1	3.0
Al$_2$O$_3$ porous	420	89.5	1.42	2.80 (5.5 GHz)

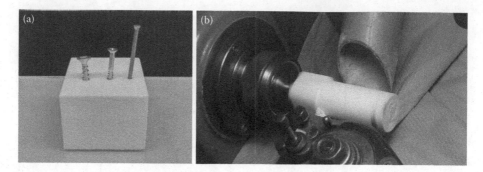

FIGURE 25.5 (a) Due to its resistance to crack formation, high porosity foam can be nailed and tolerates driving in screws. (b) Turning of 91 vol% porous α-Al$_2$O$_3$.

defined size using a band saw with hard metal coated teeth. On a reasonable machine, an accuracy of ± 0.1 mm can be achieved. Axisymmetric parts like cylinders and cones are fabricated by turning as is exemplarily shown in Figure 25.5b. Hard metal tools are, in principal, sufficient to machine 90 vol% foams. However, poly crystalline diamond cutting tools will show a longer service life.

25.5.4 ADVANCED FOAM PROCESSING

Many potential applications can be covered using a homogeneous piece of foam. The layout of porous ceramic materials for a certain application is generally a compromise between the highest possible porosity for low thermal conductivity and low weight on one hand and sufficient mechanical stability on the other hand. Through advanced foam processing, however, it is possible to integrate multiple functions into one monolithic part. One possibility is the formation of layered foam structures. An example is the manufacture of a foam part comprising a bottom layer of 80 vol% and a top layer of 92 vol% porosity. While the bottom layer is still insulating but chosen to mainly carry the mechanical load, the top layer is tuned for maximum thermal insulation. Another example is the combination of an 89 vol% α-Al$_2$O$_3$ foam part with excellent insulation properties and a nearly dense alumina top layer that effectively prevents abrasion of the underlying foam or penetration of liquids or gasses into the foam structure. A photograph of a dense porous composite sample together with electron microscopy images with focus on the interfacial zone of layered α-Al$_2$O$_3$ foam structures are presented in Figure 25.6a–c. The micrographs confirm the high quality of the interfacial layers. Under mechanical load, such structures do not fail at the interface but anywhere within the layer with the higher porosity.

25.6 INSULATION OF LIQUID METAL CENTRIFUGE SAMPLE HOLDERS

An interesting application where all the above mentioned properties of high-porosity α-Al$_2$O$_3$ come into effect is the thermal insulation of sample cavities in a liquid metal centrifuge. Figure 25.7 shows a photograph of such a centrifuge. The rotor of the machine exhibits eight symmetrically arranged cavities where sample holders can be placed. A three-dimensional visualization of the sample holder is presented in Figure 25.7b. The insulation material separates and insulates the core zone containing the liquid metal and the outer metal shell of the assembly. The demanding requirements for a suitable insulation material are listed in Table 25.3. First of all, the material needs to withstand an operation temperature of up to 1300°C at accelerations of up to 50,000 g. Secondly, the insulation properties need to be excellent. On the cold side of the insulation, the temperature should not exceed 120°C to minimize the thermal load on the metallic machine parts. A low density is crucial to reduce the stresses caused by the accelerated mass. Further, the insulation material requires to be machined with high precision to an exactly defined shape. Finally, it needs to be stable in a high vacuum of

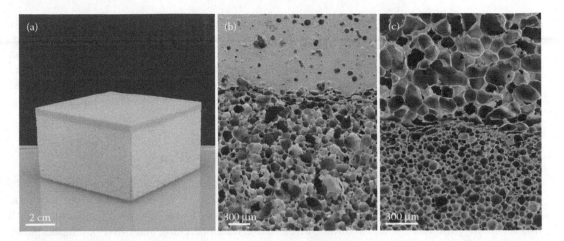

FIGURE 25.6 (a) Photograph of a 89 vol% α-Al$_2$O$_3$ foam part combined with a nearly dense top layer. (b) Electron microscopy image of the interface region of a 89 vol% porous α-Al$_2$O$_3$ foam part with a nearly dense α-Al$_2$O$_3$ top layer. (c) Electron microscopy image of the interface region of a high porosity foam part with a two layer design. The top layer consists of 92 vol% porous α-Al$_2$O$_3$, the bottom layer of 80 vol% porous α-Al$_2$O$_3$.

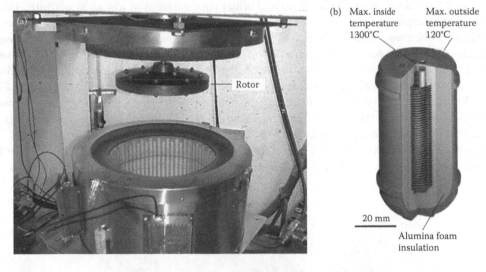

FIGURE 25.7 (a) Photograph of a liquid metal centrifuge showing the lid with spindle and rotor. The rotor contains 8 symmetrically arranged sample cavities. (Adapted from Loffler, J.F. et al. *Philos Mag* 2003; 83: 2797–2813 [13].) (b) 3D visualization of the sample holder with the shaped insulating part.

TABLE 25.3

Requirements for the Thermal Insulation Material of the Sample Cavity in a Liquid Metal Centrifuge

Requirements insulation material

Thermal stability at 1300°C

Ultralight ($\rho < 400$ kg/m^3)

Excellent thermal insulation

Machinable with high precision

High vacuum capable

at least 10^{-4} bar. The high service temperature clearly requires a ceramic material, but completely rules out silica and silica rich products. Low densities can only be achieved with fibers or foams. Fibers cannot be applied in this case because they barely take load and cannot be shaped with high accuracy. Therefore, only high porosity mineral foams can do the job. Considering the forces acting at 50,000 g, the combination of thermal and mechanical load exceeds the capability of foamed alumino silicate. Only an α-Al_2O_3 foam with a porosity of 85–88 vol% simultaneously fulfills the whole set of requirements.

25.7 HIGH POROSITY FOAM FROM FURTHER MINERALS

The direct foaming method using *in situ* hydrophobized particles as foam stabilizers is not limited to processing high quality alumina powders. In fact, a broad variety of ceramic and even some polymeric raw materials can be foamed [14]. Examples of high porosity foams that are manufactured with this technology are fused silica, alumino silicate, table ware porcelain, and β-tricalciumphosphate. Common to these materials is that they need to be sintered to transform them into monolithic, mechanically stable foam parts. The high temperature treatment is not necessary if the technology is applied to foaming hydraulic materials like ordinary Portland cement, calcium aluminate cement, reactive calcium phosphates or gypsum. The motivation to foam Portland cement is to save raw material, reduce weight, and improve the room temperature thermal insulation properties. Gypsum foams with porosities above 90 vol% exhibit thermal conductivities as low as 57 mW/mK. Because gypsum incorporates up to 23 wt% of crystal water upon hydration, the material is an interesting candidate for the manufacture of fireproof room temperature insulation boards. Porous gypsum is especially suited as an active material in fire protection components where the reduction of weight is advantageous. Such components are fire doors, insulation parts in vehicles or boards and cable channels for overhead installation. An example of a u-shaped gypsum foam part is depicted in Figure 25.8a. A further feature of the described foaming technology is that it allows for filling of complex cavities. An example is the demonstrator in Figure 25.8b that shows a high porosity gypsum fill between an inner and outer pipe assembly.

Although there are examples of mineral foam products in industry, the potential of these fiber-free materials with their fascinating combination of mechanical and insulation properties is not at

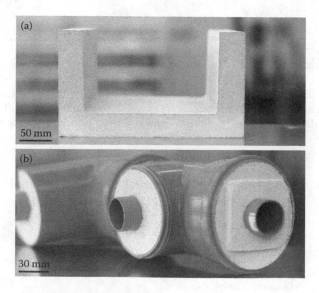

FIGURE 25.8 (a) u-shaped high porosity gypsum foam part and (b) demonstrator where the space between pipes is completely filled with high porosity gypsum foam.

all fully exploited yet. The direct foaming method using *in situ* hydrophobized particles as foam stabilizers is a versatile technology for the manufacture of high porosity foams, tailor made for the requirements of specific applications. It is versatile in respect to a broad variety of materials that can be processed. The technology allows controlling crucial properties like foam porosity and pore structure and enables the manufacture of foam parts with dimensions ranging from submillimeter coatings to bulk parts with complex, three-dimensional shapes.

REFERENCES

1. Studart, A.R., Gonzenbach, U.T., Tervoort, E., Gauckler, L.J. Processing routes to macroporous ceramics: A review. *J Am Ceram Soc* 2006; 89: 1771–1789.
2. Binks, B.P. Particles as surfactants—Similarities and differences. *Curr Opin Coll Interface Sci* 2002; 7: 21–41.
3. Kaptay, G. On the equation of the maximum capillary pressure induced by solid particles to stabilize emulsions and foams and on the emulsion stability diagrams. *Colloids Surf A, Physicochem Eng Aspects* 2006; 282: 387–401.
4. Gonzenbach, U.T., Studart, A.R., Tervoort, E., Gauckler, L.J. Tailoring the microstructure of particle-stabilized wet foams. *Langmuir* 2007; 23: 1025–1032.
5. Gonzenbach, U.T., Studart, A.R., Tervoort, E., Gauckler, L.J. Stabilization of foams with inorganic colloidal particles. *Langmuir* 2006; 22: 10983–10988.
6. Gonzenbach, U.T., Studart, A.R., Tervoort, E., Gauckler, L.J. Ultrastable particle-stabilized foams. *Angew Chem-Int Ed* 2006; 45: 3526–3530.
7. Gonzenbach, U.T., Studart, A.R., Tervoort, E., Gauckler, L.J. Macroporous ceramics from particle-stabilized wet foams. *J Am Ceram Soc* 2007; 90: 16–22.
8. Megias-Alguacil, D., Tervoort, E., Cattin, C., Gauckler, L.J. Contact angle and adsorption behavior of carboxylic acids on alpha-Al_2O_3 surfaces. *J Colloid Interface Sci* 2011; 353: 512–518.
9. Seeber, B.S.M., Gonzenbach, U.T., Gauckler, L.J. Mechanical properties of highly porous alumina foams. *J Mater Res* 2013; 28: 2281–2287.
10. Wong, C.P., Bollampally, R.S. Thermal conductivity, elastic modulus, and coefficient of thermal expansion of polymer composites filled with ceramic particles for electronic packaging. *J Appl Polym Sci* 1999; 74: 3396–3403.
11. Kingon, A.I., Maria, J.-P., Streiffer, S.K. Alternative dielectrics to silicon dioxide for memory and logic devices. *Nature* 2000; 406: 1032–1038.
12. Rajesh, S., Nisa, V.S., Murali, K.P., Ratheesh, R. Microwave dielectric properties of PTFE/rutile nanocomposites. *J Alloys Compd* 2009; 477: 677–682.
13. Loffler, J.F., Bossuyt, S., Peker, A., Johnson, W.L. Eutectic isolation in Mg-Al-Cu-Li(-Y) alloys by centrifugal processing. *Philos Mag* 2003; 83: 2797–2813.
14. Wong, J.C.H., Tervoort, E., Busato, S., Gonzenbach, U.T., Studart, A.R., Ermanni, P., Gauckler, L.J. Designing macroporous polymers from particle-stabilized foams. *J Mater Chem* 2010; 20: 5628–5640.

Index